The Nature of Diversity

D1019269

The Nature of Diversity

An Evolutionary Voyage of Discovery

Daniel R. Brooks
Deborah A. McLennan

The University of Chicago Press
Chicago and London

DANIEL R. BROOKS is professor in the Department of Zoology at the
University of Toronto. He is coauthor, with E. O. Wiley, of *Evolution
as Entropy: Toward a Unified Theory of Biology,* and, with Deborah A.
McLennan, of *Phylogeny, Ecology, and Behavior: A Research Program in
Comparative Biology* and *Parascript: Parasites and the Language of Evolution.*

DEBORAH A. MCLENNAN is associate professor in the Department of
Zoology at the University of Toronto. She is coauthor, with Daniel R.
Brooks, of *Phylogeny, Ecology, and Behavior: A Research Program in
Comparative Biology* and *Parascript: Parasites and the Language of Evolution.*

The University of Chicago Press, Chicago 60637
The University of Chicago Press, Ltd., London
© 2002 by The University of Chicago
All rights reserved. Published 2002
Printed in the United States of America
11 10 09 08 07 06 05 04 03 02   1 2 3 4 5

ISBN: 0-226-07589-3 (cloth)
ISBN: 0-226-07590-7 (paper)

Library of Congress Cataloging-in-Publication-Data

Brooks, D. R. (Daniel R.), 1951–
    The nature of diversity : an evolutionary voyage of discovery /
    Daniel R. Brooks, Deborah A. McLennan.
        p. cm.
    Includes bibliographical references and index.
    ISBN 0-226-07589-3 (cloth : alk. paper)—ISBN 0-226-07590-7
    (pbk. : alk. paper)
    1. Biological diversity.   2. Phylogeny.   3. Adaptation (Biology)
    I. McLennan, Deborah A.   II. Title.

QH541.15.B56 B76 2002
576.8'8—dc21

                                                        2001053860

The paper used in this publication meets the minimum requirements of
the American National Standard for Information Sciences—Permanence of
Paper for Printed Library Materials, ANSI Z39.48-1992.

QH
541.15
B56
B76
2002

[T]here are two factors: namely, the nature of the organism and the nature of the conditions. The former seems to be much more the important, for nearly similar variations sometimes arise under, as far as we can judge, dissimilar conditions; and, on the other hand, dissimilar variations arise under conditions which appear to be nearly uniform.

Darwin 1872:32

Contents

Preface

Scientists have odious manners unless you support their
theory; then you can borrow money off them.

Mark Twain

We wrote the predecessor to this book more than a decade
ago in an attempt to show people how the comparative phylogenetic
method could illuminate some of the dark corners of their research. At
that time, our biggest problem was finding enough studies to illustrate
all of the ideas. Over the past decade, interest in this approach has grown
tremendously. This time around we found ourselves in the enviable position of having too many studies from which to choose. Given such an
embarrassment of riches, we could not include everything we encountered during our voyage through the literature (which ended as an active
search on the eve of the new millennium). So, lest anyone feel that his
or her study or group of interest has been neglected, please understand
that our choice of examples is completely subjective and reflects our
own biases toward the organisms and questions we find interesting. We
have striven to include studies based on a wide variety of taxa published
by a geographically diverse group of researchers, but we do not claim
to have provided a true representation of either. The power of this approach is that it is infinitely malleable. All groups, all questions are
welcome!

Those of you who have read *Phylogeny, Ecology, and Behavior* will find
yourselves initially on familiar ground: an introduction, albeit updated,
to what is still a young but vigorously growing research program. There
are, however, several major differences between *Phylogeny, Ecology, and
Behavior* and *"Voyager."* The first difference results from a change in perspective. A decade ago we emphasized the extent to which adding phylogenetic history to our explanations could simplify and clarify our work.
Believing that the message was complicated, not to mention controversial, enough, we seldom ventured beyond documenting general patterns
supporting the hypothesis that a substantial degree of order in the world
around us was due to phylogenetic history. The past ten years of research
has convinced us (more than ever) that Darwin's metaphor of the tangled bank is the ultimate descriptor of biodiversity on this planet. We
now believe biologists (including ourselves) are ready to move beyond

the initial stance of "history = simplicity" to "evolution has been so historically contingent and complex that we need a phylogenetic framework if we are ever to have a hope of disentangling that complexity." The second difference is reflected in our sense that this phylogenetic framework, although necessary, is never sufficient to explain that complexity. We have found ourselves drawn more and more to research integrating phylogenetic and experimental studies in an attempt to detect the imprint of the past in the shape of extant biodiversity. Third, we believe that the research program we present in this book can benefit from connections with other independent and progressive research programs in evolutionary biology. We have attempted to highlight those connections in "Frontiers" sections at the end of each chapter. Fourth, we revisit continuously in this book the caveat that nothing should be eliminated from the "scope of the possible" unless there is strong biological *evidence* to the contrary. In other words, we do not believe that any person on this planet has a pipeline to God (or any other words you choose to associate with a sense of "truth"). Eliminating irritating data because they do not conform with what one "knows" to be the "correct" answer is simply not acceptable within the logic of scientific investigation. Another manifestation of this sentiment is our belief that virtual reality does not take precedence over empirical data. If models do not agree with the empirical data, chances are the models, not the data, should be re-evaluated. This is not an antimodel stance. A mutually reinforcing and mutually modifying dialogue between models and empirical discovery enhances progress. We are naturalists (one of us is a field biologist and the other is an experimental ethologist); therefore, our best contributions come from the accumulation of empirical knowledge about the world, which can be used both to evaluate predictions of existing models and to suggest new models.

Finally, we passionately believe that the biodiversity crisis is upon us and as such should be a real and pressing concern for all biologists. We owe a special debt of gratitude to Harry Greene and Kelly Zamudio, who (over a memorable California lunch) urged us to emphasize biodiversity science in this book. The special debt also extends, with love and gratitude, to Dan Janzen and Winnie Hallwachs, as well as the staff of the Area de Conservación Guanacaste (a World Heritage site in northwestern Costa Rica), who operate at ground zero of the biodiversity crisis every day. They have provided a living platform for integrating many of the ideas and ideals of historical ecology into biodiversity science, as well as a world standard of dedication and self-sacrifice in the dual pursuit of saving biodiversity and promoting socioeconomic development. Without the substantial intellectual input of these biologists, as well as their friendship and constant reminder of what human beings can be at their very best, one of us at least would have been tempted to give up the struggle and write murder mysteries.

A book of this scope could not have happened without the input of many people who generously helped us by discussing ideas, suggesting novel interpretations, and sharing publications. First and foremost among those are the many zoology faculty and students at Stockholm University. For nearly a decade we have been fortunate enough to lecture to undergraduates at Stockholm and to attend the annual Blod Bad conference at Tovetorp Field Station, where we have had the opportunity to air our views and to be encouraged and corrected (politely, gently, but firmly) by researchers at one of the world's foremost centers for comparative evolutionary studies. Of this large group of people, we especially wish to thank Anders Angerbjörn, Sven Jakobsson (who also introduced us to the tourist moose), Niklas Janz, Ole Leimar, Patrick Lindenfors, Sören Nylin, Hans Temrin, Birgitta Tullberg, Hans-Erik Wanntorp, Lars Werdelin, Nina Weddell, and Christer Wiklund. Tack, gode vänner. Also first in our hearts are Eric and Margaret Hoberg (U.S. National Parasite Collection and Johns Hopkins), who have shared ideas about the great evolutionary drama over good wine and good jazz for over two decades now (and still invite us back). We also thank the unwavering support and input of Diedel Kornet, Rino Zandee, Marco Van Veller, and Hubert Turner at Leiden University, Netherlands (a town of contented cats), Howard and Irene Burton, who have constantly turned our ideas upside down by questioning our basic assumptions about the origins of "life, the universe, and everything," and Rick Winterbottom, who knows more about fish (and phylogenetics) than is decent for one modest man (even if they are marine fish).

We are eternally grateful for our time spent with Douglas Causey (over too many lunches to remember), Harvard University; Michael Ryan (whose passion for those terrestrial fish we call frogs is infectious), David Hillis, James Bull, and Michael Singer, University of Texas; Gerardo Pérez Ponce de León and Virginia León-Règagnon, Universidad Nacional Autónoma de México (who both braved a Toronto winter for the mutual love of parasitos); Dalton da Souza Amorim, Universidade de São Paulo, Martin Lindsay Christoffersen, Universidade do Joao Pessoa, and Walter Boeger, Universidade do Rio Grande do Sul, Brazil; Eörs Szathmáry, Institute for Advanced Study, Budapest; Serge Morand, Université de Perpignan; Philippe Grandcolas and Louise Desutter-Grandcolas, Muséum National d'Histoire Natural, Paris (with particular gratitude for a memorable lunch during a warm day in Paris); Mats Björklund, Kåre Bremer, and Fredrik Ronquist, Uppsala University; William Wright, Colorado State University; Richard Mayden, St. Louis University (who also knows more about fish, phylogenetics, and conservation than an ordinary mortal); George Lauder, Harvard University; Bruce Lieberman and Edward Wiley, University of Kansas; Vicki Funk, U.S. National Museum of Natural History, Smithsonian Institution; Wouter Holleman, Rhodes University, Francis Thackeray, Pretoria, and Louise Coetzee, Bloemfontein, South Africa; James Dale and Susan Smith (now blissfully in Arizona, telescope and all) and Bill Presch, California State University-Fullerton; Brian Maurer, Michigan State University; John Lynch, Universidad Nacional de Colombia; John Wenzel, Ohio State University; and Geoffrey Scudder, Cyril Finnegan, and Jack Maze, University of British Columbia.

Closer to home, we greatly appreciate the support and input over numerous cups of coffee from our colleagues Doug Currie, Chris Darling, Ellen Larsen, Robert Murphy (composer of and player in the famous phylogenetic rock opera, *Rommie*), Robert Reisz, Hans-Dieter Sues, Malcolm Telford, and Polly Winsor of the University of Toronto and Royal Ontario Museum. We also thank the more than 500 University of Toronto undergraduate students who have taken our course based on *Phylogeny, Ecology, and Behavior* during the past decade (and appear, on the whole, to have enjoyed it), as well as a decade of outstanding graduate students at the University of Toronto, notably Jingzhong Fu, Hugh Griffiths, Gregory Klassen, Michelle Mattern (cofounder of the XWP lab), Randy Mooi, Brian Moore, Mark Siddall, Jon Stone, and David Zamparo. Their feedback has helped us immensely.

We have taught and spoken at numerous places during the past ten years, and we thank the organizers and participants for all those activities, as well as the Natural Sciences and Engineering Research Council (NSERC) of Canada for continuing support of our respective research programs. Finally, we burn metaphysical oxen to Athena Pronaia for inspiration while reading the first complete draft of this book—sequestered in our room at the Acropole Hotel, gazing down over a sea of olive trees to the Gulf of Itea and the beginning of the Sacred Way up to Delphi.

None of these words of gratitude, thanks, and love would have been possible without the encouragement, the belief, the fathomless capacity for understanding, the clarity, insight, intuition, and the treasured friendship of Susan Abrams. Mille grazie speciale amica!

This is an introductory book, providing a set of directions for how to begin a lifelong voyage of discovery. In the midst of striving to be clear about the basics, we may occasionally give the impression that this research is easy. It is not. Everyone on this voyage will encounter their own personal storms and sea monsters. This the price you pay for the sheer joy and wonder of discovery. To those of you who believe the voyage should be easy, cheap, and quick, we say, Put this book aside now. To those of you who understand that the voyage never ends, precisely because your passion will draw you ever onward, welcome aboard.

Chapter 1

Voyage of Discovery

"Why not go . . . and study . . . for yourself?"
Lord Dorwin raised his eyebrows and took a pinch of
snuff hurriedly. "Why, whatevah foah, my deah fellow?"
"To get the information firsthand, of course."
"But wheah's the necessity? It seems an uncommonly
woundabout and hopelessly wigmawolish method of getting
anywheahs. Look heah, now. I've got the wuhks of all the
old masters. . . . I wigh them against each othah—balance
the disagweements—analyze the conflicting statements—decide which is pwobably cowwect—and come to a conclusion.
That is the scientific method."

Isaac Asimov, *Foundation*

Isaac Asimov's Lord Dorwin is as far removed from Charles Darwin as is possible to imagine. Although he is known today primarily as a brilliant theoretician, Darwin was first and foremost a consummate naturalist. His firsthand knowledge of biological diversity in many parts of the world helped him "weigh the works of the great masters," synthesizing information from each to formulate a conception beyond any one person (Darwin 1859). With this new theory of evolution, Darwin himself became one of the great, if not the greatest, masters of all time in biological sciences. Today, nearly a century and a half later, one might think the task set out by Darwin complete, or at least that the finish line was in sight. Nothing could be further from the truth. In fact, only about 10 percent of the basic units of evolution on this planet, species, have been described and named. And we know the natural history for only a fraction of that 10 percent. It would appear that there is still a place for voyages of discovery. In this book, we will take you on one such voyage, partly with the intent of convincing younger people that many such voyages are possible and indeed necessary, and partly for the sheer joy of the journey.

Nature is complex. As sentient beings, we have always sought explanations for the origin and maintenance of that complexity, hoping that

1

somewhere during the search we would discover answers to questions about who we are and where we fit in the global biosphere. The search for those answers has been conducted from many different perspectives, from religion to sociology, art to science. Each perspective contributes a valuable piece to the puzzle, a way of "seeing the world." As biologists, we can trace the roots of our way of seeing to the earliest paintings of animals on cave walls and the emergence of myth making. All mythology is tied to an awareness of surrounding organisms and their natures. We are connected to these myths, and thus to millennia of observing, describing, and documenting the diversity of life, by our fascination with the nature of the organism. That fascination entered the formal realm of science when Darwin (building upon ideas from Lamarck and Wallace) united that vast but loosely connected network of biological information under one cohesive principle: the theory of evolution.

Darwin's original conceptual framework included two components. First, all organisms are connected by common genealogy:

> [T]he characters which naturalists consider as showing true affinity between any two or more species, are those which have been inherited from a common parent, all true classification being genealogical. (1872:346)

Second, the forms and functions of organisms are closely tied to the environments in which they live:

> [S]light modifications, which in any way favoured the individuals of any species, by better adapting them to their altered conditions, would tend to be preserved; and natural selection would have free scope for the work of improvement. (1872:59)

Many specialized research programs emerged from these two postulates over the next 100 years. Every one of those budding disciplines initially incorporated both genealogical (phylogenetic) and environmental (adaptational) factors into their explanations of evolutionary change. However, the role of phylogeny was progressively diminished in some fields, most notably in ecology, ethology, and the physiological sciences, while in other fields, most notably systematics, the role of the environment was virtually eliminated from evolutionary explanations. This, in turn, led to the emergence of markedly different worldviews even within evolutionary biology. Gareth Nelson summarized two of these perceptual differences in a discussion at a biogeography conference at the American Museum of Natural History in 1979. He told an apocryphal story of two biologists, one an ecologist and the other a systematist, who stepped into a large room together. Suspended from the ceiling by a variety of supports were thousands of balls of many different colors and sizes. All at once the supports were cut, and all the balls dropped from the ceiling, hit the floor, and began bouncing around the room. The ecologist exclaimed, "Look at the diversity!" whereupon the systematist said, "Hmm, 32 feet per second per second!"

The Darwinian revolution was founded on the concept that biological diversity evolved through a combination of genealogical and environmental processes. Although in theory the majority of biologists adhere to this proposition, until recently phylogenetic and ecological/behavioral studies have often been conducted quite independently. Is this a problem? In order to answer this question let us consider the following thought experiment. Suppose we were to pick, at random, any organism from a designated tide pool and a crab from anywhere in the world. If we then asked for a list of morphological, behavioral, and ecological characteristics of the unknown organism from a given environment and of the known organism (a crab) from an undetermined habitat, we would expect that more of the predictions would be correct for the crab than for the unknown tide-pool organism. At the same time, we would not expect to be able to predict all the ecological and behavioral features of our unknown crab without knowledge of the environment in which it normally lives. So it appears that Darwin's original intuition was correct: evolutionary explanations require reference both to phylogeny and to local environmental conditions. The answer to the preceding question is thus, Yes, the sundering of phylogenetic and ecological/behavioral studies is an important problem because the exclusion of either will weaken our overall evolutionary explanations.

This answer leads us to two new questions: First, given the conceptual framework proposed by Darwin, how did this dissociation come to be? In order to answer this question we must examine the histories of three disciplines: ethology, ecology, and evolutionary biology. This in itself has formed the central theme for numerous papers, books, and book chapters, so we will present only brief summaries.[1] Second, how can communication between ecology/behavior and systematics in evolutionary biology be reestablished? Answering this question requires the development of a

1. See, e.g., Kingsland 1985; McIntosh 1985, 1987; Lauder 1986; Hull 1988; McLennan et al. 1988; Burghardt and Gittleman 1990; Funk and Brooks 1990; Ross and Allmon 1990; Wanntorp et al. 1990; Brooks and McLennan 1991, 1993a,b; Harvey and Pagel 1991; Maurer 1991; Behrensmeyer et al. 1992; Mayden 1992a,b; Zrzavy 1992; Miles and Dunham 1993; Núñez-Farfán and Cordero 1993; Ricklefs and Schluter 1993; Shettleworth 1993; Skelton 1993a; Allmon 1994; Eggleton and Vane-Wright 1994; Grandcolas et al. 1994; Grande and Rieppel 1994; Hall 1994; Wainwright and Reilly 1994; Brooks et al. 1995; Harvey, Brown, and Smith 1995; Wagner and Funk 1995; Harvey 1996; Jablonski et al. 1996; Martins 1996; Rose and Lauder 1996; Sanderson and Hufford 1996; Begun et al. 1997; Kendrick and Crane 1997; Pérez Ponce de León 1997; Pérez Ponce de León et al. 1997; Grandcolas 1998; and Hernandez et al. 2000.

research program that will allow us to integrate ecological, behavioral, and historical information to produce a more complete picture of evolution. The remainder of this book delineates the conceptual, methodological, and empirical foundations for one such research program.

LOSING TIME IN EVOLUTIONARY BIOLOGY
The Eclipse of History in Ethology

Ethology, as a science, was founded upon a tradition of investigating behavior within an explicitly phylogenetic framework. Darwin started the ball rolling when he compared, among other things, the behavior of two species of ants within the genus *Formica* in an attempt to trace the evolution of slave making in ants. Following this example, the "founding fathers" of ethology, Oskar Heinroth and Charles O. Whitman, proposed that there were discrete behavioral patterns which, like morphological features, could be used as indicators of common ancestry. Whitman's (1899) views mirrored Darwin's: "Instinct and structure are to be studied from the common viewpoint of phyletic descent"(262). This perspective served as the focal point for a plethora of studies in the early twentieth century. Behavioral data were examined with an eye to their phylogenetic significance for birds, including anatids (ducks and their relatives) (Heinroth 1911; Herrick 1911), weaver birds (Chapin 1917), cowbirds (Friedmann 1929), and birds of paradise (Stonor 1936); and for insects and spiders, including wasps of the family Vespidae (Ducke 1913), bumblebees (Plath 1934), caddisfly larvae (Milne and Milne 1939), termites (Emerson 1938), social insects in general (Wheeler 1919), and spiders (Petrunkevitch 1926). Wheeler reiterated Darwin's and Whitman's perspective and reaffirmed the basis of ethological studies at the time: "Of late there has been considerable discussion . . . as to the precise relation of biology to history . . . and what most of us older investigators have long known seems now to be acceded, namely that biology in the broad sense and including anthropology and psychology is peculiar in being both a natural science and a department of history (phylogeny)" (1928:20).

Comparative behavioral studies flourished under the direction of Konrad Lorenz and Niko Tinbergen during the 1940s and 1950s. Both of these ethologists repeatedly emphasized two distinct but related points: behavioral patterns are as useful as morphology in assessing phylogenetic relationships, and behavior does not evolve independently of phylogeny. Lorenz stated that "all forms of life are, in a way, phylogenetic attainments whose special objects would have to remain completely obscure without the knowledge of their phylogenetic development"

(1941:3), and "every time a biologist seeks to know *why* an organism looks and acts as it does, he must resort to the comparative method" (1958:69; see also Lorenz 1950). Tinbergen outlined the comparative method:

> The naturalist . . . must resort to other methods. His main source of inspiration is comparison. Through comparison he notices both similarities between species and differences between them. Either of these can be due to one of two sources. *Similarity* can be due to affinity, to common descent; or it can be due to convergent evolution. It is the convergences which call his attention to functional problems. . . . The *differences* between species can be due to lack of affinity, or they can be found in closely related species. The student of survival value concentrates on the latter differences, because they must be due to recent adaptive radiation. (1964:421–422)

In other words, the phylogenetic relationships among species provide the platform from which explanations of processes responsible for behavioral evolution within species must be derived.

Although the comparative approach to studying behavioral evolution flourished during the 1950s and 1960s, skepticism mounted about Lorenz's assertion that species-specific behavioral characters were valuable systematic characters. By the centenary of the publication of Darwin's book, two widely divergent viewpoints had emerged:

> To assume evolutionary relationships on the basis of behavior patterns is not justifiable when such findings clearly contradict morphological considerations. The methods of morphology will therefore remain the basis for the natural system [of classification]. (Starck 1959, cited in Eibl-Eibesfeldt 1975:223)

> If there is a conflict between the evidence provided by morphological characters and that of behavior, the taxonomist is increasingly inclined to give greater weight to the ethological evidence. (Mayr 1960:345)

This difference in opinion was founded, in part, upon continuing unresolved debates among ethologists. Two questions recurred: first, how well can sequences of ancestral and derived traits be determined for attributes that left no fossil record; and second, how well can similarities due to common ancestry (homology) be distinguished from similarities due to convergent or parallel evolution (homoplasy)?[2] The question of homology was problematical because homologous characters were defined by their common origin and, at the same time, were used to reconstruct phylogenetic relationships. The inherent circularity in such a method bothered many biologists. Remane (1956) proposed a set of criteria for

2. Boyden 1947; Lorenz 1950; Tinbergen 1951, 1953; Schneirla 1952; and Michener 1953.

testing hypotheses of common origin (homology) without a priori reference to phylogeny. These were (1) similarity of position in an organ system; (2) special quality (e.g., commonalities in fine structure or development); and (3) continuity through intermediate forms. Although
authors did not agree about the universal applicability of Remane's criteria
to behavior, the majority accepted that the criterion of special quality,
studied at the level of muscle contractions (fixed action patterns), was
the fundamental tool for establishing behavioral homologies (Baerends
1958; Remane 1961; Wickler 1961; Albrecht 1966). Initial attempts to
homologize behavior in this way were admittedly vague and simplistic
when compared to the more quantitative methodology of comparative
morphology, but this reflected more the youth of the discipline than a
fundamental flaw in the behavioral traits themselves. Time and again,
phylogenies reconstructed using behavioral characters mirrored those
based solely on morphology. However, in a scathing review of the ethologists' research program, Atz made only a cursory reference to these successes when he concluded,

> The number of instances in which behavior has provided valuable
> clues to systematic relationships has continued to grow but it
> should be made clear that the establishment of detailed homolo-
> gies was seldom, if ever, necessary to accomplish this. . . . Func-
> tional, and especially behavioral, characters usually do not involve
> demonstrable homologies, but depend instead on resemblances
> that may be detailed and specific but nevertheless cannot be traced,
> except in a general way, to a common ancestor. . . . Until the time
> that behavior, like more and more physiological functions, can be
> critically associated with structure, the application of the idea of
> homology to behavior is operationally unsound and fraught with
> danger, since the history of the study of animal behavior shows
> that to think of behavior *as* structure has led to the most pernicious
> kind of oversimplification. (1970:67–69)

Lorenz had marked the beginning of the eclipse when he wrote, "I am
quite aware that biologists today (especially young ones) tend to think
of the comparative method as stuffy and old-fashioned—at best a
branch of research that has already yielded its treasures, and like a spent
gold mine no longer pays the working. I believe that this is untrue"
(1958:69). Atz's review punctuated the eclipse. For nearly 20 years, only
a few intrepid souls maintained the belief that phylogenetically relevant
behavioral homologies existed, could be studied scientifically, and were
important to explanations of the evolution of behavior.[3]

3. For example, Dunford and Davis 1975; Radinsky 1975; Drummond 1981; Greene
1983, 1986a,b, 1988; Hunt et al. 1983; Lauder 1986; Arnold 1987; McLennan et al. 1988;
Miller 1988; and Shine 1988.

Lorenz cautioned, "The similarity of a series of forms, even if the series structure arises ever so clearly from a separation according to characters, must not be considered as establishing a series of developmental stages" (1941:81). In his opinion, without reference to phylogenetic relationships, the criterion of similarity was, of itself, a dangerously misleading evolutionary marker. Unfortunately, the Gordian knot of behavioral homology drove ethologists toward a new methodology based, in direct contrast to Lorenz's warning, upon arranging behavioral characters as a "plausible series of adaptational changes that could easily follow one after the other" (Alcock 1984:432). Although intuitively pleasing, this method relies heavily on subjective, a priori assumptions concerning the temporal sequence of ethological modifications and dissociates character evolution from underlying phylogenetic relationships. This dissociation of history from behavioral evolution has had an important impact on both the nature and direction of ethological research.[4]

The Eclipse of History in Ecology

Ecology is founded upon the search for an understanding of the interactions between an individual and its environment. This simple aim masks a Herculean challenge, for the term "individual" can encompass practically all levels of biological organization, from the organism through the species to the ecosystem. The complexity of this search prompted Moore, in the opening paper of the first issue of *Ecology,* to call for an integration of ecology with other sciences.

> There have been three stages in the development of the biological sciences: first, a period of general work, when Darwin, Agassiz and others amassed and gave their knowledge of such natural phenomena as could be studied with the limited methods at hand; next, men specialized in different branches, and gradually built up the biological sciences which we know today; and now has begun the third or synthetic stage. Since the biological field has been reconnoitered and divided into its logical parts, it becomes possible to see the interrelations and to bring these related parts more closely together. Many sciences have developed to the point where . . . contact and cooperation with related sciences are essential to full development. Ecology represents the third phase. (1920:3)

Over the next 30 years, the call for integration and cooperation was answered by disciplines such as forestry and geology. Communication

---

4. Lauder 1986; McLennan et al. 1988; Brooks and McLennan 1991; Wenzel 1992; and Greene 1994a. For a discussion of nonhistorical and historical perspectives on particular topics, see Pearson et al. 1988; Mooi et al. 1989; Pearson 1990; Altaba 1991; and Vogler and Kelley 1996, 1998.

between ecologists and systematists developed more slowly, however, and this period saw only a handful of studies exploring ecological questions within a historical framework.[5] Although numerically small, this research foreshadowed the emergence of a phylogenetically based perspective in ecology at the same time that this theme was being developed in ethology. On one side of the Atlantic, Lorenz (1941), drawing on his observations of ducks and their relatives, was emphasizing the importance of phylogeny to studies of behavioral evolution. On the other side of the ocean, Bragg and his coworkers were reaching a similar conclusion from their extensive studies of the ecology and natural history of toads.

> Since variations in ecological conditions (physical or biotic) markedly effect the lives of individual organisms, and through this, of species, it follows that there is a broader line between the usual ecological emphasis upon succession of communities to the climatic or edaphic climax of a given region, on the one hand, and the taxonomic and geographic distributional emphasis of taxonomists and biogeographers on the other. The study of habits of animals, interpreted in the light of both ecology and taxonomy is, thus, an aid—indeed an absolute essential—to a complete understanding by either group of workers of the peculiar problems of either. (Bragg and Smith 1943:301)

The next 25 years were characterized by two significant changes: the appearance of papers by systematists in ecological journals, echoing this sentiment of cooperation, and a burst in the number of comparative studies.[6] The ascension of the comparative approach coincided with the appearance of the "new" evolutionary ecological perspective developed by Hutchinson and MacArthur. This research program was primarily concerned with attempting to answer the general question, Why are there so many species? and its corollary, How do these species manage to coexist? Answers to these questions had traditionally been sought within a comparative framework, an approach reinforced by MacArthur's statement, "Ecological investigations of closely-related species then are looked upon as enumerations of the diverse ways in which the resources of a community can be partitioned" (1958:617). King empha-

5. See, e.g., Baker 1927; Rau 1929, 1931; Parker 1930; Talbot 1934, 1945, 1948; Park 1945; Park and Frank 1948; and Smith and Bragg 1949.

6. Papers by systematists include Sabrosky 1950; Davidson 1952; Constance 1953; and McMillan 1954. Comparative studies include Pavan et al. 1950; Hairston 1951; Dobzhansky and da Cunha 1955; Carpenter 1956; MacArthur 1958; Kohn 1959; Cade 1963; Rand 1964; Schoener 1965, 1968a,b; Shoener and Gorman 1968; Brown 1971; Preston 1973; Laerm 1974; Roughgarden 1974; and McClure and Price 1975.

sized the importance of searching for competitive exclusion within a closely related group of organisms in his critique of MacArthur's broken stick model of species abundance.

> As realized by Darwin, the principle of competitive exclusion is most applicable to closely related sympatric species (that is, to species of high taxonomic affinity) having similar but not identical niches. This may be related to the MacArthur model since when competitive exclusion has taken place, the species of high taxonomic affinity that remain may be expected to have niches which are nonoverlapping but contiguous. Hairston suggests that tests of these species should display better fits to the MacArthur model than do tests of all species occurring in the habitat. That these predictions are valid was first indicated by the striking fits obtained by Kohn when only members of the genus *Conus* were examined. Subsequent investigations of fresh-water fishes ... reveal that in one collection from a single locality members of the class do not fit well, but when members of the same family are considered the fit is much better. (1964:723)

MacArthur set the tone for ecological studies of species coexistence and the search for correlations between changes in a species' ecology and changes in the environment. However, although evolutionary ecologists were examining experimental data within a comparative framework, few researchers were incorporating phylogenetic information into their evolutionary explanations (for a historical review, see Collins 1986). The difference between asking a question within a historical context and incorporating historical information into the answer is a critical and, at first, counterintuitive one. Consider the following simple example. Suppose you are interested in the question of species coexistence. As MacArthur noted, the best place to look for the factors involved in species coexistence is among sympatric populations of congeners. The assumption behind this recommendation is a historical one: members of the same genus should theoretically share a number of ecological, morphological, and behavioral characters in common because they are all descended from a common ancestor. The recognition that the genealogical relationships among species may influence the outcome of an experimental investigation is the first step in any evolutionary ecological study. Having discovered an appropriate group of sympatric congeners, you set about collecting a wealth of data concerning feeding behavior, habitat preference, and breeding cycles, in order to identify the way(s) in which the species are partitioning their environment. This second step in your study is primarily nonhistorical because it requires that you make assumptions about the evolutionary *past* of species' interactions, based upon characters and interactions observed in the *present* environment. What is missing here is information about the evolutionary origin and elaboration of the characters and of the associations themselves. So, when we talk about "incorporating phylogenetic data into an evolutionary explanation," we are referring to the combination of both *the history of the species and the history of the traits that characterize interactions among those species.*

The number of historically based studies began to decrease within the rapidly burgeoning field of ecology at about the same time that the comparative method was waning in ethology.[7] This trend continued through the 1980s[8] and, paradoxically, paralleled an increase in the number of studies concerned with examining ecology within a specifically evolutionary context. We cannot offer any particular explanation for this observation. Part of the answer may stem from the MacArthurian perception that historical effects, though real, would confound ecological predictions (see, e.g., Facelli and Pickett 1990). Part of the answer may simply be that the theoretical foundations for ecology were well developed by the 1970s so more ecologists turned their attention toward a rigorous examination of the assumptions underlying those theories. Although painstaking, there is no other way to test assumptions than by careful species-by-species examination. And still another part of the answer may lie in an observation by Stenseth (1984) that ecology was once the "hand-maiden" of taxonomy, but became a science on its own in the 1960s. If many ecologists felt they had been under the yoke of taxonomy, perhaps the break had more to do with desires for individual identities. If so, it would be unfortunate, because many systematists have felt the same way about the subordination of their discipline within ecology. Thus, ironically, the perception of subordination by members of each specialty has been based on mutual misapprehensions.

Whatever the reason, Ricklefs (1987) suggested that this "eclipse of history" had a profound and adverse effect on the field of community ecology. He argued that community ecology had relied mostly on local-process theories for explanations of patterns that are strongly influenced by regional processes. Local explanations rely on the action of competition, predation, and disease to explain patterns of species diversity in small areas, from hectares to square kilometers. According to this perspective, the community is maintained at a saturated equilibrium by bi-

---

7. But see Fraser 1976; Huey and Webster 1976; Huey and Pianka 1977; May 1977; Pitelka 1977; and Hubbell and Johnson 1978.

8. But see Hixon 1980; Hairston 1981; Keen 1982; Horton and Wise 1983; Kingsolver 1983; Davidson and Morton 1984; Schroder 1987; and Armbruster 1988.

otic interactions. However, independent lines of evidence from different communities suggest that regional diversity plays a strong role in structuring local communities. For example, the observations that (1) there are four to five times more mangrove species in Malaysia than in Costa Rica and four times more chaparral plant species in Israel than in California, (2) the number of cynipine wasps on a species of California oak is strongly related to the total number of cynipines recorded from the whole range of the oak species, and (3) local species richness in Caribbean birds is strongly related to total regional bird diversity cannot be explained solely by the assumption of local, saturated equilibria—otherwise similar states would be attained in systems exposed to similar environmental conditions. Ricklefs recognized the need for alternative explanations in community ecology, including phylogenetic information. Brown and Maurer (1989; also Maurer et al. 1992; Brown 1995; Maurer 1999) reinforced this conclusion with their suggestion that general statistical regularities in ecological associations occur on much larger spatial scales than previously considered. They proposed a research field, called macroecology, in which the emphasis is on large- rather than small-scale studies. Like Ricklefs, they recognized that enlarging the spatial scale of evolutionary ecological studies would increase the amount of phylogenetic influence in the systems under investigation. Given the existence of these effects, then, Brown and Maurer called for ways to incorporate them into the explanatory framework of macroecology.

The Eclipse of History in Evolutionary Biology

The centrality of comparison in biology predates the Darwinian revolution. Classification is the foundation of all biological research, and classification is inherently a comparative pursuit. Darwinism, however, required that classification add an explanatory dimension to its critical descriptive one. This did not happen immediately. The last quarter of the nineteenth century gave rise to a research program known as the "comparative method." Taking advantage of the emerging field of biostatistics spearheaded by Darwin's cousin Francis Galton, the comparative method sought to provide biology with rigorous evidence of lawlike behavior that could be ascribed to the effects of natural selection. In order to show statistically that selection acted like a natural law, each trait of each species was assumed to represent an evolutionarily independent variable, and thus an independent outcome of selection. The same or similar traits in each species could then be correlated with the same or similar environmental variables and a common lawlike cause inferred.

Boas (1896) sounded a note of concern with this new approach when he discussed limitations of the comparative method. He suggested that evolutionary biology was being studied in two different ways, which he denoted as the "evolutionist" and the "historian" perspectives (see also Bowler 1983). The evolutionists' assumption that every trait was evolutionarily independent allowed them to conform to the prevailing view of the time that the best scientific explanations were statistical in nature. The historians, on the other hand, assumed that similarities between species might be the result of common ancestry, the criterion upon which Darwin suggested biological classifications should be based. This assumption, however, lowered the number of independent origins of traits and potentially weakened statistical arguments for the lawlike behavior of selection. Boas pointed out that if all traits were independent responses to environmental selection, regardless of the history of descent (phylogeny), the comparative method risked becoming a theory of geographical determinism, a veiled reference to Lamarckism. This conundrum was not resolved until more than 70 years later, with the development of statistical methods that explicitly incorporated phylogenetic relationships.[9]

By the 1930s, the development of genetics as an experimental science had cast additional doubt on the relevance of history in evolutionary biology (Fisher 1930; Morgan 1932). In the introduction to his evolution text, Morgan (1932) wrote that evolutionary biology had freed itself from the constraints of simply recording historical sequences and had become a modern science studying contemporary processes in the laboratory. The almost simultaneous emergence of the "new synthesis" further eroded interest in historical explanations.[10] Although the new synthesis was promoted as a synthesis of genetics and paleontology (see, e.g., Mayr 1963), evolutionary history was relegated to being a passive record of the gradual accumulation of microevolutionary phenomena. The eclipse of history was complete after the evolutionary ecology revolution of the 1960s (MacArthur 1957, 1960, 1965, 1969, 1972; MacArthur and Wilson 1967), in which evolution became a synthesis of genetics and ecology, and history largely disappeared.

While we feel that the eclipse of history during much of twentieth century evolutionary biology was unfortunate, we hasten to make two

9. Clutton-Brock and Harvey 1977, 1984; Gittleman 1981; Harvey and Mace 1982; Ridley 1983; Felsenstein 1984, 1985a; Cheverud et al. 1985; Dunham and Miles 1985; Harvey and Clutton-Brock 1985; and Harvey and Pagel 1991.

10. Fisher 1930, 1958; Wright 1931; Haldane 1932, 1935, 1937; Dobzhansky (who was a post-doc in Morgan's lab) 1937; Huxley 1938, 1942, 1953; and Mayr 1942. But see Simpson 1944.

observations. First, a tremendous amount of knowledge was accumulated, both in the laboratory and in natural settings, by researchers who were not making explicit reference to phylogeny. Thus, so long as their research programs were exciting and busy, there was no reason for them to add more complications to their lives and research programs. Second, systematic biology in general failed to embark on a search for objective methods for studying phylogenetic history and rejected the efforts of those few (e.g., Hennig 1950) who did so. In fact, by the 1960s, many evolutionary biologists and systematists (e.g., Sokal and Sneath 1963) had concluded that explicit and robust protocols for inferring evolutionary history were not even possible and the search for surrogate measures of history, particularly geographic distributions, began (MacArthur 1960, 1965, 1969, 1972; MacArthur and Wilson 1967; Connor and McCoy 1979; Brown 1984).

MAKING UP FOR LOST TIME
The Past as Prologue

> [T]here are two factors: namely, the nature of the organism and the nature of the conditions. *The former seems to be much more the important,* for nearly similar variations sometimes arise under, as far as we can judge, dissimilar conditions; and, on the other hand, dissimilar variations arise under conditions which appear to be nearly uniform. (Darwin 1872, 32; emphasis added)

If Darwin was correct, and the nature of the organism is paramount, and if there was a real reason for showing a phylogenetic tree as the *only* illustration in *Origin of Species,* then clearly we need to end these eclipses and return to a more panoramic view of evolution. How do we do this? What tools should we use? First and foremost we need a robust method for reconstructing phylogeny, because *phylogeny is the embodiment of the nature of the organism through time.* We need the phylogenetic patterns to help us link real-time evolutionary processes with deep-time phenomena.

 Why use phylogenetic trees as the general reference system for the voyage of discovery?

As it is difficult to show the blood relationship between the numerous kindred of any ancient and noble family even by the aid of genealogical trees, and almost impossible to do so without this aid, we can understand the extraordinary difficulty which naturalists have experienced in describing, without the aid of a diagram, the various affinities which they perceive between the living and extinct members of the same great natural class. (Darwin 1872:410–411)

> The processes of life can be adequately displayed only in
> the course of life throughout the long ages of its exis-
> tence. (Simpson 1949:9)

A Revolution in Systematics

While evolutionary ecology and ethology were experiencing a surge of
interest in the comparative approach, the attention of systematists was
being focused in the opposite direction. The "new systematics," prompted
by the successes of the neo-Darwinian program, emphasized studies of
population variation and downplayed phylogenies. The reasons for this
shift in perspective were straightforward: systematists shared the general
concern that phylogenies could not be reconstructed in a noncircular
manner, and evolutionary biology in general was heavily influenced by
the quantum leaps occurring in population genetics. Under the influence of theoreticians such as Ronald Fisher, J. B. S. Haldane, Sewall
Wright, and Theodosius Dobzhansky, researchers sought the golden
fleece of evolution in a new arena: changes in gene frequencies within
and among populations under different environmental conditions.

Julian Huxley announced the arrival of the "new systematics" in
1940. By the late 1950s and early 1960s, systematic biology experienced
another revolutionary change, triggered as a reaction against a perceived
lack of repeatable methodology and quantitative rigor in the discipline.
Some theorists thought these problems were inherent in any attempt to
reconstruct phylogeny and suggested evolution-free systematics (Sokal
and Sneath 1963). Others believed there could be more rigor in the evolutionary approach to systematics. These researchers, however, were faced
with solving three long-standing and thorny problems: homology, levels
of generalities in similarities, and recognizing potentially informative
traits. Remane (1956, 1961), for example, proposed a set of criteria for
detailed comparisons of similarities among traits that allowed researchers to establish whether some traits that appear to be "the same" are, or
are not, "the same." However, certain traits that are homologous under
Remane's criteria could conceivably be nonhomologous evolutionarily.
This would occur, for example, if two species showing the same ancestral
polymorphism experienced similar selection pressures leading to independent fixation of the same trait. Because the fixed trait arose more
than once evolutionarily, its various manifestations among different species are not "evolutionary" homologues. What was needed, then, was a
homology criterion that would allow workers to recognize evolutionary
sequences of ancestral-to-derived traits (levels of generality) that would
not be circular.

The "evolutionary homology criterion"[11] that emerged in systematics from these methodological considerations is based on the Darwinian assumption that (evolutionarily) homologous traits all covary with phylogeny (since they are products of a single evolutionary history): "Homologous parts tend to vary in the same manner, and homologous parts tend to cohere" (Darwin 1872:158). Nonhomologous traits, on the other hand, do not covary with phylogeny and covary with each other only under special circumstances.

To implement this criterion, systematists needed a method for reconstructing phylogeny independent of assumptions about genealogical relatedness. Taxa could not be grouped according to overall similarity, because similarity embodies three different phenomena. First, there is similarity in general homologous traits—for example, humans, gorillas, and salmon all have vertebrae, and vertebrae appear to have evolved only once, but the presence of vertebrae does not help determine that humans and gorillas are more closely related to each other than either is to the salmon. Second, there is similarity due to convergent or parallel evolution (jointly termed **homoplasy**)—for example, birds and mammals have independently evolved homeothermy, which conflicts with their phylogenetic relationships. And third, there is similarity in special homologous traits—for example, among living amniotes only birds and crocodilians have submandibular fenestrae, which is taken as evidence of close phylogenetic relationship between those two groups because no other living taxa have such structures. Given this, two problems must be solved: distinguishing general from special traits and distinguishing homology from homoplasy. A solution to these problems was provided by the German entomologist Willi Hennig (1950, 1966a).

Hennig proposed that homology should be presumed whenever possible by applying criteria such as Remane's. General homology could then be distinguished from special homology by using what is now called the "outgroup criterion" (for a discussion see Wiley 1981; Wiley et al. 1991, forthcoming). Briefly, the outgroup criterion states that any trait found in one or more members of a study group that is also found in species outside the study group is a *general* similarity. Hence, the presence of vertebrae in mammals is a general trait because there are nonmammals that also have vertebrae. Those traits occurring only within the study group are *special* similarities. The members of the study group are then grouped according to their special shared traits. If there are conflicting groupings, it means that some traits presumed to be evolutionary

11. See Wiley 1981; Patterson 1982; Roth 1984, 1988, 1991, 1994; Gould 1986; Rieppel 1992; and McKitrick 1994.

homologies on the basis of nonphylogenetic criteria are actually homo-plasies. Because all evolutionary homologies covary, and homoplasies do not covary except under special circumstances, the pattern of relation-ships supported by the largest subset of special similarities is adopted as the working hypothesis of phylogenetic relationships. As more and more traits are sampled, there will be progressively more support for a single phylogenetic pattern. Traits that are inconsistent with this pattern are interpreted, post hoc, as homoplasies. Thus, the phylogenetic sys-tematic method works in the following way: (1) presume homology, a priori, whenever possible using Remane's, and other, criteria (hypothesis of homology); (2) use outgroup comparisons to distinguish general from special homologous traits; (3) group according to shared special homolo-gous traits; (4) in the event of conflicting evidence, choose the phyloge-netic relationships supported by the largest number of traits; (5) inter-pret inconsistent results, post hoc, as homoplasies (falsification of the original hypothesis that the traits were homologous). So, homologies, which indicate phylogenetic relationships, are determined *without* refer-ence to a phylogeny, while homoplasies, which are inconsistent with phylogeny, are determined as such *by* reference to the phylogeny.

The advent of phylogenetic systematics marked a return to the posi-tion advocated by Darwin: "community of descent is the hidden bond which naturalists have been unconsciously seeking, and not some un-known plan of creation, or the enunciation of general propositions and the mere putting together and separating of objects more or less alike" (1872:346; see also Wiley 1986a; de Queiroz 1988). Wiley (1981) deter-mined the minimal set of evolutionary assumptions necessary for us to proceed with attempts to determine phylogenetic patterns. We must as-sume only that evolution has occurred, that it produces phylogeny in the form of internested sets of descendant species, and that this history of phylogenetic diversification leaves its trace in the characters of spe-cies, living and fossil.[12]

Armed with a noncircular method for use in formulating, testing, and refining explicit hypotheses of phylogenetic relationships, systematists were finally in a position to begin contributing detailed information about phylogenetic effects on evolving systems of many kinds.[13] And indeed, a variety of contributions, derived from phylogenetic analyses

12. See also Felsenstein 1978, 1982, 1988a; Sober 1988a; Donoghue 1990; Swofford and Olsen 1990; and Beatty 1994.

13. For reviews see Kavanaugh 1972; Bremer and Wanntorp 1977; Cracraft 1978; Du-puis 1978, 1984; Funk and Stuessy 1978; Gaffney 1979; Eldredge and Cracraft 1980; Funk 1981; Wiley 1981, 1986a,b,c,d,e,f; Brooks 1985, 1995, 1996; Duellman 1985a; Schoch 1986; de Queiroz 1988; Brooks and McLennan 1991; Wiley et al. 1991, forthcoming; Forey

were quickly suggested.[14] By that time, however, systematists had virtually abandoned ecological and behavioral data as primary indicators of phylogenetic relationships. Their apprehensions stemmed, in part, from legitimate concerns about the dynamic nature of functional, as opposed to structural, traits. After all, we believe intuitively that verbs are more labile than nouns. These apprehensions have persisted, and until recently a vast database of ecological and behavioral characters remained virtually unexplored by systematists. In fact, despite recent successes,[15] the current state of affairs is still best summarized in a paper presented by R. D. Alexander during a symposium on the usefulness of nonmorphological data in systematic studies.

> [A]nyone with more than a passing curiosity about the study of animal behavior soon acquires the feeling that it has been neglected too frequently in many aspects of zoology, but especially among the systematists, who have almost a priority on the comparative attitude. . . . Behavioral attributes are . . . too often at the core of diverse problems in animal evolution to allow us to get by with the vague feeling that structure and physiology can be compared but behavior cannot—that a structural description is important information but that a behavioral description is a useless anecdote. (1962:70)

And so today we stand at a branching point in evolutionary studies, ecology and behavior to one side, systematics to the other. Researchers focused on ecology and ethology have increasingly turned their gaze toward evolutionary phenomena within species, whereas systematists have become preoccupied with among-species patterns. Strangely, this dichotomy has returned us, more than a century later, to Darwin's two theories of evolution, one emphasizing genealogy and the other, natural selection. What is strange is not that the disciplines have separated along these lines, but that they separated at all. Darwin's greatest contribution lay in his attempt to consolidate his two theories within a unified framework of "evolution," in which genealogy, or common history, explained

---

et al. 1992; Joysey and Friday 1992; Sundberg 1993; Amorim 1994, 1997; Christoffersen 1994; Ferris 1994; Gaffney et al. 1995; Pérez Ponce de León 1997; Pérez Ponce de León et al. 1997; Andersen 2001; and McLennan and Brooks 2001.

14. For example, Michener 1967; Ashlock 1974; Rosen 1975; Cracraft 1979; Eldredge 1979; Eldredge and Cracraft 1980; Brooks 1981; Lauder 1981; Nelson and Platnick 1981; Wiley 1981; and Fink 1982.

15. For example, McLennan et al. 1988; Prum 1990; Wenzel 1991, 1992, 1993; McLennan 1993a; de Queiroz and Wimberger 1993; Greene 1994a; Wimberger and de Queiroz 1996; Miller and Wenzel 1995; Paterson et al. 1995; Hedenäs and Kooijman 1996; Hughes 1996; Kennedy et al. 1996; Halloy et al. 1998; Slikas 1998; Stuart and Hunter 1998; Andersson 1999; Basibuyuk and Quicke 1999; Zyskowski and Prum 1999; Bostwick 2000; and McLennan and Mattern 2001.

the similarities that bind all living organisms together and natural selection explained the differences. A reunification of ethology, ecology, and systematics will return us to this multidimensional view of evolution. And, as many biologists are beginning to realize, this reunification is long overdue.

> Without taxonomy to give shape to the bricks, and systematics to tell us how to put them together, the house of biological science is a meaningless jumble. (May 1990:130)

Renewed Interest in Comparative Biology

At least three different approaches to studying comparative biology emerged in the last quarter of the twentieth century.[16] The first approach originated when Clutton-Brock and Harvey (1977), echoing Boas (1896), noted that treating each species as an independent variable in statistical analyses was tantamount to assuming that all interspecific similarities were due to convergent or parallel evolution. This could overestimate the effects of selection and possibly obscure real patterns in the data. A number of studies dedicated to disentangling variation due to common history from variation due to ongoing processes emerged in the decade following Clutton-Brock and Harvey's pioneering paper.[17] The first phase of this program's development culminated with the publication of a book by Harvey and Pagel (1991), which provided the launch point for an explosion of comparative studies using statistical methods during the last decade of the twentieth century and continuing today. Because this research program relies on statistical analysis, it has been referred to as "the comparative method," in reference to the historical roots of the statistical approach in the late nineteenth century. There is a constellation of methods available to researchers interested in this type of comparative approach,[18] but recent studies suggest that many of these

---

16. Winkler 2000. All three are represented in Martins 1996—see review by Brooks (1999).

17. For example, Clutton-Brock and Harvey 1984; Gittleman 1981; Harvey and Mace 1982; Ridley 1983; Felsenstein 1984, 1985a; Cheverud et al. 1985; Dunham and Miles 1985; and Harvey and Clutton-Brock 1985.

18. For example, Bell 1989; Garland et al. 1991, 1992; Harvey and Pagel 1991; Harvey and Purvis 1991; Lynch 1991; Martins and Garland 1991; Garland 1992; Gittleman and Luh 1992; Herrera 1992; Miles and Dunham 1992, 1993; Møller and Birkhead 1992; Pagel 1992, 1994a,b, 1999; Wickman 1992; Armstrong and Westoby 1993; Kelly and Purvis 1993; Cotgreave 1994; Peat and Fitter 1994; Fitter 1995; Harvey et. al. 1995a,b; Jordano 1995; Karlsson 1995; Leishman et al. 1995; Lord et al. 1995; McPeek 1995; Purvis et al. 1995; Purvis and Rambaut 1995; Read and Nee 1995; Westoby et al. 1995a,b; Purvis 1996; Ricklefs and Starck 1996; Martins and Hansen 1997; Ptacek and Travis 1998; and Ackerly 2000.

"methods" are variants of each other (Garland et al. 1999; Garland and Ives 2000).

The second research program in comparative biology originated in systematics. It is focused on assessing the significance of traits that evolve so slowly that they are relatively fixed in more than one species and can be treated as qualitative variables.[19] In this research program, as with the comparative method, it is important to distinguish between similar traits derived from a common ancestor (homologous traits) and those which have evolved more than once independently (homoplasious traits). This group of researchers uses a comparative approach in order to explain trait correlations. Sequences of evolutionary transformations in traits become correlated in space and time through inheritance from common ancestors. Various traits and trait combinations can co-occur on the same branch of a phylogeny, or one can precede the other. Each type of pattern can imply different questions to be asked and different explanations to be offered; for example, Do these traits evolve together? versus, Must one trait evolve in order for the other to emerge? Finding such patterns may be sufficient to refute some hypotheses and persuasive enough to encourage us to design experiments to test others.

This brings us to the third type of comparative biology, which is the focus of this book.

THE EMERGENCE OF HISTORICAL ECOLOGY AS A RESEARCH PROGRAM IN COMPARATIVE BIOLOGY

By the early 1970s some researchers had begun to focus their attention on macroevolutionary patterns of diversity. Ross (1972a,b) was particularly interested in explaining these patterns for a variety of groups within the most diverse taxonomic class on this planet, the insects. Based upon his discovery that approximately only one out of every 30 speciation events in these groups was correlated with some form of ecological diversification, Ross suggested that ecological change was consistent with, but much less frequent than, phylogenetic diversification. Furthermore, since he could not uncover any predictable patterns to explain the shifts that did occur, he proposed that ecological change manifests itself as a biological "uncertainty principle" in evolution. Within a few years, Ross's insights were corroborated by other studies. Boucot (1975a,b, 1981, 1982, 1983, 1990) reported that the majority of ecological changes leaving some trace in the fossil record occurred out of time

---

19. For example, Coddington 1988, 1990, 1992, 1994; Maddison 1990, 1994; Sillén-Tullberg 1993; Armbruster 1994; Pagel 1994b; Fitter 1995; Read and Nee 1995; Werdelin and Tullberg 1995; Crespi 1996; Hart et al. 1997; and Maddison and Maddison 2000.

phase with periods of phylogenetic diversification. Like Ross, he concluded that ecological change lags behind morphological and phylogenetic diversification, or "evolution takes place in an ecological vacuum" (Boucot 1983:1). These pioneering efforts were perhaps the first major empirical studies demonstrating that the relative importance of the nature of the organism over the nature of the conditions reveals itself on macroevolutionary time scales.

Brooks (1985) consolidated the research of authors such as Ross and Boucot, as well as the results from his own studies with parasitic organisms, into a research program he called "historical ecology." Initially, historical ecology was concerned with studying macroevolutionary components of multispecies ecological associations, especially host-parasite systems. In a previous book (Brooks and McLennan 1991), we attempted to expand the boundaries of historical ecology to include two general evolutionary processes, speciation and adaptation, and to explore the macroevolutionary effects of these processes in the production of both clades and multispecies ecological associations. To the question, What is organizing biological diversity? we answered that in part it is the cohesive influences of persistent ancestral traits. For example, some adaptations that originated in the past may become fixed and inherited relatively unchanged for long periods of time. These slowly evolving traits come to characterize genealogical groups of species, or clades, and may influence the scope of the adaptively possible at every point in the evolution of those clades.[20]

 Humans celebrate their history because they recognize that it has had a great impact on their present existence. These histories include heroes and heroic events, and it is the job of the historical ecologist to find and to explain the heroic episodes in the history of life on this planet.

Such evolutionary events may be historically unique, or at least so rare as to preclude statistical assessment. Fortunately, "beyond statistical assessment" does not mean "beyond explanation." This message struck a resonant chord with many biologists, resulting in an explosion of publications spanning so many different topics that the term "historical ecology" is no longer a sufficient descriptor of the field.[21]

20. Brundin 1972; Riedl 1978; Lauder 1982a; Brooks and Wiley 1986, 1988; and Wanntorp et al. 1990.

21. For example, McLennan and Brooks 1991, 1993; Maurer 1991; Vane-Wright and Smith 1991; Behrensmeyer et al. 1992; Mayden 1992a; Zrzavy 1992; Brooks and McLennan 1993b; Miles and Dunham 1993; Núñez-Farfán and Cordero 1993; Ricklefs and

AND NOW FOR SOMETHING COMPLETELY DIFFERENT: HISTORY AS ERROR, HOMOPLASY AS ERROR

Of course, not everyone is ecstatic over efforts to reintroduce phylogenetic information into biology. Some contemporary ecologists, following MacArthur's dictum, view phylogenetic history as an error term in a statistical model (e.g., Fitter 1995; Westoby et al. 1995a,b), one that can be "corrected" using comparative statistical approaches. Harvey, Read, and Nee (1995a,b), however, pointed out that the comparative method is not an attempt to remove historical information that may produce ambiguous explanations, but rather, "is a constructive attempt to extract information from non-experimental data which is riddled with non-independence" (1995a:535). Ackerly and Donoghue (1995) also emphasized that phylogeny per se is neither a "correction" nor a "constraint." It is a record (becoming more complete and more explicit every day) of the major events and transitions in the evolution of life on this planet. In a similar vein, some systematists have adopted the view that homoplasy is simply observational error (Kluge 1999), mistakes made by the observer that can be "corrected" once a phylogenetic tree has been generated. Both the systematically and ecologically based "correction" factors eliminate the effects of some evolutionary processes and thus place limitations the kinds of questions we ask and the kinds of evolutionary mechanisms we can seek to discover and understand. We disagree with both these perspectives because we believe

 Evolution without historical influences would be maximally complex, but this does not mean that evolutionary explanations including historical influences will necessarily be simple.

Therefore,

"we should abstain from issuing prohibitions that draw limits to the possibilities of research" (Popper 1968:250).

Given the small proportion of the world's biodiversity that has been documented, many of our generalizations are based on few examples indeed. For that reason it is important to include all available evidence

Schluter 1993; Shettleworth 1993; Skelton 1993a; Allmon 1994; Amorim 1994, 1997; Eggleton and Vane-Wright 1994; Grandcolas et al. 1994; Grande and Rieppel 1994; Greene 1994b, 1996; Hall 1994; McLennan 1994; Wainwright and Reilly 1994; Brooks et al. 1995; Harvey, Brown, and Smith 1995; Wagner and Funk 1995; Jablonski et al. 1996; Martins 1996; Rose and Lauder 1996; Sanderson and Hufford 1996; Begun et al. 1997; Kendrick and Crane 1997; Pérez Ponce de León et al. 1997; Grandcolas 1998; Mottishaw et al. 1999; Weller and Sakai 1999; and Hernandez et al. 2000.

and all available taxa in any analysis. Modern comparative biology, of which the research program we discuss in this book is a part, is based on the proposition that what we are seeking, what we should be comparing, and what questions we should be asking are not evident without a phylogeny. Phylogenetic systematics provides a logical means of discovering the unknown. This is especially true for slowly evolving traits, recurring traits, and rare evolutionary events.

We include in this book ways to begin evaluating these discoveries. All of the evaluation methods presented are connected by four guiding principles: (1) phylogenetic trees are necessary but rarely sufficient for explaining evolutionary origins and diversification; (2) we must always be responsible for well-formulated questions; (3) we must always be responsible for the quality of the data used in any level of our analyses, from generating phylogenetic hypotheses to testing general theories; and (4) everything we learn implies yet more cycles of discovery and evaluation (Kluge 1989, 1991, 1997, 1998a,b, 1999).

Phylogenies are not the end of the story, merely the end of the beginning, thus we have adopted the image of a voyage of discovery in our presentation of this research program. Like all such voyages, this one will require courage, both from the voyagers and from those who finance them. We do not promise that the journey will be easy, fast, or inexpensive. Nor do we guarantee that it will always take us closer to the "truth" in a direct manner. We do, however, believe that it will be exciting for those who embark.

Chapter 2

Tools for the Voyage

The affinities of all beings of the same class have sometimes been represented by a great tree. I believe this simile largely speaks the truth. The green and budding twigs may represent existing species; and those produced during former years may represent the long succession of extinct species. At each period of growth all the growing twigs have tried to branch out on all sides, and to overtop and kill the surrounding twigs and branches, in the same manner as species and groups of species have at all times overmastered other species in the great battle for life. The limbs divided into great branches, and these into lesser and lesser branches, were themselves once, when the tree was young, budding twigs, and this connection of the former and present buds by ramifying branches may well represent the classification of all extinct and living species in groups subordinate to groups. Of the many twigs which flourished when the tree was a mere bush, only two or three, now grown into great branches, yet survive and bear the other branches; so with the species which lived during long-past geological periods, very few have left living and modified descendants. From the first growth of the tree, many a limb and branch has decayed and dropped off; and these fallen branches of various sizes may represent those whole orders, families, and genera which have now no living representatives, and which are known to us only in a fossil state. As we here and there see a thin straggling branch springing from a fork low down in a tree, and which by some chance has been favoured and is still alive on its summit, so we occasionally see an animal like the Ornithorhynchus or Lepidosiren, which in some small degree connects by its affinities two large branches of life, and which has apparently been saved from fatal competition by having inhabited a protected station. As buds give rise by growth to fresh buds, and these, if vigorous, branch out and overtop on all sides many a feebler branch, so by generation I believe it has been with the Great Tree of Life, which fills with its dead and broken branches the crust of the earth, and covers the surface with its ever-branching and beautiful ramifications.

Darwin 1872:131–132

With these words, Darwin introduced the only illustration ever to appear in any edition of *Origin of Species*, a phylogenetic tree.

Darwin did not consider this to be merely the most appropriate pictorial metaphor for evolution, but envisioned that such depictions could be crucial elements of biological investigation and explanation: "As it is difficult to show the blood-relationship between the numerous kindred of any ancient and noble family even by the aid of genealogical trees, and almost impossible to do so without this aid, we can understand the extraordinary difficulty which naturalists have experienced in describing, without the aid of a diagram, the various affinities which they perceive between the living and extinct members of the same great natural class" (Darwin 1872:410–411).

In order to put Darwin's suggestion into practice, we must be able to produce objective hypotheses of phylogenetic relationships. Biologists have always used "similarity" as an indicator of "relationship"; however, as we discussed in the previous chapter, there are different types of similarity. Hennig (1950, 1966a) codified Darwinian-inspired approaches to reconstructing phylogeny by emphasizing that the primary goal of systematics should be the delineation of a special type of similarity (homology) which, when used to reconstruct relationships, would provide a general reference system for comparative biology (Wiley 1981, 1986a; de Queiroz 1988). Hennig reasoned that this system should be based on reconstructing phylogenetic relationships from shared homologous traits because *all homologies covary with each other and with phylogeny.* Other types of similarity (homoplasy), although evolutionarily interesting, are not phylogenetically informative because *homoplasies do not covary with phylogeny and covary with each other only under special circumstances.* Evolution, with its underlying assumption of the preeminence of genealogical ties, has been the unifying biological principle since the emergence of Darwinian ideas, so Hennig's perspective would seem conceptually unobjectionable. After all, "Homologous parts tend to vary in the same manner, and homologous parts tend to cohere" (Darwin 1872:158); homologies tend to cohere because replication rates are higher than mutation rates, because different parts of organisms are developmentally integrated, and because reproduction transmits whole organisms, even if parts of them are evolving at different rates. Response to this new perspective, however, was reserved. The reservation originated, in part, from a long-standing problem with the relationship between homology and phylogeny. Specifically, if homology is both defined by phylogeny and required to reconstruct phylogeny, then a researcher needs to adopt an Orwellian doublethink strategy in order to "know the phylogeny, obtain the homologies, and build the phylogeny." This relationship makes traditional phylogenetic reconstruction irreducibly circular.

Hennig's solution to this problem stemmed from his belief that genealogical influences in evolution are so pronounced that homologous characters will outnumber covarying homoplasious characters within any given group, a view shared by Darwin. Hennig suggested that researchers begin with the assumption that all characters which conform to *nonphylogenetic criteria for homology* are, in fact, evolutionarily homologous.[1] This assumption will sometimes lead to the incorrect, and initially undetectable, identification of homoplasious traits as homologues. When a phylogeny is reconstructed by grouping taxa according to their shared homologies, these misidentifications will be revealed because the homoplasious characters will not covary with the plurality of the other characters. These traits can then be recognized, using the phylogenetic hypothesis, as homoplasies.

The distinction between the non-Hennigian and Hennigian approaches is a subtle but vital one. Consider the following example. Suppose, while describing the behavior of four different avian taxa, you notice that all of the males perform the same type of mock-preen display during courtship. A non-Hennigian systematist would say, "Since this display looks the same in these four taxa *and* is performed by different members of a closely related group ('birds'), it is a homologous trait. We can use this trait to assess the phylogenetic relationships among these birds." A Hennigian systematist would say, "Since this display looks the same in these four taxa, it is the same (homologous). We can use this trait to assess the phylogenetic relationships among these organisms." In the first case, homology is assumed because of similarity among characters, *coupled with* presumed relatedness among the taxa bearing the characters. Hence, there is an underlying assumption of prior knowledge of phylogeny. In the second case homology is assumed solely on the basis of similarity among characters (the Wiley criterion "if it looks like a duck and quacks, it's a duck"). Hence, the approach advocated by Hennig is not circular because *homologies, which indicate phylogenetic relationships, are determined without a priori reference to a phylogeny while homoplasies, which are inconsistent with phylogeny, are determined as such by reference to the phylogeny.*

Hennig recognized that there are three types of homologous characters: (1) shared general characters, which identify a collection of taxa as a group; (2) shared special characters, which indicate relationships among taxa within the group; and (3) unique characters, which identify

1. Such criteria include things like similarity of position or similarity of structure, to name just a few ways in which two traits can appear to be "similar." For a detailed list, see Remane 1956, 1961; also de Jong 1980; Wiley 1981; Patterson 1982; Roth 1991; Haszprunar 1992; Rieppel 1992; McKitrick 1994; and Brower and Schawaroch 1996.

particular taxa within the group. In addition to these three categories of homology there is a separate category, homoplasy or "false homology," which tells us nothing about relationships among taxa. Since only shared special homologies denote particular phylogenetic relationships within a study group, characters must be assigned to one of the three homology categories before their usefulness in a phylogenetic study can be determined. Once again, the risk of circularity is high: "If two taxa are related, then a character that they share in common is a shared special homology; therefore, this character can be used to determine if the taxa are related." Hennig suggested that this determination be made by comparing the state of each character in the study group to the state of the same characters in one or more species outside the study group (outgroups). In this way, *each character is independently assigned a particular homology status (general, special, or unique) depending upon properties of species for which the phylogenetic relationships are not being assessed (the outgroups).* This "outgroup comparison" distinguishes among traits that are shared between the outgroups and at least some members of the study group (shared general traits), traits that are restricted to some members of the study group (shared special traits), and traits unique to single members of the study group (unique traits).

In the remainder of this chapter we will present a detailed discussion of the basic methods involved in phylogenetic systematics. More advanced applications of phylogenetic systematic methodology pertinent to this research program will be introduced in later chapters.

TERMINOLOGY

There is a perception that researchers must learn an inordinate number of new and specialized words before they can complete their initiation into phylogenetics. This apprehension is based, in part, on the incorporation of numerous old terms such as monophyly, ancestor, homology, and homoplasy into the field of phylogenetic systematics. Since most evolutionary biologists are familiar with these terms, this task is not really so daunting after all. There are of course some new words to learn, apomorphy and plesiomorphy being the most important and, for many researchers, the most perplexing. Hopefully by the end of this chapter and definitely by the end of this book, these words will no longer be fraught with such mystical connotations.

There are two reasons for adopting the specific terms, both the old and the new, that we will use in this book. First, any empirical field requires unambiguous terminology for constructing hypotheses and explanations. Second, because evolution is a genealogical process, it is critical that systematic data be incorporated explicitly into evolutionary explanations.

Groups of Organisms

A **taxon** is any named group of organisms. The relative position (or rank) of a taxon in the Linnaean hierarchical system of classification is indicated by the use of categories (e.g., "family," "genus"). *You should not confuse the rank of a taxon with its reality as a group.* For example, the taxon "Aves" includes exactly the same organisms whether it is ranked as a class, an order, or a family. A **natural taxon** is a group of organisms that exists as a result of evolutionary processes. There are two kinds of natural taxa: species and monophyletic groups. A **species** is a lineage, a collection of organisms that share a unique evolutionary history and are held together by the cohesive forces of reproduction and development. Every species has a unique historical origin, either by **cladogenesis**, the division of one ancestral species into new daughter species (fig. 2.1a), or **reticulate speciation**, the formation of a new species through the hybridization of two ancestral species (fig. 2.1b).

A **monophyletic group**, or **clade**, is a group of taxa encompassing an ancestral species and all of its descendants (fig. 2.2). Members of a monophyletic group are bound together by a common ancestral relationship that they do not share with any other taxa. Each monophyletic group begins as a single species, the ancestor of all subsequent members of the clade. Because of the nature of speciation, groups of species cannot give rise to other groups or to a single species. In brief, the various processes involved in speciation allow one species to give rise to another (or two species to produce a new taxon of hybrid origin), but there are

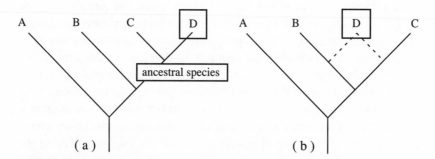

Figure 2.1. The endpoints of speciation. *Letters* = species. (*a*) Cladogenesis: species D is produced by the division of an ancestral species into two new daughter species, C and D. (*b*) Reticulate speciation: species B and C hybridize, forming species D.

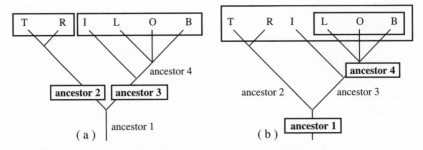

Figure 2.2. Monophyletic groups on a phylogenetic tree. *Letters* = species. Species within boxes are part of the following monophyletic groups: (*a*) ancestor 2 and all its descendants (ancestor 2 + species T + R); ancestor 3 and all its descendants (ancestor 3 + species I + ancestor 4 + species L + O + B); (*b*) ancestor 1 and all its descendants (ancestors 1 + 2 + 3 + 4 + species T + R + I + L + O + B); ancestor 4 and all its descendants (ancestor 4 + species L + O + B).

no documented processes that can produce a genus from a genus ("geni-ation") or a family from a family ("familiation"). Species are the largest units of taxic evolution; they are real, evolutionary entities (we will dis-cuss this in more detail in chapter 3). Higher-level categories, on the other hand, are artifacts of our propensity to classify our surroundings. As figments of our collective imaginations, supraspecific taxa have no evolutionary substance, whereas species, and the array of speciation pro-cesses that form them, lie at the very heart of "modification with de-scent," or evolution.

An **artificial taxon** represents an incomplete or invalid evolutionary unit. (We produce artificial taxa, evolution does not.) In a **paraphyletic group**, one or more descendants of an ancestor are excluded from the group, making the grouping an incomplete evolutionary unit (fig. 2.3). For example, most researchers think that *Homo sapiens* shares a common ancestor with the African great apes (chimpanzees and gorillas). A group within a classification comprising the African great apes plus orangutans and gibbons (the Asian great apes) while excluding humans would be paraphyletic. In a **polyphyletic group**, taxa that are separated from each other by more than two ancestors are placed together without including all the descendants of their common ancestor (fig. 2.4). Since the rela-tionship between the two taxa is so distant, this type of grouping misrep-resents the evolutionary relationships that arose from speciation events following the divergence of the shared common ancestor, making the grouping an invalid evolutionary unit. A classic example of a polyphy-letic group would be a classification that placed mammals and birds alone together in the same taxon.

An **ingroup** is any group of theoretically closely related organisms of

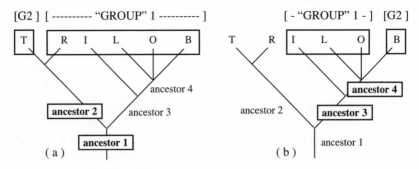

Figure 2.3. Paraphyletic groups on a phylogenetic tree. *Letters* = species. (*a*) Two groups have been distinguished: group 1 includes ancestors 1 + 2 + 3 + 4 + species R + I + L + O + B; group 2 contains species T. Species T should be included in group 1 because it shares ancestors 1 and 2 with that group. (*b*) Two groups have been distinguished: group 1 includes ancestors 3 + 4 + species I + L + O; group 2 contains species B. Species B should be included in group 1 because it shares ancestors 3 and 4 with that group.

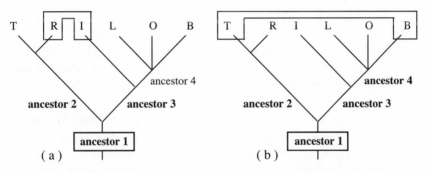

Figure 2.4. Polyphyletic groups on a phylogenetic tree. *Letters* = species. (*a*) Species R and I are grouped together because they "look the same," even though they do not share a recent common ancestor. You have to count back through two ancestors (ancestors 2 and 3) before arriving at an ancestor the taxa share (ancestor 1). (*b*) Species T and B are placed together, forcing you to count back through three ancestors before arriving at a common link between the two taxa (ancestor 1).

interest to an investigator. *Choice of an ingroup is constrained only by the rule that it must contain more than two species,* because it is impossible to determine phylogenetic relationships for only two taxa. For example, an investigator studying the genealogical relationships between tigers and gorillas can say only that, by virtue of their status as biological entities, they are related. However, by adding lions to the picture, the investigator increases the degrees of freedom from zero to three because there are four possible hypotheses of relationships: (1) they are all equally related to one another; (2) gorillas are more closely related to tigers; (3) gorillas are more closely related to lions; or (4) lions are more closely related to tigers.

A **sister group** is the taxon most closely related genealogically to the ingroup. For example, Old World monkeys are the sister group of the great apes. The ancestor of the ingroup cannot be its sister because it is a member of the ingroup. An **outgroup** is any group used for comparative purposes in a phylogenetic analysis. *Choice of an outgroup is constrained by the rule that it cannot contain any members of the ingroup.* Because genealogy is so important in evolution, it is not surprising that the most important outgroup in any study is the sister group to the taxa being investigated. So, if we were interested in studying the phylogenetic relationships within the great apes, we would use the Old World monkeys as our outgroup.

Relationships of Taxa

In phylogenetic systematics the term **relationship** refers strictly to connections based on genealogy. In other systems "relationship" may be equated with "similarity" without evolutionary implications or with the implication that taxa which are more similar to each other are more closely related evolutionarily. The latter conclusion, based on the a priori assumption that things which "look the same are the same," can lead to the recognition of polyphyletic groups because it lacks a rigorous methodology to test the validity of the assumption. *Degree of similarity is never equated with degree of relatedness in the phylogenetic system.* **Genealogical descent** at the taxic, rather than the individual, level is based on the proposition that ancestral species give rise to daughter species through speciation. A **phylogenetic tree**, or **cladogram**, is a branching diagram depicting the sequence of speciation events within a group. As a graphical representation of genealogy, *a phylogenetic tree is a hypothesis of the genealogical relationships among taxa.* Since phylogenetic trees are hypotheses and not "facts," they are dependent upon both the *quality* and *quantity* of data which support them. A tree is composed of three main parts (fig. 2.5): (1) a **branch point**, or **node**, sometimes highlighted with a circle, representing an individual speciation event; (2) a **branch**, the line connecting a branch point to a terminal taxon, representing the terminal taxon; and (3) an **internode**, the line connecting two speciation events, representing an ancestral species. The internode at the bottom of the tree is given the special term **root**.

 If you rotate the branches about a node on a phylogenetic tree, do you change the implied phylogenetic relationships? No. Phylogenetic trees can be rotated at their nodes without changing the hypothesis of relationships as long as you do not pull a taxon from the node and move it somewhere else. (Think of the node as a swivel.) For example, in figure 2.6, panels a, b, and c are the same phylogenetic tree. Panel d, however, proposes a different set of relationships and thus is not the product of simple nodal rotation.

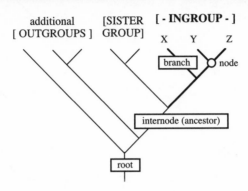

Figure 2.5. Components of a phylogenetic tree. On this diagram, the internode is the common ancestor of the sister group and the ingroup (highlighted with bold lines), the node represents the speciation event that gave rise to species Y and species Z, and the branch refers to species X. In all, there are seven nodes, six internodes, and eight branches on this particular cladogram.

Classifications

A **natural classification** contains only monophyletic groups and thus is consistent with the phylogenetic (evolutionary) relationships of the organisms. In other words, the genealogical relationships depicted on the phylogenetic tree can be reconstructed from the classification scheme. An **artificial classification** contains one or more paraphyletic or polyphyletic groups, rendering it inconsistent with the phylogeny of the organisms. In such cases the phylogenetic tree cannot be wholly reconstructed from the classification scheme. An **arrangement** is a classification of a group whose phylogenetic relationships have not yet been delineated, so its classification type, natural or artificial, is unknown. The overwhelming majority of current classifications are arrangements, serving as necessary but interim vehicles for classifying organisms until their phylogenetic relationships have been determined. Neither artificial classifications nor arrangements have been constructed via a rigorous, phylogenetic methodology. *It is therefore inappropriate to convert such classification schemes into phylogenetic trees, because you cannot assume a priori that taxonomic relationships are consistent with phylogenetic relationships.*

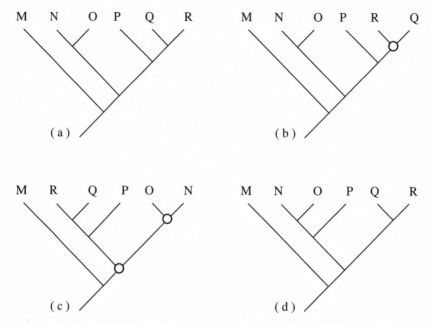

Figure 2.6. Rotating around nodes. The trees shown in (*a*)–(*c*) are identical, they have just been rotated around particular nodes (depicted as *open circles*). The tree shown in (*d*) is not the same as that shown in (*a*)–(*c*) because species P has been plucked from the node joining P + Q + R and moved to another branch. If you find this difficult to visualize, try working through the example by constructing a phylogenetic tree using toothpicks for branches, jujubes for nodes, and jelly beans for terminal taxa.

The **category** of a taxon indicates its relative place (or rank) in the hierarchy of the classification. The Linnaean hierarchy is the most common taxonomic classification system. Within this scheme, the formation of category names occupying specific places in the hierarchy is governed by rules contained in various codes of nomenclature. It is important to remember that the rank of a taxon does not affect its status in the phylogenetic system. *All monophyletic taxa are equally valid and all paraphyletic and polyphyletic taxa are equally misleading to the evolutionary biologist.*

Features of Organisms

A **character** is any observable part, or attribute, of an organism. A character has two evolutionary options: it can either remain the same and be passed on genetically from ancestor to descendant unaltered, or it can change in one species and be transmitted in the new form to its descendants. If a character changes, it is transformed from its existing (ancestral) condition into an **evolutionary novelty.** Two characters found in different taxa thus can be assigned homologous status because they are either the same character that is found in their common ancestor or they are different characters that are genealogically linked by passing through the transformation from an ancestral condition to a novel condition. The ancestral character is termed the **plesiomorphic character** (*plesio* = close to the stem; *morpho* = shape), while the descendant character is termed the **apomorphic character** (*apo* = away from the stem; *morpho* = shape). A **homoplasy** is also a character shared among taxa. Although it may appear to be "the same" in those taxa, it does not meet either of the two preceding criteria of homology, being instead a result of convergent or parallel evolution.

 Are there primitive and advanced species, just as there are primitive and advanced characters? No. Not all characters evolve at the same rate and to the same degree in different lineages. As a consequence, all species are mosaics of plesiomorphic and apomorphic traits, so it is inappropriate to speak of plesiomorphic (or "primitive") and apomorphic (or "advanced") species. You can refer to species by their relative position in a phylogenetic tree (for example, a species might be "old" or "young," basal or derived). In general, we also recommend the terms "plesiomorphic" (or ancestral) and "apomorphic" (or derived) to describe traits, rather than "primitive" and "advanced," which carry implicit value judgments.

A **transformation series** is a collection of homologous character states: two homologous character states produce a binary transformation series, while three or more homologous character states create a multicharacter or multistate transformation series. An **ordered transformation series** is a hypothesis of the particular pathway a character traveled during its evolutionary modification(s); however, without further information we cannot tell which direction the character moved along the pathway. If a transformation series is **unordered,** there are several possible routes to explain the character changes. Information about the evolutionary direction of character change is provided by polarization. For a **polarized transformation series,** the relative apomorphic and plesiomorphic status of character states has been determined, so we have a hypothesis of which character state represents the ancestral condition and which the derived condition. And finally, in an **unpolarized transformation series,** the direction of character evolution remains unspecified. There are thus four possible types of character transformation series based on the amount of information available concerning the pathway

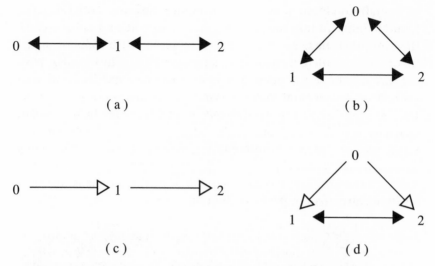

Figure 2.7. Types of character transformation series. (*a*) Ordered, unpolarized: information about pathway, no information about direction. Character modification may be either 0 to 1 to 2 or 2 to 1 to 0. (*b*) Unordered, unpolarized: no information about either direction or pathway. (*c*) Ordered, polarized: information about both pathway and direction. (*d*) Unordered, polarized: information about direction, no information about pathway. Zero is the plesiomorphic state, but we do not know whether character modification moved from 0 to 1 to 2, from 0 to 2 to 1, or whether 1 and 2 arose independently from 0.

and direction of evolutionary change: (1) ordered, unpolarized (fig. 2.7a); (2) unordered, unpolarized (fig. 2.7b); (3) ordered, polarized (fig. 2.7c); and (4) unordered, polarized (fig. 2.7d). Not surprisingly, ordering and polarizing multistate transformation series can become very complicated. Binary characters are much simpler because they are all automatically ordered (but not necessarily polarized).

Character argumentation is the logical process of determining which characters in a transformation series are plesiomorphic and which are apomorphic based on a priori deductive arguments using outgroup comparison. Frequently termed "polarizing the character states," this is the pivotal process in phylogenetic systematics. **Polarity** refers to the plesiomorphic or apomorphic status of each character state. **Character optimization** consists of a posteriori arguments concerning how particular character states should be polarized given a particular tree topology. (This is not a part of tree building, but we will return to this process in chapter 5.)

Phylogeneticists routinely use computer-assisted analysis of their data (the first quantitative implementation was proposed by Kluge and Farris (1969; see also Farris et al. 1970). Such analyses require the production of a **data matrix** composed of character transformation series and taxa. Each character in the matrix is assigned a numerical **code.** By convention, the code "0" is usually assigned to the plesiomorphic character state, while "1" is reserved for the apomorphic character state of a transformation series, if the polarity of that series has been determined (hypothesized) by outgroup comparisons. You can assign any code for the plesiomorphic and apomorphic states of your characters as long as you are consistent. If a transformation series consists of more than two character states the situation becomes more complex, as we will discuss later in this chapter.

HENNIG ARGUMENTATION: BUILDING TREES

Phylogeneticists assume that all organisms, both living and extinct, occupy a unique position on one phylogenetic tree rooted at the origin of life on this planet. Since characters are features of organisms, they each should have a place on this tree corresponding to the point at which they arose during evolutionary history. So, ultimately, we are seeking to reconstruct phylogenetic trees in which the taxa are placed in correct genealogical position and the character states are placed where they arose. For example, consider the tree for some major land-plant groups (fig. 2.8a). This diagram provides us with a hypothesis of the genealogical ties between plant groups (i.e., tracheophytes and mosses are more closely related to each other than either is to hornworts: Mishler and Churchill 1984). It also provides us with a hypothesis of the evolution of specific characters. Notice that characters are depicted on the phylogenetic tree at their hypothesized point of origin. This is the shorthand notation for stating that "the ancestor in which character x arose, and

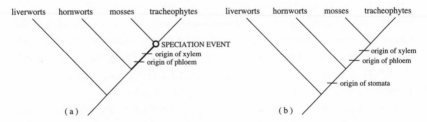

Figure 2.8.   Tracking character evolution on a phylogenetic tree for four groups of plants. (*a*) Xylem and phloem originated in the ancestor (highlighted with a bold line) of the mosses and tracheophytes. (*b*) Xylem and stomata are indeed evolutionarily independent traits; they originate on different branches of the tree. Although xylem and phloem appear on the same branch, we have additional developmental and functional evidence indicating that they are also evolutionarily independent.

all of its descendants, display character x, unless it is modified again later in the evolutionary history of the group." For example, the phylogenetic hypothesis states that xylem and phloem originated in the common ancestor of mosses and tracheophytes. In other words, these characters arose in an ancestral species between the time of origin of the hornworts and the speciation event that produced the mosses and tracheophytes. Since xylem and phloem are postulated to be homologous in all plants bearing these tissues, each trait appears only once, at the level on the tree where it is thought to have arisen as an evolutionary novelty.

Now, even without a phylogenetic tree we might suspect that all plants bearing xylem and phloem shared a common, unique ancestor, because both characters are morphologically and developmentally similar in these plants. In the phylogenetic system such detailed similarity is always considered as a priori evidence that the characters are homologous. This concept is so important that it has been termed

> **Hennig's Auxiliary Principle:** Never presume convergent or parallel evolution; always presume homology in the absence of contrary evidence.

This principle is a powerful one. Without it we could assert that all characteristics "probably" arose multiple times by convergent or parallel evolution. For example, how many times do you think the "backbone" has evolved? We say once, you say twice, another person says 42 times. Can we resolve the debate, or are all alternatives "equally valid"? Using Hennig's auxiliary principle, we can begin to address the question objectively by reasoning in the following way: all backbones in all species with backbones look remarkably similar, develop from the same embryonic tissue, and serve a similar function; therefore, all backbones are homologous. Notice that we have just used a series of nongenealogical criteria to determine homology; you don't have to know anything about the species displaying the characteristics, just some things about the characteristics themselves. The statement "all backbones are homologous" is our a priori hypothesis, it is not a statement of "fact" or "knowledge." This hypothesis will be tested during the process of Hennigian argumentation (which we will come to presently).

The empirical observation that replication rates are always higher than mutation rates justifies using this principle as the most robust (defensible) a priori criterion. Of course, just because we use Hennig's auxiliary principle does not mean we necessarily believe that homoplasies are rare or nonexistent. Homoplasy is a fact of nature and rather common in some groups. If, however, all character evolution has been driven by rampant homoplasy, then we will never be able to recover evidence of phylogenetic relationships. This is a fascinating and testable prediction. Oddly, though, the only way to test it is by attempting to reconstruct phylogenetic relationships. If rampant homoplasy overwhelms homologies, then we will have nothing but problems with that reconstruction. If, however, some similarities are in fact due to homology, not homoplasy, then we might just find that we are able to reconstruct robust phylogenetic trees. In either case, in order to pinpoint homoplasy without invoking ad hoc assumptions, you must first have a tree, and without Hennig's auxiliary principle you will never get out of the starting gate. So, whatever your perspective about the "true path" of evolution, you can evaluate that perspective only if you begin with the hypothesis that similarity = homology. This includes "homology as identity" (i.e., a finger is a finger) and "homology as evolutionary transformation" (i.e., a bat's wing is a mammalian forelimb is a coelacanth's pectoral fin). Hennig did not consider his auxiliary principle to be problematical for Darwinians. After all,

> it would in most cases be extremely rash to attribute to convergence a close and general similarity of structure in the modified descendants of widely distinct forms. The shape of a crystal is determined solely by the molecular forces and it is not surprising that dissimilar substances should sometimes assume the same form; but with organic beings we should bear in mind that the form of each depends on an infinitude of complex relations, namely on the variations that have arisen, these being due to causes far too intricate to be followed out,—on the nature of the variations that have been preserved or selected, and this depends on the surrounding physical conditions, and in a still higher degree on the surrounding organisms with which each being has come into competition,—and lastly, on inheritance (in itself a fluctuating element) from innumerable progenitors, all of which had their forms determined through equally complex relations. It is incredible that the descendants of two organisms, which had originally differed in a marked manner, should ever afterwards converge so closely as to lead to a near approach to identity throughout their whole organisation. If this had occurred, we should meet with the same form, independent of genetic connection, recurring in widely separated geological formations; and the balance of evidence is opposed to any such admission. (Darwin 1872:127–128)

When systematists accumulate character information for phylogenetic analysis (usually in the form of a data matrix), they begin with the presumption that all characters are evolving independently of one another, even though they are carried in the same genealogical flow of information, a practice stated formally as

Kluge's Auxiliary Principle: Always presume character independence in the absence of evidence to the contrary.

This is the phylogenetic equivalent of assuming panmixis or equal rates of immigration and emigration when using the Hardy-Weinberg equilibrium equation. Kluge's auxiliary principle does not assert that two characters are in fact independent, only that if you have no evidence other than the basic character description, assessments of dependence or independence cannot be made without a phylogeny.

This is a difficult point. Imagine that you are investigating the phylogenetic relationships within a group of monkeys. You have noticed that some species have bumps on the knuckles of their first two fingers, while other species have no bumps at all. Do you score this as one character, "bumps on knuckles of first two fingers: present or absent," or as two characters, "bump on knuckle of first finger: present or absent" and "bump on knuckle of second finger: present or absent"? In the absence of any other data, Kluge's auxiliary principle says, Score these characters separately and see what happens. Remember that a transformation series only depicts the evolutionary sequence of *homologous* character states. Therefore, you must place the traits in different transformation series because there is no evidence that the bump on the first finger is homologous with the bump on the second finger. You might envision a transformation series for the first finger bump being something like "bump on first finger is: small, medium, or large." There also is no evidence that the evolution of a bump on the first finger is, by necessity, coupled with the evolution of a bump on the second finger. When you analyze the data matrix phylogenetically, you will discover one of two patterns for the knuckle-bump characters: The bumps might originate on different branches of the tree, supporting the initial hypothesis of independence. Alternatively, the bumps might originate on the same branch of the tree, in which case there is still no evidence that they are acting as a single evolutionary unit. *There is, however, empirical support for questioning their independence.* The phylogenetic analysis has now revealed as much as it can. In order to investigate the problem further, you need information about the underlying genetic and developmental control of finger bumps. If those data indicate that the bumps are indeed acting as a single evolutionary unit, then you must go back to your phylogenetic analysis and recode the two characters as a single transformation series, "bumps on first and second finger: present or absent."

So, although it may sound paradoxical, the only way to demonstrate whether two or more traits are indeed evolutionarily dependent is to begin with the hypothesis that they are independent, include them in a phylogenetic analysis with as many other characters as possible, then test the hypothesis of independence a posteriori. In fact, if much of evolutionary diversification involves breaking old genetic linkage groups and forming new ones, then even finding that two or more traits are genetically linked today does not mean they have always been so in the past, so they still need to be treated as independent in a phylogenetic analysis.

With this in mind, let us return to our plant phylogeny and add another character, the presence or absence of stomata (fig. 2.8b). Stomata are epidermal pores found on some leaves, stems, floral organs, and fruits that play a role in the transfer of gases and water between the plant and the environment. Xylem also plays a role in water transportation but is derived from completely different embryonic tissue, so we score "stomata" and "xylem" as different transformation series. When we analyze the data matrix phylogenetically, we discover that stomata and xylem originated on different branches of the phylogenetic tree. Our initial invocation of Kluge's auxiliary principle was justified (the two are not acting as a single unit).

Now consider phloem. Although we place phloem in a different transformation series, which must therefore be considered independent of the transformation series that includes xylem, both traits originate on the same branch of the tree. Is this cause for alarm? Have we artificially scored xylem and phloem as two characters when, in reality, they are only one complex character ("xylem and phloem: present or absent")? In this case, the answer is no because, thanks to the work of generations of botanists, we have developmental and functional (phloem transports nutrients) evidence indicating that the two traits are evolutionarily distinct.

 Isn't mistakenly describing one "complex" trait as many independent traits a major problem for reconstructing phylogenetic relationships? Not if you are using phylogenetic systematic methodology and *artificially subdividing a homologous trait,* because all homologies covary with phylogeny. (You will get the same answer any way you slice it.) If, however, you artificially subdivide a homoplasy into so many parts that the covarying homoplasies outnumber the homologies in a data set, phylogenetic systematics (indeed all quantitative systematic methodologies) will take you very efficiently to the wrong phylogeny (Felsenstein 1978). Unfortunately, we cannot identify homologies and homoplasies a priori, so phylogenetic reconstruction must always be an open-ended process requiring constant re-examination of character descriptions against the resultant tree and constant additions of new data.

When you use the auxiliary principles, aren't you making claims about the way in which evolution works? No. Phylogeneticists do not use the auxiliary principles to make statements of "fact" (i.e., they are not formal assumptions of a model). They use them to construct a series of *hypotheses* about character evolution. The construction of these hypotheses represents the first step, not the ultimate explanation, in Hennig argumentation. Without both principles, we would have no objective basis for reconstructing phylogenies. Without Kluge's auxiliary principle, we would have no objective basis for using phylogenies, and character correlations on them, to study evolutionary mechanisms. This would leave us in the paradoxical and undesirable situation of having an evolutionary research program without a temporal dimension.

Hennig (1966a) and Brundin (1966) characterized the essence of phylogenetic analysis as the "search for the sister group." They recognized that if you could find the closest relative or close relatives of the group you are working on, you would have the basic tools for deciding which characters are apomorphic and which are plesiomorphic in a transformation series. The argument goes something like this: Say you discover that members of your study group have two different but homologous dance characters, "step to the left" and "step to the right." As a phylogeneticist, you realize that one of these characters (the apomorphic one) might provide information about relationships within your study group, but that both cannot be equally informative because *you cannot group taxa based on plesiomorphic characters*. (Does the possession of a backbone help you determine whether lions are more closely related to bats or to dogs?) If you only find left stepping in the sister group or in other closely related groups (outgroups) of the taxon you are studying, then it is fairly clear that this dance type is older than right stepping. Left stepping must be the plesiomorphic character in the transformation series. The characteristics of members from closely related groups are thus vital components to decisions regarding the generality of characters within the study group. The fundamental rule for distinguishing plesiomorphy from apomorphy is the

> **Relative Apomorphy Rule (outgroup comparison):** Homologous characters found within the members of a monophyletic group that are also found in the sister group are plesiomorphic, while homologous characters found only in the ingroup are apomorphic.

As we noted in the introduction to this chapter, phylogeneticists avoid the trap of using knowledge about relationships to determine homology, then using those homologous traits to confirm the relationships, by using outgroup comparisons to polarize characters. Once again, this principle should not be problematical for Darwinian biologists: "Mr. Waterhouse has remarked that, when a member belonging to one group of animals exhibits an affinity to a quite distinct group, this affinity in most cases is general and not special" (Darwin 1872:409).

The concept of using outgroups to polarize homologous transformation series springs from the fundamental foundation of evolutionary theory: biological evolution is a *genealogical* process (as opposed to the "evolution" of galaxies); it is descent (genealogy) with modification. This means that some homologies (like a backbone) will be shared among taxa, and it is this sharing of homologous characters that allows outgroup analysis to work. Theory aside, actual polarity arguments can be quite complex. For example, what do we do if (1) we don't know the exact sister group but have only an array of possible sister groups; (2) the sister group also has both characters; (3) either the study group or the sister group is not monophyletic; or (4) the shared character evolved in the sister group independently? Answers to these questions depend on our ability to argue character polarities using some formal rules. The best presentation of these rules was published by Maddison et al. (1984), although the issues have been discussed widely.[2] We will present the case developed by Maddison et al. for groups in which sister-group relations have already been determined. Before proceeding with this discussion, however, we need to add some more terms to our phylogenetic vocabulary.

The ingroup (IG) is the group on which we are working (fig. 2.9a). For purposes of character polarization arguments, the ancestor of the ingroup is depicted as an **ingroup node.** (Note: the IG internode = the ancestral internode; see fig. 2.5.) *We are basically on a quest to determine what the character looked like in this ancestor because it represents the plesiomorphic (or ancestral) condition of the character for the ingroup.* The **outgroup node** is the node immediately below the ingroup node. Characters, designated by lowercase letters, are placed where taxa are usually labeled. (Letters are used purely as a heuristic device to avoid connotations of "primitive" and "advanced.") The ingroup is indicated by a **polytomy** (a node connecting more than two descendant branches), since we presume that the relationships among these taxa are unknown. The relationships among outgroups (OGs) can be either **resolved** (fig. 2.9b) or **unresolved**

2. See, e.g., Ross 1974; Crisci and Stuessy 1980; de Jong 1980; Stevens 1980, 1981; Watrous and Wheeler 1981; Wiley 1981, 1986b; Farris 1982; Patterson 1982; Donoghue and Cantino 1984; Brooks and Wiley 1985; de Queiroz 1985; and Hwang 1992.

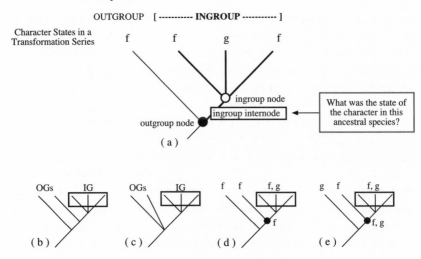

Figure 2.9. Primer for character polarization. *Lowercase letters* = character states in a transformation series. Members of the ingroup (IG) are enclosed in a box. (*a*) Examples of some general terms used in character polarization. (*b*) Outgroup (OG) relationships are known (resolved). (*c*) Outgroup relationships are unknown (unresolved). (*d*) State of the character at the outgroup node is known (decisive). (*e*) State of the character at the outgroup node is unknown (equivocal).

(fig. 2.9c). A determination of the character(s) to be placed at the outgroup node may be either **decisive**, which provides us with an unambiguous indication of the polarity of our ingroup characters (fig. 2.8d: f is plesiomorphic and g is apomorphic in the ingroup) or **equivocal**, which does not allow us to postulate the direction of the character change (fig. 2.9e: we are not sure whether f or g is plesiomorphic).

Maddison et al. (1984) began the quest for character polarities by attempting to determine the character state at the outgroup node using an optimization routine modified from earlier procedures of Farris and Fitch. They reasoned that because this node precedes the ingroup node, it will give us information about the character state in the common ancestor of the ingroup. There are two possible cases: the relationships of the outgroups are known relative to the ingroup; or the relationships of the outgroups are either unknown or only partly resolved. We will focus on the simpler, first case, using the hypothetical taxon Annidae, its sister group (P + Q + R), and other outgroups (M, N, and O).

CHARACTER POLARITY IN THE GROUP ANNIDAE

1. Draw the phylogenetic tree of the ingroup and outgroups. We cannot reconstruct the tree on the basis of the characters in table 2.1 because these characters apply to the resolution of relationships within the

Table 2.1. Data matrix for characters found within the ingroup (Annidae), its sister group (the monophyletic group P + Q + R), and other outgroups

Character TS	Taxon						
	M	N	O	P	Q	R	Annidae
1	b	a	a	b	b	a	a, b
2	b	b	a	b	b	a	a, b

NOTE: TS = transformation series.

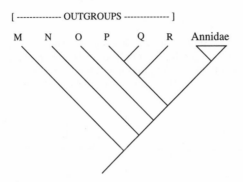

Figure 2.10. Relationships of the Annidae clade to its closest relatives.

ingroup, not to the relationships between the ingroup and the taxa in the outgroup. Presumably, then, we have access to a phylogeny for these taxa before beginning our investigation (fig. 2.10).

2. Inscribe each of the branches with its corresponding character from the first transformation series (i.e., the first row of data in table 2.1) and indicate six nodes (fig. 2.11). The lowest node on the tree is the root node, while the node connecting the Annidae to its sister group (P + Q + R) is the outgroup node.

3. Label the two nodes, other than the root node, that are farthest from the outgroup node in the following manner (fig. 2.12): (1) Label the node "a" if the two closest nodes or branches are both a or are a and a, b. Notice that "closest nodes" refers to the nodes that are adjacent to the node in question, while "closest branch" is defined as any adjacent branch at the equivalent level or lower on the tree than the node in question. (2) Label the node "b" if the two closest nodes or branches are both b or are b and a, b. (3) If the closest branches/nodes have different labels (one "a" and the other "b"), then label the node "a, b." Note that the root node is not labeled. In order for us to label the root, we would need another outgroup. Since we are primarily interested in relationships within the ingroup, we will forget about this node. So, label the farthermost two nodes according to this procedure. The node connect-

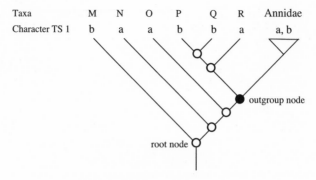

Figure 2.11.   Character states from the first transformation series (TS) in table 2.1, each listed under the appropriate taxon. In order to determine the state of the character (either a or b) at the outgroup node, we must work toward it based on the information available at other, more accessible nodes (depicted as *open circles*).

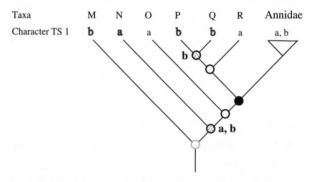

Figure 2.12.   First and second polarity decisions for character transformation series 1. *Outlined letters* = closest branches and nodes; *bold letters* = polarity decisions. The top node is labeled "b" because its closest branches are both b; the bottom node is labeled "a, b" because one of its closest branches is a and the other is b.

ing P + Q is given the notation "b," because both P and Q have character b. The node connecting N with O + P + Q + R + Annidae is given the notation "a, b" because M has character b and N has character a.

4. Continue working toward the outgroup node (fig. 2.13) in the same manner.

5. The analysis is over when we reach a decision concerning the outgroup node. In this example, the assignment is a decisive "a" (fig. 2.14).

Repeating this procedure (steps 2–5) for character transformation series 2 of the matrix eventually produces an equivocal decision for that character at the outgroup node (fig. 2.15). Notice that *each decision is made one transformation series at a time, and thus the polarization of each character occurs independently of every other character.* This does not mean

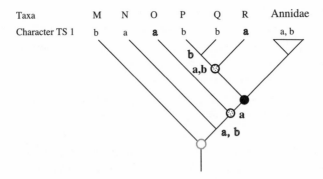

Figure 2.13. Third and fourth polarity decisions for character transformation series 1. *Outlined letters* = closest branches and nodes; *bold letters* = polarity decisions. The top node is labeled "a, b" because its closest branch is a and its closest node is b; the bottom node is labeled "a" because its closest branch is a and its closest node is a, b.

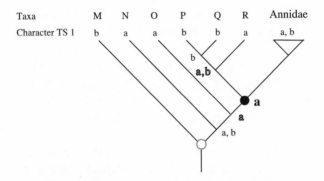

Figure 2.14. Assignment of polarity to the outgroup node for character transformation series 1. *Outlined letters* = closest branches and nodes; *bold letters* = polarity decisions. The outgroup node is labeled "a" because one of its nodes is a and the other is a, b.

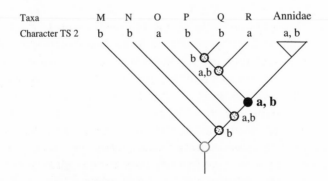

Figure 2.15. Assignment of polarity to the outgroup node for character transformation series 2. The outgroup node is labeled "a, b" because both of its closest nodes are a, b.

that equivocal decisions based on single characters examined in vacuo will remain equivocal at the end of the analysis. The final disposition of character states is subject to an overall analysis combining all of the information gleaned from each of the separate character polarizations according to a series of formal rules (which we will turn to in a moment).

A robust phylogenetic tree is based on simultaneous analysis of numerous characters, each polarized using outgroup comparisons (Wiley 1981; Clark and Curran 1986; Mayden and Wiley 1992). Any of the characters used may exhibit some degree of homoplasy, in which case the distribution of the characters on the phylogenetic tree will not correspond exactly to the sequence postulated originally when the characters were polarized (the transformation series). *Thus, it is not methodologically sound to polarize one character using an outgroup and then to assume that the postulated character transformation series represents the true phylogenetic sequence of events.* This type of error results from a subtle misunderstanding of the purpose of outgroup comparisons. While it is true that polarizing character states through outgroup comparison produces a hypothesis of the sequence of evolutionary changes in that character, it is also true that this hypothesis can only be tested by comparing it to the hypothesized evolutionary sequences of other characters, because such assessments initially assume that there has been no homoplasy. The sequence to be used for evolutionary explanations is thus the sequence that is logically consistent with the phylogenetic tree based on all the characters, and that sequence may include postulates of homoplasy. *In other words, a character transformation series does not stand on its own, independent from the underlying phylogenetic relationships of the species.*

Maddison et al. (1984) also discussed more complicated situations in which the relationships between the outgroups are either not resolved or only partly resolved. Since this watershed article is readily available, we will only mention two important observations: (1) Whatever the resolution of the outgroup relationships, the sister group is always dominant in its influence on the polarity decision. If the sister group is decisive for a particular state, no topology of outgroups farther down the tree can result in a decisively different state. (2) If you are faced with no sister group, only an unresolved polytomy below the ingroup, the frequency of a particular character among the outgroups in the polytomy has no effect on the decision at the outgroup node. For example, you could have ten possible, unresolved sister groups with character a and one with character b and the decision would still be equivocal at the outgroup node. This happens because *"common" does not equate with "plesiomorphic"* in the phylogenetic system.

> If outgroups are so important, what happens if you change the outgroup? Will you get vastly different results? It depends on the robustness of your original data set!

Let's say, for the sake of discussion, that you only used one outgroup (and that it is in fact the sister group) in your analysis of the relationships among the Annidae. A few days later, a fellow researcher comes along and uses a different outgroup to polarize your characters. Depending upon your data set, and the relationship of the new outgroup to your ingroup, one of three things will happen in this new analysis:

1. There will be no change in any transformation series polarities because both of the outgroups share the same plesiomorphic state with the ingroup for all characters. Result: The tree topology will not change.

2. Your fellow researcher will be unable to determine the plesiomorphic and apomorphic states for some characters because the outgroup he chose is too distantly related to the ingroup and thus does not share as many similarities with it as did your outgroup. Result: There will be a loss of resolution in the analysis (an increase in the number of unresolved polytomies on the tree). This produces an incomplete, rather than an incorrect, estimate of phylogeny.

3. There will be a change in the polarity of some transformation series because the new outgroup shares some homoplasious character states with the ingroup. Result: Your fellow researcher will propose an incorrect hypothesis of polarization for those characters, which should not affect the tree topology if you have an extensive and robust data set (the same outcome as in 1, above). In fact, the more resilient a tree proves to be in response to changing the outgroup, the greater confidence we have in it (Lyons-Weiler et al. 1998). Worst case scenario: If the new outgroup shares so many *homoplasious* apomorphic traits with the ingroup that the "real" underlying homologies are swamped, then you will get a different and "incorrect" topology. In order for a completely new topology to result from changing outgroups, however, you must have had either very few characters or very little support for the number of taxa you were trying to investigate in the first place. Overall, then, if you have an extensive data set, changing the outgroup may (or may not) change polarization hypotheses for some characters, and it may (or may not) cause a loss of resolution on the tree (if the new outgroup is more distantly related to the ingroup than is the original outgroup). Only under extremely unusual circumstances will changing the outgroup cause a wholesale restructuring of the tree. As you can see, the distinction is not one of finding the "correct" versus the "incorrect" outgroup, but rather

of finding the outgroups that provide the most information pertinent to polarizing characters in the ingroup. The more closely related your outgroup is to your ingroup, the more useful information you will find. Notice, however, that no one outgroup, including the sister group, can be expected to solve all of your polarization problems, because traits continue to evolve in the outgroups as well as in the ingroup. There is a simple way around this problem: *always use more than one outgroup in your analysis* (Maddison et al. 1984; Mayden and Wiley 1992).

Once you have established the relative apomorphy of your putative homologues, you may begin putting the data to work producing the most robust hypothesis of phylogenetic relationships possible. For this, you need the

> **Grouping Rule:** Only synapomorphies (shared special homologies) provide evidence of common ancestry relationships. Symplesiomorphies (shared general homologies) and homoplasies (convergences and parallelisms) are useless in this quest.

Homoplasies are useless indicators of common ancestry relationships because they evolved independently in each taxon that displays them (fig. 2.16a). The futility of using plesiomorphies in an attempt to reconstruct a particular phylogeny is more problematic. After all, plesiomorphies are homologies, so why can't they be used to denote phylogenetic relationships? In fact, they can, because *evolution is an ongoing process, so the plesiomorphic or apomorphic status of a character is a relative condition.* A symplesiomorphy cannot show phylogenetic relationships within the group you are studying because it originated earlier than any of the taxa in your study group (fig. 2.16b). However, if you increase the temporal scale of your investigation, say by examining relationships among genera within one family instead of among species within one genus, a character that is a symplesiomorphy for the original ingroup may be found to be a synapomorphy of a larger group (fig. 2.16c).

Finally, we have to consider how to combine the information from different transformation series into hypotheses of relationships for the ingroup as a whole. This is accomplished by using the

> **Inclusion/Exclusion Rule:** The information from two transformation series can be combined into a single hypothesis of relationship if that information allows for the complete inclusion or the complete exclusion of groups which were formed by the separate transformation series. Partial overlap of groupings necessarily generates two or more hypotheses of relationship.

The inclusion/exclusion rule is directly related to the concept of logical consistency. Trees that conform to the rule are logically consistent

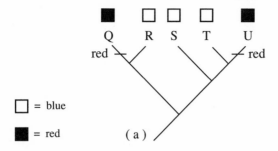

= blue

= red

(a)

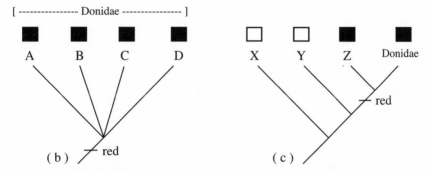

Figure 2.16.   The quest for phylogenetic relationships. *Letters* = taxa; *boxes* = the color of each taxon below. (*a*) Taxa Q and U are both red and other members of the group are blue; however, clustering Q with U would be incorrect because, according to the phylogenetic tree based on numerous other characters, red arose independently in both taxa (homoplasy). (*b*) Since all members of the Donidae are red (symplesiomorphy), that character does not tell us anything about individual relationships among taxa A, B, C, and D. (*c*) On a larger scale, Donidae shares the character red with group Z (a synapomorphy supported by many other characters on the tree); therefore, red is useful in determining sister group relationships at this level.

with each other, while trees that do not conform are logically inconsistent with each other. You can get an idea of how this rule works by studying the examples in figure 2.17. In panel a, we have four characters and three potential trees. The first tree contains no character information and thus provides no resolution of the phylogenetic relationships within the group; therefore, it is logically consistent by default with any tree that has character information. The second tree states that B, C, and D form a monophyletic group based on characters from two transformation series (1 and 2). The third tree states that C and D form a monophyletic group based on two additional characters (3 and 4). Note that the group C + D is completely included within the group B + C + D. Based

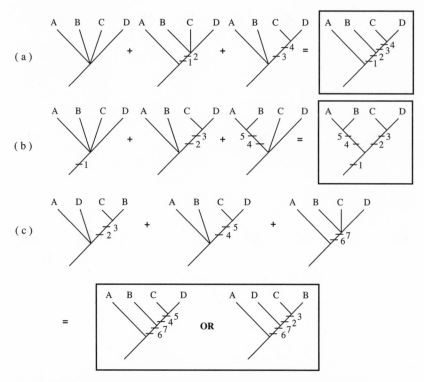

Figure 2.17.   The use of the inclusion/exclusion rule for combining information from different character transformation series into trees. (*a*) The group B + C + D completely includes the group C + D. (*b*) The group A + B + C + D completely includes the groups A + B and C + D and, within this, the group C + D completely excludes the group A + B. (*c*) The group B + C + D completely includes the groups C + D and C + B, but groups C + D and C + B overlap because they both contain C. This results in the production of two different trees.

on this distribution of characters, we hypothesize that the tree for these taxa includes C + D as a monophyletic group enclosed within a second monophyletic group, B + C + D. Panel b shows the result of the inclusion of two monophyletic groups (A + B and C + D) within a larger monophyletic group (A + B + C + D). In this case, the groupings are consistent because A + B (supported by characters 4 and 5) and C + D (supported by characters 2 and 3) completely exclude each other. Finally, the example presented in panel c violates the inclusion/exclusion rule. Although both C + B and C + D can be included within the group B + C + D, the transformation series for characters 2 and 3 group C and B and exclude D, while the transformation series for characters 4 and 5 group C and D and exclude B. Thus, the phylogenetic information gleaned from these characters conflicts, and the groups overlap (C is included in two different groups), producing two trees that are logically inconsistent with each other.

The Relationships of the ABCidae

1. Transformation Series (TS) 1 is composed of characters in the first column of table 2.2. Recall that plesiomorphies are coded as "0," synapomorphies are coded as "1," and, according to the grouping rule, relationships among taxa are reconstructed based on shared derived traits or synapomorphies. Given this, we can draw a tree with the groupings implied by the synapomorphy found in the first transformation series (fig. 2.18a). We can then repeat the process for TS 2 (fig. 2.18b). Both trees, based on the distributions of two different characters, imply the same groupings; therefore, we can say that the trees are topologically identical. Combining the trees in panels a and b according to the inclusion/exclusion rule produces the tree depicted in panel c (i.e., both characters support the group A + B + C to the complete exclusion of X). We can calculate a tree length for the tree in panel c by simply adding the number of synapomorphies that occur on it. In this case, the tree length is two steps.

2. Now, repeat this procedure for character transformation series 3 and 4. Inspection of the data matrix reveals that the synapomorphies

Table 2.2. Data matrix for determining the relationships among taxa A, B, and C

Taxon	Character Transformation Series						
	1	2	3	4	5	6	7
X	0	0	0	0	0	0	0
A	1	1	0	0	0	0	0
B	1	1	1	1	0	0	0
C	1	1	1	1	1	1	1

NOTE: X = outgroup; A + B + C = ingroup; *bold* = synapomorphies (shared special homologies).

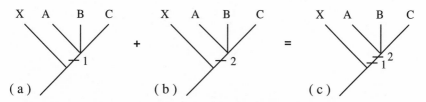

Figure 2.18. Trees for the ABCidae, based on characters 1 and 2. Trees produced by applying the grouping rule to (*a*) character transformation series 1 and (*b*) character transformation series 2. (*c*) Tree produced by applying the inclusion/exclusion rule to the information provided by both characters.

for these characters have identical distributions, implying that B and C form a monophyletic group (fig. 2.19a,b). If we combine the information from both characters, the results should look like the tree in panel c. This tree is also two steps long.

3. Only taxon C has the apomorphies listed in character TS 5, 6, and 7. Apomorphic characters which are unique to one taxon are termed **autapomorphies.** Although they can tell us nothing about relationships among different taxa (fig. 2.20), autapomorphies are useful diagnostic traits for identifying a particular taxon. For example, if we were to collect individuals displaying the autapomorphic state for characters 5, 6, and 7 (denoted by "1" in the data matrix), we would assign those individuals to taxon C. On the other hand, collecting organisms bearing the synapomorphic condition for character 4 (also denoted by "1" in the data matrix) only tells us that they are members of either taxon B or taxon C. Autapomorphies count when calculating tree length, so the length of this tree is three steps.

4. We now have three different tree topologies (figs. 2.18c, 2.19c, and 2.20). If we examine these trees more closely, we discover that although they differ, they do not contain any conflicting information. For example, since TS 5–7 only imply that taxon C is different from the other

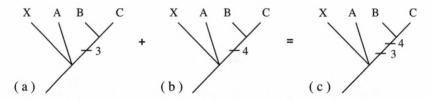

Figure 2.19. Trees for the ABCidae based on characters 3 and 4. Trees produced by applying the grouping rule to (*a*) character transformation series 3 and (*b*) character transformation series 4. (*c*) Tree produced by applying the inclusion/ exclusion rule to the information provided by both characters.

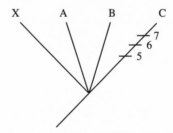

Figure 2.20. Tree for the ABCidae, based on characters 5, 6, and 7. Since autapomorphies are not useful for grouping, this tree shows no resolution of relationships among the taxa.

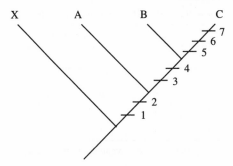

Figure 2.21. Combining all the information in the data matrix (table 2.2) produces one hypothesis (tree) of the phylogenetic relationships within the ABCidae. This is the best estimate of the relationships, based on the available data. The tree proposes that taxa B and C are sister groups bound together by the possession of synapomorphies for characters 3 and 4, and that A + B + C is a monophyletic group based on the common shared characters 1 and 2.

four taxa, this tree (fig. 2.20) does not conflict with the other two trees. Further, the distributions of character TS 1 and 2 do not conflict with the distributions of TS 3 and 4: TS 1 and 2 imply that A, B, and C form a monophyletic group, while TS 3 and 4 imply that B and C form a monophyletic group without saying anything about the relationships of A or the outgroup, X. Trees that contain different but mutually agreeable groupings are logically consistent, or **congruent.** They can be combined without changing any hypothesis of homology. When combined, the length of the resulting tree is the sum of the lengths of each subtree. For example, all of the information in the data matrix can be combined to produce one tree (fig. 2.21) with a length of seven steps, exactly the total number of subtree steps (2 + 2 + 3).

THE RELATIONSHIPS WITHIN THE RSTIDAE

1. The data matrix for TS 1 and TS 2 (table 2.3) implies that R, S, and T form a monophyletic group (fig. 2.22).

2. TS 3 and TS 4 imply that S and T form a monophyletic group (fig. 2.23).

3. Finally, TS 5, 6, and 7 imply that S and R form a monophyletic group (fig. 2.24).

4. At this point you should suspect that something has gone wrong. TS 3–4 imply a monophyletic group that includes S and T but excludes R, whereas TS 5–7 imply a monophyletic group that includes R and S but excludes T. There must be a mistake since combining these trees would violate the inclusion/exclusion rule. In such a situation we invoke the first principle of phylogenetic analysis: *there is only one true phylogeny.*

Table 2.3. Data matrix for determining the relationships among taxa R, S, and T

Taxon	Character Transformation Series						
	1	2	3	4	5	6	7
X	0	0	0	0	0	0	0
R	1	1	0	0	1	1	1
S	1	1	1	1	1	1	1
T	1	1	1	1	0	0	0

NOTE: X = outgroup; R + S + T = ingroup; *bold* = synapomorphies (shared special homologies).

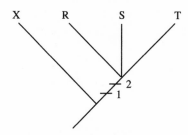

Figure 2.22. Tree for the RSTidae, based on characters 1 and 2. This tree was produced by applying the grouping rule to character transformation series 1 and 2, then combining this information via the inclusion/exclusion rule.

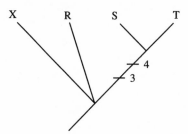

Figure 2.23. Tree for the RSTidae, based on characters 3 and 4. This tree was produced by applying the grouping rule to character transformation series 3 and 4, then combining this information via the inclusion/exclusion rule.

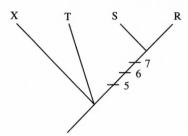

Figure 2.24. Tree for the RSTidae, based on characters 5, 6, and 7. This tree was produced by applying the grouping rule to character transformation series 5, 6, and 7, then combining this information via the inclusion/exclusion rule.

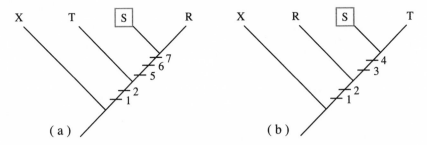

Figure 2.25. Two logically inconsistent trees produced from the information in the data matrix (table 2.3). Taxon S is the problem: characters 5, 6, and 7 place it with R, while characters 3 and 4 group it with T. Both trees cluster R + S + T together based on possession of the apomorphic form of characters 1 and 2.

Thus, one or more of our groupings must be wrong. Fortunately Hennig's auxiliary principle keeps us going until we can *demonstrate* which of the groupings is incorrect. We are now faced with the problem of trying to differentiate between two logically inconsistent trees (fig. 2.25). Note that there is some congruence between the trees based on their possession of the apomorphies from the first two transformation series (characters 1 and 2).

5. You have probably noticed by now that each of the trees in figure 2.25 is incomplete. The tree on the left lacks transformation series 3 and 4, while the one on the right lacks transformation series 5, 6, and 7. *Leaving characters out of an analysis is not acceptable.* In fact, eliminating characters that do not "fit" a hypothesis ranks among the top three heinous "crimes against phylogenetics" (the other two being grouping by symplesiomorphies and equating taxonomy with phylogeny). In order to add the missing characters into both trees, we postulate that some of the evolutionary changes within this group are due to homoplasy (fig. 2.26). There are two types of homoplasy. The first we have already discussed: a character may have arisen independently more than one time (convergent or parallel character evolution). A second type of homoplasy results from a reversion to the plesiomorphic condition (this is called a **reversal**). We must consider both types of homoplasy in this example. Recall that our first tree (fig. 2.25a) neglected to include characters 3 and 4. There are two possible ways to portray the distribution of these characters on the tree: either 3 and 4 arose independently in taxa S and T (fig. 2.26a), or 3 and 4 arose in the common ancestor of the group T + S + R and were subsequently "lost" in taxon R (reversal to the ancestral [plesiomorphic] character conditions: fig. 2.26b). Examination of the distributions of characters 5, 6, and 7, which are missing on the second tree (fig. 2.25b), produces a similar pattern of homoplasy:

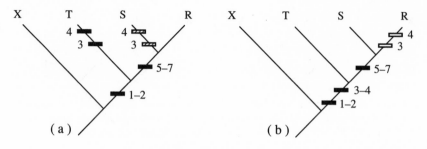

TREE TOPOLOGY 1

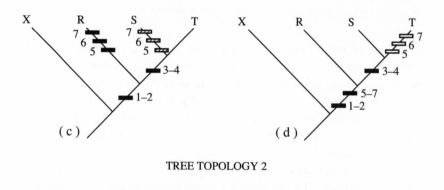

TREE TOPOLOGY 2

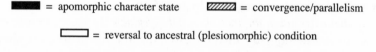

■■■■ = apomorphic character state ▨▨▨ = convergence/parallelism

▭▭ = reversal to ancestral (plesiomorphic) condition

Figure 2.26. Alternate hypotheses for the relationships of R, S, and T due to homoplasious characters. (*a*) Convergent or parallel evolution of 3 and 4 in taxa S and T. (*b*) Characters 3 and 4 revert to the plesiomorphic condition in taxon R. (*c*) Convergent or parallel evolution of 5, 6, and 7 in taxa S and R. (*d*) Characters 5, 6, and 7 revert to the plesiomorphic condition in taxon T.

either 5, 6, and 7 arose independently in taxa R and S (fig. 2.26c), or 5, 6, and 7 arose in the common ancestor of the group R + S + T, and subsequently reverted to the ancestral (plesiomorphic) character conditions in taxon T (fig. 2.26d).

6. *The question now becomes, Which of these trees should we accept?* That turns out to be a rather complicated question. If we adhere to Hennig's auxiliary principle, we should strive for a tree that includes the greatest number of homologies and the fewest number of homoplasies. Although these qualities are usually consistent with each other (i.e., the tree with the greatest number of synapomorphies is also the tree with the fewest number of homoplasies), you can find exceptions. Fortu-

nately, numbers of homologies and homoplasies are related to tree length. Before we begin counting steps, notice that the trees in panels a and b have the same topology; they are the same hypothesis (tree) of phylogenetic relationships among the RSTid taxa even though they are based on different hypotheses of character change. This occurs because *the topology of a tree is determined by synapomorphic relationships, not by the distributions of homoplasies.* The same is true for the trees in panels c and d. So, because of this logical consistency, we really have only two phylogenetic trees for the RSTidae. When you count the number of steps on each tree, you discover that tree topology 1 (fig. 2.26a,b) has nine steps while tree topology 2 (fig. 2.26c,d) has ten steps. We accept tree topology 1 as the best estimate of phylogeny because it is shorter and thus incorporates the greatest number of homology statements and the fewest number of homoplasy statements for the data set. *Note that such a conclusion can apply only to trees derived from the same data set.* Hennig's auxiliary principle coupled with the principle that there is only one true phylogeny of life carried us to this point. Given the evidence at hand, however, we cannot choose between the different sequences of character evolution postulated by the trees in figures 2.26a and 2.26b because both trees are equally parsimonious (are the same length). We will discuss this "principle of parsimony" in more detail later in this chapter. At the moment, it is important to remember (1) we invoke the principle of parsimony at the end of the analysis, not at the beginning; and (2) we use the principle of parsimony as a means of deciding which hypothesis we will use as our most defensible working hypothesis when the data support more than one alternative. When there is no homoplasy, a single phylogenetic hypothesis explains all the evidence unambiguously, and there is no need to invoke any decision-making procedure such as parsimony.

 Remember, phylogenetic systematics is a method of discovery. We do not assume homology a priori as a statement of fact. Rather, we construct individual hypotheses of homology as a starting point, hence we presume as much homology as we can at the beginning. We also presume we will make mistakes and that we need to let the data somehow tell us when we have done so. The appearance of homoplasy on the final tree is the data's way of signaling a mistake. To put it another way: the appearance of homoplasy a posteriori refutes the a priori hypothesis of homology.

Basic phylogenetic systematics, therefore, can be summarized in five steps. First, use nonphylogenetic criteria to assess similarities in traits among species and presume homology whenever possible. Second, use outgroup comparisons to distinguish shared general, shared special, and unique homologies in the context of your ingroup. Third, group according to shared special homologies. Fourth, when there are conflicts in the data, choose the most parsimonious answer. Fifth, interpret inconsistent traits, a posteriori, as instances of homoplasy (parallel or convergent evolution, or reversal to the plesiomorphic condition); that is, as cases in which the initial nonphylogenetic homology criteria were misleading. The result of this procedure is the most robust possible hypothesis of the phylogenetic relationships within your study group for the evidence you have used.

 A decade ago we wrote, "One cardinal rule in historical ecology is *never bias your analysis by using the ecological information you want to study to build your phylogenetic tree*" (Brooks and McLennan 1991:134). Although for us the critical part of the statement was "never bias your analysis" (Brooks and McLennan 1994:20–21), the wording was obviously ambiguous because many readers interpreted the statement as *"never use the ecological information you want to study to build your phylogenetic tree, otherwise you are building circularity into the analysis."*[3]

As by now you know, the issue of circularity does not arise, because we use Hennig's auxiliary principle a priori and the parsimony criterion a posteriori in phylogenetic analysis. The critical issue thus is not whether we use the characters of interest to build the tree or optimize them onto the tree after an analysis. The critical issue is one of robustness—whether or not exclusion of the trait you are studying from the phylogenetic analysis changes the genealogical hypothesis. If removing the trait changes the hypothesized relationships, then you are not beginning your study with a robust tree. Because the relationships supported by that trait are not corroborated by other traits, evolutionary explanations for that character are substantially weakened. To state our position more clearly:

 Use all available evidence to construct your phylogenetic hypothesis, but be sure that the phylogenetic tree being used to investigate the evolution of trait x depicts relationships that are maintained when trait x is excluded from the analysis.

3. Deleporte 1993; Kluge and Wolf 1993; de Queiroz 1996; Bruneau 1997; Luckow and Bruneau 1997; and Grandcolas et al. 2001.

CHARACTER CODING FOR BUILDING TREES

A **character code** is a numerical or alphabetical symbol that represents a particular character. We used character codes in the preceding section as we covered the basics of tree reconstruction using classical Hennigian argumentation and some of the approaches to determining the polarity of characters through outgroup analysis. In this section we will introduce some of the different kinds of apomorphic characters encountered in phylogenetic research, describe some of the problems associated with assigning codes to these characters, and present rules for coding these characters for the purpose of reconstructing phylogenetic relationships among taxa.

Multistate Transformation Series

All of the derived characters we have considered thus far are qualitative characters in binary transformation series. A **binary transformation series** consists of a plesiomorphy and its single derived homologue. By convention the plesiomorphy is coded as "0," while the derived homologue is coded as "1." Binary transformation series present no problem in coding: simply code each character according to the information available from outgroup argumentation and produce a matrix full of zeros and ones. Complications can arise if you encounter taxa that are **polymorphic** for a given character, exhibiting both the plesiomorphic and the apomorphic conditions. However, this problem is only critical when both conditions are found in a single species. Considerable controversy surrounds the coding of such characters, especially when molecular traits are used.[4]

Investigators working on a large group, or even a small group that has undergone considerable evolution, may discover several different homologous character states in a single transformation series. For example, if you were working on the phylogenetic relationships of fossil and recent horses, the transformation series for the number of toes on the hind foot would contain a goodly number of character states: four toes, three toes, and one toe in the ingroup, and five toes in the outgroup. Such a large transformation series, encompassing a plesiomorphic character and two or more apomorphic character states, is termed a **multistate transformation series**. As usual, Darwin recognized the significance of this:

---

4. For advanced discussions, see Buth 1984; Swofford and Berlocher 1987; Murphy 1988, 1993; Nixon and Davis 1991; Mabee and Humphries 1993; Wiens 1995; and Kornet and Turner 1999.

We have seen that the members of the same class, independently of their habits of life, resemble each other in the general plan of their organisation. This resemblance is often expressed by the term "unity of type"; or, by saying that the several parts of organs in the different species in the class are homologous. This is one of the most interesting departments of natural history and may almost be said to be its very soul. . . . What can be more curious than that the hand of a man . . . that of a mole . . . the leg of the horse, the paddle of the porpoise, and the wing of the bat, should all be constructed on the same pattern, and should include similar bones in the same relative positions? . . . Notwithstanding this similarity of pattern, it is obvious that the hind-feet of these several animals are used for as widely different purposes as it is possible to conceive. . . . . Professor Flower . . . remarks in conclusion "We may call this conformity to type, without getting much nearer to an explanation of the phenomenon"; and then he adds, "but is it not powerfully suggestive of true relationship, of inheritance from a common ancestor?" (1872:413)

Multistate transformation series can be grouped according to the amount of information available concerning the pathway and direction of evolutionary change. As discussed earlier, this produces four possible types of multistate transformation series (see fig. 2.7): (1) ordered, unpolarized; (2) unordered, unpolarized; (3) ordered, polarized; and (4) unordered, polarized. Binary characters are much simpler because they are automatically ordered (though not necessarily polarized).

A diagram of the relationships of the characters in a multistate transformation series can be termed a **character tree**. It is important to understand that a character "phylogeny" is not the same as a phylogeny of taxa. A character tree only contains information about the hypothesized relationships among character states. It is not a hypothesis of the underlying phylogenetic relationships of the taxa that exhibit the character in its various states. (At the moment we are not concerned with how to reconstruct a character tree, but rather with how to code the information presented in such a tree for use in reconstructing a phylogenetic tree. We will return to the question of just how these transformation series are derived at the end of this section.)

A LINEAR TRANSFORMATION SERIES

A **linear transformation series**, consisting of characters related to one another in a straight-line fashion, is the simplest type of multistate transformation series. Such transformation series can be coded in one of two basic ways. One way is simply to code the character states in a linear fashion, assigning a value to each character based on its position in the sequence. For example, for the character states represented in figure 2.27, we have chosen to code rounded white square as "0," white square

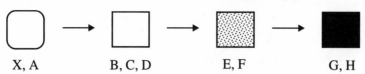

X, A   B, C, D   E, F   G, H

Figure 2.27.   A simple linear character state tree for an ordered, polarized transformation series with four states: rounded square, white square, speckled square, and black square. *Letters* = taxa displaying the character state. X = outgroup; A-H = ingroup.

Table 2.4.   Data matrix for the four character states and nine taxa in the linear transformation series in figure 2.27

Taxon	Linear Coding	Additive Binary Coding		
X	0	0	0	0
A	0	0	0	0
B	1	1	0	0
C	1	1	0	0
D	1	1	0	0
E	2	1	1	0
F	2	1	1	0
G	3	1	1	1
H	3	1	1	1

as "1," speckled square as "2," and black square as "3." Each value is placed in a data matrix (table 2.4) in a single column. Every apomorphy will contribute to the length of our forthcoming phylogenetic tree in an additive fashion. We use the term "linear" because each instance of evolutionary modification requires one step along the tree, and counting all of the steps in a straight line shows exactly how much the transformation series added to the overall tree length. Indeed, such transformation series are often termed **additive multistate characters.**

Alternatively, we could use **additive binary coding,** a method that breaks the character down into a number of binary subcharacters, each represented by its own column of information in the matrix. For example, we can consider both speckled square and black square as subsets of white square since each is ultimately derived from white square. The first additive binary column in table 2.4 reflects this fact, coding rounded white square as the plesiomorphic character state (0) and white square plus all of its descendants as apomorphic (1). The second additive binary column contains the codes for the next level of comparison, black square as a subset of speckled square. Both rounded square and white square are plesiomorphic relative to speckled square, so they re-

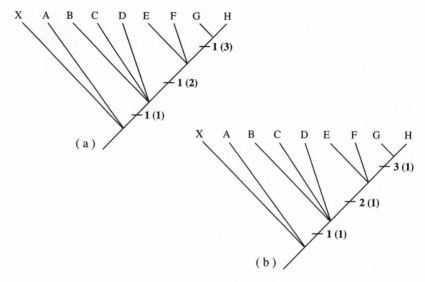

Figure 2.28. Phylogenetic tree constructed from the linear character state tree in figure 2.27. (*a*) Simple linear coding breaks the character into four parts, the plesiomorphic state (0), which need not be represented on the tree, and three apomorphic states (1, 2, and 3; one character, four states). (*b*) Additive binary coding breaks the linear series into three independent characters, then groups according to the apomorphic states (1) of those characters (three characters, two states each). Both methods of coding preserve intact the phylogenetic information in the character and thus produce the same phylogenetic tree.

ceive a "0," while speckled square and its descendant, black square, receive a "1." Finally, for the last level of comparison, black square is apomorphic relative to all the other character states in this linear series, so it is coded "1" in the third column and everything else receives a "0." Three columns now represent the transformation series. You can double-check your binary coding by adding all the ones together in each row and placing them in a single column. You should find that you have replicated the original linear coding column. Either method of coding produces exactly the same character sequence and thus contributes the same information to a phylogenetic reconstruction (fig. 2.28).

A BRANCHING TRANSFORMATION SERIES

A **branching transformation series** (also called a **nonlinear transformation series** or a **complex transformation series**) contains character states that are related to each other in a branching, rather than a straight-line fashion (fig. 2.29). Since the character states are not related in a linear fashion, simple linear coding will result in errors when the transformation series is translated into a phylogenetic tree. How can we display these complex relationships accurately? There are two basic methods: additive binary coding and mixed coding. Since we are already proficient at additive binary coding, let's turn to this method first.

We arrived at the additive binary codes in table 2.5 by working through the following steps:

1. White square is apomorphic relative to white circle and is ancestral to all other character states in the character tree; therefore, in the first column of the matrix, code white square and all of its descendants as apomorphic (1) and code white circle as plesiomorphic (0). (The relationship between white square and white circle is a simple linear one.)

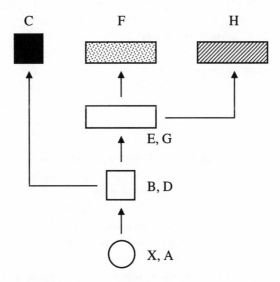

Figure 2.29. A complex branching character state tree for an ordered, polarized transformation series with six states. *Letters* = Taxa displaying each character.

Table 2.5. Data matrix for the six characters and nine taxa in the branching transformation series in figure 2.29

Taxon	Additive Binary Coding					Mixed Coding		
X	0	0	0	0	0	0	0	0
A	0	0	0	0	0	0	0	0
B	1	0	0	0	0	1	0	0
C	1	1	0	0	0	1	1	0
D	1	0	0	0	0	1	0	0
E	1	0	1	0	0	2	0	0
F	1	0	1	1	0	3	0	0
G	1	0	1	0	0	2	0	0
H	1	0	1	0	1	2	0	1

Moving to the next comparison, the relationship between white square and its most immediate modification, brings us to a branch point on the tree—both black square and white rectangle are derived directly from white square. We deal with branch points by examining the derived states one at a time.

2. Black square is derived directly from white square; therefore, in the second column of the matrix, code white square as plesiomorphic (0) and black square as apomorphic (1). Notice that only the taxa displaying either the apomorphic character or any of its modifications are assigned a code of "1"; all other taxa receive a "0." In this case, black square has no modifications and is only present in taxon C (it is an autapomorphy).

3. White rectangle is derived directly from white square; therefore, in the third column, code white square as "0" and white rectangle as "1." Again, the taxa bearing either the apomorphic condition or any of its modifications are assigned a code of "1," so taxa E and G (apomorphic state "white rectangle") and F and H (bearing the descendants of white rectangle) receive a "1"; all other taxa receive a "0."

4. Speckled rectangle is derived directly from white rectangle and is an autapomorphy for taxon F; therefore, in the fourth column, F is assigned a "1" for this character and all other taxa get a "0."

5. In the final comparison, striped rectangle is derived directly from white rectangle and is an autapomorphy for taxon H; therefore, in the fifth column, only H receives a "1" for this character.

Mixed coding, also called **nonredundant linear coding**, is a hybrid between additive binary coding and linear coding. By convention, the longest straight-line branch of the character tree is coded in a linear fashion, then branches off this linear "trunk" are coded in an additive binary fashion. Depending on the asymmetry of the character tree, this strategy can substantially reduce the number of character columns. Returning to the character tree in figure 2.29, the longest sequence of character modifications is white circle–white square–white rectangle–speckled rectangle. (Note: we could have chosen the sequence white circle–white square–white rectangle–striped rectangle. Since nodes can be freely rotated on a tree, the choice between the character states "speckled rectangle" or "striped rectangle" is completely arbitrary and does not change the outcome of the analysis.) We place the codes for this section of the tree (0–1-2–3) in the first mixed coding column of table 2.5. Each branch from this sequence, the autapomorphies black square and striped rectangle, is then assigned a code of "1" and displayed in separate columns in an additive binary coding fashion as described above.

Polarization Arguments for Multistate Transformation Series

As promised at the beginning of this section, we now turn to the methods involved in determining the evolutionary sequence of multistate characters. Sometimes the ordering of these transformation series can be determined using biological data, such as information about developmental sequences or the directions of biochemical pathways.[5] More often, though, these data are not available or are not reliable.

What can phylogenetic systematic analysis tell us in the absence of this information? For example, suppose we wish to examine a group of mammals that includes all species which do and some which do not have forelimbs modified into wings. Furthermore, suppose extensive research has revealed two types of wings (let us say "short" and "long," just for this example) within the group. Using all other mammals as outgroups reveals that the "no wings" state came first (is plesiomorphic), but from that point the evolutionary pathway of wings could have been (1) short to long, (2) long to short, or (3) short and long arising independently from the "no wings" condition. For these species, simple outgroup comparison does not resolve the polarity of the apomorphic states because wings do not occur among members of the original outgroups. In order to resolve the transformation series for apomorphic states restricted to the ingroup, it is necessary to find outgroups that possess at least one of these states. In such cases, phylogeneticists determine character polarities by looking for additional "outgroups" within the ingroup, using a method that preserves the logic of outgroup comparison. This method, developed by Watrous and Wheeler (1981), is called **functional outgroup analysis.**

FUNCTIONAL OUTGROUP ANALYSIS

Let us begin with an ingroup (taxa A–E), a set of outgroups, and three characters (table 2.6). One binary character supports the monophyly of the ingroup, but tells us nothing about relationships within that group (fig. 2.30). Two characters, one binary and the other multistate, help us resolve relationships *within* the ingroup.

1. Use the second binary character to produce a partial phylogenetic tree for the ingroup. Figure 2.30 depicts the distribution of states for this character among the outgroup and ingroup taxa.

5. For examples of ordering by developmental sequences, see Nelson 1978; Nelson and Platnick 1981; Patterson 1982; Voorzanger and van der Steen 1982; Brooks and Wiley 1985; de Queiroz 1985; Kluge 1985; O'Connor 1992; Mabee 1993, 2000; and Wenzel 1993. For examples of ordering by directions of biochemical pathways, see Seaman and Funk 1983; and Barkman 2001.

Table 2.6.   Distribution of states for one multistate and two binary characters among five members of an ingroup (A–E) and a set of outgroups

Taxon	Characters		Multistate
	Binary		
Outgroups	0	0	x
A	1	0	x
B	1	0	y
C	1	+	y
D	1	+	z
E	1	+	z

NOTE: The first binary character supports the monophyly of the ingroup.

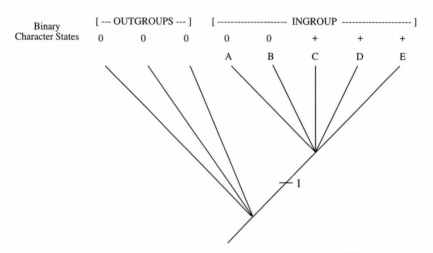

Figure 2.30.   Starting point for phylogenetic analysis of ingroup A–E using three outgroup taxa. Synapomorphy 1 supports the monophyly of the ingroup. Distribution of states (0 and +) for the second binary character are indicated at the tips of each branch.

2. State 0 is clearly supported as the plesiomorphic condition by outgroup comparison. Hence, taxa C, D, and E are united as a group within the ingroup by the shared possession of the derived (synapomorphic) trait, "+" (fig. 2.31). Because C + D + E is a putative clade, we can consider it to be a "functional ingroup" (FIG). A and B, then, form the "functional outgroups" (FOGs), because their logical relationship to C + D + E is the same as the relationship of the outgroups to A + B + C + D + E.

3. Use the original outgroups and the functional outgroup to polarize the multistate transformation series. First, place the corresponding character states x, y, and z at the tips of the branches for the outgroup and ingroup taxa (fig. 2.31). Polarizing via the outgroup taxa indicates that

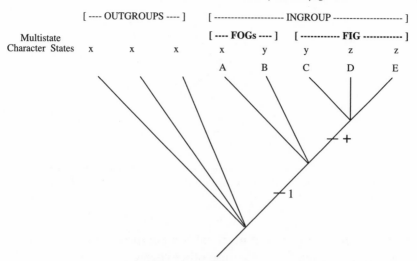

Figure 2.31. Partial resolution of phylogenetic relationships among members of ingroup A–E, based on distribution of states for one binary character. Outgroup comparisons support the interpretation that state + is apomorphic, linking taxa C, D, and E into a clade within the ingroup. This clade now serves as a functional ingroup (FIG) within the original ingroup, with the other members of the original ingroup (A and B) serving as functional outgroups (FOGs). The distribution of states (x, y, z) for a multistate character among members of the ingroup and outgroups is indicated above the taxa.

character state x is plesiomorphic to either y or z. Now, examine the functional outgroups. One of them (A) exhibits state x, while the other (B) exhibits state y. One of the functional ingroup members (C) exhibits state y, while the other two (D and E) exhibit state z. Because state y occurs in *both* the functional ingroup and the functional outgroup, we conclude that state y is plesiomorphic to state z. The original outgroups tell us that x arose first, while the functional outgroup tells us that y arose next. Consequently, state z arose last and the transformation series for this multistate character is the linear sequence x to y to z.

4. Combine the information from the binary and multistate characters. Outgroup comparisons for the binary character support the interpretation that state 0 is plesiomorphic and state + is apomorphic, linking taxa C, D, and E together as a clade. Functional outgroup analysis suggests that state x is plesiomorphic for the ingroup, that state y is apomorphic to x, linking B, C, D, and E together as a clade, and that state z is apomorphic to y, linking D and E together as a clade. Invoking the inclusion/exclusion rule results in a fully resolved phylogenetic tree (fig. 2.32).

To return to our example of winged mammals, we use some characters

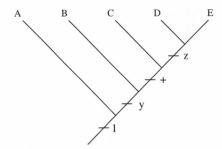

Figure 2.32. Phylogenetic tree for taxa A–E based on outgroup comparisons for two binary characters, and outgroup and functional outgroup comparisons for one multistate character.

to tell us that one group of mammals (ingroup) is distinct from other mammals (outgroup). We then use other characters to give us enough divisions among the winged members of this ingroup (functional in- and outgroups) to establish the sequence in which short and long wings evolved.

The use of functional outgroup comparisons (the logical justification of which is called "reciprocal illumination": Hennig 1966a; Wiley 1981), strikes many people as being circular. Although the distinction is a fine one, we believe that the curse of circularity is avoided because *the groupings within the original ingroup that allow us to determine polarities for the multistate character are determined a posteriori by reference to other traits.* Of course, if we do not have very many other characters, the robustness of our assessment of multistate polarities is not very great. Consequently, many phylogeneticists worry about using multistate characters at all.

 In a complementary manner, other phylogeneticists worry about problems associated with treating states of an evolutionary sequence as independent characters. For example, if we treated x, y, and z as three binary characters, we would be able to use the original outgroups to determine that "x present," "y absent," and "z absent" were plesiomorphic conditions. The states "y present" and "z present" would be interpreted as synapomorphies of B + C and of D + E, respectively. Considering "z present" as a synapomorphy of D + E would not conflict with other data (i.e., the second binary character), but considering "y present" to be a synapomorphy of B + C would conflict. Hence, in this case, breaking x, y, and z into independent characters would result in unnecessary postulates of homoplasy (violating Hennig's auxiliary principle). This is the reason phylogeneticists study each character carefully, argue transformation series as independently as possible (often considering both binary and multistate options), and constantly seek additional data in an effort to provide the best possible hypothesis at any given time.

HYBRID SPECIES AND PHYLOGENY: REPRESENTING RETICULATIONS

The assumption that all of phylogeny can be represented accurately by an internested branching diagram is refuted each time a new species is formed by the hybridization of members of two other species (called the parental species). This produces a reticulated phylogenetic history. Most zoologists consider hybrid speciation to be rare in animals, but botanists have recognized for a long time that many plant species have been formed in this manner, and the endosymbiont theory of eukaryote origin assumes that at least some major transitions in evolution are the result of hybridization.[6] For phylogeneticists, the presence of species of hybrid origin creates several problems. The first is expected homoplasy: a hybrid species may exhibit some or all of the apomorphies of both parental species. The second is expected multiple equally parsimonious phylogenetic trees: the hybrid species may exhibit traits supporting two or more different sets of relationships. The third is ambiguous classification: with which parental species should a hybrid species be classified? In this section, we will deal with the first two problems; for possible solutions to the third, see Wiley 1981 and Wiley et al. 1991 and forthcoming.

Phylogeneticists have long been aware that species of hybrid origin might pose special problems.[7] Funk (1985) first noted that when there are multiple equally parsimonious trees for a clade suspected of containing species of hybrid origins, those trees actually represented a small number of distinct topologies. Furthermore, inspection of the trees indicated that most of the homoplasy, and thus most of the ambiguity, was contributed by a small number of species. Removal of those species from the analysis produced a stable topology with reduced homoplasy.

WHEN SISTER SPECIES HYBRIDIZE

Let us consider the simplest case, one in which a species results from the hybridization of parental species that are themselves sister species (A and

6. See, for example, Schwartz and Dayhoff 1978; Margulis 1981; Markowicz and Loiseaux-de Goer 1991; Sitte 1993; and Maynard Smith and Szathmary 1995.

7. See Bremer and Wanntorp 1979; Funk 1981; Humphries 1983; Nelson 1983; Wagner 1983; and Wanntorp 1983a.

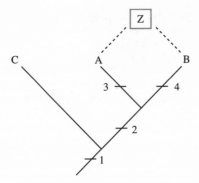

Figure 2.33. The problem with hybrid species. In this example, Z is a distinct species that arose from hybridization between two sister species, A and B.

Table 2.7. Data matrix for four characters and four taxa for the group shown in figure 2.33

Taxon	Character Transformation Series			
	1	2	3	4
A	1	1	1	0
B	1	1	0	1
C	1	0	0	0
Z	1	1	1	1

NOTE: *Bold* = synapomorphies.

B in fig. 2.33). The hybrid species (Z) exhibits the autapomorphies of both parents. (The data matrix for species A–Z is shown in table 2.7.) Phylogenetic analysis produces two distinct topologies, each with two alternative character optimizations, for a total of four trees (fig. 2.34). If we remove species Z from the analysis (temporarily), we obtain a single tree and single character optimization, with no homoplasy (fig. 2.35), from which we conclude that Z is problematic and is perhaps a hybrid species. We could not, however, postulate the parental species from this approach (given that Z is no longer part of the analysis!).

The inclusion/exclusion rule permits us to reach the same conclusions without removing species from the analysis, thus giving us an initial hypothesis of the parental species. If species Z is a hybrid of species A and species B, then Z is simultaneously the sister species of A and of B. We can represent this by treating Z as two different species in the analysis, one coded for the autapomorphy of A (Z_1 in table 2.8) and the other for the autapomorphy of B (Z_2 in table 2.8). The resulting phylogenetic analysis (fig. 2.36) depicts species Z as simultaneously the sister species of species A and B, which we interpret as an indication that Z is

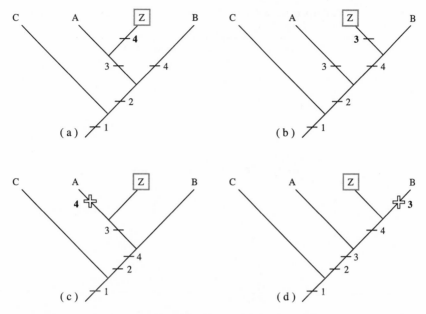

Figure 2.34. The presence of hybrid species leads to the production of more than one equally parsimonious tree. In this case there are two different topologies placing Z as either the sister group to A (the topology shown in panels *a* and *c*) or B (the topology shown in panels *b* and *d*). There are four trees because there are two equally parsimonious optimizations for the problematic characters 3 and 4 (convergent homoplasy, as shown in topologies *a* and *b;* or reversals to the plesiomorphic condition, as shown in topologies *c* and *d*).

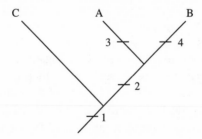

Figure 2.35. Excluding Z from the analysis removes the homoplasy, resulting in one tree.

a hybrid of A and B, the parental species. Note that the hypothesis that Z is indeed a hybrid species is refuted if the two subdivisions of Z—Z_1 and Z_2—cluster together on the tree.

WHEN DISTANT RELATIVES HYBRIDIZE

Now let us consider a more complex case. In this example, species G is a hybrid of parental species A and F, which are distantly related (fig.

Table 2.8.   Data matrix constructed by dividing the suspect taxon, Z, from figure 2.33 into two separate taxa, Z_1 and Z_2

Taxon	Character Transformation Series			
	1	2	3	4
A	1	1	1	0
B	1	1	0	1
C	1	0	0	0
Z_1	1	1	1	0
Z_2	1	1	0	1

NOTE: *Bold* = synapomorphies.

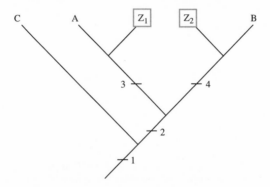

Figure 2.36.   Representing the potential hybrid species Z as two taxa in the analysis. Notice that Z_1 and Z_2 do not cluster together as sister species.

2.37). Phylogenetic analysis of species A–G (table 2.9) produces two equally parsimonious trees (fig 2.38). Once again, the ambiguity in the analysis is caused by the hybrid species, G, which is equally parsimoniously assigned as the sister species of A and of F. Coding G as two different species, one identical to A and the other identical to F (table 2.10) produces a single unambiguous result (fig. 2.39) depicting species G as simultaneously the sister species of species A and F, which we interpret as an indication that G is a hybrid of A and F. This application of the inclusion/exclusion rule leads us to the

> **Taxon Duplication Convention:** Whenever ambiguity in phylogenetic analysis disappears with the removal of particular species, it is reasonable to investigate the possibility that the species involved have reticulate histories. The most parsimonious hypothesis of reticulate history can be shown by duplicating the taxa in such a way as to remove the ambiguity without removing the taxa from the analysis.

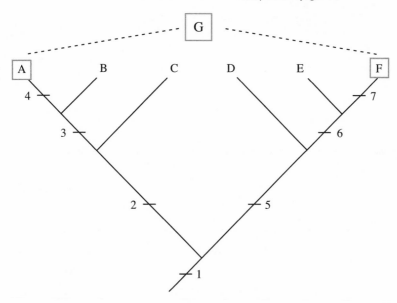

Figure 2.37. A more complicated hybrid species scenario. In this example, the parental species are not closely related. Taxon G arose via hybridization of taxa A and F.

Table 2.9. Data matrix for seven characters and seven taxa for the group shown in figure 2.37

Taxon	Character Transformation Series						
	1	2	3	4	5	6	7
A	1	1	1	1	0	0	0
B	1	1	1	0	0	0	0
C	1	1	0	0	0	0	0
D	1	0	0	0	1	0	0
E	1	0	0	0	1	1	0
F	1	0	0	0	1	1	1
G	1	1	1	1	1	1	1

NOTE: *Bold* = synapomorphies.

In practice, phylogeneticists represent species of hybrid origin by listing such taxa only a single time, with branches connecting them to both putative parental species. The result is a reticulated diagram, such as shown in figures 2.33 and 2.37 for our hypothetical examples, representing the hypothesized reticulate nature of phylogenetic diversification in the clade.

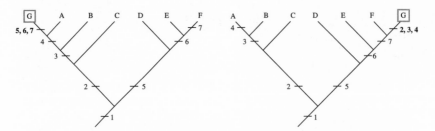

Figure 2.38.   Analysis of the data matrix shown in table 2.9 produces two equally parsimonious trees, placing G as either the sister group of A or of F.

Table 2.10.   Data matrix constructed by dividing the suspect taxon, G, in figure 2.38 into two taxa, G_1 and G_2

	Character Transformation Series						
Taxon	1	2	3	4	5	6	7
A	1	1	1	1	0	0	0
B	1	1	1	0	0	0	0
C	1	1	0	0	0	0	0
D	1	0	0	0	1	0	0
E	1	0	0	0	1	1	0
F	1	0	0	0	1	1	1
G_1	1	1	1	1	0	0	0
G_2	1	0	0	0	1	1	1

NOTE: *Bold* = synapomorphies.

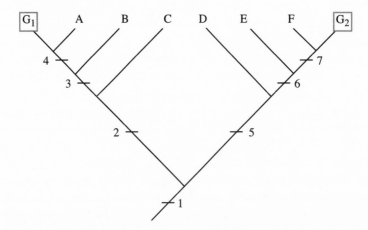

Figure 2.39.   Representing the potential hybrid species G as two taxa in the analysis removes the homoplasy because the two taxa do not cluster together.

A Caveat

This approach works when enough of the heritage of each parental species is expressed in the hybrid product(s). In some cases this means that we can actually recognize the parental genomes, segregate them as different taxa (according to the protocol discussed above), and perform phylogenetic analysis (see, e.g., Sang and Zhang 1999; Vargas, McAllister, et al. 1999; Shaw and Goffinet 2000). We expect that the more distantly related the parents, the less likely we will be able to recover the parental species, but also the less likely that hybrid speciation will occur. In addition, we often expect hybrid species to segregate phylogenetically more with one parent than another.[8] For cases in which ambiguity in phylogenetic analysis results from a small number of species, however, this approach gives us a first approximation of the possibility that the ambiguity is due to those species being hybrids.

At least some prokaryotes have an interesting ability—in times of stress they not only shut down their metabolic activities, they may actually digest the portion of their genomes associated with those metabolic activities. They are also capable of regaining genetic machinery for metabolic activities, but they do not always do so from a close relative. Recent studies of prokaryote phylogenetics have raised the possibility that this propensity on the part of prokaryotes has produced some wildly reticulated phylogenetic relationships. Some researchers have even expressed the opinion that this means we should change our view of phylogeny from a tree to a reticulated network (e.g., Doolittle 2000). Maynard Smith and Szathmary (1995, 1999), however, pointed out that genetic machinery associated with storing and transmitting information shows no similar evidence of horizontal transfer. Thus, the true phylogeny may still be treelike in form, but with substantial homoplasy in metabolic genes due to reticulated horizontal transfer. In other words, researchers interested in uncovering phylogenetic relationships for the prokaryotes should concentrate their attention on nonmetabolic genes in addition to ultrastructural traits. Even if the best depiction of phylogenetic relationships for prokaryotes is a reticulated network, we still do not have to develop fundamentally different approaches to assessing the evidence about relationships. We must simply recognize that the more horizontal transfer there has been, the more we will have to resort to the taxon duplication convention to find the source of those transfers. And, as we

8. See McDade 1990, 1992a, 1997; Mindell 1992, 1993; Zrzavy and Skala 1993; Skala and Zrzavy 1994; Maddison and McMahon 2000; Sang and Zhong 2000; and Xu 2000.

have shown here, this can be done completely within the framework established by phylogenetic systematics. If there is evidence of reticulation, the taxon duplication convention should be the way to find it parsimoniously, and then it can represented as a reticulated network if desired. In other words, we do not need a new discovery mode, though we may need a more flexible representation mode for our discoveries.

HOW ARE WE DOING?
Character Analysis: Assessing the Data

Character selection and analysis, not the use of computer programs, is the most important part of phylogenetics. It is not only a rigorous objective exercise, but also the place where one's experience and knowledge of various groups can contribute to the greatest extent. Editors of publications often limit the space allocated to character analysis, allowing only a list of character states and the data matrix. Space permitting, you should provide a discussion of each character used, a description of how the states were discriminated, and the polarity argumentation based on comparisons with multiple outgroups.

The logic of character analysis is the same for all types of data. The first step is character description, determining as many aspects as possible of the position and fine structure of any attribute to be considered as a character and also documenting the range of variation (Stevens 1992). Then criteria such as Remane's are used to assess similarity of position, similarity in parts/fine structure, and continuity through intermediate forms for similar characteristics from different species to determine character states. Third, Hennig's and Kluge's auxiliary principles (detailed similarity equals homology and all characters are independent) are invoked to produce a matrix of characters for each taxon being analyzed. Finally, character states are polarized to the greatest extent possible through comparisons with multiple outgroups.

The first applications of this logic used comparative anatomical data, the traditional form of systematic information. Systematists have mined this almost inexhaustible source of information extensively, using novel sources of **anatomical** data, **ontogenetic** data, **chromosomal** data, and **biosynthetic pathways** for secondary chemical compounds.[9] More re-

9. Examples of anatomical data include Hickey and Wolfe 1975; Brooks et al. 1985a,b; Brooks 1989a,b; Brooks, Bandoni, et al. 1989; M. Wake 1990, 1992, 1993, 1994; Raikow 1990; Winterbottom 1990, 1993; Siegel-Causey 1991; Prum 1993; Muñoz-Chápuli et al. 1994; Shubin et al. 1995; Stone 1997; Ferraguti and Erseus 1999; Fritsch 1999; Newton and de Luna 1999; Rouse, 1999; Ashe 2000; Buschbeck 2000; Fischer et al. 2000; Herendeen and Miller 2000; Katinas and Crisci 2000; Krenn and Kristensen 2000; Luter 2000; Renzaglia et al. 2000; Raikow and Bledsoe 2000; Sallum et al. 2000; Scheltema and

cently, **morphometric** data, long a bugaboo of phylogeneticists due to problems with assessing character states and homology,[10] seem to be joining the fold.[11] Even **fossils** have been willing to give up their secrets as readily as specimens representing living species,[12] often providing data critical to resolving key areas in phylogenetic analyses.[13]

In this spirit of enthusiastic exploration, and as anticipated by Mayr (1958) and Cracraft (1981a), it is becoming increasingly apparent that **ecological** and **behavioral** traits also provide suitable data for phylogenetic analysis.[14] Qualitative ecological and behavioral information is being used in two different ways in comparative biology, however. One approach groups species into a few very general functional classes (sometimes called "adaptive syndromes": e.g., "herbivory," "parental care," "mating displays," "pollinators," "parasitoids," or "parasites").[15] Phylogenetic analyses using these functional classes as if they were characters results in a substantial amount of homoplasy. The second approach views ecological and behavioral traits as characters. Those studies report substantially lower homoplasy and strong congruence with phylogenetic trees derived from other data.[16] Why the difference?

Real phylogenetic characters are historical outcomes represented in the reproductive and phylogenetic relationships among organisms and species. Membership in any class of objects, on the other hand, is based

---

Schander 2000; Staniczek 2000; Vilhelmsen 2000; Wroe et al. 2000; and Zamparo et al. 2001. Examples of otogenetic data include Blackmore et al. 1987; Klompen and O'Connor 1989; O'Connor 1992; Mabee 1993, 2000; Wenzel 1993; Buscalioni et al. 1997; Kampny and Denglers 1997; Stone 1998; and Rotheray and Gilbert 1999. Examples of chromosomal data include Farris 1978; Borowik 1995; Baker et al. 1987; and Mattern and McLennan 2000. Examples of biosynthetic pathway data include Seaman 1982; Seaman and Funk 1983; and Barkman 2001.

10. For example, Archie 1985; Bookstein et al. 1985; Goldman 1988; Farris 1990; Johnston et al. 1991; and Crowe 1994.

11. Fink and Zelditch 1995; Sosa and de Luna 1998; Swiderski et al. 1998; Zelditch and Fink 1998; and Zelditch et al. 1998, 2000.

12. For example, Eldredge 1972, 1973; Eldredge and Branisa 1980; Butler 1982; Trueb and Cloutier 1991; Norell and Novacek 1992; Novacek 1992a,b; Lieberman 1993, 1994, 1997; Gould 1995; Lieberman and Eldredge 1996; Brochu 1997; Lieberman and Kloc 1997; Lieberman et al. 1997; Sidor and Hopson 1998; Lieberman 1999, 2001; and Wang et al. 1999.

13. For example, Wiley 1976; Patterson 1981a; Doyle and Donoghue 1987; Gauthier et al. 1988; Donoghue et al 1989; Donoghue and Sanderson 1992; and Begun et al. 1997.

14. For example, Dunford and Davis 1975; Arnold 1987; McLennan et al. 1988; Prum 1990; Wenzel 1991, 1992, 1993; McLennan 1993a; de Queiroz and Wimberger 1993; Greene 1994a; Wimberger and de Queiroz 1996; Miller and Wenzel 1995; Paterson et al. 1995; Hedenäs and Kooijman 1996; Hughes 1996; Kennedy et al. 1996; Halloy et al. 1998; Slikas 1998; Stuart and Hunter 1998; Basibuyuk and Quicke 1999; Zyskowski and Prum 1999; Bostwick 2000; and McLennan and Mattern 2001.

15. Harry et al. 1996, 1998; Johnson et al. 1998; and Price and Carr 2000.

16. See references cited in note 14, above.

on possessing the defining trait(s) of the class, regardless of the origin of those trait(s). Many classes, therefore, do not owe their membership to common ancestry. A frequent mistake is to confuse the end product, or outcome, with the trait itself, which involves the mechanism by which the outcome is achieved.[17] Nuptial coloration in stickleback fishes is not the same as nuptial coloration in birds of paradise, and no competent systematist would ever code them as the same character. The resulting elevated homoplasy that would result from coding these two characters as "the same" (nuptial color present) would thus not be indicative of real convergent evolution but of confusion on the part of the researcher.

Finally, the most rapid growth in phylogenetic systematics concerns the use of **molecular,** especially nucleotide sequence, data. Biologists have been using molecular data to elucidate phylogenetic relationships for nearly half a century, but only in the past 20 years or so has the rigor of phylogenetic systematics been applied to such analyses.[18] Robust analysis of nucleotide sequence data follows the same precise logic for character analysis as any other type of data.[19] First, choose molecular data that will form phylogenetic hierarchies, for example, those that are nonrecombinant, maternally inherited alleles or fixed attributes (Davis and Nixon 1992). Second, generate sequences from those sources. Third, align the sequences based on the presumption that similarity of position equals homology (i.e., sequence similarity = sequence homology). Fourth, maximize the number of such a priori homology assumptions (e.g., Gatesy et al. 1993; Philips et al. 2000). Finally, perform phylogenetic systematic analysis on the resulting aligned sequence data.

At the present time, phylogeneticists have identified three areas of special concern for those working with sequence data. The first is that molecular systematics, despite its technological sophistication, is still in its infancy. As a consequence, we have comparable information for a very small number of species and a very small number of genes, so sampling error is a problem.[20] As we generate more sequences from more

17. For a good recent discussion, see Stuart and Hunter 1998.

18. For example, Miyamoto 1981, 1983, 1984; and Hillis 1985, 1987. For reviews and texts, see Hillis 1988, 1991, 1994; Hillis and Davis 1986; Hillis et al. 1983, 1993; Moritz et al. 1987; Hendy and Penny 1989; Hillis and Moritz 1990; Mindell and Honeycutt 1990; Cracraft and Helm-Bychowski 1991; Miyamoto and Cracraft 1991; Hillis and Huelsenbeck 1992; Soltis et al. 1992; Helm-Bychowski and Cracraft 1993; Brower et al. 1996; and Hillis et al. 1996.

19. For example, Brower and DeSalle 1994; Graybeal 1994; Hillis 1994; Mishler 1994; Brower et al. 1996; and Wägele 1996a,b.

20. For example, Wheeler 1992; Lecointre et al. 1994; Brower 1997; Wägele and Rödding 1998; Margulis et al. 1999; and Caterino et al. 2000.

genes for more species, we expect these problems to disappear. The second concern stems from the fact that each base position can be occupied by one of only four alternate bases; the range of potential evolutionary changes for any base position at any time is accordingly very small. We therefore expect routinely to encounter high degrees of "built-in" homoplasy (Brooks 1996) due to this "design constraint" (Wake 1991). Additional features, such as the complexity of secondary structure, may make alignments ambiguous.[21] The third concern is that all sequences, like all suites of morphological and behavioral traits, seem to be comprised of relatively fixed ("conserved") and highly variable ("non-conserved") regions. And like morphological and behavioral traits, the highly variable characters tend to be problematic for the reconstruction of phylogenies (Hendy and Penny 1989).

When we include non-conserved regions, our analyses may be affected by what has been termed "long branch attraction." This is an effect that may occur when the evolution of a clade includes some rapidly evolving lineages; for example, rapid speciation events early in the phylogeny of a group producing two or more ancient species lineages in which neutral mutations have been in effect long enough to "saturate" or randomize the sequences. Given the relatively high levels of "built-in" homoplasy in nucleotide sequence data overall, such saturated sequences may actually "attract" each other, resulting in two distantly related members of the ingroup being grouped together, either at the base or at the tip of a phylogenetic tree. Additionally, if two or more outgroup taxa exhibit saturated sequences relative to the ingroup, long branch attraction effects may produce radically different yet apparently robust results using each outgroup, or results differing markedly from phylogenetic analyses based on other data, including other genes.[22] These effects can be manifested at very high levels (e.g., the phylogeny of major eukaryotes: Stiller and Hall 1999; Morin 2000), intermediate levels (e.g., the phylogeny of the major lineages of the Clitellata: Martin et al. 2000), or low levels (e.g., species level analysis of some iguanid lizards: Wiens and Hollingsworth 2000).

The region in "possible tree space" affected by long branch attraction is commonly referred to as the "Felsenstein zone" (Huelsenbeck and Hillis 1993), a place you do not want to visit if you can avoid it. Fortu-

21. Lake 1991; Gatesy et al. 1993; Wägele and Stanjek 1995; Winnepenninckx and Backeljau 1996; Wägele and Rödding 1998; Hancock and Vogler 2000; and Savill et al. 2001.

22. Felsenstein 1978; Archie 1989; Hendy and Penny 1989; Swofford 1991; Huelsenbeck and Hillis 1993; Huelsenbeck 1995; and Morin 2000.

nately, we can often distinguish conserved from non-conserved regions during character analysis.[23]

 Remember, you are responsible for the quality of the data upon which your analysis is based; hence, if you are using sequence data, you must ensure that your character analysis is as rigorous as that performed on other types of data which might be at your disposal.

The most effective way to reduce the impact of long branch attraction (indeed, homoplasy in any data set) on phylogenetic analyses is to collect more data (Mitchell et al. 2000). This recommendation stems from Hennig's (1950, 1966a) proposition that any data, analyzed phylogenetically, will give the same or highly similar answers because organisms are the result of a dynamic interaction among genetic, developmental, physiological, morphological, and behavioral systems through time. There is thus no a priori reason to believe that one type of data contains more information about phylogenetic relationships than any other. Hennig termed his approach to character analysis **holomorphological.** The published record of the past 25 years suggests that he was correct—there does not seem to be any aspect of organismal biology, from nucleotide sequences to complex feeding ecologies to mating behaviors, that cannot be assessed using phylogenetic character analysis and outgroup comparisons and then used to reconstruct phylogenetic relationships. At the same time, there appears to be no type of data that is immune from homoplasy, and thus no particular type of data can be assumed a priori to be a better indicator of phylogenetic relationships than any other type of data. The good news is that morphological/functional and molecular data provide highly similar results when analyzed using phylogenetic systematic methods based on sufficient data,[24] and it even appears that the rate and degree of character change at the genotypic and phenotypic levels are similar (Olmland 1997). The take-home message here is simple: *One type of data is not inherently better for phylogenetic reconstruction than another. For particular clades and particular types of data, some characters will be more phylogenetically informative than others.* This applies to all data sets—molecular, morphological, ecological, or behavioral.

23. For example, Hillis 1991; Hillis and Huelsenbeck 1992; Lyons-Weiler and Hoelzer 1997; Wägele and Rödding 1998; Grundy and Naylor 1999; Castresana 2000; and Steel et al. 2000.

24. For example, Miyamoto 1981, 1983, 1984; McKitrick 1985; Hillis and Davis 1986; Hillis 1987, 1988; Patterson 1987; Arntzen and Sparreboom 1989; Cracraft and Mindell 1989; Bledsoe and Raikow 1990; DeSalle and Grimaldi 1991; Patterson et al. 1993; Larson 1994; Mishler 1994; Olmland 1994; Brower 1997; Harris et al. 1998; Silvain and Delobel 1998; Fang et al. 2000; and Mattern and McLennan 2000.

Assessing the Robustness of Your Results

We can evaluate the performance of each character originally coded as a synapomorphy by calculating its **consistency index** (Kluge and Farris 1969). The consistency index (CI) of a character is simply the reciprocal of the number of times that character appears on the tree; therefore, true homologues (real synapomorphies) have a CI of 1.0, or 100%. The CI is a favorite summary "statistic" (usually given as a percentage) in most computer programs, so it is worthwhile practicing some hand calculations to remove some of the computer mystique.

On the most parsimonious tree in the RSTidae example (topology 1: fig. 2.26a,b; table 2.3), the apomorphy coded "1" in transformation series 3 appears twice, so its CI = ½ = 50%. Based on this CI, we can see that this character is not really homologous in all species. Of course, our best estimate of a true homologue is an a posteriori test, because we can only calculate the consistency index after we have determined our best estimate of common ancestral relationships based on all the characters. There is no way to know in advance that a particular apomorphy will have a CI of less than 100%.

The interesting point about calculating consistency indices for individual characters is that, if you have the tree before you, it is completely unnecessary. Examination of the tree will tell you everything you need to know about the number of times a particular character has appeared during the evolution of your group. This calculation is simply a numerical convenience. More important to considerations of choices among multiple trees is the consistency index for the entire tree (Kluge and Farris 1969). Basically, this is a goodness-of-fit measure designed to indicate the degree of support for a particular tree by a particular data set. The CI ranges from 0% to 100%, with a value of 100% indicating that all characters support the tree (no homoplasy). Once again, the calculation is very simple:

$$CI = \frac{\text{\# of apomorphic characters in the data matrix}}{\text{\# of character changes on the tree}} \times 100$$

So, for the trees shown in figure 2.26, the consistency indices would be: (1) $\frac{7}{9}$ = 77.8% (fig. 2.26a,b) and (2) $\frac{7}{10}$ = 70% (fig. 2.26c,d). For a given data set, trees that have the same length have the same CI. Tree length gives an indication of the quantity of character support, whereas the CI gives an indication of the quality of character support for a given tree.

A major drawback of the CI is that its value can be inflated by the addition of autapomorphies, so most phylogeneticists now report a CI value calculated without using autapomorphies. Also, because the consistency index is affected by both the number of taxa and the number of characters in your analysis, it has been considered axiomatic that the CIs from different data sets cannot be compared directly in order to identify the best supported tree. Recent simulation studies, however, have provided methods for directly comparing the relative robustness of the CIs from different data sets (Klassen et al. 1991; Meier et al. 1991).

In the 30 or so years since the development of the consistency index, there has been an increase in the number of methods available to assess the amount of phylogenetic information, or signal, in a given data set: for example, the **retention index** and the **rescaled consistency index** (Farris 1989a,b; Naylor and Kraus 1995); **tree "rot"** (Sorenson 1999) approaches (Felsenstein 1985b; Davis et al. 1993; Bremer 1994; Davis 1995; Morgan 1997; Gatesy 2000; Wilkinson et al. 2000); **bootstrap analysis** (Felsenstein 1985b; Sanderson 1989, 1995; Hillis and Bull 1993); and **jackknife analysis** (for reviews, see Lee 2000a and Mort et al. 2000). In general these methods provide various ways to identify which portions of a tree are well supported and which portions are weak. They do not, however, provide a way to strengthen the weak portion of an analysis. That can only be accomplished by following two steps. First, take a longer look at the original descriptions and codings of the problematic characters; you may have made a mistake in your original hypothesis of homology (Kluge 1999). This is an important and often overlooked point: phylogenetic systematic analysis is an open-ended, iterative process. No matter how carefully you argue the characters the first time, you could have made a mistake. The only way to remedy this mistake is to return to the data. If all seems in order there, then move to step two: collect more data.

MULTIPLE EQUALLY PARSIMONIOUS TREES FROM THE SAME DATA SET

Phylogeneticists are often faced with multiple equally parsimonious trees even after substantial effort (e.g., Chase et al. 1993; Soltis et al. 1997). There are two options for dealing with this situation—either summarize the information contained in the trees or try to refute some of the trees (Carpenter 1988a). Multiple equally parsimonious phylogenetic trees (fig. 2.40a) can be summarized using **consensus trees.** Consensus trees may highlight groupings that are identical on all of the equally parsimonious trees (a **Nelson,** or **strict,** consensus tree: fig. 2.40b), groupings that do not conflict with each other on any tree (an **Adams** consensus tree: in this case, also fig. 2.40b), or groupings that occur in an arbitrary proportion of the trees (a **majority rule** consensus

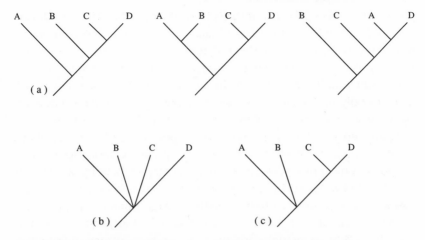

Figure 2.40. Three different types of consensus trees. (*a*) Three equally parsimonious phylogenetic trees recovered from the analysis of a single data set. (*b*) Strict and Adams consensus tree. (*c*) Majority rule consensus tree.

tree, special cases of which are jackknife and bootstrap consensus trees: fig. 2.40c).[25] As you can tell from figure 2.40b,c, all consensus trees lose some information. This should not be unexpected, since the purpose of consensus trees is to reduce the impact of conflict in the data and give a conservative estimate of phylogenetic relationships. Nevertheless, you can also see that different consensus methods may produce different results, so it is important to justify your choice of a particular type of consensus tree a priori. Remember, "consensus" is not the same as phylogenetic analysis of a data set (that analysis has already produced multiple trees). Consensus is a way to "make the best of a bad situation" until you have collected more data.

One of the reasons for getting multiple equally parsimonious trees in an analysis is the presence of polytomies. Although the computer programs will try to find all possible dichotomous representations of those polytomies, phylogenetic systematics does *not* assume dichotomous speciation always occurs (see, e.g., Wiley 1981; Wiley and Mayden 1985; Funk and Brooks 1990; Brooks and McLennan 1991). There are two major explanations for the presence of polytomies: (1) the data are insufficient or inadequate and (2) the phylogenetic history of the group includes evolutionary phenomena expected to be reflected in polytomous

25. For detailed discussion of the construction and properties of different consensus trees, see Adams 1972, 1986; Margush and McMorris 1981; Nelson and Platnick 1981; Miyamoto 1985; Hendy et al. 1988; Bremer 1990; Wiley et al. 1991, forthcoming; and Wilkinson 1994, 1996.

phylogenetic trees (Nixon and Davis 1991; Platnick et al. 1991; Maddison 1993; Wilkinson 1995). Insufficient data may produce a single, most parsimonious phylogenetic tree with polytomies indicating the places where you lack the data necessary to produce a dichotomous result. Inadequate data produce multiple equally parsimonious trees, the consensus of which includes polytomies indicative of the ambiguity in the data. These are "soft" polytomies, that is, polytomies that are artifacts of the data.

"Hard" polytomies, by contrast, may be produced either by peripheral isolates speciation, in which the ancestral species persists after the production of one or more descendant species (chapter 3), or by rapid bursts of speciation followed by long periods of relative stasis.[26] During such episodes, it is postulated that a relatively large number of species are formed in such a short period of time that few, if any, synapomorphic traits have time to evolve; consequently, the burst of new species produced will all be characterized by autapomorphies. This is thought to be especially true for nucleotide sequence data. (The effect is identified as a polytomy of many "short" branches: Avise 2000.)

The distinction between soft and hard polytomies is an important one. For example, a number of studies have been published in which a single gene or even a gene fragment constitutes the only data from which a phylogenetic hypothesis has been derived. When the analysis produces multiple equally parsimonious trees due to ambiguity at only one or a few basal nodes, this is often taken as de facto support for the hypothesis that there was a rapid burst of speciation at that point in phylogeny. Documenting such bursts of diversification, in turn, is a key component of research in phylogeography and evolutionary radiations (chapters 3, 4, and 6). If, however, the data used to support the hypothesis of a rapid burst of speciation are the only data used to build the tree, then the tree has no independent support. This means that an explanation based upon a burst of speciation cannot be distinguished from an explanation based upon inadequate or insufficient data. Such evolutionary scenarios should be treated with caution.

MULTIPLE TREES FROM DIFFERENT DATA SETS

There are two schools of thought about how we can integrate information from different data sets into one hypothesis of phylogenetic relationships. The first school supports a method called **taxonomic congruence**, which involves the search for a consensus among different tree topologies. There is an important difference between this use of "consensus" and our previous discussion of the topic. Multiple equally parsi-

---

26. For an excellent overview, see Avise 2000.

monious trees produced from the same data set have, by definition, the same amount of character support (hence the term "equally parsimonious"). Trees reconstructed from different data sets, however, need not, and indeed rarely do have, the same degree of support (hence they are called "different topologies," not "different, equally parsimonious topologies"). Researchers who advocate the adoption of taxonomic congruence believe different types of characters evolve in such different ways that they cannot be compared directly (and perhaps cannot even be analyzed using the same methods) or can be combined only with difficulty.[27]

Most researchers who disagree with the use of taxonomic congruence do so on both philosophical and empirical grounds.[28] The philosophical objection is that the principles of phylogenetic systematics are so fundamental to evolutionary thought that all phylogenetically relevant data should be capable of being, and should be, analyzed in the same manner. Therefore, *the hypothesis that different forms of data have different evolutionary properties should be tested by reference to simultaneous analysis of different types of data, not assumed a priori* (Kluge 1997, 1998a,b). The empirical objections are twofold. First, as we noted above, phylogenetic analysis of particular types of data often produces multiple equally parsimonious trees, which can be summarized in a consensus tree. Taxonomic congruence studies of such data sets make consensus trees of consensus trees. Since consensus trees are less informative as general classifications than any single, equally most parsimonious alternative tree produced by a given analysis (Miyamoto 1985), opponents of taxonomic congruence argue that such an approach may obscure information substantially.

Second, and more important, taxonomic congruence treats every tree as equally well supported. Is this a reasonable assumption? Let's say you have spent several years reconstructing a phylogenetic tree for the ABCidae based upon 20 characters (fig. 2.41a). At the same time your results are published, two other papers also appear, agreeing on a different set of relationships for the ABCidae using 6 and 13 characters, respectively (fig. 2.41b, c). A fourth researcher, looking at the two trees through the filter of taxonomic congruence using a majority rule consensus tree, concludes that the topology shown in figure 2.41b, c is better supported

27. For example, Anderberg and Tehler 1990; Bledsoe and Raikow 1990; Bull et al. 1993; Kim 1993; Chippindale and Wiens 1994; Miyamoto et al. 1994; Miyamoto and Fitch 1995; Purvis 1995; Wiens and Reeder 1995; Page 1996; and LaPointe and Cucumel 1997.

28. For example, Kluge 1989; Barrett et al. 1991; Eernisse and Kluge 1993; Friedlander et al. 1994; Olmstead and Sweere 1994; de Queiroz et al. 1995; Tehler 1995; Allard and Carpenter 1996; Nixon and Carpenter 1996; Baker and DeSalle 1997; Kluge 1997, 1998a,b; Siddall and Kluge 1997; and Wyner et al. 2000.

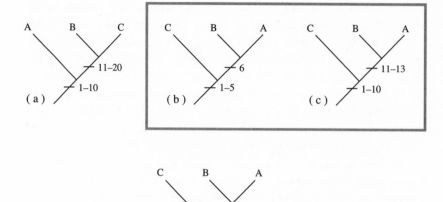

Figure 2.41. The equally valid world of taxonomic congruence. (*a*) Tree topology based upon 20 characters; (*b*) Tree topology based upon a second data set of six characters; (*c*) Tree topology based upon a third data set of 13 characters. (*d*) The topology shown in (*b*) and (*c*) is preferred by taxonomic congruence using a majority rule consensus analysis.

because two different data *sets* converged on the same result. This conclusion occurs irrespective of the fact that you have ten characters supporting the placement of B and C as sister groups, while your colleagues have only one and three characters, respectively, supporting the group B + A.

As is often the case, Darwin realized the solution to this problem more than a century ago:

> The value indeed of an aggregate of characters is very evident in natural history. Hence, as has often been remarked, a species may depart from its allies in several characters, both of high physiological importance, and of almost universal prevalence, and yet leave us in no doubt as to where it should be ranked. Hence, also, it has been found that a classification based on any single character, however important that may be, has always failed; for no part of the organisation is invariably constant. (1872:398)

As we noted above, Hennig used Darwin's perspective to support the "holomorphological" approach; it has been re-emphasized by Kluge (1989, 1997, 1998a,b, 1999) as the concept of **total evidence**, or *"use all available data from any source in your phylogenetic analysis."* This second school of thought advocates using the tree that is supported by the most data (characters), not by the largest number of data sets. So, back to our example. Combining all of the information from the three different

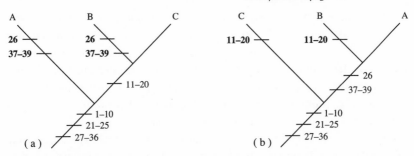

Figure 2.42. Total evidence versus taxonomic congruence. (*a*) The total evidence tree based upon combining all the characters from the three trees depicted in figure 2.41 shows four homoplasious traits (bold). (*b*) The taxonomic congruence analysis of the same data set produces ten homoplasies, none of which are the same traits as those identified by analysis of the actual characters.

studies into a single 39-character data matrix produces one tree (39 steps in the matrix/43 steps on the tree = CI of 90.7%: fig. 2.42a), placing B and C as sister groups because that is what the majority of the characters support. In the world of phylogenetic systematics, therefore, not all trees are equally valid. If they were, we would not need so many methods for assessing the robustness of our results. Because the taxonomic congruence tree is not the best supported hypothesis based upon actual data, the method not only gives us an incorrect hypothesis of relationships in the ABCidae, but it also artificially inflates the amount of homoplasy on the tree (39 steps in the matrix/49 steps on the tree = CI of 79.6%: fig. 2.42b) and highlights a different subset of characters as being homoplasious. This difference between the two methods is not trivial. Identifying homoplasious traits is an important component of many research programs in evolutionary biology (as you will discover in the remaining chapters), and as you can see from figure 2.42, the two methods will take you down dramatically different pathways.

Overall, we concur with Hennig, Kluge, and their fellow researchers: *the best way to achieve the goal of robust phylogenetic hypotheses is to analyze as many characters from as many sources as possible for any group.* We have already reported that comparisons of phylogenetic analyses using morphological and molecular data for the same taxa tend to agree more than they disagree. It should therefore be no surprise that total evidence studies tend to produce results that are more robust than analysis of any subset of the data.[29]

29. For example, Miyamoto 1985; De Sá and Hillis 1990; Barrett et al. 1991; Jones et al. 1992; Vane-Wright et al. 1992; Eernisse and Kluge 1993; Wheeler et al. 1993; Friedlander et al. 1994; Olmland 1994; Olmstead and Sweere 1994; de Queiroz et al. 1995; Tehler 1995; Wiens 1995; Harry et al. 1996; Huelsenbeck et al. 1996; Nixon and Carpenter 1996: Soltis

The answer to the question we posed a decade ago, "When do you have enough characters?" (Brooks and McLennan 1991), thus remains the same. The reconstruction of phylogeny is an open-ended process, so you never have enough characters, and total evidence studies become an imperative. The corollary of this position is, *Don't waste time trying to coerce an answer out of a poor data set. No matter what computer manipulation you perform to produce a hypothesis, if your basic data are not robust, eventually you or someone else will have to obtain more data to produce a truly robust result.* This is one of the most important rules in comparative phylogenetic research:

 Murphy's Rule: Garbage in garbage out, no matter how powerful your computer.[30]

If the only available tree is a bad one, should you use it? No (Coddington 1985). Increasing the number of homologous characters in an analysis decreases the number of equally parsimonious trees because homologous characters correspond to a single phylogenetic tree. If the number of homologies has been greater than the number of covarying homoplasies in evolution, then the most biologically robust method of dealing with a weakly supported tree (e.g., multiple equally parsimonious trees) is to collect more data.

What's wrong with equating a taxonomic classification with phylogeny? It is inappropriate to use a taxonomic classification as a phylogeny because many classifications portray paraphyletic (or polyphyletic) taxa as monophyletic groups (Lanyon 1994; Rognes 1997). Evolutionary explanations based on such classifications are misleading.[31] For example, they will overestimate the

---

et al. 1996; Baker and DeSalle 1997; Nandi et al. 1998; Wiley et al. 1998; Fredericq et al. 1999; Les et al. 1999; Light and Siddall 1999; Renner 1999; Trontelj et al. 1999; Vargas, Baldwin and Constance 1999; Wang et al. 1999; Anderberg et al. 2000; Anderson 2000; Backlund et al. 2000; Baker et al. 2000; Damgaard et al. 2000; Doyle and Endress 2000; Flores-Vilela et al. 2000; Gatesy and Arctander 2000; Ghebrehiwet et al. 2000; Giribet et al. 2000; Graham and Olmstead 2000; Lara and Patton 2000; Mattern and McLennan 2000; Miadlikowska and Lutzoni 2000; Newton et al. 2000; Phillips et al 2000; Renzaglia et al. 2000; Savolainen et al. 2000; Smith 2000; Wyner et al. 2000; Yang et al. 2000; Flores-Villela et al. 2001; Hoberg et al. 2001; Lee 2001; and McLennan and Mattern 2001. Presently underway is an impressive effort by botanical systematists to provide a massively documented total evidence phylogenetic framework for angiosperms. For current progress, see Doyle 1998; Soltis et al. 2000; Graham et al. 2000; Floyd and Friedman 2000; Kramer and Irish 2000; Mathews and Donoghue 2000; Mishler 2000; and Qiu et al. 2000.

30. See also Presch 1989; and Crother and Presch 1992, 1994.

31. See Greene 1998 for an excellent exposition of the pitfalls of teaching "paraphyletic biology."

amount of homoplasy because diagnoses for paraphyletic groups list synapomorphic traits more than once. This gives the impression that these traits are actually examples of parallel or convergent evolution. If homoplasy is as important in evolution as researchers think, then it is in our own best interest to have as robust a hypothesis of homoplasy as possible.

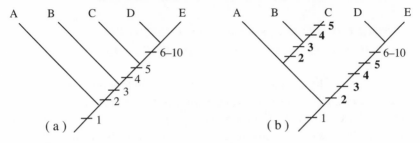

Figure 2.43. Why it may be incorrect to equate classification (or taxonomy) with phylogeny. (*a*) Phylogenetic tree reconstructed for species A–E based upon phylogenetic systematic methods. (*b*) Tree reconstructed from a taxonomic classification scheme that includes the paraphyletic group A + B + C. This tree forces us to postulate that characters 2, 3, 4, and 5 evolved twice.

Consider the following example. Figure 2.43a depicts the phylogenetic tree for a group of hypothetical taxa. Although the presence of five synapomorphic traits (characters 6–10) distinguishes species D and E from species A, B, and C, these taxa are all members of a monophyletic group characterized by the presence of traits 1–5. Now, suppose we have a classification scheme that places species D and E in one taxon because they are so distinct, and places species A, B, and C in another taxon. Reconstruction of phylogenetic relationships based on this classification will produce the tree depicted in figure 2.43b. This arrangement forces us to postulate that characters 2–5 evolved twice, overestimating the amount of homoplasy required by the data. Since most commonly accepted classifications include paraphyletic groups, they cannot serve as independent templates for estimating the origins, diversifications, and associations of characters through evolutionary time. Unfortunately, given the current dearth of available phylogenetic trees, many researchers have been forced to utilize classifications in their preliminary analyses of ecological/behavioral evolution.

> The large number of paraphyletic groups in currently accepted classifications is due to a shortage of phylogenetic analyses of groups, not to inherent shortcomings in the Linnaean system (Benton 2000; Nixon and Carpenter

2000). Potential problems rendering phylogenetic trees into Linnaean classifications stem from the fact that not all branches in a phylogenetic tree need be specified in a Linnaean classification. Protocols exist for recovering a complete phylogenetic tree from a Linnaean classification, based solely on monophyletic groups, that does not denote each branch as a distinct supraspecific taxon (Wiley 1981; Wiley et al. 1991, forthcoming).

BEATING THE HOUSE: CHARACTER WEIGHTING AND ALTERNATIVE METHODS OF PHYLOGENETIC ANALYSIS

Hypotheses of phylogenetic relationships provided by phylogenetic systematics are based on the maximum number of homology statements permitted by the data, accompanied by an effort to minimize input based on the researcher's preconceived notions of relationships. This imbues the method with a very convenient property; hypotheses produced in this manner do not depend directly on the number of homologies (assuming there is a sufficient number to resolve all sister-group relationships for the members of the ingroup), but rather on their distribution. That is, giving some homologies more significance than others (differential weighting) will not change the tree (and a priori character weighting violates Kluge's auxiliary principle). On the other hand, it is correlated homoplasy that produces ambiguous results, so differential weighting of homoplasious traits could alter a phylogenetic hypothesis substantially. And if phylogeneticists adopt the position that they do not know what the homoplasies are a priori, then

> Can't you just manipulate the data to get the answer you "want"? Of course.

In principle, it is possible to invent a weighting scheme to produce any desired phylogenetic tree (Farris 1969). The problem of data manipulation, which also includes eliminating taxa and characters from the analysis because they do not give you the "correct" answer, is not unique to either phylogenetic systematics in particular or biology in general. In fact, phylogenetics may be slightly more open to scrutiny because you have to report both the character descriptions and the codings in each publication; therefore, the original data are more accessible to re-analysis than, say, a table of p values. Manipulating character state weighting a priori renders the resultant hypothesis less testable (i.e., subject to falsification by the addition of new characters) than a tree derived from an unweighted analysis of the same data. Because testability has consistently been paramount in phylogenetic systematics and research incorporating phylogenetic information (e.g., Wiley 1975; Gaffney 1979; Kluge 1997, 1998b, 1999, 2001), most phylogeneticists do not attempt to "help" their data out by awarding some characters more significance than others a priori. They believe such character weighting places limitations on what we can discover from our data, and hence what we can learn. *We are talking here about a priori subjective weighting, not about character selection, which is highly subjective (being dependent on the interests, skills, and finances of the investigator), or about character analysis (e.g., applying Remane's criteria, outgroup comparisons), which is not subjective at all.*

And yet, the suspicion persists among evolutionary biologists that without a priori weighting, data analyzed using phylogenetic systematic methods may not produce "true" phylogenetic reconstructions (e.g., Bryant 1989). Scientists holding these views are not concerned about data that frustrate us because they produce ambiguous results, and whose influence can be overcome by including more characters in any analysis. They are concerned about data that may provide a robust result depicting "incorrect" relationships. So long as the data are analyzed using phylogenetic systematic methods, such a situation could occur *only if homologies are outnumbered by covarying homoplasies.* The possibility of this happening will increase if your characters are drawn from a restricted and functionally connected set of traits (e.g., 15 characters all drawn from wing morphology might lead you to suspect that your results could be influenced by selection operating simultaneously on all traits associated with "flight") or from a type of data that has a high rate of change in a relatively small number of options (e.g., nucleotide sequence data, with only four options for change, have a substantial amount of "built-in" homoplasy).[32]

As we noted above, phylogenetic systematic analysis of different data sets, say of morphological/behavioral/ecological data and of nucleotide sequence data, generally produce highly similar results. That's the good news. The bad news is that this is not always the case, and the growing appreciation for the significance of phylogenetic hypotheses has made us increasingly demanding about the robustness of such results. This situation has led to two parallel research programs in phylogenetic analysis. One of these is the "total evidence" approach described above. The other approach is based on the belief that data can be manipulated a priori to correct their shortcomings. This "correction" usually takes the form of an evolutionary model, either implicit or explicit, which stipu-

---

32. Felsenstein 1978; Hendy and Penny 1989; Huelsenbeck and Hillis 1993; Huelsenbeck 1995; Brooks 1996; and Milinkovitch et al. 1996.

lates differential weighting of traits. Such weighting can be particular (i.e., specific weights are assigned to specific traits) or general (e.g., probability statements covering entire classes of traits). The most popular of this second type of weighting scheme are **maximum-likelihood models** (Edwards 1972, 1996), especially applied to nucleotide sequence data for which there are concerns about non-conserved regions, saturated sites, and long-branch attraction effects.[33] The enthusiastic development of this research program during the past decade has produced a plethora of models.[34]

Not surprisingly, many phylogeneticists have reservations about the status of phylogenetic trees derived from maximum-likelihood models, *especially when the results differ from those derived using phylogenetic systematic methods.* When attempting to explain the disagreements, care must be taken not to confound the type of data with the method of analysis. The typical case is one in which maximum-likelihood analysis of sequence data conflict with a previous phylogenetic systematic analysis of morphology. Often, one or both of the data sets is not particularly robust (e.g., there were few morphological characters used, or phylogenetic systematic analysis of the sequence data produces multiple equally parsimonious trees). Advocates of phylogenetic systematic approaches fear that expressing a preference for the maximum-likelihood result over a total evidence phylogenetic systematic result places the researcher in a position of having to argue a priori knowledge of the true phylogeny, which would diminish the robustness and independence of those results with respect to subsequent testing and potential falsification. Advocates of maximum-likelihood approaches argue in return that such disagreement between the results is prima facie evidence that the characters used in the phylogenetic systematic study were ones exhibiting high degrees of correlated homoplasy, and thus should be discounted or re-analyzed.

Some researchers have found shortcomings in the assumptions of some existing maximum-likelihood models,[35] especially the notion that nucleotide sequences evolve in a neutral manner unaffected by selection (for a review, see Yang and Bielawski 2000). Others are concerned with

33. For early discussions, see Felsenstein 1973, 1978, 1981a,b,c, 1982, 1983a,b, 1988a,b,c; Kimura 1980; and DeBry and Slade 1985.

34. For example, Goldman 1990, 1993a,b; Navidi et al. 1991; Hillis et al. 1994; Sanderson 1994; Yang 1994, 1996; Hillis 1995; Huelsenbeck 1995; Swofford et al. 1996; Huelsenbeck and Crandall 1997; Huelsenbeck and Rannala 1997; Cunningham et al. 1998; Pagel 1999; Thornton and DeSalle 2000; Broughton et al. 2000; Goldman et al. 2000; Lewis 2001; Salter and Pearl 2001; and Savill et al. 2001.

35. For example, Kishino and Kasegawa 1989; Håstad and Björklund 1998; Lyons-Weiler and Hoelzer 1999; Lyons-Weiler and Takahashi 1999; Wägele 1999; Whelan and Goldman 1999; Goldman and Whelan 2000; Ota et al. 2000; Sennblad and Bremer 2000; Steel and Penny 2000; and Wyner et al. 2000.

some aspects of the methodologies associated with generating phylogenetic trees using such models. Johns and Avise (1998), for example, showed that maximum-likelihood methods are less robust than phylogenetic systematic methods when different outgroups are used. Chor et al. (2000; see also Salter and Pearl 2001) have shown that even though the programs generate a single tree, there are often multiple, equally optimal solutions under a given model for a given set of data, any one of which may give multiple trees (the early computer programs for phylogenetic systematic analysis suffered from this same deficiency).[36]

> No matter what kind of analysis has been performed, you must always be certain you know whether the results present "the tree" or "the consensus tree" for the data.

Advocates of the "total evidence" approach are concerned that we do not know enough yet to believe any of our models are universal, or even typical enough of all of phylogeny to be applied universally. If we have described only 10 percent of the species on this planet, then it is likely that we have sequence information from more than a single gene for less than 5 percent of living species.

Still other researchers believe that evolution has been so complex and historically contingent that no a priori model will be suitable for all clades, or perhaps even for all genes for a given clade. For example, Gissi et al. (2000) reported lineage-specific evolutionary rates just in mammalian mtDNA. For those researchers, statements such as "Nature is not as simple as our models. . . . But this should not be taken as a criticism of the use of a model" (Pagel 1992:425, 441) will always require responses such as "Yes, it should" (Wenzel and Carpenter 1994:101). If models are known to be simpler than reality, how can we use models of sequence evolution embodied in maximum-likelihood methods to build our trees and then consider the results to be scientific corroborations of the model?

This is the reason early advocates urged caution in applying such methods: "The weakness of the maximum likelihood approach is that it requires us to have a probabilistic model of character evolution that we can believe. The uncertainties of [probabilistic] interpretation of characters in systematics are so great that this will hardly ever be the case" (Felsenstein 1978:409). We believe this statement is important because phylogenetic reconstruction is an effort to reconstruct past events that

---

36. See Jermiin et al. 1997 for a discussion about the use of consensus trees in such cases that parallels our earlier discussion about consensus trees in phylogenetic systematic analysis.

actually happened, not to suggest what might have happened or to predict what might happen in the future. That means that if you are using maximum-likelihood models to build phylogenetic trees, you are placed in the position of using information about current evolutionary states (for which the future is still probabilistic) to assign probability statements to events (character changes) that have already happened (i.e., are no longer probabilistic). Under what circumstances is it reasonable to assume that current conditions under which traits are being maintained also explain the conditions under which those traits arose and were favored over ancestral traits? This is a recurring theme for all evolutionary research into origins, one to which we will return repeatedly in this book.

Finally, if you use a particular model to derive a phylogenetic tree from a set of data, you can assess the degree of fit of those data to that model (or to any other, for that matter), but you cannot use those data to falsify that model. (Obviously, new data can falsify a preexisting phylogenetic hypothesis regardless of the manner in which the data were analyzed.) For some, the issue of whether or not input data can falsify an a priori model is of paramount concern in phylogenetic analysis.[37] Wenzel and Siddall (1999) have suggested using the results of total evidence studies based on phylogenetic systematic methods to examine alternative maximum-likelihood models. The model that provides the best fit to (is congruent to or most consistent with) the total evidence tree is the preferred model for those data for those taxa. Should a large number of such studies indicate that a particular model typically outperforms all others, we could then entertain the possibility of a universal model of, say, sequence evolution.

The approach proposed by Wenzel and Siddall would also permit you to investigate the possibility that some of your data (specifically, sequence data in the Felsenstein zone) do not preserve phylogenetic information well. First, you must demonstrate the shortcomings of particular data by showing that in comparison with other data (i.e., in a total evidence study), they do not perform well in a phylogenetic analysis (i.e., are a source of correlated homoplasy). Then you may assess the fit of the problematic data to a variety of models, choosing the one that produces results congruent with the total evidence tree. This provides an estimate of the mode of evolution that occurred in the problematic data.

Adopting the approach advocated by Wenzel and Siddall means that you must accept the proposition that *whenever the results of a phylogenetic*

37. Wiley 1975; Platnick and Gaffney 1977, 1978a,b; Gaffney 1979; Farris 1983, 1986; Kluge 1991, 1997, 1998a,b, 1999; and Siddall and Kluge 1997.

systematic analysis and a particular maximum-likelihood analysis of the same data disagree, the maximum-likelihood model has been falsified for those particular data. That presupposes your phylogenetic systematic study is itself robust (hence our assertions about the quality and quantity of data for which you are responsible). It also leads to the following concern:

> Isn't "maximum parsimony analysis" just another model? No.

This mistaken belief has caused great confusion in evolutionary biology during the past decade. Methods of phylogenetic analysis using models to provide differential weight to the data a priori are inductive. As such, they are not used to test the assumptions of the model. We therefore have to assume that the model is either completely true or is true for situations that are typical enough that the results are a good approximation of the truth. Phylogenetic systematics, by contrast, is based on the view that evolution has been so complex that we routinely expect to find conflicts in the data requiring us to admit falsification of Hennig's auxiliary principle. The criterion we use to tell us just how many, and which, traits conflict is the principle of logical parsimony. But this principle is invoked a posteriori, not a priori. And it is invoked to identify expected departures from a priori methodological assumptions of simplicity. If evolution had been parsimonious, we would never need to invoke any decision-making criterion because the data would always agree on a single, unambiguous answer.

This is not a novel or revisionist interpretation. Early in the development of quantitative approaches to phylogenetic systematics, Farris (1973) showed how the method could be construed as a maximum-likelihood model. Farris was attempting to show that phylogenetics is based on the a priori assumption that anything is possible, and further, that such a generic model is worthless for inductive studies, but is perfect for hypothetico-deductive investigations. Felsenstein supported this characterization when he criticized Farris's formulation:

> Farris . . . establishes a general correspondence between parsimony methods and maximum likelihood methods . . . [using] the maximum likelihood method to estimate not only the parameters of the evolutionary tree, but also the states of the characters. . . . When this latter kind of maximum likelihood estimate is made, the number of parameters being estimated rises without limit as more characters are examined. (1978:408)

Felsenstein thus recognized that phylogenetics is based on an a priori presumption of complexity, not simplicity, and opted for a more parsimonious a priori view of the world embodied in a different class of maximum-likelihood models. Unfortunately, he later confused matters by stating incorrectly that parsimony-based phylogenetics assumed too simple a model of evolution (Felsenstein 1983a). Using the results of phylogenetic systematic analysis is not justified by an a priori appeal to simplicity, although it begins with a presumption of homology. Sober (1983, 1988a), for example, showed that likelihood estimates of phylogeny can be justified by simply assuming that the less variable character is the most probable character (also pointed out by Hendy and Penny 1989). This turns out to be a slightly different likelihood justification for parsimony (which Sober called the "Quack-a-Doodle" theorem) than Farris's proposal (1973).

Therefore, as counterintuitive as it may seem, "parsimony analysis" *as actually practiced by phylogeneticists* is based on the assumption that evolution is *not* parsimonious. We agree with Felsenstein (1978; contra Felsenstein 1983a) that any model asserting that true phylogenetic relationships can be reconstructed using a small number of universal probability statements, such as nucleotide substitution rates specified in maximum-likelihood models, is based on the a priori assumption that evolution *has been parsimonious*. Using the results of phylogenetic systematic analyses in general evolutionary studies is thus justified by the fact that the final results are not guaranteed to fit the investigator's preconceptions about character evolution or phylogenetic relationships.[38]

Most scientists, we believe, would agree that if we knew a particular model would provide "true" phylogenetic reconstructions more often than would phylogenetic systematic analysis, we would be crazy not to use that model in our work. The attraction of model-based research in science is thus highly understandable. In addition, since a major purpose of science is the search for general principles, and since the universe appears to be made up of many general properties, the success of model-based research in increasing scientific knowledge is also understandable. The fundamental value of having alternative approaches in any important scientific discipline is a form of reciprocal illumination. That is, we use the differences in results for a given set of data analyzed according to alternative approaches as a point of departure for criticizing and assessing each other's approach. The presumption is that such disagreements can lead to improvements in both approaches, and possibly to

---

38. For excellent reviews of this issue, including a discussion of the novel criticism that "maximum parsimony analysis" is deficient because it is *not* based on any a priori model, see Tuffley and Steel 1997 and Steel and Penny 2000.

mutual appreciation for each other's endeavors, even if advocates of each approach still retain a philosophical preference for one (for such a proposal, see Faith and Trueman 2001).

We have adopted the belief, shared by many phylogeneticists, that the utility of phylogenetic systematics in evolutionary biology stems from its relative independence from any particular evolutionary mechanism beyond descent with modification. As Wiley (1981) pointed out at the beginning of the modern era of phylogenetic analysis, the only assumptions of phylogenetics are: (1) evolution (descent with modification) has occurred, and a record of it remains in the characters of species; (2) each species is a historically unique mosaic of traits, some old, some new, some unique to it; (3) we do not know beforehand which characters are the most significant and which are the most liable to homoplasy, hence (4) we do not know a priori what the phylogenetic relationships are, nor do we know what the relative or absolute rates of divergence are. This returns us full circle to Hennig's auxiliary principle; we can now see that this principle is used to begin our voyage of discovery and not to justify its findings before we know what they are. If we are truly on a voyage of discovery, we do not want to *assume* homology (or homoplasy for that matter) a priori in the sense of a rigid assumption of a model. That is, we do not want simply to assess the degree of "fit" for our data to something we think we already know. We instead *presume* as much homology as we can at the beginning, simply as the launching point for our voyage. At the same time, we cannot *presume* that everything we have done is a mistake, or there is no reason to even contemplate a voyage of discovery. (Falling off the end of the world is not an option.) In a sense, we hope the data will help us discover when we have made a mistake (we call this falsification of Hennig's auxiliary principle "homoplasy"), in addition to helping us discover the phylogenetic relationships among the members of any group we are studying. We believe that we will discover more on our voyage if we remain, as much as possible, skeptics (Watkins 1984; Miller 1999; Kluge 2001) about what we think we already know. (No matter how convinced we are that the world is flat, we may not fall off the end of the world when we sail into the setting sun.)

 We all have powerful computers—why don't we simply analyze our data using all available methods, and choose the best answer? Because in our voyage into the unknown, we do not know what the best answer is. We rely on the data to tell us with minimal intrusion of our preconceived ideas. It is an exercise in circular reasoning to select one particular phylogenetic tree, based on one particular subset of the data, derived from one particular method simply because it agrees with previous work or with your own personal preferences.

SUMMARY

The past 25 years has witnessed an explosion of published phylogenetic studies. Sanderson et al. (1993) examined 882 phylogenetic analyses published in 76 journals during the period 1989–91. Kristoffersen (1994) reported that 2,486 strictly Hennigian analyses had been published by 1992. The listing of phylogenetic studies produced and maintained by Vicki Funk and Michael Donoghue for the Society of Systematic Biology records an increase in rate of production in the final decade of the last century with no indication of slowing.[39] This has given us a sobering glimpse of the true complexity of information contained in the phylogenetic diversity on this planet and of the scope of the task ahead (Sanderson et al. 1993). Efforts already underway to make this growing database accessible include "super tree analysis," linking together phylogenetic trees of different groups and different levels of resolution[40] using a form of the matrix representation method of Brooks (1981, 1985, 1990; also Wiley 1986d, 1988a,b),[41] coupled with dissemination of information via the Internet.[42]

The good news is that computer algorithms have been designed to help perform these laborious tasks. The bad news is that the ease with which computers perform this task may seduce us into forgetting that the results coming out are only as good as the data going in and the analysis of those data (remember Murphy's Rule). Phylogenetic systematics does not promise to provide a single, quick, simple, true answer. It certainly does not promise to force a robust answer out of bad data. Unfortunately, there is no relationship between an actual understanding of phylogenetic systematics, including the expertise needed to describe

39. http://www.utexas.edu/ftp/depts/systbiol/cladliterature.html

40. Gordon 1986; Steel 1992; Constantinescu and Sankoff 1995; LaPointe and Cucumel 1997; and Sanderson et al. 1998.

41. Baum 1992a; Ragan 1992; Purvis 1995; Ronquist 1996; Bininda-Emonds and Bryant 1998; and Sanderson et al. 1998.

42. For some examples, see the Tree of Life (http://phylogeny.arizona.edu/tree/phylogeny.html), Species 2000 (http://www.sp2000.org), the Global Taxonomy Initiative (http://research.amnh.org/biodiversity/acrobat/gti2.pdf), the International Organization for Plant Information (http://www.life.csu.edu.au/iopi/), and the TAXACOM mailing list hosted by the U.S. Organization for Biodiversity Information (http://usobi.org).

characters, and the ability to manipulate data using the various computer programs. Because of this, although the number of "phylogenetic trees" or "cladograms" is growing in the literature, not all of these diagrams are constructed using phylogenetic systematic methods, and even those that have been may not be very robust. The problem of methodological misunderstanding is not unique to phylogenetics. Papers are published on a regular basis in ecological journals castigating researchers for their misuse of statistics. The information presented in this chapter will help you critically evaluate these diagrams, for *only trees produced in accordance with phylogenetic systematic principles provide the robust estimates of genealogical relationships free of a priori assumptions about evolutionary mechanism which are the necessary precursors for our particular voyage of discovery.*

At the same time, it is important to remember that phylogenetic patterns by themselves imply, but are insufficient to explain, evolutionary transformations.[43] Phylogenetic trees are descriptions of patterns in nature; they are the outcome of a logic of discovery. We can use these patterns to falsify predictions made by evolutionary hypotheses that are contrary to the discovered phylogenetic evidence. To the extent that the goal of science is to provide explanations of how things happen (of mechanisms), phylogenetic trees on their own can never be *sufficient* for evolutionary explanation. Mechanisms must be corroborated experimentally as well. Phylogenetic patterns are, however, *necessary* for a complete evolutionary study, because evolution is a temporal process that produces systems with a past, present, and (indeterminate) future. In the rest of this book, we will highlight some of the interesting studies that can be performed using phylogenetic information in the search for evolutionary explanations.

43. Felsenstein 1978; Wiley 1981; Beatty 1982, 1994; Kluge 1985; de Queiroz 1985; Ridley 1986; Coddington 1988, 1990; Hull 1988; Sober 1988a; de Queiroz and Donoghue 1990; and Brooks and McLennan 1991, 1994.

Chapter 3

Species:
Exploring the Entities

Why is not all nature in confusion, instead of the species
being, as we see them, well defined?

Darwin 1872:161

Diversity is idiosyncratic. It is impossible to reconcile idio-
syncrasy with preconceived ideas of diversity. The search for
hidden likenesses is unlikely to yield a unifying species
concept.

Levin 1979:381

It is not uncommon for major concepts in a given field of
scientific endeavor to have ambiguous definitions. In biology, one such
concept is that of species. In this case, however, the ambiguity is particularly vexing because Darwin made species the central player in evolution. Evolution is the unifying concept of biological science; we need a
coherent and productive approach to dealing with species.

Although Darwin complicated matters, the roots of our modern dilemma predate Darwin (Stevens 1992). Initially, the "species" was simply
a category in classifications, the least inclusive group of organisms to be
recognized formally. That the biosphere was organized into an internested hierarchy of species and groups of species was thought to be self-
evident. It was simply the nature of the world that organisms grouped
naturally into discrete species, species into genera, genera into families,
and so on until all living things were included in a single classification
of life. Prior to the advent of evolutionary thinking, taxonomists, like
all natural philosophers of the day, argued that the only "real" entities
were those that had immutable spatio-temporal existence. Because of
their unchangeable nature, such bits of reality could be grouped into
"classes" defined by the fixed properties of their components. Classic
examples of such "real" entities, which at the time were called "species," are hydrogen and gold. Like these other natural kinds, biological
species existed and conformed to a single hierarchical classification

simply because it pleased the Creator for it to be so. In the absence of any causal principles explaining the taxonomic hierarchy linking species together, pre-evolutionary taxonomists relied on aesthetics, intuition, and personal judgment to make decisions they felt best mirrored the mind of God.

In 1832, Charles Lyell, the leading British geologist of the day, criticized Lamarck for postulating that biological species were mutable (we will discuss this in more detail in chapter 5). By the 1850s, however, accumulated evidence about fossils, geographic distributions of similar species on different continents, and his ongoing discussions with other natural philosophers had led Lyell to abandon his former perspective, stating that science had reached a critical state with respect to the "species question" (Wilson 1971).

Lyell's doubts were raised, in part, by a monkey wrench thrown into the system by one of his former students. Charles Darwin (1859) suggested that our ability to group organisms naturally into species and groups of species was not necessarily evidence for a Divine plan but could also be seen as evidence that biological "species" were not immutable. They could change over time, a process we now call **anagenesis**, and if subdivided they could produce new species, a process we now call **speciation**, or **cladogenesis**. Species descended from the same ancestors would naturally group together in a classification mirroring their common history of descent. Darwin's belief in both the reality and the evolutionary nature of biological species was based on several empirical observations. First, as suggested by Lamarck, comparing fossil and living species indicated that not all of the species populating the planet today are the same as those populating the planet in the past. Second, the world today is not populated by a small number of species widely distributed in many different habitats in many different parts of the world; many different species live in specific habitats in many different parts of the world. And third, all living and extinct species fit into a single hierarchical classification that looks like a genealogy, or family tree. Using perhaps the original form of the "Wiley criterion," Darwin suggested that if it looked like a genealogy, it was one. This led Darwin to his first theory, that all species on this planet were related to each other through a single, common history of descent with modification—phylogeny. As noted earlier, it was not mere happenstance that the only illustration Darwin ever placed in any edition of *Origin of Species* was a phylogenetic tree (Bowler 1996). The search for the ways in which the different species making up that tree were produced led Darwin to his second theory, the theory of natural selection. Thus, modern evolutionary biology was born. Ironically, Darwin's proposals created more than a century of debate about the nature and reality of species, and of their place in evolutionary theory.

Darwin's proposals about evolution created a dual role for species in biology. In addition to being units of classification, species were now also units of evolutionary process. From his various statements in *Origin of Species,* it is clear that Darwin recognized this duality. On the one hand, he felt that taxonomists were seeing and classifying something naturally delimited, but was not sure that all of the units recognized by them were real, that is, units of evolutionary process. On the other hand, he felt that species as units of evolutionary process were real, but were not necessarily always well delimited in nature. "Hence, in determining whether a form should be ranked as a species or a variety, the opinion of naturalists having sound judgment and wide experience seems the only guide to follow. We must, however, in many cases, decide by a majority of naturalists, for well-marked and well-known varieties can be named which have not been ranked as species by at least some competent judges" (Darwin 1872:63).

Following the publication of *Origin of Species,* taxonomists, by and large, continued to concentrate on species as units of classification and tended to ignore evolutionary principles, generally invoking them only to justify taxonomic decisions made following traditional (preevolutionary) practices (Stevens 1992). For example, at the end of the nineteenth century, the primary taxonomic guidelines for dealing with species were (1) they should be easily recognizable to laymen, (2) taxonomic nomenclature should be stable, and (3) there should not be too many names [the *Amadeus* criterion]. This led some biologists to assert that species were whatever good taxonomists said they were (Regan 1926; Dobzhansky 1937; Gilmour 1940). Such concerns still preoccupy the minds of some taxonomists (e.g., Grant 1985) and, as we will discuss in the final chapter of this book, form the basis for policy-making by virtually all private and governmental agencies dealing with conservation and biodiversity. In the second half of the twentieth century, this evolutionary taxonomy (Mayr 1969) epitomized the pre-Darwinian view that taxonomy was partly science and partly art form, with a sizeable component of personal judgment and intuition involved in trying to justify taxonomic decisions in terms of evolutionary principles. Phenetics, or numerical taxonomy (Sneath and Sokal 1973), by contrast, abandoned all efforts to provide a link between taxonomy and evolution by advocating purely empirical classification procedures identifying species as entities having no particular connection with evolutionary principles.[1] As Stevens (1992) pointed out, it is ironic indeed that even those who disavowed the reality of species, particularly taxonomic species, nonetheless used them to focus their research. Hull (1965) presciently described taxonomy as the inheritor of 2,000 years of stasis resulting from the Aristotelian notion that all units of classification must have uniquely defining (and hence unchanging) characters, an indication that the Darwinian revolution had yet to penetrate deeply into taxonomic practice.

While systematics struggled with the evolutionary nature of species, evolutionary biologists struggled with the question of whether species were real entities at all. Three of the founders of the new synthesis, Theodosius Dobzhansky (1937), Ernst Mayr (1942), and George Gaylord Simpson (e.g., 1944, 1953, 1961), argued that the species (and speciation) was the "keystone of evolution."[2] For example, Simpson advocated integration of taxonomic and evolutionary concepts of species, reasoning that without such integration, taxonomy had no scientific basis and evolutionary biology had no way to determine its units of study. As a paleontologist, Simpson was particularly interested in making certain that the development of evolutionary biology included the "deep history" evidence provided by fossils. He viewed species as "active participants" in evolution, identifying them as historical lineages that carried part of the past with them while also responding to long-term selection pressures. Approaching the question from the perspective of population genetics, Dobzhansky wrote, "[I]n nature we do not find a single greatly variable population of living beings which becomes more and more variable as time goes on; instead, the organic world is segregated into more than a million separate species, each of which possesses its own limited supply of variation which it does not share with the others. . . . The origin of species . . . constitutes a problem which is logically distinct from that of the origin of hereditary variation" (1937:119).

Despite the insights of the founders, many biologists interested in seeing "evolution in action" concentrated more and more on local inbreeding populations, or **demes**, claiming that these groups of "replicators" were the most inclusive level of evolutionary processes (e.g., Ehrlich and Raven 1969; Brown and Gibson 1983, 1998; Brown 1995). A critical step in developing this view was the emergence of what Mayr (1942, 1963, 1988) called **populational thinking.** Advocates of popula-

---

1. For example, Sokal and Crovello 1970. For a discussion anticipating this approach, see Boas 1898.

2. Mayr 1963. For excellent historical reviews see Coyne 1994 and Gould 1994.

tional thinking treat species as assemblages of organisms, held together by reproductive bonds exclusive to them, that can develop like individual organisms (but do not have to die of old age)and can "reproduce" by something analogous to binary fission. This approach allowed biologists to slip comfortably into a transformational or evolutionary mode, moving away from a static or typological view, because it treated species as collections of organisms characterized by both common and variable traits. Emphasis on populational thinking, however, led many biologists to conclude that only demes and populations were real, species being artificial constructs: "[S]pecies lack reality, cohesion, independence, and simple evolutionary or ecological roles. . . . The concept that is most operational and utilitarian . . . is a mental abstraction" (Levin 1979:381). Ironically, one of the earliest and most influential advocates of this perspective was Julian Huxley, editor of *The New Systematics* (Huxley 1940) and a descendant of Darwin's most influential and vocal advocate, Thomas Huxley.

Neo-Darwinists like Simpson, Mayr, and Dobzhansky argued that species must be real entities for two reasons. First, the collection of demes construed as representing a species often exhibits more geographical and ecological cohesion and temporal persistence than the demes themselves (demes can disappear and reform without destroying the species: see, e.g., Wiley 1981; Wiley and Brooks 1982; Brooks and Wiley 1986, 1988; Pannell and Charlesworth 1999). For example, consider a species of fish widely distributed throughout the streams and ditches in your area. If one population goes extinct because the Department of Highways comes through and scours a ditch, does the species go extinct? The answer to this question is obviously no (unless that was the last remaining population of the species). Of course, this does not mean that the ditch population was an unimportant reservoir of genetic material for the species. It does mean, however, that species as cohesive entities have properties beyond any one population; a species is more than the sum of its parts (see also Avise 1989a,b, 1992, 2000). Second, reducing evolution solely to reproductive exchange within individual demes (changes in gene frequencies in populations) does not explain the origins of the diversity among demes and collections of demes.

Proponents of the "species are real" school of thought faced an important problem: determining how species could be real without being typological or essentialistic (Hull 1965). Michael Ghiselin (1974) provided the solution to the problem by considering species as if they were individual, rather than collective, entities.[3] Biological species are real, but

---

3. See also Hull 1976, 1978, 1980; Wiley 1978, 1980a,b, 1989; Mishler and Donoghue 1982; Cracraft 1983a, 1987, 1989a,b; Donoghue 1985; McKitrick and Zink 1988; Kluge

not in the same sense that hydrogen is real. A molecule of hydrogen found anywhere, and formed at any time, in the universe would be a member of the class hydrogen. By contrast, an organism that looks like a tiger on this planet would not be part of the same species as an organism that looks like a tiger but evolved on another planet.

Simpson (1944) anticipated at least some sense of Ghiselin's view of species by emphasizing that species are historically unique lineages extending through time. Under the "species-as-individuals" view, the most important characteristic of a species is that its members are cohesive wholes, bound together primarily by unique, common ancestry as well as by unique sets of reproductive bonds or common ecology. To Simpson, the evolution of a single species was analogous to the development of a single organism; just as an organism changes its appearance without losing its identity during development, so a species can change its appearance without losing its identity during evolution. The formation of new species is analogous to asexual reproduction, in which new individuals are distinct from the old individual because they form independent evolutionary lineages. Over time, distinct historical trajectories emerge from the speciation process, each differing to some degree from its ancestor and closest relatives but retaining some of its ancestry in the form of synapomorphies. We take advantage of this historical mosaic nature of the attributes of organisms that make up species when we use synapomorphies to reconstruct phylogenetic trees. Thus, among the founders of the new synthesis, Simpson (1944, 1951, 1953, 1961) included a sense of evolutionary history in his conception of what species are. Almost 35 years after Simpson proposed the **evolutionary species concept,** E. O. Wiley (1978) linked Simpson's views on species to the empirical rigor of phylogenetic systematics (see also Eldredge 1985).

Our discussion suggests a fundamental dualism in the way biologists have thought about species. Frost and Kluge (1994; see also Kluge 1990) suggested a reason for this dualism. They contrasted *regular science,* in which the basic units are classes of entities (atom or hydrogen) that have "real," or extensional, definitions, and *historical science,* in which the basic units are the entities themselves, that is, historical individuals. Because the entities in historical science have no a priori defining characteristics, Frost and Kluge argued that historical sciences progress through cycles of discovery and evaluation (see also Kluge 1989, 1991, 1997, 1998a,b, 1999), both of which are necessary but neither of which is sufficient for complete understanding, and both of which require objective

---

1990; Löther 1990; O'Hara 1993, 1994; Frost and Kluge 1994; Mayden and Wood 1995; Mayden 1997; and Wiley and Mayden 2000a,b,c.

methods of study. Evolutionary biology is a historical science in which species play a central role.

Prior to the infusion of phylogenetic thinking into discussions of species concepts, biologists tried to define species using what Mayr (1988; see also Ereshevsky 1992; O'Hara 1994) has called **"nondimensional"** **or "relational" species concepts.** That is to say, "A is a species relative to B and C if it maintains its identity as a distinct entity relative to B and C" (Löther 1990). The prevailing nondimensional species concept for more than a quarter century was the **biological species concept** (BSC), proposed by Dobzhansky (1937, 1940, 1970, 1976) and championed most strongly by Mayr (1942, 1963, 1976, 1982, 1988; see also Ayala 1982). This concept centered on the ways in which species maintain a distinct reproductive identity; a species is a group of interbreeding or potentially interbreeding organisms reproductively isolated from all other such groups. Van Valen (1976) changed the focus of attention from reproduction to ecology, proposing that each species has its own unique niche and it is the niche, an ecological concept, that defines the species (the **ecological species concept**, or ESC). The BSC and the ESC focus on what makes species distinct, separated, or isolated from other such entities. Other researchers have focused their attention on what keeps species together (makes them cohesive entities). A highly persistent proponent of such views has been Hugh Paterson (1978, 1982, 1985, 1987), whose **recognition concept** defines species as groups of individuals held together by a common and unique **specific mate recognition system** (SMRS). We will refer to this and related proposals as **cohesion concepts of species.**[4] All nondimensional species concepts strive to provide a universally applicable definition of what species are (e.g., compare Coyne et al. 1988 with Masters and Spencer 1989 and Masters and Rayner 1996). All, however, imply prior determination of the units to which particular concepts will be applied; they do not tell us how to discover them.

 In other words, we must invoke another, more fundamental concept to discover species before we can apply any nondimensional species concepts to evaluate them.

SPECIES: HOW DO WE DISCOVER THEM?

Species function in evolutionary theory as the basic units of the origin, diversification, and extinction of biodiversity. They are produced by a

---

4. For an extensive catalog and discussion of these and numerous other nondimensional species concepts, see Mayden and Wood 1995; also Mayden 1997, 1999.

variety of mechanisms, all of which involve the *irreversible* splitting of ancestral lineages into descendant ones (Dobzhansky 1970). It is the irreversibility of speciation that produces the unique historical lineages which Simpson recognized as species. Both Darwin and Simpson postulated that those lineages were organized in a hierarchy of phylogenetic diversification. With the advent of phylogenetic systematics, which provided a strong empirical method for recovering the phylogenetic hierarchy, came the possibility of discovering the basic units of that hierarchy. We will refer to species concepts based upon phylogenetic trees as **historical concepts of species.**[5]

Historical concepts of species integrate well with Ghiselin's proposal that species be considered ontological individuals. Although there are numerous formulations of historical species concepts, they are all variations on the original evolutionary species concept formulated by Simpson and modified by Wiley. For example, Eldredge and Gould (1972) referred to species as homeostatic systems held together by the cohesion of common evolutionary history. Löther (1990) asserted that species should be considered ontological individuals because they are material supraorganismal systems, spatio-temporally organized, forming an integrated whole. Wiley (1978; see also Wiley and Brooks 1982; Brooks and Wiley 1986, 1988) suggested that any general species concept must (1) encompass species persistence through time (which makes them diagnosable taxonomic units) as well as their divergence (their evolutionary role), (2) recognize that species are cohesive wholes, and (3) recognize the inherent historicity of species. In a similar vein, Frost and Kluge (1994; see also Kluge 1990) considered species to be genealogical (phylogenetic) systems (implying cohesion) existing as lineages of reproductive connections extended through time (historical entities) preserving a unique (mutually exclusive) mixture of genetic information (genealogical reservoirs that are the most inclusive units of actual process in evolution but which, as genealogical reservoirs, are not active participants in evolutionary processes). Kornet (1993a; see also Kornet 1993b; Kornet and McAllister 1993; Kornet et al. 1995) went further, suggesting that the historically irreversible splitting of cohesive lineages provided the

---

5. Wiley 1978, 1980a; Eldredge and Cracraft 1980; Mishler and Donoghue 1982; Cracraft 1983a, 1987, 1989a,b, 1992; Donoghue 1985; Mishler and Brandon 1987; McKitrick and Zink 1988; Endler 1989; Ridley 1989; Nixon and Wheeler 1990; Sluys 1991; Baum 1992b; Davis and Nixon 1992; Williams 1992; Kornet 1993a,b; Kornet and McAllister 1993; O'Hara 1993, 1994; Baum and Donoghue 1995; Baum and Shaw 1995; Kornet et al. 1995; Mayden and Wood 1995; Maynard Smith and Szathmary 1995; Zink and McKitrick 1995; Davis 1996; Zink 1996, 1997; Mayden 1997, 1999; Packer and Taylor 1997; Adams 1999; de Pinna 1999; Brooks and McLennan 1999; Sangster et al. 1999; Sluys and Hazevoet 1999; Wheeler 1999; and Wiley and Mayden 2000a,b,c.

only universal criterion for recognizing species. She concluded that species function in evolutionary theory as mutually exclusive, nonarbitrary (discovered, not imposed *sensu* Frost and Kluge 1994) historical entities (see also O'Hara 1993). Following these authors, we therefore consider the evolutionary species concept to be the fundamental ontological species concept for evolutionary biology.[6] As an ontological concept, however, the evolutionary species concept is not **operational**; it does not specify either discovery modes or evaluation criteria. *What is needed is an operational surrogate for the evolutionary species concept.*

Dupré suggested, "Phylogenetic taxonomy aims at more direct connection with historical components of evolutionary theory, by starting from the principle that taxonomy should accurately reflect phylogeny" (1992:315). In practice, most taxonomists propose species names in ways that do not require a particular species concept. We can accept such typologically based names as scientific hypotheses, but we cannot say that they represent explicit hypotheses of evolutionary species because for the most part the naming of species by taxonomists follows pre-evolutionary practices (traditions, or historical constraints on the discipline: Stevens 1992). It is not sufficient for us to say that because we *wish* to be discovering evolutionary species, our descriptions become hypotheses of evolutionary species. Thus, until Latin binomina are subjected to a discovery mode appropriate for discerning evolutionary species, we do not know whether or not a taxonomist has discovered something that might be an evolutionary species.

 The operational surrogate for the evolutionary species concept, therefore, begins with phylogenetic analysis of the collections of organisms given names by taxonomists.

Although phylogeneticists have a common discovery method and substantial agreement on the role species play in evolutionary theory, we do not have complete agreement on what species are. This is partly because we use characters to reconstruct the phylogenetic hierarchy, and thus to point the way to species, but characters are neither the hierarchy nor are they the entities we ultimately strive to discover (Frost and Kluge 1994; Wiley and Mayden 2000a,b,c). For example, your right hand is not the descendant of your mother's right hand (Wiley and Mayden 2000a,b,c). Or, as Löther (1990) put it, characters are not "properties of membership" but "evidence of partship." Moreover, not all organisms

6. See also Wiley 1978, 1980a; Mayden and Wood 1995; Mayden 1997, 1999; Brooks and McLennan 1999; and Wiley and Mayden 2000a,b,c.

have all the "defining" characters of the species (e.g., in cases of sexual dimorphism, age-related expression of traits, or teratologies). It is important to remember that evolution is an ongoing process; "defining" characters can evolve and change. What is a "defining" character of a species today might serve only to put an organism among a large group of species 10 million years from now. Sometimes these changes are so pronounced that original definitions may not make sense. When the first tetrapods (four limbs) appeared, who could have predicted there would come a time when many tetrapods would lack limbs altogether?

Evolution requires that we use characters as approximations or extensions, not definitions, of species. Thus, it should not come as a surprise to discover that there are several different views on the ways in which a given phylogenetic tree might be subdivided into individual species. Phylogenetic trees have two components, apomorphic characters and branches. Any intellectual arena that includes two variables (A and B) can give rise to as many as three different viewpoints (only A, only B, both A and B). This is exactly what has happened with systematics. Biologists have proposed three distinct categories of historical species concepts (see also Baum and Donoghue 1995): two forms of the **phylogenetic species concept** (which we will call the PSC-1 and the PSC-2) and the **composite species concept** (CSC).[7] These views differ primarily with respect to how they subdivide a phylogenetic tree and how they deal with the issue of ancestral species.

Figure 3.1 illustrates the similarities and differences among these species concepts. In figure 3.1a, the phylogenetic tree for species A–G is supported by a single apomorphic trait for each nonterminal (internode) and terminal branch. All three species concepts recognize A–G as 7 species; in addition, all three recognize 6 ancestral species, the 6 nonterminal branches, for a total of 13 species. In figure 3.1b, species B, C, and G lack apomorphies. Once again, all three species concepts recognize A–G as separate species. The PSC-2 would again recognize the 6 nonterminal branches as separate species, for a total of 13 species. The PSC-1 and CSC, however, would recognize only ten species, since B, C, and G are not distinguishable from their common ancestors. Finally, in figure 3.1c, 4 of the 6 nonterminal branches of the phylogenetic tree have more than one apomorphic trait. As previously, all three concepts recognize A–G as separate species; the PSC-2 recognizes 6 ancestral species, for a total of 13 species; and the PSC-1 recognizes only 10 species, since B, C, and G are not distinguishable from their common ancestors. The CSC

---

7. Kornet 1993b; Kornet and McAllister 1993. See also Baum and Donoghue 1995; Mayden and Wood 1995; Mayden 1997, 1999; and Brooks and McLennan 1999.

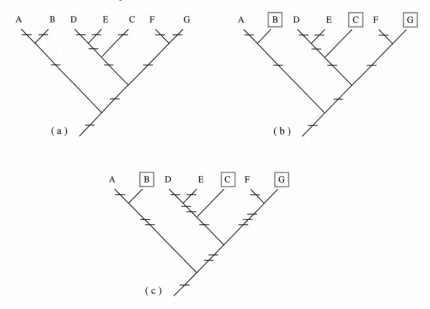

Figure 3.1. Differences among the three major historical species concepts with respect to the number of species recognized on a phylogenetic tree. (*a*) Each terminal and nonterminal branch has a single apomorphic character change; the PSC-1, PSC-2, and CSC all recognize seven terminal and six nonterminal species. (*b*) Three terminal branches lack autapomorphies; the PSC-1 and CSC consider those indistinguishable from their common ancestor. (*c*) Four nonterminal branches have more than one synapomorphy; the CSC recognizes a distinct species for each character. (Modified from Brooks and McLennan 1999.)

recognizes 15 total species, one for each apomorphic trait on the phylogenetic tree. These three examples demonstrate that the fundamental differences between the approaches has to do with how many ancestral species will be recognized. We will see later in this chapter that these distinctions are sometimes, but not always, important to research on species and speciation.

The CSC complements the PSC-1, recognizing species on the basis of evidence for character evolution and branching. It differs by assuming that each instance of character fixation is evidence of past, irreversible lineage splitting (see Goldstein and DeSalle 2000). This sets the upper limit on the number of possible species for a given set of organisms and their characteristics, providing insight into what we may be missing. The origin and fixation of each apomorphic trait, it is argued, requires permanent lineage splitting; therefore, each apomorphic trait indicates the present or prior existence of a distinct species. In this regard, the CSC differs from the PSC-1 and PSC-2, both of which permit anagenesis (evolutionary change within a single species lineage), a process that cannot occur in the CSC. Anagenesis is important for estimating times and rates of divergence and for distinguishing the reproductive and ecological roles of species (Futuyma 1989 and references therein). Consequently, the CSC would appear to represent an a priori constraint on the kinds of phenomena evolutionary biologists may study.

The PSC-2 is, in a sense, the most consistent of the historical species concepts, since it takes into account only the branching structure of the tree. It will differ from the PSC-1 only in not permitting the existence of persistent ancestors. In figure 3.1, the PSC-2 recognizes B, C, and G as different species from their ancestors on the basis of evidence of lineage splitting but in the absence of evidence of character evolution. In terms of evolutionary investigations, this species concept rules out the possibility of any speciation modes that involve descendant species "budding off" from the ancestral species, with the ancestral species continuing to persist. If we adopt the PSC-2, we would have to reject some possible modes of speciation, thereby giving weight to alternative mechanisms. Thus, despite the attractiveness of stability, we reject the PSC-2 because (as suggested previously) "we should abstain from issuing prohibitions that draw limits to the possibilities of research" (Popper 1968:250).

The PSC-1 is the most conservative approach, requiring evidence of *both lineage splitting and character evolution* to recognize a species. And yet, it does not rule out any putative evolutionary processes a priori. It can even be challenged by the postulates of additional species made by the other two concepts. For the PSC-2, the hypothesis that a particular terminal taxon is *not* a persistent ancestor can be corroborated by finding a fixed apomorphy for the terminal taxon. In such a case, the number of species recognized by the PSC-1 would increase toward the number recognized by the PSC-2. Likewise, for the CSC the hypothesis that each apomorphy on a branch was accompanied by lineage splitting will be corroborated each time we find a new species that possesses only a subset of the apomorphies of what was considered previously to be a single species. The hypothesis that there has been more than one character fixation between speciation events (anagenesis) can be tested by examining evolutionary rates and mechanisms underlying character change to establish cases in which it is reasonable to assume that character fixation occurred more often than irreversible lineage splitting. In such cases, the number of species recognized by the PSC-1 and CSC would approach each other. It therefore appears that all species recognized by the PSC-1 are recognized by the PSC-2 and CSC, and empirical corroboration of the postulates of additional species stemming from either the PSC-2 or CSC will always narrow the gap among all three. Or, to put it another way, if we say that in the absence of character differences, two species are the same, the PSC-1 and PSC-2 become the same concept. Likewise, if we say that in the absence of evidence of lineage splitting, there has been no permanent lineage splitting, the CSC becomes the PSC-1. This makes the PSC-1 the primary historical species concept, with the PSC-2 and CSC embodying challenges to any hypotheses about numbers of persistent ancestors or numbers of unrecognized, irreversible lineage splits.

All three historical species concepts recognize only those nonpersistent ancestral species for which there is evidence of character evolution. As first noted by Wiley (1980a), this is due to a limitation of phylogenetic systematic methodology, from which the assessment of species stems. Figure 3.2a shows the "true" historical sequence of two speciation events producing terminal species A, B, and C. The common ancestor of A, B, and C is indicated by an apomorphic trait; B and C exhibit autapomorphic traits. Figure 3.2b shows the phylogenetic tree for these taxa and characters resulting from phylogenetic systematic analysis. For figure 3.2b, the PSC-1 and CSC would recognize three species, A, B, and C, with A being a persistent ancestor; the PSC-2 would recognize four species, since A and the ancestor cannot be the same. For figure 3.2a, the PSC-1 and CSC would still recognize three species, A, B, and C, with A being an ancestral species that persisted without character change through two speciation events. The PSC-2 would recognize seven species, because the ancestor of B and its sister, the ancestor of C and its sister, and the ancestor of those two ancestors cannot be the same species. Later in this chapter, we will explore the ramifications of this type of speciation scenario. In such cases, the PSC-1 will recognize the correct number of species involved, even if the phylogenetic analysis fails to clearly indicate the correct number of independent speciation events.

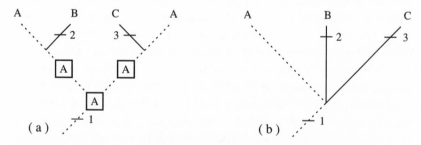

Figure 3.2. Limits of resolution of phylogenetic analysis with respect to speciation. (*a*) Depiction of two episodes of speciation in which the common ancestor does not disappear, resulting in three distinct species. (*b*) Cladogram of those three species derived from phylogenetic analysis. (Modified from Brooks and McLennan 1999.)

Stevens recently noted, "It is not clear to me that our taxonomic interests can be squared with the taxonomic species. We are interested in the whole process of evolution, not only in what 'has evolved'" (1992:311). The evolutionary species concept and its operational surrogate, the PSC-1, seem to bridge this gap. For most biologists, this breakthrough should be a cause for universal rejoicing. In the absence of guidelines provided by evolutionary theory, however, researchers making decisions about species based solely on patterns eventually resort to authoritarianism to maintain their positions. Having a phylogenetic tree is thus necessary but not sufficient for the study of speciation. In other words, we need to study *what* species are in order to understand *why* species are.

SPECIES: HOW DO WE EVALUATE THEM?

Phylogenetic systematic methodology takes us on a voyage of discovery—that is, it allows us to erect hypotheses about what historical, mutually exclusive, and diagnosable entities (identified by the preceding criteria) *might* be species. In order to determine whether these entities actually *are* species, we need to move to the next level in our analysis, that of evaluation. Examining both the ways in which new species are formed and the ways in which such new species maintain their identities once they have formed constitute valuable research programs aimed at evaluating our understanding of species and the evolution of biodiversity. Within the rubric of evaluation, there are two different general research programs. One couples phylogenetic patterns with geographical distributions in order to determine the conditions under which speciation was *initiated*. The other uses information from quantitative genetics, demographics, mate choice experiments, etc., in order to determine how speciation was *completed*. Each is necessary but, on its own, insufficient to explain speciation.

The Origin of Species: Modes of Initiating Speciation
BASIC ASSUMPTIONS
Mayr (1963) recognized three general classes of speciation. The first is **reductive speciation**, in which two existing species fuse to form a third. Harlan and DeWet (1963) proposed the term "compilo-species" for cases in which one species absorbs another; however, examples of this phenomenon have not been documented to date. The second is **phyletic speciation**, in which a progression of forms within a single lineage are assigned species status at different points in time. As noted above, we consider each individually evolving lineage to be a single species under-

going anagenesis. The CSC, however, would recognize each episode of anagenetic change as an instance of phyletic speciation. The third class, called **additive speciation** or **cladogenesis**, is characterized by an increase in the number of species. The majority of speciation models, although based on several different mechanisms, are models of additive speciation. The most important thing to remember about speciation is not that it produces species, but that it produces *sister species*. Because of this, explanations about speciation modes cannot be based on analysis of a single species; they must be based on examinations of sister species and clades.

Wiley suggested that various models of additive speciation could be studied if phylogenetic, biogeographic, and population biological data were available, and if two assumptions were made.[8] First, assume there have been no extinctions in the clade. If we are to use phylogenetic trees to study particular modes of speciation, we must have confidence that sister species are each other's closest relatives and not, in reality, more distantly related due to the extinction of one or more species. Consider the following hypothetical example. Two species of fish, demonstrated to be sister species on the basis of a phylogenetic analysis, are located on either side of a mountain range (fig. 3.3a). We might hypothesize that the disjunct distribution was caused when the upheaval of the mountains separated the ancestral species into two groups, which subsequently diverged in isolation (fig. 3.3b). Unfortunately for our theory, a group of enthusiastic paleontologists discover an abundance of fossil evidence suggesting that at least two other species fall between the extant representatives (fig. 3.3c). Hence, the current disjunction of fishes B and C (which are no longer postulated to be sister species) was probably derived through a series of speciation *and* extinction events; only one of which need have been associated with the tectonic activity (fig. 3.3d).

Second, assume that the influence of geographical separation during the evolutionary divergence of a clade has not been obscured by rampant dispersal of the descendant species (fig. 3.4). Pairs of sister species or clades that have experienced such dispersal exhibit large-scale sympatry. Uncovering such sympatry is problematic, however, because it is difficult to determine whether the current distribution pattern existed during the speciation of the group, or whether it represents widespread dispersal following speciation in isolation. This assumption does not imply that dispersal is unimportant, only that postspeciation dispersal does not ob-

8. Wiley 1981. See also Mayr 1954, 1963; Bush 1975a,b; Endler 1977; White 1978; Wright 1978a; Lande 1980, 1981; Templeton 1980, 1981, 1982, 1989a; Wiley and Mayden 1985; and Brooks and McLennan 1991.

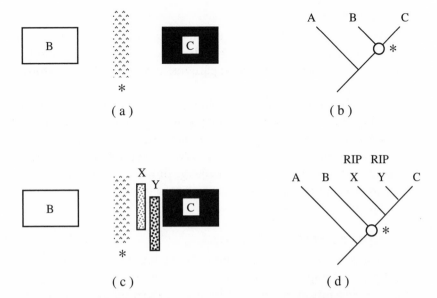

Figure 3.3. Problems arising from extinctions. (*a*) Two fishes, B and C, are located on either side of a mountain range. (*b*) Phylogenetic tree for the genus containing species B and C. *Open circle* = the speciation event; * = the upheaval of the mountains. (*c*) Fossil evidence of extinct species X and Y on the same side of the mountains as extant species C. (*d*) New phylogenetic tree incorporating fossils. According to this new hypothesis, the mountains may have played a role in the production of species B and the ancestor of the X + Y + C clade. RIP = extinct species.

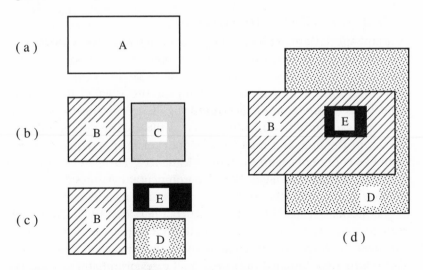

Figure 3.4. Problems arising from widespread dispersal of descendant species. (*a*) Ancestral species A. (*b*) Geographic separation of A produces descendant species B and C. (*c*) Geographic separation of C produces descendant species D and E. (*d*) Rampant dispersal of descendants B and D produces a current distribution pattern of widespread sympatry and obscures the original pattern of geographic disjunction.

scure speciation patterns. Like many biological assumptions, this is a necessary starting point. Without it we have no a priori justification for attempting to reconstruct speciation patterns, and thus no hope of studying the process.

It is probably true that for many groups one or both assumptions are not justified; however, until a larger database is established, it is impossible to determine whether these nonconformists need be accorded the status of an overwhelming majority or a confounding minority. We are confident that numerous clades will emerge in which phylogenetic patterns and distribution patterns are congruent with predictions from particular speciation modes.

ALLOPATRIC MODES OF SPECIATION

During the past 50 years, most evolutionary biologists have become convinced that geographic isolation is often crucial to species formation. **Allopatric speciation** is a generic term for speciation *initiation* due to the complete geographical separation of two or more populations of an ancestral species. There are two general types of allopatric speciation, defined by whether the ancestral species is fragmented by changes in the environment or by dispersal into a new area. Let us take a closer look at these two modes (see also Wiley 1981; Wiley and Mayden 1985; Funk and Brooks 1990; Brooks and McLennan 1991).

Allopatric Speciation Mode I (Passive Initiation)

Vicariant speciation, or **vicariance,** occurs when a species is geographically separated into two or more isolated populations (initiation of speciation) by the formation of a barrier to dispersal and gene flow, with subsequent lineage divergence occurring in the isolated descendant populations (completion of speciation) (fig. 3.5). We tend to associate large-scale geological perturbations, such as the uplifting of mountains, changes in the flow and interconnectedness of river systems, and the floating of continents, with the term "vicariance." There are, however, a number of factors associated with "environmental harshness" (Cracraft 1985a) (e.g., desertification of an area) that also can subdivide a species geographically (see also Mayr 1970; Lynch 1989; Frey 1993). Note that vicariant speciation can produce dichotomous or polytomous phylogenetic trees, depending on the circumstances under which different speciation events were initiated (fig. 3.6). In all cases of initiation of speciation by geographical isolation, the ancestral species plays a passive role (i.e., does not disperse across barriers).

Two predictions from this mode are of interest to students of specia-

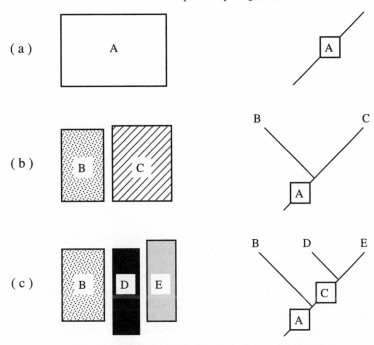

Figure 3.5.   Allopatric speciation mode I producing dichotomous phylogenetic patterns. (*a*) Species A extends throughout a geographical area. (*b*) The species is divided by the appearance of some geographical barrier preventing gene flow; the two populations continue to evolve independently, producing new species B and C. (*c*) Species C undergoes another geographical upheaval, gene flow is eliminated, and changes continue in isolation, leading to the eventual production of new species D and E. The outcome of this division of space through time is the production of three extant species (B, D, and E) and the extinction through total speciation of two ancestors (A and C).

tion. First, the points of geographical disjunction between sister species will correspond to the historical boundaries established by the geological changes. Based on this, the ancestral range may be estimated by combining the distributions of the descendant species, assuming no substantial range expansion or contraction following speciation (see fig. 3.4). Second, a multitude of ancestral species, fragmented in the same way by the same geological event, could theoretically all speciate subsequent to the event because the mechanism initiating speciation is independent of any particular biological system. Hence, we would expect to find the same biogeographic distribution pattern shared by a number of different clades. Historical biogeographical studies rely on this mode of allopatric speciation to detect episodes of parallel biological and geological evolution.

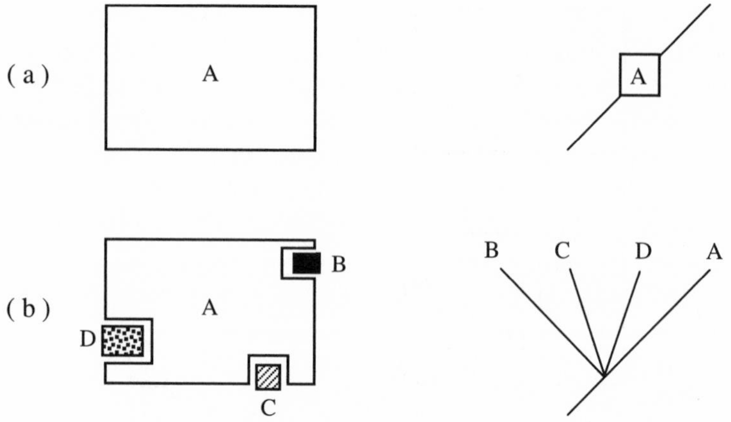

Figure 3.6. Allopatric speciation mode I producing polytomous phylogenetic patterns. (*a*) Species A extends throughout a geographical area. (*b*) Gene flow is severed by a series of rapid microvicariance events, resulting in the isolation of descendant species B, C, and D. The traits present in each species represent a combination of characters that existed prior to the isolation of the populations (symplesiomorphies) and evolutionary modifications that occurred subsequent to the isolation (autapomorphies). Since derived traits are not shared between populations under these circumstances, the sequence of the speciation events will be impossible to determine and the resultant phylogenetic pattern will be polytomous.

Allopatric Speciation Mode II (Active Initiation)

Peripheral isolates speciation, or **peripatric speciation,** postulates that a new species may arise from a small, isolated population that is usually, but not always, on the periphery of a larger ancestral population. Proponents of this speciation mode often assume that peripheral habitats are necessarily marginal in composition. Lynch (1989) pointed out the dangers in equating "different" with "marginal." This is an important distinction because it allows the potential for evolutionary change to occur in response to the presence of a new selective regime, without invoking the assumption that the habitats occupied by the central populations are somehow "better" than the habitats occupied by their peripheral counterparts. The small, isolated population results from active dispersal by members of the ancestral species into a geographical locality not previously occupied by the ancestor and separated from the rest of the ancestral species by a preexisting geographical barrier. The barrier cannot be an absolute bar to dispersal, or there would never be any founder populations. If the barrier reduces dispersal by only a small amount, occasional gene flow between the peripheral and central populations could be sufficient to maintain species cohesion, keeping novel traits from being fixed in either population to the exclusion of the other. The dispersal barrier, therefore, must be sufficient to permit the establishment of novel phenotypes in the isolated populations. Because species differ with respect to what constitutes a barrier to dispersal, we do not expect episodes of peripheral isolates speciation in one clade to mirror those in another clade (except in special circumstances, which we discuss below).

If dispersal across a barrier establishes geographic isolation from the central population, gene flow is stopped. The rate at which speciation in the peripheral isolate is completed then depends upon a complex interaction among the rate at which gene flow decreases (Mayr 1963; Hennig 1966a; Brundin 1966; García-Ramos and Kirkpatrick 1997); the size of the founding population, including its genetic composition and the effects of mutation plus drift[9]; and the strength of divergent selection, including sexual selection, in the new area.[10] This mode of speciation is generally thought to occur more rapidly than vicariant speciation if strong selection is involved (Mayr 1963; Grant 1981; Carson and Templeton 1984). The small size of the founding population, however, may also render species formed in this manner, or the founding population itself for that matter, more subject to extinction.

Two primary phylogenetic patterns arise from peripheral isolates speciation. First, if isolation is due to random settlements of individuals around the margins of the ancestor's range, we expect to find phylogenetic trees comprised of *polytomies* in which the number of terminal taxa equals the number of peripheral descendants "budded off" from the ancestor plus the ancestor itself (fig. 3.7). Frey (1993) pointed out that repeated cycles of peripheral isolation from the same ancestor, coupled with character evolution in that ancestor between cycles, could produce a completely bifurcating tree similar to that produced by vicariance (see fig. 3.5). Although the phylogenetic patterns might be similar, we should be able to distinguish between the two scenarios based upon degree of asymmetry in species ranges and the presence (or absence) of obvious geographical barriers. Second, if isolation is due to individuals from the ancestral species dispersing into a new habitat and speciating, followed by movement of some members of the new species into another peripheral area, repeating the process, we would again expect to find dichoto-

---

9. Carson 1975, 1982; Templeton 1980, 1981, 1989b, 1996; Lande 1981, 1986; Carson and Templeton 1984; Goodnight 1987; Charlesworth and Rouhani 1988; Barton 1989, 1996; Giddings et al. 1989; Stebbins 1989; Coyne 1990; Weinberg et al. 1992; Sites and Reed 1994; Whitlock and Wade 1995; Gavrilets 1996, 2000; and Rundle et al. 1998.

10. Mayr 1954, 1963, 1982; Hennig 1966a; Barton and Charlesworth 1984; Patton and Smith 1989; Møller 1993; Schluter and Price 1993; and Whitlock 1997.

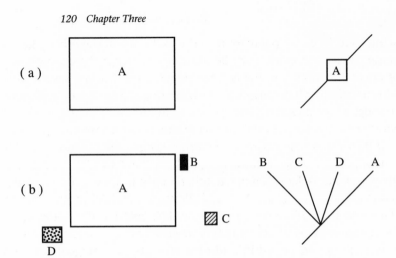

Figure 3.7. Allopatric speciation mode II producing polytomous phylogenetic patterns via random dispersal. (*a*) Species A extends throughout a geographical area. (*b*) Three additional random dispersal events by members of the ancestral species occur and gene flow is severed, producing species B, C, and D. The traits present in each species represent a combination of characters that existed prior to the isolation of the populations (symplesiomorphies) and evolutionary modifications that occurred subsequent to the isolation (autapomorphies). Since derived traits are not shared between populations under these circumstances, the sequence of speciation events will be impossible to determine and the resultant phylogenetic pattern will be polytomous.

mous phylogenetic patterns reflecting the alternating episodes of dispersal and isolation (Hennig's "progression rule"; fig. 3.8). An excellent example of this pattern might be found among organisms that have speciated repeatedly during progressive colonization of island archipelagos (stepping-stone colonization: Darlington 1938; MacArthur and Wilson 1967; Ricklefs and Cox 1972, 1978; Wagner and Funk 1995). This is a special circumstance in which we might expect multiple clades to show the same general pattern of speciation.

Differentiating Allopatric Speciation Modes I and II

In order to differentiate between the two modes of allopatric speciation, we must answer three questions:

1. Were the disjunct populations created by the actions of geological processes (passive role for the ancestor) or by the dispersal of some ancestral individuals over preexisting barriers (active role for the ancestor)?

2. Was gene flow among ancestral populations present (fig. 3.9a,b) or absent/rare (fig. 3.9c) prior to the isolating event? This is an important question because the rate of speciation will be affected by the interaction between local differentiation in response to selection, which tends to

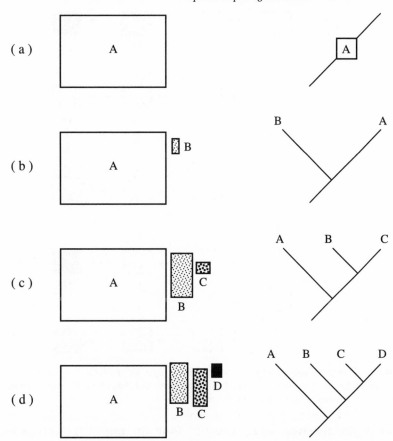

Figure 3.8. Allopatric speciation mode II producing dichotomous phylogenetic patterns via sequential dispersal. (*a*) Species A extends throughout a geographical area. (*b*) Some individuals disperse into a new area and gene flow is severed, producing species B. (*c*) Individuals from species B disperse into a new area and gene flow is severed, producing species C. (*d*) Individuals from species C disperse into a new area and gene flow is severed, producing species D. In this case, sister species can be identified by the presence of shared derived traits; therefore the resultant phylogenetic pattern will be dichotomous.

promote speciation,[11] and cohesive forces such as persistent ancestral traits, mate recognition and fertilization systems, and gene flow among demes, which tend to inhibit speciation.[12]

3. Did the ancestral population divide equally or did only a very small

11. For example, Fisher 1930; Wright 1931, 1940, 1978b; Haldane 1932; Lande 1980, 1981; Templeton 1980, 1981, 1982; and Coyne and Kreitman 1986.

12. Wiley 1978, 1981; Wiley and Brooks 1982; Brooks and Wiley 1986, 1988; and Templeton 1989a.

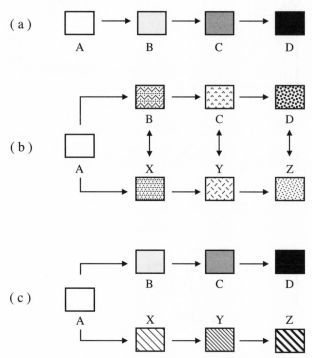

Figure 3.9. Three types of population structure. (*a*) A single deme evolving through time. (*b*) Two demes, linked by gene flow, evolving through time. (*c*) Two demes with no gene flow evolving through time.

part of the ancestral range "bud off" from the rest of the species (or something in between these two extremes)?

Differentiating the two modes based solely upon phylogeny plus geographical distributions of sister species becomes increasingly more difficult as the range of the peripheral isolate species increases. For example, Lynch (1989) suggested that identification of peripheral isolates be limited to species whose range encompassed 5 percent or less of the range of their sister species. Clearly (as Lynch pointed out), this distinction is rather arbitrary (fig. 3.10), and just as clearly we will need to incorporate other evidence for distributions that fall into that gray area between 5 and 50 percent. (We will return to this problem in chapter 4.)

Both vicariant speciation and peripheral isolates speciation may produce dichotomous or polytomous phylogenetic trees. The discovery that allopatric speciation may produce polytomies is critical. First, computer programs that force all phylogenetic trees into completely bifurcating patterns will conceal the evidence for this process and overestimate the importance of other speciation modes (Wiley 1981; Brooks and McLennan 1991; Hoelzer and Melnick 1994). Second, each of the terminal taxa

ALLOPATRIC MODE I

ALLOPATRIC MODE II

symmetrical distribution
range of A = range of B

asymmetrical distribution
range of X ≤ 5% of range of Y

Figure 3.10. Size counts. How much asymmetry in size between sister species is enough to qualify for allopatric mode II?

in the polytomy may be a different age; therefore, comparing degrees of evolutionary divergence will not necessarily give an accurate picture of evolutionary rates.

> There are no phylogenetic patterns that, by themselves, distinguish allopatric modes of speciation. Differentiating various modes requires additional evidence and new applications of our discovery method (chapter 4).

NONALLOPATRIC MODES OF SPECIATION

Nonallopatric speciation is a generic term for speciation initiation when there is incomplete geographical separation of two or more populations of an ancestral species. Unlike vicariance and microvicariance, which postulate that gene flow between populations is initially severed by factors extrinsic to the biological system, nonallopatric speciation modes require the involvement of biological processes intrinsic to the system. In this, they are similar to peripheral isolates modes. Nonallopatric modes differ from peripheral isolates speciation because speciation is postulated to progress despite substantial and continuous gene flow between populations. For this reason, nonallopatric modes (Futuyma and Mayer 1980) are often called divergence-with-gene-flow (Rice and Hostert 1993).

Nonallopatric Speciation Mode I

Parapatric speciation[13] occurs when two populations of an ancestral species differentiate into descendant species despite the maintenance of some gene flow and geographical overlap during the process (fig. 3.11). Stochastic events (e.g., drift), adaptive responses to different local selection pressures along an ecological cline, or both, initiate the differentia-

13. See Smith 1965, 1969; Endler 1977; Lande 1982; Barton and Charlesworth 1984; Gavrilets, Li, and Vose 1998, 2000; Gavrilets 2000; Jiggins and Mallet 2000; and Arthur and Kettle 2001.

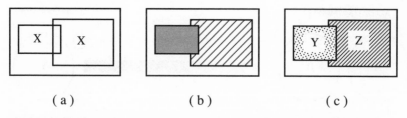

(a)   (b)   (c)

Figure 3.11. Parapatric speciation. (*a*) Overlap of two populations of ancestral species. (*b*) Differentiation of populations begins while they are still in contact. (*c*) Speciation of Y and Z is completed despite maintenance of the contact area.

tion; low vagility among members of the populations (decreasing gene flow even when overlapping), a decrease in heterozygote/hybrid fitness leading to positive assortative mating, or both, promote it. Endler (1977) presented a detailed defense of parapatric speciation in his extensive treatise on the microevolutionary aspects of geographical and clinal variation. He suggested that members of the anuran *Rana pipiens* group; the mosquito-fish genus *Gambusia;* the fruit fly genus *Drosophila;* the plant genus *Gilia;* the frogs *Hyla ewingi* and *H. verreauxi;* and the frogs *Pseudophryne dendyi, P. bibroni,* and *P. semimarmorata* might all be examples of parapatric speciation. In the case of *H. ewingi* and *H. verreauxi,* hybrids from the zone of overlap showed depressed fitness relative to hybrids from parents taken from allopatric portions of the species population—one of the conditions of parapatric speciation. What is required now is evidence that the overlapping species are indeed sister species.

Nonallopatric Speciation Mode II

Sympatric speciation[14] occurs when one or more new species arise without geographical segregation of populations (fig. 3.12). Although this was the mode originally preferred by Darwin, support for sympatric speciation wavered when quantitative geneticists demonstrated that the homogenizing effects of recombination during meiosis coupled with gene flow among populations would tend to destroy the nonrandom associations of traits (the double-variation model) arising via divergent selection (Felsenstein 1981b; see also Mayr 1963; Futuyma and Mayer 1980; Paterson 1981). If gene flow were restricted or interrupted, as in the allopatric speciation modes, the novel trait would have a better chance of becoming fixed within a deme, and the whole process would operate much more smoothly. The work of the geneticists was coupled with the

14. Darwin 1859; Fisher 1930, 1958; Bush 1966, 1969, 1974, 1975a,b, 1982; Maynard Smith 1966; Dickinson and Antonovics 1973; Tauber and Tauber 1977a,b, 1989; Rosenzweig 1978; and White 1978.

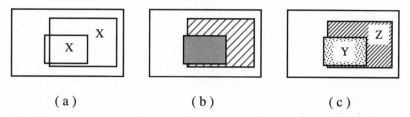

(a)   (b)   (c)

Figure 3.12. Sympatric speciation. (*a*) Extensive overlap of two populations of ancestral species X. (*b*) Differentiation of populations begins while they are still in contact. (*c*) Speciation of Y and Z is completed despite maintenance of the contact area.

earlier recognition that most "related" species (this usually meant members of the same genus) exhibited allopatric but adjacent distributions (e.g., Wallace 1855; Jordan and Kellogg 1908; Mayr 1942; Wallace 1955), and this combination provided a strong foundation for the hypothesis that most speciation was allopatric.

The lack of theoretical support notwithstanding, one of the most passionate advocates of nonallopatric speciation modes, Guy Bush, argued that "the future holds many surprises. . . . I suspect that macromutations and rapid nonallopatric mechanisms of speciation will prove to be far more important in many groups of organisms than previously imagined" (1982:128).

A revival of interest in sympatric speciation followed hard on the heels of Bush's prediction as researchers began to uncover potential roles for phenotypic plasticity, disruptive selection, ecological segregation, sexual selection, and mutation plus random drift in both theoretical and empirical investigations.[15] For example, Rice (1985) and Rice and Salt (1988, 1990) demonstrated that divergent selection for habitat preference could lead to reproductive isolation between sympatric populations of *Drosophila* if (1) selection was applied to numerous aspects of habitat preference and not restricted to a single trait, increasing the chance that reproductive isolation will evolve as a correlated response to selection via pleiotropy or genetic hitchhiking (the single-variation model) and (2) the intensity of divergent selection was high relative to the level

15. West-Eberhard 1983, 1989; Diehl and Bush 1984, 1989; Dominey 1984; Rice 1984, 1985, 1987; Wu 1985; Bush and Howard 1986; Kondrashov and Mina 1986; Butlin 1987a; Fialkowski 1988, 1992; Grant and Grant 1989; Tauber and Tauber 1989; Higgs and Derrida 1991, 1992; Todd and Miller 1991; Craig et al. 1993; Bush 1994; Turner and Burrows 1995; Johnson et al. 1996; Kawecki 1997; Gavrilets and Boake 1998; Møller and Cuervo 1998; Orr and Smith 1998; Dieckmann and Doebeli 1999; Gavrilets 1999; Seehausen, Mayhew, and van Alphen 1999; Seehausen, van Alphen, and Lande 1999; Arnqvist et al. 2000; Doebeli and Dieckmann 2000; and Via et al. 2000. For an excellent discussion and additional references, see Johnson et al. 1996.

of gene flow. In their impressive summary of evidence for the major "engines" of speciation, Rice and Hostert vindicated Bush's optimism, concluding that "the single-variation model of divergence-with-gene-flow speciation has been largely overlooked . . . because it represents a 'theoretician's nightmare,' that is, it can be completely deduced from experimental results, it is simple, and it is obvious. There just does not seem to be much that is theoretically interesting about divergence-with-gene-flow speciation via pleiotropy/hitchhiking, despite its potential [we would argue likely] importance in nature" (1993:1650).

Differentiating Nonallopatric Speciation Modes I and II

Phylogenetic trees reflecting incidents of nonallopatric speciation may be either dichotomous or polytomous, depending on how many species have been produced sympatrically or parapatrically from the same ancestor, and depending on whether or not the ancestor persists. Differentiating the two types of nonallopatric modes thus requires that we answer only one question: How much overlap exists between the sister species? At the moment, there are no suggestions about the amount of overlap required before sympatric devolves to parapatric speciation. For example, should we recognize sympatry as more than 50 percent overlap, more than 75 percent, only 100 percent? If we restrict sympatry to 100 percent overlap, then sympatric speciation suffers the same problems of identification as peripheral isolates speciation; in both cases the criterion for recognition is so restrictive (100 percent overlap, 5 percent the size of the sister species, respectively), that we might underestimate the importance of the two modes. At the moment, there have been few studies of putative nonallopatric modes that have actually investigated differences between sister species; we therefore cannot make any generalizations based upon empirical evidence about the impact of percent overlap on speciation rates. Overall, though, nonallopatric modes are postulated to occur rarely because of the initial difficulties in overcoming gene flow.

> ! Can't you just use distribution patterns to identify the products of parapatric and sympatric speciation? No. Just because two "closely-related" species have completely overlapping ranges does not mean that they are sister species. Remember, the most important thing about the speciation process is that it produces sister species. You thus need a phylogeny to determine relationships and distributions of sister species to determine possible speciation modes. In other words, do not equate sympatric/parapatric distributions with sympatric/parapatric speciation.

Both modes require observations of overlap among *sister species* that differ in some special ecological or genetic characteristics that could, in themselves, produce independent species. Identification of overlapping sister species is thus the first step in a complete study of nonallopatric modes. The second step requires that we identify the autapomorphic traits for each member of that pair. The hypothesis that any or all of those traits may have been causally involved in the speciation process sets up the third, and most difficult, step in the research program: testing the hypotheses via experimental manipulation.

SOME TEST CASES

In the following pages we shall focus our attention on examples of the first steps in a speciation study—combining phylogenetic and geographical distribution patterns to uncover the factors that initiated speciation in various groups.

Freshwater Fishes: Identifying Modes at Nodes. The freshwater stream fishes of southeastern North America represent an excellent model system for studying speciation in a comparative phylogenetic framework. Detailed distributional data are available and explicit phylogenetic hypotheses have been published and are being reinforced or upgraded on a regular basis. Although there are a plethora of examples, we will discuss only a few of these from each region based on the pioneering study by Wiley and Mayden (1985).[16]

Darters are small, bottom-dwelling percids. Although distributed in freshwater throughout the northern hemisphere, approximately 90 percent of the species are restricted to locations east of the Rocky Mountains (Moyle and Cech 1982). Killifish are small, brightly colored cyprinodontids that are both geographically and ecologically diverse. They can tolerate a wide variety of habitats, from freshwater streams and desert springs, to salt marshes and mangrove swamps. This discussion will focus on speciation patterns within one group of sand darter (*Ammocrypta*) and one group of killifish (*Fundulus*).

Examination of the disjunct distribution patterns of these fishes reveals that vicariant speciation (allopatric mode I) has been the predominant mode in these groups. Within the sand darters (fig. 3.13), geographic division of ancestor x into two populations produced *Ammocrypta clara* and its sister species, ancestor y, with some apparent extinction of *A. clara,* leaving two disjunct populations of the species today. Notice that there is some asymmetry in size between *A. clara* and

16. For even more extensive examples, see Mayden 1988, 1991.

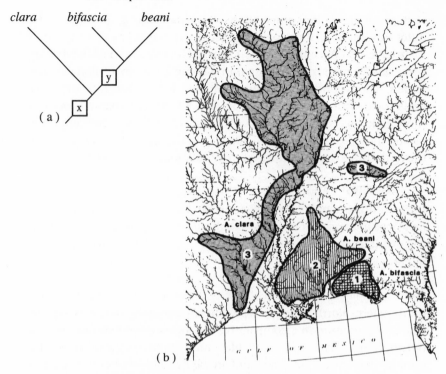

clara   *bifascia*   *beani*

(a)

(b)

Figure 3.13. Speciation of sand darters. (*a*) Phylogenetic tree for sand darters in the *Ammocrypta beani* group. *Names* = species; *letters* = ancestral species. (*b*) Distribution map for the three *Ammocrypta* species (modified from Wiley and Mayden 1985).

its sister group, species y (whose range = approximately the range of its descendants, *A. beani* + *A. bifascia*) but not enough of a difference to postulate peripheral isolates speciation of ancestor y. Geographic division of ancestor y into populations east and west of the Mobile drainage system next produced *A. beani* and *A. bifascia*. *Ammocrypta* is significant for another reason. Earlier, we emphasized that your interpretation of speciation modes may be affected by the exclusion of species. In this case, phylogenetic analyses by Simon (1992) and Near et al. (2000) have added three species to the *Ammocrypta* clade. Both of those studies agree with Wiley and Mayden that *A. beani* and *A. bifascia* are sister species, but suggest that *A. clara* is not the sister species of *A. beani* + *A. bifascia*. The origins of *A. clara* and of the common ancestor of *A. beani* + *A. bifascia*, therefore, appear to be the result of two different speciation events. Interestingly, the addition of three species between *A. clara* and *A. bifascia* + *A. beani* solves the original problem of range asymmetry.

The analysis by Near et al. indicates that the three additional speciation events in the group are also the result of vicariance.

The killifish pattern (fig. 3.14) is slightly more complicated. Production of sister species p and q from ancestor o is problematic because, since these species no longer exist, we can only obtain an estimate of their distribution by combining the ranges of their descendants. This method assumes that there has not been any widespread extinction or dispersal in the area, assumptions which are tenuous at best for these fishes. The results of such an analysis indicate that the ranges of p and q potentially overlapped near the mouth of the Mobile drainage area. Although fossil evidence of overlap would be illuminating, it would not resolve this quandary because, since p and q are extinct, we can no longer test for the presence or absence of interpopulation interactions in the parapatric area. So, for the present, this speciation event must be tentatively assigned an indeterminate status (parapatric mode?). Division of ancestor p producing *Fundulus blairae* and *F. dispar,* division of ancestor q producing *F. lineolatus* and r, and division of ancestor r into populations east and west of the Mobile drainage system producing *F. nottii* and *F. escambia* all represent apparent examples of vicariant speciation (allopatric mode I). Although suggestive of sympatric speciation, the observation that *F. nottii* is located almost entirely within the range of *F. blairae* is not important to a study of what initiated speciation, because the two fishes are not sister species. (See also the slight overlap between *F. escambia* and *F. lineolatus.*) It is important to remember that there is a fundamental difference between modes of speciation and mechanisms of speciation. Speciation modes describe different original conditions which set the stage for the subsequent divergence of ancestral populations. This divergence, in turn, may be accomplished by a variety of biological processes. For example, it could be envisioned that once ancestor r was separated on either side of the Mobile River, interactions between *F. blairae* and the population of ancestor r isolated on the west side of the basin were involved in driving the population along the pathway to "species-hood" (to becoming *F. nottii*). Nevertheless, these interactions have no effect on the original event which established the potential for the speciation of ancestor r; that is, its separation into two disjunct populations.

Examples of allopatric speciation mode I are widespread among fishes inhabiting the central areas of the Mississippi drainage from the Ozark Plateaus and Ouachita Highlands in the west to the Interior Low Plateau, Ridge and Valley Province, and Blue Ridge Province in the east. Darters are well represented in this area, as are the small, species rich, silvery cyprinids aptly referred to as shiners. During the breeding season many

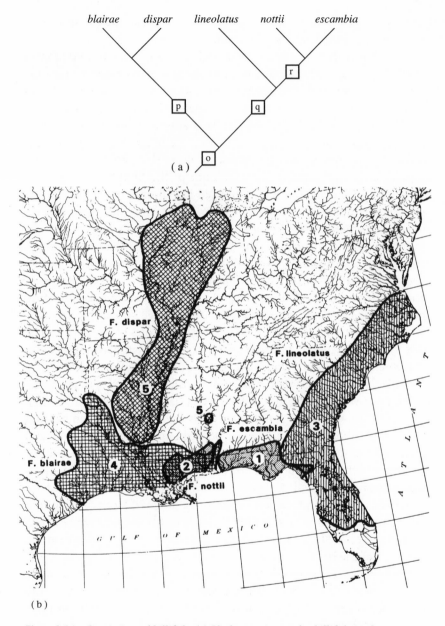

Figure 3.14. Speciation of killifish. (*a*) Phylogenetic tree for killifish in the *Fundulus nottii* group. *Names* = species; *letters* = ancestral species. (*b*) Distribution map for five *Fundulus* species (modified from Wiley and Mayden 1985).

shiner species don flamboyant red, orange, and/or yellow breeding liveries, and because of this, they have drawn the attention of many researchers. This discussion will be restricted to speciation patterns within two groups of shiners (*Notropis*) and one group of darters (*Etheostoma*).

Examination of the distribution patterns of these fishes reveals that, like the situation in the Mobile drainage system, vicariant speciation (allopatric mode I) has been the predominant speciation mode in these groups. The simplest pattern occurs in the *Notropis nubilus* clade (fig. 3.15), where two speciation events associated with geographical vicariance have been coupled with the apparent loss of one species' (*N. nubilus*) central populations.

The second *Notropis* example (fig. 3.16) is equally straightforward. The distributions and phylogenetic relationships support the proposal that the four putative speciation events within this clade have been vicariant via the geographical divisions of (1) ancestor o, possibly through a vicariance event eliminating the central populations, producing sister species p and q; (2) ancestor p, producing *N. pilsbryi* and *N. zonatus;* (3) ancestor q producing *N. cerasinus* and ancestor r; and finally (4) ancestor r, producing *N. coccogenis* and *N. zonistius.*

The pattern depicted for the darters (fig. 3.17) is slightly more problematic because the current pattern appears to have resulted from a combination of two and possibly three speciation modes. First, there is the possible asymmetrical division of (1) ancestor u, producing *Etheostoma blennius* and ancestor v; (2) ancestor v, producing *E. sellare* and ancestor w; and (3) ancestor y, producing *E. variatum* and ancestor z, all examples of allopatric speciation mode II. Second, there are two putative speciation events that indicate vicariant (allopatric mode I) speciation: (1) the division of ancestor w, associated with extinction of central populations, producing ancestors x and y; and (2) the division of ancestor x, producing *E. tetrazonum* and *E. euzonum*. Finally, there is the potential parapatric speciation of *E. osburni* and *E. kanawhae* from ancestor z. The case for parapatric speciation in this instance would be strengthened if hybrids between the two species could be documented in the region of overlap, especially if those hybrids could be shown to be at a fitness disadvantage with respect to nonhybrids of either parent outside the zone of overlap.

From Cat Food to Cats: Adding Ecology and Behavior.

> What sort of philosophers are we, who know absolutely nothing of the origin and destiny of cats?
>
> Henry David Thoreau

Human beings have been associated with cats (Felidae) in one way or another for millennia. We hunt them for their fur and medicinal proper-

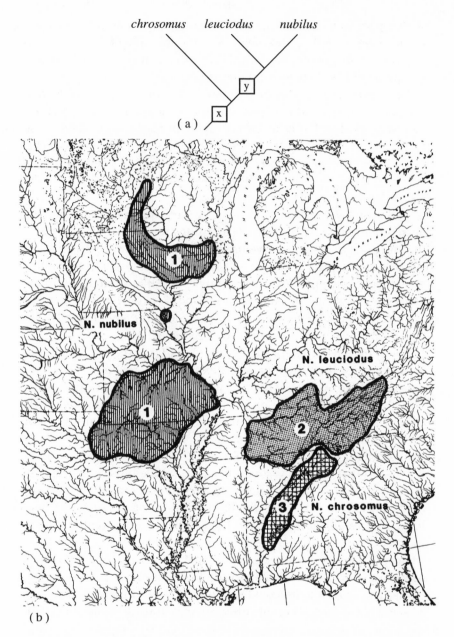

Figure 3.15. Speciation of shiners. (*a*) Phylogenetic tree for shiners in the *Notropis nubilus* group. *Names* = species; *letters* = ancestral species. (*b*) Distribution map for three *Notropis* species (modified from Wiley and Mayden 1985).

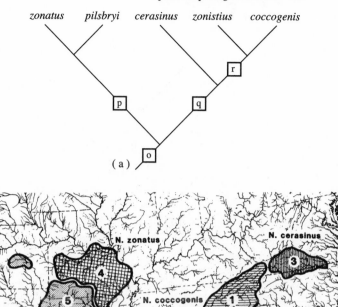

Figure 3.16. Speciation of more shiners. (*a*) Phylogenetic tree for shiners in the *Notropis zonatus-coccogenis* group. *Names* = species; *letters* = ancestral species. (*b*) Distribution map for five *Notropis* species (modified from Wiley and Mayden 1985).

ties, display them in zoos, associate them with magic and witchcraft, worship them as gods, domesticate them for pest control, and keep them as pets. Undisputed cultural value aside, cats are also prime candidates for studies of nonallopatric modes of speciation because many felid species have overlapping distributions. In order to study these modes further we need to be able to disentangle two problems: (1) sympatric speciation versus sympatric contact between congeners (a phylogeny resolves this problem); and (2) sympatric speciation versus sympatric overlap of sister species due to secondary dispersal by one or both of the newly fledged sisters after speciation was completed in allopatry. This concern

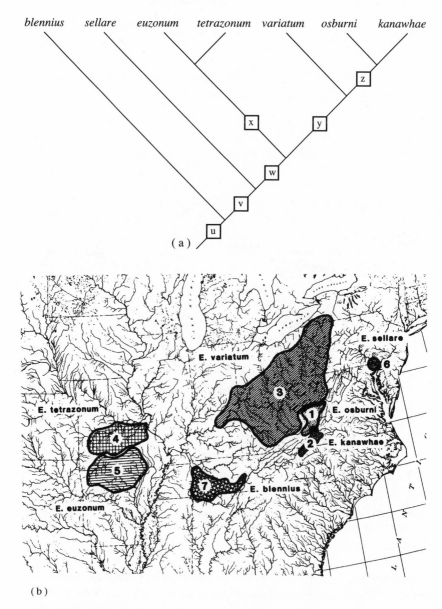

blennius   *sellare*   *euzonum*   *tetrazonum*   *variatum*   *osburni*   *kanawhae*

(a)

(b)

Figure 3.17. Speciation of more darters. (*a*) Phylogenetic tree for darters in the *Etheostoma variatum* group. *Names* = species; *letters* = ancestral species. (*b*) Distribution map for seven *Etheostoma* species (modified from Wiley and Mayden 1985).

is particularly apposite when the study species are highly mobile creatures like cats. It is also pertinent when we are considering the widespread sympatry between ancestral sister species buried deep in the phylogeny of the group, because we would expect the obfuscating effects of dispersal to increase with increasing age of the speciation event (Brown and Gibson 1983, 1998; Chesser and Zink 1994). Until we have access to a time machine, we will have to use indirect methods to distinguish primary from secondary contact. Mattern and McLennan (2000) proposed that one such indirect piece of evidence concerns ecological change between sister species. Nonallopatric modes of speciation require the evolutionary diversification of some special ecological or genetic characteristics that could, in themselves, produce independent species. So, if two sister species show a moderate amount of overlap (let the question of how much overlap wait until chapter 4) but do not show any ecological character change, then the pair can tentatively be assigned to allopatric speciation followed by dispersal and secondary contact. "Tentatively" is the operative word here, because at any one time it is unlikely that we will have documented all the important ecological/behavioral differences for members of a clade.

Mattern and McLennan investigated potential patterns of felid speciation by integrating information from phylogeny, geographical distributions, ecology, and behavior (habitat preferences, activity patterns, and estimates of body size [because changes in body size are hypothesized to be associated with different prey preferences in the felids: Kiltie 1984, 1988]). For the purposes of this chapter, we will focus on only one of the eight clades discussed by the authors, the group containing most of the small, South American cats. These cats appear to have speciated in the following manner (fig. 3.18): Node 1: the distributions indicate parapatric speciation of ancestor 1 and ancestor 2, but there are no obvious functional changes associated with this event. This node is tentatively assigned the status of vicariant speciation followed by dispersal and secondary contact. Node 2: sympatric speciation accompanied by an increase in body size (small to medium) in *Leopardus pardalis* (the ocelot) and a change in ecology (from primarily terrestrial to arboreal) in *L. wiedii* (the margay). Node 3: the distributions indicate sympatric speciation of ancestor 3 and ancestor 4, but again there are no obvious functional changes associated with this event (vicariant speciation with postspeciation dispersal). Node 4: peripheral isolates speciation of *Oncifelis guigna* (the kodkod), presumably following the movement of a small population of ancestor 3 west across the Andes into Chile. This putative speciation event was not associated with any ecological change in *O. guigna*. It is possible that the movement from one side of the Andes to the other,

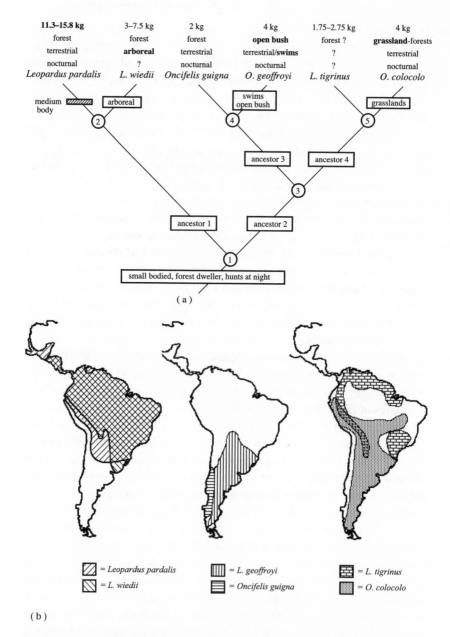

Figure 3.18. Speciation in the ocelot group. (*a*) Ecological and body-size data optimized onto the phylogeny for the ocelot group. (You will learn about character optimization in chapter 5.) *Bold* = apomorphic changes in the character. Nodes are labeled for discussion of potential speciation modes (modified from Mattern and McLennan 2000). (*b*) Distribution map for ocelot group.

a rather Herculean feat for such a small cat, was a rare event. In other words, a small population of ancestor 3 crossed the mountains into unoccupied, plesiomorphic habitat, evolved in isolation, and were never able to repeat the dispersal event successfully (classic waif dispersal: Frey 1993). While the kodkod was making its amazing journey, its sister species, *O. geoffroyi* (Geoffroy's cat), continued to evolve, moving out of the forest into more open habitat and adding swimming to its list of accomplishments. Node 5: parapatric speciation associated with an ecological shift by *O. colocolo* out of the forests into grasslands and surrounding woods.

Combining data from the entire family indicates that 7 of 27 speciation events appear to be the result of vicariance (fig. 3.19). Although five of the cases are ambiguous (accorded vicariant status solely because there were no differences in ecology between sympatrically or parapatrically distributed sister species), four of them were buried deep within the phylogenetic tree, so it is not untoward to postulate that they represent examples of postspeciation dispersal with secondary contact. Even with the ambiguous cases included, vicariant speciation only accounts for 25.9 percent of felid diversification. This is a very different picture from that reconstructed for other vertebrate groups, in which vicariance has been hypothesized to play the dominant role (think back to the North American freshwater fishes). Peripheral isolates speciation also appears to have been less important in the cats than in other clades (four episodes based upon the geographical distributions of sister species). Only two of the four putative isolates (*Uncia uncia* [the snow leopard] and *Otocolobus manul* [Pallas's cat]) displayed an apomorphic ecology compared to their sister species, supporting proposals that the small isolate should display more character change than its large sister group (due to drift, divergent selection in a novel environment, etc.). There are three possible explanations for the lack of ecological differentiation in the other two isolates (*Oncifelis guigna* [the kodkod] and *Felis nigripes* [the black footed cat]). The first, and most obvious, explanation is that differences will be found with a more detailed ecological and genetic investigation. The second, and less easily tested, hypothesis is that these putative instances of peripheral isolates speciation are in fact the remnants of a vicariant speciation event, followed by substantial range contraction in one of the sister species (Frey 1993). Of the two species, this second hypothesis could most easily apply to the black-footed cat, a small (1.5–2.75 kg) felid restricted to an equally small area in South Africa, while its sister group ranges thousands of miles to the north across northern Africa, the Middle East, and Europe. It is possible that the original vicariance event involved a widespread climate/habitat change in northern Africa (e.g., the origin and expansion of the Sahara desert) followed by

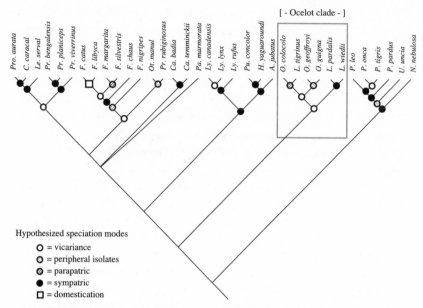

Figure 3.19. Summary of speciation in felids. The phylogeny is a strict consensus of two equally parsimonious phylogenetic trees for the Felidae based upon 1,504 morphological, molecular, and karyological characters (CI = 41.3%). The only difference between the two trees involves the placement of *Pardofelis marmorata*. Hypothesized speciation mode is identified at each node. (For a detailed discussion of these modes, see Mattern and McLennan 2000.) *A.* = *Acinonyx; C.* = *Caracal; Ca.* = *Catopuma; F.* = *Felis; H.* = *Herpailurus; L.* = *Leopardus; Le.* = *Leptailurus; Ly.* = *Lynx; N.* = *Neofelis; O. Oncifelis; Ot.* = *Otocolobus; P. Panthera; Pa.* = *Pardofelis; Pr.* = *Prionailurus; Pro.* = *Profelis; Pu.* = *Puma; U.* = *Uncia.*

long-term habitat destruction, fragmentation, and extinction of black-footed cat populations. Finally, the pattern of more autapomorphies in the sister species of the putative peripheral isolate supports the often overlooked centrifugal mode of peripheral isolates speciation. According to this model, the larger, central population is assumed to be the "principal source of evolutionary change" (Brown 1957:248) because, among other things, large populations have a higher probability of gaining new mutations and a lower probability of going extinct than do small populations. This explanation might fit the kodkod–Geoffroy's cat pair. Overall, in two cases the peripheral isolate became more highly derived ecologically compared with its more widespread sister species (the snow leopard and Pallas's cat), and in two cases the peripheral isolate remained relatively plesiomorphic compared with its more widespread sister species (the kodkod and the black-footed cat). Further research is required to determine the actual amount of differences between these sister pairs. If the preceding patterns survive the additional scrutiny, then studies about population structure/biology are needed to determine why the "principle source of evolutionary change" can vary from the larger to the smaller population within this relatively small, conservative vertebrate family.

In summary, the relatively rapid diversification of the cats appears to have been more strongly associated with ecological, than geological, separation. The partitioning of the environment by sympatric sister species provides more support for the hypothesis that adaptation via ecological character change may play a causal role in the speciation process.[17] However, macroevolutionary patterns plus distributional data, on their own, cannot tell us anything about underlying processes. For example, Toft (1985) concluded that resource partitioning in a wide variety of amphibians and reptiles resulted from a number of causes, including competition, predation, and physiological constraints, acting alone or in concert. In order to understand how these different selective forces have influenced the evolution of resource partitioning and the subsequent speciation of many felids, we need more detailed information about the physiology, ecology, and behavior of the felids in general. We also need to demonstrate that the changes in ecology are linked to the evolution of reproductive isolation between the sister pairs. Overall, though, this preliminary investigation indicates that closely related cats are able to coexist (Hutchinson 1959; MacArthur and Levins 1967; Houle 1997) because of changes in one of the three traditional resource dimension categories (Pianka 1975): displacement in habitat (space), displacement in time (stay in the same place, but hunt at different times), and displacement in energy source (hunt in the same area at the same time, but focus on different prey items). If the cats are as young as molecular evidence would indicate (8–9 million years), it is unlikely that postspeciation dispersal has clouded numerous instances of vicariance speciation. It is more likely that such movement has blurred the distinction between peripheral isolates, parapatric, and sympatric speciation.

Hawaiian Crickets: Examining Hypotheses concerning the Sequence of Speciation. The Hawaiian islands were created over a fixed hot spot in the Pacific Ocean.[18] This hot spot is caused by an upwelling of magma from deep within the earth that periodically breaks through the slowly drifting Pacific plate to form volcanoes. The highest of these volcanoes may rise

17. See excellent discussions in Losos 1990a, 1992; Bush 1994; Schluter 1996a, 1998; Taylor et al. 1996; Kawecki 1997; Losos et al. 1997; Colbourne 1998; Nagel and Schluter 1998; and Bridle and Jiggins 2000.

18. The following brief discussion is based on an excellent and much more detailed review by Carson and Clague (1995).

above the ocean's surface, forming islands. Because the volcanoes/islands are part of the Pacific plate, they are also on the move, drifting slowly northwestward, being worn down through erosion and subsidence as they go. As one island drifts off, another forms behind it over the hot spot, adding new, large islands as old, worn, and now low lying ones sink beneath the ocean. This of course is rather catastrophic for terrestrial life, but may be a boon for marine organisms since sinking islands generally acquire coral reefs (suggested by Darwin 1837; recently confirmed by Christie et al. 1992). There has been a complicated geological history of coalescence between one or more islands and the appearance and destruction of land bridges, which means that faunal exchange among islands has been possible at various times in the past.

Into this complicated story come the swordtail crickets (Trigonidiinae: *Laupala*), small, flightless insects living on the ground or in small bushes in the rain forests of the six high islands of the Hawaiian archipelago. Each species is endemic to only one island. The observation that species living on the island of Maui exhibit the greatest degree of morphological and acoustic differentiation (cricket males serenade their mates by rubbing their wings together) led Otte (1989) to propose that speciation began on Maui, with colonists dispersing to other islands at a later date. This hypothesis differs from the sequence proposed for picture-winged *Drosophila,* which has colonizers tracking islands as they emerged, from the older Kauai and Oahu to the younger Maui and Hawaii (Carson 1983). Fortunately these two hypotheses differ in their predictions of where different species should be located on a phylogeny. For example, the Maui hypothesis predicts that the Maui species should cluster together as the basal-most clade in the genus, while the Kauai hypothesis places the Maui species at a more derived (younger) position on the tree.

Shaw (1995a, 1996a) examined these different predictions by reconstructing a phylogeny for 17 of the 35 *Laupala* species based upon mtDNA characters. The consensus of the eight equally parsimonious trees is shown in figure 3.20. Differences among those eight trees involved resolutions of polytomies within monophyletic groups and thus will not affect the overall picture of island colonization in these crickets (although it will affect hypotheses of divergence on particular islands). Mapping geographic distribution above the tree uncovered a very straightforward picture of island hopping from the oldest (Kauai: 3.8–5.6 million years) to the youngest (Hawaii: 0.68 million years; McDougall 1964) islands (allopatric mode II), with at least two episodes of hopping backward to an older island (Hawaii to Maui: *L. fugax* and *L. kolea*). This sequence is congruent with the general picture-wing *Drosophila* pattern and does not

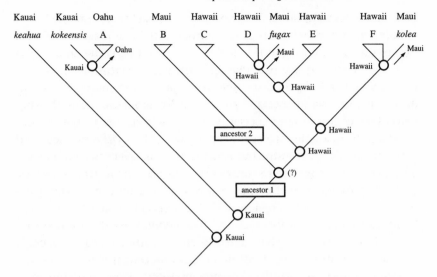

Figure 3.20. Speciation in crickets. Geographical distributions mapped onto the strict consensus of eight equally parsimonious phylogenetic trees for swordtail crickets (*Laupala*) based upon mtDNA base pairs. Monophyletic lineages on one island are represented by letters: A = *pacifica* + *tantalus* + *hapapa*; B = *verspertina* + *eukolea* + *prosea*; C = *cerasina* + *nigra*; D = *cerasina* + *pruna* + *kona* + *hualalai*; E = *cerasina* + *pruna*; F = *cersina* + *kohalaensis* + *paranigra*. The polyphyletic status of several "species" (e.g., *L. cerasina, L. pruna*) needs to be addressed with further research.

support the "Maui-as-center-of-origin" hypothesis. The cricket fauna on Maui appears to be a complicated combination of evolution *in situ* (clade B on fig. 3.20) and back colonization by species that had evolved on Hawaii. These polyphyletic origins of the fauna probably explain why the Maui assemblage is so morphologically and acoustically diverse. The study also shows that (successful) inter-island transfers have not happened very often within *Laupala*. Once a colonizer gets to an island, it tends to settle there and speciate. *Drosophila* shows a much more complicated pattern of movement between islands (Carson 1983), which is not surprising given that fruit flies fly, whereas swordtail crickets are flight-challenged.

   The study highlights enough unanswered questions about cricket speciation to generate research for decades. For example, ancestor 1 could have been enjoying itself on Maui, Kauai, or Hawaii. Is there any way to determine where the Maui lineage came from? Estimates of lineage age may throw some light on the problem. If the Maui lineage is older than 0.68 million years, its ancestor can't have come from Hawaii (there are no marine crickets). This would leave only one alternative, Maui was colonized from Kauai, either earlier (ancestor 1) or later (ancestor 2).

When it comes right down to it, how do the crickets get from one island to another, and where did they come from in the first place (see discussion in Carson and Clague 1995)? Equally interesting is the observation that there has been a substantial amount of diversification on Hawaii within a very short period of time. How have those species been produced? If inter-island speciation is driven by allopatric mode II, what percentage of intra-island speciation has also been due to that mode? For example, in a series of elegant experiments, Shaw (1996b) discovered that differences between the pulse rate of male *L. kohalensis* and *L. paranigra*'s courtship songs were controlled by at least eight genetic factors. These two (putative) sister species are separated by the northern expanses of the Mauna Kea region of Hawaii. Speciation has thus occurred in allopatry and relatively recently. Shaw (1999) then determined that discontinuities in song characters associated with distinct gene pools were also present in areas boasting at least three sympatric cricket species. In conjunction with the polygenic basis for a male courtship character, these studies hint that sexual selection may have played an important role the completion of speciation in these insects, both in allopatry and sympatry (Orr and Coyne 1992). Clearly, what is needed now is a fully resolved phylogeny for all of the species in the genus, as well as (further) detailed studies of the mechanisms responsible for the completion of speciation.

The patterns of speciation in the Hawaiian archipelago are a researcher's dream (or nightmare, depending upon how well organized you prefer the world to be). Virtually everything that can happen has happened (see studies in Wagner and Funk 1995). There are species that are as old as the island on which they reside, suggesting peripheral isolates speciation by colonization from a neighboring island soon after the species' island rose above the ocean surface. There are entire clades that have diversified by a variety of means on a single island, producing some species that are relatively old and some that are very young. There are species that have evolved on a younger island and "back-colonized" an older island. And finally, there are species from older islands that have colonized younger islands relatively recently. Thus, the biota of any given island is likely to be a historically unique and complex mosaic of very old, relatively old, relatively young, and very young species. Aloha.

Of Mice and Rings: Appearances Can Be Deceiving. Two species of pocket mice, *Perognathus amplus* and *P. longimembris,* are distributed throughout scrubby habitats in the Mojave, Sonoran, and Great Basin deserts of the southwestern United States. Although the two species are nearly indistinguishable at the northernmost extent of their ranges (in northern Arizona, separated by the mighty Colorado), they are clearly differentiated further south (*P. longimembris* is smaller than its relative). Populations of the two species form a roughly circular pattern, with *P. amplus* to the east and *P. longimembris* to the west of the Colorado River, overlapping in only one locality in western Arizona. Taken together, their morphology and geographic distributions led Hoffmeister (1986) to propose that these pocket mice formed one "ring species." According to this scenario (fig. 3.21), some members of the ancestral population of *P. longimembris* in Utah crossed the Colorado into northern Arizona and began dispersing throughout the Arizona desert, increasing in size as they did so. Other members of the ancestral population dispersed in a westerly direction along the northern side of the river and either did not change in size or actually got smaller. When the two ends of the "ring" came back into contact, they were so different morphologically that they no longer interacted with one another even though gene flow continued between populations throughout the ring.

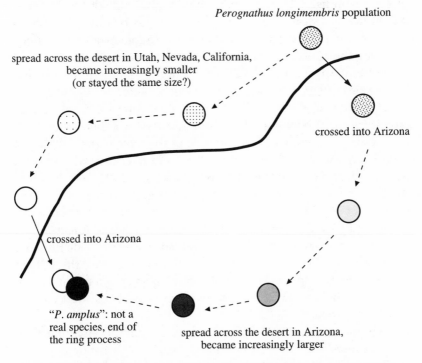

Perognathus longimembris population

spread across the desert in Utah, Nevada, California, became increasingly smaller (or stayed the same size?)

crossed into Arizona

crossed into Arizona

"*P. amplus*": not a real species, end of the ring process

spread across the desert in Arizona, became increasingly larger

Figure 3.21. Scenario proposed for production of a "ring species" in pocket mice. The story begins with ancestral populations of *P. longimembris* dispersing along two disjunct pathways south and west. By the time the two dispersal lines come back into contact (forming the end of the ring), they have diverged so much during their evolutionary and geographic travels that they no longer recognize one another as the same species.

The ring hypothesis is based upon the relationship between variability in morphology and the geographic distribution patterns of populations. The hypothesis makes two predictions that are amenable to being tested within a phylogenetic framework: (1) *Perognathus longimembris* is paraphyletic without inclusion of *P. amplus;* and (2) the northern most populations are the basal members of the group. McKnight (1995) tested these predictions by using cytochrome-b sequence data to reconstruct the phylogenetic relationships among 20 populations of pocket mice. His analysis, which produced one tree with a CI of 66% (fig. 3.22), refuted the first prediction of the ring hypothesis: according to the cytochrome-b data, *P. amplus* and *P. longimembris* are distinct species that diverged from one another approximately 5–6 million years ago (see McKnight for details of calculating lineage age). Although the tree does not support the hypothesis that the mice are part of one continuous ring species, it does support the suggestion that the group has a northern origin. More specifically, it appears that the ancestor of the two species (ancestor 1 on fig. 3.22) was subdivided on either side of the Colorado River, either by a vicariance event (allopatric mode I) or by the colonization of some members of ancestor 1 across the Colorado (allopatric mode II). During the Miocene, semidesert habitat was widespread throughout southwestern Arizona and into eastern California, southern Nevada, and Utah (Axelrod 1979, 1983). Geological evidence indicates that the Colorado changed course 5–8 million years ago (Lucchitta 1990), which may account for the subdivision of ancestor 1. Ancestor 3 then found itself stranded on the rapidly desertifying side of the river as the uplifting of California mountain ranges to the west of the river cast "a severe rain shadow . . . over the lands to the east" (McKnight 1995:824). It is possible that changing habitat created the selection arena that favored the evolution of smaller body size in *P. longimembris.* The two populations that currently overlap are from the basal-most lineage of *P. amplus* and from the most recently derived lineage of *P. longimembris.* Evidently, after several million years of independent evolution, including the shift in body size, the two species came back into contact after a population of *P. longimembris* from eastern California dispersed across the Colorado River into Arizona.

Given the time elapsed since divergence and the environmental changes experienced by the mice during that time, it is not surprising that the two species were reproductively isolated at the point of secondary contact. What is required now are detailed genetic, ecological, and behavioral studies to document what keeps the two species apart in sympatry. For example, McKnight and Lee (1992) noted that *P. amplus* possesses two pairs of chromosomes with secondary constrictions, whereas

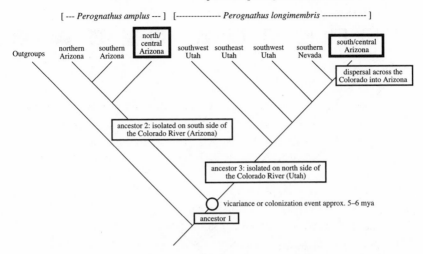

Figure 3.22. Speciation in mice continues. Phylogeny of pocket mice populations based upon cytochrome-b sequence data. Two species are identified on the tree. Populations are represented by geographical distributions on the tree (all distributions are near the Colorado River). Each distribution includes a cluster of populations for which the interpopulation sequence divergence is < 2% (for details see McKnight 1995). Populations in bold boxes are sympatric, representing the former "ends" of the ring shown in figure 3.21.

P. longimembris possesses only one constricted pair. Which of these conditions is derived (outgroup information is needed to answer this question), and could it (or other karyological differences) have been causally involved in the speciation process? Are the pocket mice ecologically segregated based upon body size and, if so, does this, or the differences in size alone, have ramifications for reproductive isolation?

South American Plants: When One Clade Isn't Enough. *Lepechinia* is a group of small, white flowered shrubs distributed mainly throughout tropical and subtropical Latin American highlands. Within this range, the section *Parviflorae*, comprising 12 weedy species living at altitudes of 1,700–3,900 meters in Andean shrub/forest zones, shows complicated latitudinal and altitudinal distribution patterns. Hart (1985a) analyzed the phylogenetic relationships in this group, using 24 morphological characters, producing a single most parsimonious tree with a consistency index of 96% (fig. 3.23a). Figure 3.23b depicts the geographic distribution of these 12 species. The six oldest species in this section are allopatrically distributed with respect to the appropriate sister species. *Lepechinia graveolens,* the sister species to the rest of the section, occurs in north Chile and southern Bolivia. *Lepechinia vesiculosa* ranges through Peru

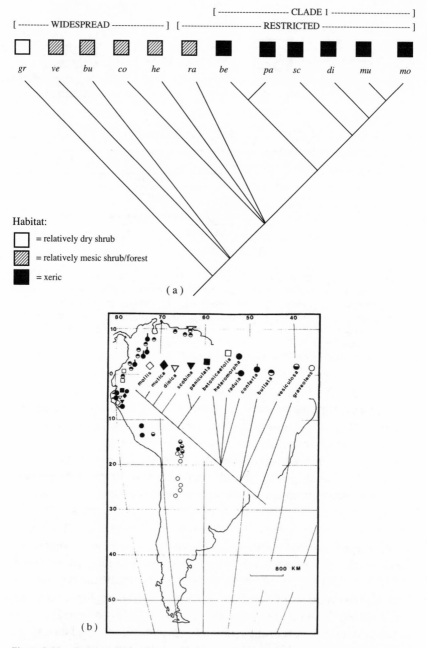

Figure 3.23. Speciation in plants. (*a*) Phylogenetic relationships within *Lepechinia* section *Parviflorae* based on 24 morphological characters. Details of the characters used are provided in Hart 1985a. *Boxes* = habitat preferences. Geographical distribution is listed as either widespread or restricted. *be* = *Lepechinia betonicaefolia; bu = L. bullata; co = L. conferta; di = L. dioica; gr = L. graveolens; he = L. heteromorpha; mo = L. mollis; mu = L. mutica; pa = L. paniculata; ra = L. radula; sc = L. scobina; ve = L. vesiculosa.* (*b*) Distribution map for 12 species of *Lepechinia* section *Parviflorae* (from Hart 1985a).

and Bolivia, while its (possible) sister species *L. bullata* is widely disjunct in Colombia and Venezuela. Finally, *L. heteromorpha* ranges from eastern Ecuador to southern Peru and Bolivia, while *L. radula* is distributed throughout southwestern Ecuador and northern Peru, and *L. conferta* is widespread throughout Colombia and Venezuela. *Lepechinia bullata* and *L. conferta* occur in moist, upper Andean forests at similar altitudes in Colombia and Venezuela, but they are not sister species, so their potential geographical overlap is not pertinent to studies of speciation initiation, although that overlap may be pertinent to studies of speciation completion.

The remaining species, *Lepechinia betonicaefolia, L. paniculata, L. scobina, L. dioica, L. mutica,* and *L. mollis,* form a clade (clade 1 on fig. 3.23a). In contrast to the widespread distributions of their higher elevation, forest-dwelling relatives (five of the six preceding species in the section and all species in the outgroup), all members of clade 1 are locally endemic with very small ranges and are found only in lower elevation, dry habitats. The small ranges, marginal distributions, and specialized habitat preferences of these six species look like peripheral isolates speciation. Hart (1985a,b), however, proposed that a series of forest expansions and contractions during the ebb and flow of the Pleistocene glacial periods produced a repeating cycle of ancestral range expansion, followed by range contraction, producing geographic isolation and vicariant speciation.

Hart's hypothesis for the evolution of clade 1 was based on a strong correlation between the phylogenetic patterns for the clade and a particular geological/environmental scenario. The members of clade 1 are located within the Huancabamba Deflection region of southern Ecuador and northern Peru. Unlike the usual north-south orientation of the Andean mountain range, this area is characterized by low mountain chains, bisected by deep, dry, east-west running valleys. These topographical differences, then, set the geographical stage on which the following scenario is hypothesized to have been enacted. Forests in the higher mountains were reduced and the eastern/western slopes separated by the lowering snow line during the cold, wet glacial periods (van der Hammen and Gonzales 1960; van der Hammen 1972; Geel and van der Hammen 1973). The Pacific side of the Andes was colder and wetter, so forests forced down the mountainsides and to the east were more likely to find refuge than their western counterparts (Hastenrath 1971a,b; Simpson 1975). Interestingly, this scenario, based upon a variety of geological data, is supported by the extant distributions of *L. vesiculosa* and *L. heteromorpha* in eastern slope forest habitats. Within the Huancabamba Deflection, the decrease in mountain size and the communication between the eastern and western slopes through valleys is postulated to have allowed a westerly expansion of the eastern slope forests. Following this expansion, the climate gradually shifted toward a drier state, driving the forests back up the mountainsides following the retreating ice fields. This movement, combined with the complex geography in the area, left populations isolated within moister refugia. Increasingly xeric conditions, small population size, and severed gene flow eventually resulted in a series of vicariant speciation events, producing the six species we see today.

The *Lepechinia* example provides an outstanding illustration of a common problem in speciation studies based on one clade: a number of conflicting scenarios can often fit the available data. In this example, we have three such conflicting scenarios. First, the clade is young and has been formed by a series of biotic expansions and contractions associated with moist/dry cycles during the Pleistocene. Speciation occurred via vicariance due to changes in environmental harshness as suggested by Cracraft (1985a). According to this scenario, each member of clade 1 independently adapted to increasing xeric conditions. Second, the clade is young, but was formed by a series of peripheral isolates speciation events associated with the dispersal by members of ancestral species from one xeric habitat to another. In this case, we would hypothesize that the ancestor of clade 1 was already adapted to living in xeric conditions before the speciation sequence began. Finally, although we know that the phylogenetic tree plus the geographic distributions correlate well with one geological/environmental scenario, how do we know that the speciation events producing the members of clade 1 occurred *at that time*? The Huancabamba Deflection was created during the uplifting of the Andes in the Miocene. It is thus possible that the six species in clade 1 are older than the Pleistocene; that the glaciations simply dissected an already established flora, leaving the distributions that we see today and possibly obscuring the evidence for the speciation events that occurred prior to glaciation (see Lynch 1986 for a discussion of the origin of Andean herpetofauna).

As you can see, then, correlation between a phylogeny plus geographic distributions and a particular geological/environmental scenario may not be sufficient for understanding modes of speciation initiation. In such cases, what can we do? For discovery-based approaches, the answer is always the same: When in doubt, get more data. As we mentioned previously, the key to differentiating between the two allopatric modes of speciation is theoretically quite simple: vicariance events should affect many clades living in the same area because the mechanism initiating speciation is independent of any particular biological system, while peripheral isolates speciation is directly dependent upon the biology of the organisms involved. Allopatric mode I should thus produce replicated patterns of speciation for multiple clades, while allopatric mode II is expected to produce clade-specific patterns. As we will show in the next chapter, historical biogeographical studies rely on this difference to distinguish vicariance and peripheral isolates speciation using multiple clades. Our answer to the *Lepechinia* problem is, therefore, Collect data from other clades living in the area and search for general versus unique patterns.

FREQUENCY OF INITIATION MODES

More than a decade ago, we concluded, "Contrary to many early theoretical predictions, the predominant mode appears to be vicariant speciation" (Brooks and McLennan 1991:124). How has that conclusion fared? As discussed in chapter 2, the number of phylogenies has increased dramatically in the literature. Unfortunately, the number of studies addressing speciation modes based upon those phylogenies has not grown at the same rate, so the database is still very small. Although that database supports our conclusion (and the conclusion of other authors: e.g., Cracraft 1982a; Wiley and Mayden 1985; Lynch 1989) that vicariance has been the dominant mode of speciation on this planet, the last ten years has witnessed a slight decline in the imposing lead we once thought vicariance held over other speciation modes. This decline (from 71.1% documented by Lynch to 65% shown in table 3.1) has been precipitated by the addition of groups such as birds, mammals, and plants to the foundation of freshwater fishes and amphibians. The new percentages shown in table 3.1 must still be taken with a large grain of salt because the table does not include all published speciation studies (in particular, it does not include many of the freshwater fish groups discussed by Wiley and Mayden 1985 and Mayden 1988, 1991) and because the methods used to determine various modes vary among researchers (e.g., the amount of asymmetry used to distinguish allopatric mode II from mode I and the amount of overlap used to identify sympatric speciation: see discussions in Lynch 1989 and Chesser and Zink 1994). We have not included an exhaustive list of speciation studies here because we believe that the search for the "dominant" initiating mode has obscured a more fundamental and interesting point: not all clades are the same. So, while it is important to begin with vicariant speciation as the null hypothesis, it is equally important to allow information from phylogeny, geographic distributions, ecology, and genetics to identify instances of peripheral isolates, parapatric, and sympatric speciation which refute the null hypothesis. Biology may be influenced by geology, but it is not completely subservient to it.

Table 3.1. A Heuristic Exercise: Assigning Frequencies of Speciation Modes in Different Vertebrate Groups

Group	N	Allopatric Mode I (%)	Allopatric Mode II (%)	Parapatric (%)	Sympatric (%)
Rana[a]	19	89.4	5.3	?	5.3
Ceratophrys[a]	6	83.3	16.7	?	0
Eleutherodactylus[a]	7	71.4	14.3	?	14.3
Heterandria[a]	8	62.5	37.5	?	0
Xiphophorus[a]	13	61.5	23.1	?	15.4
Poephila[a]	5	80	0	?	20
Various passerines[b]	30	66.7	3.3	0	30
Ammocrypta	2	100	0	0	0
Fundulus	3	100	0	0	0
Notropis	6	100	0	0	0
Etheostoma	6	33.3	50	16.7	0
Felidae	27	25.9	14.8	7.4	51.9
Lepechinia	5	?	100	?	?
Laupaula	5	?	100	?	?
Perognathus	1	100	0	0	0
Total (mean ± s.d.)	143	65 (35)	24.3 (34.1)	1.6 (4.6)	9.1 (15.2)

NOTE: Chesser and Zink (1994) argue that postspeciation dispersal has inflated the percentage for sympatric speciation and obscured the amount of allopatric speciation mode II. ? = speciation mode equivocal or not considered

[a] Data taken from Lynch 1989.

[b] Data taken from Chesser and Zink 1994.

The Completion of Speciation: Evaluating Reproductive Isolation

Speciation involves the irreversible splitting of a lineage. At the beginning of this chapter we discussed how phylogenetic patterns can be used to identify which historical, mutually exclusive, and diagnosable entities *might* be species. Geographical distribution patterns were then added to the phylogenetic information in order to determine the factors involved in the initiation of speciation (Were the populations that participated in the speciation process disjunct or overlapping?). In the third step of a speciation study, we must ask whether the divergence highlighted by character changes on the phylogeny is in fact complete (irreversible). To answer this, we must address issues of reproductive isolation. Not surprisingly, the number of ways in which reproductive isolation can occur mirrors the number of nondimensional species concepts.

Theoretical input from quantitative genetic models and empirical data from both field and laboratory studies have demonstrated that *there is no one general mechanism for completing speciation.* Completion results from a complex interaction among population size, intensity of gene flow between the diverging sister groups, drift-induced genetic differentiation, incidental pleiotropy or genetic hitchhiking as a correlated response to multifaceted divergent selection in different environments, sexual selection, and sexually antagonistic selection.[19] *Nor is there any particular combination of these factors unique to a particular speciation mode.* Thus, for example, sexual selection,[20] multifarious divergent natural selection (leading to adaptation within, but genetic incompatibilities among, populations),[21] and mutation plus random drift[22] may be associated with speciation under conditions of both allopatry and sympatry.[23] By the same token, a number of general mechanisms produce reproductively isolated groups almost instantaneously, including the production of asexual lineages from a sexually reproducing ancestral species, ploidy shifts, and hybridization.[24] In these cases, the initiation of speciation is equivalent to the completion of speciation, so these mechanisms are not unique to any particular speciation mode either (they can occur in allopatry or sympatry).

What if two species hybridize repeatedly to produce a third species, and each event produces a hybrid that "looks" identical to the other hybrids? Doesn't this mean that we have evidence for multiple, independent origins of a species? No. Sexual reproduction ensures that each offspring will be a distinct product of its parents. This applies to speciation by hybridization (and changes in ploidy) as well as individual reproduction. In other words, although the parental (ancestral) species are the same, each successful hybridization (ploidy) event poten-

19. For an excellent discussion, see Rice and Hostert 1993.

20. Lande 1982; West-Eberhardt 1983; Lande and Kirkpatrick 1988; Barraclough et al. 1995; Civetta and Singh 1998; Gleason and Ritchie 1998; Price 1998; and Higashi et al. 1999.

21. Muller 1942; Mayr 1963; Rice and Hostert 1993; Orr and Orr 1996; Schluter 1996a, 1998; Funk 1998; Orr and Smith 1998; Hatfield and Schluter 1999; and Miyatake and Shimizu 1999 and references therein.

22. Gavrilets and Hastings 1996; and Gavrilets 1997a,b, 1999, 2000 and references therein.

23. Doebeli and Dieckmann 2000; and Gavrilets, Acton, and Gravner 2000.

24. Harlan and DeWet 1975; Grant 1981; Funk 1982; Patton and Sherwood 1983; Barrett 1989; Harrison and Rand 1989; Hewitt 1989; Wake et al. 1989; Orr 1990; Werth and Windham 1991; Cannatella and de Sá 1993; Levin 1993; Dufresne and Herbert 1994; Ptacek et al. 1994; Vickery 1995; Gillespie et al. 1997; and Barrier et al. 1999.

tially produces a unique, new species.[25] This shouldn't surprise you. After all, one ancestral species can give rise to many peripheral isolates.

One of the most hotly debated issues in evolution at the moment is the importance of secondary contact between "close relatives" to the completion of speciation (speciation by reinforcement).[26] At first, this might seem counterintuitive, given the earlier resistance to parapatric and sympatric speciation (the nonallopatric modes); however, the argument is actually quite simple. Both nonallopatric modes require that speciation be initiated *and* completed while the diverging sister species remain in contact with one another (scenario 1). Classical isolation (completion of speciation) via reinforcement requires that divergence begins (is initiated) in allopatry and is completed only when the allopatric populations come into secondary contact with one another (scenario 2: Dobzhansky 1937, 1951; Key 1968). Endler (1977) called this **alloparapatric speciation**. Reinforcement happens when the two groups have diverged so much in allopatry that there is a penalty for making mating mistakes in sympatry (otherwise the two groups would merge back together, and we would have no evidence for speciation). The penalty is the occurrence of postzygotic incompatibilities between the two populations due to changes in the fertilization system that have evolved in allopatry (e.g., as a correlated response to divergent selection). These incompatibilities manifest themselves along a line of decreasing cost, from the absolute failure of fertilization to reduced hybrid fitness under certain environmental conditions. Selection for prezygotic isolating mechanisms then occurs as a response to these costs. A variety of laboratory studies on *Drosophila* have indicated that positive assortative mating does arise when two divergent populations are brought into secondary contact if (1) additive genetic variation exists for the prezygotic traits and (2) hybrids are removed from the study, mimicking the effects of strong (absolute) postzygotic penalties to breeding.[27] The third scenario proposes that speciation can be initiated *and* completed in allopatry. In this scenario, a population exposed to a new selective regime in a novel habitat responds with adaptive shifts in ecology/morphology *and* in mate-recognition characters (Paterson 1985). The linear progression from postzygotic to prezygotic mechanisms is not necessary in this type

25. For an excellent discussion of the ploidy problem, see Soltis and Soltis 1999.

26. Dobzhansky 1937; Mayr 1954. See also Liou and Price 1994; Hostert 1997; McMillan et al. 1997; and Kirkpatrick 2000.

27. For excellent reviews, see Howard 1993; Rice and Hostert 1993; Sites and Reed 1994; Hostert 1997; Wade and Goodnight 1998; and Jiggins and Mallet 2000.

of dynamic. In fact, it may be possible that both will coevolve, or that prezygotic (mate-recognition) characters will evolve in the absence of postzygotic costs.[28] Under these circumstances, sister species that reestablish contact will not recognize each other as potential mates and so will make few, if any, mating mistakes.

Of these three concepts, reinforcement is the most difficult to recognize based upon current distribution patterns, because the postulate that two currently overlapping sister species were disjunct in the past is not easy to corroborate (or to falsify for that matter). In fact, without additional data (e.g., fossil, biogeographic, or both [recall the pocket mice example]: McKnight 1995; Latta and Mitton 1999), the overlap of sister species is more parsimoniously explained as either parapatric or sympatric speciation. The situation is even more complex because:

 The mechanisms that complete speciation need not involve sister species. This is distinctly different from the initiation of speciation, which involves the production of sister species. Once speciation has been initiated, completion can be influenced by many factors regardless of the speciation mode. In other words, there are no predictable phylogenetic patterns or combinations of factors associated with the completion of speciation.

There are numerous examples from the field in which traits involved in mate recognition (prezygotic isolation) are more intense in the areas of overlap (sympatry) between congeners than in areas where the congeners are isolated from one another (allopatry: see references in Coyne and Orr 1989, 1997a,b; and Rice and Hostert 1993). The dynamic is exactly the same as that described for sister species, with the added assumption that congeners retain some plesiomorphic elements of the mating system that allow them to recognize each other as potential mates and hence make mating mistakes (Marler and Ryan 1997; McLennan and Ryan 1997, 1999). Two important points come from this observation. First, because the ability to hybridize is the ancestral condition (reproductive isolation is derived), evidence of hybridization between two species is not evidence of sister-species status (Rosen 1979; Cracraft 1983a; McKitrick and Zink 1988; Zink 1994). Second, the completion of speciation via secondary overlap (sympatry or parapatry) with a close relative *is not the same as* sympatric or parapatric speciation (which produce sister species: Cracraft 1989a,b; Brooks and McLennan 1991; Zink 1994).

The question then becomes, How closely related are the congeners,

<hr />

28. See discussions in Butlin 1987b, 1989; Butlin and Ritchie 1994; Hollocher et al. 1997; and Civetta and Singh 1998. For an interesting study, see Gleason and Ritchie 1998.

and when did the traits that are purportedly keeping them apart evolve? For example, suppose you go into the field and discover two species of "closely related" butterflies, one of which is emerald green (species EM) and the other of which is cobalt blue (species COB), that overlap over a portion of their range. After extensive observations, you also discover that (1) populations of EM are intensely green in the area of overlap and light green in allopatry, (2) females prefer conspecific males based upon wing color, (3) there is a little hybridization between the two species in the area of overlap, and (4) such hybrids are at a disadvantage because they are aquamarine and can only avoid predators by constantly hovering against a background of blue sky. Explanation: Reinforcement in the area of overlap (selection against hybridization leading to the intensification of green) has played a role in the *speciation* of EM. Now, further suppose we have a phylogenetic tree for the clade comprising EM and COB, as well as the geographical distributions (fig. 3.24a). In conjunction with the phylogeny, the distribution patterns indicate that (1) vicariance events produced A + ancestor 1, COB + ancestor 2, and B + ancestor 3 and that (2) EM is a peripheral isolate of species C (which may or may not be persistent ancestor 3). Let's begin with the scenario depicted in figure 3.24b: blue is the plesiomorphic color for butterflies in this clade and green arose in ancestor 3 (name your mechanism) in allopatry. When the peripheral isolate population budded off ancestor 3/species C and eventually moved into contact with COB, it *brought its green color with it.* Fortunately, there was sufficient variation in the production of green for selection to favor the evolution of more intensely green males in the area of overlap (avoiding the hybridization penalty). Has this dynamic completed the speciation of EM or is this a case of character evolution in the area of overlap between two closely related species (EM and COB)? In order to answer this question, we must determine whether EM has completed the speciation process by comparing it with its sister species, C, and asking what, if anything, forms the basis for reproductive isolation between the two. Given that all populations of C and EM are green and that females prefer green males, it seems unlikely that this trait will be involved in the split between the peripheral isolate and its ancestor. Overall, this type of pattern tends to support the hypothesis of *character evolution,* not the completion of speciation, in the area of overlap.

Now, suppose green is an autapomorphy for EM (fig. 3.24c). There are two possible scenarios here: (1) Green arose following strong divergent selection in the peripheral habitat (correlated with the origin of reproductive isolation between EM and C). Once again, the overlap between EM and COB represents a case of character evolution when the green

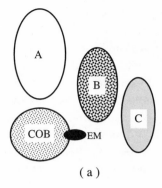

(a)

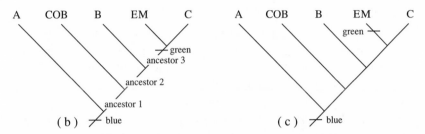

(b)   (c)

Figure 3.24.   Studying reinforcement. (*a*) Geographical distributions of the hypothetical butterflies. (*b*) Phylogenetic tree (well supported, of course) showing evolution of wing color. In this example, both EM and its sister, C, have green wings. (*c*) Alternate example (but same phylogeny) in which only EM has green wings.

species comes into contact with a blue relative, not the completion of speciation. (2) Green arose in the area of overlap between the peripheral isolate population and COB, decreased the likelihood of hybridization, and spread back through the allopatric populations by immigration. Recall that EM is descended from ancestor 3 and therefore has passed through two, possibly three, waves of reproductive isolation compared to COB (step one: isolation between ancestor 2 and its sister species COB; step 2: isolation between ancestor 3 and its sister species B; step 3: isolation to some unknown degree between EM and C). Given this, we would predict that there has been more divergence in postzygotic mechanisms between COB and EM than between EM and C because there is greater genetic divergence between COB and EM (e.g., Gleason and Ritchie 1998). In this case, reinforcement in the area of overlap with a "close relative" may indeed be responsible for the completion of EM's speciation. (For a detailed model of this process, see Liou and Price 1994.) The challenge now is to distinguish between scenarios 1 and 2. Overall, then, the hypothetical butterflies have demonstrated that (1) overlap between "close relatives" in conjunction with (2) character change in the area of overlap and (3) resultant reproductive isolation between the "close relatives" are not sufficient on their own to distinguish between character evolution and completion of speciation via reinforcement. (The special exception to this discussion will be a case in which the "species" is limited to the "population" in the area of overlap.) The critical piece of missing information in this example is the relationship between EM and its sister species, C. Are they in fact reproductively isolated? If not, then we are not studying different species, although we may be catching speciation in action.

SUMMARY

Our voyage through species and speciation has been a long one, through the stormy waters of discovery (studies of pattern, allowing us to identify and erect hypotheses about which biological units might be species) and evaluation (studies of process, allowing us to falsify or corroborate our hypotheses). This voyage represents a research program (fig. 3.25) in which discovery (historical definitions of species: phylogenetic species concept) and evaluation (nondimensional definitions of species: biological species concept) are *equally important.* This admittedly pluralistic view of species (see also Hull 1999; Reydon 2000) allows us to reconcile both statements at the beginning of this chapter:

 A great part of the living world is ordered into well-delimited species (Darwin), but there is no single attribute or type of attribute that encompasses all those species (Levin).

As we have already mentioned and will continue to emphasize throughout the rest of this book, reducing biology to an either-or choice may allow us to study one part of the system, but such simplifications come at a price: lost explanatory power and generality. A system of integration, collaboration, and mutual respect is the most effective way to increase that power (see also Hall 1996; Brooks and McLennan 1999). Toward that end, we offer

The Frontiers of Discovery in Speciation

MACRO-AND MICROSPECIES: INTEGRATING SPECIATION STUDIES AT DIFFERENT LEVELS

The integration of macroevolutionary and microevolutionary studies of speciation has been the Holy Grail for many evolutionary biologists since the beginning of the new synthesis. Phylogeography (Avise 1989a,

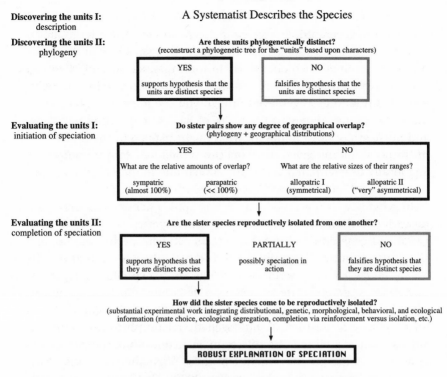

Discovering the units I: description

Discovering the units II: phylogeny

A Systematist Describes the Species

Are these units phylogenetically distinct?
(reconstruct a phylogenetic tree for the "units" based upon characters)

YES	NO
supports hypothesis that the units are distinct species	falsifies hypothesis that the units are distinct species

Evaluating the units I: initiation of speciation

Do sister pairs show any degree of geographical overlap?
(phylogeny + geographical distributions)

YES		NO	
What are the relative amounts of overlap?		What are the relative sizes of their ranges?	
sympatric (almost 100%)	parapatric (<< 100%)	allopatric I (symmetrical)	allopatric II ("very" asymmetrical)

Evaluating the units II: completion of speciation

Are the sister species reproductively isolated from one another?

YES	PARTIALLY	NO
supports hypothesis that they are distinct species	possibly speciation in action	falsifies hypothesis that they are distinct species

How did the sister species come to be reproductively isolated?
(substantial experimental work integrating distributional, genetic, morphological, behavioral, and ecological information (mate choice, ecological segregation, completion via reinforcement versus isolation, etc.)

ROBUST EXPLANATION OF SPECIATION

Figure 3.25. Flow chart showing the different steps needed to move from a description of a species to a robust explanation for the way in which that species was produced. Notice that information from both the historical and nondimensional species concepts is required in order to identify and evaluate the units we think might be species.

2000) has helped place that Grail within our sights, possibly within our grasp. Evolving species are commonly viewed as collections of populations distributed horizontally across geography (Avise 1989a, 2000; Futuyma 1989). Kornet (1993a,b; Kornet and McAllister 1993; Kornet et al. 1995) added an important dimension to this perspective by suggesting that species are collections of historical lineages subdivided into smaller historical lineages extending through time. When one of these smaller lineages experiences a permanent, or irreversible, split from the others, the relationship between that lineage and the others changes from a microevolutionary to a macroevolutionary one. The divergent lineage is now called a **macrospecies** to indicate that it has experienced a permanent split from its ancestor and may now produce its own complement of smaller lineages extending through time, which constitute its own **microspecies.**

"Microspecies" is a general term encompassing all those assemblages of conspecific organisms that we have called subspecies, differentiated populations, geographical races, or incipient species. These terms are all linked by two implications. First, the group of organisms you are studying can be distinguished objectively in some manner. Second, there is a probability that this group will become a (macro)species in its own right, but it has not done so yet, even though you can identify it objectively. Microspecies thus represent the realm of what is happening right now and the realm of possibilities for what might happen in the future. The microspecies of any given macrospecies should be relatively numerous and locally differentiated yet highly similar due to their close common history, so naturally replicated exemplars will abound.

In general, we expect a host of demographic phenomena, such as local extinctions and fusions with other microspecies as a result of dispersal and gene flow to limit the number of microspecies that become macrospecies in their own right. This is exactly the pattern that is starting to emerge from the explosion of studies in intraspecific phylogeography. Phylogeographers are interested in describing the "deployment of genetic variation within species" (Zink 1996:308). This deployment is uncovered by reconstructing phylogenetic relationships among populations then examining the effects of relatedness and geography on differences in the genetic structure of those populations (microspecies) (see, e.g., Chernoff 1982; Avise 1989a, 2000). One of the major generalizations arising from these studies is that the relationships among populations within a species are complex and reticulated, often showing only moderate to very little resolution, as indicated by the occurrence of numerous, equally parsimonious or statistically indistinguishable trees.[29] *This is exactly what we would expect if (as suggested by Mayr, Dobzhansky, and Simpson) species, and not just populations, are key evolutionary entities.* Although the relationships among microspecies are often ambiguous,[30] many phylogeographic studies do detect the effects of geographic processes on their distribution. For example, Vogler and DeSalle (1993) uncovered a basal split in populations of the North American tiger beetle, *Cicindela dorsalis,* that could be traced to a disruption in gene flow following the emergence of the Florida peninsula in the Pliocene/ Pleistocene,[31] while Klein and Brown (1994) documented complex pat-

29. See, e.g., Ellsworth et al. 1994; Klein and Brown 1994; Phillips 1994; Seutin et al. 1994; Zink 1994; Riddle 1996; Shaffer and McKnight 1996; Hedin 1997; Barrowclough et al. 1999; Lessios et al. 1999; and Avise 2000.

30. For excellent discussions of problems with the methodology, see Klein and Brown 1994; Zink 1994; and Avise 2000.

31. For a review of other taxa showing the same pattern, see Avise 1992, 2000; also Weisrock and Janzen 2000.

terns of colonization and divergence for Caribbean populations of yellow warblers that did not corroborate a simple island-hopping scenario.

Kornet's distinction between nonpermanent (reversible) and permanent (irreversible) splits provides us with a clear view of a major component differentiating micro- and macroevolution. That component is not magnitude, but quality. It embodies time, and history, because it is the difference between reversible and irreversible phenomena.[32] "A loose distinction is often drawn between microevolution and macroevolution, according to the magnitude of the morphological, physiological, or genetic changes. One may say that microevolutionary changes are possibly reversible, whereas macroevolutionary ones are not. This does not mean that there are two kinds of evolution, micro- and macro-, as was contended by some authors. . . . The distinction is quantitative, and any boundary can only be arbitrary" (Dobzhansky 1970:429). It also involves assessing the quality of the cohesion holding the microspecies together as a single "corporate entity."[33] As Futuyma (1989) suggested, population/demic-level changes are likely to have no net evolutionary impact unless they are partitioned (irreversibly) by speciation.

Adopting this view allows us to "make up for lost time" in evolutionary biology, helping to end the eclipse of history. There is nothing mysterious about the reversible/irreversible distinction. Maynard Smith and Szathmary (1995) recently suggested that biological processes become effectively irreversible whenever two or more individually improbable events affect a given system (change the nature of the organism). The odds that two improbable events would occur at the same time in the proper sequence for a specific biological system are astronomically large. History decreases those odds when the first improbable event becomes incorporated into the plesiomorphic background (retained evolutionary history) of the system. The second improbable event can now occur at any time, in any place, and the system will respond accordingly. In other words, history increases the chances that the first improbable event will still be part of the system when the second improbable event occurs.

To this we would add that both micro- and macrospecies can be affected irreversibly by extrinsic factors (changes in the nature of the conditions) such as tectonic activity or major changes in environmental conditions (the "environmental harshness" of Cracraft 1985a; the "environmental stress" of Hoffman and Hercus 2000). Such influences can be

---

32. Dobzhansky 1970; Wiley and Brooks 1982; Brooks and Wiley 1988; Maynard Smith and Szathmary 1995; and Brooks 1998a, 2000.

33. Wiley 1981; Wiley and Brooks 1982; Brooks and Wiley 1986, 1988; Templeton 1989a; and Maynard Smith and Szathmary 1995.

singular and need not be improbable a priori, though they may be rare. We conclude this chapter with a discussion of two paradigm systems for studying the interface between micro- and macrospecies.

Fish in Lakes: A Surfeit of Cichlids. Cichlids, small fishes beloved of the aquarium trade, are found in a variety of freshwater habitats throughout central-south Africa (including Madagascar), India, and the Neotropics. Both morphological (Stiassny 1987, 1990, 1991) and molecular (Streelman et al. 1998; Farías et al. 1999; Farías et al. 2000) data indicate that the family is an old one (fig. 3.26), with the major subdivisions occurring in sequence with vicariance events involving the separation of Madagascar-India from Gondwana and of South America from Africa (Storey et al. 1995). The group is characterized by a novel and complex suite of changes involving the bones, muscles, ligaments, and nerves of the jaw and associated areas on the fish's head (the pharyngeal jaw apparatus: Liem 1973, 1980; Liem and Osse 1975). This new unit is at once highly integrated, complex, and so evolutionarily flexible that even minor alterations to any of its components (e.g., change in the position of a bone or the site of muscle attachment) can produce profound differences in function. Because these changes are associated with the jaw, this flexibility permitted cichlids to explore trophic space, and explore it they did. Over the 200 million years or so since the group originated, species capable of eating practically every type of food available to them have evolved: "In sharp contrast with the condition in generalized percoids, the complex pharyngeal jaw apparatus of cichlids is an exceedingly efficient precision instrument capable of crushing mollusks . . . , manipulating fish scales into packets to facilitate deglutination, molding filamentous algae onto more manageable masses, masticating crustaceans and insect larvae, and cutting leaves of higher plants, etc." (Liem and Osse 1975:445).

The pattern of moderately well resolved relationships among major clades plus the novel jaw complex that characterizes the family as a whole sets the historical stage for what appears to have been an amazing burst of morphological differentiation among the cichlids inhabiting the large East African Rift Lakes. The scenario goes something like this:[34]

1. Formation of the Rift Lakes: Lake Tanganyika was formed approximately 12 million years ago and underwent a dramatic drop in water level, dividing it into three sublakes about 25,000 years ago (Scholz and Rosendahl 1988). Lake Malawi dates back approximately 2 million years,

---

34. For excellent overviews, see Ribbink 1994; Martens 1997; Sturmbauer 1998; and Kornfield and Smith 2000.

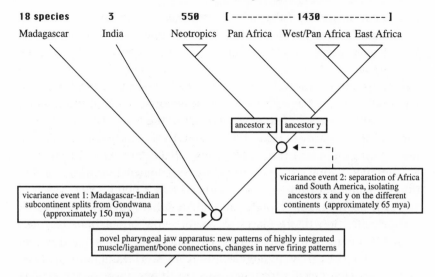

Figure 3.26. Speciation in cichlids: the big picture. Hypothesized dates of divergence for major lineages and appearance of critical characters are mapped onto the reduced phylogenetic tree (consensus tree modified from Streelman et al. 1998; see also Stiassny 1991). *Numbers* = approximate number of species per clade.

with one massive die-off of fish fauna in the shallow southern end documented at about the same time as the Tanganyika subdivision and another following a drop in water levels in the 1700s (Fryer and Iles 1972; Owen et al. 1990). Lake Victoria is the youngest of the three lakes, forming possibly 400,000 years ago and undergoing three major desiccation events since, drying completely in the last event then reforming following flooding about 14,600 years ago (Johnson et al. 2000; Stager and Johnson 2000; Talbot and Laerdal 2000).

2. Colonization of the lakes by fishes from surrounding rivers: The lakes, which provided a variety of different habitats (e.g, rocky versus sandy or muddy, deep water versus littoral), functioned as a large, heterogeneous arena where numerous competitors with and predators of the riverine cichlid species came together in one place. Once in the lakes, the novel cichlid pharyngeal jaw apparatus came face to face with a novel and complicated suite of selection pressures. Result: rapid trophic diversification of cichlid lineages, producing hundreds of ecologically and phenotypically distinct ecomorphs (Fryer 1959; Greenwood 1964, 1973; Fryer and Isles 1972). Because the lakes are not the same age, Greenwood (1965, 1984) suggested that the degree of specialization should become more marked with the age of the lake. Although most researchers have focused their attention on the relationship between food and diversification, more recent studies have added sex to the pic-

ture. Most of the Rift Lake cichlids are sexually dimorphic (males colored, females cryptic), and color patterns appear to have undergone almost as explosive a diversification as jaw structure. Experimental evidence and field observations indicate that individuals from some "species" mate assortatively based upon color, so it is possible that sexual selection also may have played a role in lacustrine cichlid speciation.[35]

3. Finally, rising and falling water levels in the lakes over time further complicates the story of cichlid speciation. How a particular species will react during high and low water levels will depend upon its dispersal abilities, population structure, ecological requirements, and the distribution of habitats within the lake (Marlier 1959; Ribbink 1994; Meyer et al. 1996). So, for example, high water might either provide opportunities for diversification (more places to go) or allow populations to intermingle more freely and thus reduce diversification. Low water levels, on the other hand, can have four possible outcomes: (1) extinction; (2) fusion of microspecies forced into close contact with one another; (3) completion of speciation, initiated in the high water days, via reinforcement during secondary contact; and (4) initiation of speciation in isolation.[36]

By now you realize that we can begin to ask questions about how species are produced in the first place *only after* we have rigorously identified the species following the research program outlined in figure 3.25. So, first we must *discover* how many evolutionarily distinct lineages (macrospecies) there are among the East African ecomorphs. In order to do this, we need a phylogenetic systematic analysis of as many of the East African ecomorphs as possible, which is no small feat given the 1,400 or so lineages that have been described. Although many researchers are still using distance-based phenograms (Mayer et al. 1998; Albertson et al. 1999) or a consensus of nonphylogenetic-and phylogenetic-based methods (Rüber et al. 1999), there are some phylogenetically based studies investigating upper-level relationships (e.g., Sültmann et al. 1995; Meyer et al. 1996; Lippitsch 1998). The results of these and previous studies hint that the "species flocks" in each of the three major lakes may be complicated assemblages of lineages from the Rift Lake plus surrounding rivers and lakes (e.g., Nishida 1991; Sturmbauer and Meyer 1993; Sültmann et al. 1995; fig. 3.27). One lake at least, Lake Tanganyika, is be-

35. The extent of that role is still controversial. See Dominey 1984; McElroy and Kornfield 1990; McKaye 1991; McKaye et al. 1993; Deutsch 1997; Seehausen et al. 1997; Galis and Metz 1998; Knight et al. 1998; Seehausen and van Alphen 1998, 1999; Seehausen et al. 1998; Seehausen, Mayhew, and van Alphen 1999; Seehausen, van Alphen, and Lande 1999; Genner, Turner, Barker and Hawkins 1999; Genner, Turner and Hawkins 1999; Bouton 2000; and Seehausen 2000.

36. Fryer and Iles 1972; Greenwood 1974; Fryer et al. 1983; Owen et al. 1990; Ribbink 1994; Rossiter 1995; and Bouton 2000.

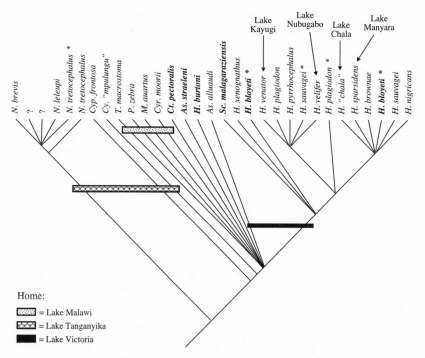

Figure 3.27. Speciation in cichlids: focusing on Africa. Mapping where the species are found onto a simplified genetic distance phenogram (modified from Mayer et al. 1998) hints that each of the lakes is a unique mosaic of cichlid lineages, assembled by a unique history of dispersal into and out of the lakes plus speciation *in situ*. Bold = species in East African rivers; * = species that are paraphyletic on the phenogram. *As.* = *Astratoreochromis; Ct.* = *Ctenochromis; Cy.* = *Cyprichromis; Cyp.* = *Cyphotilapia; Cyr.* = *Cyrtocara; H.* = *Halpochromis; M.* = *Melanochromis; N.* = *Neolamprologus; P.* = *Pseudotropheus; Sc.* = *Schwetzochromis; T.* = *Tyrannochromis.*

lieved to have polyphyletic origins! It is thus critically important to document the "big picture" of East African cichlid phylogeny in order to uncover the number and sequence of invasion by different lineages into each of the three lakes and the timing of particular bursts of speciation.

Any further discussion at this time is pure speculation, but let's speculate. Most studies show large numbers of polytomies among the East African cichlid lineages. Let us assume, for the moment, that future extensive phylogenetic analysis also produces polytomies. Where does this get us? The two speciation modes that have been proposed for the East African cichlids, allopatric mode II (peripheral isolates speciation) and sympatric speciation, will both produce polytomies if speciation occurs rapidly. One way to simplify the problem is to move to the next stage of a speciation study: ask whether each "species" identified as a unique evolutionary entity on the basis of the phylogenetic analysis does indeed represent an irreversible lineage split. In other words, we need to *evaluate* each of the lineages identified from the phylogenetic analysis. For example, mate-choice experiments in the laboratory indicate that female haplochromine cichlids from Lake Victoria prefer males of their own lineage based upon color patterns even though no postreproductive barriers (costs) to hybridization have been identified among the halpochromines in either Lake Victoria or Lake Malawi (Crapon de Caprona and Fritzsch 1984; Crapon de Caprona 1986; Stauffer et al. 1996; Seehausen et al. 1997). Does such assortative mating indicate an irreversible split? This question is (unfortunately) being answered in a massive field "experiment" even as you read this paragraph. For the past decade, human-induced eutrophication of Lake Victoria has been associated with a decrease in the number of brightly colored cichlid species. Seehausen et al. (1997) demonstrated that when females can no longer make choices based upon color because the water is too murky, the diverging lineages collapse back together. In other words, some of the "species" in Lake Victoria (and possibly Lake Malawi) are in fact microspecies, populations on a distinct evolutionary trajectory that have not yet made the irreversible jump to macrospecieshood. This is an important piece of information. It means that although we may have documented a substantial amount of character diversification to date, we still have little idea of how much of that diversification is coupled with irreversible lineage splitting.

So, the good news is that we appear to know a fair amount about the factors initiating speciation (the deployment of microspecies within macrospecies). What we lack is information about the completion of speciation: How many macrospecies are there? Macrospecies should be less numerous than microspecies, and better individuated, because they embody a history of irreversible change. Interestingly, this is exactly what you find in Lake Tanganyika. As suggested by Greenwood (1965), that lake (which as you recall is the oldest of the three by many millions of years) contains only 171 "species," but those species display more extreme morphological and behavioral specializations than do the fish in the younger lakes. (For an excellent discussion, see Sturmbauer and Meyer 1993.) The African Rift Lakes and surrounding watersheds are living evolutionary galleries showcasing every stage in the speciation process. There is a surplus of information in these lakes; the kind of information that is generally neither accessible nor abundant in most other systems. With that information in hand we may finally begin to understand the complicated relationships between micro- and macrospecies, between rates of irreversible change (speciation) and rates of character change and genetic divergence. Unfortunately, if we don't find a work-able solution to counter fishery-based species introductions and increased deforestation around the lakes (to name just a few anthropogenic activities: Kaufman 1992; Cohen et al. 1996), then they will serve one last function for evolutionary biology: they will become a gallery showcasing the complicated interaction between micro/macrospecies and increased rates of extinction.

Shape Shifters: Speciation in Three-Spined Sticklebacks (More Fish in Lakes). Three-spined sticklebacks (*Gasterosteus aculeatus*) exist in a variety of different ecomorphs on the west coast of North America. These morphs can be divided into roughly two general types: the streamlined, blue-green dorsally/silver ventrally, marine/anadromous forms (think of a very small tuna) and the smaller, deeper bodied, mottled gray-green, freshwater forms. There are, however, two additional types living in some lakes, the existence of which has sparked the interest of researchers studying speciation. "Benthic" ecomorphs are very deep bodied, robust fishes living relatively close to the lakeshore, feeding on the small invertebrates sheltering on the lake bottom among plants and in the mud. "Limnetic" ecomorphs are smaller, slender fishes found in more open waters of the lake, where they pursue a variety of planktonic prey. Pairs of benthics and limnetics are found in six lakes on three islands off the coast of southern British Columbia.[37] The interesting thing about these ecomorph pairs is that they live in lakes that were below sea level until about 12,500 years ago, when the uplifting of land following the retreat of the glaciers (isostatic rebound) raised the lakes to their present level (Mathews et al. 1970; Clague et al. 1982; Clague 1983). Approximately 2,000 years later, low-lying lakes were flooded by the sea a second time, as ocean levels were raised around the world following the melting glaciers. So, if the lakes are so young, how did these pairs of benthic and limnetic sticklebacks come to inhabit them? Did they find shelter in glacial refugia then recolonize the lakes, or did they speciate *in situ*? Why are the pairs found in only six lakes, when dozens of other lakes can boast the presence of only a single morph?

These questions have intrigued J. D. McPhail (and the legions of students he has trained) for more than 30 years now.[38] Three decades of research have established that (1) the two ecomorphs differ significantly

37. A distressing sidebar: one of the pairs (Hadley Lake, Lasqueti Island) has recently gone extinct (Schluter 1996a).

38. For an excellent review of the system, see Taylor and McPhail 1999.

in morphological traits associated with efficiency in capturing their pre-ferred prey[39]; (2) the freshwater populations have reduced genetic varia-tion compared with marine populations (Withler and McPhail 1985; Taylor and McPhail 2000) (3) there are consistent allozyme and microsa-tellite differences between the two morphs, even though some pairs are indistinguishable in terms of mtDNA restriction sites (McPhail 1984, 1994; Taylor and McPhail 1999); (4) premating isolating mechanisms exist between the two morphs (the morphs mate assortatively: Ridgeway and McPhail 1984; Nagel and Schluter 1998; Rundle and Schluter 1998); (5) the morphological differences between the two morphs have a heri-table basis (McPhail 1984, 1992; Lavin and McPhail 1987); (6) F_1 and F_2 hybrids between the two morphs show no reduction in viability, fertility (McPhail 1984, 1992), or the ability to acquire mates (Hatfield and Schluter 1996); and (7) such hybrids are at a disadvantage compared to their parental ecomorphs, demonstrating poorer foraging abilities and a reduced growth rate (McPhail 1984, 1992; Bentzen and McPhail 1984; Schluter 1993, 1994, 1995; Vamosi et al. 2000). Combining all of this information indicates that the benthic and limnetic ecomorphs may be distinct macrospecies (Hagen and McPhail 1970; Ridgeway and McPhail 1984; McPhail 1994; Schluter and Nagel 1995; but see Kraak et al. 2001), which prompted McPhail (1992, 1993) to propose two hypotheses for the route to that speciation.

Both scenarios begin with the colonization of the newly formed lakes and streams by anadromous three-spined sticklebacks (or the isolation of anadromous/marine populations in those lakes as the land uplifted). The scenarios present different possibilities for what happened next. In scenario 1, sympatric speciation produces benthic-limnetic species pairs independently in each of the six lakes. Scenario 2 is more complicated, invoking a second wave of anadromous colonists wafted in with the ma-rine incursion. According to this hypothesis, the first colonists had al-ready begun to diverge in the new habitat (remember that freshwater morphs are different from marine morphs) when the new colonists ap-peared. Competitive interactions over food between the first and second colonizers are then postulated to have caused disruptive (or divergent) selection, pushing the colonizers along two different evolutionary trajec-tories. The first colonizer was closer to the benthic form (solitary fresh-water populations tend to be deeper bodied), while the second colonizer, with its streamlined anadromous shape, was closer to the limnetic

39. Hagen and Gilbertson 1972; Larson 1976; Bentzen and McPhail 1984; McPhail 1984, 1992, 1994; Lavin and McPhail 1985, 1986; Schluter and McPhail 1992; and Schluter 1993.

morph, so natural selection had "full scope" for amplifying these differences. Scenario 2 represents an unusual case of alloparapatric (peripatric) speciation in which members of the ancestral species (the marine/anadromous fish) come back into contact with the peripheral population, which had diverged in allopatry (not the other way around, as we traditionally envision the process). Speciation was completed via reinforcement (favoring premating isolating mechanisms) caused by selection against hybrids (McPhail 1984, 1992; Rundle and Schluter 1998; Taylor and McPhail 1999; see also Kirkpatrick 2001). Because of the potential involvement of ecological character displacement due to competitive interactions among the diverging lineages (Darwin 1859; Lack 1947; Brown and Wilson 1956), the preceding two scenarios have been called "ecological speciation" (Schluter 1996a,b). It is important to understand that "ecological speciation" refers to the specific involvement of ecological factors, generally competition-or predator-driven divergence,[40] in completing speciation. Although completion always involves direct contact between diverging lineages, initiation of that divergence may occur in either allopatry or sympatry (Day 2000; Via et al. 2000).

If the benthics and limnetics are indeed different macrospecies, what type of phylogenetic patterns are predicted from McPhail's two scenarios? In the case of a double invasion by the same marine/anadromous ancestor at different points in time (fig. 3.28), we would expect to find a complete polytomy among all of the species because both waves of colonization represent multiple, independent instances of peripheral isolation from the same ancestor. Sympatric speciation in some lakes following a wave of independent colonizations would be expected to produce a polytomy with sister species identified in the appropriate lakes (fig. 3.29). In the first case the benthics and limnetics are not closely related; in the second case they are sister species. Equally important, the putative ancestor of the freshwater ecomorphs should cluster in the polytomy with its descendants because it shares common traits with those descendants that it does not share with other gasterosteids. Because all freshwater populations have at least some morphological modifications distinguishing them from the marine threespines, we have included them on the hypothetical diagram. Note that this represents the first step in a speciation analysis, identification of different "taxonomic units." The second step in such a study would require building the kind of database for the solitary freshwater populations (be they micro- or

---

40. For example, Abrams 2000a,b; Doebeli and Dieckmann 2000; Hansen et al. 2000; Schluter 2000; and Travisano and Rainey 2000.

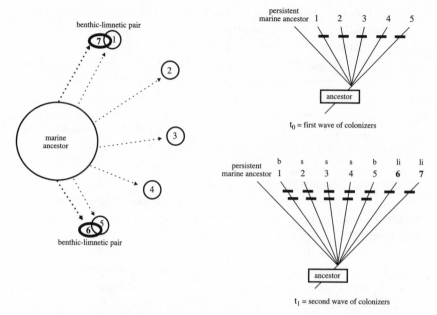

Figure 3.28. Phylogenetic pattern expected from a pure double invasion by the same marine ancestor into numerous isolated lakes and rivers. The first wave of colonization (t_0) isolates populations that begin to diverge in freshwater (black bars = autapomorphies for these populations). The second wave of colonization (t_1: bold lines and numbers) affects only a few low-lying lakes close to the ocean. Interactions between the resident and the colonizer lead to divergent selection, producing two distinct, reproductively isolated ecomorphs (species) in each lake. More autapomorphies are indicated for the residents than the colonizers under the simple assumption that the residents have been evolving *in situ* for a longer period. Given the putatively brief interval between colonization events (about 2,000 years), this difference may not represent a general pattern. In this scenario, benthics and limnetics are not sister species, and benthics are marginally older than limnetics. b = benthic; li = limnetic; s = solitary.

macrospecies) that is available for the benthic and limnetic macrospecies to determine whether the phylogenetically distinct solitary forms are indeed reproductively isolated from their relatives.[41]

Differentiating between the two speciation scenarios (and others that might occur to you) can be done only within a phylogenetic context. At the moment, there is no phylogenetic systematic analysis of the Pacific threespine complex, so we cannot give you a definitive answer to the problem. Taylor and McPhail (1999) presented a phenetic analysis of mtDNA restriction sites for 28 populations from the complex. As you recall from chapter 2, phenetic analyses are not a robust general method

41. For the beginnings of this database, see Hay and McPhail 1975 and Zuiganov et al. 1987.

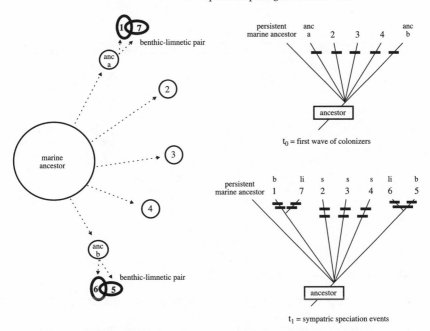

Figure 3.29. Phylogenetic pattern expected if production of species pairs is due to sympatric speciation following a single invasion by a marine ancestor into numerous isolated lakes and rivers. The first wave of colonization (t_0) isolates populations that begin to diverge in freshwater (*black bars* = autapomorphies for these populations). Sympatric speciation (possibly via divergent selection on foraging modes) in two lakes produces sister species, one benthic and the other limnetic (t_1). In this case, the benthics and the limnetics in the same lake are the same age, although there is no stipulation about when each sympatric event occurred because these speciation episodes are independent of one another. b = benthic; li = limnetic; s = solitary.

for determining genealogical relationships because they are based upon clustering organisms according to overall similarity. The results of the analysis must therefore be treated with extreme caution until a phylogenetic tree is produced for all ecomorphs. Instead of using the phenogram as a marker of phylogeny, let us use it to contrast hypotheses about genealogical relationships and speciation that can be tested with a phylogenetic tree. Examining figure 3.30 reveals the following, *if we assume the phenogram is giving us a picture of ancestor-descendant relationships:* (1) The limnetic-benthic pairs in two of the four lakes investigated are sister species (Enos Lake [L4-B4], Paxton Lake [L1-B1]). These pairs appear to be the product of sympatric speciation in those lakes. (2) The benthic-limnetic pairs are not sister species in the other two lakes (Priest Lake [L3-B3], Emily Lake [L2-B2]). This might support the double invasion hypothesis, except the almost completely bifurcating tree implies that

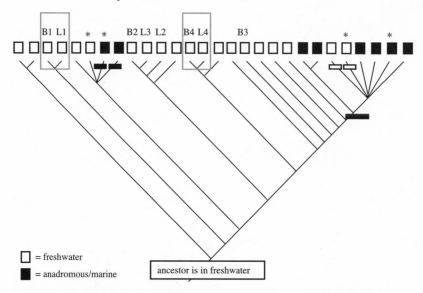

Figure 3.30. "Population phylogeny" (modified from Taylor and McPhail 1999) based on phenetic analysis of mtDNA restriction sites. If this diagram is representative of phylogeny (shows ancestor-descendant relationships), then it implies (1) two of the benthic-limnetic pairs were produced by sympatric speciation (B1-L1; B4-L4); (2) two of the pairs are the result of more complicated processes (B2-L2; B3-L3); (3) living in freshwater was ancestral for these micro-species; (4) there have been at least three independent colonizations of the ocean from freshwater (black bar) and two movements from the ocean to freshwater (white bar); and (5) sequential colonization by freshwater ancestors has been responsible for the origin of most populations in the lakes and rivers. * = anadromous and freshwater population in the cluster are in the same river; B = benthic; L = limnetic.

the freshwater populations in each lake are descended from freshwater ancestors through a series of sequential colonization events. For example, the phenogram indicates that the ancestor of the Paxton Lake species pair was itself freshwater, as was the ancestor of the Enos Lake pair, and so on. In fact, the placement of the putative marine/anadromous ancestral sources in the derived position at the tip of the diagram (and clustered with the Brannon Lake fish) suggests that the group as defined by the analysis is plesiomorphically freshwater and has independently colonized the ocean three times (with two subsequent movements back into freshwater)! Paradoxically, Taylor and McPhail reported that bootstrap resampling of "Wagner parsimony" (phylogenetic systematic analysis) results placed most species in a single polytomy and resolved only six near-terminal pairs of haplotypes more than 50 percent of the time. As mentioned previously, this kind of large-scale polytomy is the expected result of multiple, independent colonization events from a common ancestor. (We are also assuming that this places the marine populations in a polytomy with the freshwater populations, which also supports the marine colonizer/ancestor hypothesis.) This result, although less intuitively appealing because relationships are not "completely resolved," may in fact provide stronger support for the general "invader from the sea" scenario than the phenetic analysis shown in figure 3.30. At the same time, the lack of resolution may also indicate that the analysis includes a mixture of macro- and microspecies.

Only one phylogenetic pattern could refute the hypothesis that divergent selection has been responsible for the repeated and independent production of each benthic-limnetic species pair: the benthics and limnetics form separate monophyletic lineages. Of course that still would not refute the hypothesis that divergent selection had been responsible for the original speciation event producing the two ecomorphs, but it would imply that the presence of both morphs in the majority (five out of six) of the lakes is due to dispersal by already differentiated species. The patterns shown on Taylor and McPhail's phenogram refute that hypothesis. If those patterns reflect phylogeny to any degree, the analysis implies that speciation within the Pacific coast three-spined stickleback complex has itself been extraordinarily complex, possibly encompassing both of McPhail's original scenarios and perhaps even adding complications based upon dispersal and colonization by freshwater ancestors. None of this ambiguity decreases the potential role for divergent selection in the speciation of some or all of the benthic-limnetic species. It simply indicates that the story is incomplete, and the saga must continue. The stickleback system has provoked a cascade of excitement among researchers because studies using wild-caught organisms to demonstrate a direct relationship between ecology, natural selection, and speciation are relatively rare. Additionally, both hypotheses proposed by McPhail indicate that divergent selection driven by competition for resources could be responsible for extremely rapid diversification and speciation.[42] Clearly there is substantial room for even more research on these most intriguing of fishes. At the moment, the most critical missing piece of the puzzle is a phylogenetic systematic analysis (1) based upon data from as many sources as possible (including actual sequences from more than one gene), for (2) as many of the Pacific coast populations as possible, including other putative members of the Euro-North America

42. McPhail 1994; Schluter 1996a,b, 2000. For an overall review, see Orr and Smith 1998.

clade of threespines (Orti et al. 1994) from the neighboring Queen Char-lotte Islands and Alaska, using (3) members of the Trans-North Pacific clade and other gasterosteid species as outgroups.

Darwin expressed concern throughout *Origin of Species* that not every-thing systematists called species corresponded to systems affected by natural selection and that not all systems affected by natural selection were actually species. We suggest (and we are not the first to do so by any means) that in general both population biologists and systematists are recognizing real causal units of evolution. Neither one is, however, reducible to the other, and robust explanations about both the origin and the maintenance of these systems through time requires informa-tion from micro- and macroevolutionary sources. For example, both re-versible and irreversible splits are identified by some degree of character change. The question then becomes, Of all the traits on a branch that you identify by PSC-1, which ones contributed to the irreversible split-ting of the lineage and which ones represent cases of anagenetic change that did not contribute to cladogenesis (although they may have played a role in individual survival, etc.)? In practical terms, the macrospecies are necessary and sufficient for studies such as the ones presented in this book. They are not, however, the whole story. Evolutionary biologists are interested in many different properties of species, both micro and macro. We hope to explore some of the interesting research that can be undertaken using macrospecies, including assessment of the persistent historical influences on microdynamics that constitute the arena within which microspecies function. We would hope that such researchers will eventually make connections with those studying microdynamics to see if we can predict which microspecies are most likely to emerge as macro-species and, ultimately, which of those microspecies that become macro-species will be likely to continue producing more macrospecies. As O'Hara so succinctly stated,

> Because evolutionary history is something we are still in the midst of, it will not always be possible for us to determine which varie-ties—which distinctive populations in nature—are temporary and which are permanent, and so our counts of species across space and through time will always have some measure of ambiguity in them that we cannot escape. If there is any consolation in this, it must be that the very existence of this ambiguity—the very fact that some organisms in nature cannot easily be grouped into species— is itself, as Darwin recognized, one of the most important pieces of evidence for the historical process we call evolution. (1994:20)

Chapter 4

Historical Biogeography: Exploring Space

> In considering the geographic distribution of organic beings over the face of the globe . . . [w]e see . . . some deep organic bond, through space and time, over the same areas of land and water, independently of physical conditions. The naturalist must be dull who is not led to enquire what this bond is. The bond is simply inheritance, that cause which alone, as far as we positively know, produces organisms quite like each other, or, as we see in the case of varieties, nearly alike.
>
> Darwin 1872:346–347

A BRIEF HISTORY

The mid-1960s witnessed three events that energized the field of biogeography and helped change it into a dynamic discipline for studying the geographic context of speciation and the evolution of biotas. The first event was the publication of the theory of island biogeography (MacArthur and Wilson 1963, 1967), which provided an analytical and quantitative basis for **ecological biogeography**, a discipline rich in tradition and narrative accounts about the relationships between current environments and the number and relative abundance of species in particular areas. The MacArthur-Wilson model also established that additional developments in any area of biogeography needed to be rooted in quantitative, testable methodologies.

The second event was the acceptance of the theory of **plate tectonics** and continental drift by geologists (e.g., Dietz and Holden 1966; Windley 1986). This breakthrough permitted some biologists to recapture a wealth of ideas linking geographic distributions of related species to general distribution patterns on a regional or global scale, embodied in a discipline called **historical biogeography**. The assertion that such patterns existed and were important had been explained previously by reference to geographic determinism (e.g., de Candolle 1844) and then by reference to the theory of continental drift (Wegener 1912, 1924; for a precursor, see von Ihering 1891, 1902). Faced with increasing hostility from the geological community, however, historical biogeographers abandoned continental drift for land bridges (Matthew 1915, 1939) or chance dispersal across fixed geographical barriers mediated by climate (Mayr 1942; Darlington 1957; Udvardy 1969). We do not use the word "hostility" lightly; geologists' antagonism toward continental drift affected biological thought for more than half a century. Consider the example of Harold Manter, who helped establish the modern basis for evolutionary studies of host-parasite assemblages. When Manter arrived as a new faculty member at the University of Nebraska in the 1920s, he presented a seminar on the geographic distributions of marine fish parasites in light of the "Wegener hypothesis" (continental drift). Following his seminar, the chairman of the geology department told him he would lose his faculty position if he ever mentioned the Wegener hypothesis again.[1] Manter continued to document transoceanic and transcontinental distribution patterns for more than 40 years but did not refer to continental drift in print until 1967, a scant four years before his death.

The third event was the advent of phylogenetic systematics. From the beginning, Hennig (1950, 1965, 1966a,b) was impressed with the observation that superimposing phylogenetic trees on maps seemed to produce general patterns of geographic distribution. Following a seminal study by Brundin (1966) that provided the first modern evidence of trans-Gondwanian distribution patterns in extant taxa, Nelson (1969) wrote an article calling the lack of a method for documenting the generality of such patterns and the absence of an explanation for them "the problem of historical biogeography." He noted that estimating ancestral distributions by adding together the ranges of sister species implied that many modern groups were old enough to have been affected by tectonic activities predating fossil evidence for them.[2] This suggested that historically associated remnants of ancient biotas existed in areas now disjunct, but connected in the distant past.[3] Given this, Nelson concluded that combining information from phylogenetic systematics and plate tectonics might be key elements in solving the "problem." Ironically, this advancement proved more of a basis for dispute than for integration between ecological and historical biogeographers. Ecological biogeographers adopted the position that comparing phylogenies with patterns of continental drift might be useful for understanding the origins of long-

1. Correspondence between H. W. Manter and H. B. Ward, archives of the H. W. Manter Laboratory, Division of Parasitology, University of Nebraska State Museum; also M. H. Pritchard, pers. comm.

2. See also Cain 1944; Camp 1947; Wulff 1950; and Croizat 1952, 1958, 1964.

3. Cracraft 1974; Croizat et al. 1974; Rosen 1975, 1978; and Platnick and Nelson 1978.

extinct groups, such as dinosaurs (excluding birds, of course), but was irrelevant for studying modern biotas, which were thought to be too young to have been affected by tectonic activities. As a result, ecological biogeographers did not incorporate phylogeny or geological history into their explanatory framework.

Testing Nelson's hypothesis called for new applications of phylogenetic methods. In particular, researchers needed to find a way to identify the common historical biogeographical elements for multiple groups of organisms. Rosen (1975, 1985) made one of the first attempts to provide such a method. He began by taking phylogenetic trees and replacing the names of the species with their geographic distributions, calling the diagrams **unreduced area cladograms.** In order to highlight the common elements, he then removed all inconsistent branches from each tree, producing a set of **reduced area cladograms,** each of which depicted the same set of historical biogeographical relationships. These relationships were assumed to indicate a *common history of vicariant speciation* (supported in the case of Rosen's study by a scenario of geological changes in the areas occupied by the species). For some researchers, this was all the evidence needed to establish the reality of general vicariance-based patterns and to assert that the true general patterns of biogeography reflected old history more than current ecology. Platnick and Nelson (1978; see also Nelson and Platnick 1981) extended and refined Rosen's approach in a method called **component analysis.** Component analysis begins with unreduced area cladograms and labels each node according to the areas included in the distributions of the taxa forming the monophyletic groups denoted by each node. Nodes common to more than one clade, or common to a clade and an a priori hypothesis of vicariance based, for example, on geological evidence, are used to produce a consensus tree of areas, which depicts the general vicariance pattern as the dichotomous branching points. Taxa not conforming to the general pattern appear as unresolved, or ambiguous, polytomies in the consensus tree. The question thus became, Do the polytomies represent something real (modes of speciation other than vicariance), or are they artifacts of confounding variables such undetected extinctions, sampling error (all the relevant species have not yet been discovered), poor phylogenetic analysis (the relationships are incorrect), and widespread taxa (the same species occurring in more areas than the one in which it originated)? Nelson and Platnick reasoned that any one of these problems is likely to occur at some point during a study, making it even more impressive that time after time, historical biogeographical studies uncovered strong support for general patterns. They thus felt justified in proposing two assumptions designed to help investigators detect when apparent inconsistencies (indicated by polytomies) are actually part of a general pattern and not representative of alternative speciation modes. In brief, **assumption 1** permitted a researcher to postulate lineage duplications (we will explain this concept later) and extinctions to remove all ambiguities, while **assumption 2** permitted a researcher to postulate lineage duplications and extinctions, and to remove widespread taxa, to eliminate ambiguities (see also Wiley 1986d, 1988a,b; Zandee and Roos 1987).

In May 1979 a conference was held at the American Museum of Natural History ostensibly to critique the new historical biogeography, by then dubbed "vicariance biogeography" (Nelson and Rosen 1980). For ecological biogeographers, the component approach smacked of selectively pruning phylogenetic trees to achieve a desired result (e.g., Simberloff et al. 1980; also Simberloff 1987, 1988) or of selective interpretation of results (Endler 1982a,b). For those researchers, the simplifying assumptions of component analysis made it appear that there was more historical order and less ecological contingency in geographical distributions than the ecological biogeographers felt was warranted. Some even questioned the need for a historical method in biogeography, feeling that the MacArthur-Wilson model was sufficiently general to provide all the information necessary to explain the geographical distributions of species. Attempts to extend the MacArthur-Wilson model to evolutionary phenomena, however, quickly illuminated a different problem. The model is based, in part, on the assumption that all species occurring in a given area dispersed there from someplace else—one or more "source areas." Extended to evolutionary questions, this means that all species in all areas evolved "someplace else," or to turn the problem on its head, no species actually evolved anywhere. Thus, while the model remains an excellent tool for assessing patterns of immigration and emigration from particular areas, and the relationship between areas and the number of species that inhabit them, it is not capable of addressing questions about the geographic *origins* of species, and hence the origins of the geographic distributions of their closest relatives. Unfortunately, the net result of the conference was a hardening of positions by both ecological and historical biogeographers.

Meanwhile, several researchers within the systematics community began to analyze the general properties of component analysis. Nelson and Platnick's (1981) explanation of their method was laborious and the properties of the assumptions were unclear. Zandee and Roos (1987) and Wiley (1986d, 1988a,b) found both assumptions permitted a researcher to conclude that inconsistent monophyletic groups were actually paraphyletic or polyphyletic *if that helped strengthen the general pattern of vicariance*. They concluded that component analysis gave less than parsimonious results and that the simplifying assumptions represented unwarranted ad hoc mechanisms that protected the hypothesis of a single general pattern of vicariance from falsification. Once again, the specter of fiddling with the data to achieve a predetermined result was raised, this time from within the systematics community. This concern was summarized recently by Marshall and Liebherr:

> Other methods (Assumptions 1 and 2) . . . utilize information from widespread taxa differently, diminishing their relative importance in determining biotic relationships. . . . Unfortunately, with numerous widespread taxa these methods often lead to a great many equally parsimonious fundamental area cladograms . . . that when combined into a consensus tree . . . often produce less resolved patterns that are not as explanatory, thereby serving as weaker tests of other biogeographic patterns. (2000:209)

To avoid this problem, Wiley (1986d, 1988a,b) and Zandee and Roos (1987) proposed

> **Assumption 0:** You must deal with all species and all distributions in each input phylogeny without modification, and your final analysis must be logically consistent with all input data.

This is the historical biogeographical analogue of the acronym "WYSI-WYG" (what you see is what you get) used in computer applications (note the similarity with the Wiley criterion in chapter 2). In other words, you must provide a most parsimonious assessment of all the data, regardless of whether it is consistent or inconsistent with a general vicariance pattern. Wiley suggested that the most appropriate way to apply assumption 0 was to use the quantitative phylogenetic systematic methodology developed by Farris (e.g., 1970, 1982, 1988) and used first by Brooks (1981) to identify points of congruence and incongruence between host and parasite phylogenies and later for historical biogeography (Brooks 1985). Wiley called this application of standard phylogenetic systematic methodology Brooks parsimony analysis (BPA). Rather than assume that all speciation events conform to a general pattern of vicariance, researchers using BPA in historical biogeography ask, Which of the speciation events in my groups conform to a general pattern of vicariance and which could be due to other modes of speciation?

Unification of Ecology and History in Biogeography

We believe that ecological and historical biogeographers have more in common than is generally realized. Brooks (1988a) suggested that the complexity of phylogenetic influences in historical biogeography may be affected by the chosen spatial scale. Specifically, the larger the spatial scale of the study, the more likely we are to find evidence of replicated speciation events, the greater the phylogenetic effects on the diversity examined, the older the origins of the biotas studied, and the more complicated the historical explanations for the species composition of those biotas. As a result, historical biogeographers need to adopt a global perspective (see also Ricklefs 1987, 1990; Ricklefs and Schluter 1993). Choice of spatial scale also greatly influences the types of questions asked and the analytical methods used in ecological biogeography (Brown and Gibson 1983, 1998). Advocates of macroecology, for example, have suggested that researchers studying ecological associations should search for regular patterns of distribution and abundance by expanding the spatial scale of their studies (Brown and Maurer 1987, 1989; Brown 1995; Maurer 1999), recognizing this will involve phylogenetic diversification (Maurer 2000).

The conceptual agreement on the relativity of space and time in evolution provides a platform for sharing results. "Arrive in a place" need not mean only "dispersal from another place"; it can mean "arrival in the same place from a different (earlier) time." Or, in the language of our voyage, geographic distributions can be phylogenetically conservative through time and space. At the same time, there may be macroecological principles at work that influence the structure and abundance of species within biotas over long periods of time as well as over large stretches of geography. In the empirical arena, ecological biogeography calls attention to the propensity for organisms to move about, while historical biogeography reminds us that those movements may be constrained by geological history and phylogenetic history, as well as by the capabilities of the species themselves (Dial and Grismer 1992). It is therefore important to have a method that elucidates the patterns of species origin and geographic co-occurrence. Such a method will allow us to differentiate between species that occur in an area because they evolved (speciated) there, and those that occur in an area because they dispersed into it from somewhere else. (The method should even allow us to determine where that "somewhere else" was.)

It is ultimately important to determine which components of a biota evolved together *in situ* and which entered the biota by dispersal (colonization) in order to assess the temporal and environmental context in which particular interactions between and among species have emerged. We will investigate this aspect of the evolution of multispecies ecological associations in chapters 7 and 8. In this chapter, however, the possibility of shared phylogenetic histories among geographically co-occurring species is investigated without making a priori assumptions about the type or extent of their ecological interactions. In fact, beyond occupying the same general area, the study species need not interact directly. In chapter 3, we suggested that vicariance be considered the null hypothesis for speciation because it does not require any species-specific attributes. By the same reasoning, vicariance should be the null hypothesis for studies of historical biogeography because it is the only mode of speciation that will always produce replicated patterns of association by descent with respect to geographical areas (Wiley 1981; Brooks 1985; Wiley and Mayden 1985; Kluge 1988; Brooks and McLennan 1991). Other modes of speciation (e.g., sequential colonization of islands) may, but need not (e.g., Funk and Wagner 1995), produce replicated patterns.

We are also interested in asking to what extent geographical distributions of species have been promoted by the active movements of organisms and to what extent they have been promoted by the active movements of the earth. The robustness of such studies depends primarily on how many clades we are able to analyze phylogenetically and how well supported those phylogenies are. Any information we have about the geological histories of the areas inhabited and the natural history of the species involved provides additional depth to our studies.

DISCOVERING BIOGEOGRAPHIC HISTORIES

Investigating the macroevolutionary components of biotas requires some advanced applications of phylogenetic systematic methods. First, new terminology: Up to this point we have equated "branching diagram" with "phylogenetic tree" because we have been dealing with the unique historical relationships among species based on their traits. In historical biogeography, the branching diagrams derived by using the phylogenetic relationships of the species as characters and the geological areas as taxa are called **area cladograms** ("cladogram" literally means "branching diagram"). We realize that the term "cladogram" is commonly used synonymously with "phylogenetic tree"; however, we believe that it is important to distinguish between a genealogical reconstruction and a biogeographic (or coevolutionary: chapter 8) reconstruction in evolutionary biology. Therefore, in this and the following chapters, the term "phylogenetic tree" will be reserved for estimates of historical relationships among species based on characters intrinsic to them. In the following pages we will discuss applications of phylogenetic systematic principles for discovering historical associations among different clades, with respect to their geographic distributions. Our discoveries will take the form of area cladograms.

Basic Methodology

Your introduction to phylogenetic analysis of historical biogeographical relationships begins with the Amphilinidea, interesting organisms somewhat resembling a lettuce-leaf that are the sister group of the tapeworms. The eight known species of amphlinids live throughout the world in the body cavities of freshwater and estuarine ray-finned fishes and in one species of freshwater turtle.

1. The first step in the search for possible historical components in the association between these organisms and their geographical distributions is to *reconstruct the phylogenetic relationships of the organisms.* Phylogenetic systematic analysis of 46 morphological characters produced a single tree with a consistency index of 87.5% (Bandoni and Brooks 1987). Figure 4.1 depicts these relationships for five of the eight species. (For heuristic purposes, we will show you the simple pattern first and return to the more complicated picture in a few pages.)

2. The next step is to *designate the areas in which the species occur as if they were taxa.* Begin by making a list of amphilinid species and the areas in which they occur (table 4.1).

3. The phylogenetic relationships of the five amphilinid species (fig.

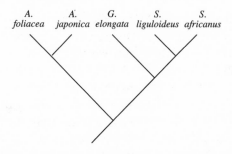

Figure 4.1. Phylogenetic tree for five species of amphilinid flatworms. *A.* = *Amphilina; G.* = *Gigantolina; S.* = *Schizochoerus.*

Table 4.1. List of Geographical Areas and Species of Amphilinid Flatworms That Inhabit Them

Area	Species	Species Name
A Eurasia	1	*Amphilina foliacea*
B North America	2	*Amphilina japonica*
C Australia	3	*Gigantolina elongata*
D South America	4	*Schizochoerus liguloideus*
E Africa	5	*Schizochoerus africanus*

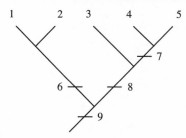

Figure 4.2. Phylogenetic tree for five species of amphilinid flatworms, with internal branches numbered for biogeographic analysis. 1 = *Amphilina foliacea;* 2 = *A. japonica;* 3 = *Gigantolina elongata;* 4 = *Schizochoerus liguloideus;* 5 = *S. africanus.*

Table 4.2. Binary Codes Representing the Phylogenetic Relationships among Five Species of Amphilinid Flatworms

Species	Binary Code
1 *Amphilina foliacea*	100001001
2 *Amphilina japonica*	010001001
3 *Gigantolina elongata*	001000011
4 *Schizochoerus liguloideus*	000100111
5 *Schizochoerus africanus*	000010111

4.1) can now be treated *as if they were a completely polarized and ordered multistate transformation series,* in which each species and each internal branch of the tree is numbered (fig. 4.2). The sequence of numbering is arbitrary, but each internal branch of the tree must have a number.

4. Each species of amphilinid now has a code that indicates its identity *and* its common ancestry. For example, the code for *Amphilina japonica* is (2, 6, 9) and the code for *Schizochoerus africanus* is (5, 7, 8, 9). These codes, in turn, can be represented in a data matrix, in which the presence of a number in the species code is listed as "1" and the absence of a number in the species code is listed as "0" (table 4.2). You should recognize this as an application of additive binary coding (chapter 2). The phylogenetic relationships of the study group are now represented by the binary codes. This can be confirmed by performing a phylogenetic systematic analysis for species 1–5 using the binary codes from table 4.2 (fig. 4.3). If all is well, the analysis will reproduce the original phylogenetic tree (shown in fig. 4.1).

5. Now, *replace the species names* in table 4.2 *with their geographic distributions* (table 4.3).

6. Construct an area cladogram based on the binary codes representing the phylogenetic relationships of the amphilinids (fig. 4.4). This pro-

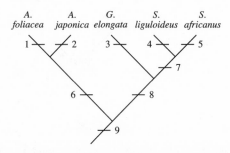

Figure 4.3. Cladogram for five species of amphilinid flatworms, based on the additive binary matrix representing the phylogenetic tree for those species. *Numbers accompanying slash marks* = codes for species and their relationships from table 4.2.

Table 4.3. Binary Codes Representing the Phylogenetic Relationships among Five Species of Amphilinid Flatworms

Area	Binary Code
A Eurasia	100001001
B North America	010001001
C Australia	001000011
D South America	000100111
E Africa	000010111

NOTE: Species are replaced by the five geographical areas in which they live.

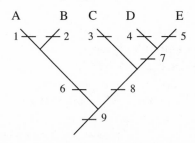

Figure 4.4. Area cladogram for five areas, based on the phylogenetic relationships of five species of amphilinid flatworms that inhabit those areas. A = Eurasia; B = North America; C = Australia; D = South America; E = Africa; *numbers accompanying slash marks* = codes for species and their relationships from table 4.2.

duces a picture of the historical involvement of areas in the evolution of those species.

7. You can now use geological evidence to produce an area cladogram showing the historical connections among the study areas independent of any species living in them (fig. 4.5). We will show later that it is not

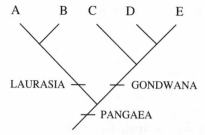

Figure 4.5.   Area cladogram for major continents on this planet, based on historical geological data. A = Eurasia; B = North America; C = Australia; D = South America; E = Africa.

necessary to have any a priori information about geological history to *describe* the biogeographic patterns, but such information does help to *explain* those patterns.

In this example the area cladogram based upon geological evidence (fig. 4.5) and the area cladogram reconstructed from the phylogenetic relationships of the taxa occurring in each region (fig. 4.4) are identical. In addition, the consistency index for the area cladogram reconstructed from the amphilinid phylogenetic relationships is 100%, indicating that all the speciation events postulated by the phylogenetic tree are congruent with the geographically based area cladogram. In the absence of any other information, it is most parsimonious to hypothesize that the occurrence of these species in these areas is a result of a long association between amphilinids and the areas in which they now occur, with speciation mediated by the geological fragmentation of those the areas. You should recognize this as vicariant speciation.

The historical relationships between species and the areas in which they live can be complex. Organisms have the ability to move around. If they did not, species would be confined to a small area somewhere on the planet where life originated, rather than being dispersed across it. Depending upon its dispersal capabilities and the changing environment in which it lives, any given species' range may expand and contract over time. The species itself may even go extinct. All of this is happening against a background of geographical evolution, as larger areas split into smaller ones and smaller areas fuse to form larger ones. Is it possible to uncover general patterns of speciation in the midst of all this complexity? To answer this question we must make some special modifications to the phylogenetic methodology described above. These modifications are designed to handle three sources of "noise in the system": (1) redundant species (more than one representative of a clade in an area); (2) widespread species (one species existing in more than one area); and (3) missing species (one or more clades do not have species in a particular area).

MORE THAN ONE MEMBER OF THE CLADE ENDEMIC TO THE SAME AREA (REDUNDANT SPECIES)

Let us return to the amphilinids and include the remaining three members of the group in the analysis.

1. The complete phylogenetic tree for the Amphilinidea is shown in figure 4.6. The "new" flatworms are *Gigantolina magna* (species 6), *Schizochoerus paragonopora* (species 7), and *Schizochoerus janickii* (species 8).

2. The three additional species of amphilinids inhabit South America (*S. janickii*) and Indo-Malaysia (*G. magna* and *S. paragonopora*). Species codes (from fig. 4.6) are converted to binary codes and listed for each area in table 4.4. (When more than one species occurs in an area, the codes are combined—we will discuss this more fully later.)

3. The area cladogram reconstructed from the data in table 4.4 is depicted in figure 4.7. The consistency index for this area cladogram is 93.8% because species 10 appears twice on the tree. This indicates that the common ancestor of species 3 and species 6 (ancestor 10 from fig. 4.6) must have occurred in both area C and area F. Identification of homoplasy falsifies the hypothesis that all species conform to a single vicariance pattern. *The problem now is to find an explanation for that homoplasy.* The occurrence of species 10 in area C coincides with the geological history of the areas (e.g., Dietz and Holden 1966, 1970), so we explain this by saying that species 3 evolved in the same place (area C) as

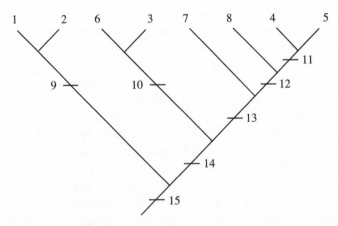

Figure 4.6. Phylogenetic tree for eight species of amphilinid flatworms, with internal branches numbered for biogeographic analysis. 1 = *Amphilina foliacea;* 2 = *A. japonica;* 3 = *Gigantolina elongata;* 4 = *Schizochoerus liguloideus;* 5 = *S. africanus;* 6 = *G. magna;* 7 = *S. paragonopora;* 8 = *S. janickii.*

Table 4.4. Primary Matrix Listing the Binary Codes Representing the Phylogenetic Relationships among Eight Species of Amphilinid Flatworms

Area	Binary Code
A Eurasia	100000001000001
B North America	010000001000001
C Australia	001000000100011
D South America	000100010011111
E Africa	000010000011111
F Indo-Malaysia	000001100100111

NOTE: Taxa are replaced by the six geographical areas in which they live.

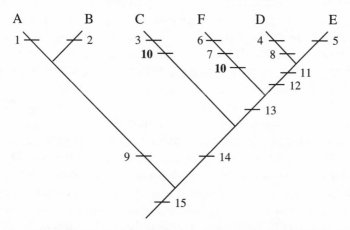

Figure 4.7. Primary area cladogram, based on phylogenetic relationships among eight species of amphilinid flatworms. A = Eurasia; B =North America; C = Australia; D = South America; E = Africa; F = Indo-Malaysia; *numbers accompanying slash marks* = codes for species and their relationships from table 4.4; *bold* = homoplasy.

its ancestor (ancestor 10). By extension, the occurrence of ancestor 10 in area F must not coincide with the same geological history of the areas. We explain this by hypothesizing that at least some members of ancestor 10 dispersed to area F, where the population evolved into species 6. Hence, the occurrence of species 7 in area F is due to common history, whereas the occurrence of species 6 in area F is due to dispersal of its ancestor into that area. If this is true, then what we have called "area F" is, from a historical perspective, two different areas for species 6 and 7. That is, area F has a *reticulate history* with respect to this clade. We can test this possibility and further examine the question of ancestor 10's dispersal by using a version of the taxon duplication convention for uncovering hybrid species that we proposed in chapter 2.

Area Duplication Convention: Whenever areas have reticulate histories with respect to the species inhabiting them, assumption 0 will be violated unless those areas are represented as separate entities for each separate historical episode. Duplicate the ambiguous areas until assumption 0 is satisfied.

The area duplication rule is thus necessary to maintain consistency with assumption 0 whenever there have been reticulated geographic histories of speciation. Hereafter, we will refer to phylogenetic analyses that do not invoke the duplication rule as **primary BPA** and those that do invoke the duplication rule as **secondary BPA** (Brooks and McLennan 2001; Brooks et al. 2001; Van Veller and Brooks 2001).

Recoding the data matrix in table 4.4, listing species 6 and species 7 in different subsections of area F (F_1 and F_2), produces the matrix shown in table 4.5. Secondary BPA using this new matrix produces the area cladogram depicted in figure 4.8. Notice that areas F_1 and F_2 are now in different parts of the area cladogram, with F_1 connected to area C (Australia) and F_2 associated with areas D (South America) and E (Africa). The placement of F_2 is in accordance with the patterns of continental drift (vicariance), but the placement of F_1 is not. Area F_1 is "misplaced" according to geological evidence, strengthening our hypothesis that ancestor 10 did some dispersing (from Australia [area C] into Indo-Malaysia [area F_1]). The separation of F_1 and F_2 confirms our suspicions that they are not the same areas historically. This conclusion is reinforced by the observations that F_1 encompasses estuarine Indo-Malaysian habitats, while F_2 represents freshwater, nuclear Indian subcontinent habitat. Finally, *Schizochoerus janickii* (species 8) and *S. liguloideus* (species 4) co-occur in South America as a result of two consecutive speciation events: ancestor 12 gave rise to *S. janickii* and ancestor 11, while ancestor 11 gave rise to *S. liguloideus* and *S. africanus*. This is an example of a lineage duplication event (to which we alluded in our earlier discussion of assumptions 1 and 2). Such events indicate either an ancient episode of sympatric speciation or (as in this case) two allopatric speciation events within an area (South America) that we cannot explain fully given the degree of geographic resolution in our study. We will discuss additional examples later in this chapter.

This example highlights some of the analytical problems that can occur when the patterns of species distribution include incidents of dispersal, either with or without speciation. In the case of the amphilinids, this movement resulted in more than one member of the group occurring in the same area. When this happens, the binary code for the area in primary BPA is a composite of the codes from all the taxa in that area. For example, the codes for taxa 6 and 7 were combined to give a

Table 4.5. Secondary Matrix Listing the Binary Codes Representing the Phylogenetic Relationships among Eight Species of Amphilinid Flatworms

Area		Binary Code
A	Eurasia	100000001000001
B	North America	010000001000001
C	Australia	001000000100011
D	South America	000100010011111
E	Africa	000010000011111
F_1	Indo-Malaysia	000001000100011
F_2	Indo-Malaysia	000000100000111

NOTE: Indo-Malaysia is listed once for species 6 (F_1) and once for species 7 (F_2).

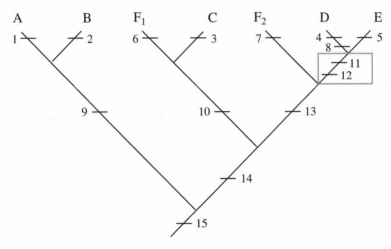

Figure 4.8. Secondary area cladogram, based on phylogenetic relationships of eight species of amphilinid flatworms, listing Indo-Malaysia (F) as two separate areas. A = Eurasia; B = North America; C = Australia; D = South America; E = Africa.

composite binary code for area F, while the codes for taxa 4 and 8 were combined to give a composite binary code for area D. This procedure is called "inclusive ORing" (Cressey et al. 1983). Fortunately, the patterns of dispersal did not override the historical patterns in the analysis, so the resulting area cladogram coincided with the geological history of the areas. The dispersal episode appeared as a homoplasy, and the two speciation events within one area (producing species 4 and 8) appeared as autapomorphies for the area. The ambiguity resulting from the occurrence of species 6 and 7 in Indo-Malaysia was resolved by assuming that Indo-Malaysia for species 6 was different from Indo-Malaysia for species 7.

Inclusive ORing produces a more serious analytical problem *when a large enough number of relatively young species evolve by dispersing into areas inhabited by older members of the clade* (see also O'Grady and Deets 1987).

1. Figure 4.9 depicts an area cladogram for hypothetical areas A–D based on geological evidence.

2. Figure 4.10 depicts the phylogenetic tree for hypothetical species 1–6 based on a combination of many morphological, behavioral, and molecular characters. (This is one robust tree.)

3. The matrix depicting the geographic distributions of species 1–6 among areas A–D is shown in table 4.6.

4. Primary BPA produces the area cladogram shown in figure 4.11. You should notice two problems with this cladogram. First, although it has a consistency index of 100%, the positions of areas B and C are reversed. The new cladogram is not congruent with the area cladogram based on geological evidence (fig. 4.9) because area B contains the two most recently evolved members of the clade, species 5 and 6; therefore, area B is assigned a highly derived status when species codes are combined. How can you tell whether your new area cladogram is congruent with geology if you do not already have a hypothesis of area relationships? This brings us to the second problem: although there is no homo-

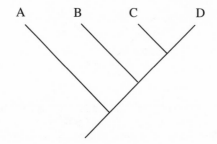

Figure 4.9. Area cladogram for hypothetical areas A–D.

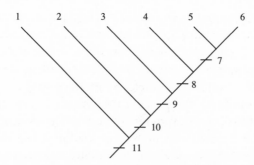

Figure 4.10. Phylogenetic tree for hypothetical species 1–6, with internal branches numbered for biogeographic analysis.

Table 4.6. Primary Matrix Listing the Geographic Distribution of Hypothetical Species 1–6 among Hypothetical Areas A–D, along with the Binary Codes Representing the Phylogenetic Relationships among Species 1–6

Area	Species	Binary Code
A	1	10000000001
B	2, 5, 6	01001111111
C	3	00100000111
D	4	00010001111

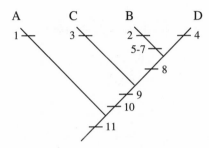

Figure 4.11. Primary area cladogram for hypothetical areas A–D, based on phylogenetic relationships of hypothetical species 1–6.

plasy in the primary area cladogram, you are alerted to the fact that there is something wrong in the historical geographical scenario because *the phylogenetic relationships of the species are not represented accurately on that cladogram.* (Compare the hypothesized series of relationships depicted on the phylogenetic tree in figure 4.10 with the relationships in figure 4.11.) For example, the primary BPA area cladogram shows species 2 originating after ancestor 8.

5. Because three species currently occur in area B, we treat area B as three areas, B_1 for species 2, B_2 for species 5, and B_3 for species 6. This treatment produces a new data matrix (table 4.7) for secondary BPA.

6. Secondary BPA produces the area cladogram depicted in figure 4.12. Maintaining the phylogenetic tree for species 1–6 requires that area B_1 and areas $B_2 + B_3$ be placed in different parts of the area cladogram. In this case, because we have a hypothesis of the geological history of the areas (fig. 4.9), we can explain the placement of areas B_1, B_2, and B_3 as one episode of vicariance (B_1) and one episode in which ancestor 7, which links areas B_2 and B_3 dispersed from area D into area B, and subsequently speciated, producing species 5 and 6. If this is true, we can consider areas B_2 and B_3 as a single area, simplifying the analysis. We will show later that this is the way we keep from postulating more area duplications than necessary to explain the data.

Table 4.7.   Secondary Matrix Listing the Geographic Distribution of Hypothetical Species 1–6 among Hypothetical Areas A–D, along with the Binary Codes Representing the Phylogenetic Relationships among Species 1–6

Area	Species	Binary Code
A	1	10000000001
B_1	2	01000000011
C	3	00100000111
D	4	00010001111
B_2	5	00001011111
B_3	6	00000111111

NOTE: B is listed as three separate areas, one each for species 2, 5, and 6.

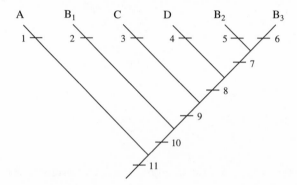

Figure 4.12.   Secondary area cladogram for hypothetical areas A–D, based on phylogenetic relationships of hypothetical species 1–6 and treating area B as if it were three separate areas historically.

WIDESPREAD SPECIES

We now turn to the ambiguity that may result from the occurrence of widespread species. Any given species may occur in more than one of the areas being studied, either because it has dispersed from its area of origin into the other areas without speciating (postspeciation dispersal) or because it has failed to speciate in response to vicariance events. Reliance on Hennig's auxiliary principle means that phylogenetic analysis will tend to treat widespread species as if their presence is homologous for all the areas inhabited. This may produce two types of ambiguity: (1) relationships supported by the area cladogram that are inconsistent with the original estimates of phylogeny; and (2) improper postulates of secondary loss (extinction) of the widespread species if it does not occur in all of the areas that are linked historically by the members of the clade to which the widespread species belongs.

1. Figure 4.13 depicts our geologically based area cladogram for hypothetical areas A–D.

2. Figure 4.14 depicts a phylogenetic tree for hypothetical species 1–4.

3. Table 4.8 lists the data matrix for the areas and the species that inhabit them (plus the codes for their phylogenetic relationships).

4. Primary BPA produces the area cladogram depicted in figure 4.15. In this instance, the area cladogram derived from biological data is congruent with the area cladogram derived from geological data (fig. 4.13). Nevertheless, something is still amiss, because interpreting the absence of species 1 in area D as a reversal, or extinction event, in that area requires placing species 1 in a position ancestral to species 2, 3, and 4. This conflicts with the phylogenetic tree for the clade, which places species

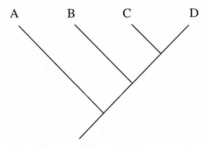

Figure 4.13.   Area cladogram for hypothetical areas A–D.

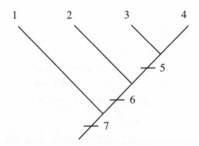

Figure 4.14.   Phylogenetic tree for hypothetical species 1–4, with internal branches numbered for biogeographic analysis.

Table 4.8.   Primary Matrix Listing the Geographic Distribution of Hypothetical Species 1–4 among Hypothetical Areas A–D, along with the Binary Codes Representing the Phylogenetic Relationships among Species 1–4

Area	Species	Binary Code
A	1	1000001
B	1, 2	1100011
C	1, 3	1010111
D	4	0001111

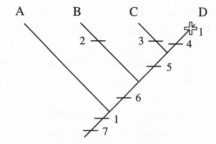

Figure 4.15. Primary area cladogram for hypothetical areas A–D, based on phylogenetic relationships of hypothetical species 1–4. *Cross* = putative extinction of species 1 in area D.

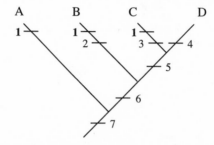

Figure 4.16. Area cladogram for hypothetical areas A–D, based on phylogenetic relationships among hypothetical species 1–4 and using an optimization rule that allows no reversals. In this case, species 1 is hypothesized to have colonized areas B and C following its origin in area A.

1 as the sister group to the remaining taxa (fig. 4.14). Hence, while the most parsimonious interpretation based simply on the occurrence of species in particular areas supports the postulate of extinction in area D, phylogenetic evidence rules against it.

There are two general strategies available for dealing with this type of problem. The first, proposed by Wiley (1988a), is to perform a phylogenetic analysis using an optimization rule that allows no reversals. For the preceding example, this produces an area cladogram showing two episodes of postspeciation dispersal by species 1 (fig. 4.16). However, there might be cases in which extinction is a better explanation than colonization, so this option should be invoked with great care. The second strategy involves using the area duplication convention again. Figure 4.17 depicts the area cladogram for areas A–D, with numbers superimposed indicating the species in clade 1–4 that occur in each area. Table 4.9 shows areas B and C duplicated, once for each species occurring in the area. Secondary BPA produces a new area cladogram (fig. 4.18) show-

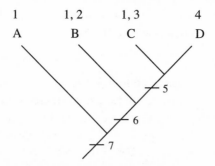

Figure 4.17. Area cladogram for hypothetical areas A–D, with numbers mapped above the areas indicating the species from clade 1–4 that reside there.

Table 4.9. Secondary Matrix Listing the Geographic Distribution of Hypothetical Species 1–4 among Hypothetical Areas A–D, along with the Binary Codes Representing the Phylogenetic Relationships among Species 1–4

Area	Species	Binary Code
A	1	1000001
B_1	2	0100011
B_2	1	1000001
C_1	3	0010111
C_2	1	1000001
D	4	0001111

NOTE: B is listed as two separate areas, one each for species 1 and 2; C is listed as two separate areas, one each for species 1 and 3.

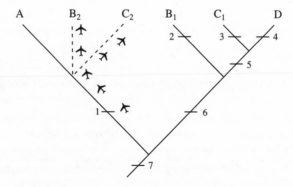

Figure 4.18. Secondary area cladogram for hypothetical areas A–D, based on phylogenetic relationships of hypothetical species 1–4 and treating areas B and C as if they were each two separate areas historically. *Airplanes* = postspeciation dispersal by species 1 into areas indicated by dotted lines.

ing areas A, B$_2$, and C$_2$ as sister taxa, supporting the explanation that species 1 dispersed, postspeciation, from area A into areas B and C.

"MISSING" SPECIES

We have just discussed methodological protocols for distinguishing species that have been *added to* (dispersed into) an area from those that have evolved *in situ*. Conversely, there are a number of reasons why certain clades might be *absent from* an area. Are we dealing with a primitive absence of the species (no member of the clade ever reached the area) or a secondary loss (a member of the clade once inhabited the area but has become extinct)? Asking questions about absence requires comparison using more than one clade. Let us begin with a simple example demonstrating how to handle two clades in the same analysis:

1. Figure 4.19 is the area cladogram based on geological evidence.

2. Table 4.10 lists the occurrence of species representing two hypothetical clades in the five areas.

3. Figure 4.20 depicts the phylogenetic trees for the clades containing species 1–5 and species 10–14.

4. Table 4.11 lists the binary codes for members of each clade for each

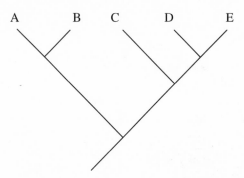

Figure 4.19.   Area cladogram for hypothetical areas A–E, based on geological evidence.

Table 4.10.   Occurrence of Species Representing Two Hypothetical Clades (1–5 and 10–14) in Five Hypothetical Areas

Area	Clade 1 species	Clade 2 species
A	1	10
B	2	11
C	3	12
D	4	13
E	5	14

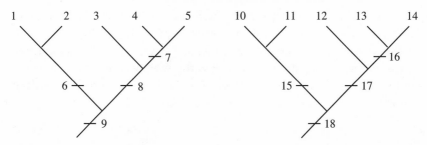

Figure 4.20. Phylogenetic trees for the clades containing species 1–5 and species 10–14, with internal branches numbered for biogeographic analysis.

Table 4.11. Primary Matrix Listing the Geographic Distribution of Hypothetical Species from Two Clades (1–5 and 10–14), along with the Binary Codes Representing the Phylogenetic Relationships among Those Species

Area	Species	Binary Code
A	1, 10	100001001100001001
B	2, 11	010001001010001001
C	3, 12	001000011001000011
D	4, 13	000100111000100111
E	5, 14	000010111000010111

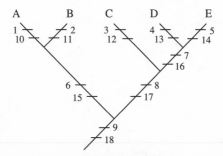

Figure 4.21. Primary area cladogram for hypothetical areas A-E, based on phylogenetic relationships of species representing clades 1–5 and 10–14. There are no exceptions to the general pattern in this example.

area. Columns 1–9 represent codes for species 1–5, while columns 10–18 represent codes for species 10–14.

5. Primary BPA produces the area cladogram in figure 4.21. In this hypothetical example, the consistency index for the area cladogram is 100%. The new area cladogram, in turn, is congruent with the geological history of the areas (fig. 4.19). The evolutionary history of the clades represents, therefore, an example of two co-occurring groups that have

speciated in response to the same vicariance episodes. Now, let us leave this perfect world and travel to a place (very similar to the real world) where two groups of species are not equally represented throughout the areas under investigation. Specifically, the members of one clade occur in five areas (A–E) and the members of the other clade occur in only four areas (A–D).

The Problem

1. Figure 4.22 depicts the phylogenetic trees for the clades containing species 1–5 (clade 1) and species 10–13 (clade 2).

2. Table 4.12 lists the binary codes for the members of clades 1 and 2 in the areas which they inhabit.

3. Primary BPA produces two equally parsimonious area cladograms, one congruent with the historical relationships among the areas (fig. 4.23a) and the other placing area E at the base of the cladogram rather than with area D (fig. 4.23b). The reversals in characters 14–16 on the first area cladogram imply that clade 2 went extinct in area E. In contrast, the second area cladogram postulates three cases of parallel dispersal by an-

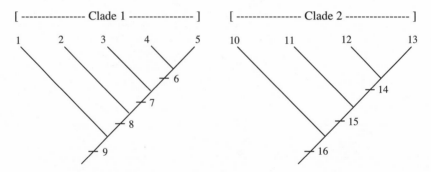

Figure 4.22.   Phylogenetic trees for hypothetical clades 1 and 2, with internal branches numbered for biogeographic analysis.

Table 4.12.   Primary Matrix Listing the Hypothetical Areas, the Species That Inhabit Them, and the Binary Codes Representing the Phylogenetic Relationships among Those Species

Area	Species	Binary Code
A	1, 10	1000000011000001
B	2, 11	0100000110100011
C	3, 12	0010001110010111
D	4, 13	0001011110001111
E	5	0000111110000000

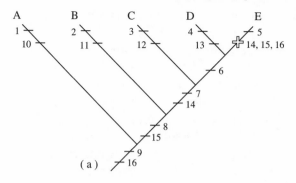

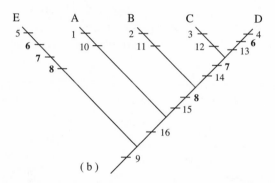

Figure 4.23. Two equally parsimonious area cladograms, based on phylogenetic relationships of species in clades 1 and 2.

cestral species 8, 7, and 6, the latter producing species 5 in area E. The second pattern, although "equally parsimonious" with the first, is inconsistent with the phylogenetic relationships of the species. This is an obvious artifact of the inclusive ORing method. Is the first pattern (extinction) also an artifact of the method, and if so, how can we tell?

Secondary loss (extinction) appears as a series of reversals in this analysis because species that are absent in an area are coded with a "0," which is equivalent to saying they were primitively absent from the area. Wiley (1988a,b) suggested that because we do not know a priori whether or not members of a clade occurred in any area where they are now absent, absence of species should be treated as "missing data" ("?") for the relevant area. The protocol for coding absence as "0" was proposed by Brooks (1981, 1985; see also Zandee and Roos 1987) because at the time computer programs for phylogenetic analysis did not accept "missing data" codes. Fortunately, contemporary computer programs for phylogenetic analysis have such an option. In this way, we should be able to distinguish postulates of extinction that are artifacts of our a priori assumptions from postulates of extinction required by the data. So, let us reanalyze the preceding example coding missing species with a question mark.

1. Table 4.13 is the matrix produced by coding absent species as "?".

2. Primary BPA produces only one area cladogram (fig. 4.24). Even though that area cladogram is congruent with figure 4.23a in terms of hypothesized area relationships, it is not identical with that cladogram. We now have no hypothesis concerning either the presence or absence of clade 2 in area E. This occurs because the absence of clade 2 from area E is not considered informative in constructing the area cladogram (i.e., relationships cannot be reconstructed from question marks, or, *it is dangerous to group on the basis of character absence*). With only these data we cannot differentiate between two equally parsimonious scenarios. If species 4 and 5 have been produced as the result of a vicariance event, then

Table 4.13. Primary Matrix Listing the Hypothetical Areas, the Species That Inhabit Them, and the Binary Codes Representing the Phylogenetic Relationships among Those Species

Area	Species	Binary Code
A	1, 10	1000000011000001
B	2, 11	0100000110100011
C	3, 12	0010001110010111
D	4, 13	0001011110001111
E	5	000011111???????

NOTE: ? = species missing from the area.

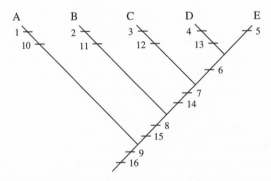

Figure 4.24. Area cladogram for hypothetical areas A–E, based on phylogenetic analysis of clades 1 and 2 using "missing data" coding option. (Note the absence of notation for clade 2 on the branch connecting area E to the rest of the area cladogram.)

it is possible that the same event fragmented clade 2, producing species 13 and its sister species (which later went extinct). On the other hand, species 5 could exist in area E because some members of ancestor 6 dispersed into that area and speciated there (allopatric mode II). In that case, clade 2 is absent from area E because its members never dispersed into the area.

> Page (1990) confused things by calling the approach of coding missing data as primitively absent "assumption 0." This usage contrasts with Wiley's original "assumption 0" (1986d; 1988a,b), which was an admonition that all available evidence must be considered; thus, even widespread taxa are seen a priori as providing evidence of area relationships. "Assumption 0 analysis," as implemented in various computer programs for component analysis, reconciled tree analysis, and three area statement analysis, is *not* BPA using Wiley's assumption 0.

> If we compare two clades *we can identify the problem* of missing taxa, *but we cannot determine a solution to that problem* because we do not know whether the ambiguity is caused by the presence or absence of species in the unusual area. In other words, we do not know which alternative represents the general pattern. Within phylogenetics the answer to, What is the general pattern? is always, Get more data.

The Solution

> **"Threes Rule" Rule:** In order to distinguish between general and special patterns of historical association based solely on the available data, in particular to determine whether absence from an area is due to secondary extinction or primitive absence, you must analyze *at least 3 co-occurring clades.*

Let us return to our example and add a third clade (species 17–21: fig. 4.25) with members in all five areas. A new matrix incorporating information from all three clades is shown in table 4.14. The new area cladogram (fig. 4.26) now indicates that the presence of species 5 in area E conforms to a general pattern (along with species 21). It is thus the absence of the sister to species 13 (clade 2) in area E that is the unusual observation, and we must invoke an episode of secondary extinction to explain that absence.

By using at least three clades, we no longer need to refer to an independently derived area cladogram (e.g., one based on geological data

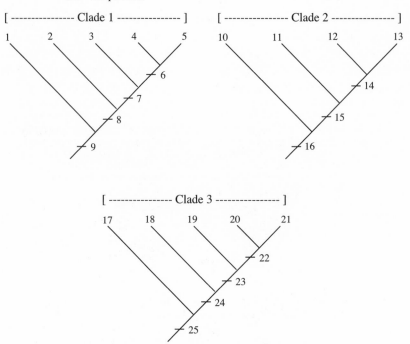

Figure 4.25. Adding a third clade to the problem (species 17–21).

Table 4.14. Primary Matrix Listing the Hypothetical Areas, the Species That Inhabit Them, and the Binary Codes Representing the Phylogenetic Relationships among Those Species

Area	Species	Binary Code
A	1, 10, 17	10000000110000011000000001
B	2, 11, 18	01000001101000110100000011
C	3, 12, 19	00100011100101110010000111
D	4, 13, 20	00010111100011110001011111
E	5, 21	000011111???????000011111

NOTE: ? = species missing from the area.

from an arbitrarily chosen time period) as a reference point to determine the general patterns. We are able to determine the general and special elements of geographic distributions based on biological evolution, and *then* ask if these elements conform to identifiable episodes in the geological history of the areas. In this manner, we can begin to understand how the earth and life have evolved together, even when each evolves independently. We do not need to assume that the history of a lineage always indicates a singular history of place (Brown 1995; Hovenkamp

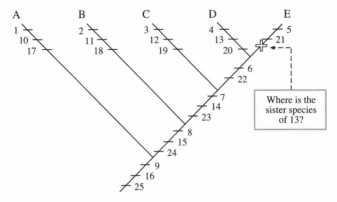

Figure 4.26. New area cladogram based upon data analysis from three clades. Adding information from a third group indicates that it is the absence of the sister to species 13 (clade 2) in area E that is unusual.

1997) because we can now identify biogeographic phenomena other than vicariance.

We are now ready to address a key methodological issue:

> Isn't the area duplication rule ad hoc and arbitrary? The area duplication convention may seem arbitrary, but it is not. Rather, it is a logical extension of phylogenetic systematic reasoning as embodied in assumption 0. The null hypothesis is that all areas have a single history with respect to all the species that inhabit them. When all available data (total evidence) do not support that null hypothesis, we duplicate areas to indicate the exact departures from (falsifications of) the null hypothesis, *but we do not duplicate areas beyond necessity (logical parsimony)* (Van Veller and Brooks 2001).

Consider four clades, each containing four species, with each species living in a single area. Figure 4.27 depicts the phylogenetic trees for hypothetical species 1–4 (clade 1), 8–11 (clade 2), 15–18 (clade 3), and 22–25 (clade 4). The matrix depicting the geographic distributions of these species among areas A–D is shown in table 4.15. The general area cladogram constructed by primary BPA (fig. 4.28) has a consistency index of 88% and is congruent with the area relationships supported by species in clades 1, 2, and 3, but differs from that supported by species in clade 4. This difference appears on the primary BPA area cladogram as the homoplasious occurrence of ancestral species 26 and 27, allowing us to reject the null hypothesis that all observed distributions are due to a single history of vicariance for each area. You can see immediately that area A has a reticulate history ("A" for clades 1, 2, and 3 is not the same

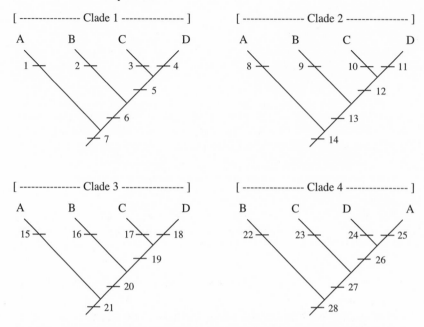

Figure 4.27. Phylogenetic trees for four hypothetical clades of organisms, with internal branches numbered for biogeographic analysis. You can tell already that one clade does not show the same pattern of area relationships as the other three clades.

Table 4.15. Primary Matrix Listing the Geographic Distribution of Four Hypothetical Clades (Species 1–4, 8–11, 15–18, 22–25) among Hypothetical Areas A–D, along with the Binary Codes Representing the Phylogenetic Relationships for All Four Clades

Area	Species	Binary Code
A	1, 8, 15, 25	100000110000011000001000111
B	2, 9, 16, 22	010001101000110100011100001
C	3, 10, 17, 23	001011100101110010111010011
D	4, 11, 18, 24	000111100011110001111001011

as "A" for clade 4). Other examples may not be so straightforward, so we are going to show you the logic behind the analysis to remove the mystique of the method:

1. Reconstruct all possible combinations of area duplications that can be applied to the general pattern. This example requires duplicating four areas (ABCD), three areas (ABC, ABD, ACD, BCD), two areas (AB, AC, AD, BC, BD, CD), and one area (A, B, C, D), for a total of 15 possible combinations of area duplications.

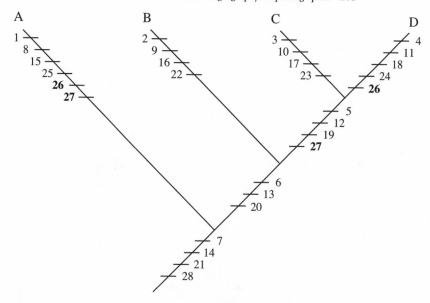

Figure 4.28. Primary area cladogram for hypothetical areas A–D, based on phylogenetic relationships of the four clades. *Bold* = homoplasy.

2. Perform a secondary BPA for each of the 15 combinations of area duplications. Whenever duplicates of the same area are connected to the same node, combine those areas (remember areas B$_2$ and B$_3$ on fig. 4.12).

3. This reduces the 15 possible combinations to four different area cladograms. Seven combinations produce the same area cladogram as the primary BPA (fig. 4.28). Figure 4.29 shows the result of duplicating areas B and C. Combinations BCD, BD, CD, B, C, and D also reduce to the primary BPA area cladogram. (We leave you to work these cladograms out for yourself.) These combinations ultimately show no area duplication because the duplicated areas cluster together, but the homoplasy is retained. Remember, we are trying to find the most parsimonious explanation for the homoplasy on the primary BPA area cladogram, so this gets us no further ahead. The remaining three area cladograms eliminate the dispersal-based homoplasy, but require different numbers of duplicated areas: (1) one combination produces an area cladogram in which the biogeography of clade 4 is viewed as completely independent of the other clades (combination ABCD: fig. 4.30); (2) three combinations produce an area cladogram in which areas D and A are duplicated (combinations ACD, ABD, and AD: fig. 4.31); and (3) four combinations produce an area cladogram in which area A is duplicated (combinations ABC, AB, AC, and A: fig. 4.32). Overall, then, the combinations depicted by the summary cladogram in figure 4.30 require four area duplications

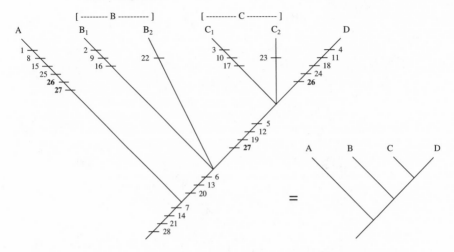

Figure 4.29. Duplicating areas B and C produces the same area cladogram as the primary analysis (duplicating BCD, BD, CD, B, C, or D also produces the primary area cladogram). In this example, areas B_1 and B_2 are attached to the same node, so they are collapsed into area "B" (the same rationale applies to C_1 and C_2 = "C"). The resultant area cladogram has no area duplications but does have homoplasy (bold).

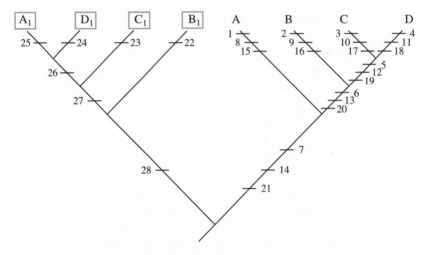

Figure 4.30. Eliminating homoplasy I: cladogram that results from duplicating all four areas. Duplicated areas are enclosed in boxes.

(A, B, C, and D), those in figure 4.31 require two area duplications (A and D), and those in figure 4.32 require only a single area duplication (A).

4. Choose the area cladogram that requires the fewest number of area duplications to explain the species that do not fit the general pattern of vicariance (the homoplasy), in this case, figure 4.32.

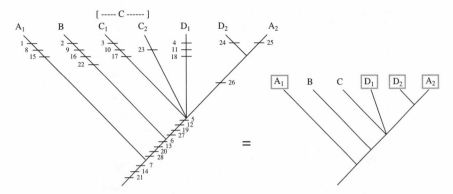

Figure 4.31. Eliminating homoplasy II: cladogram that results from duplicating areas A, C, and D (duplicating ABD and AD also produces the same area cladogram). Notice that areas C_1 and C_2 are attached to the same node, so they are collapsed into area "C." Duplicated areas A_1 and A_2 do not come from the same node, nor do areas D_1 and D_2, so they remain separate.

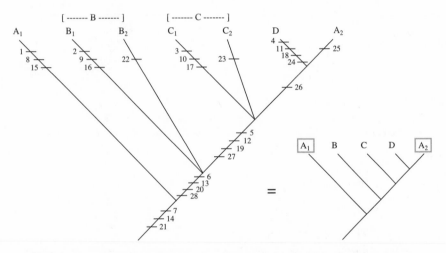

Figure 4.32. Eliminating homoplasy III: cladogram that results from duplicating areas A, B, and C (duplicating ABC, AB, AC, and A also produces the same area cladogram). Areas B_1 and B_2 are attached to the same node, so they are collapsed into area "B" (the same rationale applies to C_1 and C_2 = "C"). Duplicated areas A_1 and A_2 do not come from the same node, so they remain separate.

By now you are probably somewhat tired from drawing all 15 combinations. Fortunately, you do not actually need to perform the laborious task of trying all possibilities for all taxa and all areas. Instead, simply focus on the homoplasious portions of the primary BPA. In this case, the homoplasy uncovered by the primary BPA (remember fig. 4.28) involves area A. Duplicating area A (table 4.16 is the secondary BPA matrix) pro-

Table 4.16.   Secondary Matrix Listing the Geographic Distribution of Four Hypothetical Clades (Species 1–4, 8–11, 15–18, 22–25) among Hypothetical Areas A–D, along with the Binary Codes Representing the Phylogenetic Relationships for All Four Clades

Area	Species	Binary Code
A_1	1, 8, 15	10000011000001100001??????
A_2	25	?????????????????0001111
B	2, 9, 16, 22	0100011010001101000111000001
C	3, 10, 17, 23	0010111001011100101110100011
D	4, 11, 18, 24	0001111000111100011110010111

NOTE: A is listed as two separate areas, one for species 1, 8, and 15 and the other for species 25. ? = species missing from the area.

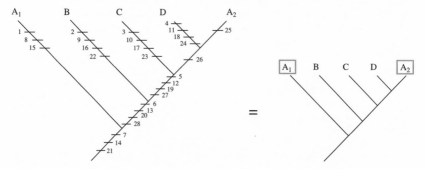

Figure 4.33.   Secondary area cladogram for hypothetical areas A–D, based on phylogenetic relationships of four clades of organisms and treating area A as if it were two separate areas historically.

duces a single area cladogram (fig. 4.33: the same as fig. 4.32) with a CI of 100%.

PUTTING IT ALL TOGETHER

Lynch (1989; see also Grady and LeGrande 1992) made the first attempt to combine phylogenetic and geographic information to assess the relative frequency of particular modes of speciation initiation. Lynch's method requires species-level phylogenetic hypotheses for groups for which there is also sufficient data to calculate species range and percent overlap or asymmetry in size between sister groups. Lynch argued, quite plausibly, that the older the speciation event, the more likely that post-speciation dispersal had obscured the original conditions surrounding that event. He thus modulated his expectations of overlap based upon the relative age of speciation. So, for example, as the age of the specia-tion event increased, he required a greater degree of overlap to recognize sympatric speciation (to a point that, if the event was very old, it was assigned an ambiguous status despite substantial overlap). When Chesser and Zink (1994) applied Lynch's method to avian clades, they felt that the 20 percent overlap criterion produced levels of sympatric speciation that were unrealistically high on theoretical grounds. They suggested that the criteria for recognizing significant amounts of overlap or asymmetry could be fine-tuned to reflect the dispersal abilities of different study groups, but cautioned that such fine-tuning would have to be based upon rigorous descriptions of those abilities and implemented a priori to avoid the allegation of data manipulation to get the "right" answer.

This discussion returns us to a problem we encountered in chapter 3: assigning values to "how much" overlap and asymmetry is required to recognize nonvicariant modes of speciation is always a somewhat arbitrary exercise. Even if vicariance is the dominant mode of speciation on this planet, it is not the only mode. Furthermore, even if a species is produced initially by vicariance, there is no guarantee that it will remain living happily within the confines of its ancestral distribution; some species do so, others expand their ranges substantially, and yet others experience a decrease in range. The specter of postspeciation dispersal or range contraction is thus always lurking in the background of any speciation study, particularly if the clades are old. Should we just abandon such studies as hopelessly vague? Green et al. (forthcoming) suggested that the answer to this question is, No, we should bring the power of BPA to bear on the question. That is, rather than searching for explanations of speciation on a clade by clade basis, we should attempt to examine the problem using as many clades as possible for a given set of areas. So, vicariant speciation can be recognized as elements conforming to general biogeographic patterns, while other modes of speciation can be recognized as departures from those patterns. *We can thus recover episodes of nonvicariant modes of speciation and postspeciation dispersal without recourse to arguments about degree of range asymmetry or overlap.*

For example, consider five more hypothetical clades containing missing, widespread, and redundant (more than one clade member in the same area) species (table 4.17; fig. 4.34). Primary BPA produces a single area cladogram (fig. 4.35), but there is considerable homoplasy on the tree. We have now falsified the hypothesis that one vicariance pattern can explain all the speciation events. Secondary BPA resolves that ambiguity by duplicating area A once and areas B and D three times (table 4.18; fig. 4.36). In order to interpret figure 4.36, you must keep one thing in mind: we are searching for general patterns among five clades. If the history of speciation for the clades has simply been one of vicariance events affecting all clades equally, then we would expect to see an area

Table 4.17.   Primary Matrix Listing the Geographic Distribution of Five Hypothetical Clades (Species 1–4, 8–11, 15–18, 22–26, 31–33) among Hypothetical Areas A–D, along with the Binary Codes Representing the Phylogenetic Relationships for All Five Clades

Area	Species	Binary Code
A	1, 8, 11, 15, 22, 31	10000011001111100000110000000110001
B	3, 9, 10, 17, 23, 25, 33	00101110110111001011101010111100111
C	2, 9, 16, 24, 32	01000110100011010001100100011101011
D	1, 4, 11, 17, 18, 26	10011110001111001111100001111?????

NOTE: ? = species missing from the area.

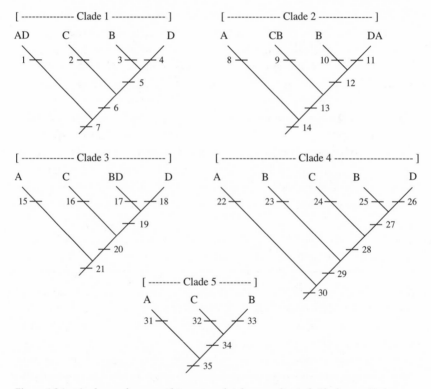

Figure 4.34.   And now for something completely complicated. Phylogenetic trees for five hypothetical clades containing widespread, redundant, and missing species.

cladogram with no homoplasy (which we have already falsified) and *with representatives of all five clades in each area.*

Bearing this in mind, let us examine figure 4.36 more closely. The most obvious misfit between the hypothesis of a single general pattern and the secondary area cladogram occurs in duplicated areas D_2, B_2, D_3,

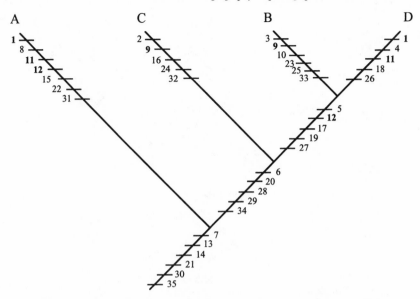

Figure 4.35. Primary area cladogram for the fiendish example. *Bold* = homoplasy.

Table 4.18. Secondary Matrix Listing the Geographic Distribution of Five Hypothetical Clades (Species 1–4, 8–11, 15–18, 22–26, 31–33) among Hypothetical Areas A–D, along with the Binary Codes Representing the Phylogenetic Relationships for All Five Clades

Area	Species	Binary Code
A_1	1, 8, 15, 22, 31	10000011000001100000110000000110001
A_2	11	???????0001111?????????????????????
B_1	3, 10, 17, 25, 33	00101110010111001011100010111100111
B_2	9	???????0100011?????????????????????
B_3	23	?????????????????????010000011?????
C	2, 9, 16, 24, 32	01000110100011010001100100011101011
D_1	4, 11, 18, 26	00011110001111000111100001111?????
D_2	1	1000001?????????????????????????????
D_3	17	?????????????0010111?????????????????

NOTE: A is listed as two separate areas, and B and D are each listed as three separate areas. ? = species missing from the area.

and A_2, which are not connected to similar areas (e.g., D_2 is connected to area A, not another part of area D). More important, none of these duplicated areas have any autapomorphies (unique species) defining them. Instead, they each have only one species which they share with their "sister area" on the secondary area cladogram. This particular pat-

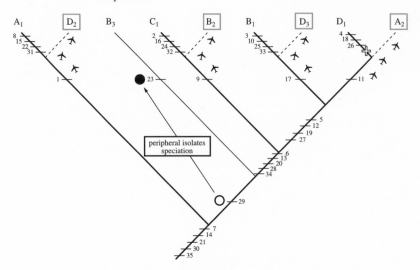

Figure 4.36.   Secondary area cladogram for the fiendish example. *Airplanes* =
postspeciation dispersal by a particular species into the area (dotted lines).
Dispersal by ancestor 29 into area B_3 represents an instance of peripheral isolates
speciation. *Cross* = extinction (the putative sister of species 33); *bold lines* = the
backbone of vicariant speciation.

tern identifies instances of **postspeciation dispersal:** in this case, the
analysis implies that species 1 evolved in area A and dispersed into area
D, and that species 9 evolved in area C and dispersed into area B. (Follow
this train of thought for species 17 and 11.) Less obviously unusual is
the duplicated area B_3, which has its own unique species (species 23).
Why is this so odd? Recall that the general pattern predicts representatives of all five clades in each area. The secondary area cladogram thus
indicates that the species in area B_3 cannot be the result of a general
vicariance event because four clades are missing from this area. We could
attempt to salvage the vicariance hypothesis by postulating that a vicariance event really did occur producing area B_3 plus new species from all
five clades in B_3, and that sometime after that event, four of five species
have subsequently gone extinct (leaving only species 23). At the moment, we do not have any evidence of four extinctions. We do, however,
have the presence of species 23 on its own, in an area unique to it. We can
attribute such unique events on the secondary area cladogram to dispersal followed by speciation (allopatric speciation mode II: peripheral
isolates speciation). In this example, the secondary BPA supports the hypothesis that some members of ancestor 29 dispersed into the new area
(B_3), where they speciated, producing species 23. Finally, if you list all
the species in each area, you will find representatives of each clade that

speciated as a result of vicariance in areas A (species 1, 8, 15, 22, and 31), C (species 2, 9, 16, 24, and 32), and B (species 3, 10, 17, 25, and 33). There are, however, only four such species in area D (4, 11, 18, and 26). The most parsimonious explanation for the absence of a representative of clade 5 in area D is that there was indeed a species from clade 5 there at one time. In other words, we hypothesize that the vicariance event that split area B and D produced five pairs of sister species: 3–4, 10–11, 17–18, 25–26, and 33 and its sister (x). Sometime subsequent to that event, species x has gone extinct. (We cannot tell when it went extinct based on the information at hand.) Once again, we are letting the general patterns indicate what happened in a particular area, then choosing the explanation for departures from that general pattern that requires the fewest number of ad hoc assumptions. In this case, one extinction is the simplest way to explain the absence of species x in area D. Overall, then, the secondary BPA indicates that speciation in these five clades has been influenced very strongly by vicariance events (allopatric mode I: highlighted with bold lines on fig. 4.36), but that the history of the groups also includes one episode of peripheral isolates speciation (allopatric mode II), four episodes of postspeciation dispersal into new areas, and one episode of extinction following a vicariance episode.

 These explanations are based solely upon two pieces of information, the phylogenetic relationships among species in the areas and the distribution patterns of those species. As is true for any scientific endeavor, we choose the simplest explanation that fits the data we currently have. That explanation then becomes a hypothesis that may be refuted at any time with the incorporation of additional information. For example, suppose an avid paleontologist were to discover fossils of four species representing the four missing clades in area B_3. In that case, we would have to redo the original phylogenetic analysis for each clade incorporating the new species, then rerun the primary and secondary BPA based on these new phylogenies. In this way, historical biogeographical studies are themselves constantly evolving with the addition of new information.

 Although secondary BPA produces as explicit a depiction of both general patterns and departures from those patterns as possible *given the data at hand*, this does not mean that we will always achieve a single most parsimonious answer. As geological and environmental factors (including periods of environmental harshness and extinction) become increasing complex, so do the patterns of biolog-

> ical diversification. There is only one way to compensate for ambiguity (homoplasy) in a phylogenetic data set: collect more data. In the case of historical biogeography, "collecting more data" means increasing the number of (well-supported) phylogenetic trees you use in your analysis.

Now, on to the world of real organisms, a much better test of whether or not phylogenetic methodology can help us discover patterns of speciation.

Some Examples from the Real World
THE ICE IS COMING

In 1846, Edward Forbes proposed that climatic upheavals during the Pleistocene fragmented previously continuous and widespread distributions of species, producing populations isolated in refuges. Wagner (1868) took this proposal a step further, suggesting that isolation in refugia could actually promote speciation in addition to fragmenting previously widespread biotas. It took almost another 100 years before Wagner's (and Forbes's) insights were formalized by Ernst Mayr in 1963, and within a decade the Pleistocene forest refugium theory (PFRT) was created, christened, and launched. The PFRT initially was based upon mainly avian distribution patterns,[4] but as the idea gained more acceptance, other animal and plant groups joined the fold.[5]

North American Freshwater Fishes: Pre- or Post-Pleistocene Origins? The North American freshwater fish fauna, containing many localized (endemic) species, is quite diverse. Gilbert (1976) noted that North America contains approximately 1,000 described and undescribed species of freshwater fishes, compared with 230 in Australia, 250 in Europe, 1,500 in Asia, 1,800 in Africa, and 2,200 in South America. This diversity has traditionally been attributed to the dispersal of fishes into North America from Europe, Asia, and South America. If this is true, then the North American fish communities are younger than their counterparts elsewhere. The patterns of this diversity vary across the continent. In general, the drainages east of the Rocky Mountains are species rich and dominated by the members of only a few genera in a small number of families (Cyprinidae,

---

4. African birds: Chapin 1932 and Moreau 1952. Australian birds: Keast 1961; Haffer 1967, 1969, 1974; Hall and Moreau 1970; and Snow 1978. For additional references, see Lynch 1988 and Cracraft and Prum 1988.

5. Vanzolini and Williams 1970; Vuilleumier 1971; Prance 1973, 1982a,b; Simpson 1975, 1983; Brown et al. 1974; Hamilton 1976; Simpson and Haffer 1978; and Simpson and Todzia 1990.

Percidae, Centrarchidae, and Ictaluridae). By contrast, the western fauna is somewhat depauperate and composed of a different assortment of fishes (distinctive cyprinid and catostomid genera, Salmonidae, Cyprinodontidae, and Cottidae). Researchers have sought explanations for these patterns by identifying the putative center of origin for a particular group, then invoking a variety of dispersals and extinctions (see Mayden 1985, 1988 for a detailed discussion). Once again, such explanations were often predicated on the hypothesis that most species had dispersed to this continent from elsewhere. The lack of explicit phylogenetic hypotheses for the evolution of individual clades has made it difficult to formulate rigorous explanations about the origins of this fauna. However, due to the recent efforts of a variety of researchers, who are building upon the taxonomic legacy bequeathed by ichthyologists such as D. S. Jordan and C. L. Hubbs, a solid foundation for such evolutionary investigations is currently under construction (e.g., Mayden 1992a).

Mayden (1988) examined the historical biogeography of fishes in seven different clades: the *Notropis leuciodus, Luxilus zonatus-coccogenis, Etheostoma variatum, Nocomis biguttatus,* and *Fundulus catenatus* species groups and the subgenera *Etheostoma (Ozarka)* and *Percina (Imostoma).* These fishes inhabit drainage systems within the Central Highland region (fig. 4.37). The Central Highland consists of three currently disjunct regions. To the west of the Mississippi River lie the Ozark and Ouachita highlands, separated by the floodplain of the Arkansas River. The western highlands, in turn, are separated from their expansive counterpart, the eastern highlands, by the floodplain of the Mississippi River. Prior to the disruptive influences of Pleistocene glaciation, these regions were continuous (see references in Mayden 1988).

Two hypotheses have been proposed to explain the diversity patterns of the freshwater fish fauna in this region. The first hypothesis is based upon observations that the same or related species often occur on the eastern and western sides of the Mississippi. Given this, several researchers proposed that much of the current diversity was produced by the fragmentation and isolation of populations during Pleistocene glaciation. According to this scenario, speciation occurred subsequent to this isolation and has been accompanied by widespread dispersal of the new species following the retreating glaciers. The second hypothesis postulates that current diversity existed before the Pleistocene glaciation. According to this scenario, the glaciers fragmented the freshwater fauna existing in the expansive and continuous highland province. While this does not rule out a role for glaciation in the production of the current disjunct distribution patterns of freshwater fishes in the Central Highland region, it minimizes glaciation's role as the stimulus for widespread

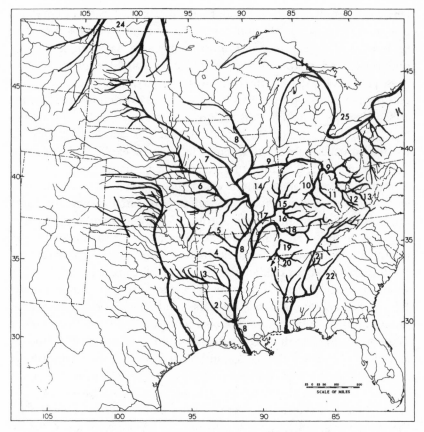

Figure 4.37. Preglacial river drainages in North America superimposed on existing drainages. 1 = Plains Stream; 2 = old Red River; 3 = old Ouachita River; 4 = old Arkansas River; 5 = White River; 6 = old Grand-Missouri River; 7 = ancestral Iowa River; 8 = old Mississippi River; 9 = old Teays-Mahomet River; 10 = old Kentucky River; 11 = old Licking River; 12 = old Big Sandy River; 13 = Kanawha River; 14 = Kaskaskia River; 15 = Wabash River; 16 = Green River; 17 = old Ohio River; 18 = old Cumberland River; 19 = old Duck River; 20 = old Tennessee River; 21 = Appalachian River; 22 = old Tallapoosa River; 23 = Mobile basin; 24 = Hudson Bay drainage; 25 = St. Lawrence River. (Redrawn from Mayden 1988.)

and recent speciation among these organisms. Phylogenetic analysis has uncovered speciation events in multiple groups of fishes that are correlated with geological events predating the Pleistocene in both glaciated and unglaciated areas, supporting the second hypothesis. Mayden focused his attention on the Central Highland drainages in an attempt to shed further light on the problem:

> For Central Highland fishes one may examine the origin of the fauna by comparing the history of the drainage basins involved

and the history of the fishes, inferred from geologic data and phylogenetic relationships, respectively. If congruence is obtained between the phylogenetic relationships and drainage relationships existing prior to the Pleistocene then one may predict that the fish groups existed prior to glaciation and the vicariance hypothesis would be supported. However, if relationships of fishes are congruent with drainage patterns developed after glaciation, then an explanation of dispersal during and after the Pleistocene glaciation may be appropriate. (1988:340)

Mayden performed a phylogenetic analysis using 34 river drainages as area taxa and the phylogenetic relationships from the seven different clades as characters (in all, 37 "characters"). This analysis produced 33 equally parsimonious area cladograms. We will base the following discussion upon a simplified version of Mayden's consensus tree for those cladograms (fig. 4.38). This area cladogram highlights four points of agreement among all of the alternate hypotheses. First, all area cladograms place the rivers of the Interior Highlands in a single clade. That clade, in turn, is the sister group to the drainages of the upper Mississippi River and the Old Teays River. Second, prior to the Nebraskan and Kansan glaciations, the Teays River flowed west into the upper Mississippi. Following these glacial advances, the course of the Teays shifted southward to flow into the Ohio River, which is the present-day (postglaciation) configuration. All 33 area cladograms place the rivers of the old

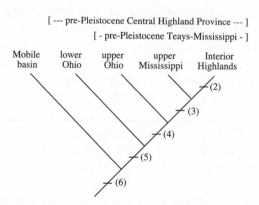

Figure 4.38. Area cladogram for the Central Highland drainage systems, reconstructed from the phylogenetic relationships of various freshwater fishes. The original area cladogram, based on 34 contemporaneous rivers, has been simplified to highlight five groups of rivers that indicate preglacial origins of the fish that inhabit them: Mobile basin (drainages 21–23 in figure 4.37); lower Ohio (drainages 15–20 in figure 4.37); upper Ohio (part of the old Teays River drainage; drainages 10–14 in figure 4.37); upper Mississippi (drainages 7–9 in figure 4.37); and Interior Highlands (drainages 2–6 in figure 4.37). The number of characters supporting each of the five groupings is listed in parentheses. (Modified from Mayden 1988.)

Teays system with the upper Mississippi drainages; the Teays system never clusters with drainages forming part of the lower Ohio River system. Third, the Mobile Basin, which was separate from the Mississippi Drainage long before the Pleistocene, retains its independent status in this analysis, clustering at the base of the area cladogram. And finally, one clade encompasses all the Mississippi offshoots thought to have formed the drainages for the pre-Pleistocene Central Highlands, while yet another clade reconstructs the pre-Pleistocene Mississippi-Teays River system.

In summary, most of these distributions coincide with *pre-Pleistocene*, rather than *post-Pleistocene* or contemporary, drainage patterns (for a detailed discussion of individual rivers see Mayden 1988). This suggests that there was a diverse and widespread Central Highlands ichthyofauna prior to the Pleistocene glaciation, corroborating the studies of speciation patterns reported by Wiley and Mayden (1985; see chapter 3). More recent studies (e.g., Mayden and Matson 1992) also support this conclusion. This does not mean that dispersal and glaciation have been unimportant in this system. For example, in seven cases, Mayden was able to demonstrate that specific instances of geographic homoplasy coincided with episodes of Pleistocene glacial alterations in river flow patterns that apparently resulted in some faunal mixing (secondary BPA would depict these instances using area duplications). Not surprisingly, then, current distributions reflect an interaction between the relatively ancient origins and diversification of the fauna and the recent effects of large-scale environmental changes. Following Mayden's groundbreaking study, similar antiquity has been reported for the speciation patterns of crustaceans in the genus *Daphnia,* one of the favorite food items of many North American freshwater fish, at least during some part of their lives (Colbourne and Hebert 1996; Witt and Hebert 2000); for North American softshell turtles (Weisrock and Janzen 2000); for North American passerine birds (Klicka and Zink 1997; Zink et al. 2000), most of which have little to do with fish; and for elements of the riverine and terrestrial biota of Europe (Hewitt 1999; Oberdorff et al. 1999). All of these studies suggest origins that predate the Pleistocene, in some cases considerably, with post-Pleistocene dispersal and differentiation of contemporary microspecies whose relationships indicate post-Pleistocene dispersal routes.

Amazonian Parrots and Toucans: What Was Happening in the South during the Pleistocene? The effects of Pleistocene climatic changes, most obvious in the glaciation of temperate areas, were global. Even the lush, warm Amazon Basin was not immune (e.g., Prance 1982a,b; Mayr and O'Hara 1986; Haffer 1987, 1993). Instead of physically intruding into the area, how-

ever, the glaciers are thought to have been responsible for a general drying of the Amazon Basin, creating relatively small "islands" of ancient
forests surrounded by "seas" of new xeric habitat.[6] PFRT predicts that
much of the Amazonian terrestrial diversity was produced via multiple
episodes of the form of vicariant speciation shown in figure 3.6 (i.e.,
isolation of ancestral populations resulting from barriers created by environmental change). We do not necessarily expect to find general patterns of speciation, even though we are dealing with vicariance, because
refuges are small and isolate only a subset of available diversity. In addition, we might expect to find many polytomies on phylogenetic trees of
clades living in the Amazon basin, reflecting the essentially simultaneous formation of the refugia.

At the same time, we expect the impact of this relatively recent environmental harshness to be superimposed upon an older backbone of
vicariance due to geological changes. For example, biologists have long
noted that many South American clades are composed of distinct species
living on different sides of the Andes (Chapman 1917, 1926), on different sides of rivers (e.g., Sclater and Salvin 1867; Wallace 1889a; Hellmayr
1910; Snethlage 1913; Sick 1967), or along an upstream-downstream gradient within the rivers (Lara and Patton 2000; Patton et al. 2000). Given
the complexity of this area, is it possible to disentangle the older vicariance events from more recent speciation (if any) due to environmental
changes in the Pleistocene?

Cracraft and Prum (1988) investigated this question by tracking the
evolutionary divergence of one clade of parrots and three clades of toucans in the Neotropics. So, what do the biogeographic patterns for the
Amazon Basin look like for these birds?

1. The phylogenetic trees for the four groups are shown in figure 4.39.

2. Primary BPA of the data in table 4.19 produces three equally parsimonious area cladograms (CI = 92.3%). The consensus of these three
cladograms is shown in figure 4.40. The homoplasy is concentrated in
area C (southeastern Brazil) because two species in the study group, *Selenidera maculirostris* (species 4), which is a relatively highly derived member of its clade, and *Pionopsitta pileata* (species 13), which is the basal
member of its clade, occur there. Southeastern Brazil is therefore divided
into two different areas, C_1 and C_2, with the binary codes adjusted accordingly (table 4.20).

3. Secondary BPA produces one area cladogram with a CI of 100%
(fig. 4.41). Recall that a cornerstone of the PFRT is that Amazonian diver-

---

6. Geologists and paleoecologists are currently examining these assumptions. See discussion and references in Brumfield and Capparella 1996.

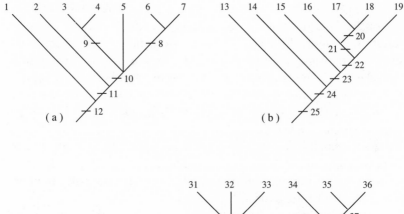

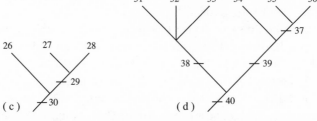

Figure 4.39. Phylogenetic trees for four clades of Neotropical birds, with internal branches numbered for biogeographic analysis. (*a*) Members of the toucan genus *Selenidera:* 1 = *S. spectabilis;* 2 = *S. culik;* 3 = *S. gouldii;* 4 = *S. maculirostris;* 5 = *S. nattereri;* 6 = *S. langsdorffi;* 7 = *S. reinwardtii.* (*b*) Members of the parrot genus *Pionopsitta:* 13 = *P. pileata;* 14 = *P. coccinicollaris + P. pulchra + P. hematotis;* 15 = *P. caica;* 16 = *P. vulturina;* 17 = *P. aurantiigena;* 18 = *P. barrabandi;* 19 = *P. pyrilia.* (*c*) Toucans in the *Pteroglossus viridis* group: 26 = *P. viridis;* 27 = *P. inscriptus;* 28 = *P. humboldti.* (*d*) Toucans in the *Pteroglossus bitorquatus* group: 31 = *P. sturmii;* 32 = *P. reichenowi;* 33 = *P. bitorquatus;* 34 = *P. azara;* 35 = *P. mariae;* 36 = *P. flavirostris.* (Modified from Cracraft and Prum 1988.)

Table 4.19. Primary Matrix Listing the Geographic Distribution of Four Avian Clades in Eight Neotropical Areas, along with the Binary Codes Representing the Phylogenetic Relationships for All Four Clades

Area	Binary Code			
A	0100000000	1100100000	0011110001	??????????
B	0010000011	1100010000	1111101011	1110000101
C	0001000011	1110000000	00001?????	??????????
D	0000100001	1100001001	1111100111	0001000011
E	0000010101	1100000101	1111100111	0000101011
F	0000001101	1100000101	1111100111	0000011011
G	1000000000	0101000000	00011?????	??????????
H	??????????	??00000010	01111?????	??????????

NOTE: A = Guyana, northeast of Amazon; B = southeast Amazon; C = southeast Brazil; D = southwest Amazon; E = northwest Amazon; F = west Amazon; G = Panama–northeast South America; H = north Colombia–Maracaibo. ? = species missing from the area.

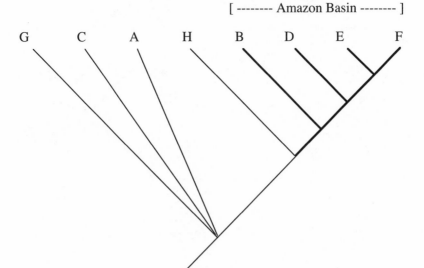

Figure 4.40. Primary area cladogram for eight Neotropical areas, based on phylogenetic relationships of one group of parrots and three groups of toucans.

Table 4.20. Secondary Matrix Listing the Geographic Distribution of Four Avian Clades in Eight Neotropical Areas, along with the Binary Codes Representing the Phylogenetic Relationships for All Four Clades

Area	Binary Code			
A	0100000000	1100100000	0011110001	??????????
B	0010000011	1100010000	1111101011	1110000101
C_1	??????????	??10000000	00001?????	??????????
C_2	0001000011	11????????	??????????	??????????
D	0000100001	1100001001	1111100111	0001000011
E	0000010101	1100000101	1111100111	0000101011
F	0000001101	1100000101	1111100111	0000011011
G	1000000000	0101000000	00011?????	??????????
H	??????????	??00000010	01111?????	??????????

NOTE: A = Guyana, northeast of Amazon; B = southeast Amazon; C_1 and C_2 = southeast Brazil; D = southwest Amazon; E = northwest Amazon; F = west Amazon; G = Panama–northeast South America; H = north Colombia–Maracaibo. ? = species missing from the area.

sity results from multiple, independent, simultaneous episodes of speciation, leading us to expect polytomous area relationships and a paucity of general patterns. These expectations are not supported, at least for these birds, the relationships of which provide strong evidence for sequential vicariant speciation in the Amazonian portion of the area cladogram (areas B, D, E, and F).

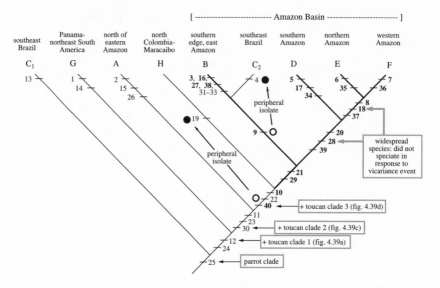

Figure 4.41. Secondary area cladogram for eight Neotropical areas, based on phylogenetic relationships of one group of parrots and three groups of toucans. *Light lines* = areas outside Amazonia; *bold lines* = areas in Amazonia in which endemic members of all four bird clades reside; *open circle* (ancestor) + *solid circle* (peripheral isolate) = instance of peripheral isolates speciation.

The area cladogram also suggests that this component of the Neotropical avifauna was *assembled* sequentially through time and space. Southeastern Brazil (area C_1) is inhabited by only one study species, *Pionopsitta pileata,* the basal parrot species in its clade. Central America plus the Choco (area G) occurs next on the area cladogram because it contains *Selenidera spectabilis,* the basal member of that toucan clade, and *Pionopsitta coccinicollaris* + *P. pulchra* + *P. hematotis,* the second basal parrot group. Finally, the Guyanan area (A), is inhabited by the basal member of the second toucan clade, the *Pteroglossus viridis* species group (*P. viridis*), the second member of *Selenidera* (*S. culik*), and the third member of *Pionopsitta* (*P. caica*). It is not until we come to the first Amazonian area (area B) that we find all four clades represented. Although the cladogram allows us detect the points at which the various clades became associated, we cannot tell whether each point of association represents an older vicariance event producing the ancestor of a group or speciation by dispersal. In order to address this question, we must add information from the sister groups of these four clades to the analysis, and perhaps from additional clades as well.

There are two curious instances of independence from the background of vicariant speciation. First, the occurrence of two sets of sister species, *Selenidera langsdorffi* (6) + *S. reinwardtii* (7) and *Pteroglossus mariae* (35) + *P. flavirostris* (36), in the sister areas E and F conforms to a hypothesis of vicariant speciation. *Pionopsitta barrabandi* (18) and *Pteroglossus humboldti* (28), however, both occur in areas E and F (they are widespread species). This suggests that the vicariance event affecting the *Selenidera* and *Pteroglossus bitorquatus* species-group representatives in those areas did not affect the members of *Pionopsitta* and the *Pteroglossus viridis* species group. Second, *Pteroglossus humboldti* also occurs in area D, suggesting that the vicariance event producing *Selenidera nattereri* (5) plus ancestral species 8, *Pionopsitta aurantiigena* (17) plus *P. barrrabandi* (18), and *Pteroglossus azara* (34) plus the ancestor of *P. mariae* plus *P. flavirostris* (37) did not affect the *Pteroglossus viridis* clade. Thus, the members of *Selenidera* and the *Pteroglossus bitorquatus* species group exhibit three instances of vicariant speciation in the Amazon, members of *Pionopsitta* exhibit two instances, and members of the *Pteroglossus viridis* species group show only one.

Differential speciation rates among members of clades living in a single area represents another form of evolutionary independence in the assembling of this Neotropical avifauna. The area along the southern edge of the eastern Amazon (B) is inhabited by birds from all four study clades. Three of these clades are represented by a single species that is widespread throughout the area, *Selenidera gouldii* (3), *Pionopsitta vulturina* (16), and *Pteroglossus inscriptus* (27). The fourth clade, the *Pteroglossus bitorquatus* species group, is represented by three species with restricted allopatric distributions, *P. sturmii* (31), *P. reichenowi* (32), and *P. bitorquatus* (33). Cracraft and Prum (1988) reported that the species status of these toucans is uncertain at the moment. However, if they are differentiated species, the *Pteroglossus bitorquatus* species group exhibits an interesting coupling of decreased distribution and increased diversity compared to its coinhabitants along the banks of the Amazon. It is possible that these three species were produced in the forest refugia during the Pleistocene.

Areas H (northern Colombia to Lake Maracaibo) and C_2 (southeastern Brazil) also represent departures from the general pattern of area relationships depicted in figure 4.41. Both of these areas are placed among the Amazonian regions on the area cladogram due to the relationship of a single species to the other members of its genus: (1) the parrot *Pionopsitta pyrilia* (species 19; area H) and (2) the toucan *Selenidera maculirostris* (species 4; area C_2). These distributions appear to be the result of two independent speciation events. If this is the case, then we have identified two additional examples of the speciation mode associated with PFRT. However, the theories were designed to explain species diversity and distribution patterns within the Amazon basin, and these birds live along the periphery of that area.

This study thus indicates that the Neotropical avifauna has been assembled by a combination of (1) the sequential addition of basal members of the clades into areas surrounding the Amazon, (2) vicariant speciation among the toucans and parrots within the Amazonian areas, (3) three possible cases of peripheral isolates speciation within Amazonian refugia, (4) two cases of peripheral isolates speciation on the periphery of the Amazon basin, and (5) factors unique to two clades that affected the form and nature of their response to the vicariant effects (e.g., species 18 and 28 did not respond to the vicariance event by speciating). Cracraft and Prum thus came to the same general conclusions as did Weitzman and Weitzman (1982), Lynch (1986, 1988, 1999), Amorim (1991), Brumfield and Capparella (1996), and Patton et al. (2000) about PFRT in the Neotropics, both highlands and lowlands: "These generalized historical patterns . . . [arose] via fragmentation (vicariance) of a widespread ancestral biota. . . . These vicariance events could have originated . . . at various times during the Cenozoic. The inference that diversification of the Neotropical biota is primarily the result of . . . isolation within Quaternary forest refugia, is unwarranted, given present data" (1988:618; see also Cracraft 1988).

The Pleistocene Forest Refugium Hypothesis: Current Status

Objections to the PFRT began to surface within 20 years of its formalization by Ernst Mayr (e.g., Endler 1982a,b; Campbell and Frailey 1984; Liu and Colvinaux 1985; Räsänen et al. 1987). Endler, for example, suggested that the number, form, and extent of contact zones between populations of some east African birds in the Guinea highlands could have been the result of present-day interference competition limiting contact. By 1997, enough evidence from various geographic regions had been accumulated to indicate that the PFRT on its own did not provide a general explanation for the current extent and distribution of species on this planet. As a general explanation, then, the PFRT has failed (Klicka and Zink 1997). And yet, like some plague victims in a Monty Python movie, it is not really dead yet, nor should it be.

A number of phylogeography studies have shown that many macrospecies predate the Pleistocene (confirming the general conclusion from Mayden's study), but that much current diversification, both geographical and genetic, of differentiated microspecies within those macrospecies, are post-Pleistocene in age.[7] The Pleistocene clearly had a global impact on biodiversity, including isolation, extinction, speciation, and stasis at the level of individual clades, and origination of new ecological associations. Phylogenetic analyses indicate, however, that the effects differed from place to place (recall the Amazonian birds) and taxon to taxon (e.g., Valentine and Jablonski 1991). For example, the general perception among biologists is that the increasingly harsh polar environments resulted in widespread extinctions during the Pleistocene glaciations. Based upon his work with the tapeworms of alcids (puffins, murres, auks, and their relatives) and pinnipeds (seals, sea lions, walruses, and their relatives), Hoberg (1986, 1987, 1992; Hoberg and Adams 1992, 2000) suggested an intriguing alternative: the dramatic environmental fluctuations characterizing the Pliocene-Pleistocene glaciation promoted the diversification, as well as the extinction, of Arctic and boreal organisms (the **boreal refugium hypothesis**). In the decade since Hoberg's proposal, data have begun to accumulate documenting speciation in boreal refugia for a variety of other organisms, including shags in the western North Pacific Ocean (Siegel-Causey 1991), members of the *Daphnia pulex* complex in northern Canada (Weider et al. 1999), and species in the genus *Maudheimia,* a clade of oribatid mites living in the soil of Antarctica (Marshall and Coetzee 2000). So, rather than just being the places where clades went to die, the extreme environments at the planet's poles may also have been among the few places where evolutionary diversification took place during the last glacial period.[8] Clearly, the evolutionary impacts of the Pleistocene were so complex and varied that we need extensive and detailed phylogenetic analyses of many groups, and more geological and paleoecological information, to tell the story properly.

Let us now leave discussions of ice and snow and head to a warmer area to search for general patterns of speciation.

The Birds Are Coming: The Australian Avifauna. Many birds possess the potential for widespread dispersal, so it seems counterintuitive to expect avian speciation patterns to produce much in the way of congruent area cladograms. Nonetheless, it is also true that even a champion flier such as the

7. For example, Avise 1992, 2000; Vogler and DeSalle 1993; Ellsworth et al. 1994; Klein and Brown 1994; Phillips 1994; Seutin et al. 1994; Zink 1994, 1996; Riddle 1996; Shaffer and McKnight 1996; Hedin 1997; Rosenzweig 1997; Barrowclough et al. 1999; and Lessios et al. 1999.

8. For a recent assessment of Antarctic evolutionary biology, see Bargelloni et al. 2000; Brandt 2000; Clarke 2000; Eastman 2000; Eastman and McCune 2000; Pisano and Ozouf-Costaz 2000; Rodhouse et al. 2000; and Vincent 2000.

ruby throated hummingbird is not found in every area of the world, so it is possible that some remnants of historical associations are retained even in the most highly vagile organisms. In a series of detailed studies, Cracraft (1983b, 1986, 1988, 1994) examined the biogeography of eight avian groups (finches, pigeons, fairy wrens, rifle-birds, flycatchers, emu-wrens, and quail thrushes) occurring in 14 different parts of Australia and in New Guinea. In this example, we follow Cracraft's area descriptions with two modifications (fig. 4.42). In 1986, Cracraft did not identify any isolating barrier between the southeastern corner and the eastern forest, which lie along the eastern and southern regions of Australia, so we combined those two areas into one region (eastern forest). In 1994, however, he postulated that a subsection of the southeastern forest, the Eyre Peninsula + Adelaide, was a separate area. This is consistent with his 1986 description that the peninsula and surrounding lands, with their shrub-steppe and *Eucalyptus* scrubland, formed an ecological barrier between the moist forests to the east and west along the southern margins of Australia. We have therefore added "Eyre Peninsula + Adelaide" to the study and coded them as "AD."

If, as intuition suggests, postspeciation dispersal obscures vicariance patterns (or if birds are more likely to speciate via peripheral isolates speciation because of their dispersal abilities), we would expect to find no trace of a general vicariance pattern when avian phylogenies are matched with areas. Astonishingly, Cracraft documented substantial support for two general vicariance patterns, one involving north/east Australia plus New Guinea and the other comprising central and southern Australia. At the time of the original studies, secondary BPA had not been developed, so it was difficult to identify possible exceptions to those patterns. Let us reexamine the data in light of the new methodology.

Primary BPA of all eight clades produces 245 equally parsimonious area cladograms, which agree only on a grouping of the Atherton Plateau (AP) as the sister area of the eastern forest (EF). The consensus of these MPTs is a large polytomy, with only (AP, EF) showing well-supported relationships. The lack of resolution is partly due to missing data; even though there are eight clades in the study, none of them contain species that occur in all of the areas, and most comprise species that occur in fewer than half the areas. Removing the areas that contain the fewest representatives of the eight clades (CN, WN, EN, and ND) reduces the number of MPTs from 245 to 16, but the consensus tree remains the same. Most of the ambiguity resides in the only four areas that link all eight clades together: the Cape York Peninsula (CY), the Atherton Plateau (AP), the western desert (WD), and the eastern forest (EF). The lack

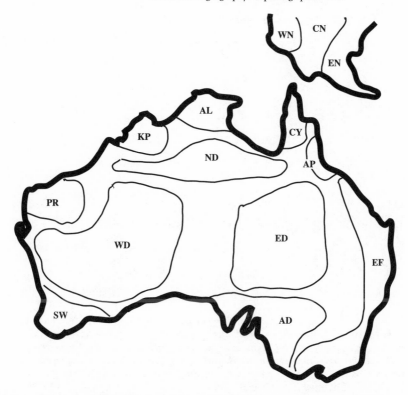

Figure 4.42. Map of Australia and New Guinea highlighting the 14 areas used in the analysis of Australian bird biogeography. AD = Adelaide + Eyre Peninsula; AL = Arnhem Land; AP = Atherton Plateau; CN = central New Guinea; CY = Cape York Peninsula; ED = eastern desert; EF = eastern rainforest; EN = eastern New Guinea; KP = Kimberley Plateau; ND = northern desert; PR = Pilbara region; SW = southwestern corner; WD = western desert; WN = western New Guinea.

of resolution, even after correcting for areas contributing substantial missing data, indicates that there is something wrong with the initial assumption that the history of speciation for these organisms can be reduced to one general vicariance pattern. In his 1986 study, Cracraft suggested that two separate historical patterns accounted for the historical biogeography of these birds. Following that suggestion, we next analyzed each grouping separately.

The first grouping comprises three clades (fig. 4.43), *Psophodes, Stipiturus,* and *Cinclosoma,* distributed in central-southeastern coastal Australia. Primary BPA of these data (table 4.21) produces one most parsimonious area cladogram (fig. 4.44) with a consistency index of 93%, indicating some homoplasy. Secondary BPA (table 4.22) produces a

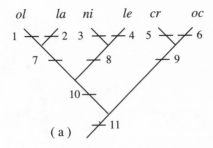

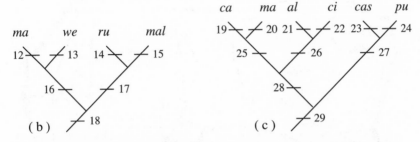

Figure 4.43. Phylogenetic trees for three clades of Australian birds numbered for biogeographic analysis. (*a*) *Psophodes: cr = cristatus; la = lateralis; le = leucogaster; ni = nigrogularis; oc = occidentalis; ol = olivaceus.* (*b*) *Stipiturus: ma = malachurus; mal = mallee; ru = ruficeps; we = westernensis.* (*c*) *Cinclosoma: al = alisteri; ca = castanoethorax; cas = castanotum; ci = cinnamomeum; ma = marginatum; pu = punctatum.* (Modified from Cracraft 1986.)

Table 4.21. Primary Matrix Listing the Geographic Distributions and Binary Codes Representing the Phylogenetic Relationships for Three Clades of Australian Birds (Species 1–6, 12–15, 19–24)

Area	Species	Binary Code		
AP	2	0100001001	1?????????	?????????
EF	1, 12, 24	1000001001	1100010100	000100101
SW	3, 13, 23	0010000101	1010010100	001000101
AD	4, 12, 23, 24	0001000101	1100010100	001100101
ED	5, 15, 19, 22, 23	0000100010	1000101110	011011111
WD	6, 14, 20, 21, 23	0000010010	1001001101	101011111
PR	6, 14	0000010010	10010011??	?????????

NOTE: AD = Eyre Peninsula + Adelaide; AP = Atherton Plateau; ED = eastern desert; EF = eastern forest; PR = Pilbara region; SW = southwestern corner; WD = western desert. ? = species missing from the area.

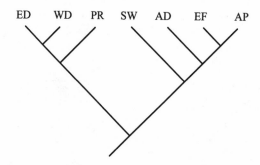

Figure 4.44. Area cladogram produced by primary analysis of three Australian bird clades. AD = Adelaide + Eyre Peninsula; AP = Atherton Plateau; ED = eastern desert; EF = eastern forest; PR = Pilbara region; SW = southwestern corner; WD = western desert.

Table 4.22. Secondary Matrix Listing the Geographic Distributions and Binary Codes Representing the Phylogenetic Relationships for Three Clades of Australian Birds (Species 1–11, 12–18, 19–29)

Area	Species		Binary Code	
AP	2	0100001001	1??????????	??????????
EF	1, 12, 24	1000001001	1100010100	000100101
SW	3, 13, 23	0010000101	1010010100	001000101
AD$_1$	4	0001000101	1??????????	??????????
AD$_2$	12, 24	??????????	?100010100	000100101
AD$_3$	23	??????????	???????00	001000101
ED$_1$	5, 15, 19, 22	0000100010	1000101110	010011011
ED$_2$	23	??????????	???????00	001000101
WD$_1$	6, 14, 20, 21	0000010010	1001001101	100011011
WD$_2$	23	??????????	???????00	001000101
PR	6, 14	0000010010	10010011??	??????????

NOTE: AD is listed as three separate areas, and ED and WD are each listed as two separate areas. AD = Eyre Peninsula + Adelaide; AP = Atherton Plateau; ED = eastern desert; EF = eastern forest; PR = Pilbara region; SW = southwestern corner; WD = western desert. ? = species missing from the area.

single most parsimonious area cladogram (fig. 4.45), which is consistent with Cracraft's vicariance hypothesis, but also suggests that there have been seven instances of postspeciation dispersal: *S. malachurus* and *C. punctatum* from EF to AD, *C. castanotum* from SW to WD, ED, and AD, and *P. occidentalis* and *S. ruficeps* from WD to PR. There are two cases of peripheral isolates speciation: (1) A population of ancestor 7 dispersed into AP producing *P. lateralis* (species 2). That dispersal was most likely

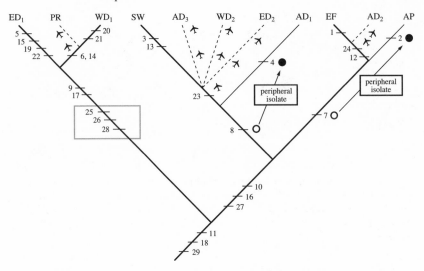

Figure 4.45. Secondary area cladogram for three Australian bird clades with three duplicated areas. *Airplanes* = postspeciation dispersal into the areas indicated with dotted lines. Dispersal by ancestor 8 into AD_1 and ancestor 7 into area AP represent instances of peripheral isolates speciation. *Dashed box* = lineage duplication; *bold lines* = the backbone of vicariant speciation; *numbers accompanying slash marks* = species codes (from fig. 4.43).

from EF which is adjacent to AP (fig. 4.42), and where *P. olivaceus* (species 1), the sister species of *P. lateralis,* occurs. This raises the possibility that *P. olivaceus* is a persistent ancestor (i.e., that *P. olivaceus* is ancestor 7). (2) A population of ancestor 8 (the sister species of ancestor 7) dispersed into AD producing *P. leucogaster* (species 4). Finally, the analysis uncovers a potential episode of sympatric speciation: ancestor 28 gave rise to ancestors 25 and 26 in the central desert. Both of these ancestors speciated following the vicariance event separating the desert into eastern and western segments, ancestor 25 giving rise to *C. castaneothorax* (species 19) + *C. marginatum* (species 20) and ancestor 26 producing *C. cinnamomeum* (species 22) + *C. alisteri* (species 21).

Now for something far more complicated (for those of you who like a challenge). The second grouping comprises five clades (fig. 4.46), *Poephila, Malurus, Petrophasa, Ptiloris,* and *Tregellasia,* distributed in the Cape York peninsula (CY), New Guinea [central (CN), eastern (EN), and western (WN)], the Atherton Plateau (AP), the eastern forest (EF), Arnhem Land (AL), the Kimberly Plateau (KP), the northern desert (ND), and a coastal area south of Arnhem Land called the Pilbara region (PR). Primary BPA of those data (table 4.23) produces 6 equally parsimonious

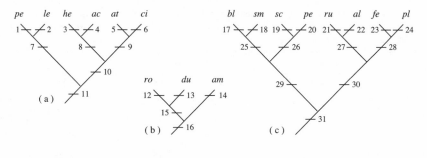

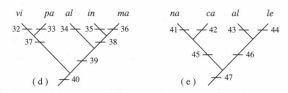

Figure 4.46. Phylogenetic trees for five clades of Australian birds numbered for biogeographic analysis. (*a*) *Poephila: ac = acuticauda; at = atropygialis; ci = cincta; he = hecki; le = leucotis; pe = personata.* (*b*) *Malurus: am = amabilis; du = dulcis; ro = rogersi.* (*c*) *Petrophasa: al = albipennis; bl = blaauwi; fe = ferruginea; pe = peninsulae; pl = plumifera; ru = rufipennis; sc = scripta; sm = smithii.* (*d*) *Ptiloris: al = alberti; in = intercedens; ma = magnificus; pa = paradiseus; vi = victoriae.* (*e*) *Tregellasia: al = albigularis; ca = capito; le = leucops* (species complex); *na = nana.*

area cladograms. Once again, this is an artifact of having considerable missing data in the data matrix because five areas (CN, EN, WN, ND, and PR) have only one species each in them. If we (temporarily) remove those unique elements, we obtain a single most parsimonious area cladogram (fig. 4.47a) congruent with Cracraft's vicariance scenario. If we add CN back into the analysis we still get one most parsimonious tree (fig. 4.47b), congruent with the vicariance scenario. Adding any of the remaining areas increases the ambiguity substantially due to problems with both missing data and widespread species. The sister species *Petrophasa ferruginea* (species 23 in the Pilbara region) and *P. plumifera* (species 24 in the northern desert) also occur in the Kimberly Plateau, suggesting that some secondary dispersal occurred. Was that dispersal from the Kimberly Plateau to the other areas or the reverse? To answer this question, we duplicate the Kimberly Plateau for each occurrence in the data matrix.

Secondary BPA of that data set (table 4.24) produces 6 equally parsimonious area cladograms. Excluding eastern and western New Guinea (again, only temporarily) produces two equally parsimonious area clado-

Table 4.23.   Primary Matrix Listing the Geographic Distributions and Binary Codes Representing the Phylogenetic Relationships for Five Clades of Australian Birds (Species 1–6, 12–14, 17–24, 32–36, 41–44)

Area	Species	Binary Code				
KP	1, 4, 12, 17, 22, 23, 24	1001001101	1100111000	0111101111	1?????????	???????
AL	1, 3, 13, 18, 21	1010001101	1010110100	1000101011	1?????????	???????
PR	23	??????????	??????0000	0010000101	1?????????	???????
EN	35	??????????	??????????	??????????	?000100111	???????
WN	36	??????????	??????????	??????????	?000010111	???????
CN	44	??????????	??????????	??????????	??????????	0001011
CY	2, 5, 14, 20, 34, 43	0100101011	1001010001	0000010010	1001000011	0010011
AP	6, 19, 32, 41	0000010011	1?????0010	0000010010	1100001001	1000101
EF	33, 42	??????????	??????????	??????????	?010001001	0100101
ND	24	??????????	??????0000	0001000101	1?????????	???????

NOTE: AL = Arnhem Land; AP = Atherton Plateau; CN = central New Guinea; CY = Cape York Peninsula; EF = eastern forest; EN = eastern New Guinea; KP = Kimberley Plateau; ND = northern desert; PR = Pilbara region; WN = western New Guinea. ? = species missing from the area.

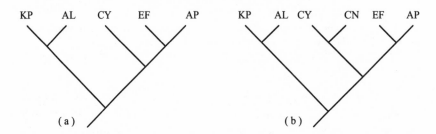

Figure 4.47.   Area cladograms from primary analysis of five Australian bird clades. (*a*) Area cladogram recovered after the five unique elements have been removed. (*b*) Area cladogram retrieved after one unique element, CN, is added back into the analysis.

grams (fig. 4.48a,b). The two area cladograms agree on the relationships within three general units (1, 2, and 3 on fig. 4.48). The disagreement centers on how those units are related to one another: specifically, is unit 1 more closely connected historically to unit 2 or unit 3? The disagreement is due solely to the speciation event producing *Petrophasa ferruginea* (species 23) and *P. plumifera* (species 24), which is the only support for unit 2 in the analysis. At this point it is important to remember that the two area cladograms are equally parsimonious *only* if we do not account for missing data. Taking the missing data into account, we find that placing unit 2 as the sister area of units 1 and 3 requires us to postu-

Table 4.24. Secondary Matrix Listing the Geographic Distributions and Binary Codes Representing the Phylogenetic Relationships for Five Clades of Australian Birds (Species 1–6, 12–14, 17–24, 32–36, 41–44)

Area	Species	Binary Code				
KP_1	1, 4, 12, 17, 22	1001001101	1100111000	0100101011	1?????????	???????
KP_2	23	??????????	??????0000	0010000101	1?????????	???????
KP_3	24	??????????	??????0000	0001000101	1?????????	???????
AL	1, 3, 13, 18, 21	1010001101	1010110100	1000101011	1?????????	???????
PR	23	??????????	??????0000	0010000101	1?????????	???????
EN	35	??????????	??????????	??????????	?000100111	???????
WN	36	??????????	??????????	??????????	?000010111	???????
CN	44	??????????	??????????	??????????	??????????	0001011
CY	2, 5, 14, 20, 34, 43	0100101011	1001010001	0000010010	1001000011	0010011
AP	6, 19, 32, 41	0000010011	1?????0010	0000010010	1100001001	1000101
EF	33, 42	??????????	??????????	??????????	?010001001	0100101
ND	24	??????????	??????0000	0001000101	1?????????	???????

NOTE: KP is listed as three separate areas. AL = Arnhem Land; AP = Atherton Plateau; CN = central New Guinea; CY = Cape York Peninsula; EF = eastern forest; EN = eastern New Guinea; KP = Kimberley Plateau; ND = northern desert; PR = Pilbara region; WN = western New Guinea. ? = species missing from the area.

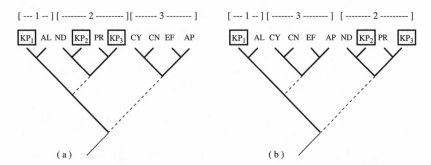

Figure 4.48. Secondary area cladograms excluding the eastern and western New Guinea areas and duplicating the Kimberley Plateau area. *Bold lines* = points of congruence between the two area cladograms; AL = Arnhem Land, AP = Atherton Plateau, CN = central New Guinea, CY = Cape York Peninsula, EF = eastern forest; KP = Kimberley Plateau, ND = northern desert; PR = Pilbara region.

late an extinction event (i.e., no sister species of ancestor 30 in unit 3). No such additional explanation of extinction is required for grouping unit 2 as the sister of unit 1 (fig. 4.48a). It is thus actually a more parsimonious explanation for the data, and therefore forms the basis for our discussion of speciation modes.

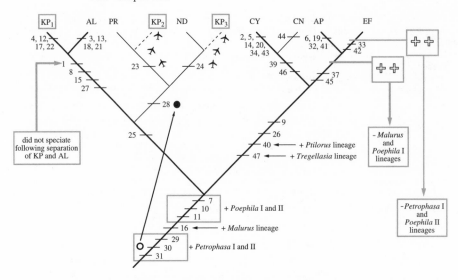

Figure 4.49. Explanations for departures from the vicariance backbone (still excluding eastern and western New Guinea). *Airplanes* = postspeciation dispersal by a particular species into the area (dotted lines). Dispersal by ancestor 30 into northwestern Australia represents an instance of peripheral isolates speciation. *Dashed boxes* = lineage duplications; *crosses* = extinctions; *bold lines* = the backbone of vicariant speciation; AL = Arnhem Land, AP = Atherton Plateau, CN = central New Guinea, CY = Cape York Peninsula; EF = eastern forest; KP = Kimberley Plateau, ND = northern desert; PR = Pilbara region; *numbers accompanying slash marks* = species codes (from fig. 4.46).

Examining this pattern (fig. 4.49) in more detail reveals substantial biogeographic complexity. First, the data indicate that *Poephila personata* (species 1) did not speciate following the vicariance event separating Arnhem Land (AL) and the Kimberly Plateau (KP). (This is an example of a "nonresponse" to a vicariance event.) Second, *P. ferruginea* (species 23) and *P. plumifera* (species 24) evolved in the Pilbara region (PR) and in the northern desert (ND), respectively, so their occurrence in the Kimberly Plateau (KP) is the result of postspeciation dispersal. The evolution of ancestor 28, the common ancestor of *P. ferruginea* and *P. plumifera,* is best explained as an episode of peripheral isolates speciation followed possibly by vicariance to produce *P. ferruginea* and *P. plumifera,* which are allopatric and have similar-sized ranges (although their unique distribution pattern might suggest an additional episode of peripheral isolates speciation). Third, there have been two episodes of lineage duplication (i.e., possible sympatric speciation). The distribution of *Poephila* species suggests an episode of lineage duplication in the speciation event that saw ancestor 11 subdivided into ancestors 10 and 7, which is responsible for the co-occurrence of species *P. personata* (species

1) with *P. hecki* (species 3) + *P. acuticauda* (species 4) in Arnhem Land and the Kimberly Plateau. The distribution of *Petrophasa* species suggests an episode of lineage duplication in the speciation event that subdivided ancestor 31 into ancestors 30 and 29, which is responsible for the co-occurrence of the nonsister species *P. blaauwi* (species 17)/*P. albipennis* (species 22) in the Kimberly Plateau and *P. smithii* (species 18)/*P. rufipennis* (species 21) in Arnhem Land. These ancient lineage duplications may represent old episodes of sympatric speciation or indicate allopatric speciation events that occurred prior to the time these clades became associated historically. Fourth, the analysis has uncovered evidence of potential extinction events in four of the five lineages. EF and AP lack members of *Malurus* and *Poephila* I. Because there are representatives of six clades in CY, it is most parsimonious to explain this absence as two extinction events. The *Petrophasa* I and *Poephila* II lineages are missing from EF. It is equally parsimonious to suggest that EF has experienced either two extinction events associated with the absence of those clades or two episodes of peripheral isolates speciation (producing *Ptiloris paradiseus* [species 33] and *Tregellasia capito* [species 42]). We prefer the extinction explanation, because CY, AP, and EF are part of a larger vicariance pattern, and because *P. paradiseus* and *T. capito* exhibit geographic ranges at least as large as those of their sister species.

Now, let's return to those pesky areas in New Guinea (fig. 4.50). Both *Tregellasia* and *Ptiloris* support a grouping of ((AP, EF) (CY)). *Tregellasia leucops* (species 44) in central New Guinea (CN) is the sister species of *T. albigularis* (species 43), endemic to the Cape York Plateau (CY). This is a unique element for the data set, implying an episode of peripheral isolates speciation (movement by a population of ancestor 46 from Cape York into New Guinea). *Ptiloris intercedens* (species 35) and *P. magnificus* (species 36) in eastern and western New Guinea (EN and WN) are sister species, the common ancestor (species 38) of which was the sister species of *P. alberti* (species 34), endemic to the Cape York Plateau (CY). This is also a unique element for the data set, implying yet another episode of peripheral isolates speciation in which a population of the ancestor of the *P. alberti* + *P. intercedens* + *P. magnificus* clade (species 39) moved from Cape York to New Guinea. These two unique speciation events, with the more general events associated with the Atherton Plateau and the eastern forest, are all linked historically with the Cape York Plateau.

The ambiguity in primary and secondary BPA associated with these areas is thus an artifact of using computer programs that attempt to guess at missing data (Nixon and Davis 1991; Platnick et al. 1991; Maddison 1993; Wilkinson 1995) rather than simply implementing the inclusion/exclusion rule to the fullest extent possible, then stopping.

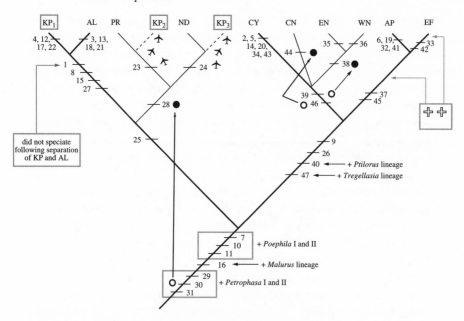

Figure 4.50. Finally, the big picture (including eastern and western New Guinea). The explanations are exactly the same as those depicted in figure 4.49, except for two additional instances of peripheral isolates speciation. AL = Arnhem Land, AP = Atherton Plateau, CN = central New Guinea, CY = Cape York Peninsula, EF = eastern forest; EN = eastern New Guinea; KP = Kimberley Plateau; ND = northern desert; PR = Pilbara region; WD = western desert; WN = western New Guinea; *numbers accompanying slash marks* = species codes (from fig. 4.46).

The only program we know of that does not "guess" is called CAFCA, produced by Rino Zandee of Leiden University. CAFCA implements group component compatibility analysis (CCA: Zandee and Roos 1987), which is based on strict application of the inclusion/exclusion rule and differs from primary BPA only in the way missing data are coded (Van Veller et al. 1999, 2000, 2001a,b). At present, no computer program implements secondary BPA.

Our analysis indicates that the biogeographic history of the Australian avifauna, represented by these eight clades, has been much more complex than initially thought. In addition to the strong vicariance backbone reported by Cracraft, we have found evidence for one nonresponse to vicariance, five instances of peripheral isolates speciation, four episodes of extinction, nine cases of postspeciation dispersal, and three lineage duplications that could indicate ancient episodes of sympatric speciation. Despite this, in only one case, *Petrophasa ferruginea* and *P. plumifera,* do sister species co-occur. Both species come together in the

Kimberly Plateau (KP), having dispersed there from their vicariant origins in the northern desert (ND) and the Pilbara region (PR).

Crisp et al. (1995) suggested that plant distributions demonstrated a double tropical invasion of Australia. Because Cracraft's original analyses did not demonstrate the same double tropical invasion, they felt that the historical patterns uncovered for the birds could not serve, on their own, as a general explanation for the evolution of Australian terrestrial ecosystems. Given that most birds eat either seeds or insects that feed on plants, and nest and sleep in vegetation, this seems a reasonable assumption. On the other hand, it is possible that the current flora of Australia is older than the portions of the avifauna Cracraft studied. We would therefore cast the question in a slightly different manner and ask if the apparent differences in biogeographic patterns are due to method of analysis (Crisp et al. used reconciled tree analysis) or to real differences in the biogeographic history of the plants and the birds. Our analysis supports Cracraft's argument that the data for these eight clades of birds indicates two different elements of biogeographic history. Are those two elements compatible with a double tropical invasion hypothesis? We cannot provide a definitive answer to this question, because none of the avian data link the two biogeographic elements shown in figures 4.45 and 4.50 into a single pattern. (Recall that when we tried to run all eight clades together, we uncovered 245 equally parsimonious trees. Most of that ambiguity was due to missing data for some clades in one of the two major areas, and to the fact that the root of the area cladogram is not supported by any characters.) If each of the two elements represents a separate tropical invasion from the north, we would expect to find representatives in the Cape York region, but we do not. At the same time, species currently in the Cape York region may have displaced earlier arrivals southward, or the earlier residents kept the newer arrivals from moving inland. Thus, it is possible that at least some of the plants and birds of Australia arrived and speciated together. It is interesting to note that the reanalysis of Cracraft's data uncovered two novel events from the perspective of the double invasion scenario: it appears that two species of birds, *Tregellasia leucops* and the ancestor of *Ptiloris intercedens* and *P. magnificus*, actually moved in the opposite direction, from northern Australia further northward into New Guinea. Evidently, Australia has enjoyed a dynamic geological and biological history (see also, e.g., Wilson and Johnson 1999; Matthews 2000). In order to investigate this complicated problem further (from the perspective of the birds), we need to increase both the number and size of avian clades in the analysis.

Range asymmetry revisited: Explaining markedly asymmetrical ranges of sister species has become one of the most contentious issues in the study of speciation modes. In particular, how do you differentiate between peripheral isolates speciation and microvicariance? (Recall the *Lepechinia* example in chapter 3.) We have already provided you with a partial answer to this question: microvicariance conforms to the general pattern in a historical biogeographical analysis. The sizes of each of the sister pairs' ranges are irrelevant in terms of identifying the general pattern *because there is no rule stating that all clades in an area will be subdivided in the same way by a vicariance event.* So, in figure 4.51a we see the amazing situation in which the uplifting of mountains has partitioned three very different clades in a similar manner, producing species C, F, and R by microvicariance because the mountains rose on the extreme edge of each ancestor's range. Figure 4.51b shows a more complicated situation in which the vicariance event bisected ancestors 1 and 2 equally (producing sister pairs B + C and E + F on either side of the mountains with approximately symmetrical ranges), but isolated just a small portion of ancestor 3 (producing species Q on one side with a large range and its sister species, R, on the other side with a small range). In both examples, vicariance is the initiating force in the speciation of B + C, E + F, and Q + R. Now, examine figure 4.51c closely, and you will see that something is amiss. Species R is on the same side of the mountain as its sister species, Q. This positioning of R does not conform to the general vicariance pattern indicated by the other two clades (B + C, E + F). The presence of this unique element in the biogeographical analysis indicates that species R was produced by peripheral isolates speciation (allopatric mode II). As we discussed in chapter 3, there is one important exception to the expectation that peripheral isolation events will produce only unique biogeographic elements: repeated biotic dispersal through temporary corridors (e.g., island hopping; Endler 1982b). Explanations invoking convergent episodes of dispersal for numerous clades require evidence about the timing of geological events and the ages of the clades under study (see, e.g., studies in Wagner and Funk 1995). Bear in mind that the debate about peripheral isolates versus vicariance concerns the mechanisms *initiating* speciation. Once the populations have been isolated, the same mechanisms for *completing* speciation may occur in either case. This is particularly true for the microvicariance-peripheral isolates distinction since both initiating modes produce small, isolated descendant species.

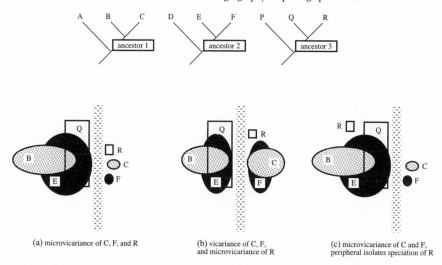

Figure 4.51. Hypothetical example showing how range asymmetry does not give you a complete picture of speciation modes. Three clades of very different organisms are subdivided in different ways by the uplifting of a mountain range. Species R has the same sized range in all cases, but in (*a*) and (*b*) it is hypothesized to be the product of microvicariance, while in (*c*) it is the product of peripheral isolates speciation.

SUMMARY

Hovenkamp (1997) noted that methods of historical biogeography seemed to comprise two areas of focus: the history of the areas in which species live (area history methods) and the history of the species living in the areas (species history methods). Van Veller et al. (1999, 2000, 2001a,b) referred to species-history methods as a posteriori methods (the "parsimony group": Brooks parsimony analysis [Brooks 1981, 1985, 1990; Brooks et al. 2001] and component compatibility analysis [Zandee and Roos 1987]) and to area-history methods as a priori methods (the "component group": component analysis [Nelson and Platnick 1981; Page 1988, 1990], reconciled tree analysis [Page 1993, 1994], and three-area statement analysis [Nelson and Ladiges 1991a,b]). Both groups have always believed that phylogenetic and geographic distribution data of taxa are informative for the reconstruction of the historical relationships among their areas of distribution and that common phylogenetic patterns for different taxa most often indicate a common history of vicariant speciation. Past that point, however, the "component" versus "parsimony" groups diverge, because the methods are founded upon two very different ways of looking at the world. Those differences are embodied in the way the null hypothesis is treated (Andersson 1996; Hovenkamp 1997; Van Veller at al. 1999, 2000, 2001a,b; Van Veller and Brooks 2001).

If the null hypothesis embodies your belief about the way the world is structured (your ontology), you may attempt to bolster (or verify) that null hypothesis by searching for supporting data. A priori methods are based on an **ontology of simplicity**, that is, a simple model (vicariance) explains all the data, or at least all the relevant data (contrary data being analogous to experimental error). Any data that conflict with the null hypothesis must be flawed. This is the reason Nelson and Platnick (1981:417) referred to "items of error" (i.e., the data are faulty) rather than "items of falsification" (i.e., the null hypothesis has been falsified by the data) in describing component analysis. A priori approaches are based upon the belief that all patterns of speciation are due to common episodes of vicariant speciation in given areas. Page (1987) was the first to acknowledge explicitly that within the component ontology the "best" phylogenetic trees were the ones that showed the largest number of matches (common nodes) with an a priori general area cladogram, regardless of how many additional ad hoc assertions are needed to explain incongruent data. In the original formulations of component methodologies, the belief of common cause could not be tested because incongruent data were eliminated by either postulating as many extinctions as were needed to generate the same pattern of vicariance for all clades, by deleting widespread taxa from the analysis (a process called **tree reconciliation:** Page 1994), or both. Component analysis over the years has undergone several modifications specifically designed to allow some instances of dispersal in the explanatory framework (see e.g., maximum cospeciation analysis: Page and Charleston 1998); however, the basis for the method is still one in which the preferred explanation maximizes the number of hypotheses of historical congruence (vicariance) and then minimizes the number of ad hoc assertions of lineage duplications and extinctions necessary to *maintain* that hypothesis, regardless of the original input data.

For a posteriori approaches, the null hypothesis serves only as a convenient beginning point and default explanation. Users of these methods believe that when data conflict with the null hypothesis, the null hypothesis, not the data, is flawed. It is such falsifications that provide objective (i.e., a posteriori) support for one's ontology (Popper 1968). For all a posteriori methods, that ontology is thus an **ontology of complexity,** which may be encapsulated in the following way: (1) speciation produces evolutionarily independent units, so the mode of speciation for any given speciation event does not predetermine or predispose the next mode of speciation in the clade; (2) geological and biological evolution are causally independent; (3) geological and biological evolution are nonetheless syntopic and synchronic (they take place at the same place and time); (4) geological and biological evolution will thus be historically correlated; but (5) these historical correlations *need not be simple.*

In historical biogeographical studies, falsifications of the null hypothesis are interpreted as evidence for modes of speciation other than vicariance (peripheral isolates speciation producing unique biogeographic elements, sympatric speciation producing lineage duplications), for postspeciation dispersal, for failure to speciate in response to a vicariance event, and for extinction.[9] Any of these may also occur in the context of areas having reticulate histories of evolution with respect to the species being studied. This is the reason that logical parsimony plays such an important role in a posteriori methods; it requires that you postulate the minimum number of ad hoc falsifications of your null hypothesis necessary to explain all the data if you are to discover the maximum amount about the world around us.[10] Some authors have suggested that a priori methods are more parsimonious than a posteriori methods (e.g., Crisci et al. 1991) because they are based on an assumption of simplicity. This places practitioners of a priori methods in an odd position, because if the world is indeed a complex place, then they must invoke more and more ad hoc assumptions in order to maintain their assertion of simplicity (see, e.g., Platnick et al. 1996). Explanatory parsimony, on the other hand, allows us to identify exceptions to our hypothesis of one general vicariance pattern, then get on with the job of uncovering additional data to explain those exceptions.[11] As Marshall and Liebherr put it,

> Various methods of creating general biogeographic patterns exist, differing in how, or if they incorporate incongruent pieces of information. The analysis adopted here [BPA and CCA] treats each piece of evidence (congruent or incongruent) as equal and searches for a pattern that is maximally congruent (most parsimonious) with all the data. Consensus techniques also yield general patterns . . . but do not truly incorporate incongruent information, since consensus techniques treat incongruence as ambiguity (e.g., unresolved nodes). (2000:216; see also Carpenter 1988a; Barrett et al. 1991)

In addition to the examples we have discussed in this chapter, there are a number of other studies that have used some form of BPA or CCA

---

9. See also Cracraft 1988, 1989a, 1994; Brooks 1990; Axelius 1991; Polhemus 1996; Hovenkamp 1997; Ronquist 1997; and Van Soest and Hajdu 1997.

10. For example, Wiley 1975; Platnick and Gaffney 1977, 1978a,b; Gaffney 1979; Sober 1988b; and Kluge 1997, 1998b.

11. For an excellent series of detailed discussions about this distinction, see Van Veller et al. 1999, 2000, 2001a,b.

in historical biogeography.[12] Like the studies we have discussed in detail, these analyses tell us that there is substantial historical structure, but also substantial complexity, in the history of geographic assemblage of biotas. Phylogeographic studies have also demonstrated that a high degree of historical biogeographical ordering may be evident at the scale of both microspecies and macrospecies.[13] Many ecologists, however, "remain skeptical about the extent to which the reconstructed history of lineage can be used to reconstruct history of place" (Brown 1995:190).

The phylogenetic studies discussed in this chapter provide the evidence needed to address Brown's skepticism in two ways. First, the data suggest that there is almost always an old vicariant signature in biotas, suggesting that Nelson was correct in asserting that many biotic components are older than we have thought. Thus, the history of lineages, especially of multiple lineages, *can* tell us about the history of place. This does not mean, however, that the history of any single lineage necessarily tells us about a *unique* history of place. That is, most biotas are a mosaic of dispersal, extinction, and vicariance, so the geographic areas in which they are localized often have reticulated histories with respect to the species that live in them. Three major phenomena create mosaic areas, or reticulated biogeographic histories: (1) successive vicariance events (e.g., the Andean biota); (2) successive invasions of the same areas, generally resulting from rises and falls in sea levels or the evolution of island arcs or coral reef systems (e.g. Indo-Malaysian areas, Hawaii, Indo-Pacific coral reefs, the Galapagos Islands) or both; and (3) expansion of one or more ancient biotas into the same region (e.g., the Antilles). Secondary BPA suggests that less than half, often fewer than one-third, of the areas considered are responsible for the reticulated histories, but as we add more species to our studies, it may turn out that the majority of areas are historical mosaics. The greatest diversity occurs clumped in areas that have had both an extremely complex geological history, producing many endemic species *in situ,* as well as repeated dispersal events mediated by one or both of these processes. Brown (1995)

---

12. For example, Indo-Malaysian and Wallacean distributions of plants (Baas et al. 1990; van Welzen 1990; Ridder-Numan 1995, 1996; Turner 1995, 1996) and shrews in the genus *Crocidura* (Ruedi 1996); the herpetofauna of Mexico (Flores-Villela 1991); southern South American plants (Crisci et al. 1991); Australasian corals (Pandolfi 1992); plants and animals of the Hawaiian Islands (Wagner and Funk 1995); marine nemerteans (Harlin 1996); the biota of the Caribbean (Liebherr 1988; Crother and Guyer 1996); avian clades in northwestern South America (Brumfield and Capparella 1996); Rhombognathine mites worldwide (Abe 1998); the terrestrial biota of the western Mediterranean (De Jong 1998); and 33 clades of insects, fish, and reptiles for Mesoamerica (Marshall and Liebherr 2000).

13. For example, Avise et al. 1987; Avise 1989a, 1992, 2000; Zink 1996; Johns and Avise 1998; Weider et al. 1999; Riddle et al. 2000; and Zink et al. 2000.

was also concerned that marked historical structuring might be a feature only of particular kinds of assemblages, such as communities of parasites associated with particular hosts. This is clearly not the case—only one of the studies we have discussed in detail or listed above is a study of parasites. We do note, however, that studies using parasites have found patterns similar to those documented by studies of free-living organisms (see Brooks and McLennan 1993b).

The biogeographic data collected to date refute a priori models based either on extreme amounts of vicariance (all of nature is reduced to a single specifiable area cladogram in which each area appears once) or of dispersal (all of nature is reduced to a simple counting of the number of species and their relative abundance per unit area). In its place, we have historical biogeography, based upon a posteriori primary and secondary BPA incorporating elements of both the vicariance biogeography and the ecological biogeography research programs. Such a fusion of perspectives is a necessary prerequisite to understanding the history of biotic evolution, because that history has involved complex patterns of speciation, extinction, environmental change, and geological evolution. We will thus return to historical biogeography in chapter 7, when we consider the evolution of communities.

The Frontiers of Discovery in Historical Biogeography
LONG AGO AND FAR AWAY: CYCLES OF BIOTIC EXPANSION AND VICARIANCE IN THE FOSSIL RECORD

If vicariance has played a major role in the diversification of life on this planet, there must have been episodes of biotic expansion between episodes of habitat fragmentation leading to vicariant speciation. If the complexities caused by the interaction between dispersal and vicariance aren't enough for you, consider the fact that dispersal has two components. The first, and most obvious, component involves the active movement of organisms. The second, and more subtle, component involves why the organisms are dispersing. Are they simply continuing a process of range expansion or has something changed geologically that permits them to move into previously inaccessible habitat? This latter process, called geodispersal (the fusing of areas due to drops in sea level, tectonic fusion, or both) is the analogue of area fragmentation via a vicariance event. Given that cycles of vicariance and geodispersal tend to occur on large geographic and time scales, can we ever hope to disentangle their various effects on the production of biodiversity? This question is being addressed by Bruce Lieberman, a young paleontologist studying the evolution of trilobites (one of the most charismatic deceased faunal groups). Lieberman's method begins with optimizing geographic distributions on existing phylogenetic trees for different clades. (We explain phylogenetic character optimization in the next chapter.) Two matrices are then created based upon the optimizations. One matrix (vicariance), comprising all taxa and nodes for which the area optimization indicates range reduction, provides information about the general history of speciation for the groups. The other matrix (geodispersal), comprising all those taxa and nodes for which the area optimization indicates range expansion, provides information about the history of biotic expansion. Finally, area cladograms are reconstructed using the vicariance and the geodispersal matrices. The points at which the two area cladograms agree are interpreted as cases in which geodispersal set the stage directly for subsequent vicariance. Lieberman and Eldredge (1996; Lieberman 1997, 2000) have taken advantage of the temporally and geographically dense fossil record and excellent phylogenetic resolution for many trilobite groups (e.g., Lieberman 1993, 1994, 1997; Lieberman and Kloc 1997; Lieberman et al. 1997) to assess the efficacy of their method. In the Middle Devonian, for example, there is a strong correlation between both kinds of area cladograms for a variety of trilobite groups (Lieberman and Eldredge 1996), suggesting a close cycling of biotic expansion and vicariance mediated by tectonic fusion and splitting. The story appears to be different during the "Cambrian Explosion," a period characterized by the first appearance of almost all modern animal groups (Lieberman 1997, 2000). Here the vicariance area cladogram is well resolved but the geodispersal area cladogram is not, suggesting a period dominated by habitat fragmentation with little corresponding habitat fusion. Current geological evidence supports this interpretation, indicating that the Cambrian was a time of active continental fragmentation.

THE COMPLEMENTARY ROLES OF DISCOVERY-BASED AND EVENT-BASED METHODS

Event-based methods (e.g., Bremer 1992; Ronquist 1994, 1995, 1996, 1997a,b, 1998) allow the researcher to specify certain "costs" (weights) for various possible phenomena (e.g., vicariance, dispersal, extinction), then search for the solution that provides the minimum overall cost for a data set. In essence, evaluation is based on the most parsimonious solution given the preset constraints (weights or costs). In chapter 2, we referred to a study by Wenzel and Siddall (1999) suggesting that alternative maximum-likelihood models could be evaluated using the results of phylogenetic systematic analysis. Here we suggest that in an analogous way, phylogenetic systematic analysis (BPA) can be used to evaluate the results of alternative event-based models with the basic discovery information in historical biogeography.

Still More Birds: North American Passerines. Zink et al. (2000) recently discussed the historical biogeography of warblers (*Vermivora ruficapilla* group), scaled quail (*Callipepla squamata* group), brown towhees (*Pipilo fuscus* group), black-tailed gnatcatchers (*Polioptila melanura* group), and thrashers (*Toxostoma curvirostre* and *T. lecontei* groups) living in seven areas of the southwestern United States, Mexico, and northern Central America. Using an event-based method called DIVA (**dispersal-vicariance analysis:** Ronquist 1996, 1997a,b), Zink et al. uncovered as many as 14 and as few as 6 dispersal events, depending upon the value assigned to the maximum number of ancestral source areas (two [favoring dispersal] versus unconstrained [favoring vicariance]). Based upon this, they concluded that (1) although vicariance was the dominant speciation mode (approximately 75 percent), a simple vicariance scenario could not explain all of the speciation in three, and possibly four of the clades; (2) DIVA could not unambiguously identify the source areas for the ancestral species of each clade; and (3) the data provided strong support for a vicariance event linking only two of the seven areas, California-Baja/California and the Sonoran Desert (A + B). Even then, comparing genetic distance data for the pairs of sister species occurring in those two areas showed unequal rates of divergence, possibly indicating noncontemporaneous speciation events. How do those results compare with BPA of the same data?

Primary BPA for the six clades (fig. 4.52a–f; table 4.25) produces two equally parsimonious area cladograms with a CI of 88% (fig. 4.53). Note that the two area cladograms differ only in their placement of area E (the high plateau of southern Mexico). When areas F (northern Central America) and G (eastern United States) are removed (as in the DIVA analysis), these area cladograms are congruent with the two equally parsimonious reduced trees from the DIVA analysis. The primary BPA provides strong support for a vicariance relationship between areas A (Baja California/California) and B (the Sonoran Desert), moderate support for a vicariance relationship between areas (A + B) and (C + D), and weak support for a vicariance relationship between areas C (the Chihuahuan Desert) and D (the Sinaloan shrubland of western Mexico). These three episodes account for 19 (68 percent) of the 28 contemporaneous species in the six clades (*Vermivora* "ruficapilla west" [1], *V. luciae* [2], *V. virginiae* [3], *Callipepla californica* [10], *C. gambelii* [11], *C. squamata* [12], *C. douglasii* [13], *Pipilo crissalis* [17], *P. aberti* [18], *P. fuscus* 1 [19], *Polioptila calif-*

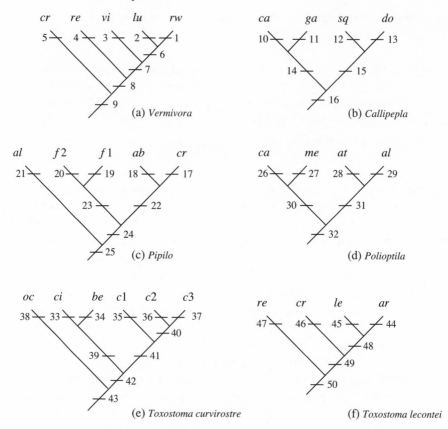

Figure 4.52. Phylogenetic trees for members of six avian clades living in the southwestern United States, Mexico, and northern Central America, numbered for BPA. (*a*) *Vermivora ruficapilla* group of warblers: *cr = crissalis; lu = luciae; re = "ruficapilla east"; rw = "ruficapilla west"; vi = virginiae.* (*b*) *Callipepla squamata* group of scaled quail: *ca = californica; do = douglasii; ga = gambelii; sq = squamata.* (*c*) *Pipilo fuscus* brown towhees: *ab = aberti; al = albicollis; cr = crissalis; f1 = fuscus 1; f2 = fuscus 2.* (*d*) *Polioptila melanura* group of black-tailed gnatcatchers: *al = albiloris; at = atriceps; ca = californica; me = melanura.* (*e*) *Toxostoma curvirostre* group of thrashers: *be = bendirei; ci = cinereum; c1 = curvirostre 1; c2 = curvirostre 2; c3 = curvirostre 3; oc = ocellatum.* (*f*) *Toxostoma lecontei* group of thrashers: *ar = arenicola; cr = crissale; le = lecontei; re = redivivum.*

ornica [26], *P. melanura* [27], *P. atriceps* [28], *Toxostoma cinereum* [33], *T. bendirei* [34], *T. curvirostre 3* [37], *T. arenicola* [44], *T. lecontei* [45], and *T. crissale* [46]) and 14 (64 percent) of the 22 speciation nodes on the six phylogenetic trees.

If we examine the inconsistencies in the primary BPA area cladograms (bold numbers on fig. 4.53), we find that they involve areas A (one or two cases), B (three or four cases), C (three or four cases), and E (one or two cases). When these areas are duplicated for secondary BPA, we dis-

Table 4.25. Primary Matrix Listing the Geographic Distribution of Six Clades of North/Central American Birds, along with the Binary Codes Representing the Phylogenetic Relationships for All Six Clades

Area	Binary Code				
A	1000011111	0001011000	0101110001	0110000010	0111001111
B	0100011110	1001010101	0111101001	0101100010	1110110111
C	0010101110	0100110010	0011101001	0100001001	1110010011
D	?????????0	001011????	?????00100	1100001001	111???????
E	??????????	??????0000	10001?????	??00010101	111???????
F	??????????	??????????	?????00010	11????????	??????????
G	000100011?	??????????	??????????	??????????	??????????

NOTE: A = Baja California + California; B = Sonoran Desert; C = Chihuahuan Desert; D = Sinaloan scrublands; E = southern Mexican plateau; F = northern Central America; G = eastern United States. ? = species missing from the area.

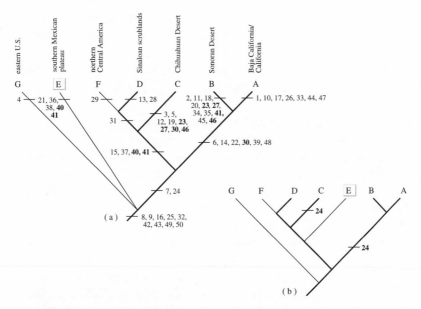

Figure 4.53. Two equally parsimonious area cladograms produced from primary BPA of six North American bird clades. *Bold lines* = points of congruence between the two area cladograms; A = California + Baja California; B = Sonoran Desert; C = Chihuahua Desert; D = Sinaloan shrublands; E = highlands of southern Mexico; F = northern Central America; G = eastern United States; *numbers accompanying slash marks* = species codes (from fig. 4.52); *bold* = homoplasy.

Table 4.26.   Secondary Matrix Listing the Geographic Distribution of Six Clades of North/Central American Birds, along with the Binary Codes Representing the Phylogenetic Relationships for All Six Clades

Area	Binary Code				
A	1000011111	0001011000	0101110001	0110000010	0111000111
A$_1$	??????????	??????????	??????????	??????????	???0001001
B	0100011110	1001010100	0101101001	0101000010	0110100111
B$_1$	??????????	??????????	??????????	??00100000	111???????
B$_2$	??????????	??????0001	00111?????	??????????	???0010011
C	0010001110	0100110010	00111?????	??00001001	1110010011
C$_1$	000010001?	??????????	??????????	??????????	??????????
C$_2$	??????????	??????????	?????01001	01???????	??????????
D	????????0	001011????	????100100	1100001001	111???????
E	??????????	??????0000	10001?????	??00000100	001???????
E$_1$	??????????	??????????	??????????	??00010001	111???????
F	??????????	??????????	?????00010	11???????	??????????
G	000100011?	??????????	??????????	??????????	??????????

NOTE: A = Baja California + California; B = Sonoran Desert; C = Chihuahuan Desert; D = Sinaloan scrublands; E = southern Mexican plateau; F = northern Central America; G = eastern United States. ? = species missing from the area.

cover that a single additional replicate of areas A and E, plus two additional replicates of areas B and C, are required to fully explain the inconsistency. Analysis of the data in table 4.26 produces 80 equally parsimonious area cladograms with a CI of 100%. These cladograms, in turn, collapse to a single area cladogram (fig. 4.54) when we eliminate "soft polytomies," which are an artifact of guesses about missing data made by the computer program. The unresolved polytomy at the base of the single area cladogram accounts for 5 of the 28 contemporaneous species: *Toxostoma redivivum* (47), *Vermivora crissalis* (5), *Pipilo albicollis* (21), *Toxostoma ocellatum* (38), and *Vermivora* "*ruficapilla* east" (4). Neither BPA nor DIVA can resolve this polytomy because the basal members of the clades live in so many different areas. We may be seeing evidence that the ancestral species of the various clades which participated in the three vicariance events discussed above dispersed into the relevant ancestral area from different source areas where, perhaps, they were elements of yet other vicariance events. Alternatively we may be seeing the remnants of an explosive burst of speciation prior to the three episodes of vicariance that make up the backbone of the upper portion of the area cladogram. Resolving this issue would require adding more basal species to the phylogenetic trees for the six clades, thereby expanding the spatial and temporal scope of the study. This means that, like DIVA,

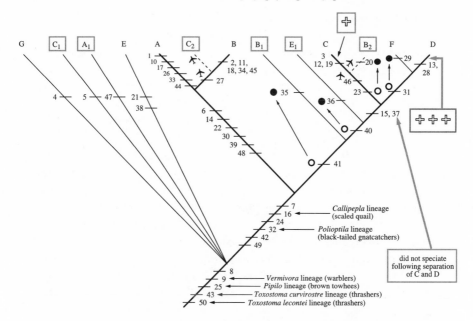

Figure 4.54. Secondary area cladogram showing the big speciation picture for the North American birds. *Bold lines* = the backbone of vicariant speciation; *airplanes* = postspeciation dispersal into the areas indicated with dotted lines; *open circle* (ancestor) + *solid circle* (peripheral isolate) = instance of peripheral isolates speciation; *crosses* = extinctions; *numbers accompanying slash marks* = species codes (from fig. 4.52); A = California + Baja California; B = Sonoran Desert; C = Chihuahua Desert; D = Sinaloan shrublands; E = highlands of southern Mexico; F = northern Central America; G = eastern United States.

BPA cannot determine how many ancestral, or source, areas were involved in producing the vicariant associations shown by these six clades.

The secondary analysis produces a similar result to the primary BPA and DIVA analyses: strong support for a vicariance relationship between areas A and B, moderate support for a vicariance relationship between areas (A + B) and (C + D), and weak support for a vicariance relationship between areas C and D. DIVA does not support the relationship between areas C (Chihuahuan Desert) and D (the Sinaloan shrubland of western Mexico) strongly because there are large amounts of nonvicariant complexity in the evolutionary histories of species living in the those areas. DIVA combines that complexity into one "dispersal" category. Secondary BPA, however, allows us to examine the details of the interaction between dispersal and speciation in more detail. For example, secondary BPA suggests that four species in the C + D portion of the area cladogram constitute unique biogeographic elements indicating episodes of peripheral isolates speciation (*Toxostoma curvirostre* 1 [35 in area B₁], *T.*

curvirostre 2 [36 in E$_1$], *Polioptila albiloris* [29 in F], and *Pipilo fuscus* 2 [20 in B$_2$]). *Polioptila melanura* (27) and *Toxostoma crissale* (46) represent instances of *postspeciation dispersal* from area B to area C$_2$ and from area C to area B$_2$, respectively. *Toxostoma curvirostre* 3 (37) represents a *nonresponse to vicariance,* occupying both areas C and D. In other words, even though these species appeared as "dispersal" in DIVA, they clearly are not all the products of the same type of event, nor do they produce the same kind of historical biogeographical signal.

Our analysis indicates that there are six endemic sister-species pairs (one representative from each clade) in areas A and B, all produced via a vicariant speciation event. Interestingly, when Zink et al. used genetic distance data to determine the approximate age of the speciation events in the birds, they found that not all of the six pairs exhibited the same degree of divergence (which would be the expected result if all speciation events were the product of one vicariance episode). This paradoxical result highlights two important points about the assumptions we frequently make in historical biogeographical analyses. First, as we discussed in chapter 3, although we tend to associate the term "vicariance" with large-scale geological perturbations, a number of factors associated with "environmental harshness" can also subdivide a species geographically. In this case, the expansion and contraction of glacial refugia and cycles of desertification during the Pleistocene may have resulted in episodes of vicariant speciation in the same area at different times. Such episodes will appear as one general pattern in a BPA even though some of the sister pairs are different ages. The take-home message from this problem is simple—the study does not end with the area cladogram. We need additional data in order to interpret the results of the analysis in the most rigorous way possible. The use of genetic distance data to estimate time of speciation is one potential source of additional data. This, however, leads us to the second assumption: the molecular clock is constant and, once calibrated, can be used to estimate the timing of speciation across many lineages.

Examination of the genetic distance analysis of speciation ages depicted by Zink et al. (their figure 3) indicates that three of the sister-species pairs in areas A and B appear to be approximately the same age (*Vermivora ruficapilla* west [1] + *V. luciae* [2]; *Callipepla californica* [10] + *C. gambelii* [11]; and *Toxostoma cinereum* [33] + *T. bendirei* [34]), supporting the hypothesis that they are the product of the same vicariance event. The other three pairs appear to be older (in order of origin from oldest to youngest: *Polioptila californica* [26] + *P. melanura* [27]; *Toxostoma arenicola* [44] + *T. lecontei* [45]; and *Pipilo crissalis* [17] + *P. aberti* [18]). It is interesting to note that all of these older pairs have other influences acting on them besides simple fragmentation via a vicariance event. In two of the pairs, at least one member is sympatric with a congener: *Pipilo aberti* (18: vicariant endemic) and *P. fuscus* 2 (20: peripheral isolate endemic) in area B; *Toxostoma arenicola* (44: vicariant endemic) and *T. redivivum* (47: vicariant or peripheral isolates endemic) in area A; and *Toxostoma lecontei* (45: endemic) and *T. crissale* (46: postspeciation dispersal from area C) in area B. The third pair has one member, *Polioptila melanura* (27), in area B that has undergone postspeciation range expansion into area C.

These data indicate a possible alternative explanation for the differences in age noted by Zink et al. Perhaps range expansion and reinforcement (overlapping of congeners) influence divergence rates and thus distort age estimates based upon the assumption of a constant clock. This alternative explanation, of course, requires that we collect data about, for example, the genetic structure of widely dispersed versus nondispersed sister groups, interactions between sympatric congeners, and differences in the genetic structure of a species sympatric with a close relative and its isolated sister species. Even if further research does detect an influence for such factors on divergence rates, the combined results of this and Zink et al.'s studies indicate that this influence is not a general "rule." One of the sister pairs within the group of three that are roughly the same age (*Toxostoma cinereum* [33] + *T. bendirei* [34]) also shows the potential for reinforcement (the vicariant endemic *T. bendirei* is sympatric with the peripheral isolates endemic *T. curvirostre* 1 [35] in area B). In this case, reinforcement appears not to have influenced the estimated age of divergence *if all the vicariant speciation in area A + B is the result of one vicariance event*. Evidently the effect, if any, of factors like secondary contact between close relatives and range expansion on rates of divergence may be historically contingent, differing among clades or perhaps even among speciation events within clades. Clearly then, the interpretation of any historical biogeographical study must be a cautious one, based upon integration of data from as many clades as possible (to disentangle general from special elements), molecular and geological evidence (to produce estimates of age), and experimental and field data (to uncover any additional factors highlighted as unique elements in the BPA analysis that might be influencing those age estimates).

Because secondary BPA can differentiate among nonvicariant factors, it provides crucial data for researchers interested in studying speciation mechanisms. Researchers wanting to study the effect of reinforcement on the completion of speciation might focus their attention on the coexistence of the three pairs of congeners in the Sonoran Desert (area B). Each of these pairs, as discussed above, comprises a species produced by

vicariance and a colonizer; in all three cases, the colonizer came originally from the environmentally similar and geographically adjacent
Chihuahuan Desert (area C): (1) *Pipilo aberti* (18: vicariant endemic) and
Pipilo fuscus 2 (20); (2) *Toxostoma bendirei* (34: vicariant endemic) and *T.
curvirostre* 1 (35); and (3) *Toxostoma lecontei* (45: vicariant endemic) and
T. crissale (46). For pairs 1 and 2, the colonizing species is a product of
peripheral isolates speciation and is endemic to area B; for pair 3, however, the colonizing species is the result of postspeciation dispersal, having been produced by vicariance in area C. None of these pairs of species
are sister species; in fact, there are no cases of sympatric sister species in
this data set. Nonetheless, some of these co-occurring species may be
closely related enough to make a mating mistake. Have any of the interactions between the colonizer and the resident species for these three
pairs been responsible for the completion of speciation for either the
vicariant species or the colonizer? If reinforcement has played a role in
completing speciation, then we would predict that the overlapping congeneric pairs will show both post- and premating isolating mechanisms.
Do they?

Finally, *Polioptila melanura* (27) colonized the Chihuahuan Desert (C)
from its vicariant origin in the Sonoran Desert (B). This was a potentially
significant event in the history of the *Polioptila* clade because there is no
endemic member of that clade in area C even though a strictly vicariant
scenario would lead us to expect one (fig. 4.54). The BPA suggests that
this absence is a result of secondary extinction. Did that extinction permit *P. melanura* to colonize the Chihuahuan Desert, or did *P. melanura*
cause the extinction when it appeared (the ghost of competition past)?
It might be possible to begin addressing this question by comparing the
ecological requirements of the colonizer with those of *P. atriceps* (the
species living in area D), the putative sister species to the extinct gnatcatcher. If those requirements overlap enough to cause competition between the two species, then this provides some support for the "competition causing extinction" hypothesis. If, on the other hand, these two
species do not show broadly overlapping ecological requirements, then
this provides some support for the "colonized the area after the extinction" hypothesis.

The results of comparative phylogenetic studies, like those in other
scientific research programs, are only as robust as the underlying database. In studies of historical biogeography, that database comes in the
form of phylogenetic trees, just as the database for population genetics
might come in the form of experimental manipulations of *Drosophila*
wing shape. Zink et al. indicated that some of the trees used in their
initial study were not well supported, which might (or might not)

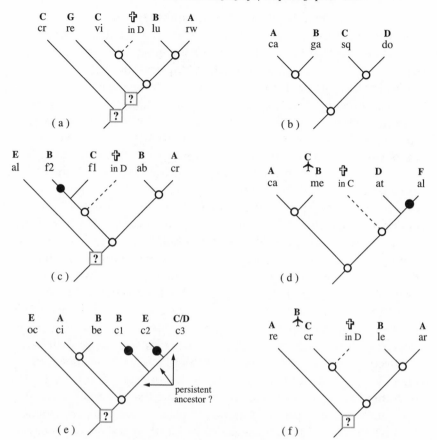

Figure 4.55. History of speciation for members of six avian clades inferred from secondary BPA. *Open circle* = vicariance; *solid circle* = peripheral isolates speciation; *dotted line* + *cross* = vicariance plus extinction of one sister species; *airplane* = postspeciation dispersal from the area to the right of the airplane in to the area above the airplane; *question mark* = speciation mode ambiguous at this node; A = California + Baja California; B = Sonoran Desert; C = Chihuahua Desert; D = Sinaloan shrublands; E = highlands of southern Mexico; F = northern Central America; G = eastern United States. For identities of clades and their included species, see figure 4.52.

change some of the hypothesized elements in this study. Bearing this caveat in mind, we mapped all of the hypothesized vicariance and "dispersal" events revealed by the secondary BPA onto the phylogenetic trees for the six avian clades (fig. 4.55). There is a total of 26 nodes on these trees, 5 of which are ambiguous basal nodes requiring additional information in order to resolve the associated speciation mode. Of the remaining 21 nodes, 13 (62 percent) indicate the effects of vicariance, 4 are hypothesized to represent vicariance plus extinction (19 percent), and 4 represent peripheral isolates speciation (19 percent). There has been a total of six dispersal events, four of which resulted in the isolation and subsequent speciation of the dispersing group, the remaining two representing episodes of postspeciation dispersal increasing a species' range. For this study at least then, DIVA and BPA produce similar results in terms of highlighting the vicariance backbone. The BPA thus corroborates the a priori weighting scheme (costs) used by DIVA to identify vicariance. At the same time, it appears that DIVA is not sensitive to the various outcomes of dispersal. Secondary BPA, on the other hand, allowed us to delve into the black box of "dispersal" and identify instances of peripheral isolates speciation, postspeciation dispersal, nonresponse to a vicariance event, secondary contact between congeners as a result of dispersal (and the potential for reinforcement completing speciation in these cases), and potential extinction resulting from competitive exclusion following dispersal. Together these two approaches corroborate the growing appreciation for the complex nature of the interaction between vicariance and dispersal in the production of biodiversity.

In historical biogeography, then, you have two choices based upon the structure of the world in which you are most comfortable. If you are interested in studying the effects of geology on speciation, then both discovery-based and event-based (evaluation) methods will allow you to identify the vicariance backbone for any set of clades and trees. If, on the other hand, you are interested in studying the exceptions to vicariance, such as the impact of peripheral isolates or nonallopatric modes of speciation, or of postspeciation dispersal, then the discovery-based method we have discussed in this chapter represents the best place to begin your search. Historical biogeography involves a lifelong voyage of discovery. If you begin that voyage with a ticket to a specific destination, you will miss many interesting places during the journey. Methods based in phylogenetic systematic principles do not rule out any evolutionary mechanism beforehand, so all destinations are possible.

Chapter 5

Functions:
Exploring Options

Adaptation is something that can no longer be safely assumed by evolutionary or other biologists. Indeed, the
more one examines the concept, the more it comes to resemble a newly landed fish: slippery, slimy, obstreperous, but
glittering with potential. There it is, flapping about, full
of energy, but the significance of all that commotion is not
clear. Perhaps the solution of some evolutionary biologists
is best—just throw the damn thing back in the water. But
of course our authors have not chosen that course of action,
and we are left with the problem of what to do with the
fish.

Rose and Lauder 1996:9

People have always noticed two things about the world. The
first and most obvious observation is that we are surrounded by an abundance of biological diversity. Explanations for this phenomenon are generally sought by attempting to uncover the origins of species. The second perception is slightly more subtle: organisms appear to have what
they need to survive. In order to explain this, we must drop our level of
investigation from species to their characters, from questions about the
origin of species to questions about the origin of adaptations. The concept of adaptation is an old one, spanning several millennia and a variety of cultures. In the following discussion we shall touch only briefly
on a few of the key players in the adaptation drama. It is not our purpose
to review the history of the concept (for this, see the excellent paper by
Amundson 1996), only to demonstrate that the concept has taken many
different guises during its evolutionary history.

A VERY QUICK HISTORY

For almost 2,000 years, explanations of biological origins were founded
upon the concept of stasis. To most medieval European naturalists, adaptations were evidence of God's design, the perfection of God's mind re-

flected in the perfection of each organism's fit to its environment. William Paley formalized this perspective in 1802 in a book called *Natural Theology*. There Paley laid out his "argument from design," which goes something like this: suppose you are out walking one day, and you find a watch lying on the ground. With great interest (and no avarice in your heart) you pick up the watch. After some minutes of detailed examination, you conclude that this novel item is an intricate and complex thing. It is inconceivable that such a marvel could have appeared by a random conjunction of various metals, so you conclude that the existence of the watch implies the existence of a watchmaker. So far, so good. Now substitute "white cat" for "watch" and work your way through the sequence, and you will have the bare bones of Paley's argument: just as the intricate watch requires a designer, so the complex cat requires a creator (but see Schrödinger for debates about the existence of the cat).

The theory of divine creation was coupled with the idea of organic stasis: in effect, a creator would not make mistakes, so the diversity we see around us today, a reflection of the creator's perfection, is the way things have always been. This idea was seriously challenged by the discovery of the fossilized remains of creatures that clearly no longer existed. Debates raged around the question of whether such fossils were inorganic deposits or the remains of once-living creatures. Naturalists taking the latter position placed themselves in a logical paradox: if the fossils were, as they maintained, once life forms, then there had to be living representatives of those forms still on the earth if the theory of stasis was to be upheld. In other words, fossils had to represent dead individuals, not dead species (Ruse 1979).

Evolutionary Biology and Adaptation: The First Generation

Into the fray came a French systematist, Jean Baptiste de Lamarck. During the early stages of his career he agreed with the prevailing theory of stasis, a position that changed following his appointment as a professor at the Natural History Museum in Paris. That appointment granted Lamarck unlimited access to the museum's extensive invertebrate collection, where he spent years describing the number of ways in which organisms could defend themselves from a hostile world. Given such protection, he argued, how could an entire species possibly go extinct? From this starting point, Lamarck developed his theory of evolution, which he published in his extensive book, *Philosophie Zoologique* (1809). In a nutshell, his argument went something like this:

Many fossils represent previously living forms that are no longer present on the earth. Given this, we must reject the theory of stasis.

+

Extant species have many characteristics that allow them to survive in a harsh world. Given this we must reject the theory of extinction.

Question: If species do not go extinct, what has happened to the forms represented by fossils?

Answer: There are no living representatives of many fossil species *because those species evolved into something else.*

The proposition that species are not immutable was the first big intuitive leap in the evolutionary drama.

Having argued for the existence of biological evolution, Lamarck provided a detailed mechanism by which species could gradually change from one form to another. This mechanism rested upon his belief that there was a connection between an organism and its environment, a connection so intimate that a change in the environment could *drive* a change in the organism (fig. 5.1). Lamarck saw the history of biological forms on this planet as evidence for the triumph of adaptation. He thought this mechanism so complete that nothing could ever go extinct because of a failure to adapt. Although his mechanism was eventually falsified, his belief that adaptation is the outcome of an interaction between an organism and its environment remains with us today.

Lamarck formulated his ideas of the evolutionary mechanism underlying adaptation in the midst of yet another heated debate, as biologists fragmented along form versus function (Russell 1916), teleological versus antiteleological (Ospovat 1978, 1981), and structuralist versus adaptationist lines (Amundson 1996). Structuralists advocated the supremacy

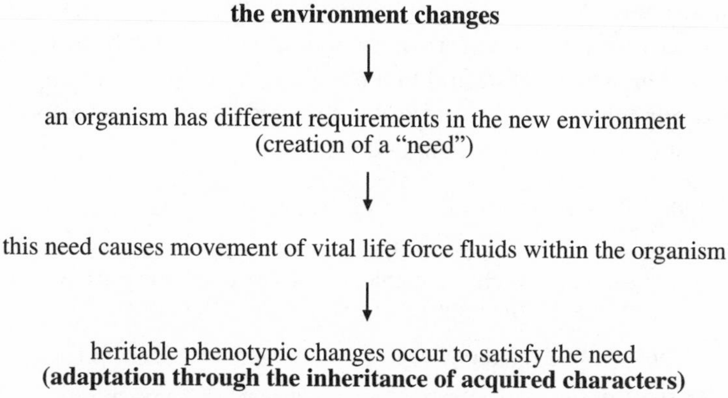

the environment changes

↓

an organism has different requirements in the new environment
(creation of a "need")

↓

this need causes movement of vital life force fluids within the organism

↓

heritable phenotypic changes occur to satisfy the need
(adaptation through the inheritance of acquired characters)

Figure 5.1.   Lamarckian evolution in a nutshell.

of form. These researchers were the keepers of information about homology. They argued that, because similar structures could be identified in different organisms irrespective of the environment and the function of those structures, the homologous structures existed before the origin of their divergent functions. Adaptationists were advocates of the supremacy of function. They argued that the need for internal functional integrity or for an external functional fit between the organism and the environment was so strong, it was probable that an adaptive reason existed for every structure. In their worldview, functional needs shaped form; similarity of form simply reflected similarity of adaptation, which was not nearly as interesting as divergence of form and function. And thus was born the fractious debate about pattern versus process, a debate stilled only briefly by the inspired synthesis presented in the *Origin of Species* (Darwin 1859).

Evolutionary Biology and Adaptation: The Second Generation

Darwin's England was a very different world from Lamarck's France. By the early part of the nineteenth century, English society was in a state of violent turmoil created, in part, by the imposition of derived technologies (the industrial revolution) upon an ancestral social system, coupled with a rapidly exploding urban population. In 1826, the Reverend T. R. Malthus published *An Essay on the Principle of Population*. In this essay, Malthus argued that because human population increases at a geometric rate and crops increase at an arithmetic rate, we will never be able to feed everyone (scientific "fact"). Charity, therefore, is not only a waste of time but also dangerous because, by fostering overpopulation, it increased competition for limited food supplies. This argument eventually led to the repeal of the Poor Laws (in the 1830s) and the construction of the massive workhouses that so impressed writers like Charles Dickens.

Into this turmoil came Darwin, who combined ideas from Lamarck, Malthus, the structuralists and functionalists, and debates among geologists about the age and evolution of the earth, into a new astounding theory of evolution. Darwin argued thus:

THE RULE OF GEOMETRIC INCREASE

There is no exception to the rule that every organic being naturally increases at so high a rate, that, if not destroyed, the earth would soon be covered by the progeny of a single pair. (Darwin 1872:77)

Observation: This obviously has not happened. Why not?

Hypothesis: Not all offspring are surviving. Some are being eliminated.

THE STRUGGLE FOR EXISTENCE

> In looking at Nature it is most necessary to keep the foregoing considerations always in mind—never to forget that every single organic being may be said to be striving to the utmost to increase in numbers; that each lives by a struggle at some period of its life. . . . Lighten any check, mitigate the destruction ever so little, and the number of the species will almost instantaneously increase to any amount. (Darwin 1872:78–79)

Question: What causes the struggle?

Answer: The struggle is caused by the interaction of two factors:

> **Environmental Uniformism:** Although the environment changes gradually over time, so organisms are faced with new challenges, the challenges are not so catastrophic that organisms cannot meet them.

> **The Law of Offspring Variability:** Offspring are not all the same.

Outcome: Some individuals in a population will be able to meet the problem posed by the new environment (to **adapt**) better than others.

Increase + Environment + Variation + Adaptation = **Natural Selection**

> If such [*variations*] do occur, can we doubt (remembering that many more individuals are born than can possibly survive) that individuals having any advantage, however slight, over others, would have the best chance of surviving and procreating their kind? On the other hand, we may feel sure that any variation in the least degree injurious would be rigidly destroyed. This preservation of favourable individual differences and variations, and the destruction of those which are injurious, I have called Natural Selection. (Darwin 1872:89)

DESCENT WITH MODIFICATION (EVOLUTION)

> The outcome of differential survival and reproduction is that the next generation is different in some, but not all, ways from its ancestors.

Darwin thus presented two theories in *Origin*. His first theory was that evolution was *descent* with modification, that species were descended from common ancestors. He used evidence from the structuralists' research to support this theory, which he depicted as a branching diagram or family tree. Darwin's second theory involved the delineation of a mechanism, natural selection, by which *modification* could occur. He drew upon evidence from the adaptationists' research program to support this theory. It is important to note that adaptation precedes natural selection in the Darwinian formulation. Adaptation to Darwin was the logical outcome of an interaction between a changing environment and a variable population. Those individuals that met the new environmental challenge better (had "some advantage, however slight, over others") were *better* adapted. These individuals had a better "chance of surviving and procreating their kind" than did their less well adapted brethren (natural selection). The whole evolutionary process thus revolved around two concepts, variation and conservatism: variability among individuals in their ability to meet environmental challenges over their lifetime leading to variability among individuals in their ability to survive and reproduce, locked within the constraining influences of genealogy, producing a new generation which is different in some, *but not all,* ways from its ancestors (descent with modification, or evolution). This was Darwin's great insight (one of many, actually): much of evolution (although by no means all) has resulted from the interplay between inheritance (descent) and natural selection (with modification). Pattern and process became part of a unified whole, each explaining one part of biological diversity and each weaker standing alone.

Reactions to Darwin's new theory were generally enthusiastic within the scientific community. Outside of that community, however, many people found Darwinism a harsh philosophy. For example, Samuel Butler wrote, "To state this doctrine is to arouse instinctive loathing; it is my fortunate task to maintain that such a nightmare of waste and death is as baseless as it is repulsive" (from *The Deadlock in Darwinism,* 1890). A much more passionate denunciation of the emerging evolutionary perspective came in the form of a poem written almost a decade before Darwin published *Origin,* by a contemporary of Darwin's at Cambridge, Alfred Lord Tennyson, in memory of Arthur Hallam, a young poet who had just committed suicide:

> Are God and Nature then at strife,
> That Nature lends such evil dreams?
> So careful of the type she seems,
> So careless of the single life;
> That I, considering everywhere
> Her secret meaning in her deeds,
> And finding that of fifty seeds
> She often brings but one to bear,
>
> . . . .
> . . . "So careful of the type?" but no.
> From scarped cliff and quarried stone
> She cries, "A thousand types are gone:
> I care for nothing, all shall go.
>
> . . . .
> . . . Man, her last work, who seem'd so fair,
> Such splendid purpose in his eyes,
> Who roll'd the psalm to wintry skies,
> Who built him fanes of fruitless prayer,
> Who trusted God was love indeed
> And love Creation's final law—

> Tho' Nature, red in tooth and claw
> With ravine, shriek'd against his creed—
> ("In Memoriam," 1850)

After the blatant optimism of Lamarckism, the Darwinist perspective that not everything could survive ("Nature, red in tooth and claw"), that many individuals and indeed species would go extinct because of a failure to adapt, looked bleak indeed. That perceived harshness was to be amplified many times over in the next evolutionary incarnation.

Evolutionary Biology and Adaptation: The Third Generation

During the period spanning the 1890s to the 1940s, Darwinism was challenged by a variety of alternative evolutionary theories, including a resurgence of Lamarckism (Bowler 1983). One of these theories was a new perspective called "orthogenesis." The orthogenetic movement developed as a response to the perception that Darwinism overemphasized the role of the environment in evolution. Many biologists believed that they could distinguish trends in development, morphology, and the fossil record that followed a linear pattern from simple to increasingly complex to progressively degenerate (either secondarily simplified and increasingly specialized or increasingly bizarre) to extinct. Orthogeneticists sought a mechanism that could explain this trend. One of the more widely accepted proposals (although by no means the only one) among scientists of the time was that evolution is driven by internal "rules of growth." Internal forces constantly drive variation along rigidly determined lines, so the environment has no (or little) influence. Adaptation is thus generally irrelevant in the face of the overwhelming and built-in drive toward extinction. This perspective represented the ultimate in pessimism, a far cry from the exuberant optimism of Lamarck.[1]

Scientific theories are not conceived by detached individuals. All scientists are influenced, to varying degrees, by the state of the world around them. It is not so surprising that Darwin's views would be influenced by images of overpopulation, starvation, and the outcomes of competition in a rapidly expanding industrial sector, while the orthogeneticists responded to a world seemingly spinning out of control through 50 years of crippling wars.

Overall, then, by the time the new synthesis emerged in the late 1930s and early 1940s, there were three competing major ideas about

---

1. For an excellent discussion of the history of this complex movement, see Bowler 1983.

just what constituted an "adaptation" and where adaptation fit into the greater panorama of evolution:

> **Neo-Lamarckism:** The interaction between the external environment and internal rules of development and growth leads to adaptive change. Adaptive change is the cause of all variation among organisms, and that variation is ordered according to internal rules in a progressive fashion. *Every day, in every way, we're getting better and better.*

> **Darwinism:** The interaction between the external environment and already existing variability among organisms might or might not lead to adaptation. Variation is random with respect to the environment. Adaptation and the changing environment do not cause each other nor does either cause variation. The environment is always changing in unpredictable ways; therefore, adaptation, when it occurs, does not show any long-term trends. Evolution is branching, not linear, and is contingent, not progressive, over time. *It's a dog-eat-dog world . . . Nature, red in tooth and claw . . . but there is hope.*

> Darwinism is summarized by the *Jagger principle:* You can't always get what you want, but if you try sometimes, you just might find you get what you need.

> **Orthogenesis:** The interaction between the external environment and the organism is generally (or always) immaterial. Variation is driven by internal rules of growth and development. Evolution follows a rigid, predetermined, linear sequence that inevitably works its way toward overspecialization, degeneration, and extinction. *All hope abandon, ye who enter here.*

Although Darwinism eventually prevailed as the dominant theory of evolution, what we call Neo-Darwinism today retains vestiges of all three approaches. For example, the widely held belief that parasites are "secondarily simplified, highly specialized and degenerate" is taken directly from orthogenesis (Brooks and McLennan 1993b), while statements like "the first land creatures developed lungs *so* they could breathe air" are Lamarckian (Acot 1997). Even terms like "optimality" and phrases like "select for" as opposed to "select against" embody the upbeat perspective of Lamarck. The complicated new vistas now being revealed by molecular geneticists have led some researchers to propose that a complete theory of evolution will eventually require both Darwinian and Lamarckian processes (discussed in Barker 1993).

Even in its "pure form," Darwinian evolution is rarely simple. Character evolution represents the outcome of a complex interaction among numerous factors, including mutation, genetic drift, selection, alterations in relative rate or timing of developmental events, additions and

deletions/arrests of developmental stages, and changes in the spatial patterning of ontogenetic processes.[2] If you adopt the position that all of
evolutionary diversity is structured predominantly by only one of these
forces, you are making an a priori judgment, invoking the ontology of
simplicity; that is, attempting to force a complex world into a simplistic
framework. It is entirely possible that evolution may be reducible to one
general process. It is also possible that explanations of character evolution will generally be clade specific, varying from group to group. Until
we understand substantially more about the intricacies of development;
the complexities of numerous selection vectors acting in different directions on one character or complex of characters correlated through pleiotropy or identity/linkage disequilibrium (see, e.g., discussion and references in Rose and Lauder 1996; Kelly 1999); and the patterns of response
to those vectors in the field, as opposed to in the laboratory or the virtual
world of the computer, it is better to err on the side of conservatism.
Our response to the question of whether "adaptation" is or is not a critical part of character evolution is thus, Embark on a voyage of discovery;
"go and study for yourself."

In the rest of this chapter, we will present one part of a much larger
comprehensive research program that can give us the tools, and we hope
as much objectivity as possible, to help us "get the information firsthand." In so doing, we hope to expand upon the longstanding tradition
of examining adaptation within a comparative framework, advocated
and illustrated by numerous researchers "ever since Darwin." Before we
begin, however, let us reemphasize and expand upon some general
methodological points from chapter 2.

CHARACTER OPTIMIZATION: HOW TO INTERPRET CHARACTERS ON A PHYLOGENETIC TREE

We must begin our explorations with two pieces of information, a phylogenetic hypothesis and the relevant functional data. The best-supported
(most parsimonious) sequences of evolutionary transformations for the
characters, be they binary or multistate, can be determined by reference
to the phylogenetic tree through a process called character optimization
(not to be confused with optimality theory). The following discussion

---

2. For example, Simpson 1953; Gould 1977; Lauder 1980, 1982a,b, 1983a,b, 1985a,b,
1988, 1990, 1991a,b; Cracraft 1981a; Wake 1982; Maynard Smith et al. 1985; McNamara
1986; Schaefer and Lauder 1986, 1996; Lauder and Liem 1989; Lauder et al. 1989; Westneat and Wainwright 1989; King 1991; Hall 1992; Hufford 1995, 1996; Raff 1996; Zelditch
and Fink 1996; Reilly et al. 1997; Rice 1997; Friedman and Carmichael 1998; Guralnick
and Lindberg 1999; and Wagner and Schwenk 1999.

presents the original form of phylogenetic character optimization, developed by Farris (1970).

UNAMBIGUOUS BINARY CHARACTER

1. Figure 5.2 depicts a phylogenetic tree for a hypothetical group of taxa. The distribution of a binary functional character (exhibiting states a and b) is mapped at the ends of the branches.

2. "Generalizing" down the tree (fig. 5.3): Label the two nodes that are farthest from the ingroup node in the following manner: (1) Label the node "a" if the two closest nodes or branches are either both a or are a and a, b. (2) Label the node "b" if the two closest nodes or branches are either both b or are b and a, b. (3) If the closest branches/nodes have different labels (one a and the other b), then label the node "a, b."

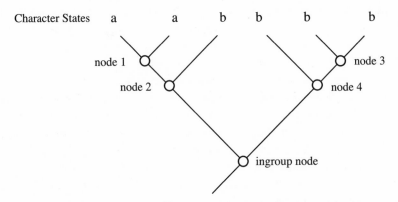

Figure 5.2.   Phylogenetic tree showing the distribution of functional character states a and b.

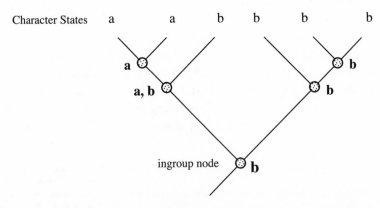

Figure 5.3.   Step 1 in Farris optimization of a binary character: generalize down the tree.

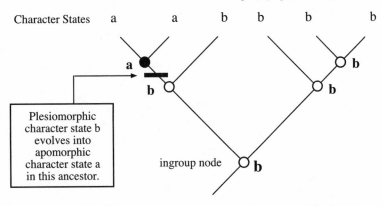

Character States   a   a   b   b   b   b

Plesiomorphic character state b evolves into apomorphic character state a in this ancestor.

ingroup node

Figure 5.4. Step 2 in Farris optimization of a binary character: predict up the tree.

Continue working toward the ingroup node in this manner. (Does this sound familiar?) Let's start on the left side of the tree. Node 1 connects two species exhibiting character a, so we generalize that the nodal (ancestral) state is a. This node, in turn, has a sister-group relationship with a taxon exhibiting state b; therefore, the state at node 2 is ambiguous for a or b. On the right side of the tree, nodes 3 and 4 are both labeled "b" because all three taxa exhibit that character state. Finally, we assign a value of b to the ingroup node because the two nodes directly above it have states a, b and b, so b "wins out" over a by majority vote (the principle of parsimony). Any ambiguity at the ingroup node may be resolved by reference to outgroups. (Starting to sound familiar yet? If not, refer back to the discussion of resolving the outgroup node presented in chapter 2.)

3. "Predicting" up the tree (fig. 5.4): Move from the ingroup node up the tree, resolving any ambiguity by comparing the value of the ambiguous node with the value of the node directly below it. In this example only node 2, designated a, b in figure 5.3, is ambiguous. Since the value of the node below it is b, node 2 is reassigned state b. All nodal states have now been unambiguously resolved on the tree. Farris optimization thus provides us with the following evolutionary hypothesis for this binary character: (1) State b is a persistent ancestral condition in four of the six terminal taxa (plesiomorphy). (2) There was a change from b to a in the ancestor of the two species that now exhibit state a (apomorphy).

UNAMBIGUOUS MULTISTATE CHARACTER

1. Figure 5.5 depicts a phylogenetic tree for a hypothetical group of taxa. The distribution of a multistate functional character (exhibiting states x, y, and z) is mapped at the ends of the branches.

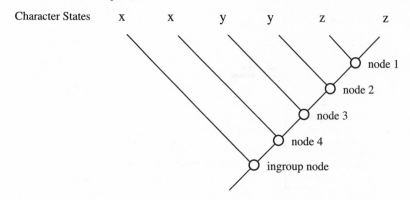

Figure 5.5. Phylogenetic tree showing the distribution of functional character states x and y and z.

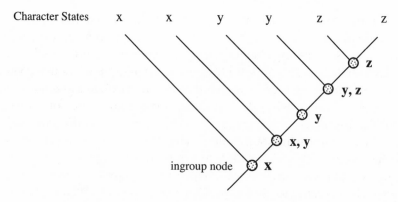

Figure 5.6. Step 1 in Farris optimization of a multistate character: generalize down the tree.

2. Generalizing down the tree (fig. 5.6): Using the same logic as that described for binary characters, node 1 is labeled with state z because both taxa connected by the node exhibit state z; node 2 is y, z because it connects a z node with a y branch (taxon); node 3 is y because it connects a y, z node with a y branch; node 4 is x, y because it connects a y node with an x branch; and finally, the ingroup node has a value of x because it connects an x, y node with an x branch.

3. Predicting up the tree (fig. 5.7): Nodes 2 and 4 are ambiguous in this example. As discussed for the binary character, resolution of each ambiguous node is dependent upon the character state of the node immediately below it. Accordingly, node 4 is reassigned a value of x and node 2 is labeled y. All nodal states have now been unambiguously resolved on the tree. Farris optimization thus provides us with the following evolutionary hypothesis for this multistate character: (1) State x, occurring in

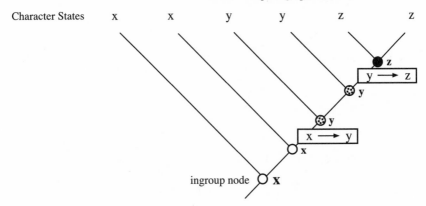

Figure 5.7. Step 2 in Farris optimization of a multistate character: predict up the tree.

the oldest two taxa, is plesiomorphic for the clade. (2) There was a change from x to y in the common ancestor of the remaining four taxa (apomorphy). (3) There was a change from y to z in the common ancestor of the two terminal taxa (apomorphy). All the functional traits are persistent ancestral conditions in the species that are extant today (and there is no reason to believe that any of them evolved more than once).

Like many aspects of the natural world, optimization does not always produce such unambiguous results. In the remainder of this book we hope to show you that the real world is a mixture of cases, some readily interpretable and others not. So first, let us consider an example that cannot be unambiguously resolved.

AMBIGUOUS BINARY CHARACTER

1. Figure 5.8 depicts a phylogenetic tree for yet another hypothetical group of taxa. The distribution of a binary functional character (exhibiting states t and s) is mapped at the ends of the branches.

2. The first stage of Farris optimization proceeds as previously described (fig. 5.9).

3. The fun begins when we proceed to the second stage of the optimization procedure. Predicting up the tree produces the solution presented in figure 5.10. According to this solution, (1) s is plesiomorphic, (2) t evolved from s twice (convergence), and (3) s evolved from t once (reversal). This implies that not all the species exhibiting either t or s inherited that trait from the same ancestral source.

4. If we examine this particular example more closely, however, we realize that the result of Farris optimization is not the only hypothesis that explains the distribution of t and s among these species. An equally

Character States

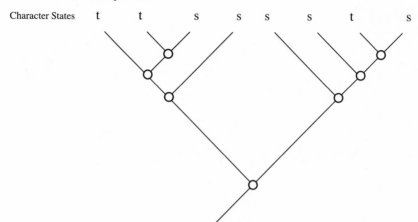

Figure 5.8. Phylogenetic tree with the distribution of functional character states t and s.

Character States

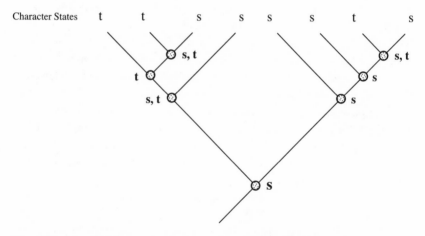

Figure 5.9. Step 1 in Farris optimization of an ambiguous binary character: generalize down the tree.

parsimonious interpretation of these data is that (1) s is plesiomorphic and (2) there have been three transitions from s to t (convergence) in extant species (fig. 5.11). This implies that all species displaying s inherited that trait from the same ancestral source. Identification of alternative hypotheses for character evolution (when present) is as far as we can go with phylogenetic optimization. We must incorporate additional biological information to resolve that ambiguity (e.g., the genetic, developmental, and physiological control of character production). In other words, we need to understand the "how" of character evolution, rather than just investigating the end-products of that evolution.

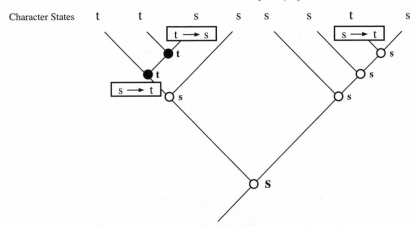

Figure 5.10. Step 2 in Farris optimization of an ambiguous binary character: predict up the tree. This type of optimization is called "ACCTRAN," for accelerated transformation. This optimization procedure places the origin of the apomorphic state (t) as early as possible on the tree. If there is homoplasy in the character, ACCTRAN favors reversal, in this case, t to s (see Wiley et al. forthcoming).

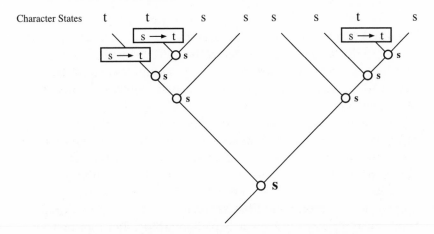

Figure 5.11. An equally parsimonious interpretation of the evolution of the binary character in figure 5.10. This type of optimization is called "DELTRAN," for delayed transformation. This optimization procedure places the origin of the apomorphic state (t) as late as possible on the tree. If there is homoplasy in the character, DELTRAN favors convergence (see Wiley et al. forthcoming).

As you can imagine, the potential for alternative explanations increases when your tree has polytomies, or when multistate or continuous variable characters are considered.[3] Donoghue (1989) noted addi-

---

3. For example, Swofford and Maddison 1987, 1992; Maddison 1989, 1991; Mickevich and Weller 1990; Hillis et al. 1992; Maddison and Maddison 2000; and Oakley and Cunningham 2000 and references therein.

tional theoretical and empirical examples in which optimization failed to resolve completely some sequences of character evolution. Although we would like nature to provide us with a perfect, and unambiguous, record of evolution, this does not always happen. In such cases, we expect that *phylogenetic optimization will tell us exactly where the ambiguity lies, even if it cannot provide an unambiguous interpretation.*

Methodological Caveats for the Comparative Phylogeneticist

 At this point let us reiterate a point we made way back in chapter 2: More than a decade ago, we wrote, "One cardinal rule in historical ecology is *never bias your analysis by using the ecological information you want to study to build your phylogenetic tree.*" Several authors took us to task for this statement (both in press and privately) because we seemed to be implying that it was circular to use the character of interest to build the tree, then ask questions about the patterns of origin and diversification of that character by reference to the tree.[4]

So, to clarify: it is not circular to use the features you wish to study as characters to build your tree because, as discussed in chapter 2, phylogenetic systematics is not circular. *But,* if the traits you wish to study represent the only character support for certain groupings in your tree, you have a weak, and therefore potentially biased, hypothesis (Brooks and McLennan 1994).

For example, suppose you are investigating the evolution of habitat preference in a group of sea urchins. Some species in your group prefer sand and some prefer rocks. Including the character "habitat preference" in your analysis of the relationships among these sea urchins (table 5.1) gives you the tree shown in figure 5.12a. Based upon this tree, your hypothesis for the evolution of habitat preference in this group would be "preference for a sandy habitat is plesiomorphic, and that preference changed to rocks in the ancestor of species D + E." Now, suppose you remove the trait "habitat preference" (character 7 in the data matrix) from your analysis. All of a sudden, you have two equally parsimonious phylogenetic hypotheses for your group. One of these trees is the same as the tree in figure 5.12a. The second tree, however, gives you a much more complicated (and ambiguous) picture of the evolution of habitat preference in the sea urchins. Both trees support the hypothesis that preference for sand is plesiomorphic, but the second topology indicates

4. Deleporte 1993; Kluge and Wolf 1993; de Queiroz 1996; Bruneau 1997; and Luckow and Bruneau 1997.

Table 5.1. Data Matrix for Hypothetical Sea Urchin Clade

	Characters											
Taxa	1	2	3	4	5	6	7	8	9	10	11	12
X	0	0	0	0	0	0	0	0	0	0	0	0
A	1	1	0	0	1	1	0	0	0	1	0	0
B	1	1	0	0	1	1	0	0	1	1	0	1
C	1	1	1	1	1	1	0	1	0	1	0	1
D	1	1	1	1	1	1	1	1	1	1	1	1
E	1	1	1	1	1	1	1	0	1	1	1	1

NOTE: Character 7 refers to habitat preference (0 = sand, 1 = rocks). X = outgroups.

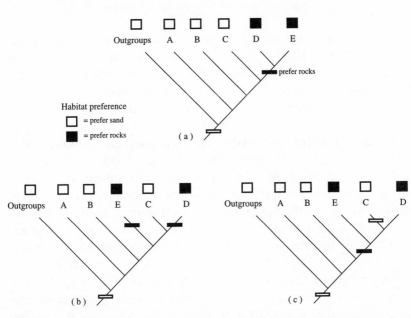

Figure 5.12. Should you use the characters you are studying to build your tree? (a) Phylogeny for hypothetical group of sea urchins based upon all of the characters in table 5.1 (CI = 85.7%). States for character 7 (habitat preference) are mapped above the tree, and evolution of the character highlighted on the tree. Eliminating character 7 from the phylogenetic analysis produces two equally parsimonious tree topologies (CI = 84.6%). One topology is the same as that shown in (a). The other topology, shown in (b) and (c) uncovers two new equally parsimonious explanations for the evolution of habitat preference. Note: the bar for habitat preference state on the root of the tree is only included to show the plesiomorphic condition for the character (in this case, preference for sandy habitats). This bar does not imply that the plesiomorphic state originated here. (We would need to expand the phylogenetic scope of our study to pinpoint the origin of that character state.)

that either (1) preference for rocks originated independently in species E and D (in this case the preference for sand in species C is plesiomorphic: fig. 5.12b) or (2) preference for rocks originated once in the ancestor of species E + C + D (in which case preference for sand represents a reversal in species C: fig. 5.12c). Removing this one character from your analysis has increased the number of potential evolutionary explanations from one to three.

> A better statement of this cardinal rule: *Don't waste time and energy exploring character evolution on a poorly supported tree.* Your resources would be better spent in improving the reliability of the phylogenetic hypothesis before asking specific evolutionary questions.

In some cases, the only way to be sure that you have the most parsimonious optimization for your character is to include information from outgroups in your analysis, *even if the optimization within the ingroup produces what appears to be an unambiguous result.* For example, suppose you are interested in studying the evolution of infanticide in a group of primates. Optimizing the trait "infanticide present or absent" onto your (well-supported) tree (fig. 5.13a) for the six ingroup species indicates that infanticide originated in the ancestor of C + D + E + F and was secondarily lost in species E. According to this scenario, infanticide is plesiomorphically absent in species A and B. The implication here is that the ability to become infanticidal simply did not show up until well into the evolution of the ingroup. Now, suppose you conduct further research and discover that the closest relatives of the ingroup also do not display infanticide. This strengthens your hypothesis of character evolution shown in figure 5.13a and is cause for serious celebration. Suppose, however, the opposite occurs: you discover that the next five closest relatives of the ingroup do display infanticidal behavior. When you include this information in your analysis, you get a somewhat different evolutionary picture for the ebb and flow of infanticidal tendencies over the history of the ingroup. You now have two equally parsimonious optimizations for the character. According to the first optimization, infanticide is an old trait within the larger group. It was lost in the ancestor of the ingroup, then reappeared in the ancestor of C + D + E + F, only to be lost again in species E. This scenario is not that dissimilar from your original hypothesis, but there is one important point of divergence. In the first case (fig. 5.13a) you might have been tempted to conclude that the absence of infanticide in species A and B occurred because the appropriate mutation (or whatever the underlying mechanism required for the emergence of the novel behavior) had not yet appeared in the group. The

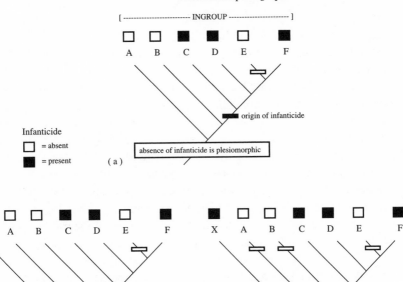

Figure 5.13. Why outgroups might be important in optimizing ingroup traits. (*a*) The presence or absence of infanticide mapped above the phylogenetic tree for the ingroup. In this particular example, including outgroups in the analysis changes the optimization shown in (*a*) even though the phylogenetic relationships of the ingroup do not change. (*b*) ACCTRAN optimization for infanticide. (*c*) DELTRAN optimization for infanticide.

patterns shown in figure 5.13b, however, indicate that the ability to show infanticide was present ancestrally and was somehow lost or suppressed in the ancestor of the ingroup. These are two very different explanations for character "absence" in species A and B. By contrast, figure 5.13c indicates that the presence of infanticide in species C, D, and F is plesiomorphic, with three independent (convergent) losses of the character in species A, B, and E. This is a substantially different picture from your original hypothesis (fig. 5.13a). Overall, the only point of congruence between your original explanation and the two new optimizations is the secondary loss of infanticide in species E.

> It is not methodologically sound to polarize one character using an outgroup and then to assume that the postulated character transformation series represents the "true" phylogenetic sequence of events.

This type of error results from a subtle misunderstanding of the purpose of outgroup comparisons. While it is true that polarizing (binary) character states through outgroup comparison produces a hypothesis of the sequence of evolutionary changes in that character, it is also true that this hypothesis can only be tested by comparing it to the hypothesized evolutionary sequences of other characters in the context of a phylogenetic analysis of all characters. Remember that a robust phylogenetic tree is based on analysis of many characters, each polarized using outgroup comparisons. Any of the characters used in that analysis may exhibit some degree of homoplasy, in which case the distribution of the characters on the phylogenetic tree will not correspond exactly to the sequence postulated when the characters were polarized (fig. 5.14). In other words, a character transformation series does not stand on its own, independent from the underlying phylogenetic relationships of the species. Remember also that outgroup comparison by itself can never specify the evolutionary sequence for a multistate character. A final point:

> ! An adaptive explanation can be weakened if some members of the study clade are removed from the analysis.

The world of phylogenetic reconstruction is fraught with enough difficulties in terms of sampling procedures and the incorporation of fossil

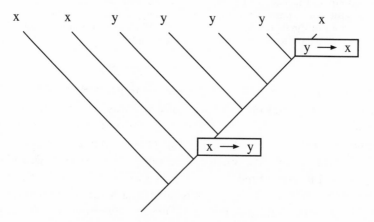

Figure 5.14. Dangers inherent in assuming that the character transformation series for one character represents the true evolutionary sequence of character change. In this example, the outgroup state for the character is x. Since both the outgroup and some members of the ingroup possess the character state x, the character transformation series is hypothesized to be x to y. However, when the character is optimized on a phylogenetic tree constructed from a number of different characters, the actual phylogenetic transformation is discovered to have been x to y to x.

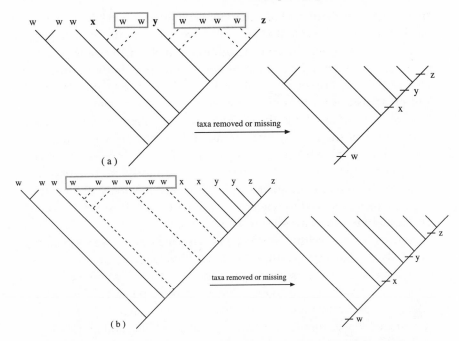

Figure 5.15. Example demonstrating the problems that can arise when adaptive hypotheses are formulated within an incomplete phylogenetic framework. (*a*) A nonlinear evolutionary transformation of traits could appear linear when taxa are removed (taxa in boxes are eliminated from the analysis). (*b*) The amount of phylogenetic conservatism is greatly underestimated by removing taxa from consideration.

evidence (Doyle and Donoghue 1986; Donoghue et al. 1989) without exacerbating the problem by eliminating available information from the analysis. For example, consider a monophyletic group of fishes displaying the nest-building characters w, x, y, and z (fig. 5.15). Failure to incorporate some of the taxa bearing trait w into the analysis will produce the incorrect postulate of a linear sequence of character change from w to x, x to y, and y to z (fig. 5.15a). When all the members of the group are considered, the evolutionary sequence is demonstrated to be the nonlinear w to x, w to y, and w to z. (You should be able to work out this optimization.) The situation depicted in figure 5.15b is slightly different. Here, although the hypothesized sequence from w to x to y to z is correct, this analysis will grossly underestimate the degree of stasis throughout the evolution of the character. In this clade, 60 percent of the taxa bear the ancestral character state w. Eliminating some taxa bearing the plesiomorphic condition from the analysis artificially inflates the proportion of taxa displaying a derived character state, thus making it appear that the character has been more evolutionarily labile. Unfortunately it is often the case that you will have either an incomplete phylogeny for your group, incomplete information about the traits you are interested in studying for all members of the group, or both. In this case your initial conclusions are weakened by the absence of information, and no statistical analysis or computer algorithm can compensate for that weakness. If anything, the preliminary study might help you determine which of the missing taxa are the most critical to your study. In this book we sometimes use the terms "simplified" and "reduced" when referring to phylogenetic trees. These terms are shorthand notation for trees that have had monophyletic groups whose members all have the same character reduced to one branch on the tree. This is not the same as eliminating information from the analysis.

Now, on to the exciting world of character evolution. Like speciation, the study of character evolution within a phylogenetic framework has undergone a dramatic radiation over the last decade, so we can only present a fraction of the studies that have been published. For additional, and generally more complicated, examples, interested readers are encouraged to see (among others) the evolution of: diet and morphology in thaidid gastropods (Kool 1987); endothermy in fast-swimming, scombroid fishes (Block 1991a,b); almost all aspects of ecology and behavior of water striders (Andersen 1982, 1991, 1993, 1994, 1995, 1996, 1997a,b, 1999); social behavior in sweat bees (Packer 1991, 1998); concealed ovulation and mating systems in anthropoid primates (Sillén-Tullberg and Møller 1993); lekking behavior in manakins (Prum 1994); sand-diving behavior in lizards (Arnold 1995; Halloy et al. 1998); mating systems in freshwater aquatic plants (Barrett 1995; Graham and Barrett 1995); breeding systems in some Hawaiian flowering plants (Weller et al. 1995); myrmecophagy in poison frogs (Caldwell 1996); habitat preference in crocodilians (Jackson et al. 1996); feeding and nonfeeding larvae in echinoids (Wray 1996); pheromone-based communication in bark beetles (Cognato et al. 1997); stridulatory apparatus in crickets (Desutter-Grandicolas 1997); sexual dichromatism in tanagers (Burns 1998); gall shapes (Crespi and Worobey 1998); color patterns in carnivores (Ortolani 1999); eusociality in sponge-dwelling shrimps (Duffy et al. 2000); silk decorations of orb-weaving spiders (Herberstein et al. 2000); postcopulatory displays in dabbling ducks (Johnson et al. 2000); cooperative and pair breeding in thornbills (Nicholls et al. 2000); interaction between percent starch in pollen and pollinator type (Roulston and Buchmann 2000); and social behavior of macaques (Thierry et al. 2000).

FOCUSING YOUR RESEARCH BEAM

> The one merit that is claimed [for comparative studies] is
> that it suggests new ways of looking at facts and new sorts of
> fact to look for.
>
> Simpson 1944:xviii

There are two components to character evolution: origin and maintenance.[5] Microevolutionary studies concentrate on the maintenance of
traits in current environments where the processes shaping the interactions between the organism and its environment can be observed and
measured directly. Having untangled this complicated web, researchers
attempt to extrapolate backward to the processes involved in the character's initial appearance in, and subsequent spread through, the ancestral
species. The research we are discussing in this chapter complements
these studies by providing direct estimates of phylogeny that can be
used as templates for reconstructing the historical patterns of character
origin and diversification. Such templates can help specialists focus their
search for the processes underlying character evolution. For example,
suppose you are an enthusiastic ornithologist interested in exploring the
evolution of that enigmatic avian species, the green jacana (*Jacanus ver-
tus*). You discover that the green jacana is green (hence the name), is
polyandrous, and that all of the individuals in the species have a long
beak. Based upon this information alone, would you ask why is the
green jacana green, why does it have a long beak, or why is it polyandrous? In order to answer this question, you must decide whether you
are interested in studying the factors involved in the maintenance of a
character or the factors involved in the origin of that character. Once
you have decided this, the only way to choose an appropriate character
is to examine the evolutionary patterns of character origin and diversification within a phylogenetic context.

So, back to the jacanas. You have decided that you are interested in
origins. Which of the three characters should you pick? Clearly it is impossible to answer this question by looking at just one species. Fortunately, while you were pondering the issues of origin versus maintenance, other researchers were collecting data about your three characters
for all of the species in the genus *Jacanus,* and a friendly systematist was
reconstructing a robust phylogenetic tree for the group. Mapping the
three traits onto that phylogeny reveals that you should focus your at-

5. Ridley 1983; Wanntorp 1983b; Brooks 1985; Dobson 1985; Greene 1986a; Brooks
and Wiley 1988; Coddington 1988; and Brooks and McLennan 1991.

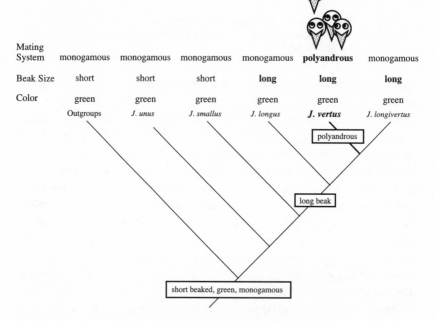

Figure 5.16. Focusing your research beam. Searching for the difference between origin and maintenance of three characters in a hypothetical group of jacanas. Character states are mapped above each species and optimized onto the phylogeny.

tention on the evolution of jacana mating systems (fig. 5.16). If you ask, Why is the green jacana polyandrous? you will have the best chance of uncovering the factors influencing the original success of the character because polyandry is an autapomorphy of the green jacana. This means that you have the best chance of uncovering information about the origin of the trait by assessing its function in its current environment. The next stage in your study might include searching for other autapomorphic changes in the green jacana that could be causally involved in the success of polyandry once it appeared (e.g., changes in operational sex ratio, clutch size, predation pressure, etc.).

Without the phylogeny you might have decided to ask, Why is the green jacana green? If "why" implies origin, the answer to this question is simple: The green jacana is green because it inherited green body color from its ancestors. It is an ancestral legacy. This does not mean that green is uninteresting. In fact, green is your best choice if you are passionately interested in studying character maintenance (What factors have been maintaining green body in this clade through so many speciation events?) because of all three traits it has persisted unaltered (plesio-morphy) through the largest number of speciation events. Each species displaying green body represents an opportunity to test your hypothesis that factor "x" is causally involved in the maintenance of green body under current environmental conditions. Is green maintained because all jacana species are subjected to high levels of predation, and green functions as camouflage in the birds' habitat (in this case you may have evidence for stabilizing selection)? Or is it "maintained" because there is no underlying genetic variation upon which selection can work? If you want to directly address why green was originally successful, you need two additional pieces of information: an expanded phylogenetic analysis incorporating close relatives of the jacanas to establish the evo-lutionary "point of origin" of green and an assessment of that *ancestor's* biology in *its* environment. This requires an excursion into the distant past, both genealogical and environmental—a more difficult quest than your original single-species perspective would have led you to believe.

 All of the jacanas are green. Is this an example of a phy-logenetic constraint? We prefer to use the term phyloge-netic stasis to indicate a lack of change in a particular character across two or more speciation events because this term does not imply that we know anything about the underlying mechanisms responsible for that lack of change. It is simply a pattern-based observation. The term "phylogenetic constraint" (or "phylogenetic iner-tia"), on the other hand, implies that something is ac-tively impeding the evolution of a particular character. Because so many different mechanisms (e.g., lack of suf-ficient genetic variation upon which natural selection can work, mechanical/architectural limitations on de-sign, the dynamics of development and physiology, stabilizing selection[6]) theoretically can produce the same outcome (observed stasis in a character over macro-evolutionary time), it is critical that researchers use the term "constraint" only if they have empirical evidence to support the existence and the limiting effects of one of those mechanisms. The key word here is "limiting," for this distinguishes between the *maintenance* of a char-

6. Alberch 1982; Wake 1982; Maynard Smith et al. 1985; Gould 1989; Lauder et al. 1989; Antonovics and van Tierderen 1991; Carrier 1991; Arnold 1992; Kirkpatrick and Lofsvold 1992; Perrin and Travis 1992, 1994; Björklund and Merilä 1993; McKitrick 1993; O'Neil and Schmitt 1993; Hall 1994; Losos and Miles 1994; Moran 1994; van Tierderen and Antonovics 1994; Schwenk 1995a; Cheverud 1996; Hansen 1997; Hayes 1997; Pigli-ucci et al. 1998; Wagner and Schwenk 1999; Witte and Döring 1999; Getty 2000; Morales 2000; and Pigliucci and Kaplan 2000a,b.

acter in its plesiomorphic state (for example, by stabilizing selection) and a character's *inability to change* during
a given interval.[7]

Recognition of phylogenetic stasis is important in itself precisely
because it highlights areas for future research into mechanisms. For
example, are there similarities in the ecological conditions, life history
parameters, or social structure of the jacanas retaining the symplesiomorphic feather color that could be operating as a mechanism for stabilizing selection in these birds? Is there any genetic variability in green at
all among these birds? Alternatively, is a portion of the observed stasis
an artifact of missing information, or incorrect character coding? In order to address the issue of accuracy in character description and coding,
we would need to delve further into the issue of, What is green, really?
For this, we would have to understand how green is made (pigments
and pigment cell types), controlled (neural, hormonal, developmental
information), and inherited (genetic data). By now you begin to understand the complex data collection required to formulate more complete
hypotheses of evolution, hypotheses that can explain both the origin,
maintenance (stasis), and divergence of characters, species, and ecosystems.

Overall, then, a phylogenetic tree, *in conjunction with* extensive
process-based studies (e.g., cost/benefit analyses, experimental manipulations), can help us disentangle the factors involved in the origin of
a character from those involved in its current maintenance. There is a
desperate need for detailed studies of this type on all levels, from genetic
to phylogenetic, for a variety of groups so that we can begin to ask questions about the interaction between "origin" and "maintenance." In
some cases we would expect to discover that the processes influencing a
trait's origin are very different from the processes currently responsible
for its maintenance. In other cases, it might be possible that the same
processes are responsible for both origin and maintenance. Because the
connection between these two components of character evolution is
currently a black box in evolutionary biology (and is thus a promising
research area), it is critical that researchers be very explicit about which
component they are studying. We believe such studies will ultimately
tell us that we should not be asking, Is it stasis *or* adaptation/diversifica-
tion? but rather, How much of each contributed to the evolution of a
character or clade? (see, e.g., Price and Carr 2000).

---

7. For excellent reviews of the problem see Schwenk 1995a; and Wagner and
Schwenk 1999.

Boldly Going Where No One Has Gone Before

Many of the macroevolutionary studies that have been published in the last decade have been concerned with reconstructing the patterns of character origin and elaboration. These studies generally represent the first stage in their authors' evolutionary odyssey: discovery. The most important thing about this stage is that it shows you where to go next!

The Secret Lives of Leeches. Leeches have a bad reputation in human folklore, bringing to mind for many the famous scene from *The African Queen* in which Katherine Hepburn, with a perceptible shiver, brushes hundreds of the blood-sucking creatures off Humphrey Bogart's back. Hollywood aside, leeches are not just restricted to the role of freshwater exsanguinators. Some species are marine predators, others, terrestrial ones. Most leeches, like their earthworm cousins, secrete a cocoon into which fertilized eggs are deposited, then abandoned to their fate. Some leeches, however, are dedicated parents, covering and fanning the cocoon with the ventral surface of their bodies and even remaining with the young for some time after hatching (Sawyer 1971, 1986). Clearly, these creatures are more ecologically and behaviorally diverse than popular conceptions would have us believe.

Siddall and Burreson (1995, 1996) were interested in uncovering how that diversity had been assembled across the millennia. They began by reconstructing a phylogenetic tree for members of the Euhirudinea (true leeches). Mapping the different ecological and behavioral traits onto that tree highlighted a cornucopia of ideas for future research projects (fig. 5.17). The evolution of parental care is unambiguous for the group. The plesiomorphic condition (cocoon with a hard cover, attached to substrate then abandoned) has changed twice, independently, to (1) secretion of hard-cover cocoon in soil or vegetation (loss of attaching behavior) before abandoning (in ancestor 4 on fig. 5.17) and (2) loss of a hard covering coupled with appearance of parental care in the ancestor of *Theromyzon* + *Glossiphonia* + *Haementeria* (ancestor 2). Did parental care appear before or after the loss of the hard covering on the cocoon? When two or more characters co-originate on a branch of a tree, we cannot determine the sequence of their origination just by looking at the macroevolutionary patterns. In the case of our leeches, we might address the issue of which came first, the fragile cocoon or parental care, by removing parents from their watches over the fragile cocoons and determining whether rates of predation on the potentially tasty hors d'oeuvres bags increased dramatically. If so, then it is difficult to envision a situation in which the loss of the protective cocoon cover could

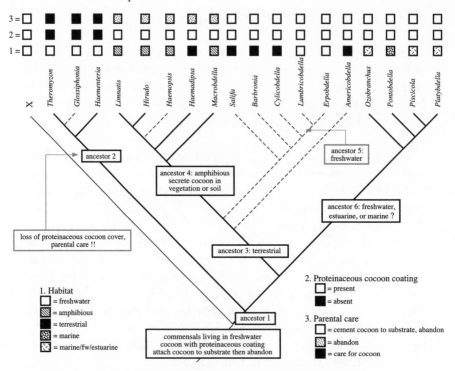

Figure 5.17. Ecological and life history traits mapped onto the reduced phylogenetic tree for the true leeches. Relationships are a composite of Siddall and Burreson (1995) and Apakupakul et al. (1999: combined morphology and molecular data; one tree, CI = 60.3%). Not all of the same taxa were used in both analyses. Width of line represents the ACCTRAN optimization for feeding mode: *thin line* = commensal; *dotted line* = predatory; *bold line* = sanguivorous. (There is one other equally parsimonious way to optimize feeding mode.) X = outgroups, including *Tubifex* and *Lumbricus* (the "earthworm").

have persisted as an autapomorphic change for any length of time because individuals showing the novel character would suffer decreased fitness. On the other hand, if parental care originated first, this would create a background in which the subsequent loss of the hardened covering would be *allowed* once it originated because the hypothesized anti-predation function of the cocoon covering had been replaced by increased parental vigilance.

There are still unanswered questions to be addressed here. Our initial set of experiments is based upon the assumption that a hardened cocoon cover provides some degree of protection for the developing eggs. Does that protection apply to predation (as we have assumed)? Is the loss of the hardened covering "adaptive" or just something that happened by chance and persisted because it was not selected against in the new evolutionary playing field? This second question might seem trivial, but it is important in the context of Darwinian versus Lamarckian explanations. A Lamarckian mechanism predicts that costly traits will always be lost once they are no longer needed. The loss of unnecessary traits can only occur in a Darwinian framework, however, if there is some variability in the trait that is available to natural selection and if losing the trait is less costly than maintaining it (one could imagine, for example, that the loss of the trait might have deleterious effects on other traits). So, in this example we would have to postulate that the ability to secrete the proteinaceous coating was variable in ancestor 2. Once parental care appeared, selection would act against individuals that produced thick cocoon coats if that production was energetically costly. Because the descendants of ancestor 2 no longer produce the proteinaceous coat, the only way to examine the "variable ancestor" hypothesis is to examine the state of that character in ancestor 2's sister group. If we discover that individuals in the basal members of the sister group (always the easiest place to start) produce cocoons with varying degrees of hardening, we have evidence to support the variable ancestor hypothesis, and to support, by extension, the hypothesis that selection acted against individuals that produced thick cocoon coats once parental care had evolved. Of course, that evidence is weak on its own, but it is a beginning. The next steps in our study should involve a delineation of the mechanisms underlying the production of the proteinaceous coating, identification of the step that has been lost or suppressed in *Theromyzon* + *Glossiphonia* + *Haementeria,* and demonstration that production of the coating is (or is not) energetically costly.

The evolution of habitat preference has been slightly more complicated than the evolution of the cocoon: Ancestor 3 made a direct transition from living in freshwater to living on land. Ancestor 4 made a partial movement back to the water, a shift concurrent with the evolution of abandoning the cocoon in vegetation or soil and representing a type of amphibious lifestyle in which the adults spend most of their time in water, wriggling their way onto land to deposit their eggs. The ancestor of *Lumbricobdella* + *Erpobdella* seems to have made a complete transition from the land back into freshwater. Researchers interested in uncovering the processes responsible for such (apparently) radical habitat shifts should focus their attention here. For example, are there differences between species of *Lumbricobdella* + *Erpobdella* and their terrestrial sister group(s) in terms of preferred prey or levels of predation? Did traits allowing the movement into freshwater appear before the transition, or concurrently with it? If the former, was the movement into freshwater correlated with any widespread environmental changes that might have favored the survival of species in an aquatic habitat (in other words, was freshwater a refuge of some sort)? The tree also shows an equivocal number of transitions among freshwater, estuarine, and marine habitats. Aside from the general indication that ancestor 6 was aquatic, we cannot say anything more about the evolution of habitat preferences in this clade (freshwater versus estuarine versus marine) without further information. The first thing we should do is to increase the phylogenetic detail of the tree to include species-level relationships for *Ozobranchus*, *Piscicola*, and *Platybdella*. So (for example), if we discover that the basal members of these genera are all marine, then we would hypothesize that ancestor 6 showed an apomorphic movement from freshwater into the marine environment. According to this scenario, there would be nothing unusual about the marine genus *Pontobdella* (its presence in the marine environment is plesiomorphic). Without these phylogenetic relationships, we can continue to propose any number of possible pathways, but such an exercise is rather futile. (Our time would be better spent taking on the more difficult job of working out the phylogenies for these groups.) Once resolved, we could begin to ask questions about the mechanisms underlying osmoregulation in leeches—differences in osmoregulatory problems posed by freshwater versus terrestrial versus marine systems are not trivial, yet leeches appear to have made these transitions a number of times. How has that happened?

Finally, the ancestor of the leeches appears to have been a sanguivore (Apakupakul et al. 1999). Past that point, the optimization is somewhat ambiguous, sanguivory has either been replaced by carnivory four times (dotted lines on fig. 5.17) or was replaced by carnivory in ancestor 3 (and in *Glossiphonia* and *Haemopsis*), then reappeared in ancestor 4. Siddall and Burreson (1995) noted that there are two different ways to be sanguivorous: species in the lineage descended from ancestor 4 make incisions through their host's skins, then suck up the blood as it oozes out of the cut (these are the medicinal leeches), whereas the descendants of ancestors 6 and 2 actually insert their proboscises into their host. Unfortunately the presence of two different methods for blood feeding is not a priori evidence that the trait evolved twice. These two characters could also be part of a homologous transformation series for sanguivory. At this stage there is no way to choose between the single- (transformation series), and double-origin hypotheses. In order to do so, we would need information about the genetic/physiological mechanisms that are responsible for the development of blood-feeding behavior. Absence of those mechanisms in the predatory descendants of ancestor 3 would favor the double-origin hypothesis. Presence and suppression of those mechanisms in the predatory species would favor the single-origin hypothesis for blood feeding (as shown in fig. 5.17) and might even give us some insights into the mechanisms underlying the appearance of predatory behavior. For example, members of *Haemopsis,* a predatory species nestled within the medicinal leech clade, shred their prey with large teeth, a condition not so far removed from their sanguivorous relatives' "slash then lap" behavior (Apakupakul et al. 1999). These are difficult but critical paths to travel if we ever want to fully understand the evolution of trophic interactions.

Although the optimization of foraging modes is ambiguous at the moment, the different patterns do agree on one interesting point: there have been two independent transitions from sanguivory to predation in *Haemopsis* and *Glossiphonia.* Researchers have traditionally looked upon the convergent appearance of a trait as cause for celebration because such independent origins of a trait are assumed to provide us with an increased number of (independent) tests of a particular adaptive hypothesis (Coddington 1988; Arnold 1990, 1995; Brooks and McLennan 1991). This assumption presupposes that everything else is equal; that is, there are no other differences among the taxa showing the convergent trait that could be influencing our analysis of the evolution of that character. Now, let's take a closer look at our two predators. *Glossiphonia* is a freshwater group that shows the derived state of parental care + loss of hard cocoon cover. *Haemopsis,* on the other hand, contains amphibious species that leave the water to secrete their hard-covered cocoons in the soil, then return to the water. A substantial amount of evolutionary change has occurred before the convergent appearance of predatory behavior in these two genera. Is it therefore plausible to assume that identical forces will have shaped the evolutionary shift from sanguivory to predation in *Haemopsis* and *Glossiphonia?*

Crickets in Caves. The Amphiacustinae is a monophyletic group of long-legged, Neotropical crickets distributed throughout the Caribbean, Mexico, and Central America as far south as Panama. This clade is nestled within a larger group of crickets (Phalangopsidae) that are generally diversified throughout the understories of tropical rainforests and are very sensitive to dry conditions. Most members of the Amphiacustinae forage during the night in the leaf litter on the forest floor then hide in hollow trees, burrows, crevices, or in cave entrances during the day. Some species, however, are completely troglodytic, living deep within the dark, moist recesses of caves. The origin of a cave-dwelling lifestyle has fascinated researchers for centuries. Some authors have seen the movement into caves as a response to periods of climatic stress (in which case the cave is a refuge: e.g., Vandel 1964; Barr 1968), while others have suggested that such movement represents the exploitation of a new resource (in which case the cave is a conquest: e.g., Matile 1970; Howarth 1987, 1991). In either case, refuge or conquest, the question remains, How do "normal" epigean species become "adapted" for cave living?

Desutter-Grandcolas (1993, 1994, 1995) approached this problem by collecting 22 morphological traits (predominantly of male genitalia) for the nine amphiacustine genera (many of which she had described herself) and two outgroups. Phylogenetic analysis of the data matrix produced one tree with a CI of 92%. She then optimized a variety of ecological traits (preferred habitat for hiding, preferred habitat for foraging, diurnal versus nocturnal behavior) onto that tree, with fascinating results (fig. 5.18). The first, and most obvious, discovery is that cave dwelling has originated at least twice in this family, once in the ancestor of *Mayagryllus* + *Arachnopsita* + *Prolonguripes* + *Longuripes* (ancestor 1 on fig. 5.18) and at least once within the genus *Noctivox*. At the moment, the patterns are ambiguous for *Noctivox* because we do not know whether the cave-dwelling species in this genus form a monophyletic group. Whatever that ultimate number of cave dwellers, in each case they are descended from nocturnally active ancestors, which foraged amid the forest-floor leaf litter at night and hid themselves in crevices, burrows, hollow logs, and even at the mouths of caves during the day. The habitat transition into caves was thus a gradual one that evolved from hiding in the leaf litter (symplesiomorphic) during the day then emerging to forage in the leaf litter at night (symplesiomorphic), to moving into shallow cavities at ground level during the day (synapomorphy for the subfamily) then emerging at night, to remaining within deep cavities during the day and night. This transition supports previous hypotheses that taxa living in dark, humid places were most likely to give rise to cave dwellers (e.g., Vandel 1964; Juberthie 1984). Not surprisingly, the movement into caves was associated with the loss of foraging in the leaf litter. (How much leaf litter do you find deep within a cave?) This apparent "loss" does not tell us whether cave crickets do not forage in leaf litter because there is none any available in their environment, or whether they have actually lost the ability to forage in that way. Both explanations beg the question of exactly where (along the cave bottoms, high up in rocks, etc.) and when (in the absence of visual cues, did the crickets change their circadian rhythms?) the cave dwellers are foraging.

The reduction and loss of ocelli associated with the origin of true troglodytes is also fascinating. As mentioned previously, there is no requirement within a Darwinian framework that "unused" traits be lost, only that if they are costly and are lost, there will no evolutionary penalty and may possibly be an evolutionary benefit. In order to examine the

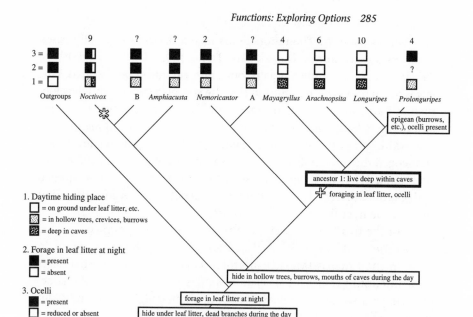

Figure 5.18. The evolution of cave dwelling in a group of crickets (Amphiacustinae). Characters are optimized onto the phylogenetic tree for the family reconstructed from morphological characters (Desutter-Grandcolas 1993). Groups A and B are undescribed genera. *Numbers above boxes* = number of species in each genus; *asterisk* = evolution of cave dwelling, with associated reduction of ocelli and loss of foraging in leaf litter in some species of *Noctivox; cross* = loss of particular traits.

mechanism(s) responsible for this loss, we need to determine the genetic/developmental basis for ocellular production (see, e.g., Jones et al. 1992). For example, how much variability in trait development is present in the sister groups to the ancestors in which cave dwelling originated? How much variation remains in species with "reduced" ocelli? Whatever that mechanism, it seems likely that the "ability to make ocelli" was suppressed, not completely lost, in the ancestor of the troglodytic species, because there has been a reversal from cave dwelling to a surface-based epigean lifestyle in the ancestor of *Prolonguripes*. Such suppression may form the basis for many documented instances of reversal to the plesiomorphic condition. If research on the evolution of troglodytic lifestyles in other groups indicates that there have never been reversals back to the surface, it would be interesting to document whether those cases involved the actual loss of a component critical to the production of "eyes," rather than just a suppression of trait expression. It is important to note here that the instance of reversal could only be identified because we did not place any restrictions on the direction of character evolution. The most common restriction that biologists employ is irreversibility. The rationale for such a decision is usually based upon an application of Dollo's Law (Dollo 1893, 1922; Gould 1970), which states that complex organs, once lost, can never be regained. This law is based upon the assumption that the complex genetic and developmental foundations for such organs "decay" once the organs are lost, and that it is highly improbable that these foundations will reappear de novo (Muller 1939). Studies at the genetic and developmental level have challenged the universality of these assumptions (e.g., Lande 1978; Marshall et al. 1994; Raff 1996; Li 1997), so Dollo's law should be applied with caution, preferably only after empirical data have been collected to support the hypothesis of irreversibility.[8]

All of the troglodytic species in the monophyletic group comprising *Mayagryllus* + *Arachnopsita* + *Prolonguripes* + *Longuripes* are found in caves in the Chiapas (Mexico)–northern Guatemala region. This raises some interesting issues for speciation studies. Once the series of character changes associated with cave dwelling originated, how did individuals disperse to (and speciate in) different areas and caves? Some authors have proposed that cave dwellers in this area could have moved during the wet phases of the Late Mesozoic (Peck 1978, 1990); troglodytic species, however, are rarely seen moving outside of their caves today (Howarth 1983). Adding species numbers to the phylogenetic patterns, indicates that cave-dwelling clades are not depauperate compared to epigean clades. In fact, the cave-dwelling *Longuripes* is 2.5 times as species rich as its secondarily epigean sister group (*Prolonguripes*). The refuge and the conquest hypotheses make different predictions about why epigean species would enter caves in the first place. Neither hypothesis, however, makes specific predictions about what will happen to those species once they make the transition. Certainly the relative success of the cave-dwelling clades (in terms of species number) would support the proposal that the cave habitat represents a novel resource that, once accessed, does not have a negative impact on speciation and may even promote it under certain conditions. This at least refutes the proposition that troglodytic species are descendants of epigean species forced into caves during periods of environmental harshness, where they wound up as "evolutionary blind alleys" (Papp 1982).

Diet Changes in the Acanthuroidei. The Acanthuroidei is a suborder of marine fishes that includes many familiar coral-reef inhabitants (e.g., surgeonfishes, unicorn fishes) as well as some less well known species like

8. For a superb discussion of this problem, see Lee and Shine's study (1998) on the evolution of viviparity in reptilomorphs.

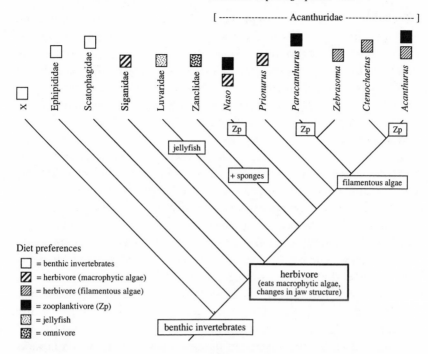

Figure 5.19. The evolution of foraging preferences in acanthuroid fishes. Dietary information is mapped above the phylogenetic tree for the Acanthuroidei (one tree, CI = 92.5%: Winterbottom 1993). X = the outgroup, *Drepane*. (Modified from Winterbottom and McLennan 1993.)

the open ocean *Luvarus*. There is a wide variety of dietary preferences in the group: some species dine on zooplankton, others eat only algae, and still others disdain all foodstuffs but jellyfish. How have those preferences evolved? Winterbottom and McLennan (1993) investigated the problem by mapping diet onto a phylogenetic tree for the acanthuroids (fig. 5.19). They discovered that foraging on benthic invertebrates is plesiomorphic for the suborder, while foraging on macrophytic (leafy/fleshy) algae is derived, a rather curious sequence of events given the traditional view of energy flow from detritivores to herbivores to carnivores. The origin of herbivory was associated with morphological changes reducing the protrusibility of the upper jaw and limiting its movement primarily to rotation. Since a protrusible premaxilla is thought to increase a fish's ability to seize moving prey, this change to a more rigid jaw may be causally, as well as phylogenetically, associated with the appearance of herbivory.[9] Once herbivory appeared, it was re-

9. For discussions of feeding mechanisms, see Lauder 1982b; Tyler et al. 1989; and Westneat 1994.

tained through most speciation events in the clade; however, a moderate amount of flexibility in foraging mode is superimposed upon this ancestral background. For example, there has been a transition from eating macrophytic algae to eating filamentous algae in the ancestor of the acanthurins, from herbivory to jellyfish in *Luvarus,* from herbivory to omnivory by the addition of sponges in *Zanclus,* and from herbivory to zooplankton picking in *Paracanthurus* and in some species of both *Naso* and *Acanthurus.*

Having identified the origins of diet switching, can we identify any correlated changes in morphology, the environment, or both that initially would support an adaptive explanation for those switches? There is only one species of *Luvarus.* The change in diet from macrophytic algae to jellyfish is associated phylogenetically with a number of autapomorphies, including a habitat shift from the coral reef to the open ocean, and extreme modifications to the caudal fin supporting elements and vertebrae that permit prolonged bouts of propulsion by slow caudal sculling. This swimming mode may allow the fish to forage more efficiently (Tyler et al. 1989; Carpenter 1990). Since there are few if any macrophytic algae globally present in the open sea, the diet shift to jellyfish allowed *Luvarus* to survive in its new environment. So, in *Luvarus* we have a correlation between the origins of a novel habitat preference, a novel diet preference, and a novel foraging mode (swimming ability). In order to determine whether these changes are indeed adaptive, rigorous experimental investigation of the hypothesized performance advantage for each autapomorphy is required.

The evolution of zooplanktivory is more difficult to study, but the patterns can direct our attention to areas for future research (fig. 5.20). Given the phylogenetic relationships among the genera, the most parsimonious explanation for the origin of zooplanktivory is that it has evolved independently at least three times from an herbivorous ancestor in *Naso, Acanthurus,* and *Paracanthurus.* Without a resolved phylogeny for *Naso* and *Acanthurus,* we cannot tell how many times eating zooplankton has evolved in those genera, nor can we tell whether there have been any reversals to herbivory. The evolution of zooplanktivory in *Naso, Acanthurus,* and *Paracanthurus* has involved a dramatic shift from feeding on medium-sized food items (that don't swim away) to feeding on extremely small prey. A variety of morphological characters have been postulated to influence a fish's ability to detect and engulf small prey items (snout length, eye size relative to head size, and the angle between the snout and eye), to digest food (relative intestine length), and to avoid predation (degree of streamlining). Mapping mean values for these traits above the genera indicates that there have been two gen-

	Naso	Prionurus	Zebrasoma	Paracanthurus	Ctenochaetus	Acanthurus
GL/SL	2.9 (1.6)	2.3	3.7	2.3	3.5	4.6 (3.1)
Eye/M	35° (27°)	50°	46°	48°	52°	56° (40°)
Sn/Eye	2.8 (1.7)	2.3	3.2	3.1	3.5	3.1 (1.7)
SL/Sn	4.9 (5.2)	5.2	4.4	5.4	4.2	4.4 (8.0)
SL/BD	2.8 (3.3)	1.8	1.8	2.3	1.9	2.0 (2.3)

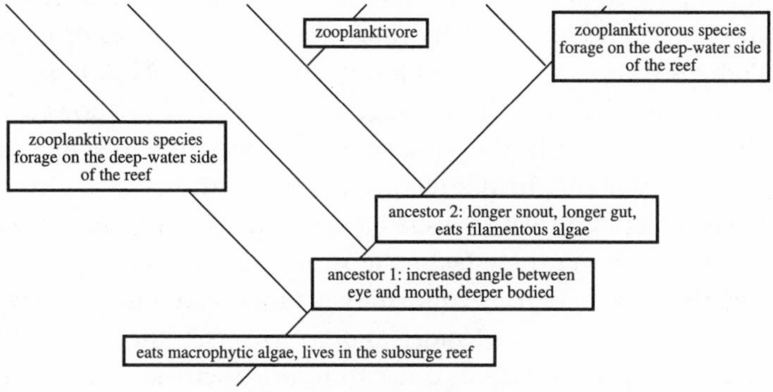

zooplanktivore

zooplanktivorous species forage on the deep-water side of the reef

zooplanktivorous species forage on the deep-water side of the reef

ancestor 2: longer snout, longer gut, eats filamentous algae

ancestor 1: increased angle between eye and mouth, deeper bodied

eats macrophytic algae, lives in the subsurge reef

Figure 5.20. The evolution of foraging preferences in acanthurids. Mean values for morphological traits hypothesized to play a role in foraging efficiency are mapped above the tree: GL/SL = gut length/standard length; Eye/M = angle between eye and mouth; Sn/Eye = snout length/eye diameter (the larger the value, the smaller the eye relative to the snout); SL/Sn = standard length/snout length (the larger the value, the shorter the snout relative to body size); SL/BD = standard length/ body depth (the larger the value, the more streamlined the fish). *Bold values* = zooplanktivorous species only. (Modified from Winterbottom and McLennan 1993.)

eral modifications in body shape along the ancestral backbone of "herbivorous species living on the inland side of the reef": the shift to a deeper bodied fish with a mouth well below eye level in ancestor 1, followed by the addition of a longer snout (and hence a larger snout/eye ratio) and a longer intestine (associated with a dietary shift from leafy/ fleshy algae to filamentous algae) in ancestor 2. This continued evolution of body shape means that, minimally, the zooplanktivores in *Paracanthurus* and *Acanthurus* evolved from a different ancestral morphological background than did the zooplanktivores in *Naso*. Additionally, the zooplanktivores in *Acanthurus* and *Naso* evolved in a different habitat (deep-water side of the reef) than did the zooplanktivorous *Paracanthurus* species (subsurge reef). Given this difference in habitat, it is striking that, relative to the herbivorous members of each genus, pursuit of small zooplankters in more open waters appears to have been associated with body streamlining, gut shortening, snout shortening (and hence increasing relative eye size), and a decreasing angle between the eye and mouth in both *Acanthurus* and *Naso*. All of these changes have been hypothesized to increase the efficiency of foraging and predator avoidance in coral reef zooplanktivores (Jones 1968; Hobson 1991). Unlike the zoo-planktivores in *Naso* and *Acanthurus, Paracanthurus* has retained the ple-siomorphic habitat, angle between the mouth and eye, and relative (small) eye size. If an increase in relative eye size and a decrease in the angle between the mouth and eye allow a fish to detect and capture small prey items more effectively, could the retention of these plesio-morphic states explain why there is only one species of *Paracanthurus*? Perhaps *Paracanthurus* is not a very effective zooplanktivore!

In general, then, the phylogenetic analysis has allowed us to trace the evolution of feeding preferences in this suborder of fishes and to tentatively highlight the relative points in history at which changes in behavior/ecology were associated with changes in morphology. The rela-tionships among these factors must now be studied experimentally, us-ing the power of the comparative functional morphology approach developed by George Lauder and coworkers.[10] Increased phylogenetic resolution is necessary in order to reconstruct the details of diet evolu-tion in these fish. For example, *Acanthurus olivaceus* grazes upon diatoms and detritus in sand patches within coral reefs. Although classified as an herbivore, this species displays a more elongated body than, and a simi-lar angle between the eye and mouth to, the zooplanktivore *A. thomp-soni*. If these two species are sister species, or if *A. olivaceus* is the sister species to a clade of zooplanktivores, then two putative adaptations for hunting zooplankton, "streamlined body" and "large angle between the eye and mouth," predated the appearance of zooplankton picking. This, of course, begs the question, Is *A. olivaceus* morphologically unique com-pared with other closely related herbivores, and if so, why? Finally, we might ask, Are the general modifications of body shape in ancestors 1 and 2 functionally involved in the shift from macrophytic to filamen-tous algae? Potential details of foraging evolution and directions for fu-ture research aside, this study has demonstrated one very important point: the ability of different species to solve the same problem is influ-enced by their underlying history. In this case, because of the different morphological backgrounds, the origin of zooplanktivory in *Naso, Para-canthurus,* and *Acanthurus* is associated with similarities in the direction of change for some characters, but not in the magnitude of those changes. At the end of the day (so far), the zooplanktivorous species in the three different genera are similar to one another, but they are not identical.

10. Lauder 1981, 1982a,b, 1983a,b, 1985a,b, 1986, 1989a,b, 1990, 2000; Lauder and Shaffer 1985, 1993; Lauder and Liem 1989; Lauder and Wainwright 1992; Lauder et al. 1995; Zweers and Vanden Berge 1997; Aerts et al. 2000; Liem and Summers 2000; and Van Leeuwen et al. 2000.

From Fish to Fish Food: *Daphnia* Ecology. If you have ever spent summer afternoons gazing into freshwater ponds and marshes, you have probably seen swarms of what appear to be small, highly animated, (usually) reddish dots. Chances are those dots belong to a species in the genus *Daphnia*, perhaps the most famous of the crustaceans inhabiting freshwaters around the world. Besides being coveted by aquarists for fish food, daphniids are admirable targets for evolutionary studies. During periods of relative environmental stability, female daphniids are parthenogenetic, producing diploid eggs that develop into daughters. When conditions change in a way that depresses their metabolic rate, however, females switch to the production of both male and female offspring. Once the sex ratio changes in the population, females begin producing haploid eggs that require fertilization to order to develop. These eggs are encased in a protective structure (the ephippium) that resists desiccation and freezing and are thus well suited to surviving harsh conditions or acting as agents of dispersal (Hebert 1987). Not surprisingly, daphniids have invaded a wide variety of habitats which, in combination with their large population sizes, rapid reproductive rates, and ancient origin (fossilized ephippia are known from the early Cretaceous: Fryer 1991), should promote substantial diversification. Given all of these characteristics, why then are daphniids notorious for their morphological stasis (Frey 1987)? Is it possible that the perceived ecological diversification is a taxonomic artifact? In other words, perhaps there has been so little morphological change in *Daphnia* because species have not invaded new habitats all that often. If this is true, then we would expect to find that species inhabiting the same type of environment are most closely related to one another (part of the same monophyletic group) than they are to species living in different environments.

Colbourne (1998) investigated this question by reconstructing the phylogenetic relationships among the 33 species of North American daphniids. The first analysis, based upon 270 variable characters of the 12S rRNA gene, produced seven equally parsimonious trees, which differed only in the relationships among the six species of the *D. pulex* complex (Colbourne and Herbert 1996). Those relationships were further resolved using the 498 nt sequence of the ND5 gene (Colbourne et al. 1998), a simplified version of which is shown in figure 5.21a. Subdividing habitat preferences into three categories (ephemeral pond, permanent pond/small lake, or larger lakes) produces four equally parsimonious optimizations for the diversification of those preferences (fig. 5.21b). All scenarios agree that living in ephemeral ponds is the plesiomorphic condition for North American daphniids and that there have been (1) three transitions from ephemeral ponds to permanent ponds/

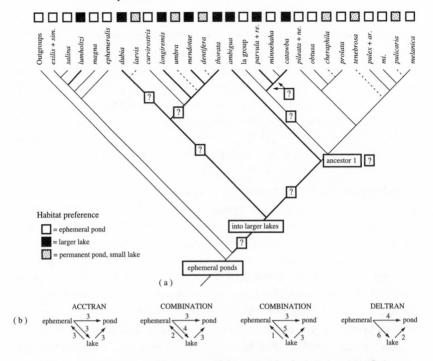

Figure 5.21. The evolution of habitat preferences in North American *Daphnia* species. (*a*) Characters optimized onto a reduced phylogenetic tree for the clade (Colbourne et al. 1998). *Thin lines* = ephemeral habitats; *dotted lines* = permanent ponds/small lakes; *thick lines* = larger lakes. Outgroup = *Simocephalus; ar* = *arenata;* la group = (*latispina* + *villosa*) + *oregonensis; mi* = *middendorffiana; ne* = *neo-obtusa; re* = *retrocurva, sim* = *similis.* *Question mark* = state ambiguous. Only one of many optimizations is shown on this figure. (*b*) The number of possible transitions in habitat preference under different optimization schemes.

small lakes in *Daphnia pulicaria, D. tenebrosa,* and *D. cheraphila,* (2) one transition from ephemeral ponds to larger lakes in *D. lumholtzi,* and (3) no transitions from permanent ponds/small lakes back to ephemeral habitats. Researchers interested in studying the forces influencing the evolution of habitat shifts in this clade should focus their attention on the preceding four species. For example, do *D. pulicaria, D. tenebrosa,* and *D. cheraphila* show any convergent morphological or ecological traits that might be associated with the apomorphic movement into the "same" habitat? Past that point, the optimizations indicate there have been as many as six and as few as three independent invasions of lakes from the plesiomorphic backbone of ephemerality. Almost all of the ambiguity in this group is centered around the number of times that the reverse transition, from large lakes to ephemeral habitats, has occurred.

So, do *D. curvirostris, D. minnehaha,* and most of the other descendants of ancestor 1 live in ephemeral habitats because their ancestors did, or because they show a secondary reversal to the plesiomorphic condition? In general, the patterns do indicate that even if the transition from ephemeral to permanent water is reversible, it does not happen very often. This observation is not trivial. Are there some genetic, developmental, physiological, or ecological factors opposing that movement? Is it possible that both transitions are equally likely, but that ephemeral species go extinct much more often, thus obscuring the patterns of habitat evolution in the group?

Overall, the analysis indicates that North American daphniids are capable of, and have indulged in, a fair amount of habitat shifting. Colbourne expanded his study to investigate the evolutionary interactions among habitat, antipredator, filtering, and ephippial morphology, melanization, obligate versus cyclic parthenogenesis, and chromosome number. Although the details of that analysis are too complicated to discuss here, there is one conclusion of particular relevance to studies of character evolution.[11] The purported morphological stasis in the clade is, in fact, the result of three factors: (1) a burst of morphological innovation in the mid-Mesozoic, which established the basic phenotypes of the three modern subgenera, followed by (2) a long period of stasis ("pools trod by dinosaurs undoubtedly contained a daphniid fauna similar to extant forms": Colbourne 1998:76), punctuated by (3) the convergent appearance of various morphological and ecological characters following repeated invasions of similar habitats (Taylor et al. 1996; Weider et al. 1999). Only by examining these changes with a phylogenetic framework could the contributions of plesiomorphic and convergent characters to the overall picture of "stasis" in the group be disentangled. The rapid phenotypic diversification may have represented the reemergence of suppressed characters under similar selection regimes. For example, back mutations, recombinations, and gene conversions can lead to the recovery of a lost transcription factor binding site, reactivating a previously silenced gene (Li 1997). If daphniids are indeed "predisposed to evolve particular traits under certain selection regimes" (Colbourne 1998: 111), this implies that developmental pathways lying dormant for up to 100 million years are still capable of being reactivated in these crustaceans.

---

11. If you are interested in a template for conducting comparative phylogenetic studies on a grand scale, see Colbourne 1998.

Increasing Your Objectivity

As we mentioned previously, it is difficult to completely remove the effects of social conditioning from the ways in which we see the world and consequently frame our hypotheses. How can a phylogenetic tree help? Consider the following example. You are an avid ichthyologist who has discovered a species of fish (*Biggus fishus*) that shows very marked sexual dimorphism for body size (males > females: fig. 5.22a) in contrast to all the other members of the family, which appear to be sexually monomorphic. In the past, most researchers would have looked at *B. fishus,* focused their attention on the male, asked, Why did the males get bigger? and collected data about male territoriality, dominance, swimming endurance, etc. Now, let's assume that you have a robust phylogeny for the *Biggus* clade, and the phylogenetic relationships indicate that *B. fishus* is one of the most recently derived members of the group. This tree supports your hypothesis that sexual dimorphism in *B. fishus* is unusual (derived) within the group. Depending upon the actual size of the males and females in all of the *Biggus* species, however, there are

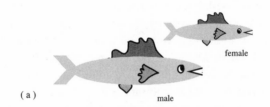

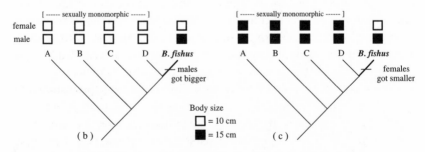

Figure 5.22.   Increasing your objectivity when framing an evolutionary hypothesis. (*a*) The newly discovered, sexually dimorphic *Biggus fishus*. According to the phylogenetic tree, *B. fishus* is the unusual member of a sexually monomorphic clade. The route to sexual dimorphism is different, however, depending upon whether (*b*) the male got bigger and the female stayed the same size or (*c*) the female got smaller and the male stayed the same size.

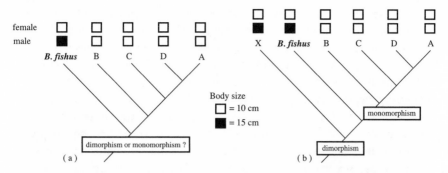

Figure 5.23. Increasing your objectivity continued. (*a*) If *B. fishus* is the basal member of the group, then we cannot determine which character state is unusual (monomorphism or dimorphism) in the clade without additional information from outgroups. (*b*) All of the outgroups (X) are sexually dimorphic, indicating that monomorphism is derived in the *Biggus* clade. If you had decided that monomorphism was plesiomorphic based upon the assumption that common = primitive, you would have been wrong.

three possible explanations for the origin of that dimorphism: (1) the males got bigger (fig. 5.22b), (2) the females got smaller (fig. 5.22c), or (3) the males got bigger and the females got smaller (we leave this for you to work out).

Now suppose *Biggus fishus* is in fact the most basal member of the *Biggus* clade (fig. 5.23). Without information from outgroups to the clade, we cannot tell whether it is the sexual dimorphism in *B. fishus* or the sexual monomorphism in the remaining species that is unusual (derived) within the group (fig. 5.23a). For example, if the closest relatives of *B. fishus* are all sexually dimorphic, then the relevant question for this clade is, Why are the remaining *Biggus* species monomorphic? (fig. 5.23b). In this case, there is nothing markedly unusual about *B. fishus* at all, despite all the initial excitement about discovering a species that seemed so "different" from other members of its group.

 This example illustrates a very common mistake many biologists make: the assumption that "common (or widespread) = primitive (or ancestral)." This mistake is not trivial. If the situation shown in figure 5.23b gives us the correct picture of character evolution, then someone assuming that sexual dimorphism is an autapomorphy for *B. fishus* could potentially waste a substantial amount of time and energy looking for correlations between extant environmental conditions and a character that may have evolved millions of years ago under very different conditions.

Big Females, Little Males: The Wonderful World of Spiders. Female spiders are often larger than their suitors. In one group of orb-weaving spiders the
dimorphism reaches astonishing proportions, with females approximately 12 times larger than the (very respectful) males. This extreme
dimorphism has traditionally been examined by searching for the forces
favoring the evolution of male dwarfism. For example, Vollrath and Parker (1992, 1997) hypothesized that male dwarfism was an evolutionary
strategy that evolved in response to a complex feedback among low population densities, predation on males searching for mates, and biased
operational sex ratios. Although the interactions among the various
components of the model were complicated, the overall picture was
simple: one general model could explain the presence and distribution
of sexual dimorphism in spiders. Bearing *Biggus fishus* in mind, let us
examine whether the model is, in fact, supported by the data.

Hormiga et al. (2000; see also Hormiga et al. 1995; Coddington et al.
1997) collected such data for 79 genera of orb-weaving spiders (9 of the
12 families in the Orbiculariae), plus representative species from the sister group (*Dictyna*). As per previous discussions in the literature, species
were scored as sexually dimorphic if one sex was at least twice as large
as the other sex. Mapping the ratios onto the composite tree revealed
that sexual dimorphism has originated four times in the orb weavers
(fig. 5.24). Contrary to the male dwarfism hypothesis, all four origins of
sexual dimorphism involved an increase in *female* body size. Two of
those increases were coupled with a decrease in male size (in the ancestors of *Tidarren* and of *Kaira*), while the other two increases occurred
without any appreciable change in the males (in the ancestors of *Nephila*
+ *Herennia* + *Nephilengys* and of the argiopoids). Researchers who attempt to explain the evolution of sexual dimorphism in these groups by
delineating the selective advantages accrued to males that "get small"
will thus be either "digging in the wrong place" (Lucas et al. 1981) or
digging in the right place, but uncovering only part of the picture. Once
again, a parsimony-based analysis has refuted the hypothesis that the
world is simple; clearly there is more than one explanation for the evolution of sexual dimorphism in these spiders.

The analysis also revealed that the ancestral background of sexual dimorphism (caused by an increase in female size) has reverted to monomorphism at least seven times within the argiopoid clade (fig. 5.25). Let
us assume that we have a selection-based hypothesis for the evolution
of monomorphism from dimorphism. The phylogenetic analysis has
identified seven independent tests of our hypothesis. After modest celebrations, should we rush out and begin collecting data? The answer to
this question is, Not right away. We have just determined that dimor-

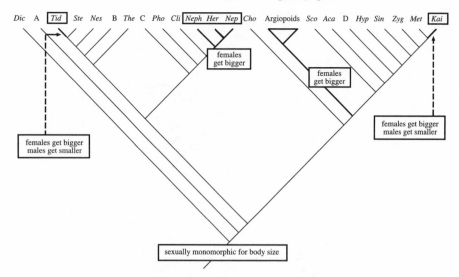

Figure 5.24. Did female spiders get bigger, or did the males get smaller? The tree is a composite of several studies and is thus only a tentative hypothesis of relationships within the orb-weaving spiders (Hormiga et al. 1995; Griswold et al. 1998; Scharff and Coddington 1997). *Aca* = *Acanthepeira; Cho* = *Chorizopes; Cli* = *Clitaetra; Dic* = *Dictyna; Her* = *Herennia; Hyp* = *Hypsosinga; Kai* = *Kaira; Met* = *Metepeira; Nep* = *Nephilengys; Neph* = *Nephila; Nes* = *Nesticus; Pho* = *Phonognatha; Sco* = *Scoloderus; Sin* = *Singa; Ste* = *Steatoda; The* = *Theridiosoma; Tid* = *Tidarren; Zyg* = *Zygiella;* A = *Deinopoidea;* B = *Linyphia* + *Pimoa;* C = other Tetragnathidae + *Azilia* + *Dolichognatha;* D = other Araneinae. *Thick lines* = sexually dimorphic for body size. (Modified from Hormiga et al. 2000.)

phism can be achieved in two different ways in these spiders, so we need to look more closely at the details of the "reversals" before assigning them all to the same potential mechanism. In other words, the first thing you should do when you have identified homoplasy a posteriori is go back and reexamine the way the character was coded in the first place. In this example, closer examination of our homoplasious character indicates that there are four different pathways to monomorphism in this group. Specifically, monomorphism has been achieved via (1) an increase in male/no change in female size (three times), (2) a decrease in female/no change in male size (two times), (3) an increase in both sexes (with a greater increase in male than female size: one time), and (4) a decrease in both sexes (with a greater decrease in female than in male size: one time). Clearly "reversal to sexual monomorphism" is not the same in all groups. Any attempt to use each of the seven reversals as an independent test of a particular hypothesis about the evolution of monomorphism would thus be doomed to failure. If you are interested in replicated tests of an evolutionary hypothesis, the best question to

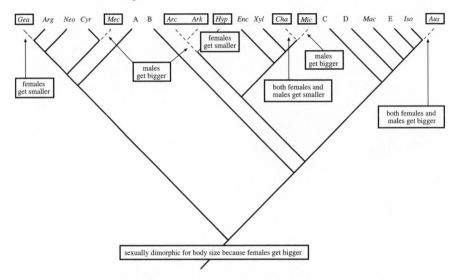

Figure 5.25. Examining changes in the size of male and female spider within the argiopoid clade (see fig. 5.24 for the position of this clade within the larger phylogeny). *Arc = Archemorus; Arg = Argiope; Ark = Arkys; Aus = Austracantha; Cha = Chaetacis; Cyr = Cyrtophora; Enc = Encyosaccus; Hyp = Hypognatha; Iso = Isoxya; Mac = Macracantha; Mec = Mecynogea; Mic = Micrathena; Neo = Neogea; Xyl = Xylethrus; A = Arachnura + Witica; B = Mastophora + Cyrtarachne + Pasilobus; C = Aspidolasius + Caerostris; D = Gastroxya + Augusta; E = Togacantha + Gasteracantha + Aetrocantha. Dotted lines* = sexually monomorphic for body size.

ask might be, What factors have influenced the evolution of increased male size in argiopoids? (potentially three independent tests).

It is likely that additional pathways to sexual monomorphism-dimorphism, including instances of male dwarfism in the absence of changes in female size, will be uncovered as more taxa are added to the analysis, and the relationships within and among genera are uncovered in more detail (although resolving the relationships among 10,000 species will take more than one person working diligently over an entire career!). In all, though, this fine-tuning will only heighten our appreciation for the "complex and rich tapestry" (Hormiga et al. 2000) of size evolution in spiders.

Myth Busting: Parasites Are Morphologically Simple and Degenerate. **Parasites are generally thought to demonstrate unusually high degrees of reductive evolution, signified by wholesale loss of characters. In free-living organisms, such as smelts and eels among fish, caecilians among amphibians, and snakes among amniotes, secondary loss of characters and the resulting simplification in morphology is explained as evolutionary spe-**

cialization; in parasites it is often called "degenerate evolution." Some authors have dispensed with the value-laden term "degenerate," stating only that parasites are extremely simplified; however, this perspective is also problematic because it is generally based upon a comparison between a parasite and its host. Clearly, a tapeworm is not as complex as a cat. Just as clearly, this is a specious comparison. If you want to demonstrate that parasites have experienced an evolutionary drive toward simplification, the appropriate comparison is between a parasite group and its free-living sister group(s). These statements about parasite evolution owe their origin directly to orthogenesis (for a more detailed discussion, see Brooks and McLennan 1993b). Although recent authors have attempted to recast the statements within a Darwinian framework, most researchers never stop to ask whether the generalizations about parasite biology are in fact supported by evidence. For example, Rogers (1962) stated that "the loss of sense organs is a common feature of parasitism." Rohde (1989), however, discovered in a single species of endoparasitic flatworm eight types of sense receptors not reported for their free-living relatives, opening the door for a reconsideration of our beliefs about parasite evolution.

So, let us examine the evidence (Brooks and McLennan 1993b,c) for one group, the parasitic platyhelminths (Neodermata: approximately 15,000 described species, including the much maligned liver flukes, beef tapeworms, and blood flukes: fig. 5.26). Of the 1,882 morphological

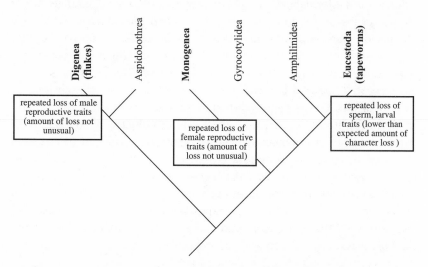

Figure 5.26. Examining our preconceived ideas. The amount of character loss is not unusual within the major groups of the parasitic Platyhelminthes. There is a difference among the three groups in the types of characters that are repeatedly lost.

Table 5.2.   Character Loss for Four Different Character Categories among Major Groups
of Parasitic Flatworms

Category	Character Loss				Total[a]
	Neodermata	Digenea	Monogenea	Eucestoda	
Male	4/23 (17%)	31/130 (23%)	27/188 (14%)	5/74 (5%)	67/430 (15%)
Female	0/24	6/230 (3%)	14/98 (14%)	2/109 (2%)	22/491 (4%)
Adult Nonreproductive	4/52 (8%)	31/222 (14%)	22/263 (8%)	6/128 (5%)	63/687 (9%)
Larval	7/58 (12%)	20/157 (13%)	10/38 (26%)	7/19 (37%)	44/274 (16%)
Total	15/157 (10%)	88/739 (12%)	73/587 (12%)	20/330 (6%)	196/1882 (10.4%)

[a]Information from the Aspidobothrea, Gyrocotylidea, and Amphilinidea included.

character changes described for the Neodermata, only 10.4 percent represents derived losses (table 5.2). In other words, character *innovation* (89.6 percent) far outstrips character loss in these organisms. Because we have little information about proportions of losses in free-living organisms (but see Begle 1991), in particular the free-living platyhelminths (see Lundin 2000 for a beginning), we do not know if the proportion of losses exhibited by the parasitic flatworms is unusually high. Nevertheless, this result casts serious doubt on the traditional perspective that parasites are highly simplified through secondary loss of characters.

The total amount of character loss is not distributed equally among the three major parasite groups (table 5.2). Character loss in the Digenea (flukes) and Monogenea occurs in proportion to total change, while character loss is disproportionately lower than total change within the Eucestoda (tapeworms). Once again, this is counterintuitive because tapeworms are often cited as one of the best examples of degenerate evolution. About half of the character loss in the Neodermata is due to the repeated loss of a few characters. In the flukes, the highest level of homoplasious loss is found in male reproductive (the genital sac and male intromittent organs) and adult nonreproductive characters (e.g., the ventral sucker has been lost at least eight times), and the lowest level is found in female reproductive characters. This pattern is reversed in the Monogenea, which show the highest level of homoplasious loss in female characters (the vagina and egg filaments) and the lowest level in male and larval traits. Finally, the tapeworms show very few homoplasious losses. Of these, the loss of an axoneme in the sperm's tail and the loss of cilia on hexacanths and tails on plerocercoids (larval) occur more than once.

Most theories of parasite evolution are based upon the assumption that the absence of a trait generally represents a secondary loss. Because it is impossible to differentiate between secondary loss and plesiomorphic (ancestral) absence outside of a phylogenetic framework, orthogenetic-based theories of parasite evolution could not be tested until a sufficient database and a method for reconstructing patterns of character origin and elaboration had been developed. Now that both exist, we see that parasites do not require special theories to explain their evolution. They are neither unusually secondarily simplified nor degenerate and thus are as amenable as any other creatures to being investigated within a Darwinian paradigm. The phylogenetic perspective also highlights several fascinating avenues for future research in life history evolution. For example, under what conditions can the male intromittent organ (or the vagina for that matter) be lost (e.g., are all these species self-fertilizing)? So, besides emphasizing the (obviously) fascinating nature of these (unfairly) maligned creatures, the take home message from this example is simple: *Beware of untested assumptions.*

The Temporal Sequence of Evolutionary Change

Lorenz (1941:81) cautioned "The similarity of a series of forms even if the series structure arises ever so clearly from a separation according to characters, must not be considered as establishing a series of developmental stages." In his opinion, without reference to phylogenetic relationships, the criterion of similarity was, of itself, a dangerously misleading evolutionary marker. Forty years later a "new" methodology had developed based, in direct contrast to Lorenz's warning, upon arranging characters as a "plausible series of adaptational changes that could easily follow one after the other" (Alcock 1984:432). Although intuitively pleasing, this method relies heavily on subjective, a priori assumptions concerning the temporal sequence of evolutionary modifications and dissociates character evolution from underlying phylogenetic relationships. In many cases, the intuition-based pathways follow a "simple to complex" sequence reminiscent of orthogenetic reasoning. The re-introduction of "history" into our evolutionary perspective has prompted researchers to seek alternate methods for uncovering the direction and sequence of character change and to test our assumptions about the generality of linear character evolution.[12]

The comparative phylogenetic approach also allows us to test preexisting hypotheses of character evolution that have been constructed based upon rigorous mathematical models or functional analyses. For example, Kaneshiro (1976, 1980) postulated that speciation based upon

12. See Eldredge and Cracraft 1980; Greene 1986a; Carpenter 1989; Donoghue 1989; Ross and Carpenter 1991a,b; and Packer 1997.

small founding populations (peripheral isolates speciation) or bottle-necks would be characterized by destabilization of the mate-recognition system in the founding population (Templeton 1979; Giddings and Templeton 1983). His model, which is generally applied to *Drosophila*, involved two critical steps: (1) the loss of some elements of the male courtship sequence as a consequence of genetic drift and (2) selection for increased permissiveness in female mating preferences due to the low availability of mates, leading eventually to a simplification of mating requirements by females in the derived taxa. Given these two changes, Kaneshiro predicted that females from ancestral populations should discriminate against males from derived populations, while derived females should accept the courtship overtures of males from both ancestral and derived stock. Indeed, derived females might even show a preference for the more complex or vigorous courtship of ancestral males over the simplified overtures of conspecifics. Kaneshiro's proposal that the direction of the mating asymmetry could be used to differentiate basal from more recently derived species (i.e., that it was an unambiguous indicator of phylogeny) has been hotly debated on a variety of theoretical grounds (Kaneshiro and Giddings 1987; deSalle and Templeton 1987; Ehrman and Wasserman 1987; Arnold et al. 1996). From a comparative phylogenetic perspective, resolution of the debate seems fairly straightforward. First, we need a robust phylogeny for the *Drosophila* clade of interest. Now, let's examine the scenario in which the two study species are not sisters. The Kaneshiro hypothesis is supported if the species with the most complex courtship is basal and the species with simplified male courtship is more recently derived on the tree (fig. 5.27a). This, of course, does not "prove" that the hypothesis is correct, only that it cannot be ruled out on phylogenetic grounds. If, on the other hand, the species with the simplified courtship falls out as basal to the species with the more complicated courtship, then the Kaneshiro hypothesis is falsified (fig. 5.27b). This second pattern would indicate that we need to propose other explanations for the observed asymmetry in interspecific mate choice.

How do we resolve the problem if the two species are sisters? If the species with the simplified male courtship is indeed the peripheral isolate and the species with the more complicated courtship its persistent ancestor, then we need a method that will allow us to differentiate between the two. You should recognize this as a job for historical biogeography.

Locks, Plugs, and Predation: Copulatory Behavior in Rodents. Of all living creatures on this planet, we probably know more about the copulatory behavior

Observations:

Males from species D have a simple courtship repertoire.
Males from species A have a complex courtship repertoire.

Results of interspecific versus conspecific mate choice tests: Asymmetry in mate choice.

Females from species D either show no choice or actually prefer males from species A over conspecific males.
Females from species A prefer conspecific males over males from species D.

Phylogenetic prediction based on Kaneshiro hypothesis:

Species A is more basal within the clade than species D.

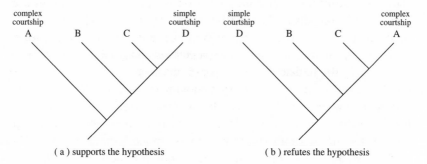

(a) supports the hypothesis   (b) refutes the hypothesis

Figure 5.27.   Testing existing evolutionary hypotheses. Phylogenetic patterns that will (*a*) support or (*b*) refute the Kaneshiro hypothesis.

of rodents than of any other creatures (excluding *Homo sapiens*). One trait of particular interest to psychologists and ethologists alike has been the evolution of a mechanical connection between the male and female that holds the pair together during mating. This "lock" may be so strong that one partner can actually drag the other about by moving away at an inappropriate time. Dewsbury (1975) proposed that the evolution of locked from unlocked copulation was driven by movement of species from dangerous to relatively safe habitats. He envisioned the following scenario: (1) A basal species is subject to high levels of predation or harassment by conspecifics during mating. If the copulatory lock originates, the costs of increased predation or harassment (lock prevents rapid escape) will outweigh the benefits from increased paternity assurance, etc., and the lock will not persist. (2) The ancestral species moves into a safer environment (lower levels of predation, lower population densities, and decreased competition for mates). If the copulatory lock originates here, the benefits will outweigh the costs, and the lock will spread throughout the population, eventually becoming a fixed feature of the ancestral species and its descendants. Although this was based upon the assumption that the sequence of locked copulation evolution went from absent to present in all groups showing the behavior, there is no necessary prohibition to the sequence being reversed. The more interesting question is whether the sequence is irreversibly locked in time; are any species that show an absence of locked copulation demonstrating a secondary loss? If the trait can be lost, are there any other changes associated with the disappearance of what seems to be a fairly important part of the mating system?

Langtimm and Dewsbury (1991) tackled these questions by investigating the origin and diversification of copulatory behaviors in a monophyletic group of New World rodents. They analyzed the distribution of (1) a copulatory lock; (2) multiple intromissions: a cycle of mounting, penetration, and dismounting that eventually ends in ejaculation; (3) pelvic thrusting following intromission; (4) rapid or ballistic dismount following ejaculation (rather than ejaculation followed by passive male dismount as the female moves away); (5) postejaculatory intromissions: continuation of the mount-mate-dismount cycle after ejaculation has occurred (Dewsbury 1972, 1975). Mapping these traits onto a phylogeny for the clade derived from morphological traits revealed a number of fascinating things about the rodent male-female mating dialogue (fig. 5.28). First, and most surprisingly, the proposed hypothesis for the evolution of the copulatory lock was reversed in this group: the presence of the lock is plesiomorphic for the group and was (apparently) lost twice, in ancestor 1 and in *Neotoma lepida*. Although the sequence of character origin is reversed, this does not necessarily negate the original adaptive hypothesis, because that hypothesis was not itself specifically directional. So, in this group of rodents it is possible that locked copulation was retained in relatively safe environments for all of the reasons discussed above. The subsequent loss of the behavior permitted *Neotoma lepida* and ancestor 1 (+ its subsequent descendants) to disperse into more dangerous areas, in essence, to take up "unsafe" lifestyles by practicing very safe sex (from a predator avoidance point of view). In order to test this hypothesis in detail, we would need to compare predation pressures in lockless species with predation pressures in their close relatives. Fortunately, there are two opportunities to examine this question: *Onychomys* versus *Podomys* (this comparison is not confounded by the elaboration of novel mating behaviors found in other descendants of ancestor 1) and *Neotoma lepida* versus its sister species.

The copulatory locking story gets even more fascinating with the addition of more characters. For example, both losses of the lock were coupled with the appearance of a copulatory plug. This gives additional weight to Dewsbury's hypothesis because, in essence, the fitness benefits of the lock are retained in the plug (and may even be increased with the addition of such functions as prevention of sperm loss or preventing the female from mating with another male for a period of time: Voss 1979;

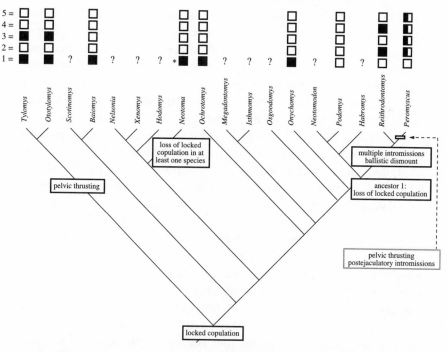

Figure 5.28. Mating behaviors mapped onto a simplified phylogenetic tree, based upon morphological characters for murid rodents (CI = 71%; Carleton 1980). Numbers refer to characters: 1 = locked copulation; 2 = multiple intromissions; 3 = pelvic thrusting; 4 = ballistic dismount; 5 = postejaculatory intromissions. *White boxes* = absent; *black boxes* = present; *asterisk* = absent in one species (*N. lepida*); *question mark* = behavior unknown.

Dewsbury 1988), without the associated costs in terms of survival. This, of course, begs the questions, How did this happen? and What are the mechanisms underlying the production of the mechanical lock and copulatory plug? Further examination of male genitalia indicates that species with the lock have a (relatively) thick glans penis with large spines and incomplete accessory glands (Arata 1964; Dewsbury 1975), whereas the loss of locking is correlated with a decrease in penile width, fewer spines, and elaborated glands. It is thus possible that changes in penis structure affect the presence or absence of the lock, whereas changes in secretions from the accessory glands affect the presence or absence of a plug. In order to determine whether these two characters are always coupled, we would need to broaden the evolutionary perspective of the study to include other murid lineages.

Finally, loss of the copulatory lock in ancestor 1 was followed by the evolution of elaborate copulatory behaviors such as multiple intromissions and ballistic dismounts. Prior to that point, mating behaviors had been fairly conservative for a long period of time (locked copulation with one quick intromission and a passive separation of partners following ejaculation). After the loss of locked copulation, however, there was a relatively rapid diversification of mating behaviors in the *Reithrodontomys* + *Peromyscus* lineage, which hints at female choice for increased mating stimuli (Eberhard 1985). One way to examine this hypothesis might be to offer female *Podomys* a choice between their own males (no lock, no novel behaviors) and males from their more elaborate, lockless relatives (e.g., *Reithrodontomys* [multiple intromissions + ballistic dismount] or *Peromyscus haplomylomys* [all four novel behaviors]). In order to study this diversification further, we would need to collect more behavioral data and expand the phylogenetic resolution within *Peromyscus*.

The Evolution of Eusocial Behavior in Wasps. Eusociality has attained a special place in evolutionary biology because it is one of the systems that perturbed Darwin the most (along with the presence of exaggerated traits that appeared to hamper male survival). His initial response to the problem was that it posed "one special difficulty, which at first appeared to me insuperable, and actually fatal to my whole theory" (1859:236). West-Eberhard (1978) proposed that the evolution of eusociality followed this pathway: (stage I) solitary: female seals an egg and food into a nest cell then provides no further care; (stage II) casteless polygynous family group: females, including mothers and daughters, share nests, a situation favored possibly because appropriate nest sites are limited; (stage III) rudimentary caste-containing group: differences in reproductive success among females sharing a nest site leads to temporary or incomplete (and still facultative) division of reproductive labor; (stage IV) eusocial stage: increased reproductive and social dominance leads to permanent division of labor, and "loser" females always fail to reproduce. Eusociality itself could pass through various stages (fig. 5.29), depending upon the length of time reproduction was monopolized by individual queens (either shorter or longer than the average egg-adult developmental time) and the number of queens in the colony (monogyny: single queen; polygyny: more than one queen at a time: see discussion in Ross and Carpenter 1991a).

Carpenter (1989, 1991; see also Ross and Carpenter 1991a,b; Hunt 1999) examined West-Eberhard's hypothesis using one of the best-studied hymenopteran groups, the vespid wasps. Three of the six vespid subfamilies are solitary (the Euparagiinae, Masarinae [pollen wasps], and Eumeninae [potter wasps]: Carpenter and Cumming 1985), two are eusocial (the Polistinae [paper wasps] and the Vespinae [yellowjackets and hornets]), and one subfamily, the Stenogastrinae (hover wasps), shows

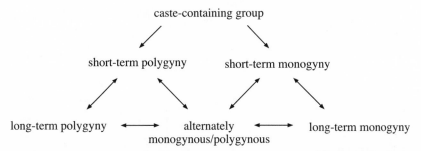

Figure 5.29. Transitions among different eusocial states depending upon number of queens in the colony (monogyny versus polygyny), and the length of time individual queens monopolize reproduction. (Redrawn from West-Eberhard 1978, Carpenter 1991.)

a more complicated diversity of behaviors. Carpenter (1988b) reported that three additional characters relevant to the study of eusociality were present in all hover wasps that had been investigated in detail: (1) females showing extended parental care (tending offspring to the pupal stage), (2) facultative nest-sharing (adult females were added to the nest site by either emergence and persistence of daughters in the area, by less closely related females joining the group, or both), and (3) helping behavior (newly emerged females assisted in the care of other females' broods until they themselves left to begin their own nest). Based upon these data, he classified the ancestral state for the Stenogastrinae as a rudimentary caste-containing group, with the caveat that this hypothesis could change as additional behavioral data were collected.

Mapping the relevant behavioral characters onto the phylogeny for the Vespidae (fig. 5.30) indicates that the evolutionary pathway to eusociality goes from solitary (stage I) to nest sharing with a rudimentary division of reproductive labor (stage III) to true eusociality (stage IV). The evolutionary transition from stage III to IV also appears to have occurred independently within the Stenogastrinae (a permanently sterile worker caste has been reported from one species). The patterns thus corroborate all of West-Eberhard's basic model except for one component: there is no phylogenetic evidence for the occurrence of the casteless, nest-sharing stage II along the main evolutionary line, although that stage has evidently arisen independently from solitary stage I in a few species of Masarinae and Eumeninae (Carpenter 1991). West-Eberhard proposed that this stage might be more transitory than the others if worker-like behavior was expressed facultatively at first, so the absence of stage II in the deep history of the wasps is not problematic for her model.

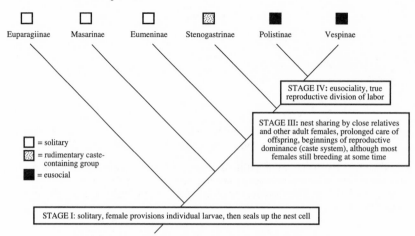

Figure 5.30. Major stages in the pathway to eusociality mapped onto the phylogenetic tree for the Vespidae (one tree, based upon of 48 morphological and six behavioral characters: Carpenter 1982, 1987, 1988).

Carpenter then investigated the predicted transitions among the different eusocial states by mapping characters associated with the transitions onto phylogenies for the Polistinae and the Vespinae. Within the Polistinae (fig. 5.31), short-term, behaviorally enforced monogyny is found in two genera; long-term polygyny is displayed by 24 genera; both short- and long-term monogyny are found in two genera (*Polistes* and *Mischocyttarus*); and short-term monogyny and long-term polygyny are both present in *Ropalidia*. None of the long-term monogynous species are the basal members of either *Polistes* or *Mischocyttarus,* so those groups are scored as displaying ancestral "short-term monogyny." Mapping these traits onto the phylogeny supports two of West-Eberhard's proposed transitions, from short- to long-term monogyny (at least twice within *Polistes* and *Mischocyttarus*) and from short-term monogyny to long-term (swarm-founding) polygyny (three times in *Polybioides, Ropalidia,* and the ancestor of the remaining 23 genera). Within the Vespinae (fig. 5.32), all groups but one (*Vespa*) have a long-term monogynous system. Two clades, the *Vespula vulgaris* and *V. squamosa* species groups, also contain long-term polygynous species but, once again, these species appear to be derived within their respective clades. Unfortunately the behavior and phylogenetic relationships among species of *Vespa* are poorly known. This is somewhat frustrating because that genus is the basal member of the Vespinae, which makes it critical for determining the ancestral condition for eusociality in the subfamily, even though the remaining vespine groups all display the same character state. This

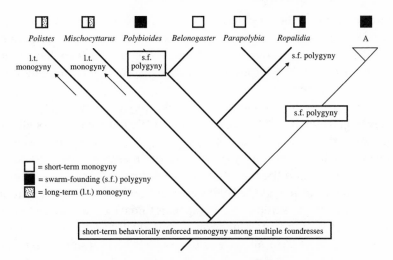

Figure 5.31. Different states of eusociality mapped onto the phylogeny for the Polistinae (paper wasps: Wenzel and Carpenter 1994; Wenzel 1998). The tree, based upon 97 morphological and nest-architectural characters (Carpenter 1991; Wenzel 1991), is a consensus of 47 equally parsimonious trees differing only in the relationships among taxa in group A (the remaining 23 polistine species). Changes in A do not affect the optimization. *Bold lines* = ancestral (symplesiomorphic) condition of short-term monogyny.

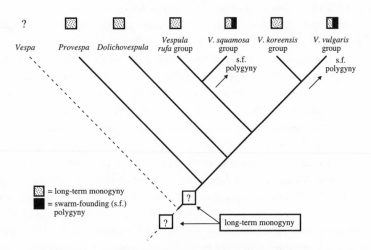

Figure 5.32. Different states of eusociality mapped onto the phylogeny for the Vespinae (hornets and yellowjackets). Phylogenetic analysis of 32 morphological and ten behavioral characters produced one tree, CI = 76% (Carpenter 1987). *Bold lines* = plesiomorphic condition of long-term monogyny.

ambiguity aside, the phylogenetic patterns for the hornets and yellowjackets provide unambiguous support for at least two transitions
from long-term monogyny to long-term polygyny. Until we have more
information from *Vespa,* the character state preceding long-term monogyny remains a mystery.

Overall, then, Carpenter's comparative phylogenetic analysis provides support for all but two components of West-Eberhard's model. The
only steps that were not discovered were the transition from a casteless
polygynous family group (stage II) to a rudimentary caste-containing
group (stage III) and the transition from a rudimentary caste-containing
group directly to polygyny. Pending resolution of the eusocial state in
Vespa, the major evolutionary path appears to have been solitary to nest
sharing with a rudimentary division of reproductive labor to eusociality
based upon short-term monogyny to eusociality based upon long-term
monogyny, with numerous independent origins of polygyny from monogyny. Additionally, it appears that some of the eusocial stages depicted as being reversible in the model may in fact be unidirectional. For
example, once long-term polygyny appears, it does not revert to either
long- or short-term monogyny, and once long-term monogyny appears,
it does not revert to short-term monogyny. This suggestion is only tentative pending further analysis of the species-level relationships within the
eusocial groups, but if that analysis supports the apparent irreversibility
of these stages, the model needs to be adjusted to incorporate explanations for why such a pattern might be selectively advantageous (if at all).

There is a further wrinkle to the social wasp story. Schmitz and Moritz
(1998) performed a phylogenetic systematic analysis for the wasps but
included honey bees (Apinae) in the study. The analysis produced a
single tree (CI = 62.1%: fig. 5.33) that placed the Stenogastrinae (hover
wasps) as the sister group to a clade comprising the honey bees + Eumeninae (potter wasps) + Polistinae (paper wasps) + Vespinae (yellowjackets and hornets). Unfortunately, the two solitary subfamilies, Euparagiinae and Masarinae [pollen wasps], were not included in this new
analysis, so we cannot trace the evolution of sociality with confidence.
For example, if we assume that the pollen wasps remain as the basal
members of the clade, then optimization of social behavior could still
follow the pathway described by Carpenter (fig. 5.30), depending upon
the state of sociality in the Apinae. As mentioned above, Schmitz and
Moritz used honey bees in the their analysis, and honey bees are eusocial
(Ross and Carpenter 1991a). This would move the transition from stage
III to stage IV to an older point in the history of the Vespidae and would
also imply that the potter wasps have secondarily lost social behavior. Is
it possible to lose such a complicated system? As we have discovered

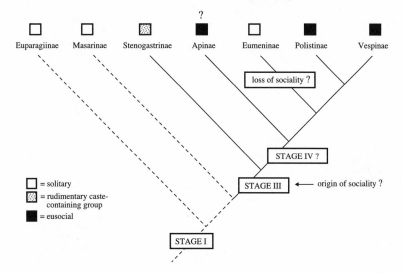

Figure 5.33. A new phylogenetic tree, based upon the inclusion of honey bees in the analysis (Schmitz and Moritz 1998). The basal two vespid subfamilies were not included in this study. *Question mark* = data insufficient at the moment, but the sequence of social evolution might still follow the pathway detected by Carpenter (see fig. 5.30) if the characters and relationships shown in this heuristic diagram are confirmed by future research.

before, discussions of character evolution are only as robust as the underlying phylogeny, so clearly there is ample room for additional research here. In particular, we need a total evidence phylogeny for all members of the Vespidae. That analysis must also include a detailed look at the rest of the Apidae (stingless bees + honey bees + bumble bees + orchid bees), because one of the tribes within the family, the orchid bees, is solitary.

Associations between Characters through Time

> We shall presently see that simple inheritance often gives the
> false appearance of correlation.
>
> Darwin 1872:141

There are at least three ways in which characters may be functionally related (fig. 5.34): (1) Traits x and y may be controlled by the same genetic system (genetic coupling) that is (for example) expressed differently in males and females. We would expect to see perfect patterns of co-origination and codivergence for traits of this type (fig. 5.34a). (2) Traits x and y may be controlled by independent genetic systems that have become linked through time (coevolution, reviewed by Butlin and

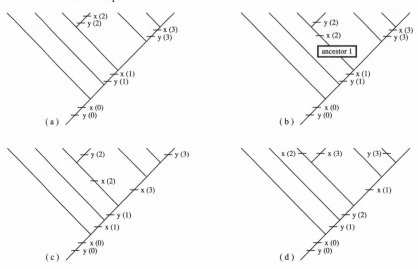

Figure 5.34. Using macroevolutionary patterns of origin and diversification to determine whether two traits might be correlated. (*a*) All three explanations for character correlation are supported. Notice that the apparent coupling of x (0) and y (0) is not relevant to the discussion because this is the plesiomorphic state for the characters. If the plesiomorphic states are both "absent," then that state has no bearing on the discussion of character correlation. If, however, the plesiomorphic condition involves character state presence (e.g., "x (0) = red dots" and "y (0) = bad taste"), then we would have to expand the scope of our study to pinpoint the actual origins of those character states before we could include them in our analysis. (*b*) The hypothesis that x and y are controlled by the same genetic system is falsified. The hypothesis that the two traits may have been genetically linked (the linkage was broken in ancestor 1) or that the evolution of character x is causally involved in the evolution of character y is supported. The hypothesis that the evolution of character y is causally involved in the evolution of character x is falsified. (*c*) The hypothesis that the two traits are controlled by the same genetic system or are genetically linked is falsified. The hypothesis that a modification of x is a necessary prerequisite for a modification of y (x is causally involved in the evolution of y) is supported. The hypothesis that the evolution of character y is causally involved in the evolution of character x is falsified. (*d*) The hypothesis of any direct interaction between the two characters is falsified.

Ritchie 1989). Genetic linkage should also produce tightly coupled patterns of origination and divergence between the two traits, however, linkage is sensitive to disruption (O'Neil and Schmitt 1993), so we might detect a break in that pattern over time (fig. 5.34b). (3) A change in x does not cause a change in y, but rather allows or promotes the change when the appropriate variant appears. Given that mutations producing variation in x and y are random, it is unlikely that the two traits will show perfect co-origination and codivergence (fig. 5.34c). Although this may produce macroevolutionary patterns of staggered origin and divergence, these patterns must still fall within boundaries specified by the biological interaction between the two characters in order to support a hypothesis of causality. In other words, if the two characters do not show a pattern of co-origination and codivergence, they must *originate and diverge in a predictable sequence* in order to support the hypothesis that they are causally related. Any hypothesis of functional relatedness that is based upon phylogenetic patterns thus must also include information about the biological functions and interactions of the characters (Sillén-Tullberg 1993; Jacobs and Wray 1995). Because both completely congruent and (apparently) decoupled patterns can support a hypothesis that two characters are interacting with one another, the patterns on their own are only incomplete guideposts to underlying mechanisms.

Prey Detection and Prey Capture: A Tale of Tongues in Lizards and Snakes. The animals we commonly think of as "lizards" form a paraphyletic assemblage that includes pet store favorites like skinks, geckos, chameleons, and anoles, the legendary (in evolutionary circles anyway) marine iguana, and the fierce varanids of Hollywood fame. In order to turn the "lizards" from a paraphyletic assemblage to a monophyletic group, we must add the snakes (Serpentes) to form a larger group called the Squamata. Of all their varied behaviors, the trait that has most fascinated people about squamates is their tongues (think of chameleons and snakes). Fascination aside, what do we actually know about the squamate tongue's function?

There are a number of ways to answer this question. Let us begin with the relationship between one aspect of tongue function, prey detection (olfaction), and foraging behavior. Active foragers tongue-flick with (relative) abandon. Sit-and-wait foragers, on the other hand, only tongue-flick while they are moving from one ambush site to another, not while they are actually sitting quietly, waiting for dinner to walk, crawl, hop, slither, or fly by. A number of adaptive hypotheses have been proposed to explain the evolutionary coupling of tongue flicking and foraging mode. For example, an ambush forager, as the name suggests, relies heavily upon the element of surprise to capture its meal. Tongue flicking has an obvious negative impact on a strategy of not being seen. Herbivores need not fear detection by their prey (or, to put it another way, fruit cannot escape even if it does detect the iguana). Active foragers may have already been seen, so one additional small movement more or less is not going to affect their hunting success. Based upon the data collected so far, the correlation between foraging mode and tongue flicking (present or absent) in squamates appears to be perfect, so you do not need a phylogeny to tell you that the two characters are intertwined. In order to determine whether that twining is causal or casual, you need information about the genetic, developmental, ecological and functional basis for the linkage. You do, however, need a phylogeny if you want to determine (1) which character complex came first, ambushing + no tongue flicking or active foraging + tongue flicking, (2) how many times these complexes have changed, if at all, and (3) whether the diversification of the complex has been correlated with any other character changes that could be causally involved in the evolution of foraging modes in this clade.

William Cooper has been collecting data about the foraging behavior of squamates for more than a decade (Cooper 1989a,b, 1994, 1995, 2000). In 1995 he wondered whether the hypothesis about the relationship between foraging mode and tongue flicking behavior would stand up under rigorous phylogenetic scrutiny. He optimized the two complexes onto a reduced phylogenetic tree for the squamates and concluded that the patterns did in fact provide support for the hypothesis (fig. 5.35). The optimization rests, in part, upon our confidence in the foraging mode assigned to the sister group of the squamates, the

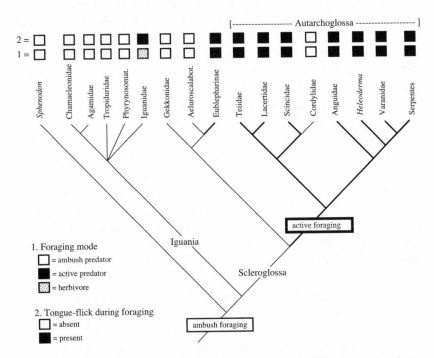

Figure 5.35. The evolution of foraging modes in squamates ("lizards" and snakes). This is a symplified phylogenetic tree based upon resolution from two studies (Caldwell 1999; Lee 2000b), so the results are tentative pending a complete phylogenetic analysis for the groups. Aeluroscalabot. = Aeluroscalabotinae; Phyrynosomat. = Phyrynosomatidae. (Modified from Cooper 1995.)

Sphenodontida (tuatara). All of the available data for the two species of tuatara indicate that they are ambush foragers that do not tongue-flick. In concert with data available for one of the basal squamate lineages, the Iguania (iguanas, chameleons, frill-neck and dragon lizards, anoles, basilisks), this indicates that ambush foraging is plesiomorphic for the clade. From that ancestral background there have been two changes (the double-origin hypothesis) to active foraging + tongue flicking, in the ancestor of the Autarchoglossa and in the ancestor of the Eublepharinae (e.g., the leopard geckos), and one reversal from active to ambush foraging with the associated loss of tongue flicking in the ancestor of the Cordylidae (girdle-tailed lizards). In addition to asking questions about the adaptive value of each strategy on its own, the patterns indicate that we should also be asking questions about the factors that influenced the diversification of those strategies. For example, are there any ecological/ morphological autapomorphies associated with the evolution of active foraging in the Eublepharinae or with the reversal to ambush foraging in the Cordylidae? For that matter, do cordylids hunt in the same way as their plesiomorphically ambushing relatives? Is the convergent appearance of active foraging associated with similar selection pressures in the Autarchoglossa and the Eublepharinae? Is active foraging "the same" in these two clades?

Having established the preliminary patterns of foraging evolution in the squamates, let us turn our attention to the tongue itself and examine its two functions, prey detection and prey capture, in more detail.[13] The generalized squamate tongue consists of (1) a complex muscular structure (the tongue proper) covered with a mucous-producing epithelium, (2) a skeletal component (the hyoid) suspended by a system of straplike muscles attached at one end to the lower jaw and at the other end to the sternum and shoulder, and (3) a small, thin bone (the entoglossal process) off the hyoid and into the tongue itself. Tongue projection is accomplished by moving the entire hyoid apparatus forward, by sliding the tongue along the entoglossal process, or both. Lingual (tongue) feeding involves three basic steps: (1) project the tongue forward, (2) curl the tip downward (allowing the sticky dorsal surface to make contact with the prey item), and (3) retract the tongue, along with the tasty item, into the mouth. Most squamates that use lingual feeding to capture food do not project their tongues very far. There are, however, two additional stages in the evolution of lingual feeding that result in increased tongue projection. Stage 1, shown by some agamid species, involves extending

13. The following discussion is based upon the research of Kurt Schwenk and coworkers; see Schwenk 1994, 1995b for additional references.

a substantial portion of the tongue downward and wrapping it around the anterior end of the entoglossal process, coupled with increased activity in the *M. verticalis* muscle, which pushes the entoglossal process out of the mouth. These two changes allow the lizard to project its tongue further, and most important, the anterior portion of the tongue, which is about to make contact with something reasonably hard, like a fly, forms a rigid block of muscle. Stage 2 involves ratcheting the action of the *M. verticalis* up another notch, so that the anterior part of the tongue (which now forms a muscular, glandular pad) continues moving forward beyond the entoglossal process. This second change allows the dramatic feats of tongue extension shown by chameleons, many of which can project their tongues as much as one and a half times their body length at a speed exceeding 5 meters per second.[14] As fascinating as lingual feeding is, it is not the only way squamates gather food. Many species grasp prey with their jaws and use their tongues only to manipulate food once it is inside their mouths.

The tongue also plays an important role in the chemosensory system of squamates. The chemical world around us can be loosely partitioned into two sensory systems, gustatory (taste buds on the tongue) and olfactory. In squamates, olfaction is itself subdivided into the nasal olfactory system (NOS: sensory cells in the nasal cavities transmit information via the olfactory nerve to the main olfactory bulb of the brain) and the vomeronasal system (VNS: sensory cells in an outpocketing of the nasal cavity called the vomeronasal organ transmit information via the accessory olfactory nerve to [wait for it] the accessory olfactory bulb). Volatile chemicals are detected as air is inhaled through the nose and are delivered to the VNS via air tongue-flicks. Tongue flicking appears to be mediated by complex hydrostatic elongation-retraction sequences, rather than the actual movement of the hyoid-tongue apparatus as described for feeding. The two behaviors, feeding and flicking, are thus mechanically as well as functionally distinct. How the tongue actually detects chemicals is a still a mystery, as are the precise mechanisms for delivering those cues from the tongue to the VNS. The tongue-VNS system is further extended to direct contact with the object (tongue touches) in some lineages, which contributes information about nonvolatile chemicals to the overall picture of the chemical world being developed by the other systems. The discriminatory abilities of tongue touching have been sharpened by the evolution of deeply bifurcate, or forked, tongues. In essence a two-tined system allows an animal to take simultaneous samples

14. For a more detailed description of lingual feeding, see Schwenk and Bell 1988. For an amazing picture of a feeding chameleon, see Cogger and Zweifel 1998.

at two different locations, compare the concentrations of the chemical cues at those two different locations, and turn toward the point of highest concentration (tropotaxis). Remarkably, observations of forked-tongue species show that the tines are spread as wide apart as possible just before the tongue makes contact with the surface that is being sampled. Such simultaneous comparisons are a marked improvement over sequential sampling, which involves a more time-consuming sequence of touch tongue to ground, retract tongue to mouth, send cue from tongue to VNS, clean tongue, move head, body, or both, repeat sequence (klinotaxis). Putting all the individual pieces together produces a hierarchical pathway for reconstructing the chemical profile of the environment: detection of volatiles by the NOS sends information to the central nervous system and stimulates air tongue-flicks. The information gathered by air tongue-flicks fills in some of the finer details about the stimuli and directs additional investigation via touch tongue-flicks. These flicks collect additional information about nonvolatiles and direct movement toward or away from the source of the stimuli. The presence of a forked tongue allows the animal to make decisions about that movement more rapidly and to follow scent trails more efficiently.

So, let us integrate this additional phylogenetic, functional morphological, and behavioral information (Schwenk 1988, 1993, 1994, 1995b; Schwenk and Bell 1988; Schwenk and Throckmorton 1989) into the comparative phylogenetic picture of foraging mode evolution developed by Cooper. Preliminary analysis indicates that lingual feeding and ambush foraging are plesiomorphic for the group (fig. 5.36). Tuatara do not have connections between their vomeronasal organs and their mouths, and they do not tongue-flick. In other words, there is no correlation between ambush foraging and the absence of tongue flicking during hunting at this point because tongue flicking has not yet evolved. Tongue flicking originated within the plesiomorphic ambush foraging background in the ancestor of the squamates, as did the connection between the vomeronasal organ and the mouth. It is at this point that prey detection was added to the tongue's ancestral role in prey capture. Detection was limited to klinotaxic responses (sampling the environment sequentially), which involves a fair amount of movement around an area. Within the iguanian lineage, the tongue became increasingly important in prey capture (fig. 5.36). The amplification of this function may have been coupled with a reduction in the tongue's role in prey detection. For example, the masters of tongue tossing, the chameleons, show a decrease in the importance of the VNS system (coupled with an increase in visual acuity). Now, recall that a hypothesis for the absence of tongue flicking during prey ambushing was that it would render the

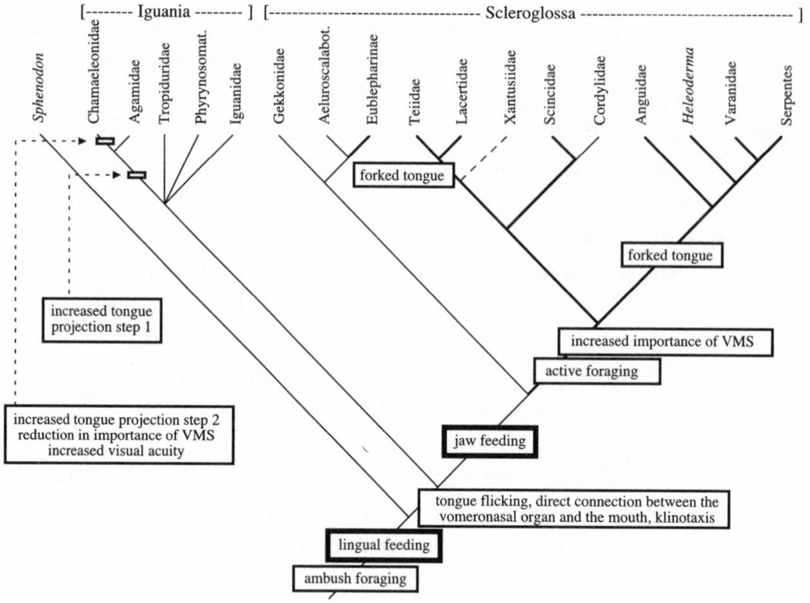

[-------- Iguania --------]   [--------------------------------- Scleroglossa ---------------------------------------]

Sphenodon
Chamaeleonidae
Agamidae
Tropiduridae
Phyrynosomat.
Iguanidae
Gekkonidae
Aeluroscalabot.
Eublepharinae
Teiidae
Lacertidae
Xantusiidae
Scincidae
Cordylidae
Anguidae
Heloderma
Varanidae
Serpentes

forked tongue

forked tongue

increased tongue
projection step 1

increased importance of VMS

active foraging

increased tongue projection step 2
reduction in importance of VMS
increased visual acuity

jaw feeding

tongue flicking, direct connection between the
vomeronasal organ and the mouth, klinotaxis

lingual feeding

ambush foraging

Figure 5.36.   More than just a tongue: increasing the resolution of foraging evolution in squamates.

hunter more obvious to the huntee (ruin the advantage of surprise). The functional analysis indicates a second hypothesis: ambush foragers do not tongue-flick while they are waiting in ambush because they are holding their tongue-hyoid structure in readiness for prey capture. In other words, the tongue's role in capture takes precedence over its role in detection (which may be primarily visual). In this case, we need to look for the selective factors responsible for the evolutionary amplification of lingual feeding in ambush foragers. It is possible that selection for lingual feeding has been responsible for the decrease in tongue flicking associated with ambush foraging, and not selection for immobility or crypsis. Conflicting hypotheses aside, one of the most exciting outcomes of this research is that it has highlighted a "natural" group that may be useful for studying the effects of many selection vectors acting in concert and in opposition on the evolution of a particular system. In order to examine these complicated relationships further we would need a resolved phylogeny for the Iguania, detailed studies of the importance of various sensory systems in prey detection, and even more data about lingual feeding in the clade.

The tongue's evolution followed a different pathway in the scleroglossan (hard tongue) lineage. Jaw feeding originated in the ancestor of the Scleroglossa (fig. 5.36). Evidently freeing the tongue from its role in prey capture set the stage for an amplification of its role in prey detection (Wagner and Schwenk 1999). That amplification was eventually associated with the increased sophistication of the VNS and the convergent appearance of forked tongues (and associated tropotaxis). Once again, those changes are not completely correlated with foraging strategy. If, as indicated by Cooper's initial study, active foraging showed up in the ancestor of the Autarchoglossa, then the divergence of the VNS occurred after the origin of jaw feeding. This is not problematic in one sense because a change in the prey-capture function of the tongue does not require an immediate change in its prey-detection role unless the two complexes are genetically coupled in some way. It is more likely that the origin of jaw feeding permitted the amplification of the tongue's importance in chemodetection once the novel changes in the chemosensory system appeared. The staggering of a novel capture mode and a novel foraging strategy, however, does have profound implications for the lizard clade, the Gekkota, situated between those two origins. Many gekkotans retain the plesiomorphic hunting strategy of ambush foraging but use the novel prey capture technique of grabbing food with their jaws. Perhaps this lineage is the place to examine the hypothesis that selection for predator crypsis has acted against tongue flicking during ambush. One of the suggested mechanisms for delivery of chemical cues from the tongue to the vomeronasal organs involves movement of sublingual plicae in the floor of the mouth (Young 1993). Interestingly, sublingual plicae have been secondarily lost in gekkotans, and the importance of the nasal olfactory system in prey detection has increased (Schwenk 1988, 1993). Does this mean that gekkotans don't tongue-flick during ambush because the functional morphology of prey detection via tongue flicking has changed? If so, what could be the reason for that shift? For example, if an ambush predator has to get closer to its prey to capture it with jaws rather than with a tongue, it is possible that selection would favor a shift in prey detection from the tongue to the nose because this would increase available information about incoming food while maintaining absolute stillness. Although this scenario harkens back to the original hypothesis about selection against tongue-flicks in ambush predators, it incorporates a number of different and more complicated interactions, including the decrease in one sensory system coupled with a rise to prominence in another sensory system within the historical backdrop of a novel prey capture mode.

Cooper expanded the scope of this research even further by examining the differences in behavior between congeneric species that have different foraging strategies. Previous experimental studies across larger phylogenetic groups had demonstrated that only active foragers can identify the chemical cues from prey items using their tongues (Cooper 1995, 1997). Cooper asked whether this apparently tight correlation between tongue flicking and prey discrimination ability would appear when the phylogenetic scale of the study was reduced (in other words, how quickly can evolution modify these traits?). A genus of lizards within the lacertid clade (*Acanthodactylus*) contains a species, *A. scutellatus*, that is classified on the borderline between active and ambush foraging. Another member of the genus, *A. boskianus*, displays the plesiomorphic active foraging mode. Cooper (1999; see also Cooper 2000 for another study with scincids) rubbed a cotton-tipped swab across a cricket (considered to be tasty morsels by both species), positioned the swab 2 cm in front of the lizard's nose, and recorded the ensuing number of tongue-flicks and bites. (Mennen Skin Bracer was used as one control.) Although both species were attracted to the potential prey item based on chemical cues, the actively foraging *A. boskianus* displayed stronger discriminatory abilities (slightly more tongue-flicks, more bites, shorter time to biting) than did its ambush-hunting relative. (Both species appeared unimpressed with the aftershave lotion.) There are two possible explanations for these interspecific differences: (1) we may have caught evolution in action as selection acts to reduce the ambusher's ability to detect prey items chemically (in which case, further reductions are predicted until the tongue is no longer used in this manner) or (2) perhaps the moderate ability to detect prey items by tongue flicking reflects a compromise in *A. scutellatus*, a lizard that moves slightly more than the usual ambusher. In either case, it would appear that chemosensory morphology and behavior can be modified very quickly in response to different selection pressures in these lizards. This is a fascinating discovery because it indicates that some characters, which appear conservative on a macroevolutionary time scale, may retain enough variability to respond quickly when the selection arena is changed. You simply see different patterns on different levels of analysis.

Many of the preceding scenarios are provisional pending further phylogenetic resolution within the Squamata and the addition of information from missing taxa. For example, some gekkotans other than the Eublepharinae are active foragers (Webb and Shine 1994; Werner et al. 1997a,b). These observations may ultimately change the double-origin hypothesis for active foraging shown in figure 5.35 if the actively foraging species are the basal members of the group. Also of interest for the double-origin hypothesis is the placement of the Xantusiidae, a small clade of ambush hunting, primarily nocturnal, New World lizards (commonly called, not surprisingly, night lizards). Although they share some characters in common with geckos (including their foraging ecologies), they are generally grouped within the Scincomorpha basal to the Teiidae + Lacertidae clade (see, e.g., Caldwell 1999). If this pattern is robust, then the evolution of active foraging within the Autarchoglossa may be more ambiguous than the initial study indicates. If, on the other hand, xantusiids are more closely related to geckos (as suggested by Lee [2000b]), then the patterns shown in figures 5.35 and 5.36 are maintained. In fact, the double-origin hypothesis is primarily meant to reflect changes on a higher phylogenetic level. As we have seen from Cooper's experimental work, foraging mode evolution is more labile within some families. So many lizards, so little time. This is especially true for gathering information from endangered species, like the tuatara, found only on islands near the northern end of the south island of New Zealand, and the enigmatic *Aeluroscalabotes*, inhabitants of areas in Borneo and the Malay Peninsula that are being rapidly deforested (Cooper 1995).

Statistical Approaches

Behavioral ecologists interested in the correlated evolution of continuous characters have been plagued by the observation that data from related taxa are not statistically independent ever since Boas raised the issue in 1896. While some researchers attempt to identify and remove the confounding effects of history from their analyses (e.g., Cheverud et al. 1985; Gittleman and Kot 1990; Gittleman and Luh 1992), others are interested in analyzing the relationship between quantitative characters within a phylogenetic context. Two general methods, independent contrasts (Felsenstein 1985a, 1988b; Burt 1989; Björklund 1990, 1994) and squared-change parsimony (Maddison 1991) are used, both of which examine associations by partitioning character variance among taxa then comparing sister groups.[15] Like all methods involved in character reconstruction, these tests are sensitive to missing data. Simplified phylogenies (e.g., trees depicting, say, 9 of 23 taxa) will produce *at best* extremely ambiguous results, particularly if branch lengths from molecular data are being used to infer amounts of morphological change, which in turn are used to search for character correlations. Excellent studies in this area include: push-up displays in *Scleroporus* lizards (Martins 1993), physiology and behavior in desert beetles (Ward and Seely 1996), morphology and life history in maples (Ackerly and Donoghue 1998), sexual size dimorphism in horned lizards (*Phrynosoma:* Zamudio 1998), and sprint speed and nocturnal behavior in geckos (Autumn et al. 1999).

15. Huey and Bennett 1987; Swofford and Maddison 1987; Harvey and Pagel 1991; Martins and Garland 1991; Garland et al. 1992; Pagel 1992; and Diaz-Uriarte and Garland 1996. For an excellent discussion, see Oakley and Cunningham 2000 and references therein.

Although there are relatively few methods designed to test whether changes in two discrete characters are independent,[16] they have been used in a number of interesting studies, including investigations of lekking and sexual dimorphism in birds (Höglund and Sillén-Tullberg 1994), avian mating systems (Sillén-Tullberg and Temrin 1994; Temrin and Sillén-Tullberg 1994, 1995), larval gregariousness and warning coloration in lepidopterans (Tullberg and Hunter 1996), and the evolution of primate behavior of many kinds (Lindenfors and Tullberg 1998; Ah-King and Tullberg 2000). For an excellent study demonstrating the different methods see Westneat 1995a,b.

Integrating the Experimental and Phylogenetic Approaches

As discussed previously, the power of macroevolutionary patterns lies in their ability to falsify hypotheses of character evolution. For example, a proposal that the evolution of character x was influenced by character y is falsified if x evolved before y. However, failure to falsify a hypothesis of character evolution based upon macroevolutionary patterns provides only weak support for the hypothesis. This is true whether you are investigating unique events or analyzing correlated patterns of character origin in different lineages statistically, because neither macroevolutionary patterns nor statistical analyses are direct tests of mechanisms. The only way to unambiguously demonstrate that mechanism x has had an effect upon the evolution of character y is to perform the appropriate experiment. The comparative phylogenetic approach is thus a powerful tool because it allows us to construct a set of predictions based upon macroevolutionary patterns that can serve as the focus for experimental investigation.[17] Although we tend to think of a comparative study as a linear process, the following examples indicate that such studies are more aptly described as feedback loops. The loop begins with the reconstruction of the phylogenetic tree for the study group and proceeds with the optimization of specific characters onto the tree, delineation of the relationship between the origin and diversification of the characters, construction of a hypothesis about the forces that have shaped the observed patterns, use of the hypothesis to make a series of predictions that can be tested experimentally, and experimental investigation of the predictions. During the course of the experimental analysis, it is not unusual to discover information that may either reinforce or change the

16. For example, Donoghue 1989; Maddison 1990, 1994; Sillén-Tullberg 1993; Wenzel and Carpenter 1994; Werdelin and Tullberg 1995; and Maddison and Maddison 2000.

17. McLennan 1991. For an excellent discussion of a detailed protocol for experimental studies of adaptation within a phylogenetic framework, see Arnold 1994, 1995.

original phylogenetic hypothesis. Such feedback highlights one of the certainties of biology: the situation is usually more complex than originally thought. A comparative study is thus an ongoing process, the successive sweeps of which we hope resemble the ever shortening strokes of a pendulum moving closer and closer to the central attractor representing "truth."

Sex: The Evolution of Male Courtship Calling in Frogs. When Darwin (1859) first published his theory of evolution by natural selection, the presence of male characteristics like elaborate plumage, bright colors, and large antlers was a thorn in the side of the newly created research program. Fortunately, that thorn was removed by Darwin's (1871) own suggestion that the survival-threatening costs of such traits are more than balanced by the benefits in attracting or sequestering females (mates), that individual "fitness" represents a compromise between survival and reproduction. Darwin's suggestion was not greeted enthusiastically by his supporters (e.g., Wallace 1905; Huxley 1938; see discussion in Ryan 1990a,b). As empirical studies documenting the existence of female choice began to accumulate, however, the initial controversy was replaced with debates about underlying mechanisms. Two hypotheses were originally proposed, one based upon a genetic correlation between the female preference and the male trait (**runaway sexual selection:** Fisher 1958; O'Donald 1962, 1967; Lande 1981; Kirkpatrick 1982) and the other upon the benefits accrued to females that mate with "genetically superior" males (**good genes:** Trivers 1972; Zahavi 1975; Kodric-Brown and Brown 1984). Under both scenarios, the male trait and the female preference are involved in a tight coevolutionary dynamic.

Michael Ryan (1990a) approached the problem from a different angle. He proposed that intersexual selection favored male traits that exploited preexisting biases in the female's sensory system (**sensory exploitation**).[18] This hypothesis decouples the evolution of the male and female traits *at their point of origin* because the character creating the "bias" in the female's sensory system is influenced by evolutionary pressures that predate the appearance of the sexual selection dynamic. Once that dynamic is established in the population, however, female preference may continue to be favored through the simple benefit to the female of acquiring a mate in an often unpredictable world, or through the establishment of either a runaway or a good genes coevolutionary interaction between the female preference and the male trait, or both (Ryan 1997).

18. Ryan 1990a,b; Ryan and Rand 1990; Ryan et al. 1990; see discussion in Ryan and Keddy-Hector 1992 of the various incarnations of this hypothesis that preceded its explicit formulation by Ryan in 1990.

The sensory exploitation hypothesis thus makes one powerful prediction that distinguishes it from both the runaway and the good genes hypotheses: the origin of the female "preference" predates the origin of the preferred male trait. *This prediction can only be tested directly within a phylogenetic framework.*

One such test involved a group of Neotropical frogs in the genus *Physalaemus*. The most famous member of the group, the túngara frog (*P. pustulosus*), and its sister species *P. petersi* are restricted to middle America and South America east of the Andes, while the sister clade is found in South America west of the Andes. Although they are generally small frogs (less than 30 mm long), the males are mighty singers. All species in the clade produce an advertisement call that consists, minimally, of a long whine. The túngara frog also adds a series of shorts chucks to the end of the whine, producing a more complex call. Research with the túngara frog indicated that the dominant frequency of the whine matches the most sensitive frequency of the amphibian papillae (AP) in the frog's inner ear, while the dominant frequency of the chuck is close to the most sensitive frequency of the basilar papillae (BP) (reviewed in Ryan 1991). The whine alone is necessary and sufficient for species recognition (Ryan and Rand 1993a,b), but once a conspecific has been identified, females prefer complex calls (calls with chucks) to simple whines (Rand and Ryan 1981; Ryan 1983, 1985; Ryan and Rand 1990). These results corroborate the hypothesis that sexual selection has influenced the evolution of call complexity in this group but do not allow us to differentiate among sexual selection mechanisms. In order to do this, Ryan and Rand (1993c) expanded the phylogenetic scope of the research to include the chuckless species *P. coloradorum,* with compelling results (fig. 5.37). When given a choice between regular calls from conspecific males and conspecific calls with the added feature of a chuck at the end, the females of *P. coloradorum* preferred the chucked call, even though their own males had never produced such a sound. The tuning in the BP of *P. coloradorum* is the same as that for *P. pustulosus* (Ryan et al. 1990), which explains why females from that species were able to respond to a sound they presumably had never heard before. This suggests that the neurophysiological characteristics of the basilar papilla evolved for processing information other than male calls. This is the one macroevolutionary pattern that most strongly falsifies the hypothesis that either a runaway sexual selection or a good genes dynamic was involved in the *origin* of the female preference, while supporting the third option, sensory exploitation. In other words, the female response was originally shaped under an earlier selective regime and was later co-opted to serve a role in the evolution of male-female communication when the male

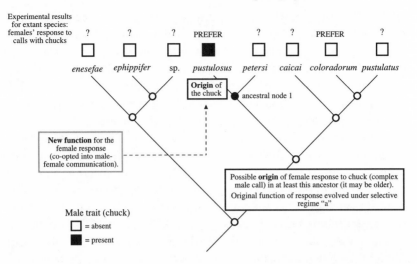

Figure 5.37. Male advertisement songs and female responses in the túngara frog clade. Which came first, the female preference or the male "chuck"? *Question mark* = data yet to be collected; *circles* = nodes for which ancestral male calls were reconstructed; *Sp.* = undescribed species. (Phylogenetic tree in Cannatella et al. 1998.)

character appeared. In order to disentangle the factors that make up that earlier selection regime, we would have to (1) pinpoint when that particular response originated (increase the phylogenetic scope of the study to determine the preferences of *P. petersi, P. caicai,* and *P. pustulatus,* and at least one outgroup species) and (2) identify the other functions of the female's response.

Ryan and Rand (1995, 1999) then carried the experiment one step further into the realm of deep history. Using seven models of evolution incorporating different assumptions about, among other things, the tempo of evolution and tree topology, they reconstructed the male's call at the ancestral nodes. The models produced different reconstructions for ancestral calls about 50 percent of the time, with the deepest nodes the most difficult to characterize. They then recorded female túngara frogs' responses to synthesized versions of all those hypothetical ancestral calls. (For anyone who has never waited patiently for a frog, fish, or bird to respond in an experimental setting, rest assured, this is no small feat.) Ryan and Rand were interested in documenting how females perceive variability in the male call. At what point does variation in a signal become meaningful to the female (cause her to alter her behavior in some fashion)? If females and males are locked together in a tightly coevolving "preference-call" unit, we would not expect females to respond to calls that are too different from the calls produced by their own males.

Surprisingly, the females responded to about half of the proffered ancestral variability. This astonishing result means that females are capable of perceiving variability in the call that is beyond anything produced by conspecific males. The ability to *recognize* various calls as potentially interesting, however, did not confuse the females. In most cases, they discriminated in favor of the conspecific song over the hypothetical ancestral songs, paralleling the experiments with extant species.

There was, however, one intriguing exception: females did not discriminate in favor of their own males when offered the call from their immediate ancestor (node 1 on fig. 5.37); in fact, depending upon the method used to reconstruct the node, females showed a weak bias toward the ancestor. This anomaly confirms the results from the recognition tests: although there is an interaction between the two components of the male-female dialogue, that interaction is looser than previously thought. In general, females appear to have a fairly broad preference for a "complex" male call, providing the call includes the species specific whine coupled with something else. That "something else" may be a series of chucks, white noise, or the amplitude modulated prefix used by *Physalaemus pustulatus* males (Ryan and Rand 1990, 1993a,b). Interestingly, the preference flexibility occurs at the level of individual females (Kime et al. 1998). Does this sound familiar? Think back to Cooper's experiments with squamate foraging behavior, tongue flicking, and the ability to recognize prey based on chemical cues. In both systems there is a moderate amount of variation (ability to detect and respond to chemical or acoustic cues) underlying what appears, on the surface, to be a conservative macroevolutionary pattern (the evolution of foraging mode, the evolution of female preference for a "complex" male call). In true Darwinian fashion, this underlying variation sets the stage for periods of rapid evolutionary change, following many different pathways, should the "nature of the conditions" change.

In the most recent chapter of the túngara story, Phelps and Ryan (1998, 2000; Phelps 2001) used artificial neural networks to examine how perceptual biases might evolve within a clade. They discovered that there are many different routes to preference for the chuck. Each of those routes has its own unique history, so although the "prefer chuck" component of the response was similar, responses to novel stimuli varied across networks. The network that incorporated the most information about the hypothesized history of the túngara frog sender-receiver system (recall the extensive experiments detailing extant call components and ancestral reconstructions) was the pathway that most closely approximated the behavior of real túngara females. Evidently the past does leave detectable traces of its presence in the patterns of character origin and diversification. This is the sense in which "history" is an important part of the present, as well as the future, of evolving systems.

 Sensory exploitation has provoked a number of responses that highlight common misunderstandings about the comparative phylogenetic approach. For example, the presence of the chuck in túngara frogs increases predation risks because it is attractive to frog-eating bats (*Trachops cirrhosus:* Tuttle and Ryan 1981; Ryan et al. 1982). Some researchers have voiced concerns that predation could therefore be a selective force operating against the presence of the chuck in other *Physalaemus* species (see Shaw 1995b for a review). According to this argument, chuckless species might actually have secondarily lost the trait, leading unwary researchers to misinterpret the actual widespread nature of the character and hence incorrectly estimate its point of origin. This argument is based upon the important recognition that character evolution is generally the outcome of numerous selection vectors, some of which may reinforce, and others of which may oppose, one another. It is thus possible that any change in the selection arena that tips the scales in favor of one selection vector over another may have a dramatic effect on character evolution. Although this dynamic interaction between selection and character evolution is important, it is not really the point here. At issue is the evidence required to refute a particular hypothesis. Bearing this in mind, let's take another look at the argument.

We begin with two facts: male túngara frogs produce a chuck; that chuck is attractive to frog-eating bats (documented selection pressure + presence of a trait). If we want to hypothesize that increased predation pressure has been responsible for the loss of the character in species that do not display the trait, what evidence can we use to refute that hypothesis? In other words, how do we differentiate between plesiomorphic absence and secondary loss? Minimally, we would have to demonstrate that (1) predators responding to the chuck were present in higher concentrations around chuckless species than around the túngara frog (otherwise how can we explain the presence of the chuck in the túngara frog?) and (2) the chuckless species retain the ability (the genetic framework) to produce a chuck under appropriate conditions, but that ability is being suppressed by the presence of the predators. Of these two conditions, the second is the more critical to the argument that absence of a character indicates secondary loss. An inability to uncover the genetic basis for that character in species that do not show the trait must be accepted as refutation of the secondary loss hypothesis, otherwise there is no way to falsify that hypothesis. To paraphrase a scientific dictum: absence of evidence is not evidence of (secondary) absence.

More Sex: The Nuptial Signal in Sticklebacks. Sticklebacks have been part of the charismatic microfauna of comparative ethology for over 60 years. Because of the painstaking descriptive and experimental work of Niko Tinbergen and his coworkers (see Wootton 1976, 1984 for hundreds of references), we probably know more about the behavior, ecology, and evolution of these diminutive fishes than of any other group of vertebrates. Much of that collective knowledge concerns one species, the three-spined stickleback, *Gasterosteus aculeatus;* one sex, the male; and one event, the male's striking transformation during the breeding season from an inconspicuous, gray-green fish to an individual sporting nuptial dress of scarlet red and flashing, cerulean blue. Hypotheses about the evolution of this ornate signal generally were based upon the assumption that the origin and elaboration of male nuptial coloration occurred in an ancestral *G. aculeatus* population. As we have discovered throughout this chapter, however, studies of a single species can, at best, only address questions concerning character maintenance in that species. In order to explore the origins of traits, we must expand the scope of our analysis.

Mapping the presence or absence of male ventrolateral nuptial coloration onto a phylogenetic tree for the Gasterosteidae indicates that it is a relatively old trait (originating in ancestor 2 in fig. 5.38). There are five equally parsimonious optimizations for color at its point of origin, given the current phylogenetic tree. Resolution of the complicated relationships within the *Gasterosteus* species complex (there are black microspecies within that complex that probably could be identified and named as distinct macrospecies), as well as more detailed descriptions of the exact pigment systems involved in producing the nuptial signal, may eventually help us resolve this problem (see discussion in McLennan 2000). That resolution will not change the main point of the optimization: the male color signal originated long ago (and possibly far away).

Three processes—intersexual selection, intrasexual selection (during territory acquisition), and natural selection (egg and fry guarding)— have been postulated to play a role in the evolution of the male color signal in these fishes. It is difficult to determine the relative influence of each of these processes on the trait's original success and subsequent diversification, however, because those events occurred in ancestors and environments no longer available for investigation. One way out of this dilemma may be to compare the patterns of origin and diversification of behavioral characters associated with particular processes (e.g., male-

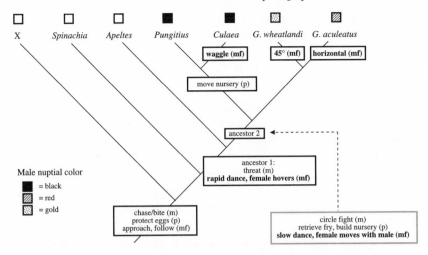

Figure 5.38.   The evolution of male nuptial coloration in sticklebacks mapped onto a total evidence tree for the Gasterosteidae (behavior: McLennan et al. 1988; McLennan 1993; behavior and morphology: McLennan and Mattern 2001]; one tree, CI = 80.9%). Behaviors used to identify particular processes are labeled: m = traits involving territory acquisition (threat and fight); p = traits involved in paternal care (egg/fry guard, build/move nursery); mf = traits involved in male-female courtship. Male ventrolateral body color is mapped above the genera (*open box* = nuptial color absent). Note that male color continues to diversify in the descendants of ancestor 2, as does the male courtship dance. Parental care changes very little, and male-male agonistic interactions do not change at all in those descendants.

female courtship and intersexual selection, male-male agonistic interactions and intrasexual selection, paternal care and natural selection) with patterns of origin and diversification of the male nuptial signal. In order to satisfy the hypothesis that a particular process has played a role in the evolution of the male color signal, two criteria must be met: (1) the color signal and the behavior must co-originate, or the origin of the behavior must predate the origin of the color signal and (2) changes in color must be accompanied by or be predated by changes in the behavior (although the degree of divergence need not be absolute, since it is unlikely that color and any of the behaviors are genetically coupled).

Optimizing color and behavior onto the phylogenetic tree (fig. 5.38) reveals three different patterns of association (for a detailed discussion and references, see McLennan 1996): (1) Color co-originated with circle fighting in ancestor 2 (intrasexual selection). Although it is difficult to envision a direct role for color in fighting, the sequence "origin of threat behavior followed by origin of color," indicates that color and threat may have become functionally coupled in ancestor 2. On its own, this association could not have been responsible for the evolution of the color signal, which continued to diversify in the descendants of ancestor 2 while threat (indeed all agonistic behaviors) remained conservative. (2) The origin of male nuptial color may have been associated with an increase in paternal care (natural selection) in ancestor 2, possibly through color's role as a threat signal directed toward intruding conspecifics of both sexes. Color and parental care change in the predicted sequence only one more time in the descendants of ancestor 2. (3) The origin of male nuptial coloration is preceded by the origin of the male zigzag dance and a change in female behavior from "follow a courting male" to "stop and hover in a head-up position" in ancestor 1 (intersexual selection). At this point, the courtship interchange became more complicated and required more time as the female stopped and waited while the male danced rapidly around her. Male and female behaviors changed again, in conjunction with the appearance of male nuptial coloration, in ancestor 2. Males now performed their zigzag dance more slowly and in front of the female, and females, holding in the head-up position, moved with (tracked) the dancing male. Both of these behavioral changes increased the female's exposure to the male's nuptial signal, providing her with ample opportunity to assess the intensity of that signal. Male courtship (orientation component) continued to diversify along with male color (hue component) in the descendants of ancestor 2, although that coupling is not perfect.

If, as the patterns suggest, the origin and diversification of male body color in gasterosteids was influenced by intersexual selection, natural selection, and intrasexual selection during territory establishment in decreasing order of importance, then we would expect to find three results: (1) The intensity of male color will be at its lowest level during territory establishment and nest building, will peak during courtship, and will reach a second, but lower peak during fry guarding. (2) The degree to which a male increases the intensity of his body color will not differ based upon the sex of an intruding conspecific during either the territorial or the egg/fry guarding stages of the breeding cycle, but will differ based upon the sex of an intruding conspecific during the courtship stage. Because of the difficulty in disentangling intersexual and intrasexual selection during the actual courtship phase in these fishes, it is particularly important to document the difference in the intensity of a sexually motivated male's response to a courting female versus a rival male. If intersexual selection has played the dominant role in the evolution of the color signal, then a courting male should develop a significantly more intense color signal in the presence of the female than in the presence of the rival male. (3) Females will discriminate among males based upon differences in the intensity of male nuptial coloration. The first two predictions are a novel byproduct of the phylogenetic comparative analysis. The third prediction, although not novel, follows directly from the analysis.

To date, experimental investigations of changes in male color across the breeding cycle have corroborated the first and second predictions for two descendants of ancestor 2, the three-spined stickleback, *Gasterosteus aculeatus* (McLennan and McPhail 1989), and the brook stickleback, *Culaea inconstans* (McLennan 1993b). The final prediction, that females will discriminate among males based upon the intensity of male nuptial coloration, has been confirmed for various populations of threespines.[19] The details of the female choice dynamic are steadily being revealed through the elegant research program of T. C. M. Bakker. Of particular interest has been the suggestion that there is additive genetic variation in the male trait and the demonstration that the male trait and the female preference are genetically correlated (Bakker 1993, 1999), that females become less selective in their choice of spawning partner as they invest more time and energy searching for an acceptable mate (Milinski and Bakker 1992), that this choice is influenced by the color of the male last seen by the female (Bakker and Milinski 1991), and that more intensely colored males receive more eggs in the field than their duller counterparts (Bakker and Mundwiler 1994). Little by little, then, information from experimental and field studies are adding support to the hypothesis based upon macroevolutionary patterns that intersexual selection has played a major role in shaping the male nuptial signal in the gasterosteids.

Although important, the role for color in the male-female interchange is not the only factor influencing the evolution of color. The macroevolutionary patterns also indicate that intersexual selection has been reinforced by the role of color as a threat signal during the courtship stage (tells intruding males to stay away: intrasexual selection) and the parental care stage (tells all intruders to keep out: natural selection) of the breeding cycle.[20]

Optimization of experimental results onto the phylogenetic tree suggests that the pattern of color cycling predicted from the macroevolutionary hypothesis is plesiomorphic for the *Gasterosteus* + *Pungitius* + *Culaea* clade (fig. 5.39). This suggestion will be strengthened if other descendants of ancestor 2, like *Pungitius pungitius* and *Gasterosteus*

19. McLennan and McPhail 1990; Milinski and Bakker 1990, 1992; Bakker and Mundwiler 1994; Baube et al. 1995; McKinnon 1995; Rowland et al. 1995b; and Kraak et al. 1999 (the most comprehensive and complex study to date).

20. For yet another elegant research program, see studies by Rowland (1982, 1984, 1988, 1989; Rowland et al. 1995a) documenting the threat value of the color signal.

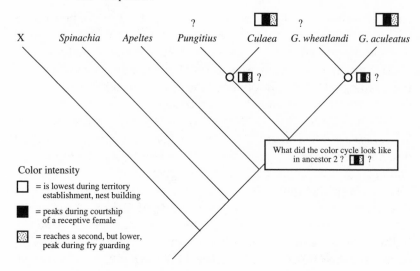

Figure 5.39. The feedback between the macroevolutionary patterns of character evolution and experimental studies. Cycling of male color is mapped above two taxa. Bearing in mind the caveat that we still need information from the other descendants of ancestor 2, we can tentatively call "low/high/medium" the plesiomorphic color-cycling pattern.

wheatlandi, show the same general pattern of color cycling. In essence, we are extrapolating backward from the processes that are currently maintaining male nuptial color in species to the processes that were responsible for the original success of male color at its point of origin. Given the complexity of the processes influencing the evolution of this one signal, is it probable that all populations will display the same cycling of male color over the breeding cycle? Intuition answers no. The most rigorous way to step beyond intuition is to make the plesiomorphic pattern shown in figure 5.39 our evolutionary "null hypothesis" of the way color should cycle, then use that pattern to highlight anomalies (derived conditions) within particular populations. Once such anomalies have been identified, we need to search for corresponding changes in, for example, environmental parameters influencing the transmission of color or changes in either the intensity of predation or the type of predator (see discussion in Reimchen 1989). The take-home message is simple: macroevolutionary patterns are not the end of an investigation, they are the beginning; they are not truth, they are simply guideposts. Any attempts to understand the forces that have shaped the evolution of a particular character must be based upon a continual feedback loop between patterns and processes.

Sexual Aids: Birds That Build Bowers. Bowerbirds (Ptilonorhynchidae) are small passerines related to catbirds, distributed across New Guinea and Australia. Males from 14 of the 19 species build stick structures called "bowers," which range from a simple avenue between two walls to an elaborate hut with a central pole.[21] They decorate their bowers with a variety of objects, including fruits, berries, flowers, bits of bone, feathers, snail shells, small stones, and even glass fragments. The existence of such a diverse array of bower structures leads us to three general questions: How many times did bower building originate in this group? What is the evolutionary sequence of bower types? and What is the function of the bower? This last question is particularly intriguing because in answering it we also must address the issue of whether that function can be responsible for the diversification of the bower structures.

Gerald Borgia and coworkers have been addressing these questions for almost two decades (e.g., Borgia 1985a,b; Uy and Borgia 2000). Borgia was initially interested in documenting the function of the bower, which required painstaking observations of bower construction and male-male, male-female interactions, coupled with equally meticulous experimental manipulations and yet more observations. By 1995 the research group had logged over 160,000 hours of videotaped and direct observation of ten species (Humphries and Ruxton 1999)! While details of the behavioral system were being investigated, blood was collected from the birds, mitochondrial cytochrome *b* sequences extracted, and a phylogenetic tree reconstructed for 14 species (Kusmierski et al. 1993, 1997). This tree then provided the phylogenetic framework for organizing the vast amount of behavioral data in order to begin answering the preceding three questions.

So, let us begin with the question of origin. There is some ambiguity about the number of times bower building arose in the family (fig. 5.40). This ambiguity is caused by the placement of the non–bower-building, monogamous catbirds (*Ailuroedus*) as the basal lineage in the family and by the placement of the toothbill bowerbird, *Scenopoeetes dentirostris,* as the basal member of the maypole-building clade. Toothbills form aggregations in which each male calls vigorously from an area that he has cleared and decorated with large objects such as freshly gathered leaves turned upside down. Has this species lost the ability to make bowers (and reverted to a simpler procedure), in which case bower building arose in ancestor 1, or did bower building arise independently in the

---

21. For the definitive discussion of bowerbird behavior and excellent diagrams of different bower types, see Borgia 1986, 1995c.

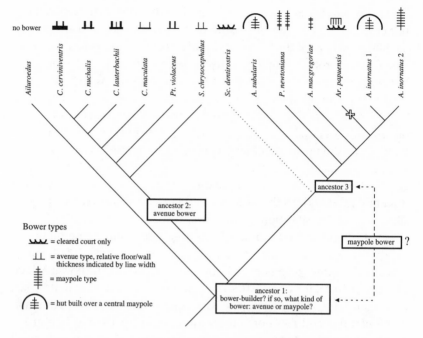

Figure 5.40. The evolution of bower building. Stylized bower types are mapped above the phylogenetic tree for the bowerbirds reconstructed from cytochrome *b* sequence data (Kusmierski et al. 1997). *A.* = *Amblyornis; Ar.* = *Archboldia; C.* = *Chlamydera; P.* = *Prionodura; Pt.* = *Ptilonorhynchus; S.* = *Sericulus; Sc.* = *Scenopoeetes.* *Cross* = loss of bower building; *question mark* = equally parsimonious placement of maypole-building origin in ancestor 1 or 3. *Ailuroedus* is represented by two catbird species, *Ai. melanotis* and *Ai. crassirostris.*

two building lineages? When we are faced with ambiguity at a node, we generally turn to the sister group for additional information. Unfortunately male catbirds neither clear display courts nor build bowers, so we are still no further ahead in determining the condition of bowers, if any, in ancestor 1. Unlike the situation described for the toothbill, all optimization procedures agree that *Archboldia papuensis* males have secondarily lost bower building. These males line a cleared court with a mat of dried ferns and feathers, decorate the court with snail shells, dark fruits, feathers, and beetle wings, then subdivide that court by draping orchid vines from overhanging tree branches to form a set of crisscrossing curtains.

The macroevolutionary analysis highlights both conservative (all species clear a display court, all species use decorations of some kind, avenue builders and maypole builders cluster in separate clades, all but one species build their bowers using sticks) and diverse (no two species build the same type of bower) elements in the bower-building repertoire (Kusmierski et al. 1997). It is this diversity, and the observation that there are two bower-building lineages, that leads us to the second question: What was the sequence of bower-type evolution? Once again, there is no unambiguous answer to this question. If bower building did evolve only once, we cannot determine what kind of bower ancestor 1 built. If the addition of the remaining five species and information from the sister group to the Ptilonorhynchidae does not resolve this ambiguity, then phylogenetic information cannot shed any more light on the problem given the current data set. The patterns do indicate that there has been a transition from a simple to a raised avenue along the avenue-building lineage (different widths of avenue floors on figure 5.40), with a reduction in the height of the avenue in *Chlamydera nuchalis* and an apomorphic avenue doubling in *C. lauterbachii*. The transitions are somewhat more complex in the maypole-building clade. If we break the bower into its component parts (presence or absence of a hut; one maypole or two maypoles), we obtain the optimization shown in figure 5.41a: the single maypole is plesiomorphic, with two independent additions of a hut, one change to a double maypole, and a loss of bower building (five steps). Coding the bower as a single unit produces the pattern shown in figure 5.41b. In this analysis, the hut type is plesiomorphic with two independent changes to a single maypole, one change to a double maypole, and a loss of building (also five steps). Closer examination of this latter scenario, however, indicates that the bower is not really acting as a cohesive entity. The transition from the hut to the single maypole bower type

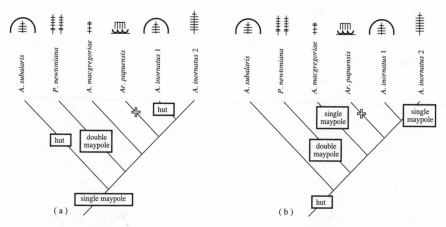

Figure 5.41. Pathways of evolution in the maypole-building lineage. Initially it appears there are two equally parsimonious optimizations depending upon whether you (*a*) score the maypole and the hut as separate characters or (*b*) score the hut + maypole as one complex character. Problems with the second scenario are discussed in the text. *A.* = *Amblyornis*; *Ar.* = *Archboldia*; *P.* = *Prionodura*. Cross = loss of bower building.

involves the loss of ability to build a hut over a single maypole, while retaining the ability to build the maypole. If we don't code "build a hut" and "build a maypole" as two distinct characters, we have to postulate that the ability to make the complete hut bower was lost and replaced with the ability to build a maypole, which actually requires two evolutionary changes, not one. The same argument holds for transition to a double maypole; either you go from building one to two maypoles (one step), or you lose the ability to make the hut and gain the ability to make two maypoles (two steps). Overall, it would appear that coding the bower components as different characters produces the most biologically defensible sequence of bower evolution in the maypole clade: a single maypole is plesiomorphic, with two forays into building a hut over the maypole (convergence) and one exuberant addition of another maypole (not to mention one loss of the maypole altogether).

Recent field observations have provided additional support for the scenario shown in figure 5.41a. The two hut builders do not construct their huts in the same way; *Amblyornis inornatus* builds from the top down, whereas *A. subalaris* begins on the ground and builds upward. In other words, closer examination of the actual behaviors involved in building causes us to reexamine the initial hypothesis that "hut + one maypole" was "the same" (homologous) in these two species.[22]

The third, and final, question is the most difficult of all to address. What processes were responsible for the success of bower building at its point of origin(s), and what factors are responsible for the continued diversification of bower structures past that point? Bowerbirds are relatively long-lived animals (for passerines). A male's building prowess increases with practice (age); older, dominant males even steal decorations from their neighbors and destroy the bowers of younger males. Not surprisingly, all of this focus on the bowers is important; in every species examined experimentally to date, a male's mating success is strongly influenced by the quality of his bower (e.g., symmetry, number and color of decorations).[23] Collecting data from extant species, then extrapolating back to the bower's point(s) of origin indicates that bower building was a successful evolutionary innovation because females preferred males that built bowers (sexual selection). Today, that preference is more finely tuned to differences among available bowers and their builders. The "preference" itself represents the outcome of an interaction among many factors. For example, bower quality may be a reliable indicator of

22. For a discussion about the dangers inherent in coding the endproduct of behavior rather than the behaviors themselves as characters, see Stuart and Hunter 1998.

23. Reviewed in Humphries and Ruxton 1999 and Uy and Borgia 2000.

male age and dominance status; some bowers may protect females from forced copulation attempts during the assessment stages of courtship (Borgia 1995b,c; Borgia and Presgraves 1998); and some decorations around the bower might serve as passive attractors, drawing the female to the area (Borgia 1995a).

Why, then, the diversity of structure? Borgia (1995c) postulated that the diversity reflected responses to ecological and biological parameters unique to each species superimposed upon the complex selection arena representing "female preference" for bowers in general. For example, the spotted bowerbird, *Chlamydera maculata,* builds a bower with a wide avenue delimited by thin, straw walls. Rather than watching males display at the bower entrance, which is the plesiomorphic condition for the avenue builders, female spotted bowerbirds gaze tremulously at their prospective mates through the semitransparent walls. These prospective mates, for their part, have extremely intense displays, including loud calls and vigorous charges toward the viewing female (see Borgia 1995c for a description). On a macroevolutionary scale, it appears that an apomorphic change in levels of male aggression may have been coupled with a change in bower construction that diffused the negative effects of this aggression. To complicate matters further, the bowers of spotted bowerbirds are widely separated, and males show lower levels of bower destruction and decoration stealing than displayed by their distant relatives, the satin bowerbirds. So, which came first, the increase in levels of male aggression or the increase in bower spacing? Are bower sites more limited in this species than in its sister group, or other members of its clade, leading to an increased intensity of competition for those sites? Or perhaps bower sites are abundant but widely spaced, leading to a decrease in male-male contact, weakening the reliability of bower variables as a marker of male dominance and vigor and indirectly selecting for an increase in courtship intensity as an indicator of male quality?

As is usually the case, the union of phylogenetic and experimental information has answered some questions and highlighted many possible avenues for future research. For example, Borgia and Coleman (2000) examined the hypothesis that autapomorphies in bower structure displayed by members of the *Chlamydera* clade functioned to diffuse a perceived increase in the intensity of male courtship. Mapping the male skrraa call onto the bowerbird phylogeny supported that hypothesis: the call functioned originally in male-male aggressive interactions and was later co-opted into the courtship repertoire of *Chlamydera* males, increasing the perceived "level of male-female aggression." The story does not stop here. Skrraa calls are only one component of a very complex courtship ritual, involving numerous visual and acoustic cues in addition to the presence of cleared courts and bowers. It took 160,000+ hours of video analysis to get to this point. We eagerly anticipate the results of the next 160,000 hours.

And Now for Something Completely Different: The Shocking Truth about Sea Slug Learning. Most people, when asked to name animals used in operant and classical conditioning experiments would reply "rats or dogs," a few might also add "zebra finches and goldfish," and very few might throw in "octopi." But it is the rare individual indeed who would add sea slugs to her list. Sea slugs (and sea hares) belong to a clade of opisthobranch mollusks that appeared approximately 340 million years ago (Moore et al. 1952), then underwent a substantial diversification about 150 million years later (Knight et al. 1960) in the Jurassic (although representatives of the group were strangely absent in the movie *Jurassic Park*). They are marine planktivores living in tide pools and kelp beds (Carefoot 1987). Researchers have been particularly interested in the learning mechanisms underlying "defensive withdrawal" reflexes in the sea hare *Aplysia californica.* "Defensive withdrawal" means exactly that: if an animal encounters an aversive stimulus in its habitat (like a predator or an electric shock, depending upon the environment), it withdraws from the stimulus (see the simplified pathway depicted in fig. 5.42). Facilitatory interneurons outside the reflex pathway respond to the first occurrence of an aversive stimulus by releasing serotonin, which in turn increases the magnitude of the sensory neuron's response (excitability and spike broadening) to the next tactile stimulus. The defensive withdrawal reflex thus allows the animal to deal with normal, nonthreatening tactile stimuli on a day-to-day basis, while stimulation of the facilitatory interneurons primes the animal to respond intensely to a potentially lethal stimulus (like a bite from a predator). Changes in the neural firing properties of the tail sensory neurons are involved in various forms of learning (including habituation, dishabituation, short- and long-term sensitization, and classical conditioning), which affect the performance of the withdrawal response.

Wright, Kirchman, et al. (1996 and references therein) were the first researchers to ask how cellular mechanisms underlying learning and memory might change across macroevolutionary time. They began their investigation by combing the literature for characters that might be useful in reconstructing the relationships among six species in the sea slug group. They discovered a total of 20 morphological traits, which produced one tree with a CI of 100%. The next step in the study involved delineating a set of morphological and physiological criteria (location, size, resting physiology, sensorin expression, proximal neuroanatomy, distal neuroanatomy, and receptive field) that could be used to identify the tail-sensory neuronal homologues among the six species. (Note: this is one of the best examples available of the painstaking research that is often required to erect a priori hypotheses of homology.) Once identified, the neurons were challenged with an application of serotonin, and the excitability and spike broadening responses recorded. The results of these intricate experiments are shown in table 5.3. If you were to interpret these results without reference to the phylogeny, you might be tempted to conclude that (1) *Dolabella* + *Bursatella* + *Phyllaplysia* form

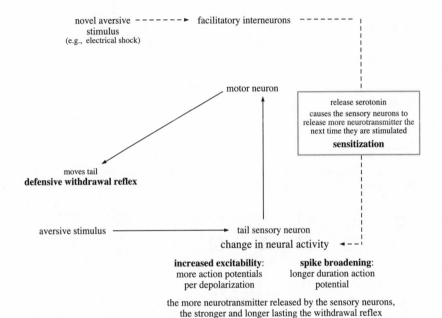

Figure 5.42. Simplified pathway for tail-withdrawal reflex in the sea hare (for detailed discussions, see Wright et al. 1995).

Table 5.3. Responses of Tail Sensory Neurons to Serotonin

Species	Excitability	Spike Broadening
Dolabrifera	−	−
Bulla	+	−
Akera	+	−
Phyllaplysia	+	+
Dolabella	+	+
Bursatella	+	+
Aplysia	++	+

a group and *Bulla* + *Akera* form a group, (2) *Dolabrifera* is the most "primitive" species because it shows neither response, and (3) the sequence of trait evolution follows the patterns shown in figure 5.43.

Bearing these nonhistorical hypotheses in mind, let's examine the patterns of character evolution uncovered by Wright et al. when they optimized the results of their experiments onto their phylogeny (fig. 5.44). Although the macroevolutionary patterns of neuronal response to serotonin are straightforward, those patterns differ for the two components of sensitization. An increase in neuronal excitability (1) is plesio-

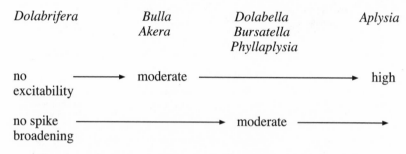

Figure 5.43. Nonhistorical hypotheses of the sequence of character evolution for the six species from the sea hare clade.

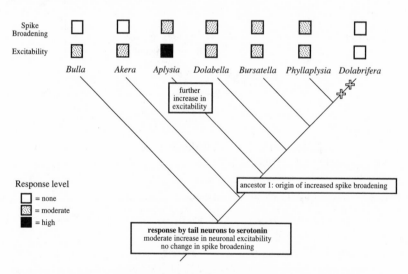

Figure 5.44. Serotonin-induced changes in spike duration and excitability and spike broadening mapped onto the phylogenetic tree for the *Aplysiidae* + *Akeridae*. The relationships shown in this figure are congruent with a more recent study combining information from three mitochondrial DNA genes for 12 species (one MPT, CI = 69.4%: Medina and Walsh 2000). *Cross* = loss of character.

morphic for the group (in order to determine how old this response actually is, we would have to expand the phylogenetic scope of the study), (2) has been increased in *Aplysia,* and (3) has been lost in *Dolabrifera.* The spike-broadening response is a much younger part of the sensitization process. Its absence is plesiomorphic, followed by the origin of the response in ancestor 1 and its secondary loss in *Dolabrifera.* These sequences differ from the nonhistorical hypotheses shown in figure 5.43 at two critical points: the presumptive state of the ancestral condition for excitability (moderate versus absent) and the reason for the absence of a response in *Dolabrifera* (secondarily lost versus ancestrally absent). This second point is not trivial. Suppose you are interested in documenting the effects of selection on the evolution of learning mechanisms. Which situation is more likely to interest you, a species that does not show a response because it is descended from ancestors that never showed that response (the character had not evolved yet) or from ancestors that did show the response (and thus may be amenable to studies on the potential selective pressures leading to the loss of the response)? Clearly, the second scenario is more interesting from an adaptationist perspective. Without the phylogenetic patterns, however, we could not have identified *Dolabrifera* as the species of choice for such research, and we might even have overlooked it as interesting at all.

The discovery that learning mechanisms can be lost evolutionarily is an intriguing one that takes us down a number of different new research pathways. For example, how did such losses occur? Is there something derived about the serotonin receptors or titers of cAMP in *Dolabrifera* compared with other members of the group (see extensive discussion and references in Wright, McCance, and Carew 1996; Wright 2000)? Do those changes represent a true evolutionary reversal to the condition displayed by species for which the absence of a response is plesiomorphic (e.g., *Bulla*), or do the changes represent a novel way to short-circuit the reflex pathway? Is there an adaptive advantage to losing a form of learning that appears to be widespread throughout the clade? In order to answer this last question, we would have to document the putative adaptive advantage of sensitization. Does it help sea slugs avoid physical harm from aversive stimuli such as predators (Carefoot 1987; Pennings 1990, 1994) or water contaminants? If so, what has changed in *Dolabrifera*'s environment that might support the loss of such useful abilities? What is the relationship between spike broadening and excitability of sensory neurons to different forms of learning? In the past, researchers would have investigated that question by pharmacologically eliminating one or other of these responses in *Aplysia.* Wright (1998) suggested that question be addressed from another angle: identifying species on the tree that could be used as naturally occurring "phylogenetic lesions." For example, drug-based blocking of the spike-broadening response in *Aplysia* eliminated short-term synaptic facilitation without affecting long-term facilitation. Based upon this result, *Bulla* and *Akera* (no spike broadening + moderate excitability) should show the "no short-term + long-term" synaptic facilitation pattern if the underlying assumptions about the importance of the two response pathways on these different types of learning are correct.

SUMMARY

Biologists and psychologists alike have very generally clung tenaciously to the idea that instincts, in part at least, have been derived from habits and intelligence; and the main effort has been to discover how an instinct could become gradually stamped into organization by long-continued uniform reactions to environmental influences. The central question has been: How can intelligence and natural selection, or natural selection alone, initiate action and convert it successively into habit, automatism, and congenital instinct? In other words, the genealogical history of the structural basis being completely ignored, how can the instinct be mechanically rubbed into the ready-made organism? Involution instead of evolution; mechanization instead of organization; improvisation rather than organic growth; specific *versus* phyletic origin. This inversion, or rather perversion, of the genealogical order leads to a very short-focused vision. The pouting instinct is supposed to have arisen *de novo*, as an anomalous behavior, and with it a new race of pigeons. . . . The incubation instinct was supposed to have arisen *after* the birds had arrived and laid their eggs, which would have been left to rot had not some birds just blundered into "cuddling" over them and thus rescued the line from sudden extinction. How long this blunder-miracle had to be repeated before it happened all the time does not matter. Purely imaginary things can happen on demand.

Whitman 1899:323

"Adaptation" is an important word in biology, one that is routinely applied to many aspects of biological function at many levels of organization. As with our discussion of species concepts, we find that what is or is not a "valid" definition of adaptation often depends on one's own research program, which generally follows the aspect of evolution about which one is most fascinated.[24] For our voyage of discovery and evaluation, we have adopted a historically based working definition of adaptation: a trait whose *origin* is associated with increased functional effi-

---

24. For discussions about valid definitions, see, e.g., Coddington 1988, 1990; Reeve and Sherman 1993; Frumhoff and Reeve 1994; Lauder 1996; and Martins 1996, 2000.

ciency that was favored by natural selection (Gould and Vrba 1982; Greene 1986a; Coddington 1988, 1990; Lauder 1996).

To evaluate the traits we have discovered that might be adaptations, we need to delineate the factors involved in the original success of a trait. In some cases, it might be possible to use fossil evidence to determine whether we can simply extrapolate from current utility to function at the time of origin (van Valkenburgh 1994). For example, bivalve mollusks deposit an energetically expensive, complex matrix of calcium carbonate and protein-rich layers inside their shells (conchiolin). Various researchers have suggested that conchiolin is adaptive because its presence deters predatory snails from drilling through the shell to dine on its owner (Lewy and Samtleben 1979; Harper 1994). Kardon (1998) investigated this hypothesis using members of the family Corbulidae by (1) demonstrating experimentally that conchiolin did indeed deter drilling, (2) examining fossil corbulids and sister groups to pinpoint the origin of conchiolin, and (3) asking whether there was any evidence for an interaction between the corbulids and predatory snails at that point in time. She discovered that conchiolin was an autapomorphy for the Corbulidae, originating in the Jurassic. Drilling snail predators, however, did not become widespread until the early Cretaceous (at least 50 million years after the origin of the Corbulidae and conchiolin), and there were no records of drillholes in Jurassic mollusks. These data clearly indicate that conchiolin did not serve an antidrilling function *at its time of origin* (this does not preclude a role in deterring other types of predation [e.g., crushing]), although it was later co-opted for that function following the origin of the drilling predators.

We also might be able to investigate the relationship between present and past utility by using the macroevolutionary patterns of character origin and diversification. Cast your thoughts back to leeches (fig. 5.17) and our hypothesis that the hardened cocoon was originally favored by natural selection because of its role as a deterrent to aquatic predators (for the purposes of this discussion, we will assume that the hardened covering originated in an aquatic environment). Looking at the tree, we might next hypothesize that the plesiomorphic presence of a cocoon with a hardened cover permitted the movement from an aquatic to a terrestrial existence because the cover protected the eggs against desiccation in the new environment. Let us assume for the moment that we have experimental evidence supporting both the antipredator function of the hardened covering in the water and the antidesiccation function of that covering on land. If an adaptation is a trait whose *origin* is associated with increased functional efficiency that was favored by natural selection, then we can say that the hardened cover is an adaptation for depositing eggs in the water. Is it also an adaptation for depositing eggs on land? According to the preceding definition, the answer to this question is No, because selection did not shape the cover for that role originally. It is just sheer happenstance that the cover is functionally useful in a completely different environment. But, is the hardened cover important in allowing leeches to escape the water? The answer to this question is equally clearly Yes (remember, we have experimental evidence showing that eggs in cocoons without hardened covers die on land).

The existence of characters that retain the plesiomorphic form while taking on a derived function (**exaptations:** Gould and Vrba 1982; Baum and Larson 1991; Arnold 1995; Armbruster 1997) are critical to Darwinism. Without these traits it is impossible to explain movements into different environments without invoking Lamarckian mechanisms. For example, consider traditional stories telling of small fishlike creatures hauling themselves laboriously out of the water into the slime and mud, where they miraculously evolved lungs in order to breathe and limbs in order to walk. Extensive analysis of fossil evidence in a phylogenetic framework, however, gives us a much more plausible, if much less heroic-sounding, tale, one in which lungs and limbs originated and were modified gradually as an adaptation to shallow, weed-tangled, ephemeral freshwater habitats (Ahlberg and Milner 1994). Once again, did these characters evolve as an adaptation to living on land? The answer is No. Were they a critical prerequisite to that quantum leap? The answer is Yes. There is a subtle but crucial difference between "the environment changed, so the organism changed" and "the environment changed, and those organisms that *already possessed* traits allowing them to survive in the new environment flourished." As a final thought experiment, let us assume that the new antidesiccation function of the cocoon covering is modified at some point in time after the transition to land, and that modification increases the efficiency of the covering (for example, suppose the coating is laid down in a double layer). Is this an adaptation to living on land? Clearly the answer to this question would be Yes. So, if we trace the evolutionary history of the cocoon covering (fig. 5.45), we find a complex pattern involving origin and adaptation to environment x, which then allowed movement into environment y when the opportunity presented itself, followed by further modification in (adaptation to) environment y.

 Researchers interested in uncovering the relationship between functional efficiency and differences in the form of a particular trait usually collect data within a particular species. For example, you might look for a correlation between the degree of bill curvature and the time it takes to crack open small nuts in species X, concluding that individuals with strongly curved beaks have an advantage (are better adapted to nut cracking) over individuals with mildly curved beaks. Elevating your investigation of "better adapted" to the species, rather than individual, level is much more complex. For example, say species A retains the plesiomorphic condition "mildly curved beak" while its sister species, B, displays the apomorphic condition "strongly curved beak." Hypothesis: At some time during the evolutionary history of B, strongly curved beaks replaced mildly curved beaks; that is, the adaptive benefit of the novel beak shape was so high that the plesiomorphic condition was eliminated in the species. If this is true, why don't individuals from species B simply take over the territories of all the species having the plesiomorphic, and less efficient, beak shape? In other words, *if we find closely related species in which the plesiomorphic form or the apomorphic form are fixed, and if the apomorphic form is better adapted than the plesiomorphic form, why is the species with the plesiomorphic condition still around?*

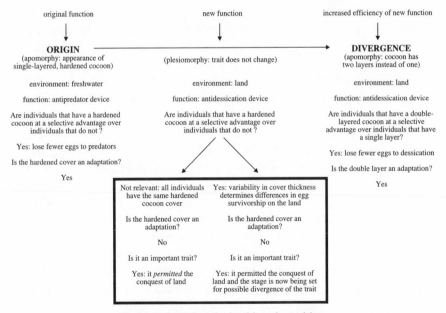

The debate about adaptation is a debate about origins.
Functional innovation need not require the evolution of a new trait.

Figure 5.45. Hypothetical sequence mapping the interaction between the origin and diversification of a trait (modification of structure), changes in the function of the trait, and changes in the environment (or, the introduction of a problem to be solved by the trait).

This is an interesting paradox, from which, fortunately, there are a number of very different escape routes (more than the four listed below may occur to you). First, species A and B may be allopatric, so they never come into contact with one another. If the apomorphic beak shape is indeed more effective than the plesiomorphic shape, then we would predict that if A and B ever do come into contact with one another (either through the vagaries of geology and time or, more likely, via human intervention), competition for the same food source will affect A adversely.

Second, it is possible that A and B will overlap without adverse effects on A for a very long time if they are sharing an abundant food source. Until and unless the environmental conditions change in such a way that only the best solution (apomorphic beak shape of species B) can survive, all adequate versions (plesiomorphic beak shape of species A) will persist (survival of the adequate: Brooks and Wiley 1986, 1988). If the environment does change in such a manner however (e.g., the food source becomes limiting), A will suffer from its interaction with B. In the latter two examples, we would expect to find a competitive interaction between species A and B at some point because they both share the plesiomorphic function of the beak (eat nuts) but have different beak structures (plesiomorphic versus apomorphic beak shapes). In these examples, the microevolutionary dynamic is played out on the macroevolutionary stage, leading to interspecific competition.

It is also possible that the apomorphic form of the trait in species B will have a new (apomorphic) function. For example, species B may be characterized by a strongly curved beak because a small ancestral population was originally isolated on an island full of banana trees, and individuals with more strongly curved beaks were better at peeling and eating bananas (the only food source). Peripheral isolates speciation now complete, individuals from species B only eat bananas. If they were ever to come into contact with their sister species again, we would predict that there would be no adverse interaction between the two because A and B would not be competing for the same resource base.

Finally, it is unlikely that two species will differ by only a single trait (beak shape). Other differences may have an impact on the net effectiveness of the novel trait you are studying. So, back to species B. This time B evolved on an island where it continued to eat nuts (individuals with a more strongly curved beak had an advantage), but where it was also exposed to an increased level of predation by snakes. After millions of years of isolation, species B has evolved a novel suite of antisnake behaviors, one of which is "fly straight upward as fast as possible whenever a snake comes into view." Now, suppose B comes back into contact with its sister species, A, on the mainland. Initially A finds itself at a disadvantage as B out-competes it for nuts (and experiments in the laboratory confirm this interspecific competition). But, hawks are the main predators of nut-eating birds on the mainland. So, every time an individual from species B sees a snake, it flies straight up into the air, where it is neatly seized by a waiting hawk. The success of B versus A is now dependent upon two different sets of selection pressures, competition for food (B is more efficient than A) and escaping from hawks (A is more effective than B). The outcome of this interaction is not easily predicted.

The Frontiers of Discovery in Exploring Options
EXPLAINING HOMOPLASY

Homoplasy affords evolutionary biologists special opportunities to study the independent origins of the same or highly similar traits, under similar or dissimilar conditions. The usual approach to studying homoplasy has been to look for genetic correlations and measure genetic variance for similar traits in similar selection arenas. The goal is to find the genetic basis for regularities in the phenotype/environment interface (for an excellent review of this approach, see Cody 1973). Only recently have we begun to ask questions about how the genetic basis is translated into the phenotype, and whether or not that influences the phenotype/environment interface. This began to change in the last decade of the twentieth century with the renewed recognition that in Darwinian evolution both homologies and homoplasies may have arisen either as adaptations or as design (or "intrinsic" or "inherent") constraints. Consequently, we cannot assume a priori that homoplasies are adaptive traits, or that adaptive traits will always be homoplasious. Therefore,

 Assessments of homology and homoplasy must be made independent of assessments of adaptation and design constraints.[25]

Phylogenetic systematics, through the use of the auxiliary principles, is a powerful tool for discovering homoplasy. The a priori presumption of homology in Hennig's auxiliary principle means that each instance of homoplasy represents a specific falsification of that a priori presumption. We therefore need not resort to a priori postulates of which traits are "likely to be" or are "probably" homoplasious—we can find the most defensible hypotheses of homoplasy right on the phylogenetic trees. In the same way, the a priori presumption that each character is evolu-

25. For example, Faith 1989; Sanderson and Donoghue 1989; Sluys 1989; Sanderson 1991; Brooks 1996; and Sanderson and Hufford 1996.

tionarily independent embodied in Kluge's auxiliary principle permits us to discover which characters may form part of a single trait complex that evolves as a unit (an **evolutionary module:** Maynard Smith and Szathmary 1995; an **evolutionarily stable configuration:** Wagner and Schwenk 1999), and whether or not those complexes represent correlated homoplasy. Once again, we need not resort to a priori postulates of what traits are "likely to be" or are "probably" dependent or independent.

This approach allows you to focus your experimental studies on the functional evolutionary significance of such homoplasies and to search for correlations between any given homoplasious traits and other traits. Among other things, this permits you to ask if the homoplasious traits evolved for the same reason each time they appeared. Winkler (2000) suggested that this is a very important question for those using the "comparative method." A relatively new research program called **integrative biology** seeks to provide rigorous explanations for the origin, persistence (stasis), and diversification of the very characteristics that provide the foundation for complex evolutionary systems. David Wake summarized the conceptual framework for integrative biology thus: "[A]n understanding of the evolution of biological form—morphology—[is] unlikely unless one combine[s] two distinct and independent approaches: neo-Darwinian functionalism and biological structuralism, in a context of rigorous phylogenetic analysis" (1991:543).

This proposal stems from the core of Darwin's view that evolution is the outcome of complex interactions between the nature of the organism and the nature of the conditions. Integrative biology already involves efforts in diverse fields including a quartet of "**morphologies**" (comparative, functional, evolutionary, and eco-), **physiology**, and **developmental biology.**[26] The last area (often called "evo-devo"), is the

26. Morphology: for example, Lauder 1980, 1981, 1982a,b, 1983a,b, 1985a,b, 1986, 1988, 1989a,b, 1990, 1991a,b, 2000; Cracraft 1981a,b; D. Wake 1982, 1991; Bramble and Wake 1985; Foreman et al. 1985; Lauder and Shaffer 1985, 1993; Liem and Wake 1985; Bemis and Lauder 1986; Brooks and Wiley 1986, 1988; Schaefer and Lauder 1986, 1996; Wainwright and Lauder 1986, 1992; Bemis 1987; Motta 1988, 1989; Lauder and Liem 1989; Lauder et al. 1989; Losos and Sinervo 1989; Schwenk and Throckmorton 1989; Wainwright 1989, 1996; Wainwright et al. 1989; Wake and Roth 1989; Westneat and Wainwright 1989; Losos 1990b,c,d, 1994; M. Wake 1990, 1992, 1993, 1994; Baum and Larson 1991; Wenzel, 1991, 1993; Lauder and Wainwright 1992; Reilly and Lauder 1992; Arnold 1994, 1995; Wainwright and Reilly 1994; Hufford 1995, 1996; Lauder et al. 1995; Zelditch and Fink 1996; Glossip and Losos 1997; Lee and Doughty 1997; Reilly et al. 1997; Rice 1997; Zweers and Vanden Berge 1997; Beuttell and Losos 1999; Guralnick and Lindberg 1999; Wagner and Schwenk 1999; Aerts et al. 2000; Iwaniuk et al. 2000; Liem and Summers 2000; Van Leeuwen et al. 2000; and Wilga et al. 2000. Physiology: Huey 1987; Lauder 1988; Schwenk 1993, 1995a; Barclay 1994; Wright, Kirchman et al. 1996;

focus for studies of "developmental constraints" in evolution. We will consider two areas of evo-devo to illustrate the potential importance of integrative approaches to studying homoplasy.

Heterochrony (changes in the timing of development that produce changes in morphology) is thought to be a source of evolutionary innovation, including homoplasy.[27] Alberch et al. (1979) developed a general model of heterochrony (adapted for phylogenetic analysis in Fink 1982) based upon representing the development of any part of an organism with a positive trajectory having a starting point (α), an endpoint (β), and a rate of growth (k) for changes in shape (γ) or size (S). Heterochrony is determined by plotting values for α and β on the x-axis (time) and values for γ or S on the y-axis (morphology).

There are two general categories of heterochrony. **Paedomorphosis** results in the production of a descendant adult morphology that is less complex than that of its immediate ancestor. This does not necessarily imply that the descendant will have a morphology comparable to that of a juvenile or even a larval ancestor; it may resemble a less developed ancestral adult (McKinney 1988). Paedomorphosis can be accomplished in three ways: growth onset can be delayed (**postdisplacement**), growth can be terminated earlier (**progenesis**), or development can proceed at a slower rate (**neoteny**). Paedomorphic phenomena can in some cases produce morphological changes in apomorphic traits that harken back to their plesiomorphic roots, manifested on phylogenetic trees as the form of homoplasy called evolutionary reversals. **Peramorphosis**, by contrast, results in the production of a descendant adult morphology that is more complex than that of its immediate ancestor. Since this produces a morphological trait in an organism that passes beyond the condition found in its ancestor, the result will be a recapitulation of the ancestral ontogeny during development. Not surprisingly, there are three ways in which peramorphosis can happen: growth onset can begin earlier (**predisplacement**), growth can continue for a longer period (**hy-**

Ritchie et al. 1997; Irschick and Losos 1998; Ketterson and Nolan 1999; Maier 1999; Mottishaw et al. 1999; Weller and Sakai 1999; Buschbeck 2000; Weiblen et al. 2000; and Wright 2000. Developmental biology: for example, Gould 1977, 1982a; Alberch et al. 1979; Alberch 1980, 1985; Alberch and Alberch 1981; Bonner 1982a,b; Fink 1982; Maderson 1982; Maynard Smith et al. 1985; McNamara 1986; Raff and Raff 1987; King 1991; Hall 1992; Stiassny 1992; Mabee 1993, 2000; Raff 1996; Albert et al. 1998; Friedman and Carmichael 1998; Holland 1999; Collazo 2000; Floyd and Friedman 2000; Kellogg 2000; Kramer and Irish 2000; Levitan 2000; and Wray and Lowe 2000.

27. For example, Gould 1977; Alberch et al. 1979; Bonner 1982a,b; Fink 1982; McNamara 1982; Shea 1983; Emerson 1986; McKinney 1986, 1988; Raff and Raff 1987; Reiss 1989; Mooi 1990; Strauss 1990; Winterbottom 1990; Stiassny 1992; Fink and Zelditch 1995; Friedman and Carmichael 1998; and Smith 2001.

permorphosis), or development can proceed at a faster rate (**accelera-tion:** see also McNamara 1986). Recurring peramorphic phenomena pro-duce homoplasious apomorphic changes.

McNamara (1986) suggested that any modification of a develop-mental sequence, whether by addition or deletion of stages, may there-fore be interpreted as an outcome of heterochrony. Based upon this, he extended paedomorphosis to include a descendant passing through fewer ontogenetic stages than the ancestor, and peramorphosis to in-clude a descendant passing through more ontogenetic stages than the ancestor. For example, the plesiomorphic ontogenetic sequence for di-geneans (the parasitic platyhelminths also known as "flukes") includes three larval stages—the miracidium, the mother sporocyst, and the cer-caria. Early in the phylogeny of digeneans, a fourth stage, the redia, was intercalated between the mother sporocyst and the cercaria. This would be an example of peramorphosis. Additionally, not all descendants of the ancestral digenean in which the redia arose possess a redial stage. Many exhibit a stage termed a daughter sporocyst that occurs, like the redia, between the mother sporocyst and the cercaria. Rediae and daugh-ter sporocysts are both derived from mother sporocyst germinal tissue, both have a birth pore, and both give rise to cercariae having (plesiomor-phically and commonly) pharynges and guts. In addition, there are no experimentally confirmed cases in which rediae and daughter sporocysts occur in the same species. These observations, in the context of Rema-ne's criteria (chapter 2), lead to the hypothesis that rediae and daughter sporocysts form a homologous transformation series. Phylogenetic anal-ysis suggests that daughter sporocysts are apomorphic, and furthermore that daughter sporocysts have evolved twice within the Digenea (Brooks et al. 1985a, 1989; Brooks and McLennan 1993b). The difference be-tween a redia and a daughter sporocyst is morphological; a redia has a pharynx and saccate gut while a daughter sporocyst has neither. This suggests the presence of paedomorphic phenomena leading to the later than expected expression of the pharynx and gut during ontogeny in those species having "daughter sporocysts" (remember that cercariae have pharynges and guts). In the absence of experimental studies, we cannot tell which of the three categories of paedomorphosis may have been involved, or whether the same process was responsible for each evolutionary origin of "daughter sporocysts," but we can say that daugh-ter sporocysts appear to be rediae that are paedomorphic for the expres-sion of the pharynx and intestine. Now we can ask, What does this mean, if anything?

Saether (1977, 1979a,b,c, 1983), following Brundin (1972), suggested that "inherited" factors termed **canalized evolutionary potential** could also contribute to developmental constraints, particularly in the production of homoplasy. He linked this discussion with the formulations by Tuomikoski (1967), who defined the concept of **underlying synapomorphy** as a plesiomorphic capacity that made it relatively easy ("cheap") to evolve similar traits (see also Crampton 1929). Simply put, this means that among members of a clade, and within that clade only, certain homoplasious changes occur repeatedly. The appearance of these traits conforms to a phylogenetic pattern of homoplasy, but the evolutionary capacity to produce them actually evolved only one time, in the common ancestor of the clade containing all the species exhibiting the trait. It is that capacity which is named the "underlying synapomorphy." Underlying synapomorphies are intriguing because macroevolutionary patterns exist that are supportive of the concept, however, the nature of, and process(es) producing, canalized evolutionary potential remain unclear (e.g., Maynard Smith and Szathmary 1995; Raff 1996).

The pelvic girdle of stickleback fishes provides an example of an underlying synapomorphy. Within the clade comprising the genera *Culaea*, *Pungitius*, and *Gasterosteus*, populations of at least four species contain members lacking pelvic girdles. Each appearance is thought to represent an independent (homoplasious) origin. No other members of the Gasterosteidae lack pelvic girdles. The restriction of this trait to one clade, coupled with its repeated appearance within the clade, corresponds to the phylogenetic pattern expected for an underlying synapomorphy. Studies of *Pungitius occidentalis* indicate that the trait is heritable, at least in part (Blouw and Boyd 1992). Developmental studies indicate two different pathways by which the loss of the pelvic girdle is manifested, at least one of which involves heterochronic processes (Bell and Foster 1994).

> Can't you just recode all homoplasies as separate character states and obtain a tree with no homoplasy? Yes, but you may gain false robustness and lose the ability to test your hypothesis objectively. In fact, adopting the view that (1) all homoplasies are coding errors *or* (2) correlated homoplasy is rampant, producing incorrect estimates of phylogeny, are the two sides of the same bad coin. Both limit what we may discover for ourselves. We must justify all our decisions by reference to what we have discovered, using a minimum of assumptions.

Distinguishing homoplasy from homology is dependent on the production of a robust phylogenetic hypothesis. That requires many characters, including as many as possible that do not show homoplasious or homoplasious-like evolution. Once we have such a hypothesis, it does not matter whether two traits are true homoplasies (identical) or only very similar, because they may still have the same functional significance evolutionarily. What is important is (1) to understand how many different pathways can be used to make similar traits; then (2) to ask if some pathways are more difficult (costly) to achieve than others; and finally, (3) to ask if the traits/pathways that are cheapest are the ones that show up the most often. If the answer to the third question is no, the traits, although easy to make, may require special conditions to be successful. This is where our particular voyage of discovery enters the picture.

For example, given the apparent evolutionary propensity for loss of the pelvic girdle to occur in some sticklebacks, we might ask, Why does the pelvic girdle persist in all species within the clade displaying the underlying synapomorphy? Answering this question leads to studies of environmental selection, including questions about the interaction between predation and the function of the pelvic girdle, or possible influences of available calcium levels and the development of the girdle (Nelson and Atton 1971; Reist 1980; Bell et al. 1993). Alternatively, having established a mechanistic basis (paedomorphosis) for the homoplasious appearance of "daughter sporocysts," researchers may now ask questions about the possible functional significance of this evolutionary innovation. First, because "daughter sporocysts" have evolved twice during digenean phylogeny, we must ask if the same process was responsible in each case, and if not, document the salient differences between them. Then we must ask, Are rediae and "daughter sporocysts" functionally equivalent, or are daughter sporocysts functionally superior in some manner (do they produce more infective larvae)? If superior, why have daughter sporocysts evolved only twice? At the moment, we do not know the answers to any of these questions. We do, however, have a protocol for beginning the search. Our voyage of discovery and integrative biology emphasize that combining information about "pattern" with information about "process" is essential for robust evolutionary explanations, and both provide the means by which you can "go and study for yourself." Just remember,

 Phylogenetic trees are necessary for discovery but not sufficient for explanations of the origin and diversification of phenotypic traits, structural and functional.

Chapter 6

Evolutionary Radiations: Exploring Time

> In adaptive radiation and in every part of the whole, wonderful history of life, all the modes and all the factors of evolution are inextricably interwoven. The total process cannot be made simple, but it can be analyzed in part. It is not understood in all its appalling intricacy, but some understanding is in our grasp, and we may trust our own powers to obtain more.
>
> Simpson 1953:393

"Adaptive radiation" is a relatively old macroevolutionary concept that is presently undergoing its own personal renaissance after decades of general neglect within mainstream evolutionary biology. That renaissance has itself undergone a rapid cladogenesis, as some authors focus their attention on the "adaptive" and others upon the "radiation" components of the concept. As a result of this divergence, two very different research programs now describe themselves with the same title, causing substantial confusion in the literature. In the next few pages, we will discover just how we got to this state of confusion in the first place, then we will attempt to find a way out of the conceptual and methodological maze created by that confusion.

The historical roots of adaptive radiation have, to a large extent, been buried in its resurrection. The story of the discovery and dissemination of the concept is dominated by two key figures: its originator, Henry Fairfield Osborn, and its champion, George Gaylord Simpson.

THE ORIGINS OF THE CONCEPTS OF ADAPTIVE RADIATIONS AND ADAPTIVE ZONES

Henry Fairfield Osborn was one of the key leaders of the orthogenetic movement that swept through North America and Europe in the early to mid-1900s. Although adaptation was generally not a key component of orthogenetic explanations, it was an important component of Osborn's research (albeit not in the Darwinian sense). For example, he described five different "lines of specialization" in mammalian teeth, diet, and limbs and feet (fitting their possessors for different habitats and types of locomotion), then combined characteristics from each line to reconstruct various mammalian orders. From this, Osborn proposed the following:

> [I]t is a well-known zoölogical principle that an isolated region, if large and sufficiently varied in its topography, soil, climate, and vegetation, will give rise to a diversified fauna according to the law of adaptive radiation from primitive and central types. Branches will spring off in all directions to take advantage of every possible opportunity of securing food. The modifications animals undergo in this adaptive radiation are largely of mechanical nature, they are limited in number and kind by hereditary, stirp, or germinal influences, and thus result in the independent evolution of similar types in widely separated regions under the law of parallelism or homoplasy. This law causes the independent origin not only of similar genera but of similar families and even of similar orders. Nature thus repeats herself on a vast scale, but the similarity is never complete or exact. Osborn 1910:355–356

According to Osborn (1910, 1934), each unique set of characteristics was an *adaptive mode,* and all species exhibiting the same adaptive mode occupied an *adaptive zone.* Because orthogenesis was founded upon the assumption that evolution is an internally driven phenomenon, Osborn believed that every lineage would travel through various adaptive zones (increasing ecological specialization), becoming so specialized that it eventually went extinct.

One of the founders of the new synthesis, George Gaylord Simpson, took Osborn's concept of adaptive radiation and tried to recast it within a Darwinian framework. In brief, Simpson argued that the environment could be divided into "a finite and more or less clearly delimited set of zones or areas" (Simpson 1944:189). He envisioned adaptive radiation as a three-step phenomenon, beginning slowly as small, isolated populations accumulated mutations (evolutionary novelties) that would gradually make them less adapted to their environment (fig. 6.1). This accumulation would either drive a population to extinction or allow it to make a "quantum leap" into a new adaptive zone. In essence, the novel traits would "pre-adapt" the population for survival in the new zone (or for extinction, if no new zone were available). Once in the new zone, the leaping ancestor and its descendants would diversify along two different pathways (fig. 6.2): (1) geographic expansion + speciation, producing many species with the same *plesiomorphic ecology* (differing in the "where you live" component of the Hutchinsonian fundamental niche of the spe-

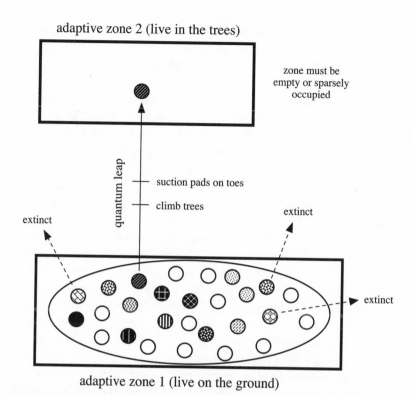

Figure 6.1. Stage 1 in an adaptive radiation à la Simpson. *Circles* = populations (different patterns = different mutations); *boxes* = adaptive zones.

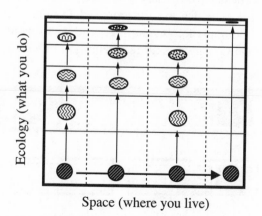

Figure 6.2. Stages 2 and 3 in an adaptive radiation à la Simpson. The population in the new zone speciates geographically. Once in the new areas, speciation occurs along ecological lines, filling up niches. In this case the niches would represent areas within the tree adaptive zone (trunk, leaves, terminal twigs, etc.).

cies) and (2) ecological diversification + speciation, producing many species with *apomorphic ecologies* (differing in the "what you do" component of the Hutchinsonian fundamental niche of the species). Simpson was developing his ideas about adaptive radiations at the same time physicists were developing theories of quantum mechanics. Quantum mechanics, as you all recall from first-year physics, is based upon the proposition that the double life of photons (energy/wave versus matter/particle) introduces a fundamental duality into the universe. Simpson built a phylogeny/ecology duality into his adaptive zones (fig. 6.3), which could be specified by the group that occupied them (the "felid zone") and by the adaptive trait that matched the environment (the "carnivore zone"). Different zones could thus have different degrees of environmental and taxonomic complexity, which made it difficult to determine exactly what constituted a zone. Although Simpson's ideas were later modified by a variety of authors, including himself, his basic components remained: an initial major change (a **key innovation:** Miller 1949) moved a group into a new zone, where the group diversified (radiation of species and adaptations) to an extent limited only by the amount of available space (unoccupied niches) in the zone (Huxley 1942; Wright 1949; Mayr 1963).

These various formulations of "adaptive radiation" had three major problems. First, as mentioned previously, the term "zone" was not easily defined. Was it an area, an ecology, or a taxon? Second, although taxonomic arrangements based upon phylogenetic zones generally were based on clades (e.g., felids), grouping organisms according to their membership in a particular ecologically based zone often created a paraphyletic group (fig. 6.4), which in turn produced an inflated assessment of the amount of homoplasious evolution occurring in nature (remember chapter 2?). Finally, there was no real definition of what qualified as a

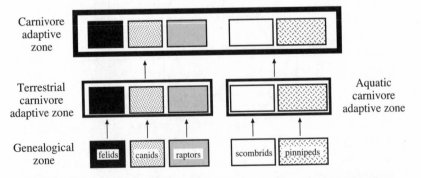

Figure 6.3. Complexities associated with defining the term "zone." For example, felids (cats) are simultaneously part of the felid genealogical zone, the terrestrial carnivore zone, and the carnivore zone.

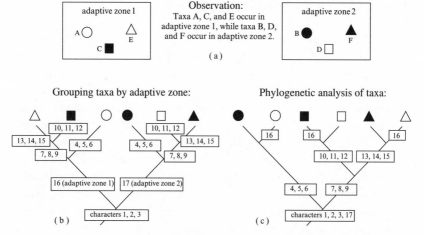

Figure 6.4. Problems with reconstructing genealogical relationships based upon membership in adaptive zones. (*a*) Six taxa (A–F) are distributed in two "adaptive zones." (*b*) Placing the taxa into two distinct groups *using the adaptive zone as the deciding character* (adaptive zone 1 = character 16; adaptive zone 2 = character 17) creates the impression that there has been substantial homoplasious evolution throughout the history of the two groups (tree is 29 steps long). (*c*) Phylogenetic analysis based upon all characters, including the traits associated with the "adaptive zone," indicates that the previous groupings were paraphyletic and that there is substantially less homoplasy than originally thought (tree is 19 steps long). According to this topology, the trait associated with adaptive zone 1 (character 16) has arisen convergently, while character 17 (adaptive zone 2) is plesiomorphic. (Note: it is equally parsimonious to hypothesize that character 16 is plesiomorphic and character 17 is convergently derived.)

key innovation or how that innovation caused a population to "leap" from one zone to another. Traditionally, a key innovation was considered to be a novel feature characterizing a group so ecologically or morphologically distinct that it could be placed in its own "higher" taxon (e.g., order or class: Simpson 1944, 1953, 1960; Miller 1949; Mayr 1963; van Valen 1971). There was no necessary link between the key innovation and speciation; only between the key innovation and the degree of distinctiveness, which in turn was related to the distinctive nature of the new "adaptive zone." If, however, many of the higher taxa grouped in such a manner were paraphyletic assemblages, then their purported "distinctiveness" might be nothing more than an illusion. And, if the groupings were illusory, then so was the importance of the key innovations.[1]

1. For an excellent review of the key innovation concept, then and now, see Hunter 1998.

ADAPTIVE RADIATIONS TODAY

"Adaptive radiation" as originally conceived was always complicated and often ambiguous. Not surprisingly, that ambiguity led to the eventual fragmentation of research along the two main ideas embodied within the term, "adaptive" (character evolution) and "radiation" (speciation). Unfortunately, the fracturing of one research program along such dramatically different lines has produced new ambiguity because both descendant programs have retained the term "adaptive radiation" to describe their research interests. This in turn has caused some friction between the groups, based more (we believe) upon misunderstandings of what each is attempting to accomplish than upon actual criticisms of the research itself. In order to avoid these misunderstandings, it is imperative that everyone be explicit about what they are studying.

From our perspective, there are two major divisions, one of which has two subdivisions, among the academic descendants of Osborn and Simpson (fig. 6.5). Researchers in one division focus their attention on the *radiation of adaptations* associated with invasion of the new zone ("so many characters, so little time").[2] Although speciation may be incorporated into this program, it is not a critical component. The only requirement is that some unit of biological diversity (microspecies, macrospecies, genus, family, etc.) show substantial adaptive diversification.

Researchers in the other division focus on the *radiation of species* ("so many species, so little time"). This research is triggered by the observation that some clades appear to be unusually species rich compared with other clades. The observation that some clades are asymmetrical with respect to species number is an old one. Rosa (1918, 1931) proposed that sister groups often contained two different "stems," the "precoce" (or smaller stem) and the "retarded" (or larger stem). Using orthogenetic reasoning, he argued that the smaller stem would be more phylogenetically conservative (archaic), would reach its apogee of large and exaggerated forms earlier, and go extinct sooner than its more species-rich sister. This theme was later adopted in modified form by Hennig (the progression rule: 1966a), who proposed that the asymmetry was due to the repeated dispersal by members from one lineage into new areas, where they were subject to different selection pressures. Result: rapid peripheral isolates speciation and differentiation, producing a species-rich group

2. See, e.g., Huxley 1942; Grant 1963, 1977; Dodson and Dodson 1985; Futuyma 1986, 1998; Avers 1989; Baldwin et al. 1990; McKinney 1993; Schluter 1996b; and Orr and Smith 1998.

Evolutionary Radiation Research Program

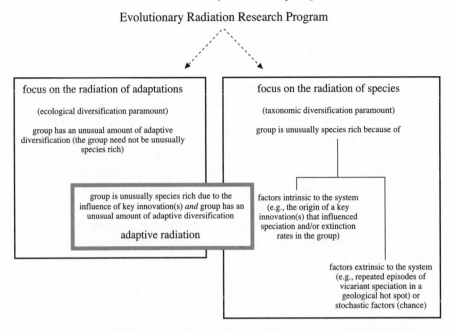

Figure 6.5. The study of evolutionary radiations in their many manifestations.

possessing many derived characteristics. Today, the question of clade asymmetry falls within the domain of the "taxic approach" to macroevolutionary studies, an approach concerned with the analysis of phylogenetic patterns resulting from processes controlling the rates of speciation and extinction (Eldredge and Cracraft 1980; Cracraft 1981a, 1982a,b, 1985a,b). The goal of this research is to identify clades of "unusually high" or "unusually low" species number, then try to uncover the processes responsible for the asymmetry.

Differences in species number between sister groups can arise for at least three different reasons. First, it is possible that asymmetry develops stochastically rather than due to the influence of any forces, extrinsic or intrinsic to the biological system.[3] For example, say you have two sister species, A and B, existing quite contentedly for eons. Then one day, purely by chance, A is subdivided and speciates, producing daughter species C and D. *If speciation occurs at random,* then the clade comprising C

3. Raup et al. 1973; Anderson 1974; Raup and Gould 1974; Anderson and Anderson 1975; Flessa and Levinton 1975; Schopf et al. 1975; Farris 1976; Gould et al. 1977; Raup 1977, 1981, 1985; Rosen 1978; Sepkoski 1978; Schopf 1979; Templeton 1979; Stanley et al. 1981; Dial and Marzluff 1989; Maddison and Slatkin 1991; and Kirkpatrick and Slatkin 1993.

and D has a higher probability of being subdivided again than does its monotypic sister group (species B), and so on. In other words, the more bets you place on the speciation roulette wheel, the greater your chance of winning.

It is also possible that asymmetry could arise due to the operation of factors extrinsic to any particular group. Cracraft (1985a) suggested that the distribution of biodiversity is affected by two major environmental factors that contribute to the asymmetric distribution of species in clades. The first is what he termed "environmental harshness" (the "environmental stress" of Hoffman and Hercus 2000). The observation that diversity in the tropics is higher than diversity in temperate or arctic regions is often attributed to differences in speciation rates. It is possible, however, that extinction rates in temperate to arctic habitats have been higher than extinction rates in the tropics due to historical increases in environmental harshness in the colder areas (see also Wallace 1878; Peterson and Black 1988; Parsons 1993), which limited the number of species living at higher latitudes. (Note that this generalization holds true only for terrestrial/freshwater ecosystems—some of the most species-rich marine ecosystems occur at high latitudes.) Cracraft's second environmental factor affecting biodiversity is the history of geological change and geologically mediated speciation. Biological diversity tends to be clumped in "hot spots" corresponding to areas with historically high rates of geological change, rather than being uniformly distributed across a given habitat or zone. For example, tropical diversity is clumped in South America, the Indo-Malaysian region, Madagascar, and the Rift Lakes of Africa, areas whose geological histories are extremely complicated. In other words, a clade may be unusually species rich because it exists in a geological hot spot, where it has been subdivided on a regular basis, producing many species via repeated episodes of vicariant (including microvicariant) speciation. Although character evolution is obviously a part of the speciation process in these examples, *no one character or set of characters is causally involved in promoting speciation.*

Finally, it is possible that asymmetry is due to some factor(s) intrinsic to the biological system, which brings us back to the concept of a key innovation. Shifting the evolutionary focus from degree of distinctiveness to the number of species in a group (also called "net diversity," the number of speciation events minus the number of extinction events: Stanley 1979) allowed the key innovation concept to be reintroduced to the study of radiations. Both changes in competition pressure by movement into a new, unoccupied space (either ecological or geographical) and improved competitive ability within the plesiomorphic space have been cited as possible mechanisms by which the origin of a key innovation might have a positive effect on net diversity.[4] It is important to emphasize here that using phrases like "reduced competition" and "ecological release" (e.g., release from competition or predation pressure) as an *explanation* for the effect of a key innovation on *patterns of radiation* is too imprecise. Both of these terms imply "absence." If movement into a new area (zone/niche) removes competition, what causes the population to change? (If the population remains static in the new "zone," you have evidence for a range expansion but not for a "radiation.") Surely the evolution (speciation, radiation) of the population in the new "zone" does not occur because of the absence of competition or predation, but because the population has encountered a new selective arena (in addition to the more stochastic effects of genetic drift: see Masters and Rayner 1993 for a thought-provoking discussion of this problem).

A key innovation may exercise its influence on net diversity in two different, but not necessarily mutually exclusive, ways. On one hand, the character may increase the likelihood that members of a clade will undergo episodes of rapid speciation (e.g., Mayr 1954; Newman et al. 1985). A variety of factors, including the adoption of a specialist foraging mode[5] or changes in breeding systems (sexual selection) and in population structure,[6] have been postulated to have a positive effect on speciation rates.[7] On the other hand, the key innovation may influence net diversity by increasing the population size, expanding the range occupied by the species, or both, decreasing its chance of going extinct. The interaction between increased rates of speciation and decreased rates of extinction is often difficult to disentangle. For example, Mitter et al. (1988) compared species richness between phytophagous (plant-eating) insect groups and their nonphytophagous sisters. In 11 of the 13 comparisons, the phytophagous lineage was more species rich, leading to the conclusion that adoption of phytophagy was significantly associated with an increase in net diversity (one-tailed sign test: $p < 0.02$). How that net diversity was produced, however, and whether only a single factor was involved, remained a mystery.

We believe that the intersection of the two "radiation" programs (see fig. 6.5) constitutes the most robust study of an "adaptive radiation" as

4. Liem 1973; Larson et al. 1981; Cracraft 1982a,b; Stanley 1987; Pimm et al. 1988; Bengtsson 1989; McKinney 1993; and Heard and Hauser 1995.

5. Eldredge 1976; Greene and Burghardt 1978; Eldredge and Cracraft 1980; Vrba 1980, 1984a,b; Cracraft 1984; Novacek 1984; Mitter et al. 1988; Ferrarezzi and Gimenez 1996; Butterfield 1997; Bond and Opell 1998; and Jordal et al. 2000.

6. Spieth 1974; Wilson et al. 1975; Carson and Kaneshiro 1976; Ringo 1977; Templeton 1979; Gilinsky 1981; West-Eberhard 1983; Barton and Charlesworth 1984; and Carson and Templeton 1984.

7. For reviews, see Ricklefs 1989 and Caillaud and Via 2000.

it was originally described by G. G. Simpson. Simpson (1944) believed that adaptive radiations resulted from *diversification* (species component) accelerated by ecological opportunity such as dispersal into new territory, extinction of competitors, or adoption of a new way of life (i.e., an *adaptive change* in ecology or behavior). That perspective was echoed more than 20 years later by Grant:

> When a species succeeds in establishing itself in a new territory or new habitat, it gains an ecological opportunity for expansion and diversification. The original species may respond to this opportunity by giving rise to an array of daughter species adapted to different niches within the territory or habitat. These daughter species become the ancestors of a series of branch lines when they, in turn, produce new daughter species. The group enters its second phase of development, the phase of proliferation. . . . Adaptive radiation is the pattern of evolution in this phase of proliferation. And speciation is the dominant mode of evolution in adaptive radiation. The first generation of speciational events gives rise to the primary branches in the now growing phylogenetic tree. These primary branches correspond to different primary ecological niches. Second, third, and later generations of speciation in each primary branch then parcel out the available ecological niches in the territory or habitat into a larger number of more highly specialized species. (1977:309)

While we agree with the basic idea behind this perspective (a causal coupling of species richness and adaptive diversification), we believe the mechanism outlined should be viewed with some caution. First, the idea that speciation produces species that progressively partition the environment into smaller and more specialized subunits comes from orthogenesis (e.g., Cope 1885, 1896). It was this aspect of Osborn's adaptive radiation scenario that Simpson, and many later authors, were never able to successfully remove from their research programs. The proposal is, of course, eminently testable. If adaptive radiations follow the pathway described above, then we should repeatedly uncover an evolutionary movement from generalist to specialist within clades. (We will return to this question in chapter 8 when we discuss coevolution.) At the moment, there are few examples that rigorously document a causal relationship between rapid levels of adaptive diversification and speciation (Schluter 1996b; Funk 1998; Orr and Smith 1998; de Queiroz 1999). For example, Rainey and Travisano (1998) demonstrated that the aerobic bacterium *Pseudomonas fluorescens* diversified rapidly from the smooth morphotype into two additional, spatially segregated morphs (wrinkly and fuzzy) when an ancestral population was raised in a heterogeneous environment. When smooth populations were raised in a homogeneous environment, no such diversification occurred, indicating that environmental opportunity (the availability of unoccupied ecological and geographical space [from the perspective of a bacterium, a petri dish has geography]) played an important role in the adaptive diversification. The questions still remain, however, Is that diversification coupled with irreversible lineage splitting (speciation)? or Does that diversification simply reflect an extremely flexible response to the proffered environmental opportunities that, although rapid and repeatable, is reversible? If the former, then we might have evidence of an adaptive radiation (assuming the sister group does not show the same abilities and has significantly fewer species); if the latter, then we have evidence for a radiation of adaptations.

Second, the concept that the environment is somehow divided into predetermined niches in the absence of the organisms that inhabit those niches suffers from the same problems as did the concept of "adaptive zones." Organisms create their own niches; they exist where they are and do what they do because of their particular natures and history (Maynard Smith 1976; Lewontin 1978, 1983; Brooks and Wiley 1986, 1988; Odling-Schmee et al. 1996). Changes in the biological system create new "niches"; unoccupied niches do not "pull" organisms into them (otherwise we would exist in a Lamarckian world). This subtle distinction was eloquently outlined by Lewontin, who wrote that the "error is to suppose that because organisms construct their environments they can construct them arbitrarily in the manner of a science fiction writer constructing an imaginary world. . . . Where there is strong convergence is in certain marsupial-placental pairs, and this should be taken as evidence about the nature of constraints on development and physical relations, rather than as evidence for pre-existing niches" (1983:283). In other words, a niche is a property of the organism. Think of the evolutionary sequence in this way: Say you are a bird living on the mainland eating small nuts, and you (and some conspecifics) fly to an island that has only big nuts available. If you have absolutely no ability to crack the nuts open, you will either leave the island or starve. The big nut niche does not "pull" you into it, creating a new beak shape de novo. But, say you have a much larger beak than needed for cracking small nuts (perhaps females prefer big-beaked males), and you find that you can eat the big nuts as well. In this case, you stay on the island and enter a new selection regime, created by different abilities among individuals to handle and open big nuts based on the amount of genetic variance in the population (which is a historical legacy). It is perfectly acceptable to call big nuts lying about with no one to eat them a "new adaptive zone," an "unoccupied niche," or an "unexploited resource." This is the sense in which unoccupied niches are "real" entities. And it is reasonable to postulate that the big nuts may at act as an agent of selection, but that will only occur when an organism already possessing the necessary genetic variation (the material on which selection works) discovers the nuts. That may never happen, no matter how long the nuts lie about. In other words, first the organism moves to the zone/niche/resource because of its own unique biological properties, then the interaction between the new zone/niche/resource/selective regime and the biological system begins.

DISCOVERING AN EVOLUTIONARY RADIATION

In order to study any radiation more rigorously, we must have a method for identifying, a priori, groups that are either "unusually species rich" or groups that have an "unusual amount of adaptive change." Let's begin with, Is this group unusually species rich? The methods for answering this question have been the result of a fertile interaction between systematists and modelers. There are four general steps in this program.[8] As of yet, these steps have not been used in studying the question of whether one group is unusually "adaptation rich" (shows a radiation of adaptations). To reconstruct how such a program might work, try inserting "adaptation rich" for "species rich."

The Path to Enlightenment

You, an indefatigable entomologist, have described two genera of butterflies, one of which (genus O) has 400 species, and the other of which (genus S) has only four species. Should you rush out and begin studying the radiation of O, or will you be "digging in the wrong place"?

Step 1. You begin your study by reconstructing the phylogenetic tree for the family (Mariposidae) containing your two new genera, using as many different characters as possible. Much to your satisfaction, you discover that O and S are indeed sister groups (fig. 6.6). Having done extensive reading (while the members of your large and well-funded research lab were sequencing various genes and describing morphological traits for the entire family), you know that the first step in any evolutionary radiation study is the identification of sister groups, because *sister groups are the descendants of a common and unique speciation event and hence must be of equal age* (Hennig 1966a; Cracraft 1981b; Mayden 1986, 1987a,b). *Comparing sister groups eliminates the possibility that one group*

---

8. Jensen 1990; Brooks and McLennan 1991, 1993d; Doyle and Donoghue 1993; Skelton 1993a,b; and Sanderson and Donoghue 1994, 1996.

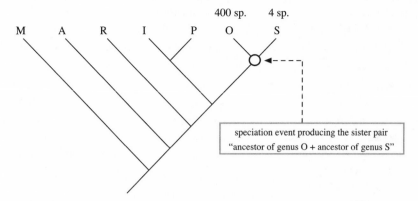

Figure 6.6. Phylogenetic relationships of the hypothetical family of butterflies, the Mariposidae.

might be more species rich (adaptation rich) than the other simply because it is older. For example, if you compared the number of ray-finned fishes (approximately 25,000 species) with the number of snakes (2,700 species), you would have to conclude there are more fish than snakes, which is not so surprising given that ray-finned fish have been speciating for more than 410 million years, while snakes have been enjoying life for a mere 95 (to 140) million years (Greene 1997).

Step 2. Looking at your tree, you confirm your initial intuition that O is more species rich than S. How do you determine whether you have identified something potentially significant? Fortunately, the topic of sister-group asymmetry has been thoroughly discussed for years, so there is an abundance of mathematically based models available to test the hypothesis that one clade contains significantly more species than its sister group.[9] For our purposes, we will use as a general rule of thumb the conclusion by Guyer and Slowinski (1993) that one group must contain more than 90 percent of the total diversity in the two sister groups in order for there to be a significant difference in species number between the two ($p = 2(N - r)/(N - 1)$, where N = total number of species in the study group [in this case, S + O] and r = number of species in the larger clade [O]: equation developed in Slowinski and Guyer 1989).This is an *extremely* conservative test. For example, when we apply this equation to the data sets from previous studies in which significant species richness was detected using other statistical approaches, only 4 of 15 compari-

9. Slowinski and Guyer 1989, 1993; Sanderson 1990; Slowinski 1990; Guyer and Slowinski 1991, 1993, 1995; Hey 1992; Sanderson and Bharathan 1993; Sanderson and Donoghue 1994, 1996; Losos and Adler 1995; Nee et al. 1996; Cook and Lessa 1998; and Chan and Moore 1999.

Table 6.1.   Some Test Studies Reanalyzed Using the Conservative Formula for Determining Significant Differences in Species Numbers

Sister-Group Comparison	Species Number	Prior Result	New Result[a]
Phytophagous insects[b]			
Tingidae vs. Joppeicidae	1,815 vs. 1	+	$p < 0.001$
Miridae vs. Isometopidae	10,000 vs. 60	+	$p < 0.012$
Trichophora vs. Aradidae	5,000 vs. 1,000	+	NS ($p = 0.33$)
Elateridae vs. Cerophytidae	9,170 vs. 1,397	+	NS ($p = 0.26$)
Scarabaeidae vs. Geotrupinae	14,000 vs. 400	+	NS ($p = 0.06$)
(or vs. Aphodiinae)	(14,000 vs. 3,200)	(+)	NS ($p = 0.37$)
Languriinae vs. Lobarinae	410 vs. 1,750	NS	NS ($p = 0.38$)
Epilachninae vs. Coccinellini	700 vs. 250	+	NS ($p = 0.53$)
Phytophagous vs. nonphytophagous Cucujoidea	130,000 vs. 10,000	+	NS ($p = 0.14$)
Symphyta vs. Panorpida	10,000 vs. 80,000	NS	NS ($p = 0.22$)
Lepidoptera vs. Trichoptera	140,000 vs. 7,000	+	NS ($p = 0.10$)
Chloropidae vs. Siphonellopsinae	2,210 vs. 80	+	NS ($p = 0.07$)
Tephritidae vs. nonphytophagous Tephritidae	4,000 vs. 1,733	+	NS ($p = 0.60$)
Agromyzidae vs. Clusiidae	2,000 vs. 200	+	NS ($p = 0.18$)
Parasitic platyhelminthes[c]			
Digenea vs. Aspidobothrea	5,000 vs. 50	+	$p < .02$
Monogenea vs. Cestodaria	5,000 vs. 3,018	+	NS ($p = 0.75$)
Eucestoda vs. Amphilinidea	2,500 vs. 10	+	$p < 0.008$

NOTE: + = significant difference; NS = no significant difference.

[a] Using formula $2(N - r)/(N - 1)$; r = number of species in the larger clade, N = total number of species in the study group.

[b] Data from Mitter et al. 1988.

[c] Data from Brooks and McLennan 1993d

sons retain a significant difference (table 6.1). Why might this have happened, especially for the phytophagous insects, a group often considered the poster child for adaptive radiations (e.g., Ehrlich and Raven 1964; Price 1980; Futuyma and Slatkin 1983; Mitter et al., 1988, 1991)?

One possible explanation is that the putative key innovation, in this case phytophagy, has a complex evolutionary history. In a few groups phytophagy has been replaced by another feeding mode (parasitism) within the "phytophagous lineage" (fig. 6.7a), while in others the ancestrally nonphytophagous sister group contains some phytophagous species (fig. 6.7b). That the key innovation might change is not surprising, given that evolution is an ongoing process, but it is methodologically frustrating. How many species should you include in the putatively

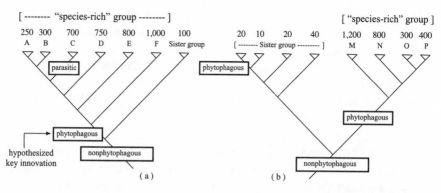

Figure 6.7. Problems that can occur when the key innovation continues to evolve within your study group.

species-rich group shown in figure 6.7a, 3,800 or 3,100? How many species in the sister group shown in figure 6.7b, 90 or 70? Mitter et al. (1988) suggested that, minimally, the problem taxa should be subtracted from the overall species total for the group. In this example, removing the problem taxa changes both tests from marginally significant to not significant (try the math on your own), so the problem is not a trivial one. In this chapter, we will discuss examples that are relatively straightforward. (If you have a taste for something more complicated, see the examples in Chan and Moore 1999.)

> A less stringent, but still quite conservative, equation for determining whether two groups differ in species number was proposed by Slowinski and Guyer in 1993: level of significance $(p) = (N - r)/(N - 1)$. Depending upon how strict you wish to be in your analysis, you can choose either equation. In this chapter, we retain the ultra-conservative $p = 2(N - r)/(N - 1)$. You can examine how the conclusions of each study might change if the requirements for determining significant differences are loosened somewhat using the 1993 formulation.

This leads us to reemphasize the importance of beginning your study with as robust a phylogenetic tree as possible. Do not be mesmerized into thinking that the application of a mathematical equation somehow magically compensates for weakly supported hypotheses of sister-group relationships. No matter how well argued the equation, its ultimate utility depends upon the quality of the data; in this case, "the data" are the phylogenetic relationships (see, e.g, Cunningham 1995; Guyer and Slowinski 1995). You undoubtedly recognize that this has been a common theme throughout this book: if you begin your evolutionary study with a poorly supported tree, then any conclusions you draw will be weak at best, and at worst may represent a significant waste of time, energy, and research funds (remember Murphy's rule from chapter 2).

Now, back to the butterflies. Inserting species numbers from your groups into the equation for determining significant differences in asymmetry indicates that O is indeed different from S ($p = 2 [404 - 400]/ [404 - 1] = 0.02$). Celebrations all round!

Step 3. Your time spent reading has been time well spent, for you realize that the next critical step in a radiation study is to *determine whether it is the species-rich or the species-poor member of the sister group that is unusual*. This requires increasing the phylogenetic scope of the study to include, minimally, the next two sister clades. Simple optimization of species richness onto the expanded phylogeny may be able to resolve the problem. For example, if you uncover the pattern shown in figure 6.8a, you would continue to focus your attention upon the (significantly) species-rich genus O. If, however, you were to uncover the pattern shown in figure 6.8b, the more interesting question for the Mariposidae would be, Why is genus S so species-poor? In general, the world is more complicated than the perfectly designed Mariposidae, so identification of the unusual clade would require recourse to more complicated analyses (see Chan and Moore 1999). For the purposes of this example, we will stay with the simpler scenario.

Step 4. Having identified O as a *significantly* and *unusually* species-rich group, you examine the synapomorphies for the clade in order to identify any putative key innovations. Recall that a key innovation is hypothesized to enhance speciation rates either directly or indirectly by allowing species to persist long enough to undergo numerous episodes

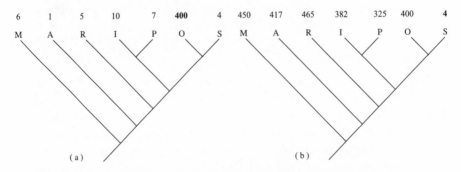

Figure 6.8. Using sister-group relationships and phylogenetic optimization of species richness to determine if (*a*) species richness or (*b*) species poverty is evolutionarily significant within a clade. *Numbers* = total number of species in the taxon.

of speciation without going extinct. In order to test this hypothesis, you need a method for determining whether or not speciation *rates* increased following the origin of the innovation.[10] Focusing on rates of diversification, rather than just numbers of species, brings you full circle, to one of George Gaylord Simpson's main messages: time, and by extension, history, must not be removed from evolutionary explanations.

Simpson's first book, *Tempo and Mode in Evolution* (1944), is enjoying a resurgence of interest as more and more biologists are becoming fascinated with documenting the tempo of speciation and extinction. In the best of all possible worlds, you would have information about both the amount and the timing of speciation and extinction events for your group (a good fossil record). In the absence of this information, analyzing molecular data phylogenetically and comparing the resultant tree topology to one constructed using maximum-likelihood models that constrain evolutionary rates to fit the assumptions of various molecular clocks will allow you to investigate the putative differences in rates of speciation more rigorously. Recall from chapter 2 that using most parsimonious trees to find the best-fitting maximum-likelihood model for your (genetic) data eliminates the circularity inherent in building a tree using molecular clock–based models, then using that tree topology to determine timing.[11] Although using molecular data in this way can give you a better estimate of divergence rates, it cannot shed much light on extinctions in the clade in the absence of fossil evidence.

This step is complicated by a number of factors. First, each ancestral branch on a phylogenetic tree may be characterized by more than one apomorphic trait. Asserting that every autapomorphy is potentially responsible for an observed radiation is of limited value at best and circular at worst (Cracraft 1990; Guyer and Slowinski 1993; Brooks and McLennan 1993d). One way around this problem is to delineate the relationship between an innovation (say, small size) and net diversification *prior* to beginning your study (Mitter et al. 1988; Lauder and Liem 1989; Slowinski and Guyer 1993). Unfortunately this is not always practicable because a researcher often knows what characters define her species-rich group, so it makes no difference whether she develops a hypothesis of how the trait and species richness are connected before or during the study.

Second, and more important, it is possible that the key innovation

10. See, e.g., Hey 1992; Sanderson and Bharathan 1993; Sanderson 1994; Sanderson and Donoghue 1994, 1996; and Chan and Moore 1999.

11. For an excellent discussion of this problem, see Peterson 1992 and Sanderson and Donoghue 1996.

may be, in fact, a series of traits (a key "complex") which together had an impact on net diversification (Simpson 1953; Lauder and Liem 1989; Brooks and McLennan 1993d) or that the innovation did not exert its influence until environmental conditions changed (the key innovation equivalent of an exaptation). In either case, the acceleration of speciation rates might be delayed compared to the origin of the key innovation (Sanderson and Donoghue 1996).

Finally, it is possible, indeed probable, that as the age of the putative key innovation in question increases, so also do the chances that its original effects on net diversification (if any) have been obscured by the evolution of other characters, which themselves may have influenced speciation or extinction rates. For example, de Queiroz (1999) investigated the possible key innovation status of image-forming eyes by comparing numbers of species in 12 sister pairs (one member with and the other without such eyes). His analysis revealed no association between the origin of the trait and increased diversification. Recognizing that the effects of eyes might have been obscured over such long evolutionary time spans, he altered the analysis to examine the effects of image-forming eyes on net diversification of sister pairs that originated during the Paleozoic ("old") versus pairs that originated after the Paleozoic (relatively "young"). There was still no significant effect. De Queiroz cautioned that this approach was crude at best but noted that it was a starting point, pending further information about rates of extinction and speciation in the clades under investigation. The age caveat is the evolutionary radiation equivalent of warnings about equating one gene with one protein or one character with one selection vector. Evolutionary systems are complex and often resist biologists' attempts to reduce them to that simplistic world in which ideal gases actually exist.

Speaking of perfect worlds, in such a universe a key innovation would be homoplasious, which would allow us to subject our hypothesis that the trait was indeed causally, rather than casually, involved in the increased net diversification to statistical analysis (Ridley 1983; Mitter et al. 1988; Cracraft 1990; Lydeard 1993). Our world, however, is seldom perfect. Because we expect a trait's influence on speciation to be itself affected by the biological context in which the trait originates, it is possible that a character might be causally involved in increased speciation rates in one clade, but "constrained" from causing the same effect in another clade. It is intuitively more plausible that a trait originating numerous times within a monophyletic group (say, mammals) might be a better test of the key innovation hypothesis than a trait that appears homoplasiously across distantly related groups (say, in mammals, crocodiles, and snakes). This, of course, depends upon your level of analysis (after all, mammals, crocodiles, and snakes are part of a larger monophyletic group, the Amniota). The critical point here involves the amount of evolutionary change between the groups displaying the key innovation and the biological background required to allow that innovation to influence the speciation process. An observation that a putative key innovation was associated with increased speciation in clade A, but not in clade B or clade C, does not eliminate the possibility that the trait had a very real impact on speciation in clade A because *the trait may not have originated under equivalent conditions in all clades.* (Hence the statistical test is flawed because it is predicated on the assumption that all other things are equal [*ceteris paribus*]. As the evolutionary distance increases between two clades, the likelihood that the *ceteris paribus* assumption will hold decreases.)

It is also possible that a key innovation will be a trait or trait complex that has originated only once in the history of life on this planet (consider traits such as meiotic sex, the cell membrane, gastrulation, the amniotic egg, the vertebral column). Although absence of homoplasy makes the hypothesis that the character was causally associated with the initial burst of speciation *at its point of origin* more difficult to test from a statistical perspective, this does not mean that historically unique events are unimportant, just harder to study (Coddington 1994; Cooper 1997). One way around the problem is to search for a reversal of the putative key innovation to the plesiomorphic condition within the species-rich group, then ask whether that reversal is associated with a decrease in net diversification. If the loss has no effect on speciation rates, then this weakens the hypothesis (Lieberman 1993; Brooks and McLennan 1993d). If the trait originates only once and is never lost (like the amniotic egg, for example), then the best we can do is demonstrate a correlation between character origin and net diversification. Experimental evidence demonstrating possible ways in which the character might have had an impact on net diversification gives additional weight to the hypothesis of causal relationship, but that is often difficult to achieve. Like the process itself, studying evolution can be very complex.

Finally, it is important to determine whether you have anything unusual to explain when offered a trait as a key innovation. For example, an elegant study by Connor and Taverner (1997) examined the hypothesis that the leaf-mining habit displayed by many groups of insects was responsible for an adaptive radiation of those groups. The hypothesis was based upon a number of suggested, and intuitively plausible, adaptive benefits to leaf mining (protection from attack by predators, parasites, and pathogens and buffering against environmental stressors such as wind, UV radiation, and desiccation) that might have a positive effect on net diversification by opening up new resources for exploitation and thus decreasing extinction rates. Connor and Taverner began their study by examining the "adaptive" portion of the radiation. An extensive survey of the literature, combined with their own experimental data, revealed both benefits (pathogen avoidance and higher feeding efficiencies) *and* costs (higher mortality rates due to parasitism and premature destruction of the leaf in which they are living, smaller body size [possibly due to constraints of tunnel living] and thus lower fecundity) of leaf mining versus external feeding. They then compared the number of species in leaf-mining clades versus their sister groups and discovered no significant difference between the two; if anything, there was a slight (but not significant) trend toward reduced species richness in the leaf miners. Connor and Taverner cautioned that many clades of leaf miners had to be eliminated from the phylogenetic statistical analysis because either the sister group or the number of species (or both) were unknown. That caveat notwithstanding, this is a fascinating study because it illustrates the power of combining information from phylogenetic and cost-benefit analyses in order to understand why evolution has taken a particular pathway. In this case, it appears that the advantages of leaf mining are sufficient to ensure that it persists under certain circumstances, but are not so overwhelming when matched against the costs to have a positive effect on net diversification.

SOME EXAMPLES FROM THE REAL (AND IMPERFECT) WORLD
Radiations of Adaptations

As mentioned previously, the major problem that needs to be addressed here is how to determine whether a clade displays significantly more adaptive change than its sister group. The first step in this program is relatively straightforward. Just as in the "determining species richness" program, you must compare sister groups in order to identify whether one group has undergone an "unusual" amount of adaptive change. Sister-group comparisons are the evolutionary equivalent of a control in experimental manipulations. Using this type of control decreases the probability of uncovering a significant difference because the two groups you are comparing are of different ages, or have been traveling such separate evolutionary pathways for so long that differences in their accumulated history (genetic, morphological, ecological, behavioral repertoires) confound attempts at comparison. Sister groups, on the other hand, are descended from the same most recent common ancestor and thus are the same age and are expected to share many aspects of their basic phenotypic "groundplan." If you find a difference between these two groups, you have evidence for differential rates of diversification in one group over another and can begin asking questions about underlying processes. In other words, it is more evolutionarily relevant to compare oranges with lemons than oranges with sponges. Finally, using sister groups completely eliminates any investigator-based bias in the choice of groups for comparison. For example, say you are interested in the amount of adaptive change demonstrated by group Y. You choose distantly related group Z for comparative purposes and determine, using the most rigorous morphometric techniques available, that they are distinctly different (Y shows more change than Z). Then, much to your consternation, a paper appears using equally robust morphometric analyses demonstrating that Y's sister group, X, shows much more change than group Y (and concluding that group Y is not unusual). Which of the two studies is giving you a more realistic picture of evolutionary change in group Y?

Having taken the first step by identifying your sister groups, how do you determine which of the two is showing an unusual amount of change (if either)? For example, consider the group of sponges shown in figure 6.9. Clade A shows two cases of adaptive character change (we will assume that the adaptive value of the changes has been demon-

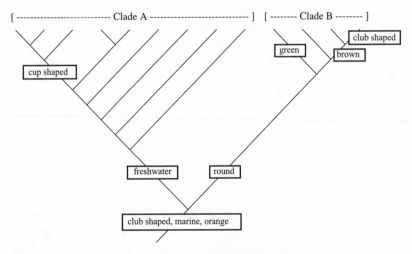

Figure 6.9. Which clade of sponges shows the greater amount of adaptive change—clade A, with two ecological character changes and nine species, or clade B, with four ecological character changes and four species? How can you test your hypothesis?

strated), while the sister group, clade B, has undergone four adaptive changes. Is the difference between the two groups great enough to be statistically significant? If so, how do we know whether it is an *increase* in the rate of adaptive change in clade B or a *decrease* in the rate of change in clade A that is unusual? Other questions, which must be addressed to inject the objective rigor of statistics into this research, will occur to you (think back to the discussions about identifying an unusual radiation of species numbers). For the time being, however, it is enough to recognize that there is substantial room for the input of mathematical models (and hence work for modelers) in the radiation of adaptations research program.

These problems aside, there are a number of examples under the rubric of "radiation of adaptations" that, in the absence of statistical approaches, are excellent examples of progress in this area. Of these, the paradigm study represents over a decade of research on the evolution of Caribbean anoles by Jonathan Losos and his colleagues.

Lizards on Islands. Anoles are (generally) small, insectivorous iguanine lizards residing in the forests and grasslands of the New World, from Florida to South America. Students of behavior, and pet enthusiasts, know them primarily for the brightly colored dewlap (throat flap) of the males, which is raised and lowered during territorial disputes and courtship. The majority of *Anolis* species are sit-and-wait predators and thus tend to be strongly territorial, moving only occasionally with rapid bursts of speed in pursuit of a tasty prey item (Stamps 1977). It is possible that the general "stay at home" tendencies of sit-and-wait foragers (the plesiomorphic condition for *Anolis*) laid the foundation for the extraordinary radiation of habitat specializations found within the group.

Losos began his study of that radiation by focusing his attention on two of the larger Caribbean islands, Jamaica and Puerto Rico. Using a principal components analysis of several morphological characters (e.g., snout-vent length, mass, tail length, forelimb and hindlimb length), he was able to subdivide species reliably into six distinct ecomorphs (see also Collette 1961; Rand and Williams 1969; Williams 1972, 1983): (1) slender, short-legged, short-tailed lizards living on twigs, (2) long-bodied, short-tailed, crown giants, (3) intermediate trunk and (4) trunk-crown morphs, (5) stocky, long-legged, trunk-ground dwellers, and (6) slender, long-tailed, grass-bush types (Losos 1990a, 1992, Glossip and Losos 1997; Losos and de Queiroz 1997; Beuttell and Losos 1999). Extensive morphometric and experimental investigation further revealed that some (but not all) differences in morphology were related to differences in functional capabilities (e.g., sprinting capability is correlated with

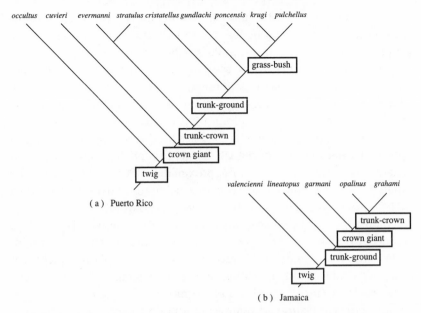

Figure 6.10. Ecomorph type optimized onto branching diagrams for anoles on two Caribbean islands. (*a*) The topology is a "best estimate" of relationships from an analysis of allozyme data, which produced a "large number" of equally parsimonious trees (Gorman et al. 1983). (*b*) The topology is a combination of an analysis by Hedges and Burnell (1990) and additional morphological data from Losos 1992 (modified from Losos 1992).

hindlimb length) and habitat use (e.g., fast moving, long-legged species tend to live in more open habitats).[12] Losos was interested in documenting whether the ecomorphs had arisen via a repeated radiation of adaptations on each island, or whether each ecomorph originated once and spread from its point of origin to neighboring islands. Mapping the ecomorphs onto trees for the anoles on both islands uncovered a similar, but not identical, pattern of habitat radiation (fig. 6.10). Losos (1992, 1994) interpreted this similarity as evidence for competition-driven habitat partitioning that occurred in a predictable sequence of diversification.[13] Once the ancestral population had moved into the new "habitat space" (Simpson's quantum leap), a variety of minor changes in climatic and prey-size preferences occurred within that space (diversification within the new space, or niche partitioning).

12. Schoener 1968b; Moermond 1979a,b; Estrada and Silva Rodríguez 1984; Pounds 1988; Losos and Sinervo 1989; Losos 1990a,b,c,d; Losos and Irschick 1996; Glossip and Losos 1997; Irschick et al. 1997; and Irschick and Losos 1998.

13. Perkins 1903, 1913; and Lack 1947. See Grant 2000 for a discussion of the important contributions made by R. C. L. Perkins to the study of island radiations.

Losos cautioned that his analysis was preliminary because some species on Jamaica were eliminated due to lack of information and, more
important, because the lizards in Puerto Rico did not represent a monophyletic group. In other words, the pattern depicted in figure 6.10a does
not represent a series of speciation events (since speciation events produce sister species) coupled with rapid diversification, but rather a combination of speciation and colonization events. Jackman et al. (1999)
filled some of the holes in the data set by combining morphological
and molecular (mitochondrial DNA) data in a phylogenetic analysis of
relationships among 53 (out of approximately 400) *Anolis* species. The
analysis, which produced four equally parsimonious trees with a low CI
(21%), provided resolution within *Anolis* subgroups but did not resolve
relationships among those groups very well (in other words, the deep
branches were not well supported). Mapping habitat preference and locality onto that tree (fig. 6.11) reveals a far more complicated picture of
adaptive change than the initial branching diagrams indicated. Some
"sister pairs" (remember, this is not a complete tree) share the same ecomorph type (speciation without ecomorph divergence: *A. coelestinus* +
A. aliniger, A. marcanoi + *A. strahmi,* and *A. distichus* + *A. brevirostris*), while
others show widespread divergence associated with speciation (e.g., the
lizards on Jamaica). Does this overall pattern represent a significant (or
unusual) radiation of adaptations? As noted, it impossible to tell at the
moment because there is so much missing data, which can have a profound influence on hypotheses of character change. For example, let us
focus our attention on Hispaniola Island. If we prune the tree to include
only the species on that island (making this a purely heuristic exercise),
there appears to have been a moderate amount of diversification (fig.
6.12a). Some of the diversification occurred on the island itself, and
some was transported in by colonizers. Adding to this picture species
that were included in a previous phylogenetic analysis (Irschick et al.
1997) increases the amount of speciation on Hispaniola that appears to
have occurred within plesiomorphic ecomorph space (fig. 6.12b). If that
speciation follows the pattern suggested previously, we would expect to
find that either these species show minor diversifications in microcli-
matic preferences or prey-size preferences (or both) or that they are allopatric to one another on Hispaniola.

There are two important take-home messages in this detailed series of
studies. The first message is methodological: you cannot undertake a robust statistical analysis contrasting the amount of character change between sister groups *if a substantial number of species are missing from that
analysis* (nor can you use character optimization procedures to reconstruct character evolution with any degree of confidence under these

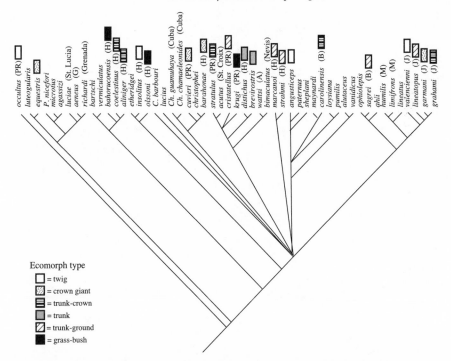

Figure 6.11. Examining ecomorph radiation on a larger phylogenetic tree for *Anolis*. The phylogeny was reconstructed by combining 16 morphological characters (most from Etheridge 1959) and 866 phylogenetically informative base pairs from mitochondrial DNA. This topology is the strict consensus of four equally parsimonious trees, each with a CI of 21% (Jackman et al. 1999). *Polychrus acutirostris* and *Diplolaemus darwinii* were used as outgroups (not shown on this figure). *P.* = *Phenacosaurus; C.* = *Chamaelinorops; Ch.* = *Chamaeleolis. Letters in parentheses* = locality: A = Antigua; B = Bahamas; G = Grenada; H = Hispaniola; J = Jamaica; M = mainland; PR = Puerto Rico.

conditions if the evolutionary picture is a complex one). If the missing species display the plesiomorphic condition, you will artificially inflate the estimate of character diversification by eliminating those species from the analysis, while failure to include taxa with derived ecologies will artificially inflate the estimate of stasis. The second message is that, missing data aside, there has clearly been a fair amount of ecomorphological diversification among the Caribbean-dwelling anoles (Losos et al. 1998).

Documenting the extent of that radiation of adaptations will require more information about the phylogenetic relationships within the large and successful *Anolis* clade as a whole, as well as detailed descriptions of the ecology and morphology of its species. This in itself will take the concerted efforts of many research teams over a long period. Once that

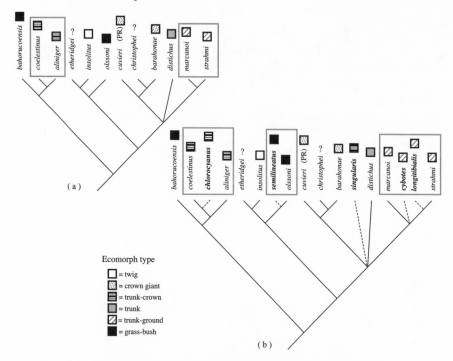

Figure 6.12. Pruning the tree (examining the evolution of ecomorphs on Hispaniola Island). (*a*) Figure 6.11 redrawn to include only the species on Hispaniola. Sister pairs that have diversified on the island within the plesiomorphic ecomorph space for the pair are highlighted in boxes. (*b*) A composite tree that includes additional species (*bold, with dotted lines*) from a previous phylogenetic analysis (Irschick et al. 1997). This is only a heuristic representation of relationships. If a comprehensive phylogenetic analysis of all species preserves this composite topology, then there has been more speciation within ancestral ecomorph space on Hispaniola (enclosed within the boxes) than the tree in (*a*) indicates.

database is in place, we can ask the more difficult question: Is the radiation of adaptations within Caribbean *Anolis* unusual? If the radiation of adaptations is confined to a subset of *Anolis,* then we must make sister-group comparisons within the genus in order to answer this question. For example, Irschick et al. (1997) reported that the relationship between morphology and ecology was very different for mainland anoles than that described for island dwellers (and, indeed, that not all island dwellers fit comfortably into the six ecomorph categories). If the rate of adaptive change varies substantially between these two general areas (mainland versus island), then we would expect to see that difference highlighted following an ancestral population's movement from one general area to another (e.g., increased rate of ecomorph diversification following movement from the mainland to an island). It is also possible that the amount or rate of radiation of adaptations is similar between the mainland and the islands, just manifested in different traits. Given this result, the next step would be to compare the amount of diversification in *Anolis* as a whole with the amount of diversification in its sister group. Finally, if there has been an unusual amount of adaptive change in this genus, is that change causally related to species richness? If *Anolis* is indeed significantly more species rich than its sister group, and if we can demonstrate a role for adaptive diversification in that speciation, this example might well make a quantum leap from the paradigm case of a "radiation of adaptations" to the paradigm case of an "adaptive radiation."[14]

Food and the Leaf-Nosed Bats. When you think about bats, probably the first thing that springs to mind is echolocation. Hard on the heels of that thought come the incredible freeze-frame images of bats grasping insects in mid-flight. Although most of the 900 or so microchiropteran bat species are insectivores, one group, the Neotropical leaf-nosed bats (Phyllostomidae), includes species that dine on pollen, fruit, nectar, blood, insects, and even the occasional small vertebrate. The Phyllostomidae is also one of the largest (approximately 142 species) of the 17 extant microchiropteran families, so it clearly represents an interesting place to examine questions of evolutionary radiations.

Ferrarezzi and Gimenenz (1996) were interested in documenting whether the apparent success of the leaf-nosed bats was related to their feeding strategies. Mapping general feeding habits onto a larger phylogeny for the order Chiroptera (all bats) unambiguously indicated that insectivory is the plesiomorphic condition for the Phyllostomidae. Although feeding-preference evolution within the phyllostomids is more ambiguous, due in part to the large number of polytomies on the composite tree, the main points of the study are clear (fig. 6.13): either strict or predominant insectivory is plesiomorphic for the leaf-nosed bats, with subsequent changes to sanguivory, carnivory, nectarivory, and frugivory occurring once each within the group. Freeman (2000) extended the analysis by examining the relationship between diet and a variety of morphological variables (e.g., body size, width of face, snout length, size and shape of teeth) for all microchiropteran bats. She discovered that insectivores have changed very little over the 60 million or so years since the origin of the Microchiroptera. Having achieved the top of an adap-

14. For other interesting lizard-based studies addressing similar questions, see Jaksic et al. 1980; Petren and Case 1997; Radtkey et al. 1997; and Vanhooydonck and van Damme 1999.

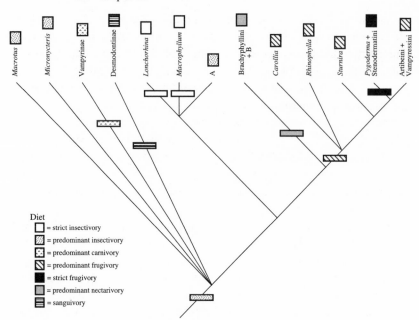

Figure 6.13.   One of two equally parsimonious optimizations for diet on a composite phylogenetic tree (Honeycutt and Sarich 1987; Baker et al. 1989; Lim 1993; Owen 1993; Gimenez et al. 1996) for the leaf-nosed bats. (Modified from Ferrarezzi and Gimenez 1996.) A = (*Mimon, Tonatia* (*Phylloderma* + *Phyllostomus*)) ; B = Phyllonycterini (Lonchophyllini + Glossophagini). "Strict" means that no other type of food is eaten, while "predominant" means that the food is preferred but supplemented with other food types.

tive peak, they are now limited to that (albeit moderately large and wide) peak with no apparent escape *as an insectivore.* Insectivorous species of leaf-nosed bats fit quite comfortably into that general morphospace. Within the leaf-nosed bats, however, four escape routes from the peak had been taken, involving correlated changes in morphology and ecology that moved evolving lineages into carnivore (increased tooth size relative to the palate), sanguivore (molars reduced to a cutting edge, incisors and canines greatly enlarged), nectarivore (lengthening of snout, decrease in tooth size, specialized tongue), or frugivore (shortening of snout, diverse specialized teeth) eco-morphospace. Given the observation that noninsectivorous phyllostomids represent only approximately 40 percent of all Neotropical bat diversity but 86–89 percent of Neotropical bat biomass (LaVal and Fitch 1977; Bonaccorso 1979; Timm et al. 1989), Freeman concluded that the various escapes from insectivore morphospace had been effective ones, particularly for the frugivores and nectarivores.

Continuing along the pathway established by the preceding two studies, Wetterer et al. (2000) reconstructed a more detailed phylogeny for
the leaf-nosed bats. Their total evidence analysis of 150 morphological,
karyological, and molecular characters produced 18 equally parsimonious trees, the consensus of which is shown in figure 6.14. The analysis
revealed a more complicated pattern of diet evolution in the leaf-nosed
bats, with a potential increase in the number of times that carnivory
(1–2 times) and frugivory (once, with ambiguous interactions between
predominant and strict frugivory) appeared. The expanded analysis provides stronger evidence for a radiation of diet preferences within the
leaf-nosed bats compared with its strictly insectivorous sister group (due
to an increase in homoplasy: all we need now is a statistical method to
determine whether this radiation of food preferences is significant). The
study also demystifies the seemingly abrupt transition into frugivore

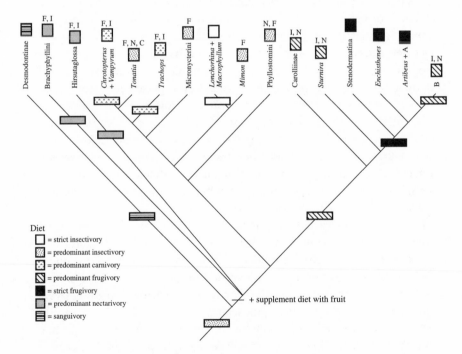

Figure 6.14. Another look at diet and leaf-nosed bats, using a simplified version of
the consensus tree (CI = 46.3%, RI = 76.5%) from Wetterer et al. 2000. The
dominant diet preference is mapped in a box above each group. *Letters =*
supplemental food items: F = fruit; N = nectar; I = insects; C = small vertebrates.
The ancestral state for diet preference is equivocal. (Only one possible
optimization is shown here. Others will occur to you.) A = *Koopmania, Dermanura;*
B = the large clade comprising *Uroderma, Platyrrhinus, Vampyrodes, Chiroderma,
Vampyressa,* and *Ectophylla.* The Hisutaglossa includes the monophyletic
subfamilies Phyllonycterinae, Lonchophyllinae, and Glossaphaginae.

eco-morphospace. Evidently the frugivore lineage is descended from ancestors that supplemented their diets with fruit and thus already had some of the characteristics in place that would later allow them to exploit that resource more fully. For example, the ability to locate fruit is based upon novel uses of older olfactory or echolocation abilities, or both (Kalko and Condon 1998 and references therein). Once potential prey items have been located, they are manipulated by a flexible tongue, one important component of which (a "free-floating hyoid") is symplesiomorphic for the Phyllostomidae (Griffiths et al. 1992; Freeman 1995, 1998). Clearly the morphological stage was set for new feeding strategies long before the ancestor of the phyllostomids appeared.

Having demonstrated that there has been a radiation of adaptations within the leaf-nosed bats, the next question to ask is whether any of those dietary changes have been associated with the putative evolutionary success of those bats (recall that Phyllostomidae is one of the largest microchiropteran families). There is no obvious relationship between escaping into any particular new eco-morphospace and an increase in species richness (fig. 6.15). For example, the nectarivorous Hirsutoglossa are not more species rich than their Phyllostominae + Carolliinae + Stenodermatinae sister group (34 species versus 103 species: $p = 0.500$), nor have the strictly/predominately frugivorous Carolliinae + Stenodermatinae outstripped their Phyllostominae relatives (70 species versus 33 species: $p = 0.65$). Expanding the analysis to include the next closest relatives of the leaf-nosed bats, the strictly insectivorous mormoopids, noctilionids, and most members of taxon X, indicates that there is no significant difference in species number between the leaf-nosed bats and their sister group (the Mormoopidae + Noctilionidae: $p = 0.13$), even though the disparity of 10 versus 142 looks suggestive. More important, there is definitely no difference between the larger sister comparison of clade X versus the Mormoopidae + Noctilionidae + Phyllostomidae ($p = 0.52$). In this case, if there had been a difference, the balance would have been tipped in favor of the strictly insectivorous (with the exception of the omnivorous mystacinids, all two of them) clade X (410 versus 152). These comparisons provide further support for the hypothesis that the phyllostomid exploration of new eco-morphospace did not cause a change in rates of net diversification in these bats. From the perspective of species number, expanding into new foraging "niches" is not better, it is simply different (or, if you prefer, equally valid). If anything, the analysis indicates that moving into one of those new niches, sanguinivory, is a specialized "dead end." Only one group of bats, the Desmodontinae, are blood feeders, and their numbers are significantly reduced compared with their sister group (3 versus 139: $p < 0.05$).

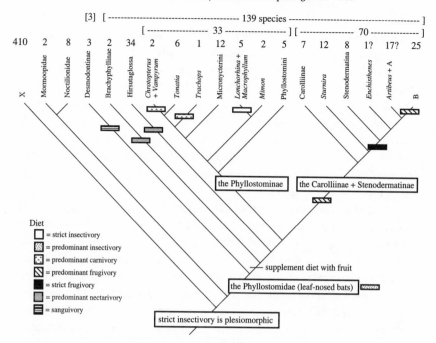

Figure 6.15. The relationship between diet and species number in leaf-nosed bats and their closest relatives. Number of species per group (from Wilson and Reeder 1993) is mapped above each lineage. This tree depicts one of two equally parsimonious placements for the Brachyphyllinae (versus the sister group to the Phyllostominae + Carolliinae + Stenodermatinae). See figure 6.14 for descriptions of taxa A and B. X = the large monophyletic clade comprising the Mystacinidae, Molossidae, Antrozoidae, Myzopodidae, Thyropteridae, Furipteridae, Natalidae, and Vespertilionidae.

Radiations of Species

Oh, What Tangled Webs We Weave. Human beings are fascinated by spiders. This fascination is somewhat surprising, given that our attention is usually drawn to organisms that are either beautiful or dangerous, or both. With the exception of the Australian funnel web, brown recluse, widow, and wolf spiders, spiders do not pose a serious threat to our health (science-fiction films notwithstanding), nor would most people rank them high on a scale of one to ten for overall beauty. Many spiders do, however, produce something beautiful: a web. We are particularly drawn to the orb weavers (the Orbiculariae: see Coddington 1986, 1989 for a discussion of orb-weaver monophyly), the spiders responsible for weaving the circular silken webs that produce the dew bejeweled effects so loved by photographers. Members of this group have also been used to test the effects of various drugs on behavior (most notably, LSD and valium: Witt et al. 1968) and have done duty as small astronauts so researchers could investigate their ability to build webs in zero-gravity conditions (Witt et al. 1976). Sometime in the early Cretaceous (or perhaps earlier), the orb-weaving clade bifurcated, producing the Deinopoidea (306 species) and the "advanced" or "derived" orb weavers, the Araneoidea (9,749 species). This seemingly obvious disparity in species number prompted Craig et al. (1994) to look for possible character changes that might be correlated with the origin of the species-rich araneoids. They focused their attention on two characters, the spectral properties of spider-produced silk and the spectral properties of the spider's foraging environment.

Craig et al. measured the spectral properties of silks collected from 25 species, representing 15 families in the order Araneae. (This order contains all spiders, more than 34,000 species assigned to 105 families: Coddington and Levi 1991.) Spider silk is a complicated mixture of proteins laid down in crystalline (strength) and noncrystalline (elasticity: Gosline et al. 1984, 1992) matrices that reflect different parts of the photoelectric spectrum. Spectral data were assigned to three general categories based on the composition of that reflected light: flat (reflects approximately 80–100 percent of all wavelengths), and disproportionately low and disproportionately high reflectance in the short wavelength (blue to ultraviolet) part of the spectrum. Data about light environment during maximum foraging activity were combined to produce three broad categories: dark (spider active at night), dim (spider active during the day in the forest), and bright (spider active during the day in nonforested habitats). Mapping these two characters on a reduced phylogenetic tree for Araneae families (fig. 6.16) reveals a complicated pattern of character divergence. For the purposes of this example, let us focus our attention on changes associated with the putatively species-rich Araneoidea. (Try reconstructing the total optimization for each character on your own.) The most obvious change involves a movement from hunting at night to hunting in the dimly lit forest during the day, either in the ancestor of the araneoids (if the basal araneids also hunt in dim environments) or independently within the Araneidae and in its sister group. The evolutionary pathway taken by the spectral qualities of silk is far more difficult to untangle; the shift from high UV reflecting properties to a flat reflectance may have occurred as far back as the ancestor of the Salticidae + Lycosidae + orb weavers, or as recently as the ancestor of the araneoids. What possible impact could differences in the composition of light reflected from a web have on a spider's hunting success? The answer is as complicated as the optimization for reflectance evolution because

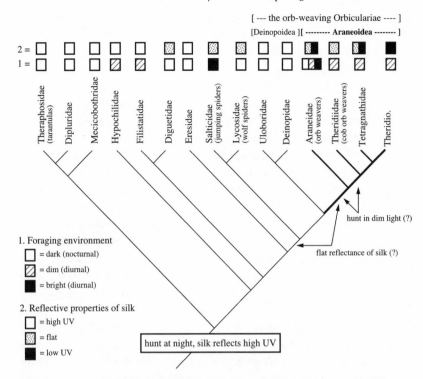

Figure 6.16. Light-reflecting properties of spider silk and foraging environment mapped onto a reduced phylogenetic tree for the spider order Araneae. The tree is a composite of several studies and is thus only a tentative hypothesis of relationships (for details, see Coddington and Levi 1991). The group of interest, the derived, orb-weaving Araneoidea, is highlighted with bold lines. Theriodio. = Theridiosomatidae.

it depends on a third variable, the visual sensitivities of the prey items being hunted by particular spiders. Depending upon those abilities, individual webs will be either invisible (advantage obvious) or visible to a potential prey item. Oddly, a visible web may actually be attractive. For example, *Drosophila* will avoid webs illuminated by light composed of moderate to long wavelengths (they can see them), but will fly straight into the same web illuminated by the full spectrum of light (the web is one that reflects a disproportionately high amount of short wavelengths: Craig and Bernard 1990). In this case, the spider has taken advantage of the fly's attraction to UV radiation, which presumably serves some function to the fly that is more important than avoiding spider predation. Craig et al. tentatively concluded that the diversification of the derived orb weavers was the result of movement into a new foraging habitat (from darkness to the dimly lit daytime in the forest) in the context of a flat, rather than high UV, reflecting web (either a symplesiomorphy or synapomorphy for the araneoids).

Bond and Opell (1998; see also Opell and Bond 2001) expanded the analysis to include two new web-associated characters: web orientation and the degree of silk thread stickiness. Previous research indicated that a vertically oriented orb web captures more insects and holds those insects for a longer period than does the same web oriented along a horizontal axis (Chacón and Eberhard 1980; Eberhard 1989). Spiders achieve web stickiness in two different ways. Some species increase the adhesive properties of the web by producing thousands of extremely fine looped fibrils laid down in repeating silken puffs along the main capture threads of the web. Others deposit a sticky secretion in regularly spaced droplets along the capture threads. Intuitively, stickiness of any sort should be functionally superior to being nonsticky; after all, a spider's web must not only capture a prey item, but hold that struggling organism long enough for the webmaster to deliver the coup de grace. Presumably the increased cost of producing a sticky web is balanced by an increased prey capture rate. Within sticky states, droplets appear to be functionally superior to thread puffs. They are less energetically costly to produce (more stickiness for your dollar) and enhance the elasticity of the capture threads, increasing the web's ability to absorb the force generated when an insect flies into it.[15] Were either of these new traits key innovations for the araneoids?

Before going any further, Bond and Opell asked whether the Araneoidea was indeed significantly more diverse than its sister group, the Deinopoidea. Applying the more conservative species asymmetry equation produced marginal results, the hypothesis that there is a significant difference in species richness between the two groups is supported only at $p < 0.061$. In general, this should mean that we reject the hypothesis of a difference between the two groups. Bond and Opell noted, however, that the basal clade in the Araneoidea, the Neotropical Araneidae, was in the process of being revised and might in fact contain substantially more species (3,400–4,400 more species: Coddington and Levi 1991) than currently recognized. If you experiment with changing the N and r values in the equation, you will discover that the number of described araneids (or indeed, any other members of the Araneoidea) will have to increase by approximately 2,500 for the comparison to be significant. At the moment, the described, as opposed to hypothesized, number of ara-

15. Chacón and Eberhard 1980; Eberhard 1989; Vollrath and Edmonds 1989; Craig and Bernard 1990; Köhler and Vollrath 1995; Lin et al. 1995; and Opell 1997a,b, 1998, 1999.

neid species rests at 2,600 (Griswold et al. 1998). For the purposes of this example, however, let us assume that more species will eventually be added and continue with the story.

Species-rich groups fascinate us because we associate numbers with "success," a word that carries its own conceptual baggage (Raikow 1988). From a research perspective, one of the costs in pursuing that fascination is that researchers rarely have thorough documentation of ecological/ behavioral diversity in the group. The orb-weaving spiders are no exception to this problem, and as you can see in figure 6.17, a number of taxa had to be excised from the analysis because of missing data. Given the number of question marks on the tree, the results are still tentative, but nonetheless intriguing. The analysis confirmed the proposal by Craig et al. that the origin of the araneoids may have been associated with a movement from a dark to a dimly lit forest environment (resolution still depends upon the plesiomorphic condition for the Araneidae), within either the plesiomorphic or apomorphic context of a flat reflecting web (the optimization remains ambiguous). In addition, two new possible key innovations were added to this character complex: the change from a horizontal to a vertical orientation of the orb web, and sticky droplets rather than thread puffs along the capture threads. Taken together, it appears that the possession of a flat-reflecting web allowed araneoids to shift their foraging activities to a time when more prey items were (possibly) available (during the day) and to capture those tasty morsels more efficiently (vertical orientation + sticky drops).

The fact that the web structure itself continued to evolve within the araneoids complicates the story somewhat (see fig. 6.17). For example, Griswold et al. (1998) postulated that the shift from vertical orbs to web sheets (which tend to be in more closed, protected spaces) may have been influenced by a variety of factors, including exploitation of a new resource (prey walking, jumping, or flying from below rather than from above), escape from aerial predators, and as a consequence of a decrease in body size. This continued evolution within the putative species-rich group confounds the hypothesis that "a vertically oriented orb web" was a key innovation for the Araneoidea because less than half of the species in the group possess that character. In fact, there appears to be a second pulse of diversification following the shift from an orb web to a sheet web, although the difference in species number (335 versus 5,914) is not significant using the most conservative test ($p = 0.107$). The other two putative innovations, sticky drops and foraging in the dim (and eventually bright) environment with a flat (and eventually low) UV–reflecting web, remain effective changes coupled with the new web structures. Although tenuous, these studies support the proposal that a "key innova-

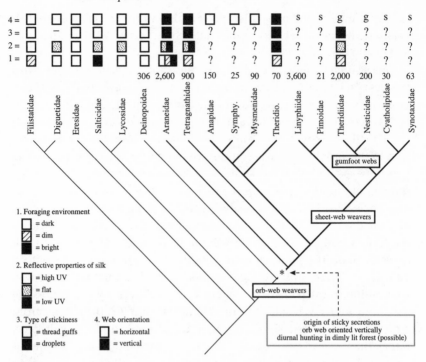

Figure 6.17.   Web-related characters and species number mapped onto a reduced phylogenetic tree for the spider order Araneae. The phylogenetic hypothesis for the clade of interest to this study (the derived, orb-weaving Aranoidea, highlighted with bold lines) is based upon 93 morphological and behavioral characters (one most parsimonious tree, CI = 64%, RI = 81%: Griswold et al. 1998). The large amount of missing data makes any optimization extremely tentative. Symphy. = Symphytognathidae; Theriodio. = Theridiosomatidae. *Letters* = the continued evolution of the orb web, past its point of origin: s = sheet web; g = gumfoot web.

tion" may often represent a suite of apomorphic changes (Lauder and Liem 1989; Cracraft 1990; Brooks and McLennan 1991, 1993d; Sanderson and Donoghue 1994), some of which may have evolved in the ancestor of the species-rich clade (synapomorphy: sticky drops) and others of which may be apomorphies within the species-rich group (possibly foraging in dim rather than dark environments). The importance of the biological background in which these apomorphies arise is also highlighted; the prior appearance of a flat reflecting web set the stage for spiders to shift their activity patterns from nocturnal to diurnal once that novel trait originated.

Assuming that new araneoid species are eventually discovered (to make the comparison between the Aranoidea and Deinopoidea significant), we still have not identified whether it is the species-rich or species-poor group that is "unusual." In order to begin this investigation, we need to expand the scope of the study to include the non–orb-web builders. Mapping approximate species number (from Platnick 1989; Coddington and Levi 1991; and Bond and Opell 1998) above the tree for the order confirms the proposition that the species-rich Araneoidea are unusual within spiders as a whole (fig. 6.18). The large-scale perspective, however, indicates that the major burst of speciation may have been initiated before the Araneoidea originated, in the ancestor of the RTA clade + the Orbiculariae (we'll call this larger group "clade X"). The disparity in diversity is immense here: approximately 28,055 species for clade X versus 348 species in its sister ($p < 0.025$). This asymmetry is mirrored in the comparison between the exuberant Araneae (all spiders) and its putative sister group, the Pedipalpi (about 34,000 versus 200 species), although, as you probably know by now, we cannot determine which of the two is unusual without expanding the phylogenetic scope

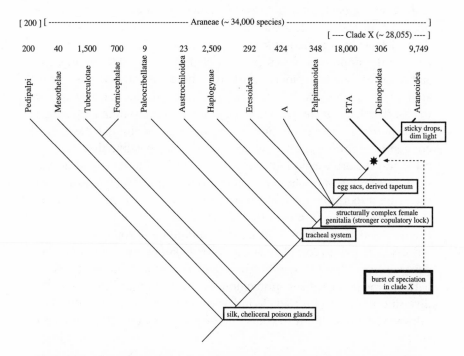

Figure 6.18. Examining spider diversification on a larger phylogenetic scale. The tree is a composite of several studies and is thus only a tentative hypothesis of relationships (for details, see Coddington and Levi 1991; Scharff and Coddington 1997). A = "other entelegynes" (an assemblage of families of which the monophyly or placement within the larger context [or both] is unknown); RTA = Dictynoidea + Dionycha + Amaurobiidae + "other amaurobioids" + Tengellidae + Lycosoidea.

even further. For the time being, we will concentrate on diversification within spiders.

Because there is no significant difference in species number between the RTA clade and its sister group, the Orbiculariae (18,000 versus 10,055: $p = 0.717$), the next question to ask is whether the initial burst of speciation was driven by a key innovation in the ancestor of the RTA clade + the Orbiculariae. We have mapped a variety of important character changes on the tree (discussed in detail by Coddington and Levi [1991]) in order to give you an idea of how complicated the search for the "key innovation" that could explain the initial burst of speciation in clade X might be. Consider the importance of the extremely old characters silk and poison glands to all spiders' hunting abilities, then add to those traits (1) the transition from posterior book lungs to a tracheal system (higher metabolic rate [increased activity levels], more efficient water conservation), (2) complicated female genitalia (grooves and projections that may help the male anchor his palpal bulb during mating and possibly may increase efficiency of sperm transfer), (3) the origin of an egg sac (decreases parasitism and desiccation), and (4) a derived tapetum (influences optical performance; further modified to provide the impressive visual prowess of the largest spider family, the Salticidae, or jumping spiders). Curiously, there is no obvious "key" synapomorphy for clade X. It is possible that this character may not appear to be a "key" change on its own, but the origin of this trait in the context of the previous changes denoted on the tree (any or all) may have been enough to begin the speciation cascade. It is also possible that the RTA clade and the orb weavers followed different routes to increased net diversification. We have already outlined some of the characteristics that might be responsible for the diversification of the orb weavers. Future studies might concentrate on the RTA clade (amongst which are nested the species-rich, cursorial hunting wolf and jumping spider groups). In any case, in order to complete our study of evolutionary radiation in spiders, we would have to demonstrate that there is a causal relationship between the proposed key innovation, or series of innovations, and net diversification, both for clade X and, within that clade, for the advanced orb weavers (in addition to collecting missing data and describing new species).

Of Pollinators and Herbs: Why Are There So Many Flowering Plants? The flowering plants are one of evolution's big success stories. According to the fossil record, it took only 25–27 million years for all the major angiosperm lineages to appear on, and then to dominate, the terrestrial landscape (Crane et al. 1995; Wing and Boucher 1998). There is a huge (and significant: $p < 0.001$) difference in species number between the Angiosperma (more than 250,000 species) and its sister group, the Gnetales (about 69 species). As an aside here, it is interesting to note that Gnetales appears to be less diverse now than it was in the Mesozoic (having suffered a rapid decline in the late Cretaceous: Crane and Lidgard 1989). Although adding the number of extinct species to the extant species of Gnetales does not change the conclusion that the angiosperms are more species rich than the Gnetales, it might very well do so in cases where the asymmetry between sister groups is not so profound. Whenever possible, therefore, data from extinct species within both sister groups must be included in the total species count in order to create a more complete data set. After all, evolution in all its forms is not defined by whatever species happen to exist on this planet at the time you choose to do a study. Now, back to the flowering plants.

A number of apomorphic changes associated with the origin of angiosperms—flowers spring to mind, also increased growth rate, seeds developing within the carpel ("ovary"), the presence of endosperm (polyploid, postfertilization, embryo-nourishing tissue)—have been identified as the key innovation responsible for this particular success story (see, e.g., Stebbins 1981; Doyle and Donoghue 1986; Bond 1989; Eriksson and Bremer 1992). If these traits are indeed the innovations that directly sparked the diversification, then we would expect to find evidence for increased net diversity immediately following their origin (remember the spider example, in which the most basal derived orb weaver, the Araneidae, is in fact very diverse). Based upon a series of models incorporating information about species number, hypothesized time of origin (fossil data) for various groups, and phylogeny, Sanderson and Donoghue (1994) concluded that diversification rates did not change within the angiosperms until *after* the basal split producing the aquatic group Ceratophyllaceae and the ancestor of the remaining angiosperms (fig. 6.19). Their analysis could not pinpoint where those changes occurred (because only three taxa were used in the models), but it did reject the hypothesis that any synapomorphies for the angiosperms (putative key innovations) were directly responsible for an increase in net diversification *at their point of origin.*

If not these synapomorphies, what other characteristics could be associated with diversification within the group? Three characteristics of flowering plants have been proposed: pollination mode (biotic versus abiotic: Grant 1949, 1981; Stebbins 1971, 1981; Crepet 1984), seed dispersal mode (biotic versus abiotic), and growth form (herbaceous versus woody). Biotic pollination decreases the probability of extinction: assuming that a given animal pollinator visits the same plant species,

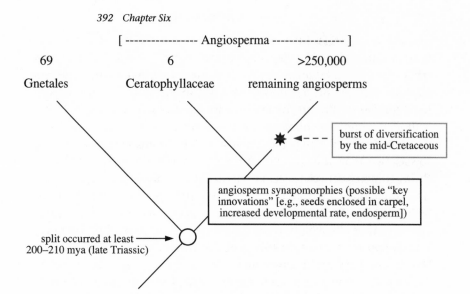

Figure 6.19. When was the burst of diversification in the angiosperms? Species numbers (Sanderson and Donoghue 1994) and diversification dates (Crane 1985; Crane et al. 1995; Kendrick and Crane 1997) are rough estimates. The burst did not occur until some time after the basal split within the flowering plants.

plants that pass their precious pollen granules to that animal have more control over where those granules are deposited than do plants entrusting them to the vagaries of the wind (Raven 1977). Biotic seed dispersal enhances the probability of speciation: animals disperse seeds over a larger area, increasing the chances for the establishment of a peripherally isolated population (Snow 1981; Tiffney 1984, 1986), increasing the effective species range and thus the chances of the species being subdivided by a vicariance event, or both. Species range aside, herbaceous lineages, with shorter generation times and less stable habitats than their longer-lived, woody counterparts, may have a higher probability of speciation (Levin 1984; Tiffney 1986; Eriksson and Bremer 1991, 1992). Not surprisingly for such a large, old group, diversity is distributed unevenly within the angiosperms, so it is possible that these three traits alone or in combination could have interacted to produce the ebb and flow of angiosperm radiation.[16]

Dodd et al. (1999) examined this possibility by optimizing the three traits onto a massive composite phylogeny for 356 angiosperm families. All three traits displayed some homoplasy, allowing the hypothesis of a causal relationship between a particular character state and changes in species number to be tested numerous times. Those tests revealed only

16. See Ricklefs and Renner 1994 for a discussion of the dynamics of that interaction.

a borderline interaction between dispersal mode and net diversity. A much stronger relationship was detected between pollination mode and species number (fig. 6. 20). One of two equally parsimonious optimizations for this trait placed the origin of animal-based pollination in a fascinating spot, on the "right" side of the first major asymmetrical split on the angiosperm tree (a synapomorphy for what Sanderson and Donoghue called "the remaining angiosperms"). As we discussed previously, the correlation between a unique event and increased net diversification can be investigated further by looking for underlying processes linking the two causally (which would require a substantial investment of time and resources) or by looking for points in time when the purported key innovation shows a reversal to the plesiomorphic condition. If the relationship between the key innovation and diversity is a causal one, we would expect to see a significant decrease in net diversification associated with each reversal (*ceteris paribus,* of course). This is almost exactly what Dodd et al. discovered in their study, with a few exceptions. The larger-scale

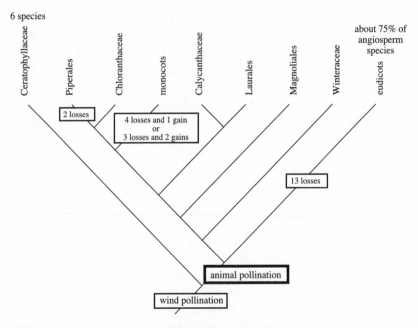

Figure 6.20. Evolution of pollination modes in the angiosperms. The shift from wind-driven to animal-based pollination may have occurred in the sister group of the wind-pollinated Ceratophyllaceae as shown here, or may be older (depending upon pollination modes in the Gnetales). For exact details of losses (reversal to wind pollination) and gains of animal-based pollination, see Dodd et al. 1998 (This is a simplified version of the composite phylogeny comprising 356 families from Nandi et al. 1998, with grafts in specific areas from other studies).

statistical analysis indicated that the predicted direction of change in net diversification following the origin/loss of animal pollinators (mainly losses, as per fig. 6.20) does occur more often than simple chance would predict ($p < 0.0001$ using the global analysis described in Slowinski and Guyer 1993). What we need now are specific explanations for the absence of that relationship in a few particular groups. For example, the "sister group" to the species-rich wind-pollinated grasses (approximately 10,000 species) contains only two species, even though it has regained animal pollinators. The analysis also supported the proposal that herbaceous lineages should be more species rich than their woody sisters. Unlike the optimization outlined for animal pollinators, the majority of transitions for this trait were independent origins of herbaceousness (24 gains of herbaceousness versus 8 losses, or 27 gains versus 5 losses). Once again, the predicted effect of gains and losses on net diversity did not hold in all cases, but the global perspective overwhelmingly supported the hypothesis ($p < 0.0001$).

Overall, then, it appears that no one "key innovation" is responsible for the tremendous angiosperm radiation. At least two (and possibly more) traits—the use of animal pollinators (decreases probability of extinction) and the assumption of an herbaceous lifestyle (increases probability of speciation)—have played a role. Because these traits are homoplasious, we would expect to find pulses of net diversification tracking their appearance and disappearance through time. This is a much more dynamic picture than our original description of one key innovation per group leading to a basal increase in net diversification.[17] The independent pulses within the flowering plants have produced such distantly related species-rich groups as the Asteraceae (approximately 21,000 species), the Cyperacea (sedges: approximately 3,600 species), and the Fabaceae (approximately 11,300 species), each of which may be characterized by a different combination of "key innovations." It seems likely that these combinations will be limited to relatively few important traits (e.g., pollination system, growth form, mating system) and continued investigation within a phylogenetic context will gradually begin to untangle these effects. For example, the trend within the flowering plants appears to be toward repeated loss of animal pollinators (decreased net diversity) and repeated gains of herbaceousness (increased net diversity). How do these two characteristics interact if they occur in the same group? Conversely, the patterns would indicate that being an animal-

17. The suggestion that many traits may be interacting in complex ways to affect net diversity has also been made for another charismatic, species-rich group: those small, fecund, opportunistic, vocal virtuosos, the passerines, or song birds (Baptista and Trail 1992).

pollinated weed is the best of all possible worlds. Are all the significantly species-rich groups within the angiosperms animal-pollinated weeds?

One of the major stumbling blocks in this study is the absence of a robust phylogenetic hypothesis for the angiosperms. Recent molecular-based studies have suggested a different arrangement for the basal angiosperm lineages (including moving Ceratophyllaceae further into the clade, which is of particular relevance to our search for possible "key innovations") and have altered some of the upper-level relationships shown in figure 6.20.[18] The issue of how the new hypotheses alter the statistical investigation of sister-group diversity is eagerly awaiting the attention of researchers. Problems with the robustness of the phylogeny aside, there is an even more difficult question to answer: Were these combinations of key innovations critical arbiters of net diversification *on their own,* or were their influences enhanced by the presence of general angiosperm synapomorphies (like the presence of closed carpels, rapid growth rates, or endosperm).[19] Recall that angiosperm synapomorphies were not directly responsible for the group's radiation because, as Sanderson and Donoghue determined, the rate of diversification did not increase following the origin of those traits. Nevertheless, those synapomorphies may have been an important prerequisite to the functioning of key innovations once the innovations appeared.

For example, there is some evidence that *Gnetum* evolved a polyploid, postfertilization, embryo-nourishing tissue equivalent to the angiosperm's "endosperm" via a different developmental pathway (Friedman and Carmichael 1998). That arrival may have occurred within the plesiomorphic background of insect-based pollination (Thien et al. 2000 and references therein). Mapping species number above the expanded tree for Gnetales + Angiosperma (fig. 6.21) indicates that the convergent appearance of a postfertilization tissue is not associated with a dramatic increase in net diversification in *Gnetum.* Notice that if you limit the comparison to *Gnetum* and its sister group, you will conclude that *Gnetum* is unusually species rich. Now, expand your horizons one step to include *Ephedra,* and you can see that it is the wind-pollinated *Welwitschia,* not *Gnetum,* that is the unusual member of the sister pair (which is not really surprising given that *Welwitschia mirabilis* has been described as "an extraordinary plant monster" [Martens 1971]). Could this unusually low diversity be linked to the appearance of wind pollination within the ple-

18. Barkman et al. 2000; Doyle and Endress 2000; Graham et al. 2000; Mathews and Donoghue 2000; Qiu et al. 2000; and Savolainen et al. 2000.

19. For exceptional studies, see Friedman and Carmichael 1998; Floyd and Friedman 2000; and Kellogg 2000.

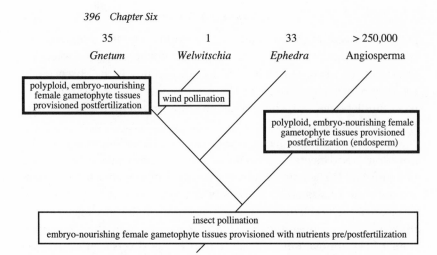

Figure 6.21. Expanding the phylogeny to include the sister group of the flowering plants. At least one trait, polyploid, postfertilization embryo-nourishing tissue, has evolved convergently in the angiosperms and *Gnetum*. Notice that this expanded perspective resolves the optimization problems for pollination mode discussed in the text: based upon this additional information, animal pollination is plesiomorphic for the angiosperms.

siomorphic embryo-nourishing background? Overall, it appears that neither endosperm nor insect pollination *on its own or the two in combination* is sufficient to explain the exuberant diversification of the flowering plants; otherwise we would expect to see a similar diversification in *Gnetum*. This does not imply that these traits were not involved in the flowering plant radiation, only that their effects (if any) appeared through a synergistic interaction with other traits apomorphic for the angiosperms (e.g., "insect pollinator in seed plants with closed carpels" or "insect pollinator attracted by floral fragrance and heat": Thien et al. 2000). In other words, the whole concept of "key innovation" is complicated by the fact that characters do not evolve in a vacuum; they operate within the context of the underlying biological system.[20]

Parasitic Flatworms: Changing the Reproductive Game. Price wrote that "the most extraordinary adaptive radiations on earth have been among parasitic organisms" (1980:3). The parasitic Platyhelminthes is an excellent candidate for Price's "extraordinary adaptive radiation." They are extremely diverse (including the familiar tapeworms and flukes: remember chapter 5) and display a substantial amount of lifecycle evolution. Although dy-

20. See Bateman et al. 1998 for a similar discussion about the search for the key innovations driving the ancient Silurian-Devonian radiation of land plants.

namic, that evolution has not been chaotic. It occurred within a relatively conservative phylogenetic framework, as life cycles were assembled in a historically coherent sequence over the course of millions of
years. The plesiomorphic pattern for the parasitic flatworms (fig. 6.22) is
one in which an arthropod is used as the only host by an ectoparasitic
species. The pattern became more complicated in the ancestor of the
Neodermata (the subgroup containing the tapeworms and flukes, among
others) as a vertebrate host was added (the arthropod was retained) and
the adult parasites became endoparasitic (Brooks et al. 1985b; Brooks
1989b; Brooks and McLennan 1993b; Zamparo et al. 2001). At this stage
in the investigation, a researcher might wonder whether these two apomorphic changes are key innovations. They certainly conform to Simpson's original scenario for a radiation, because they involve movement
into and diversification in a new habitat, both in terms of new hosts
(geography) and new places to live within the host ("niche space"). At
this level of analysis, all we can say is that the origin of these two
habitat-shifting traits is associated with a significant change in net diversification (Fecampiidae [less than 20 species] versus Neodermata [ap-
proximately 13,078 species]: $p < 0.003$). If we were to leave the story
here, our conclusions would be interesting and suggestive, but no more.
Think back to the flowering plants: many hypotheses about putative key
innovations, an obvious disparity in diversity, and no way to rigorously
test the hypotheses until Sanderson and Donoghue broke the investigation wide open by adding phylogenetic detail to the study. Fortunately,
a robust phylogenetic hypothesis exists for members of the Neodermata

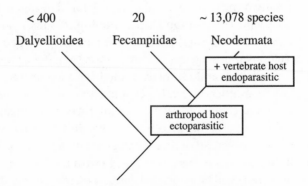

Figure 6.22. Why are there so many parasitic Platyhelminthes I? Searching for key
innovations on a tree showing relationships among large groups. The phylogenetic
analysis is based upon 159 morphological characters, which produced one tree
with a CI = 93% (Zamparo et al. 2001). The Neodermata is diagnosed by two
changes in habitat: the addition of a vertebrate host (creating a complex two-host
lifecycle pattern) and the movement from living outside to living inside that host.

(Zamparo et al. 2001 and references therein), so let us take this investigation one step further.

Expanding our phylogenetic resolution (fig. 6.23) allows us to refute the hypothesis that there is something inherently "key" about being endoparasitic because the reversal of that character to the plesiomorphic condition (ectoparasitic) in the Mongenea has not been associated with a decrease in net diversity. By the same token, the hypothesis that there is something "key" about adding a vertebrate host *on its own* is also refuted, because the Aspidobothrea, Gyrocotylidea, and Amphilinidea all have very few species, despite having a vertebrate host. Where to now? The most obvious observation at this increased level of resolution is that diversity is not evenly distributed within the Neodermata. This asymmetric distribution mirrors the patterns we uncovered for other very old groups like spiders and flowering plants. The next step in our research, then, is to search for possible key innovations within the neodermatans.

From a parasite's perspective, hosts are resource patches generally clumped in space and time. To make matters worse, parasites are usually less mobile than their hosts, so it is difficult for them to move from patch to patch. Rogers (1962) proposed that this relative immobility could be countered by three strategies: (1) direct development or autoinfection (or both); (2) production or amplification of dispersing larval or juvenile stages; and (3) increased reproductive output. The benefits of autoinfection are obvious; there is a substantial decrease in the risk that your offspring will not find a suitable host. The benefits of producing or amplifying larval or juvenile dispersal stages are also obvious. Such stages, often aided by their overwhelming numbers (asexual clones) or their host-seeking or host behavior–altering capabilities, function to (1) decrease intraspecific competition by partitioning resources among different developmental stages, (2) partition the parasite gene pool among different developmental stages in different hosts to buffer against problems with transmission, and (3) increase the likelihood of transmission from one host to another. Increasing reproductive output is the sexual equivalent of amplification of larval or juvenile stages via asexual cloning. Like its counterpart, this increases the likelihood of transmission from one host to another. All of these strategies have a direct, positive effect on the fitness of their bearers. Let us begin with the hypothesis then, that all of these traits will be associated with an increase in net diversity.

The most obvious asymmetry within the neodermatans is between the Eucestoda (tapeworms) and the Amphilinidea (3,000 versus 10 species: $p < 0.007$). The eucestodes have solved the patchiness problem by using Rogers's strategy 3 above; most adults are segmented and each segment (proglottid) contains testes, ovaries, and a uterus. The consequent

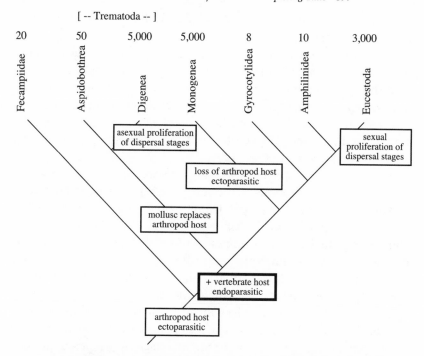

Figure 6.23. Why are there so many parasitic Platyhelminthes II? Expanding the phylogenetic resolution. The putative key innovations from figure 6.22 (the plesiomorphic lifecycle pattern for the species-rich Neodermata) are highlighted in a bold box. The gyrocotylids, amphilinids, and tapeworms have retained that general pattern. The trematodes display one variation: a molluscan host was substituted for an arthropod host in their ancestor. The Monogenea have a secondarily simplified life cycle (a return to the plesiomorphic state "ectoparasite with only one host," although that host is a vertebrate, not an arthropod). Both the Digenea and the Eucestoda have independently evolved mechanisms to proliferate larval dispersal stages. New species are being described on a regular basis within the three species-rich groups.

increase in reproductive output for the tapeworms is dramatic. The average daily ouput of 100 eggs per worm in the Monogenea, and even the exceptional 24,000 eggs per worm in some of the larger digeneans (Tinsley 1983), pale by comparison to the 720,000 eggs expelled per day by the human beef tapeworm, *Taenia saginata* (Crompton and Joyner 1980). The lifetime fecundity of a tapeworm is thus much greater than the lifetime fecundity of any amphilinid species (the sister group bearing the plesiomorphic condition "lack of repeating reproductive units"). The proliferation of dispersal stages created by the new strategy increases the likelihood that a larva will find a suitable host, decreasing the chance that a given population will go extinct (Moore 1981 and references therein). This, in turn, has an indirect effect on speciation because it allows a species to persist long enough to encounter a variety of speciation-causing factors.

Support for the positive effect of an increase in dispersal stages on net diversification can be found by investigating groups within the Eucestoda that have secondarily lost the putative key innovation "many repeating reproductive units." One such group is the Aporidae, parasites of ducks and swans. As predicted, the aporids are significantly less diverse than their sister group within the Cyclophyllidea (6 species versus approximately 300 species: $p < 0.04$).

The second-most obvious asymmetry is between the Digenea and the Aspidobothrea (5,000 versus 50 species: $p < 0.02$). Both aspidobothreans and digeneans share an ancestral lifecycle pattern involving a molluscan and a vertebrate host. In the aspidobothreans, larvae hatch from eggs and develop directly into juveniles in the molluscan host and are ingested by a mollusc-eating vertebrate, where they develop to adulthood. Hence, each aspidobothrean embryo can potentially give rise to only a single adult. Digeneans, on the other hand, are characterized by a series of complex developmental stages in the molluscan host, at least one (and usually two) of which produce a large number of cloned larvae or juveniles (depending on the species and the stage). The reproductive potential of a single digenean embryo ranges from 1,000 to 10,000 adults. This is significant; Wilson et al. (1982) estimated that of 100,000 *Fasciola hepatica* (the liver fluke) eggs deposited in a pasture from the middle of winter to the end of summer, only 17 would hatch. The reproductive potential of those 17 can compensate for the 99,983 that did not hatch, if each of the 17 is capable of producing 5,883 cercariae asexually. Like the eucestodes then, the digeneans have solved the patchiness problem by increasing the number of dispersing larvae, but they have arrived at that functional solution by a different pathway (Rogers's strategy 2). The effects of increasing the dispersal stages on net diversification should be the same as those described for the tapeworms; however, it is difficult to test this hypothesis because, unlike in the tapeworms, the putative key innovation has arisen only once and has never been lost.

Finally, we are left with Rogers's first suggested strategy to reduce the risk associated with finding a host: the development of autoinfection. Autoinfection occurs in only one clade, the Monogenea, and that clade is not significantly more species rich than its sister group. Should we reject the hypothesis that autoinfection is a possible key innovation in the monogeneans? When we investigate evolutionary radiations using statistical analyses, we are assuming that the only difference between two sister groups is the presence of a key innovation (or innovations) in one group and the absence of that innovation in the other. This assumption is the basis for the conclusion that a significant difference in species number between the two groups is related to the presence of the innovation. There are, however, two very different explanations for the absence of a significant asymmetry in species number following the origin of a putative innovation. The first, and most straightforward, explanation is that our hypothesis of a causal relationship between the innovation and net diversification is, in fact, incorrect. The second explanation is more subtle and thus more difficult to approach within the framework that we have been developing to this point: it is possible that the positive effect of a key innovation in one group was countered by the positive effect of a different key innovation in the sister group. For example, the sister group of the monogeneans contains a species-rich clade (the Eucestoda). Let us suppose, for the sake of illustration, that sexual amplification was the key force behind the diversification of the tapeworms. If the eucestodes had not developed that trait, then we would not expect them to be significantly different (in terms of species number) from their sister group. Let us use the basal members of the Eucestoda, the Caryophyllidea (Hoberg et. al 1997, 2001), as our guide. There are only about 111 species of these tapeworms, which are plesiomorphically unsegmented (lack the putative "key innovation" of multiple reproductive units). In our alternate universe, then, there are only 111 tapeworm species (and much general rejoicing amongst vertebrates), and the comparison between the Monogenea and their sister group is now significant (5,000 species versus 129 species: $p < 0.05$). Back in our universe, we believe that it is a useful exercise to reduce the confounding effects of a species-rich clade (in this case, the tapeworms) in the sister group. This manipulation should be used cautiously, however, or the entire exercise of comparing sister-group species number will be weakened by a series of ad hoc arguments for removing groups until a significant effect is found. For the moment, however, let us pursue the hypothesis that the evolution of autoinfection has driven the diversification of the monogeneans (and is undetectable by comparing sister groups statistically only because it was offset by the evolution of sexual amplification in tapeworms).

Studies have demonstrated that autoinfection has a positive effect upon the reproductive success of the bearer (see Tinsley 1983; Kearn 1986 and references therein). Is the association between the monogenean radiation and the developmental change more than just a spurious correlation? From a theoretical perspective, the answer to this is yes, because the subsequent changes in life history characteristics have a direct effect on deme structure. Parasite demes have a tendency to be ephemeral. In the vast majority of cases, they must be reassembled each generation by the relatively random immigration of larval or juvenile stages, usually from a much larger gene pool (the populations occurring in multiple hosts) than that represented by the members of the original deme. Parasites that autoinfect their individual hosts and inhabit long-lived hosts can produce more than one generation on or in the same host organism. This reduces the ephemerality of the deme structure, increasing the potential that differences appearing within a deme could be maintained by inbreeding. Tinsley hypothesized that such inbreeding could "promote rapid increase in gene frequency and potential fixing of beneficial variation" (1983:174). The monogeneans are thus characterized by the appearance of a novel reproductive strategy that increases the opportunity for relatively rapid speciation to occur under a variety of conditions. The hypothesis that the evolution of autoinfection influences the rates of speciation is supported by an additional piece of evidence. Within the monogeneans, one of the most species-rich groups, the gyrodactylids, takes autoinfection one step further. Instead of laying eggs on their hosts, they give birth to live young, some of which may even be born "pregnant" themselves.

Overall, then, we needed to increase the phylogenetic resolution of our study in order to begin investigating characters that may have played a role in the diversification of the parasitic flatworms. That diversification, as was the case for the flowering plants, has not been distributed evenly within the larger group, but rather has occurred in separate pulses within that group. If, as suggested, those pulses were driven by the origin of different "key innovations," this complicates the use of sister-group comparisons, because it is possible that two sisters will each contain species-rich groups that diversified for different reasons.

Low Diversity Groups Are Valid, Too

For some reason, our attention is generally drawn toward the flamboyance of species-rich groups. Even our choice of terminology, species *rich* versus species *poor*, is a biased one. There is obviously something inherently satisfying in "250,000 species of flowering plants" versus "one species of *Welwitschia*." Fortunately for the species-poor groups, not all researchers are so dazzled by numbers. Simpson (1944) was among the first modern evolutionary biologists to consider general explanations for groups of unusually low species numbers. He considered all such groups relicts of one form or another and postulated that different processes could produce different kinds of relictual groups. We will be concerned with two major types of relicts. **Numerical relicts** are the surviving members of once more species rich groups, the ranks of which have been depleted by extinction. **Phylogenetic relicts** are "living fossils," members of groups that have existed for a long time without speciating very much.

Such low speciation rates could be a simple byproduct of the group not evolving any of the key innovations that caused the species richness of its sister group (in which case the focus of attention should be on the species-rich sister). This explanation appears to apply to the Aspidobothrea (see fig. 6.23), an old lineage inhabiting a reasonably diverse host group, which possesses the characteristics necessary to persist over a very long period of time but does not have the traits that created the potential for increasing net diversification in its sister group, the Digenea. It is also possible that a depauperate group may have an autapomorphy (or autapomorphies) that actually limited its diversification, in which case that group should also be the focus of attention. For example, amphilinideans (see fig. 6.23) occur as adults in the body cavity of their, primarily, freshwater actinopterygian hosts, rather than inhabiting the intestine (the plesiomorphic condition for the Neodermata). Perhaps living in the body cavity of fishes has kept amphilinidean diversity low, since amphilinideans are relatively large and thus restricted to relatively large hosts. Additionally, amphilinidean eggs cannot reach the outside environment from the body cavity of all fishes, further restricting their range of suitable hosts. The gyrocotylideans (see fig. 6.23), on the other hand, are ecologically conservative, being entirely restricted as adults to the spiral intestines of chimaerid fishes, and are found in association with hosts that are themselves phylogenetic relicts. Like their hosts, the gyrocotylideans are less species rich than their sister group (in this case, the amphilinideans plus the tapeworms). This suggests the possibility that the interaction between developmental conservatism and a conservative and highly specialized ecology limited diversity in the group (see also Brooks and Bandoni 1988).

Establishing a group's relictual status first requires methods for determining that the group is old enough to be highly diverse. There are a number of methods available for estimating the ages of clades, including molecular clock criteria, paleontological data, and biogeographical analysis (see chapter 4). Second, it must be determined that the group is in fact unusually depauperate (the reverse of demonstrating unusual species richness, but using the same techniques). For example, Crocodylia (crocodilians) is the extant sister group of the species-rich clade Aves (birds). Living crocodilians, numbering about 22 species, inhabit a variety of estuarine to freshwater habitats throughout the tropical and subtropical regions of the world. They prey on a wide variety of vertebrates and some invertebrates. The fossil record indicates that crocodilians were once a more speciose group, including many fully marine species, and even some bipedal species; in addition, the earliest known crocodilian fossils suggest a terrestrial origin for the group. Hence, the current diversity of crocodilians represents only a fraction of the species number and ecological diversity once encompassed by the group, so we consider crocodilians to be numerical relicts. Now, consider the Rynchocephalia (tuatara), the sister group of the Squamata (lizards and snakes). There are two species of tuatara, compared with approximately 6,850 species of squamates. Tuatara are restricted to small islands off the coast of the South Island of New Zealand, where they dine upon insects (and the occasional egg or small bird when they can find them). Although tuatara have existed for a considerable period, they have never been highly diverse or widespread, although the fossil record indicates they have suffered a range contraction (at one time they were present on the main islands of New Zealand). Both the fossil evidence and the ecological homogeneity of contemporaneous species suggest that tuatara are phylogenetic relicts.

SUMMARY

The study of evolutionary radiations within a phylogenetic context is just beginning. Because we tend to focus on the "species-rich" side of the asymmetry coin, we are faced with the additional problem of determining robust phylogenetic relationships within an often extremely large group before we can have any confidence in our conclusions. After all, comparing the amount of diversification and evolutionary innovation between sister groups lies at the heart of this research program. The less confidence you have in your hypothesis of sister-group relationships, the less confidence you have in your overall interpretation.

What evidence is available at present suggests that many evolutionary radiations have been the result of a causal interaction between the origin of a key innovation(s) and an increase in net diversity. Details of the exact mechanism by which diversity is affected, either through an increased rate of speciation, a decreased rate of extinction, or a combination of the two, remain obscure for most systems. Once again, this is an exciting prospect for those who are attracted by a voyage of discovery into the unknown.

The Frontiers of Discovery in Radiations
DISENTANGLING THE EBB AND FLOW OF DIVERSIFICATION AND EXTINCTION

Paleontologists have been attempting to document the changing patterns of diversification and extinction on this planet for decades. The field is huge one, and a comprehensive summary of all its methodologies and proposals is beyond the scope of this book.[21] We are primarily interested in answering two questions: Have the rates of speciation and extinction been constant on this planet (not including anthropogenetic causes for extinction)? and, If not, are bursts of speciation and extinction causally linked? By now you realize that the answer to the first question is no (otherwise the research program based upon studying unusually species-rich or species-poor groups would have been out of business long ago). Answering the second question is far more difficult and contentious. Part of the problem stems from the way diversity is measured. Even a cursory glance through the literature will reveal hundreds of graphs depicting the origin and diversification of higher taxa (usually families) over the last 3.5 billion years (since the origin of life, give or take a few million years). Depicting diversity in this way is one of the last vestiges of evolutionary taxonomy in our modern system. Evolutionary taxonomists were fascinated by the origins of "major groups" and higher taxa, in part because such groups were assumed to be models of evolutionary success. Evolutionary taxonomists placed substantial emphasis on degree of similarity and grouped organisms according to that emphasis. As discussed in chapter 2, this often led to the production of paraphyletic groups. Although evolutionary taxonomy is no longer considered a valid basis for reconstructing phylogenies, some of the basic ideas from that discipline are retained within evolutionary biology. One of those ideas is that major groups (higher taxa) are in some way real and important evolutionary units.

Let us examine this proposition further. The most obvious problem with describing higher taxa as "real" is that all higher taxa, from genera to kingdoms, are created by biologists. For example, consider the phylogeny for species A–K shown in figure 6.24. One systematist places all the species in one family, the AKidae. Another divides the species into two monophyletic families, ACidae and DKidae, while still another prefers three families. (How many different families can be erected?) Even though all three biologists are recognizing monophyletic groups, this does not solve the problem, How many families are there? In fact, there are no objective criteria for recognizing higher taxa—it is simply a matter of personal preference (some of us are splitters, others are lumpers). If higher taxa are figments of our own personal biases, how can we possibly study their evolution?

Now let us consider the difference between equating diversity with either the number of higher taxa (genera, families, classes, orders, etc.)

21. If you are interested in this area, see Novacek and Wheeler 1992; Jablonski et al. 1996; and Conway Morris 1998.

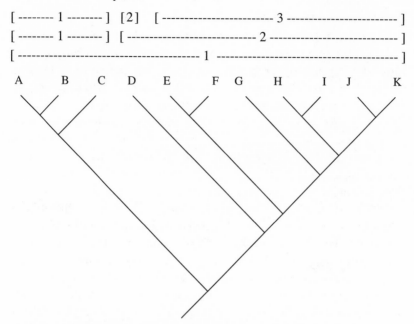

Figure 6.24. Higher-level taxa are not "real" evolutionary entities; they are artifacts of our propensity to arrange items in a hierarchical framework.

or the number of species when we examine changes in diversity across time. You are an avid paleontologist interested in studying the evolution of carnivorous mammals. Using the number of families as your marker (based upon work by a friend of yours at the museum associated with your university), you conclude that diversity has cycled over the millennia, peaking approximately 4 and 10 million years ago and crashing 7 and 11 million years ago (fig. 6.25a). While you are at the library searching for environmental correlates of the crash (perhaps a large comet or a prolonged period of drought), one of your colleagues plots number of species over the same period and uncovers exactly the opposite pattern: diversity reached its lowest points 4 and 10 million years ago (fig. 6.25b). Meanwhile, you have published a hypothesis proposing that changes in climate are correlated with the cycling of diversity (high diversity during warm periods, low diversity during cooling trends). Your colleague writes a reply to your paper, pointing out that the number of species actually decreased during the warm periods and increased during the cooler periods. Which pattern gives us a more realistic picture of biological diversity? More important, which of these two studies would you use to develop predictions about the effect of climate change on this particular group of carnivores?

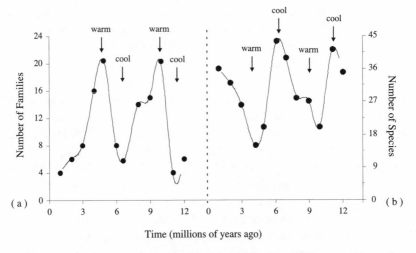

Time (millions of years ago)

Figure 6.25. Cycles in diversity over 12 million years revealed by using (*a*) number of families and (*b*) number of species as the marker of diversity.

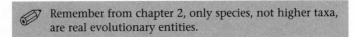

Remember from chapter 2, only species, not higher taxa, are real evolutionary entities.

By now you should be feeling vaguely uncomfortable about the whole concept of radiations. If it is incorrect to measure changes in diversity by examining differences in number of families through time, why is it any better to examine differences in diversity between two families at the same point in time? Is not the family, at any time, still an artificial construct? The quick answer to this question is yes. You can, if you wish, make the entire question of evolutionary radiations disappear by changing the taxonomy so that no one group contains a significantly different number of species from its sister. For example, we can solve the problem of why there so many passerines simply by subdividing the passerines into numerous, smaller clades and eliminating the group "passerine" completely. Perhaps radiations do not exist in the real world, but are simply an artifact of the way human beings see the world. An ant might subdivide birds into very different groupings based upon olfactory cues, rather than visual (small versus large size) or acoustic (singing versus silent) ones (although we would expect the ant to produce the same hypotheses of phylogenetic relationships as we do). After spending a considerable amount of time reading this chapter, should you now just breathe a sigh of relief and think to yourself, "Well, at least I don't have to know this for an examination"? No such luck!

One way around the problem is to approach it from the perspective of individual characters, or purported key innovations (see, e.g., Slowinski and Guyer 1993). For example, we could hypothesize that becoming an herbivore is an important innovation because it opens up a widespread resource base, allowing populations to expand their ranges and decreasing the chance that they will go extinct. Given these factors, we would predict that every time herbivory appears, it will be accompanied by an increase in net diversification. That prediction can then be tested by searching for the origins of herbivory on numerous phylogenies and using those points of origin to define the group and its sister for comparisons of net diversity (see, e.g., Archibald 1993; Reisz and Sues 1998; Sues 2000). As you can tell, this is simply a restatement of the research program described throughout the previous pages, except that we begin with a novel character or characters, then move to measuring diversity, rather than arguing backward from the species-rich group to the character. Arguing from "major innovation" to "unusual group" has probably always been a subconscious part of the way we subdivide the world. We tend to recognize and name large groups based upon *shared* distinctiveness. So, evolutionary taxonomists were on the right track all along. They simply lacked a rigorous method to test their hypotheses.

In practice, this method is difficult to employ. We have been examining the question of radiation for so long that the two components, the characters and the species-rich groups, are inexorably linked in our collective psyches. Let's face it, we all know passerines are small. Using "small" as a character, then testing our hypothesis by pinpointing the origin of "small" in the ancestor of the passerines is hardly a convincing demonstration. Discovering a correlation between increased size and decreased net diversity within the passerines would provide more support for the hypothesis, as would replicated tests for the association between size and diversity within all birds. Does this mean, then, that we should no longer be asking questions like, Why are there so many passerines, so many beetles . . . ? Not really. It just means that we cannot answer the question completely by restricting our level of analysis to the passerines or beetles alone.

So, now that we have convinced ourselves (and you as well, we hope) that the study of asymmetry in species number (or adaptations for that matter) is a worthwhile pursuit, let us return to the question of long-term changes in global diversity, particularly to the thorny problem of extinctions. True extinction involves the irreversible and permanent loss of biological information as one or more species disappear. (This differs from "pseudo-extinction," which includes the splitting and subsequent speciation of an ancestral species but does not involve the loss of biological information, which is conserved in descendant lineages [Brooks and

McLennan 1991; Brooks et al. 1992a and references therein].) Numerous authors have used the power of mathematical modeling to investigate the factors that determine the vulnerability of populations to extinction.[22] In general, there are three major players in the extinction arena at this level of analysis: (1) the strength and direction of selection (i.e., rate of environmental change: Is change stochastic, or does it show a long-term trend? Is change minor, moderate, or catastrophic?); (2) demographics: population size, carrying capacity of the environment, generation time, viability; and (3) the amount of genetic variance in the population (the historical legacy that determines the population's ability to respond to future challenges: factors that center around reproduction, including mutation/recombination rates, sex ratio, and mating system). The results of each model differ depending upon the limits placed on the variables (e.g., Is the population genetically monomorphic, or does genetic variance fluctuate? Are individuals monogamous, or do they mate randomly?).

Lande (1993), for example, examined the interaction among demographic stochasticity (random fluctuations in individual reproduction/ death), environmental stochasticity (small or moderate perturbations that affect the birth/death rate of all individuals), and random catastrophies. He concluded that the ability of a population to survive environmental perturbations is a function not only of its size, but also its growth rate. Being "large" is not enough to buffer a population from extinction if it has a negative growth rate. Bürger and Lynch (1995) reported more rapid extinction rates when environmental change was directional rather than random (e.g., global warming) and was allowed to interact with demographic and genetic factors. In essence, the synergy among these three factors lowered the critical threshold for rate of environmental change above which populations went extinct. Not all populations performed equally in the simulations: populations with a random mating system (for example, broadcast spawning fishes), narrow specialists, and broad generalists were the most at risk. Overall, then, microevolutionary analysis has demonstrated two things: (1) there is a rate of environmental change above which no population can respond, refuting Lamarck's belief that nothing ever goes extinct because of a failure to adapt and (2) the "probability of surviving extinction" is the end-product of a complicated interaction among many factors; there are many routes to extinction (or survival). Given our general lack of infor-

22. Ludwig 1976; Leigh 1981; Shaffer 1981, 1987; Tier and Hanson 1981; Ewens et al. 1987; Goodman 1987; Pease et al. 1989; Lande 1993; Lynch and Lande 1993; and Huey and Kingsolver 1993.

mation about how these factors might interact, it is thus almost impossible to rank populations on a scale of "likely to survive," with the possible exception of populations at the extreme low end of the scale.

The microevolutionary models are concerned with predicting the probability of extinction for any given population or species at a given time. The macroevolutionary models, on the other hand, investigate the patterns of extinction for many groups over long periods. The variables here are time between episodes of extinction and number of taxa in particular groups that survive during that interval (or that go extinct, depending upon your perspective). These data, collected from the fossil record, interact to produce a "kill curve," an S-shaped curve depicting the rate of extinction over time for the groups under investigation (Raup 1991). The kill curve (which is based upon actual fossil data) is used to predict what the tempo of extinction would look like over a particular period by choosing a set interval and sampling extinction intensity randomly from the curve at those intervals. For example, Raup (1996) used the kill curve for marine vertebrate and invertebrate taxa from the Phanerozoic to develop a picture of extinction rates during that period. Sampling every 10,000 years for 20 million years produced a random series of extinction spikes. (The spikes looked more like a pen tracing seismic activity on a piece of paper than either an EKG [predictable cycling] or a flatline [constant and nonfluctuating].) Although "random," there were three "bits" of information in the pattern (just as there is structure in chaos). First, there were many intervals of a million years or more during which extinction levels were extremely low (fewer than 5 percent of species). These periods of faunal stability are called **coordinated stasis** (Morris et al. 1995). Second, there were relatively few, but highly lethal, spikes during which extinction levels exceeded 30–60 percent of species. These lethal events correspond with what we call **mass extinctions.** Third, and more interesting, each of these high spikes was always closely preceded by a another major spike, extending the time of faunal die-offs and blurring the boundaries somewhat for our recognition of the "mass extinction" (Erwin 1990). Overall, the simulations showed a series of irregular and unpredictable extinction pulses over time resembling "a forest of small events punctuated by a smaller number of high spikes" (Raup 1996:428).

The most contentious part of this research program at the moment is the interpretation of what this irregular pattern means. Are there any common factors that could be influencing the ebb and flow of species-level extinction over time? Darwin (1859) argued that interspecific competition drives these patterns, with differences in fitness between species being responsible for one species completely replacing another (**compet-**

itive displacement). One of the most famous examples of the competitive displacement process is the "replacement" of the South American marsupial fauna by invading North America eutherians ("placentals") following the formation of the Isthmus of Panama. The scenario is a straightforward one: the northern eutherians out-competed the southern marsupials, driving most of the marsupial lineages to extinction (Simpson 1950, 1980; Webb 1976, 1985; Bakker 1986).

Lessa and Fariña (1996) investigated this hypothesis by looking for a relationship between the probability of extinction in 120 mammalian orders and body mass, niche (a measure of the ecological equivalence between the invaders and the residents), place of origin (North or South America), and lineage (marsupial versus eutherian). Regression analysis revealed that only body size was linked with susceptibility to extinction (larger-bodied species were more vulnerable than smaller ones: see also Martin and Klein 1984; Benton 1990; Norris 1991; Raup 1993). Once the effects of body size were removed from the analysis, there was no difference in extinction rates between the two groups. In other words, there was no support for the hypothesis that marsupials died out because they were competitively inferior to eutherians. Yet, the fact remains that the faunal composition did change. Lessa and Fariña suggested a subtle but intriguing alternative to the competition-extinction hypothesis: the northern invaders came to dominate the southern continent because they had higher rates of speciation, not lower rates of extinction. This alternative is interesting because it implies a decoupling of diversification and extinction. At the moment, the role for interspecific competition in shaping diversity patterns remains a contentious, but nevertheless dynamic, area of research.[23]

Competition aside, what other factors could play a role in susceptibility to extinction? Erwin (1996) noted that species-poor groups of gastropods (snails, winkles, whelks, limpets, etc.) were more likely to go extinct than species-rich groups during the mid-Permian to mid-Triassic. Is this simply a statistical artifact of differences in group size at the beginning of the extinction period (if you are a genus with only one species, one random, bad luck event can end your line, while the same event reduces your species-rich sister negligibly from 100 to 99 species), or does the species-rich group possess some characteristics that are actually buffering it during an extinction period? Recall our discussion of the possibility that a key innovation might increase the density or range of a species and thus decrease its likelihood of extinction. Some of these

---

23. For detailed discussions of this extensive field, see Masters and Rayner 1993 and Benton 1996.

effects (in addition to the effects of body size, as discussed above) have been demonstrated in the fossil record. For example, numerous authors have concluded that geographically widespread taxa appear to survive extinction events better than taxa with more limited distributions. Perhaps widespread taxa have greater dispersal abilities (and thus can move away from the area of stress to seek out a safe haven or refuge, then disperse back into available areas when the stress passes), or perhaps they are more ecologically diverse (and thus more likely to exploit a wider variety of resources, not all of which will be affected equally by the environmental stressor). It is also possible that being widespread simply decreases the likelihood that all of your populations will be unlucky enough to be in the wrong place at the wrong time.[24]

While all of the preceding characteristics may help species survive periods of low extinction intensity (background extinction levels), it is debatable whether they have any impact during the major catastrophies (Jablonski 1986a, 1989; Flessa and Jablonski 1996). Erwin (1996) differentiated between two types of extinction: **pulse** (or mass) **extinctions**, during which the environmental stressor is so intense and rapid that species simply do not have time to adapt, and **press extinctions**, prolonged periods of heightened environmental stress during which species have time to respond evolutionarily (by moving to refugia, etc.). In the latter case, biological factors such as dispersal abilities, population size, body size, ecological flexibility, and so on will play a large role in determining which species survive and which do not. The concept of pulse versus press is the macroevolutionary equivalent of the factor "rate of environmental change" that is incorporated into the microevolutionary models (recall that some levels of change are simply too high for any lineage to cope with). Surviving a press extinction does not appear to be connected with the probability of surviving a pulse extinction, emphasizing once again the importance of biological factors in the one event and the predominance of external factors and catastrophic stressors in the other. In fact, some authors have argued that some biological characteristics that impart an advantage under periods of low extinction levels may be disadvantageous during catastrophic extinctions (e.g., large body size: Johst and Brandl 1997). If mass extinctions do indeed "change the rules of the game," then patterns of diversification on this planet cannot be explained solely by microevolutionary processes. In other words, there are forces operating on a macroevolutionary time scale that are

---

24. Bretsky 1973; Jablonski 1986b,c, 1989, 1991, 1995; Erwin 1989a,b, 1993, 1996; and Johnson 1998. See Johnson et al. 1995 for a fascinating discussion of the relationship between colony size and susceptibility to extinction in Caribbean reef coral species.

different from, and not reducible to, forces operating on a microevolutionary level.[25]

Let us assume that a species does manage to survive a mass extinction. What happens next? Just as we currently understand little about the forces that cause species-level extinction, we understand even less about what happens after major extinction events. Researchers are fascinated by the concept of "adaptive radiation" à la Simpson because it addresses this problem, at least on an intuitive level. If adaptive zones are portions of the environment waiting to be occupied by organisms, and if there are a finite number of zones available (this is the macroevolutionary equivalent of the ecological term "carrying capacity"), then they will eventually get "filled up." In other words, the only way to stop diversification from slowing and eventually grinding to a halt is to clear the zones occasionally. Mass extinctions can thus be viewed as a kind of global spring cleaning that opens up new ecological-geological space and permits the radiation of surviving groups. Even if we dispense with the concept of zones, it is still clear that the planet can only support a finite amount of biomass, so extinction, mass or otherwise, is a vital part of evolutionary diversification. Darwin recognized this over a century ago. The picture being developed slowly by painstaking reconstruction of the fossil record confirms Darwin's intuition, indicating that some episodes of speciation and extinction follow a very general cycle: (1) major extinction event, (2) lag time, during which surviving lineages get on with the business of surviving ("macroevolutionary lag": Jablonski and Bottjer, 1990a; or "survival period": Erwin 1996), (3) burst of speciation in some, but not all, surviving lineages (radiations during the "recovery period"), which eventually slows down, passing into (4) a period of coordinated stasis, punctuated by random, low-level extinction spikes until the next lethal event. The time to recovery is long on a human scale. For example, reef communities, which appear to be extraordinarily susceptible to extinction events, may take 5–10 million years to reassemble (Fagerstrom 1978; Sheehan 1985; Pfefferkorn 1999). Although the pattern may be a general one, it is important to realize that the timing and length of the survival and recovery periods, as well as the degree of stability during the period of stasis, varies across ecosystems, localities, and through time.[26] Of course, not all radiations follow on the heels of a mass extinction. For example, Butterfield (1997; see also Stanley 1973)

25. Gould 1985, 1995; Jablonski 1986b; Kitchell et al. 1986; Vrba and Gould 1986; and Vrba 1989.

26. For an excellent discussion of the intricacies of these interactions, see Conway Morris 1995.

postulated that the extensive radiation of phytoplankton in the early Cambrian (after more than 1,300 million years of relative stasis) occurred as a response to the origin of herbivorous metazoans, which created a new selective regime (changed the rules). In one way or another, however, life always comes surging back. Given this, the central questions that must be addressed by current and future generations of researchers interested in studying radiations, adaptive or otherwise, are: What causes the surge?[27] Why do some, but not all, groups participate in the radiation? and Is there any evidence that groups which have participated in one radiation will do so again in the future (once a species-rich group, always a species-rich group, and vice versa)?

One of the most fascinating sidebars to the ebb and flow question is the demonstration that rates of speciation and extinction have not been constant. In fact, as discussed above, the general trend seems to be periods of stasis occasionally interrupted by bursts of speciation and extinction. This macroevolutionary pattern has been replicated at the microevolutionary level. In a very elegant series of experiments, Lenski and Travisano (1994) propagated 12 populations of the infamous anaerobic bacterium *Escherichia coli* in identical environments for 10,000 generations. All 12 populations were progeny of the same ancestral asexual clone. Following movement into the new environment (the experimental regime), all populations underwent rapid diversification in both morphology (cell size) and fitness for the first 2,000 generations. The rate of change then slowed, becoming virtually static between generations 5,000 and 10,000 as the populations continued to reproduce in the constant environment. More unexpectedly, the initially cloned populations diverged from each other in both morphology and mean fitness, even though they were living under identical environmental conditions. The authors interpreted these results in the following way: (1) a random sequence of mutation in each population plus the same selection regime from the new environment acting on the different historical background of accumulated mutations in each population leads to (2) an initial period of rapid diversification pushing populations up separate adaptive peaks (divergence) at which point (3) each population "runs out of ways" to become better adapted on the peak (recall the insectivorous bats) and remains there. Overall, then, the rapid change, diversification, and stasis result from an interplay between selection, chance (the stochastic effects of mutation and drift), and history.

Patterns on both macroevolutionary and microevolutionary time

---

27. For an excellent discussion of possible explanations, see Jablonski and Bottjer 1990a,b.

scales may thus support a model of episodic rather than constant diversification, forcing us to rethink our whole framework for the evolutionary
process.[28] For example, let us assume the hypothesis that molecules
evolve at a regular, clock-like rate is correct. How do we move from the
nicely balanced tick-tocking molecular level to the (apparently) randomly staggered bursts of diversification and extinction at the level of
entire clades and ecosystems or, for that matter, the patterns of diversification and stasis within a single species on a short time scale? If, on
the other hand, the assumption of an underlying clock is not a viable
one for all clades, as the fossil record suggests, then researchers who are
examining questions about rates of character change, phylogenetic relationships, and ages of lineages based on that assumption may spend
substantial time and effort digging in the wrong place (Hillis 1987; Peterson 1992; Sanderson and Donoghue 1996; Wray et al. 1996).

One of the major lessons to be learned by all biologists is that, unlike
the situation rejoiced in by physicists, it is difficult to generalize in biology. Biological systems are simply too complex to be reduced to a small
number of easily defined variables that can be reconstructed in any combination to predict the characteristics, fates, and tendencies of any species. In other words, once again there are no rules stating that either a
gradual and constant rate of speciation-extinction or an irregular cycle
of stasis punctuated by bursts of speciation-extinction is the preferred
mode for all clades (see, e.g., Erwin 1989b; Mindell et al. 1989). To us,
the more interesting questions are whether particular clades are characterized by only one of the two modes (or show a combination of the
two across long spans of time) and, if the former, whether clades that
display only one mode have any general characteristics in common.

We began this chapter with a brief discussion of how George Gaylord
Simpson attempted to take the orthogenetic concept of adaptive zones
and radiations and recast that concept in Darwinian terms. That the
study of radiations has enjoyed such a resurgence over the past decade
is a testament to the importance of Simpson's work. His understanding
of the fossil record, his fascination with unraveling the mysteries of what
he called the "tempo and mode" of evolution, lay at the heart of his
writings about radiations. As a paleontologist, he understood that the
fossil record provides us with a unique source of information because it
gives us access to individuals frozen in time. Using that information, we
can determine the patterns of extinction episodes, which types of taxa

28. Eldredge and Gould 1972; Gould and Eldredge 1977, 1993; Boucot 1978; Fagerstrom 1978; Gould 1980, 1982a,b; Cheetham 1986; Levinton 1988; Jackson and Cheetham
1990, 1994; and Kellogg 2000.

have been historically the most vulnerable during those episodes (e.g., widespread versus restricted, tropical versus temperate), and how long it has taken evolution to "repair the damage." Microevolutionary investigations involving experimental analysis and rigorous models, like the ones discussed at the beginning of this section, tell us how that repair occurs and why the damage was done in the first place. For example, all "true" extinctions, whether at the individual, population, or species level, are caused by a "failure to adapt," if "failure to adapt" = "failure to solve the problem posed by the environment." The more important question is whether that failure is due to "bad genes," "bad luck" (Raup 1993), or a combination of the two (which seems more likely). Integrating macro- and microevolutionary data is the critical next step for addressing contemporary biodiversity issues. If, as the fossil record suggests, it takes millions of years for ecosystems to return to pre-extinction levels of diversity, then repairing the damage caused by human activities will not only take longer than the destruction took, but it will take place on macroevolutionary time scales.

Chapter 7

Community Evolution: Exploring the Space-Time Continuum

> [C]onsideration of phylogeny and biogeography should be a major concern of researchers studying comparative community evolution.
>
> Cadle and Greene 1993:261

One of the most problematic aspects of studying the evolution of communities is that biologists hold widely divergent views about just what exactly a community is (Ricklefs 1990). For example, some researchers view communities as associations of species so strongly tied together by their ecological and behavioral interactions (**synecology**) that they are almost super-organisms (e.g., Wilson 1980, 1983). Other researchers feel that at least some communities are arbitrary assemblages of species that just happen to be in the same place at the same time (e.g., Strong et al. 1984). In many ways, the debate over what constitutes a community resembles the discussions among systematists about the definition of the term "species" (chapter 3) and among evolutionary biologists about the terms "adaptation" (chapter 5) and "adaptive radiation" (chapter 6). As a consequence, we are not going to champion one particular community concept or any particular evolutionary process, such as competition, that might be responsible for shaping community structure; nor are we going to summarize and categorize all of the literature in this area.[1] Rather, we will assume that a variety of community types exist on this planet, and thus almost all concepts of community will be valid for particular cases. We will begin with hypotheses about history and about origins, then use those findings as platforms for research into the processes that might be underlying those patterns. The one thing that we do expect is that we will find both historical and nonhistorical components in community structure. The interesting question

---

1. For an excellent summary, see Ricklefs and Schluter 1993.

then becomes, Are all communities influenced to the same degree by these components, or has each community traveled along a unique evolutionary pathway? In other words, we are searching, once again, for general patterns in a complex and historically unique mosaic. We must therefore be very careful about our interpretations of such patterns.

STUDYING COMMUNITIES WITHOUT PHYLOGENY

Documenting and explaining the complexity of multispecies associations has been approached by ecologists on two levels: description (discovery) and explanation (evaluation). Descriptions of diversity are concerned with the total number of species and the relative abundance of species in a given area. The value of these descriptions depends upon our ability to perceive, collect, and classify organisms and to depict the data mathematically, for example, in the form of diversity indices. Explanation, or the search for processes underlying observed biological diversity, is a complicated process that has blazed a trail of controversy through the ecological literature. For example, ecosystems ecologists have argued that community structure is an emergent property of generalized energy flow patterns rather than of specific interactions among particular organisms or species.[2] By moving the selection arena from the individual to the ecosystem (Oksanen 1988), these researchers are able to examine communities without "detailed information about individual species" (Loehle and Pechmann 1988). Evolutionary ecologists, by contrast, originally attempted to incorporate both population-level and community-level factors into their explanations of ecological structure. They examined biotic interactions based on traits that were fixed within each species but variable among species. This permitted researchers to filter out the confounding influence of intraspecific variability and thus formulate hypotheses concerning the influence of species composition on differences in interspecific interactions. Researchers then sought general rules for the production and structuring of diversity by investigating (1) specific components of the system, such as patterns of colonization and extinction of species, the age, productivity, structural complexity, and stability of the system under investigation, or predation and interspecific competition[3]; (2) organism and environment as an inseparable whole (Patten 1978, 1982); and (3) the impact of "stochastic factors"

2. Lindeman 1942; Odum 1969; Brown 1981; Wright 1983; and Glazier 1987.

3. Wallace 1878; Dobzhansky 1951; Hutchinson 1959; Hairston et al. 1960; Connell and Orias 1964; Williams 1964; MacArthur 1965; Paine 1966; MacArthur and Wilson 1967; Slobodkin and Sanders 1969; Parrish and Saila 1970; Brown 1973; Connell 1978; Pianka 1980; Lawton and Strong 1981; Strong et al. 1984; and Houle 1997.

(Simberloff 1971). All approaches have contributed some pieces to the intricate puzzle of how biological communities evolve.[4] They have also established a strong empirical basis for assuming that important features of communities may be deeply embedded in the evolutionary histories of the participating species.

Although evolutionary ecologists have performed experiments within a comparative framework on a regular basis, few have incorporated phylogenetic information into the explanations derived from those comparisons. For example, suppose you are interested in species coexistence. As MacArthur (1972) noted, the best place to look for the factors influencing species coexistence is among sympatric congeners. The assumption behind this recommendation is actually a historical one: members of the same genus should theoretically share a number of ecological, morphological, and behavioral characters, because they are all descended from a relatively recent common ancestor. The recognition that the genealogical relationships among species may influence the outcome of an experimental investigation provides the historical context in any evolutionary ecological study.

Having discovered an appropriate group of sympatric congeners, you set about collecting a wealth of data concerning feeding behavior, habitat preference, and breeding cycles in order to identify the way(s) the species are partitioning their environment, and you use that information to explain the evolution of the diversity you find. Although the choice of congeners has a genealogical basis, this approach is primarily nonhistorical because making the transition from description to explanation requires assumptions about the evolutionary *past* of species' interactions based solely upon characters and interactions observed in the *present* environment.

Such nonphylogenetic approaches are manifested in two general perspectives. The first perspective ("history is irrelevant") is characterized by the statement, "I only want to know what organisms are doing, not how they came to be doing it." If biodiversity has been produced by evolutionary processes, then this perspective limits your science to description without explanation. The most sophisticated extension of the "history is irrelevant" perspective invokes an anti-auxiliary principle: "similarity of functional components in two or more different associations = convergent evolution of those components in common environments." Employing this antiprinciple permits individual researchers to accept arbitrarily high amounts of homoplasy within and among communities.

---

4. For excellent reviews, see Brown 1981, 1995; McIntosh 1987; Oksanen 1988; and Maurer 1999.

Biologists traveling this route must be eternally vigilant, because the arbitrary nature of the antiprinciple may push their research down the slippery slope of geographic determinism. As we discussed in chapters 1, 4, and 5, the distinction between Lamarckian geographical determinism and Darwinian selectionism was well recognized more than a century ago in precisely this context (Boas 1896; also Kellogg 1906).

The second perspective holds that history is an important, but complicating, variable. For example, if species X is a keystone species in the tropical wet forest of Guanacaste, Costa Rica, but not in the same habitat in Panama, then the discrepancy is due to the "confounding effects of history." MacArthur (1972) advised researchers to avoid such effects by reducing the spatial scale of their studies. The use of spatial scale in this context is of course subject to the same caveat as is reconstructing ancestral distributions in the historical biogeography program: today's distributions are assumed to be similar to the distributions at the point of origin. The second assumption in the "history is important but confounding" perspective is that it makes no difference if you assign all specimens of closely related species to one species, as long as they are functionally equivalent. For example, you might designate all ranid frogs as "*Rana* spp." Recall that Clutton-Brock and Harvey (1977) suggested that all members of a monophyletic group having the same ecology be reduced to a single data point in a statistical analysis of character evolution. Reducing all congeners to a single species in a study of community evolution is not the same process, however, because we are studying interactions among species. It is highly unlikely that two coexisting congeners will have exactly the same set of functions and resource requirements within a given community.

By now you will recognize that we are back to the old problems of origin versus maintenance and stasis versus diversification: in this case, very few researchers have focused their attention on either the origin and elaboration of the characters involved in the association, or the origin of the community itself. Reducing the spatial scale of a study in conjunction with the assumption that there are no (or limited) historical effects on such scales, has led many researchers to make two, interconnected assumptions: (1) all traits exhibited by all species in a given area evolved *in situ;* therefore, (2) if similarities are discovered among communities in different areas, those similarities are due to convergence (the anti-auxiliary principle). Such a priori assumptions are always erected when a discipline is young—it is the only way to simplify a complex problem enough to begin collecting data (see, e.g., Schluter and Ricklefs 1993). We have been collecting such data about the ways communities are assembled for decades, so the research program should be poised to move to the next stage of a maturing discipline: detailed examination of its assumptions with subsequent modification if necessary. In other words, we need to incorporate phylogenetic (historical) information into our studies of communities.

STUDYING COMMUNITIES WITH PHYLOGENY: HOW DO WE GO ABOUT DOING IT?

There are two general components to phylogenetic studies of communities. The first component involves asking questions about the origins of the species themselves. How did the species in a community come to be where they are? Did they all evolve where we find them today, did they all arrive from somewhere else, or is the community the result of both influences? We can recover this kind of information from speciation and historical biogeographical studies (chapters 3 and 4). Studies of this type have begun to accumulate evidence indicating that the historical sequence of colonization has an impact on community structure.[5] The second component involves investigating the ways community members partition resources (**autecology**) and interact with each other (synecology).[6] How did those attributes and interspecific associations within the community arise? Did the species involved bring their traits and interactions into the area, or did those characteristics evolve *in situ*? Are the interacting species maintaining ancestral associations, or have they evolved new ways of interacting? We can recover this kind of information from phylogenetically based studies of character evolution and evolutionary radiations (chapters 5 and 6).

Consider the following hypothetical example of the interaction between past and present at the community level (elaborated from Mayden 1987a). Imagine discovering that two fish species in a large lake do not overlap ecologically; say, for example, one is a benthic forager (species D) and one is limnetic (species Z). A possible explanation for such habitat separation is that it represents the effects of competition between these two species or their ancestors in the past. Is this a reasonable hypothesis? The short answer to this question is, We do not know. The longer answer is, Without a phylogeny for the fishes and a record of their relatives' interactions with each other, it is impossible to ascertain whether the association in our research lake is a result of either interactions between Z and D or a historical legacy of interactions between

5. For example, Drake 1990; Abrams 1996; McPeek 1996; Losos 1996; Belyea and Lancaster 1999; and Brokaw and Busing 2000.

6. See Walter and Hengeveld 2000 for a recent discussion of the importance of considering both autecological and synecological factors.

their ancestors. So, you return to the field and, after extensive field work, discover species C (benthic) and Y (limnetic) in a second community, species X (demonstrates both foraging modes) and B (benthic) in community 3, and species W (demonstrates both foraging modes) and A (benthic) in community 4 (fig. 7.1a). As luck would have it, phylogenies exist for the two clades. When the foraging modes are optimized on the phylogenetic trees, you discover that foraging on the benthos is plesiomorphic for all members of the A + B + C + D clade (fig. 7.1b). These species have not changed their foraging habits, interactions with members of the other clade notwithstanding. Conversely, foraging on both benthic and limnetic prey was plesiomorphic for the W + X + Y + Z clade, but something happened during the interaction between the ancestor of Y + Z and the ancestor of C + D, and the former moved out of the benthic, into the limnetic, realm (fig. 7.1c). While this does rule out a role for interspecific competition between past populations of species D and Z in shaping the current foraging modes in these fishes, it does not rule out the possibilities that competition was involved in the habitat shift in the appropriate ancestors or is maintaining the divergent foraging habits today (see also Brooks 1980).

Both ecosystems ecologists and evolutionary ecologists have discovered valuable clues in their quest to solve the mystery of community evolution. We believe, however, that the mystery will remain unresolved if we do not incorporate the vital information contained at the macro-evolutionary level into our explanatory framework. The step from understanding the evolution of a single behavioral or ecological character or suite of characters to understanding the evolution of communities is a rather daunting one (as you can tell from the preceding example). Such a step requires that we collect information from both character evolution and species formation for many coexisting species. To complicate matters further, studies to date indicate that there may be few generalizations that can be made about the way ecological associations are assembled beyond the expectation that phylogenetic history will play a role in explaining both species composition and community structure.

At the moment, little of the research in community ecology has utilized phylogenetic information; therefore, this chapter will serve more to indicate future research possibilities than to present a database from which generalizations can be derived. In the following pages, we will discuss ideas about how we can uncover that information and, working with a coalition of ecologists and systematists, begin to solve the mystery of community evolution. At first glance, this may seem like an exercise in tiresome backtracking, but we are not suggesting that all current investigations be halted until we have better estimates of community

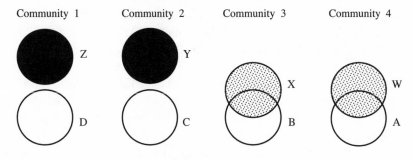

(a) Distribution of Fish Species in the Water Column

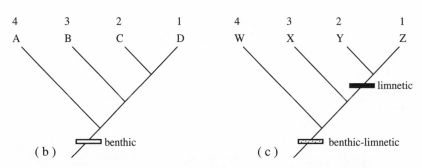

Figure 7.1.   The influence of history at the community level. (*a*) Distribution of foraging preferences. *White circle* = species foraging in the benthos; *dotted circle* = species foraging in both the benthos and the limnos; *black circle* = species foraging in the limnos. (*b*) Phylogenetic relationships for clade A–D. (*c*) Phylogenetic relationships for clade W–Z. *Numbers* = community where species is found.

origins. Clearly, useful and significant information has been and is being generated by ongoing studies lacking phylogenetic contexts. We believe, however, that incorporating phylogenetic information into those studies will resolve some confusion and add extra dimensions to already interesting perspectives.

Putting It All Together: Who, Where, What, and Why?

All species within a community contain information about their origin with respect to the biota (spatial allocation) and the origin of their individual traits relevant to the association—that is, traits that characterize a species' interactions with other species and with the environment (resource allocation). Within any given community there are two types of species: those that speciated in the community (**residents**) and those that speciated somewhere else and dispersed into the community (**colo-**

nizers). The distinction between residents and colonizers is thus based upon the species' place of origin (inside or outside the community under examination). Similarly, the ecology and behavior of any given species in a community may reflect the presence of persistent ancestral traits or of recently evolved, autapomorphic traits. Starting to sound familiar? Under the guise of spatial and resource allocation, colonization, extinction, and competition, we have returned full circle to the evolutionary processes of speciation and character evolution discussed for individual clades in chapters 3–6. We expect that all communities will be mosaics of residents and colonizers demonstrating plesiomorphic and apomorphic interactions. The relevant question is, How much of that mosaic can be uncovered by a comparative phylogenetic study? In the following discussion we will focus our attention upon what the macroevolutionary patterns can tell us about the origins of a community.

The Historical Backbone

The conservative portion of any community is composed of species that evolved *in situ* through the persistence of ancestral associations (congruent portions of phylogenies in a historical biogeographical analysis). Such species display the plesiomorphic condition for characters involved in interactions with other community members and with the environment (fig. 7.2). Since this section of the community is characterized by a stable relationship (**niche conservatism:** see Holt 1996) across evolutionary time, it may act as a stabilizing selection force on other members of the community by resisting the colonization of competitors (Brown 1995; Maurer 1999; see also "priority effects" and "invasion re-

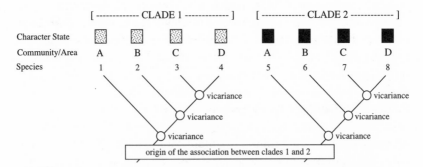

Figure 7.2.   The historical backbone of community structure. In this situation, both the origin of the association between community members and the origins of the characters involved in the interaction between those members are old. Extant community structure in areas A, B, C, and D thus represents *the persistence of an ancestral association, coupled with the persistence of ancestral interaction traits, in both members of the community.*

sistance" in Drake 1990). There is a marked similarity in many ecological associations around the world, suggesting phylogenetic influences in resource utilization/partitioning as well as in species composition.[7] For example, Hill and Smith (1984) discovered similar roosting patterns between six species of bats in a Tanzanian cave and 14 species of bats in a cave on New Ireland Island off the northeastern coast of New Guinea. In both communities, species of the genus *Hipposideros* lived near the rear of the cave in association with species of *Rhinolophus,* while species of *Rousettus* roosted just inside the first major overhang (fig. 7.3). More recently, McPeek and Brown (2000) showed that a relatively small number of parallel shifts in niche preference during speciation by North American damselflies (*Enallagma*) permitted complex patterns of allopatry and co-occurrence in the regional species pool for communities; Zimmerman and Simberloff (1996) and Vitt et al. (1999) concluded that phylogenetic history played an important role in the structure of frog and lizard communities in Amazonia; Kelt (1999) supported Brown and Zheng's (1989) suggestion that desert small mammal communities exhibit pronounced historical structuring; and Webb (2000) found a high degree of phylogenetic relatedness for members of different Bornean tree communities.

Indirect Effects within the Historical Backbone

Even in the absence of colonization, we expect that communities will continue to evolve (change) over time. For example, a fish may shift its foraging mode because of the indirect effects of strong sexual selection for a large body size. In this scenario, evolutionary changes in ecological characters occur within a phylogenetically conservative framework of species composition. Species contributing to this portion of the community structure can be recognized by their historical association with other community members, coupled with the presence of apomorphic traits characterizing their interactions with other species and with the environment (fig. 7.4). The longer a community exists below equilibrium numbers, the greater the possibility that resident species will experience this sort of evolutionary change, which may in some cases represent a type of stochastic wandering through modifications allowed by the existing community structure (e.g., Grandcolas 1993).

7. McCoy and Heck 1976; Brooks 1980, 1985; Boucot 1982, 1983, 1990; Mitter and Brooks 1983; Ross 1986; Fox 1987; Ricklefs 1987; Robinson and Dickerson 1987; Westoby 1988; Hughes 1989; Drake 1990; Foster et al. 1990; Brooks and McLennan 1991, 1992a, 1993a,b, 1994; Grandcolas 1993, 1998; Ricklefs and Schluter 1993; Futuyma and Mitter 1996; Losos 1996; Richman 1996; Wainwright 1996; Brown et al. 2000; Gauld and Wahl 2000; Leschen 2000; Lindeman 2000; Price et al. 2000; and Silvertown et al. 2001.

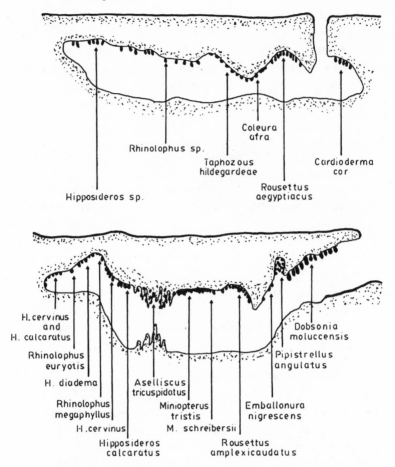

Figure 7.3. Roosting positions of several bats species in Tanzanian (*top*) and New Ireland (*bottom*) caves. (From Brooks and Wiley 1988; modified from Hill and Smith 1984.)

The Effects of Colonization

As you recall from chapter 4, a colonizing species can be recognized because its phylogenetic history is incongruent with the histories of community members that evolved *in situ*. There are a number of different pathways open to a species entering a community. First, the colonizer may fail to establish itself, in which case we will have no record of the attempt. Although such attempts may play a role in the overall dynamics of community maintenance, they are not available for analysis on a macroevolutionary level. So, let's turn our attention to the successful colonizer. Such a species may simply merge into the community because its members possess traits (plesiomorphies) that do not conflict with the

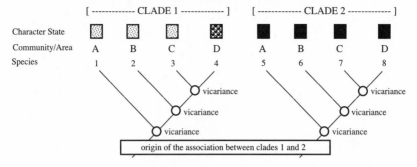

Figure 7.4. The effects of continued evolution within the community. In this situation, the origin of the association between community members is plesiomorphic; however, while species 8 retains the ancestral interaction character (plesiomorphic), the character has been modified in species 4 (apomorphic). Extant community structure in areas A, B, and C thus represents the persistence of an ancestral association, coupled with the persistence of ancestral interaction traits in both members of the community, while extant community structure in area D represents *the persistence of an ancestral association, coupled with the appearance of an autapomorphic trait in one member of the community.*

existing community structure. This scenario postulates that there is no competition between colonizing individuals and resident members of the community, or that competition is not sufficiently strong to cause a change in the newcomer or the resident. In either case, we would expect to see a colonizing event, coupled with no modifications in the traits displayed by the colonizer or by the residents with which it might potentially interact (fig. 7.5). Erwin's (1985) taxon pulse model is an example of this kind of influence in the evolution of biotic diversity, a sort of macroevolutionary succession. According to this model, a group of species begins with an ancestor that displays a certain ecological propensity (e.g., eats leaves). As time progresses, the ancestor and its descendants spread over a larger and larger geographical area, with descendant species fulfilling the same or very similar ecological roles in different locations. Subsequent to this first wave of dispersal, a new ecological trait arises in one of the descendant species in one of the localities (e.g., eats seeds). The species bearing this novel trait then undergoes widespread dissemination and a new pulse of diversity occurs, producing a new set of descendant species, all performing similar functions in different locations. Diverse and highly structured communities could be formed in many different areas in this manner, with every community containing a member of each of the "pulses" (see also the taxon cycle of Roughgarden and Pacala 1989; also Roughgarden 1992). This may be one explanation for the observation that diversity seems to increase when environmental disturbance is low (e.g., Grandcolas 1993).

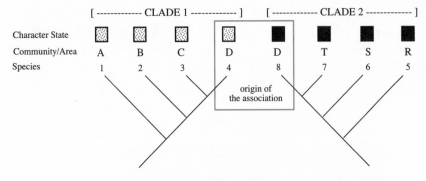

Figure 7.5. Effect of a colonizer on community structure I. Species 8, having colonized area D and now sharing space with resident species 4, is either not interacting competitively with species 4 (species 4 may be a fish that eats algae, and species 8 may be a fish that eats zooplankton) or has established a direct interaction without changing any traits (species 8 may be a bird that builds elaborate nests, and species 4 may be a tree). In this situation, the origins of the traits are plesiomorphic (old), while the origin of the association between community members is apomorphic (recent). Extant community structure in area D thus represents *the appearance of a colonizing species, coupled with the persistence of plesiomorphic interaction traits in both members of the community.*

It is also possible that the colonizer will have a direct effect on a resident, which will correspond to one of the following phylogenetic patterns: (1) The colonizer does not change, the resident goes extinct (**displacive competition:** Hallam 1989). Since the original resident is no longer a member of the existing community, is there any way to study the ghost of displacive competition past (Connell 1980)? Fortunately, the answer is yes (but it will take considerable work). As you recall from chapter 4, we can use the power of secondary BPA to uncover *instances of colonization (dispersal) by one member of a clade that are coupled with the extinction of a species from another clade in the same area.* The hypothesis that the colonization and extinction events are causally coupled will be supported by the demonstration that other members of the extinct species' clade have similar resource requirements to the colonizer. (2) The colonizing species will change (fig. 7.6a). (3) The resident species will change (fig. 7.6b). (4) Both the resident and the colonizer will change (fig. 7.6c). Except in instances of displacive competition then, direct interactions between the colonizer and the resident produce a macroevolutionary pattern in which the colonizer, the resident, or both exhibit an apomorphic condition of the traits relevant to the competitive interaction.[8]

8. For example, Drake 1990; Losos 1990a, 1992, 1994, 1996; McPhail 1993, 1994; Schluter and Price 1993; and Schluter 1996b, 1998.

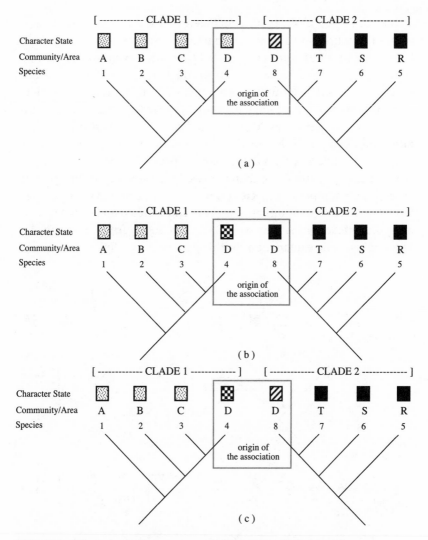

Figure 7.6. Effect of colonizer on community structure II. Species 8 has successfully colonized area D and is now interacting with species 4; therefore the origin of the association is apomorphic (recent). (*a*) The origin of the trait is plesiomorphic (old) in the resident species and apomorphic (recent) in the colonizer. (*b*) The origin of the trait is plesiomorphic in the colonizer and apomorphic in the resident species. (*c*) The origin of the trait is apomorphic in both the colonizing and resident species. In all cases, extant community structure in area D thus represents *the appearance of a colonizing species followed by modifications in the trait(s) involved in the interaction between resident and interloper.*

Boucot (1982, 1983, 1990) concluded that the fossil record demonstrates the conservative nature of community structure throughout evolutionary history. He further suggested that when evolutionary changes do occur, they tend to reverberate through most of the community structure (see also Conway Morris 1998). Comparative phylogenetic methods can provide a complementary approach to these studies. Consider the hypothetical case depicted in figure 7.7. Five areas (A–E) contain biotas composed of a member from each of three clades (clade 1: species 1–5; clade 2: species 6–10; clade 3: species 11–15). The communities in areas A, B, and C are characterized by plesiomorphic interactions among their component species (1, 6, 11; 2, 7, 12; and 3, 8, 13, respectively), whereas communities in areas D and E are characterized by apomorphic interactions among species (4, 9, 14 and 5, 10, 15, respectively). The phylogenetic explanation for this pattern is as follows: The correla-

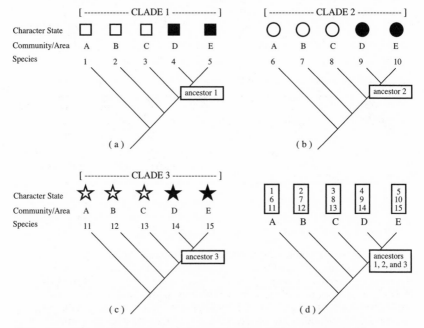

Figure 7.7. The evolution of multiple communities. *Letters* = communities; *numbers* = species from three different clades that inhabit those communities. (*a–c*) The phylogenetic trees for the members of the three clades. *White symbols* = plesiomorphic trait; *black symbols* = apomorphic trait. (*d*) Area cladogram depicting the historical relationships among the communities, based on the phylogenetic relationships of the species that occur in them. The shift from plesiomorphic to apomorphic traits in each clade occurred in ancestors 1, 2, and 3, leading to the emergence of an ecological structure in communities D and E that differs from the ecological structure in communities A–C.

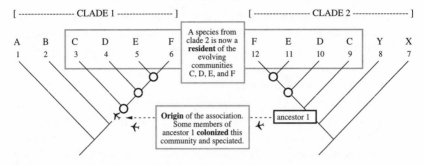

Figure 7.8. Complexities that can occur when a member of the community begins as a colonizer (origin of the association), then becomes integrated into the community and continues to speciate there as a resident.

tion among the plesiomorphic interaction traits is the result of historical conservatism in the evolution of these communities. Novel interactions among these three clades evolved in the common ancestor of species 4 + 5 (ancestor 1), the common ancestor of species 9 + 10 (ancestor 2), and the common ancestor of species 14 + 15 (ancestor 3). These clades show a pattern of biogeographic congruence; therefore, the evolutionary changes in interactions co-originated in co-occurring species in the same (ancestral) biota (the biota in which ancestors 1, 2, and 3 lived). We can thus hypothesize that a common cause is responsible for this suite of ecological changes in those ancestors.

So far, we have been discussing the origins of particular communities. Of course, once a colonizer becomes established in a community, it may continue to evolve there, becoming part of the historical backbone. So, just as a trait may be an autapomorphy on one level and a plesiomorphy on another, a species in a community may be a colonizer on one level (origin in the association) and a resident on another (fig. 7.8). There are any number of combinations involving character stasis or change and movements of lineages through resident to colonizer to resident (etc.). We are not going to delineate all possible components of the mosaic we call a "community" in this section. (Try this yourself on a blackboard with lots of chalk.) We are, however, going to illustrate how comparative phylogenetic analysis can help us identify some of those components. With that in mind, let's move from the world of hypothetical (and well-behaving) clades to the real world with all of its complexity.

HOW SPECIES COME TO BE IN A COMMUNITY

The Australian Birds Revisited. Let us return to the eight clades of Australian birds discussed in chapter 4. We had to break our analysis into two elements in order to disentangle the complexity of speciation events producing the members of the eight clades. Some of the areas, however, are occupied by species from both elements, so it is necessary to reintegrate all eight clades in order to discuss the historical assemblage of the communities to which they belong. As you might guess, we can use the historical biogeographical analysis to determine whether a given species is a resident or a colonizer in a particular area. Seven of the ten areas contain from five to seven species, at least four of which are residents (table 7.1). Only residents live in Cape York (six species), Arnhem Land (five species), and the Atherton Plateau (five species). The eastern desert and western desert both contain five species (four residents and one colonizer), and the "species-rich" Kimberly Plateau boasts seven species (five residents and two colonizers). Perhaps the most interesting areas for researchers interested in the dynamics of community evolution are the eastern forest and Adelaide/Eyre Peninsula regions. The eastern forest has five species, all residents, but three of those residents show vicariance patterns with the southwestern corner and eastern desert + western desert (*Psophodes olivaceus, Stipiturus malachurus,* and *Cinclosoma punctatum*), while two are linked by vicariance with the Atherton Plateau and Cape York areas (*Ptiloris paradiseus* and *Tregellasia capito*). This would appear to be a case in which two different sets of residents have combined to form a larger community. The Adelaide/Eyre Peninsula region contains four species from the eight clades, but all of them are colonizers on some level. *Psophodes leucogaster* is endemic to the Adelaide/Eyre Peninsula area, but evolved as a result of peripheral isolate speciation (as did *P. lateralis* in the Atherton Plateau), while *Stipiturus malachurus, Cinclosoma castanotum,* and *C. punctatum* occur in the region as a result of postspeciation dispersal. Despite having similar numbers of species then, these communities were clearly assembled in different manners historically.

Seven species fulfill the role of residents in their communities of origin and of colonizers in other communities: *Cinclosoma castanotum* originated in the southwestern corner of Australia and dispersed north and east into the western desert, the eastern desert, and the Adelaide/Eyre Peninsula region; *Stipiturus malachurus* and *C. punctatum* arose in the eastern forest and dispersed westward into the Adelaide/Eyre Peninsula region; *Petrophasa ferruginea* and *P. plumifera* dispersed northwest from their vicariant origins in the Pilbara region and the northern desert to the Kimberly Plateau; and *Psophodes occidentalis* and *S. ruficeps* moved from the western desert into the Pilbara region. Additionally, two species highlight the complicated interaction that can occur when a species disperses into a community then continues to speciate there. For example, ancestor 39 dispersed from the Cape York Peninsula into New Guinea, where it underwent peripheral isolate speciation to become ancestor 38

Table 7.1. Distribution of Resident and Colonizing Species in the Ten Australian Avian Communities That Contain More Than One Species

Area	Residents	Colonizers	Total
Kimberley Plateau	5: *Poephila acuticauda, P. personata, Malurus rogersi, Petrophasa albipennis, Pe. blaauwi*	2: *Petrophasa ferruginea, Pe. plumifera*	7
Pilbara region	1: *Petrophasa ferruginea*	2: *Psophodes occidentalis, Stipiturus ruficeps*	3
Arnhem Land	5: *Poephila hecki, P. personata, Malurus dulcis, Petrophasa rufipennis, Pe. smithii*	0	5
Cape York Peninsula	6: *Poephila atropygialis, P. leucotis, Malurus amabilis, Petrophasa peninsulae, Ptiloris alberti, Tregellasia albigularis*	0	6
Eastern forest	5: *Psophodes olivaceus, Stipiturus malachurus, Cinclosoma punctatum, Ptilorus paradiseus, Tregellasia capito*	0	5
Atherton Plateau	5: *Poephila cincta, Petrophasa scripta, Ptilorus victoriae, Tregellasia nana, Psophodes lateralis*	0	5
Eastern desert	4: *Psophodes cristatus, Stipiturus mallee, Cinclosoma castanoethorax, C. cinnamomeum*	1: *Cinclosoma castanotum*	5
Western desert	4: *Psophodes occidentalis, Stipiturus ruficeps, Cinclosoma alisteri, C. marginatum*	1: *Cinclosoma castanotum*	5
Eyre Peninsula + Adelaide	1: *Psophodes leucogaster*	3: *Stipiturus malachurus, Cinclosoma castanotum, C. punctatum*	4
Southwestern corner	3: *Psophodes nigrogularis, Stipiturus westernensis, Cinclosoma castanotum*	0	3

(see fig. 4.50). This colonization/speciation event marked the origin of the *Ptiloris* (riflebird) clade in the community. This new community itself experienced a vicariance event that separated eastern and western areas of New Guinea. Ancestor 38 was fragmented by this event and subsequently speciated (allopatric mode I) to produce *P. intercedens* and *P. magnificus.* Each of these two species is thus a resident part of the vicariance backbone of its community. A similar story can be told for the movement of ancestor 30 into the central-western corner of Australia, establishing the *Petrophasa* (pigeon) clade in that area (ancestor 28). Subsequent separation of ancestor 28 in the Pilbara region and northern desert produced two new residents of those communities, *P. ferruginea* and *P. plumifera.* Both of the preceding cases are unique events at the moment, so we cannot conclude with confidence that the sister pairs were ultimately produced by vicariance. In order to study this, we would need more clades of birds from those particular areas. Nevertheless, the point is clear: a species may begin its life in a community as a colonizer, then continue on as a resident relative to new colonizers if it speciates in the community it colonized.

Finally, as we mentioned in chapter 4, there is only one instance in which sister species are in the same community (*Petrophasa ferruginea* and *P. plumifera* in the Kimberley Plateau), the result of evolution in separate areas followed by postspeciation dispersal into the same region. Closer examination of the distribution maps in Cracraft 1986, however, reveals that the two species live on opposite sides of the plateau. It would be interesting to examine the distribution patterns of these two species in more detail in order to determine whether they interact at all within the community. There is, in general, very little overlap among congeners in these avian communities. As you can see from table 7.1, the Kimberley Plateau is an exception to this general pattern. Four of the eight *Petrophasa* species and two of the six *Poephila* species are present in that community; four evolved there (residents: *Petrophasa blaauwi, P. albipennis,* and *Poephila acuticauda, P. personata* [present because of a nonresponse to vicariance]) and two colonized from elsewhere (*Petrophasa ferruginea, P. plumifera*) (fig. 7.9). If you are interested in studying the ways in which close relatives partition resources within a community, you should book a seat on the next plane, train, or boat leaving for the Kimberley Plateau. Once there, you need to document a variety of potential interaction traits (foraging preferences, nest site preferences, mating preferences, etc.), then map those traits onto the phylogenetic tree for the clades. Having done this, you realize that the most obvious question to ask is whether the two colonizing species show a modification in any of the preceding traits in the Kimberley Plateau compared with the condition of those traits in allopatry (especially relative to populations in each colonizer's area of origin, which necessitates further travel to the northern desert and Pilbara region). You also notice that the majority of overlap among congeners occurs in two clades, *Petrophasa* and *Poephila*. This observation provokes additional questions, including, Is there something unique about the Kimberley Plateau that permits substantial overlap among congeners (perhaps it is unusually resource rich or perhaps the current level of overlap is an artifact of human-induced decreases in suitable habitat that has forced congeners into contact)? Is there something unusual about *Petrophasa* or *Poephila* (or both) that permits congeners to interact in the same community?

This study highlights the impact of geology (vicariance) and dispersal/colonization on producing the historically unique assemblages of species that are the building blocks of community structure. In the next example,

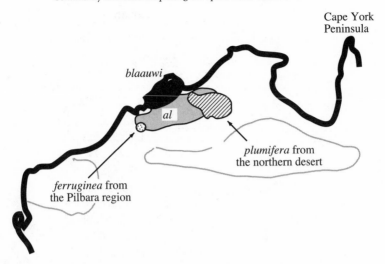

Figure 7.9. Distributions of four *Petrophasa* species in the Kimberley Plateau area. *Petrophasa blaauwi* and *P. albipennis* are residents of the community. *Petrophasa ferruginea* and *P. plumifera* evolved in the Pilbara region and the northern desert, respectively, and dispersed into the Kimberly Plateau.

we will consider another part of the picture—how coexisting related species actually partition resources.

HOW SPECIES INTERACTIONS COME TO BE WITHIN COMMUNITIES

Caribbean Anolis: How Did Each Island Get Its Guild of Ecomorphs? As you will recall from our discussion of evolutionary radiations in chapter 6, there has been a substantial amount of research on the structures of Caribbean anole communities. You will also recall that there is still a formidable amount of missing data, so any conclusions about community evolution are tentative at best. Nevertheless, we can detect some fascinating patterns by examining the three islands for which the data are the richest: Jamaica, Puerto Rico, and Hispaniola. The community on Jamaica has undergone widespread diversification *in situ*. Although we need more detailed information about the series of sister groups of the Jamaican clade, the phylogenetic tree hints that the island was colonized by a population of anoles from the mainland (fig. 7.10). Wherever that ancestral population came from, once there, it underwent substantial resource partitioning coupled with speciation, producing the ecomorph-rich assemblage we see today.

Puerto Rico is somewhat more complicated because the *Anolis* species there do not form a monophyletic group. In other words, the community on this island is a mosaic of different colonization events coupled with speciation *in situ*. Although incomplete, the phylogenetic tree

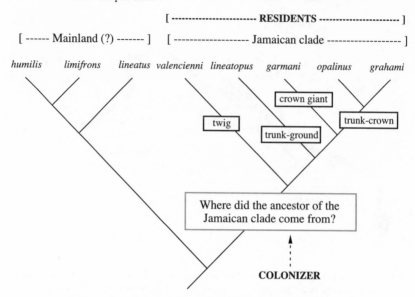

[------------------------ **RESIDENTS** ---------------------]

[------ Mainland (?) -------]   [------------------ Jamaican clade -------------------]

humilis   *limifrons*   *lineatus* *valencienni* *lineatopus*   *garmani*   *opalinus*   *grahami*

crown giant

twig

trunk-crown

trunk-ground

Where did the ancestor of the
Jamaican clade come from?

COLONIZER

Figure 7.10.   Anoles on Jamaica. The phylogeny shows the diversification of
ecomorphs following colonization by the ancestor of the group.

shown in figure 7.11 indicates that Puerto Rico has been colonized by at
least three separate lineages of *Anolis*. According to this data set, *A. occultus* was the first anole to arrive. We cannot, of course, tell where its
ancestor was living (nor what it was doing there) until we uncover the
sister group to these *Anolis* species; we only know that a population from
that ancestral species arrived on Puerto Rico and speciated (allopatric
mode II). At this stage, then, the island may have been populated by
slender, short-legged, short-tailed lizards living on twigs. Puerto Rico was
colonized at least two more times, but we cannot tell the sequence of
colonization from the phylogenetic tree because the relevant lineages
are part of a massive polytomy in the middle of that tree. Minimally it
appears that the ancestor of *A. cuvieri* arrived on the island as a crown
giant and speciated there (allopatric mode II). Apparently this colonizer
was not affected by an interaction with *A. occultus,* because it retains the
plesiomorphic ecomorph type for its clade (assuming that *A. cristophei* is
also a crown giant). Because we cannot hypothesize about the ecomorph
of the first colonizer, we don't know whether the second colonization
event was responsible for driving *A. occultus* out onto the twigs or
whether *A. occultus* was already there, in which case there was no interaction between the two species. Finally, a third colonization event by the
ancestor of *A. krugi* + *A. cristatellus* + *A. acutus* + *A. stratulus* (ancestor 1
on fig. 7.12; allopatric mode II) was followed by a rapid divergence of

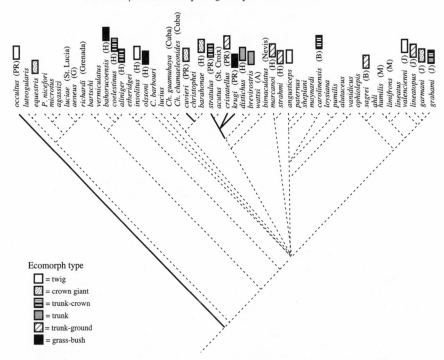

Figure 7.11. Anoles on Puerto Rico. Partial phylogenetic tree for *Anolis* with the species living on Puerto Rico highlighted in bold (for details, see fig. 6.11). Five of the nine members of the Puerto Rican community are included on this tree.

ecomorph types accompanied by speciation *in situ*. This part of the scenario most strongly resembles the picture uncovered for Jamaica. If you look back to our original discussion in chapter 6, you will discover a phylogenetic pattern for the Puerto Rican anoles that contains more species in the "ancestor 1 + descendants" clade (see fig. 6.10) than are depicted in figure 7.11. If that representation is corroborated in future phylogenetic analyses, then we also have evidence for some speciation *in situ* on Puerto Rico without habitat partitioning (fig. 7.12). Overall, it appears that Puerto Rico has experienced at least three separate colonization events, only one of which (the "ancestor 1" group) produced a lineage that continued to speciate on the island. If the twig and crown giant species were already in place when this ancestor arrived, then the final sequence of ecomorph evolution may have followed a trunk-ground to grass-bush and trunk-crown pathway (just to name one route). These residents occupy different niches than the twig and crown giant colonizers, producing an endpoint similar to that shown in Jamaica (fig. 7.10), but resulting from a different sequence of events.

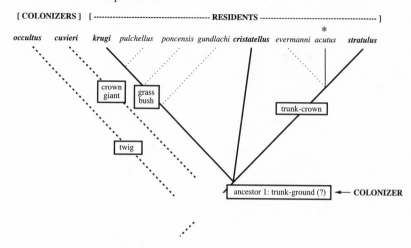

Figure 7.12. Heuristic diagram showing the three separate colonization events of Puerto Rico. This branching diagram does not represent a monophyletic assemblage. The species from the large phylogenetic analysis are highlighted in bold (for details, see fig. 6.11). *Bold, dotted lines* = separate lineages colonizing the island. *Dotted lines* = the remaining four anole species on the island provisionally added according to relationships from an earlier analysis (see Losos 1992). In order to corroborate this heuristic diagram, these four species need to be added to the larger phylogenetic analysis depicted in figure 7.11. *Asterisk* = member of the clade found on St. Croix, not Puerto Rico.

Hispaniola Island presents us with the most complicated pattern of the three communities (fig. 7.13), involving as many as seven and as few as five colonization events. It is impossible to detect the sequence of community structuring with the current database. Some colonizers brought their plesiomorphic ecology with them (*A. distichus* and *A. barahonae*), while others continued to speciate on the island without changing ecomorph type (*A. marcanoi* and *A. strahmi; A. coelestinus* and *A. aliniger*). Unlike the previous two islands, there is no lineage on Hispaniola that shows marked habitat diversification *in situ,* but that observation may be an artifact of the absence of data. There are more than 40 anole species on this island, only 9 of which are represented on this tree. The composite phylogeny of Irschick et al. (1997) added 5 more species to the picture: *A. chlorocyanus* as part of the *A. coelestinus* + *A. aliniger* clade (also a trunk-crown specialist), *A. cybotes* and *A. longitibialis* as part of the *A. marcanoi* + *A. strahmi* clade (also trunk-ground ecomorph), *A. semilineatus* as the sister to *A. olsonni* (also a grass-bush ecomorph), and *A. singularis* (trunk-crown) on its own in a big polytomy with numerous other species (see fig. 6.12b for a pared-down pattern depicting all of these species). Adding these five species strengthens the perception that there has been relatively little habitat diversification linked with *in situ*

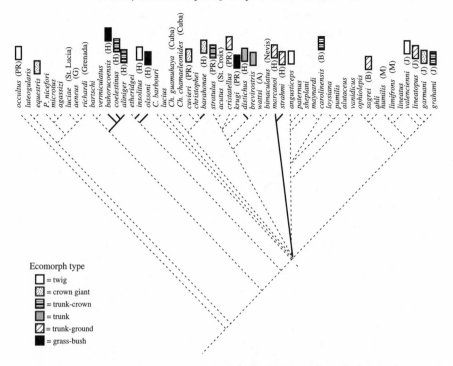

Figure 7.13. Anoles on Hispaniola Island. Partial phylogenetic tree for *Anolis* with the species living on Hispaniola highlighted in bold (for details, see fig. 6.11). Only nine of the more than 40 members of the Hispaniolan community are included on this tree.

speciation on Hispaniola Island. In order to test that perception further, we now need to collect data from the remaining 26 (or more) species.

Overall, then, the research to date has uncovered both convergence of ecomorph types among islands and what appears to be rapid diversification of those types within a lineage (but not all lineages) on two of the three islands. Although the endpoints of colonization and diversification among the anole communities often look similar, the current data set indicates that they have been reached in different ways. Each community is a historically unique mosaic. In other words, there do not appear to be any general rules about the *sequence* in which ecomorph types partition the habitat, which is precisely what we would expect to see in a Darwinian world. Expectations that we would uncover exactly the same sequence of diversification arising convergently on each island are grounded in traditional orthogenesis, with its emphasis on internally driven, predetermined, and historically repeated patterns.

Studies have also demonstrated that, on two islands at least, there has been a moderate amount of *in situ* speciation by residents, producing sister species that appear to be doing the same thing in the community (Hispaniola in particular). This begs the question of whether those islands are large enough to have more than one *Anolis* community; intra-island vicariance events may have produced isolated communities with sister species that retain the plesiomorphic ecology. If the sister species are indeed part of the same community, then we need to examine these fascinating examples of co-occurring congeners that ostensibly should be competing for the same resources (if, as suggested, ecomorph type is a direct reflection of habitat partitioning). Have they partitioned those resources even more finely within the community, or do they simply coexist because the resource base permits such coexistence? For example, researchers have discovered large dietary overlaps among closely related cichlids in Lake Malawi (Genner, Turner, Barker, and Hawkins 1999; Genner, Turner, and Hawkins 1999), Lake Tanganyika (Hori 1983; Sturm-bauer et al. 1992), and Lake Victoria (Seehausen 1996; Bouton et al. 1997). In these lakes, the initial stages of speciation by some residents are not accompanied by ecological segregation, so the community begins to build a larger and larger base of coexisting species "doing the same thing." This base of plesiomorphic ecology is coupled with wide-spread ecological divergence in other lineages to build the complex cichlid communities found in the Rift Lakes.

Most researchers have focused their attention on the role for competition-driven resource partitioning in structuring these anole communities. How can we explain the observation that those communities appear to have been assembled in different ways? The most obvious answer is that stochastic factors (what the colonizers were, when and in what sequence they arrived) have influenced the arena in which that competition is played out. The more subtle answer is that there are other forces influencing the structuring of these communities, and these forces vary from island to island. So, for example, community ecologists are renewing a long-established interest in the effects of predation on the evolution of community structure.[9] In order to examine this idea further, we will need to begin adding comparative phylogenetic analyses of predator communities to our studies. Fortunately, one such study already exists for five xenodontine snakes inhabiting Hispaniola Island (Henderson et al. 1988).

The study by Henderson et al. indicates that there have been at least three separate colonizations of Hispaniola (fig. 7.14): by *Alsophis cantherigerus* (or its ancestor); by *Hypsirhynchus ferox* (or its ancestor); and by

---

9. For example, Hairston et al. 1960; Paine 1966; MacArthur and Wilson 1967; Parrish and Saila 1970; Wong et al. 1998; and Abrams 2000a,b.

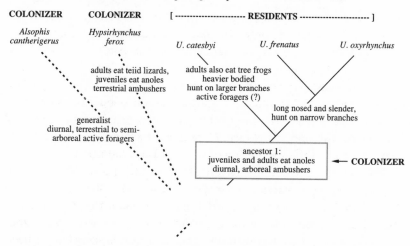

Figure 7.14. Heuristic diagram showing the three separate colonization events by some xenodontine snakes on Hispaniola Island. *Bold, dotted lines* = separate lineages colonizing the island. This branching diagram does not represent a monophyletic assemblage (for the larger tree of West Indian xenodontines, see Vidal et al. 2000). *Uromacer* speciated on the island.

the ancestor of the *Uromacer* clade (vine snakes), which continued to diversify on the island. All of these snakes eat anoles at some time in their life; some, like *U. frenatus* and *U. oxyrhynchus,* feed almost exclusively on these lizards. The story (from the anole's perspective) gets worse. Various other snakes on Hispaniola also find anoles delectable, including *Tropidophis* (dwarf boas) *haetianus* (nocturnal, terrestrial-arboreal), *Ialtris dorsalis* (terrestrial, generalist), *Epicrates* (Greater Antillean boas) *striatus* (nocturnal, terrestrial-arboreal, takes *Anolis* when young), *E. gracilis* (nocturnal, arboreal, feeds on *Anolis* almost exclusively), *E. fordi* (nocturnal, terrestrial-arboreal, feeds mainly on *Anolis*), and *Antillophis parvifrons* (leaf litter, diurnal) (data from Henderson 1982; Schwartz and Henderson 1991). Clearly there are few (if any) places on this island where anoles can find a refuge from anole-eating snakes. Was the evolution of anole community structure influenced by the presence of these snakes? The most recent phylogenetic analysis of the xenodontines places *Uromacer* at the base of a monophyletic West Indian clade (Vidal et al. 2000). *Alsophis* (paraphyletic), *Hypsirhynchus, Ialtris,* and *Antillophis* speciate later in the group. The colonization of the West Indies from South America is postulated to have occurred in the mid-Cenozoic (Cadle 1985; Hedges 1996a,b), so anole-eating snakes have been in the area for a substantial period. There have also been at least five colonizations of the West Indies by other xenodontines, so the patterns of interaction between the anoles and snakes will be complicated. When did these organisms colonize Hispaniola and in what sequence? How are the snakes partitioning their anole prey? The patterns shown for the snakes in figure 7.14 hint that the snakes might specialize on particular ecomorphs, or clusters of ecomorphs. For example, *Uromacer catesbyi* forages on larger branches, whereas its relatives, *U. frenatus* and *U. oxyrhynchus,* show the apomorphic ability to hunt anoles on slender branches. Is it possible that a coevolutionary interaction between predator and prey (diffuse coevolution, arms race, or resource tracking: see chapter 8) is playing a role in the structuring of the lizard communities?

Resolution of the entire *Anolis* tree (no mean feat) and additional ecological data for many more *Anolis* species (and their potential predators) are required before we can say anything more about the structuring of these lizard communities. What we really need is a detailed biogeographical analysis for these lizards (and snakes). For example, we have been assuming that unique lineages on an island represent instances of colonization followed by peripheral isolates speciation. Could some of these events actually represent vicariance coupled with extinction? (Think back to the Australian and North American bird examples in chapter 4.) Many of the smaller islands were once part of a larger mainland (see discussion in Losos 1994), so it is possible that marine ingression and subsidence may have fragmented an older and more continuous biota, leaving extinctions in its wake. On the other hand, if the Caribbean communities have been assembled via a substantial amount of colonization, we would expect to find that the phylogenetic tree for the clade contained numerous polytomies, reflecting those colonization events (which is precisely what fig. 7.11 shows). The important thing to determine is whether those polytomies are telling us something "real" about speciation and community structure, or whether they are artifacts of incomplete data. Besides collecting more data for the *Anolis* clade, it would be useful to combine information from other groups of organisms on these islands into a larger historical biogeographical analysis. Obviously the evolution of *Anolis* communities has been a dynamic and complicated process, one that has, not surprisingly, captivated more than one researcher for his or her entire career, and will continue to captivate the next generation of researchers (and their descendants).

HOW SPECIES COME TO INTERACT AS THEY DO IN COMMUNITIES

Demystifying Mimicry. Where we live in southern Ontario, monarch butterflies are harbingers of autumn. These beautiful, large butterflies, with their wings of orange and black stripes and borders of white dots, mass along the northern shores of the Great Lakes, feeding on late summer flowers. We have seen such a massing, thousands of butterflies congregating on a single tree, turning the bark into a kaleidoscope of orange, black, and white. Once gorged with nutrients, the monarchs begin their annual migration to forests north of Mexico City, where they spend the winter. Interestingly, these butterflies mass and migrate at precisely the same time and in the same places as many insectivorous birds. How are they able to survive in the presence of so many potential predators? The short answer is their mothers preferentially placed their eggs on milkweeds (Thompson 1988a), and when the caterpillars fed on those plants, they obtained both nutrients and high concentrations of cardiac glycosides. Avian predators learn very quickly that a monarch will make them gag and vomit; usually only a few experiences are enough to convince a bird never to try monarch again. The autapomorphic trait "concentrate cardiac glycosides in your tissues without harm to yourself" is clearly adaptive for the monarch, and being able to associate the color pattern of the butterfly with distastefulness is clearly adaptive for the birds.

The story does not end here, though, because there is a third player waiting in the wings: the viceroy butterfly, which looks reasonably similar to the monarch but is not capable of concentrating cardiac glycosides in its tissues. The rest of the evolutionary play is determined by two avian qualities. First, birds are not particularly good taxonomists. Distinguishing the viceroy from the monarch is one of the first practical lessons in taxonomy that many North American children learn. It appears that to a bird, however, all "orange-and-black-striped + white-spotted" butterflies look pretty much the same. The second avian attribute of importance is that they seem to live by the philosophy "Better safe than sorry." This means that once a bird has had a few experiences with a monarch, any butterfly in the area that looks remotely like a monarch is safe. Thus, wherever the monarch goes (and being safe from avian predators allows it to go a long way indeed), perfectly palatable butterflies with similar wing color patterns, such as the viceroy, benefit. Those benefits, direct and indirect, lead to strong stabilizing selection on the critical elements of the color pattern in the monarch and all similar species with which it comes into regular contact. The strength of this stabilizing selection maintains this **mimicry system.**

Mimicry was first described in detail by H. W. Bates. Based upon ten years of observations on butterflies and their predators in the Amazon, Bates (1862) concluded that *Dismorpha* butterflies, which are delicious, look and behave like ithomiine and *Heliconius* butterflies, which are toxic. Almost a generation later, F. Müller (1879) invoked another form of mimicry to explain why so many distasteful butterflies have similar color patterns. In the first system (**Batesian mimicry**), there is an interaction between a toxic model and a palatable mimic, while in the second system (**Müllerian mimicry**) both the model and mimics are toxic. Although the details of the systems are different, convergent evolution is the process most often invoked to explain the existence of similar behaviors and color patterns between the model and the mimics. Convergence is favored under both systems because any signal that minimizes the number of mistakes (eating a noxious prey item) made by predators will be advantageous to both predator and prey. All similarly appearing and acting prey species, palatable or unpalatable, dangerous or not, will thus benefit from living together with the species (predator and prey) that were the original focus of evolution. Additional conditions are necessary for the evolution of Batesian mimicry because the mimic is practicing a form of deception. In order for that deception to be successful, the density of the mimic relative to the model must be low enough so as not to interfere with the predator's avoidance learning process (for an extensive discussion, see Turner 1977; Mallet and Joron 1999).

Thus far we have concentrated on how mimics benefit from coexisting with models in the evasion of common enemy–predators. There is still another dimension to this problem, involving the likelihood that mimics and models will compete amongst themselves for access to resources. In a Müllerian system, there is no expectation of differential abundance of model or mimic(s) based on dynamics of the mimicry system itself, because the model and all the mimics are reliably advertising their distastefulness. If the model and mimic(s) are close relatives, they may well compete with each other, resulting in reduced abundance of one, both, or all members of the system relative to what each could achieve in allopatry. The question in this system then becomes, How do the costs of competition (if any) compare with the benefits of decreased predation? The situation is more complex for Batesian systems, in which we expect to find differential abundances of models and mimics. In these cases, the antipredator benefits of being noxious and obvious may give the model a competitive advantage over co-occurring relatives ("mimics") until the mimic becomes rare, at which point the probability of a predator finding a mimic before experiencing the joys of, say, cardiac glycosides is low. We might thus expect to find a point at which the antipredator benefit of having a model around is balanced by the cost of interspecific competition for resources. Gilbert (1983, 1991) proposed that this complicated interaction might be one way many rare species are maintained in species-rich ecosystems such as the tropics.

Let's now take this complicated biological interaction and place it within a community and a phylogenetic context. Mimicry refers to a functional state, and therefore does not on its own tell us anything about the origin of that state because evolution can produce mimicry systems through a number of different pathways. For example, the two traits involved in the monarch-viceroy Batesian mimicry system could represent an autapomorphy in the model (concentrating cardiac glyco-sides), coupled with the plesiomorphic color pattern in the model and the mimic (fig. 7.15a). Any species with wing color patterns similar to the monarch that colonizes a community inhabited by monarchs (or vice versa) will benefit immediately from the establishment of a mim-icry-like dynamic. Is this true mimicry? If the definition of "true" mim-icry requires the convergent modification of color patterns in the model and the mimic, then the answer to this question is no. Is this plesiomor-phy an important part of butterfly community structure? Clearly the answer to this question is yes, because the colonizer can establish itself with limited evolutionary change. Convergence aside, the underlying benefits to the "model" and the "mimic" stay the same, so the system can be maintained readily after the relatively small expenditure of en-ergy required to establish it in the first place. Interestingly, although there is no selection for a modification of color pattern in this situation, presumably there is some selection for a change in life history parame-ters, because the "mimics" must maintain a lower population density than the "model." Researchers interested in studying the effects of selec-tion on population demographics might consider this interaction as

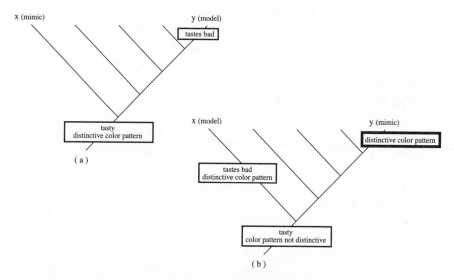

Figure 7.15. Heuristic example showing two different pathways producing a Batesian mimicry system. (*a*) The color pattern is plesiomorphic for both model and mimic and the ability to concentrate noxious chemicals in the tissues is derived for the model. (*b*) The color pattern has arisen independently in the model and the mimic and noxious taste is derived for the model.

their model system. It might also be fruitful to study the population densities of the closest relatives to the mimics, in order to determine whether the colonizer has indeed changed. Perhaps all species with the plesiomorphic color pattern have relatively low densities, and it is the model, with its new noxious taste, that has experienced a derived increase in population size.

Alternatively, both the color pattern and the ability to accumulate cardiac glycosides may be autapomorphies for the monarch. Perhaps the fitness advantage that comes with making predators sick at the sight of you allowed the monarch to expand geographically. As it colonized new communities, the monarch would come into contact with some butterfly species that possessed genetic variability in color pattern characters. Individuals in these species that most closely resembled the toxic monarchs would presumably survive longer, leave more offspring, or both, which over the course of many generations would lead to the convergence of the resident species on the colonizer's color pattern (fig. 7.15b). This represents "true" Batesian mimicry. From a community perspective, the resident has changed as a result of the interaction with the colonizer. Could such a dynamic establish itself if the future mimic were the colonizer in a community inhabited by monarchs? Presumably under certain conditions this could occur, but it would require that the colonizing species bring enough variability in color pattern to enable the model-mimic dynamic to become established in the first place. Similar arguments apply to the Müllerian system, in which two species may owe their similarity to (1) common ancestry (fig. 7.16a), (2) one of the traits being plesiomorphic in their common ancestor, the other trait arising convergently in model and mimics (fig. 7.16b), or (3) both traits being autapomorphies in the model and evolving together convergently in the mimics (fig. 7.16c).

Figures 7.15 and 7.16 suggest two general categories of mechanisms by which the traits underlying mimicry systems can evolve. The first of these involves relatively conservative traits and a minimum of convergent evolution (e.g., figs. 7.15a and 7.16a,b: see Vane-Wright and Smith 1991; Zrzavy 1994; Zrzavy and Nedved 1999). The second category involves substantial amounts of homoplasy (e.g., figs. 7.15b and 7.16c), which may be produced by developmental plasticity (Nijhout 1991, 1994), the recurring origin of similar traits from a common ancestral genetic architecture (e.g., Jiggins and McMillen 1997), or persistent ancestral polymorphisms that may be co-opted for various functions in different ecological contexts (e.g., Brower 1996; Joron and Mallet 1998; Zrzavy and Nedved 1999). Because, as we have repeatedly noted throughout this book, nature is complex, it is possible that both categories of pro-

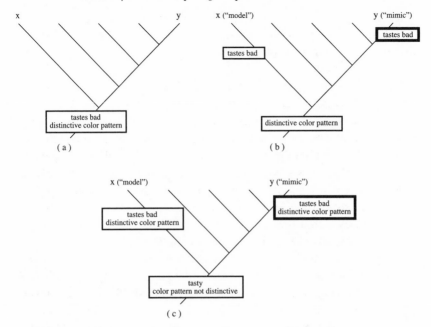

Figure 7.16. Heuristic example showing different pathways producing a Müllerian mimicry system. (*a*) The color pattern and noxious taste are both plesiomorphic. In this case there is no model or mimic in the system. (*b*) One of the traits is plesiomorphic, the other is independently derived (color pattern plesiomorphic and noxious taste derived, as shown, or vice versa). (*c*) Both traits arise autapomorphically (and convergently) in both species. There is no way to tell which species changed first ("model") and which followed ("mimic") based solely on the phylogenetic patterns.

cesses may operate in the same system. For example, Nijhout (1991, 1994) demonstrated that divergent classes of butterfly color patterns can be generated using only a few developmental rules. Making variations on a common theme (i.e., one of the rule sets) may actually be relatively cheap. Nijhout also showed, however, than many of those rule sets are plesiomorphic through entire groups. It thus may be relatively expensive to change from one rule set to another (limiting true convergent mimicry). From a phylogenetic perspective, this means that we would expect to find more mimicry systems involving close relatives than distantly related groups.

So, let us move from the theoretical world to the real world of poisonous bugs.

Cotton-Stainer Bugs of the Neotropics. The vast majority of studies of mimicry have focused on lepidopterans. Recently, however, Zrzavy and coworkers have begun to document the potentially rich source of mimicry systems (they use the term "mimicry rings") offered by bugs in the genus *Dysdercus*. Also known as cotton-stainer bugs because of their bright colors, these heteropteran hemipterans live in the Neo- and Afrotropics. The Neotropics contains a clade [the subgenus *Dysdercus* (*Dysdercus*)] comprising (at the moment) 43 nominal species (Zrzavy 1997; Zrzavy and Nedved 1997). All members of the genus are inferred to be distasteful, and their coloration aposematic, based on five observations: (1) they feed on plants known to be toxic to vertebrates (Ahmad and Schaefer 1987), and related hemipterans that are also brightly colored (Lygaeidae and Rhopalidae) are known to store toxic compounds derived from their food plants (Scudder et al. 1986); (2) they synthesize and secrete defensive chemicals, both as larvae and as adults (Farine et al. 1992); (3) they are gregarious, a trait that may enhance the effects of aposematic coloration;[10] (4) visually oriented vertebrate predators eat them rarely, if at all (van Doesburg 1968); and (5) the bright coloration of *Lygaeus equestris* (a member of the closely related hemipteran family Lygaeidae) deters avian predators (Sillén-Tullberg et al. 1982; Sillén-Tullberg 1985). Taken together, these pieces of information support the hypothesis that the coloration of *Dysdercus* species reflects a Müllerian mimicry dynamic (Van Doesburg 1968).

Zrzavy and Nedved (1997, 1999) followed a three-step protocol, now familiar to readers of this book. First, they produced a total evidence phylogenetic analysis for the 43 members of *Dysdercus* (*Dysdercus*). Second, they optimized the various components of the color patterns onto their tree to determine the sequence of evolution for the presumptive aposematic coloration. Third, they performed a historical biogeographical analysis (for their single clade) to provide an assessment of each species' geographic origin. Armed with this information, they began to explain the existence of similar color patterns among co-occurring species. The authors took a conservative position in their study: any co-occurring species having the same or similar *plesiomorphic* coloration patterns were not considered to be true mimics. They noted, as we did above, that such co-occurring species might well be engaged in a mimicry dynamic, but in the absence of convergent evolution by one or more co-occurring species with respect to their color patterns, we cannot say there had been any special evolution *for* the *origin* of mimicry. In other words, in some cases species may be functioning as mimics but may not have evolved to be mimics. Zrzavy and Nedved (from Zrzavy 1994) postulated that similarities among species could have been driven by a Müllerian mimicry dy-

---

10. See Sillén-Tullberg 1988; Gamberale and Tullberg 1996a,b; and Tullberg and Hunter 1996 for a discussion and examples using lepidopterans.

namic only if (1) no more than one member of the association exhibited the plesiomorphic coloration; (2) the associated species did not form a monophyletic group; and (3) the associated species were sympatric within the geographical range of potential predators.

Six assemblages satisfied these criteria. Two of those they named "model-mimic rings," assemblages in which one member (the model) possesses a persistent plesiomorphic color pattern, while the mimics exhibit convergently evolved similar patterns. These two assemblages reflect the initial phylogenetic divergence of the clade (fig. 7.17), separating a primarily South American group from a primarily Mesoamerican and Caribbean group. The basal members of each of these clades exhibit distinctive and different color patterns, "yellow" species in South America (the southern clade) and "yellow-black" species in Mesoamerica, the Caribbean, and northwest South America (the northern clade). Five species from the northern clade (plesiomorphically yellow-black) mimic the southern clade's plesiomorphic (yellow) color pattern: *D. rusticus, D. collaris, D. imitator, D. mimus, and D. luteus.*

Dysdercus rusticus (from Central American stock) has colonized northwestern South America, where it is sympatric with six yellow species from clade 1. The colonization and subsequent speciation of *D. rusticus* was associated with an apomorphic shift from a yellow-black to a yellow color pattern (convergent evolution). In a Batesian system, we predicted that the evolution of mimics by colonization would be extremely rare. In a Müllerian system, however, this type of change is easier to accomplish because the invading species really does taste bad. Because the ancestor of *D. rusticus* was yellow-black, it already displayed part of the signal warning predators in the community to stay away. This partial signal, if effective, may have allowed the population to persist long enough to speciate and converge on the more effective warning signal in the new community. In order to unravel the details of this evolution, we need to test the effectiveness of the yellow-black versus yellow signals in that community. Does yellow-black offer any protection from predators at all? We also need to collect data about how the colonizer and the residents are interacting with one another. Do they compete for resources, or did the colonizer bring a different ecology with it? Is there a difference in population density between the residents and the colonizer?

Populations of *D. collaris, D. imitator,* and *D. mimus* have also dispersed into communities housing members of clade 1 in northwest South America and the Andean region. Once again, the color signal of the colonizers has converged on yellow in all cases. That these three colonizers are part of a four-species monophyletic group is interesting. Is there something different about the biology of members of this clade

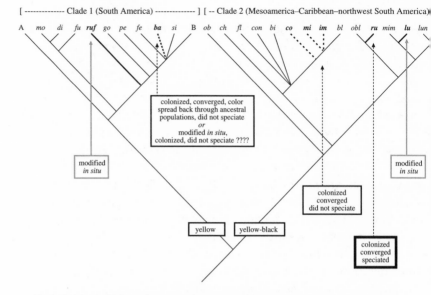

Figure 7.17. Searching for true mimics (model-mimic systems) in Neotropical cotton-stainer bugs *Dysdercus* (*Dysdercus*). The phylogenetic tree is the strict consensus of 36 equally parsimonious trees (Zrzavy and Nedved 1999). Character changes in one clade that appear identical to the plesiomorphic pattern in the other clade are indicated with bold lines and bold letters. *Dotted bold lines* = dispersal without speciation. *ba* = *basialbus*; *bi* = *bimaculatus*; *bl* = *bloetei*; *ch* = *chiriquinus*; *co* = *collaris*; *con* = *concinnus*; *di* = *discolor*; *fe* = *fernaldi*; *fl* = *flavolimbatus*; *fu* = *fulvoniger*; *go* = *goyanus*; *im* = *imitator*; *lu* = *luteus*; *lun* = *lunulatus*; *mi* = *mimus*; *mim* = *mimulus*; *mo* = *modestus*; *ob* = *obscuratus*; *obl* = *obliquus*; *pe* = *peruvianus*; *ru* = *rusticus*; *ruf* = *ruficollis*; *si* = *silaceus*; A = *chaquensis* + *maurus* + *wilhelminae* + *cordillerensis* + *honestus* + *longirostris*; B = *immarginatus* + *urbahni*; C = *sanguinarius* + *neglectus* + *andreae* + *mimuloides* + *suturellus* + *capitatus* + *ruficeps* + *albomaculatus* + *albofasciatus* + *jamaicensis* + *ocreatus* + *fervidus*.

that predisposes them to dispersal and successful colonization? Why doesn't the fourth member of the group, *D. bloetei*, show the same patterns of dispersal/color change?

The fifth species to show an apomorphic change to the yellow color pattern is the aptly named *D. luteus*. This species is a resident of western Mexico (not a colonizer into clade 1's area) and is allopatric to other cotton-stainer bugs. In order to identify this species as a mimic, we would have to postulate that a yellow species from clade 1 colonized western Mexico and that the *resident* species changed to mimic the colonizer, which subsequently went extinct. This seems unnecessarily ad hoc. It is more likely that yellow serves a different (nonmimetic) function in this community. Perhaps *D. luteus* is interacting with predators that are extremely sensitive to yellow, in which case it is the dynamic between the predator and the model that is driving the evolutionary modification of the warning signal.

There is only one possible case of mimicry in a south to north direction: *D. basialbus* overlaps yellow-black species from clade 2 in Central America. This bug, however, is also found in northwestern South America, where it presumably evolved, so it appears to have dispersed into Central America without speciating. Given that the yellow-black color pattern is mimetic in the northern extent of its range, how do we explain the presence of this color pattern in the South American communities? Was the color modified in the areas of overlap, and did it then spread back into the central populations, or did the color pattern originate in South America (for reasons other than mimicry) and thus allow members of *D. basialbus* to colonize Central American communities more easily (because they gained an immediate mimicry-like advantage)? The other member of the southern clade with a northern pattern is *D. ruficollis*, which resides in eastern South America and is not sympatric with any of its northern brethren. Once again, we must look elsewhere for an explanation of the change in the warning signal from yellow to yellow-black in this species. Perhaps the factors responsible for that modification will give us an insight into the convergent appearance of yellow-black in *D. basialbus*. If both yellow-black patterns originated *in situ*, then we have an interesting example of convergence on that pattern within clade 1 that is not driven by a model-mimic dynamic.

The remaining four mimicry systems are what Zrzavy and Nedved refer to as "co-mimics." They use this term when it is difficult to determine which of the sympatric, similar-appearing species is the model and which is the mimic. In order for co-mimicry to be a real system, and not just an artifact of missing data, all co-mimic species must have converged upon the same color pattern when they came together. For example, let's examine the evolution of the warning pattern "red with transverse black bars or spots." Six species and a population of a polymorphic species display this color pattern (fig. 7.18). All of these bugs overlap to some degree in the Antilles (of anole and snake fame). If the assemblage is indeed acting as a co-mimic "ring," then we have to assume that either (1) the population of *D. discolor*, ancestor 1, the ancestor of *D. andreae*, and ancestor 2 all colonized the islands at approximately the same time, then converged in sympatry, or (2) these populations arrived at different times but did not begin to converge on the similar red-black color pattern until some threshold for number of individuals was reached that favored the warning color change. It is also

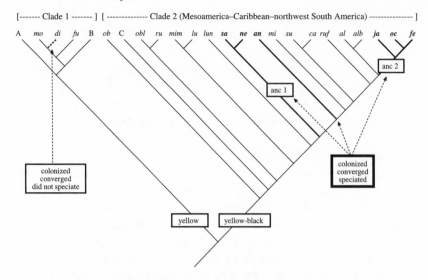

Figure 7.18. Searching for co-mimic systems in Neotropical cotton-stainer bugs. The phylogenetic tree is the strict consensus of 36 equally parsimonious trees (Zrzavy and Nedved 1999). Character changes in one clade that appear identical to changes in another clade are indicated with bold lines and bold letters. *Dotted bold lines* = dispersal without speciation. *al* = *albomaculatus; alb* = *albofasciatus; an* = *andreae; ca* = *capitatus; di* = *discolor; fe* = *fervidus; fu* = *fulvoniger; ja* = *jamaicensis; lu* = *luteus; lun* = *lunulatus; mi* = *mimuloides; mim* = *mimulus; mo* = *modestus; ne* = *neglectus; ob* = *obscuratus; obl* = *obliquus; oc* = *ocreatus; ru* = *rusticus; ruf* = *ruficeps; sa* = *sanguinarius;* A = *chaquensis + maurus + wilhelminae + cordillerensis + honestus + longirostris;* B = *ruficollis + goyanus + peruvianus + fernaldi + basialbus + silaceus + immarginatus + urbahni;* C = *chiriquinus + bimaculatus + concinnus + flavolimbatus + collaris + mimus + imitator + bloetei.*

possible, of course, that one of the populations, say, ancestor 1, colonized the islands first, speciated, and changed color (novel predator-prey dynamic in the new community) and that all subsequent colonizers were mimics of that model. The patterns cannot help us differentiate among these scenarios. We need additional information: specifically, biogeographic data, estimates of the age of each group, and possible timing of colonization events. What does appear to have happened is that ancestors 1 and 2 colonized the Antilles, convergently evolved the red-black pattern (either when overlapping or at different times) and continued to speciate in the community. This produced an assemblage of five resident species, all with similar color patterns inherited from their common ancestors. (Remember, the origin of the color patterns was convergent in ancestors 1 and 2, but its presence in their descendants is due to homology.) In none of these species did red-black warning color evolve as a case of mimicry, although the mimicry-like dynamic is most certainly responsible for the continued maintenance of that pattern. This color pattern, then, shows a complicated interaction among possible co-mimic/co-mimic, mimic/model, and plesiomorphic components.

Zrzavy and Nedved (1999) noted that many of the species involved in these mimicry systems are themselves polymorphic. This adds another level of complexity to the story because populations from the same species may be participating in more than one mimicry system. The champion mimic is *D. obscuratus,* the basal species in the northern clade (see fig. 7.18). *Dysdercus obscuratus* lives in northern South America where the bulk of the sympatric *Dysdercus (Dysdercus)* species occur and is the most polymorphic of the known species. Not surprisingly, we find it playing a role as co-mimic or model in five of the six mimicry systems and also functioning in mimicry-like systems with congeners sharing the plesiomorphic coloration pattern for the northern clade. Given the relative antiquity of *D. obscuratus,* this would seem to be a good test case for theories concerning the role of mimicry in maintaining ancient polymorphisms. The widespread occurrence of polymorphic species within this group makes it an excellent system for studying the relationship between mimicry and the evolution of microspecies, with an eye toward explaining the production of some macrospecies via mimicry-induced completion of speciation (Turner 1983; McMillan et al. 1997).

Mimicry has been seen both as a prime example of the power of natural selection[11] and as a problem for Darwinian perspectives.[12] We think that the reason for these divergent views stems from the widespread belief that the origin of mimetic systems requires convergent evolution of a form so specific that it stretches some people's credulity with respect to the power of simple chance, appearing almost Lamarckian. We hope we have shown that there are many avenues available for the evolution of mimicry systems, not all of which require seemingly unlikely amounts of convergent evolution. In addition, we would like to second Zrzavy's (1994, 1997; Zrzavy and Nedved 1999) proposal that understanding mimicry is fundamentally a comparative phylogenetic question. It is true that we do not need a phylogeny to distinguish a Müllerian from a Batesian system. Beyond that, however, we cannot reconstruct the sequence of origin and correlation of traits, the mode and timing of speciation, and the sequence of occupation of communities by models and

11. For example, Wallace 1889b; Fisher 1930; Ford 1971; Turner 1971, 1977, 1984, 1987; Gilbert 1983; Sillén-Tullberg 1988; Owen et al. 1993; Brower 1996; Joron and Mallet 1998; and Mallet and Joron 1999.

12. For example, Guilford 1990; Zrzavy 1994; and Zrzavy and Nedved 1999.

mimics without reference to phylogenetic information. This said, the next challenge to students of mimicry involves the more bizarre examples of animals apparently mimicking plants (looking like twigs or leaves) or inanimate objects and plants apparently mimicking animals (or at least their genitalia). Research is currently underway (John Wenzel, pers. comm.) documenting the timing and sequence of character modifications involved in the ultimate production of such strange apparitions.

SPECIES COMPOSITION PLUS SPECIES INTERACTIONS: PUTTING IT ALL TOGETHER

Snakes Again: Size, Diet, and Community Structure in the Neotropics. Imagine you are walking (quietly) through a rainforest in South America. Being a rugged individualist, you have decamped from the tour group and its mission to identify birds and are instead in search of snakes. Your first glimpse of a bright emerald green tree boa resting quietly entwined atop a branch seals your fate: you will spend the rest of your life studying serpent communities in the South American rain forests. Several decades later, you have a fairly comprehensive list of those communities. For example, one rainforest (fig. 7.19) contains a variety of small, slight-bodied, long-tailed snakes that live in the trees, foraging on frogs and lizards or slugs at night; small, slight-bodied, short-tailed snakes that burrow in the earth eating earthworms (also at night); slightly larger snakes that also burrow in the earth, eating other snakes and amphisbaenians (limbless "lizards") during the day; moderate-sized aquatic snakes dining on amphibians and fish; several large-bodied terrestrial species that forage during the day and feed on lizards; a moderate-bodied constrictor that is also terrestrial and feeds on small mammals at night; and a moderate-sized snake that lives in the leaf litter eating frogs and a larger snake that lives in the leaf litter snacking on arthropods. You are a little overwhelmed by all the variety, but are generally ecstatic about all of the evidence for habitat and resource partitioning that you have uncovered. "Clearly," you write, "there has been substantial ecological diversification within the serpentine community." A few years later, you attend a symposium on historical biogeography, and you begin to wonder whether the phylogenetic history of your snakes could have played any role in the structuring of your communities. Fortunately, you discover the following study by Cadle and Greene (1993) that sets you off on a whole new voyage of discovery.

Cadle and Greene (1993) were interested in documenting the imprint of history in communities of colubrid snakes in Central and South America. The Colubridae is a large family, including such memorable species as the viviparous garter snakes (*Thamnophis*), the aggressive but

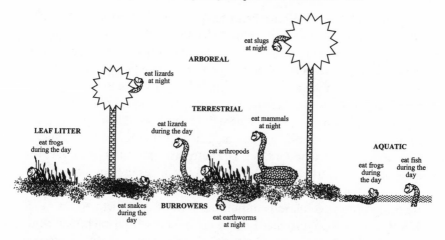

Figure 7.19. Distribution of colubrid species in a hypothetical South American rainforest.

fortunately nonvenomous water snakes (Natricines), hognosed snakes (*Heterodon*), which eat toads, rat snakes (guess what they eat), and the brightly colored milk and king snakes (no, that is *not* what they eat). The Neotropics boasts representatives from three major colubrid groups: the South American xenodontines, which evolved in South America, the Central America xenodontines, which diversified in Central America (Savage 1982; Cadle 1985), and the colubrines (a large, globally distributed group, which appear to have moved into the Neotropics following the divergence and radiation of the two major xenodontine lineages (Cadle 1984a,b,c). Xenodontine snakes began dispersing back and forth between Central and South America throughout the Tertiary. The exchange of fauna was asymmetrical; more snakes moved from Central to South America than vice versa. Based upon this biogeographic scenario, you can already catch glimpses of the rich and complex history of Neotropical colubrid communities. That history involves speciation *in situ* (producing residents), dispersal by some of those residents southward and northward, integration of some of those colonists into new communities and their subsequent evolution along with those communities, followed by the colonization of the Neotropics by a third colubrid lineage (and yet another wave of colonization, integration, and evolution). There are currently more than 600 colubrid species in the Neotropics. Aside from the biogeographic hypothesis presented above (which alerts you to the fact that communities have been assembled in complex ways), is it possible to detect additional influences of phylogeny on the structuring of communities with such a large data set?

Cadle and Greene approached this question by examining a variety of factors associated with foraging behavior (body size, habitat preference, prey preference, activity time, prey immobilization strategy) in light of the phylogenetic relationships among members of the three lineages. As you can see from figure 7.20, colubrines tend to be large (relative to other Neotropical colubrids) snakes that pursue frogs and lizards during the day on the ground. There has been a trend toward a decrease in body size, culminating in the small, slender (lighter for their length), Central American xenodontines. The patterns depicted in figure 7.20 are general ones. Clearly the specific details of changes in body size, habitat preference, and prey preference will be important to a complete understanding of how colubrid communities have been structured. In other words, we need phylogenies for specific groups within each major lineage and details of their ecology and biogeography in order to reconstruct the historical patterns of speciation, dispersal, and character evolution for all members of each community (see, e.g., Cadle 1984a,b,c) For the moment, however, let us see what we can discover on this level of analysis.

Return to your diagram of the South American snake community (fig. 7.19) and recall that you were amazed to find that so much resource partitioning had evolved within that community. Let us match that diagram with Cadle and Greene's phylogenetic perspective (fig. 7.20). When you pay closer attention to the *actual species* within your community, you discover the following (fig. 7.21):

1. The slender, arboreal, nocturnal lizard-eater is descended from a Central American colonizer. Those colonists brought their ancestral ecology with them when they entered the South American community. A few species of South American xenodontines are also arboreal lizard-eaters, but they are much less common. Have you documented an instance of interspecific competition between a resident and a colonizer, both of which have the same ecological requirements? (Note: some of those requirements are part of a plesiomorphic ecology for both the resident and the colonizer, not the result of convergent evolution.) In order to answer this question, you need to ask whether the South American species live in any of the communities inhabited by the Central American colonizers, and if so, whether the descendants of those colonizers are still out-competing them for access to lizards in trees (in which case the residents may be pushed into marginal habitats, have smaller population sizes, etc.). You can approach the question from another angle: Are the South American species more successful in communities that do not have any Central American colonizers? Finally, you need to document whether differences in activity times (the colonizer is plesiomor-

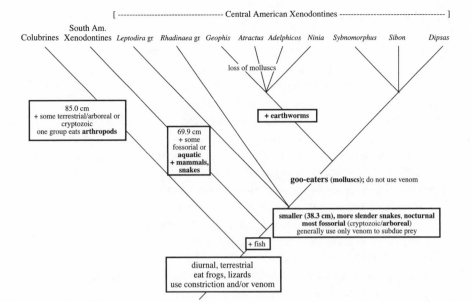

[---------------------------------- Central American Xenodontines ----------------------------------]

South Am.
Colubrines Xenodontines *Leptodira* gr *Rhadinaea* gr *Geophis* *Atractus* *Adelphicos* *Ninia* *Sybnomorphus* *Sibon* *Dipsas*

loss of molluscs

85.0 cm
+ some terrestrial/arboreal or
cryptozoic
one group eats **arthropods**

+ **earthworms**

69.9 cm
+ some
fossorial or
aquatic
+ **mammals,
snakes**

goo-eaters (molluscs); do not use venom

smaller (38.3 cm), more slender snakes, nocturnal
most fossorial (cryptozoic/**arboreal**)
generally use only venom to subdue prey

+ fish

diurnal, terrestrial
eat frogs, lizards
use constriction and/or venom

Figure 7.20. Upper-level phylogeny showing relationships among three colubrid lineages (Cadle 1984a,b,c, 1985; Vidal et al. 2000). General body sizes, diet/habitat preferences, and activity times are mapped onto each lineage. This exercise is a heuristic one, meant to identify the putative plesiomorphic states for each lineage and to show general differences among the three groups. The pathways of body size and diet evolution can only be reconstructed on a more detailed phylogenetic tree because shifts from the plesiomorphic background have happened numerous times within each lineage. Those details will eventually be necessary for a more finely tuned investigation of colubrid snake community evolution, but are not needed for this initial study. (Modified from Cadle and Greene 1993.)

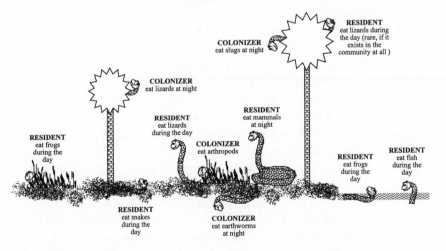

Figure 7.21. Adding history to your explanation for the evolution of colubrid snake community structure in the South American rainforest.

phically nocturnal, the resident, possibly diurnal) allow the resident and the colonizer to hunt different species of lizards, in which case there would never have been any competition for prey items between the two species. There might, however, be competition for resting sites in the trees (after all, there is more to life than food), yet another line of research that would be fascinating to pursue.

2. The arboreal, nocturnal slug-eater is also descended from a Central American colonizer. That colonizer brought its plesiomorphic ecology with it, but in this case there were no residents with which to (possibly) compete. The colonizers simply took advantage of an unused resource and settled happily in the community, snacking on slugs in trees.

3. The fossorial (ground-burrowing) earthworm-eater is also the descendant of Central American colonizers (which brought their plesiomorphic ecology with them yet again). In this case, there was already a resident burrowing throughout the substrate of the community. That resident, however, was a predator on other snakes (a prey preference that is unique to the South American xenodontines), so an interspecific competition dynamic was not established between the resident and colonizer (in terms of food, anyway). This still leaves open a number of intriguing questions, the most obvious of which is whether the colonizer discovered an unused resource base (bonus benefit), but then itself became a new source of food for the resident (unanticipated cost). Possibly the plesiomorphically different activity patterns kept the two apart and allowed the colonizer to become established in the community.

4. Both aquatic snakes are long-time residents of the community. More detailed phylogenetic and biogeographical analyses are required to pinpoint the origins of those associations and to determine the history of the resource partitioning (one eats frogs, the other, fish). For example, do all members of a particular genus eat the same thing, and does each community contain one species from several genera, or did the partitioning arise more recently due to direct interactions between the two aquatic species in this community (and other more complicated scenarios for the evolution of prey preferences)?

5. The cryptozoic (leaf litter) arthropod-eater is the descendant of a colubrine colonizer. Colubrines are the only lineage with members that display the unusual preference for arthropods, so once again the colonizer was able to add itself quietly into the community by taking advantage of an unused resource. Note that when we say "unused" it is only with reference to other colubrid snakes. Clearly other species also eat arthropods (insectivorous mammals, birds, and lizards spring to mind, not to mention other arthropods such as spiders), so it is possible, and

indeed probable, that the success of the colonizer was dependent upon interactions with some of those residents. This is "diffuse coevolution" (Futuyma and Slatkin 1983; chapter 8) at its most complex. The other cryptozoic species, the diurnal frog-eater, is a resident. Once again, although it does not compete with the colonizer for food, other areas of potential overlap between the two should be investigated.

6. And finally, both terrestrial community members are residents. As with the aquatic species, a more detailed analysis is needed in order to determine their origins in the association, and the history of resource partitioning (prey preference) evolution in those two lineages. Overall, then, adding phylogeny and biogeography to the study has given you a somewhat different perspective than your initial assumption that all of the habitat partitioning had occurred *in situ* in your community.

After repeating the above process for 15 communities, Cadle and Greene concluded that each community had its own unique history of assembly. They also noted the following points about resource partitioning in those communities:

1. Only colubrines eat arthropods (see fig. 7.20); no xenodontine considers an arthropod to be an appropriate prey item. Despite the fact that the Neotropics are crawling with arthropods, no member of the colubrid community will take advantage of that resource unless an arthropod-eating colubrine species colonizes the area. In a similar vein, the majority of Central American xenodontines do not prey significantly upon mammals. Once again, although there are many mammals inhabiting the same community with these snakes, they will generally be safe from colubrid predation unless that community is colonized by a mammal-eating South American xenodontine. In other words, as we discussed in chapter 6, "open niches" or "adaptive zones" do not act like ecological vacuums and suck lineages into them. The inability of a lineage to take advantage of such open resource bases is a fascinating area of research (returning us to our discussion of "constraints" in chapter 5). For example, consider the Central American xenodontines. As you can see from figure 7.20, they are much smaller and more slight of body than their South American relatives (or the more distantly related colubrines) and, with the exception of one genus, have lost the ability to subdue a struggling prey item via constriction (which requires the advantage of size and strength). The novel morphology gave the Central American xenodontines access to new habitats and resources, which they evidently took advantage of as they diversified into the leaf litter and up into the trees; however, that diversification had a price tag attached as doors were closed to previously accessible resources. Central American

xenodontines are simply too small to capture and subdue mammals (see, e.g., Greene and Burghardt 1978; Reynolds and Scott 1982; Greene 1983, 1997; Pough and Groves 1983).

2. A large portion of the resource partitioning seen in these communities is the result of plesiomorphic ecologies being retained by residents and their descendants or being imported by colonizers. So, for example, many colubrids eat frogs and lizards because this is the plesiomorphic prey preference among colubrids. If there are numerous species within one community feasting on frogs and lizards, this means that the resource base is large enough to support many species eating generally the same thing. Rather than postulating numerous instances of convergent evolution on "frog or lizard eating," it would be interesting to document the fine points of the interactions among the species with the plesiomorphic ecologies. The most obvious question is whether different genera of colubrids specialize on different clades of frogs and lizards, many of which are characterized by phylogenetically conservative, and distinct, ecologies of their own. It is possible that some foraging preferences and strategies have evolved convergently among members of various communities, but this hypothesis can only be tested when we have detailed phylogenies for all of the members of those communities.

When Cadle and Greene expanded their analysis to include 15 snake assemblages from southern Veracruz to southern Brazil and Argentina, they found that the communities along that more or less continuous gradient of tropical forest did not evolve due to either multiple episodes of vicariance splitting a single ancestral community or repeated episodes of convergent evolution based on resource availability, predation, and competition. Instead, regardless of species richness, each of those communities displayed a historically unique combination of species from the three colubrine lineages. This is the same picture shown by the *Anolis* communities on the Antilles (see above), and by a recent study of Central American bird communities (Price et al. 2000). Hence, both the underlying regularities and complexities in ecological interactions and community membership are due, in part, to historical influences.

SUMMARY

Each of the preceding studies, although preliminary, has illustrated three important points. First, historical influences on community composition and structure are so ubiquitous that it is possible to combine phylogenetic, biogeographic, ecological, and behavioral data to ask questions about the influence of history on the spatial and resource allocation patterns in communities (Brooks 1980, 1985; Brooks and McLennan 1991,

1992a, 1993a, 1994; Zink et al. 2000). Second, patterns of evolution in communities are generally not the result of a single influence, but rather the outcome of a complex interaction between speciation (colonization of, versus speciation in, the community) and the evolution of characters involved in the association (plesiomorphic versus apomorphic ecologies: Brooks 1985; Brooks and McLennan 1991, 1992a, 1993a, 1994; Haydon et al. 1993). These two components are the historical warp and weft of the community tapestry. Third, each community is a unique tapestry. This does not mean that each community is assembled de novo, but rather that each community will be woven of ancestral and derived threads. The ancestral threads represent historical processes and connections with other communities. So, for example, sister species of Australian birds tend to be found in sister communities. This is not really surprising because most sister species tend to be allopatric, although near each other (Wallace 1855; Jordan and Kellogg 1908; chapters 3 and 4). The derived threads represent the history of colonization and the history of character change, if any, following the appearance of such colonizers. Both the ancestral and derived threads may be affected directly by evolutionary forces such as interspecific competition for resources or predation. We expect the influence of such forces to vary from community to community and from clade to clade (or, as Darwin expressed it, according to the "nature of the organism and the nature of the conditions"). Finally, the tapestry of any community will be affected by an assortment of other factors, including the chance sequence of colonization or the evolution of characters within the community that have no direct bearing on the interaction but may cause a cascade of indirect effects. In all, the final outcome, the entity that we call the community, reflects the interaction of many influences. For this reason, simple models of species-area relationships (e.g., MacArthur and Wilson 1967; Losos and Schluter 2000) can tell us, for example, that large islands have more species than small islands, but cannot tell us why this might be so in each case.

Now that we have a method for distinguishing species exhibiting apomorphic characteristics from species exhibiting plesiomorphic ones in the community, we can begin to ask questions about the evolutionary influence of residents on colonizers and of colonizers on residents. If, for example, resident species tend to retain their ancestral ecological characteristics and colonizers tend to change, we could say that phylogenetic history exerts a cohesive and conservative influence on the evolution of community structure. Combining the results from all these studies will provide us with a more direct estimate of the relationship among the processes underlying the origin, divergence, and maintenance of biological communities. There are still so many questions to be addressed. How long has a species been in a community? What is the historical sequence of species addition to the community? Are core species usually residents, and are satellite species usually colonizers? Is there a relationship between a species' length of residence in a community and the stability of that community? Are residents or colonizers more susceptible to disturbance? Are old communities stable? Are stable communities old? Do stable communities persist longer? Answering these questions will add a macroevolutionary component to the concept of keystone species and will help refine predictions about ecosystem responses to perturbations and identify the evolutionarily most vulnerable members of a community. There may be no startling answers to these questions, but since we have never asked them, we do not know what to expect.

 This is an area where our voyage of discovery has barely left the dock.

The Frontiers of Discovery in Community Ecology
JOINING FORCES WITH MACROECOLOGY ON OUR VOYAGE

> In essence, we pose the inverse of the question asked by Brown and Maurer . . . , who sought to identify ecological processes affecting the diversification of clades on large land masses. We ask whether there are evolutionary patterns and processes within the clades that affect continental patterns of community organization. These, obviously, are two perspectives on the same question: What are the determinants of community organization in space and time?
>
> Cadle and Greene 1993:261

Macroecology views communities as dynamic, energy-processing macro-states.[13] This means that communities, like macrospecies, do not participate directly in any microevolutionary processes. Unlike macrospecies, communities are not genealogically individuated; they are open systems both energetically and informationally.[14] Like macrospecies, communities may be seen as statistical summations of their constituent parts (species) and their dynamics. We expect components of macroecological systems to exhibit diffuse to pronounced coevolutionary interactions, as well as the geographic dynamics of local immigration and emigration from a common regional pool of species produced by speciation *in situ* (Brooks 1985; Ricklefs 1987; Brooks and McLennan 1991). It is this phy-

13. Brown and Maurer 1987, 1989; Brooks and Wiley 1988; Maurer and Brown 1988; Maurer 1989, 1999; and Brown 1995.

14. Brooks and Wiley 1986, 1988; Ulanowicz 1986, 1997; Brown 1995; Losos 1996; and Maurer 1999, 2000.

logenetically cohesive core of resident species that maintains communities over macroevolutionary time,[15] even though most biotas are mosaics of dispersal, extinction, and vicariance that create reticulated "histories of place" for those communities.

We will show in the next chapter that any of these coevolutionary interactions may be phylogenetically conservative. The statistical summation of these historical effects, manifested as stability, persistence, and resilience, should exert a strong net cohesive influence on communities at all spatial scales, without sacrificing any dynamic behavior at the microevolutionary level. In this way, macroecological systems establish the arena within which a variety of microevolutionary dynamics occur, affecting individual species in particular places.

On macroevolutionary time scales, the more historically self-stabilized community structure is, the more substantial will be the environmental changes necessary to produce irreversible changes in that structure (i.e., to override the constraints of massive internal stabilizing selection). That is, we would expect such major transitions to occur only episodically throughout evolutionary history. At the same time, however, we would not be surprised to find that such events might result in widespread changes throughout many communities (described as "community level punctuated equilibrium" in the fossil record by Boucot 1975a,b, 1978, 1982, 1983, 1990). Phylogenetic conservatism (persistent history) lowers the cost of establishing community structure by reducing the number of instances in which that structure need be (re)invented, permitting relatively rapid reemergence of a (historically) predictable stable structure after a disturbance. In chapter 6, we noted that the fossil record supports an interpretation of a delayed but rapid increase in rates of speciation following a major extinction event, with those rates slowing asymptotically over time. The low diversity communities surviving in the aftermath of a major extinction event would likely exhibit a minimum of (self-)stabilizing selection, but as more and more new species are formed, communities would accumulate persistent plesiomorphic traits. This would, among other things, increase the amount of functional redundancy in the system, thereby increasing the amount of stabilizing selection exerted on each member of the community by all other members of the community (see, e.g., Behrensmeyer et al. 1992; Jablonski and Sepkoski 1996; Walker et al. 1999). Thus, in agreement with Stenseth and Maynard Smith (1984), we expect communities to be more in

15. Brooks 1985; Cracraft 1986, 1988, 1994; Cracraft and Prum 1988; Mayden 1988, 1992a,b; Brooks and McLennan 1991, 1992a, 1993a, 1994; Cadle and Greene 1993; Haydon et al. 1993; Marshall and Liebherr 2000; Price et al. 2000; and Zink et al. 2000.

"dynamic stasis" than in "Red Queen" mode. This means that for most periods of time, community structure will be stable enough, as a result of phylogenetic conservatism in membership and interactions, for density-dependent factors to operate as the macroecological determinants of community structure over microevolutionary time scales.[16] We therefore endorse wholeheartedly Brian Maurer's recent call for increasing integration of phylogenetic information in macroecological studies (Maurer 2000).

16. Brown and Maurer 1987, 1989; Brooks and Wiley 1988; Menge and Olsen 1990; Brown 1995; and Maurer 1999.

Chapter 8

Coevolution: Exploring
Personal Relationships

> It is interesting to contemplate a tangled bank, clothed with
> many plants of many kinds, with birds singing on the
> bushes, with various insects flitting about, and with worms
> crawling through the damp earth, and to reflect that these
> elaborately constructed forms, so different from each other,
> and dependent upon each other in so complex a manner,
> have all been produced by laws acting around us.
>
> Darwin 1872:459

Darwin's metaphor of the tangled bank, evoking a complex mosaic of evolved life, can never fail to rouse even the most reductionist of biologists. Species on this planet do not exist in isolation. They form associations, within which each species experiences a wide range of interactions with other organisms representing both close and distant relatives. For example, in the Area de Conservación Guanacaste in northwestern Costa Rica, the mouse *Liomys salvini* and the frogs *Rana forreri* and *Rana vaillanti* each host more than 20 species of eukaryotic parasites. Two specimens of the white-tailed deer *Odocoileus virginianus* contained 15 species of metazoan parasites (Carreno et al. 2001), and the large herbivorous iguana *Ctenosaura similis* has seven different species of nematodes all living together in the same part of the intestine. To complicate matters even more, the cuticles of some of those nematodes are covered with filamentous bacteria, hanging on with what appear to be miniature suction pads.

In fact, because every species' environment is composed primarily of other species (Stenseth and Maynard Smith 1984), all interspecific associations evolve in a context of **diffuse coevolution** (Futuyma and Slatkin 1983). Diffuse coevolution is the equivalent of saying that so many forces (selection vectors) are influencing an association that we cannot understand its evolution by examining just one of those forces (Bakker 1983). For example, a particular clade of beetles may dine on a particular clade of plants, and those plants may produce chemicals to discourage such dining. At the same time, the beetles themselves may be eaten by a bird, a shrew, and a lizard. The beetles, the plants, and the vertebrates may all be hosting a variety of parasitic species (including viruses), and all may be competing with conspecifics for access to mates. In other words, the beetle-plant association is more than just a debate over fine dining. We agree with Futuyma and Slatkin's proposal that a broad-based coevolutionary paradigm holds the key to unifying studies of interspecific associations. The outcome of such interactions should be evolutionary changes that ameliorate "conflicts of interest" arising from the independently evolved "nature of the organism" for each species involved in an association (Maynard Smith 1982; Maynard Smith and Szathmary 1995).

For the majority of biologists, coevolution is an intimate affair, involving two species face to face on an evolutionary dueling ground. At stake is the survival of at least one, possibly both. Understanding the complexity of studies in coevolution is complicated by the fact that there are two long-standing traditions for studying the phenomenon.

TRADITION I: TRACKING THE HOST SPECIES

The historical record of studies in coevolution concentrating on the host and associate species has its roots in the nineteenth century with von Ihering's (1891) observations about the similarities between some temnocephalidean (flatworm) parasites inhabiting freshwater crayfish in New Zealand and those inhabiting freshwater crayfish in the mountains of Argentina. Based on his assumption that crayfish and their parasites shared such a close ecological association that one would never be found without the other, von Ihering postulated that the species in the two disjunct areas were derived from ancestral crayfish and flatworms that were themselves associated. Hence, he argued, South America and New Zealand must at one time have been connected by fresh water. Fahrenholz (1913) examined the relationships between catarrhine (Old World monkeys) and hominid (great apes) primates based on their blood-sucking lice, with an eye on the possibility that the associations might indicate something about phylogenetic relationships of the hosts. He postulated that the information derived from the occurrence of related lice on different primates demonstrated that the catarrhines were more closely related to hominids than to any other primates, a conclusion which remains the consensus view today. In general, most parasitologists of the time believed the patterns they were uncovering demonstrated that the evolution of parasite species is irreducibly tied to the evolution of their host species.[1]

These beliefs, which became known as the "von Ihering approach" and the "Fahrenholz approach," inspired generations of parasitologists to formulate rules about coevolution based on orthogenetic, rather than Darwinian, reasoning (summarized in Brooks and McLennan 1993b). As you recall from chapter 5, the orthogenesis movement developed as a response to what many scientists saw as an overemphasis by Darwinians on the role of the environment in evolution (Bowler 1983). As an alternative to Darwinism, orthogeneticists proposed that evolutionary change resulted from an internal drive toward progressive specialization, dependence on other species, a loss of evolutionary independence, and finally self-imposed extinction. The ubiquity of parasitism was taken as prima facie support for orthogenesis. The most influential proponent of the orthogenetic view of coevolution was Wolfdietrich Eichler, a German entomologist studying Mallophaga (biting lice).[2] Eichler (1941a) set the tone for his career with an early attack on the suggestion made by Kéler (1938, 1939) that at least some biting lice species inhabited more than one host species. Eichler argued that the presence of the same parasite on different host species could be explained in one of two ways: the parasite had inhabited the common ancestor of the current host species (à la Fahrenholz) or, more likely, there had been a mistake in species-level identification of the parasite. Claiming that many of the examples Kéler discussed fell into the second category, Eichler concluded that one must base parasite taxonomy on an understanding of the parasite-host relationship (i.e., the presumed doctrine of one parasite species : one host species) rather than on morphological or other characteristics of the parasites themselves. This entrenched the perspective that the host species determined parasite evolution (see also Andrews 1927; Osche 1958, 1960, 1963).

By the late 1930s, the orthogeneticists had created their own tautological albatross: parasites are highly host specific, so they coevolve with their hosts, and because they coevolve with their hosts, they become highly host specific. To this albatross was attached the specter of unfalsi-

1. See, e.g., Kellogg (1896, 1913), who studied birds and their associated biting lice, Harrison (1914, 1915a,b, 1916, 1922, 1924, 1926, 1928a,b, 1929) and his associate Johnston (1912, 1914, 1916), who studied a variety of vertebrates and their helminth parasites, and Metcalf (1920, 1922, 1923a,b, 1929, 1940), who studied the opalinid protists inhabiting frogs.

2. Eichler 1941a,b, 1942, 1948a,b, 1966, 1973, 1982; see also Baer 1948; Dougherty 1949; and Stammer 1955, 1957.

fiability: since host specificity was the *cause* of coevolution, rather than a function of the ecological and phylogenetic interaction between lineages, any conflicting or inconsistent observations were erroneous or irrelevant because they failed to conform to the orthogenetic view of coevolution. For most of the twentieth century, parasitologists continued to study coevolution within an orthogenetic framework despite calls to integrate with mainstream evolutionary biology (e.g., Mayr 1957; Manter 1966; Janzen 1985a; Brooks and McLennan 1993b). Even today, many assumptions about coevolution held by contemporary parasitologists are vestiges of orthogenetic thinking (for a discussion, see Brooks and McLennan 1993b). This is especially true for the assumptions that evolution always produces increasing specialization, and parasites are the ultimate endpoint of specialization (e.g., Adamson and Caira 1994) and that hosts and parasites "ought" to have congruent phylogenies.[3] Outside of parasitology, similar sentiments about evolutionary specialization being a form of "dead end evolution" (e.g., Stanley 1973; Hayami 1978; Moran 1988; Thompson 1994) also represent unintended remnants of orthogenetic thinking (see Kuris and Norton 1985 for a discussion of the adaptive value of overspecialization in certain circumstances).

TRADITION II: TRACKING THE HOST RESOURCE

The second tradition in studying coevolution has historical roots almost as deep as the first. Kellogg (1896, 1913) suggested that while some host-parasite systems might well show strong phylogenetic associations, there were substantial cases of what he termed "straggling" (what we today call "host-switching"). Kellogg cast his observations about the relationship between birds and their biting lice in a Darwinian framework, which helped initiate a different approach to coevolution. This perspective was adopted primarily by researchers interested in studying the interactions between plants and phytophagous insects (e.g., Verschaffelt 1910; Brues 1920, 1924). Because those associations often showed no clear phylogenetic component with respect to host species (though they often were extremely specific), this tradition tended to develop without the influence of systematic information beyond making use of existing taxonomies for identification purposes. Instead, researchers were interested in uncovering the ecological ties between organisms, particularly the cues insects used to locate their host plants. That is, the host *species* was not the cue, it only *possessed* the cue.

---

3. For example, Hafner and Nadler 1988, 1990; Humphery-Smith 1989; Page 1990, 1994; Charleston 1998; and Hugot 1999.

The ultimate synthesis of this line of research was the classic article by Ehrlich and Raven (1964), who, following a mathematical model proposed by Mode (1958), hypothesized that the evolutionary diversification of plants and insects had been fueled by complex coevolutionary interactions involving mutual modification. This idea caught on quickly and was soon expanded to include any herbivore-crop, predator-prey, or parasite-host system.[4] Such coevolutionary dynamics might have a general phylogenetic context, but the fine details need not parallel the evolutionary history of the specific taxa involved. The distribution of insects among plants followed the evolution of host resources and the evolution of insects' abilities to utilize those resources, rather than the evolution of host species themselves. Ehrlich and Raven's views about coevolution mirrored the view that ecological specialization is the result of evolutionary diversification driven by biotic interactions such as competition (chapter 7). Thus, for most of the second half of the twentieth century, coevolutionary research involved studies of "two major groups of organisms with close and evident ecological interactions" (Ehrlich and Raven 1964:586) that were not just associated with each other, but actually modified each other evolutionarily.

Dan Janzen (1968, 1973a,b, 1980, 1981, 1983, 1985a,b) was one of the first ecologists to recognize the importance of history in coevolutionary explanations. He noted that the appearance of tight ecological tracking between two or more species in a particular habitat today may be misleading in the absence of information about the time and place of origin for that association. No matter where a given species evolved in the first place, its inherited functional abilities may allow it to survive in a variety of places under a variety of conditions through arbitrary amounts of time. In other words, species and their phylogenetically conservative traits may disperse through time and space. Janzen (1985b) called this interaction between the past history of the species and their present day associations **ecological fitting.**

Ecological fitting has different macroevolutionary manifestations. For example, any given species may be a resource specialist but may also share that specialist trait with one or more close relatives. That is, specialization on a particular resource can be plesiomorphic. Phylogenetic studies have uncovered ample evidence of conservative resource specialization.[5] Related species that share the same resource preferences are

4. For example, Pimentel and Al-Hafidh 1965; Pimentel and Stone 1968; Pimentel and Soans 1972; Lomnicki 1974; Benson et al. 1975; Lawlor and Maynard Smith 1976; Vermeij and Covich 1978; Anderson and May 1982.

5. Ross 1972a,b; McCoy and Heck 1976; Boucot 1982, 1983, 1990; Mitter and Brooks 1983; Hill and Smith 1984; Brooks 1985; Duellman 1985b, 1993; Toft 1985; Ross 1986;

called **ecological homologues** (DeBach and Sundby 1963). (Recall the mimicry systems we discussed in chapter 7.) In a complementary manner, the resources themselves may be very specific and yet still be taxonomically and geographically widespread, if they are persistent plesiomorphic traits of the hosts. This may be a key element in explaining some paradoxical elements of host and site specificity, which we will explore further in this chapter.

Such persistent plesiomorphic traits may be co-opted to perform novel functions (exaptation: Gould and Vrba 1982; Arnold 1994; Armbruster 1996b, 1997; Kardon 1998). This may be a widespread phenomenon. For example, Eggleton and Belshaw (1993) invoked exaptations to explain the different ways in which dipterans, hymenopterans, and coleopterans have become parasitoids; Armbruster (1997) documented extensive exaptations in the evolution of plant-herbivore and plant-pollinator interactions in *Dalechampia* (Euphorbiaceae); Jensen (1997) suggested that the ancestral, and now extinct, food of *Sacoglossa* (a genus of opisthobranch mollusks) established dietary preferences affecting the choice of food by present-day species; and Trouvé et al. (1998) reported that the life history traits of parasitic flatworms do not differ from the life history traits of their free-living relatives, indicating that these species do not have a "parasitic mode of life" but rather a "platyhelminth mode of life" which has been co-opted to function in a host-parasite context.

Persistent ancestral traits may also be "anachronisms," traits that evolved in a coevolutionary context that no longer exists. For example, Janzen and Martin (1982) suggested that the large thorns produced by many plant species growing in Central America today originally functioned to deter large herbivorous mammals that are now extinct. These thorns persist even though the original herbivores have not been replaced by other large plant eaters (presumably the plants could defend themselves against the current crop of herbivores with much smaller spines). These plants thus exhibit a plesiomorphic trait (enormous thorns) in an apomorphic ecological context (no big herbivores) that no longer confers the original benefit. While it might be tempting to attri-

---

Fox 1987; Ricklefs 1987; Robinson and Dickerson 1987; Westoby 1988; Brown and Zheng 1989; Hughes 1989; Sundberg 1989a,b; Drake 1990; Foster et al. 1990; Taber and Pease 1990; Brooks and McLennan 1991, 1992a, 1993a,b, 1994; Gorman 1992; Losos 1992, 1996; Pellmyr and Thompson 1992; Anderson 1993; Cadle and Greene 1993; Hedenäs and Kooijman 1996; McPeek 1996; Miller and Travis 1996; Rosenberg 1996; Wainwright 1996; Zimmerman and Simberloff 1996; Zweers and Vanden Berge 1997; Grandcolas 1998; Kelt 1999; Vitt et al. 1999; Wilkinson and Nussbaum 1999; Lindeman 2000; and Webb 2000.

bute some significance to the present-day fauna and the evolution of thorns, that inference would be unwarranted.[6]

 Why do evolutionary anachronisms persist? Maynard Smith and Szathmary (1995) suggested a number of answers to this question. For example, there may not have been enough time for the trait to disappear or be modified into something more functional. Alternatively, although it may be costly to produce the anachronistic trait, it may be even more costly or difficult to stop making it now that it has originated and been incorporated into the developmental background or genetic architecture.

The concept of ecological fitting as originally proposed by Janzen suggested novel and historically based explanations for the *functions* of complex interspecific ecological associations that exist today. Our research program extends those suggestions to phylogenetically based explanations for the *origin and diversification* of such associations. The macroevolutionary manifestations of ecological fitting tell us that the evolutionary basis for such fitting is deceptively simple, yet powerful. If specific environmental cues/resources are widespread or if traits can have multiple functions (or both), then the stage is set for the appearance of ecological specialization and close (co)evolutionary tracking, without losing the ability to establish novel associations. We think this is another manifestation of Darwin's recognition that the origin of species was an outcome of interactions between the nature of the organism and the nature of the conditions.

PHYLOGENETIC STUDIES OF COEVOLUTION TODAY

In the past 25 years, three different approaches to studying coevolution have emerged. Ecological fitting emphasizes understanding the evolution of the natural history of species involved in the association (Janzen 1985b,c). **Cospeciation** focuses on the evolutionary history of the species themselves.[7] **Diffuse coevolution** combines the preceding two perspectives into one research program emphasizing the complex nature of

6. See also persistent thorns and heterophylly in Hawaiian *Cyanea*, which may have originally functioned to deter herbivory by the large flightless birds that now have all been eaten by humans (Givnish et al. 1994), and possible diet shifts among Pleistocene-Holocene mammalian megafauna in southern South America (Fariña 1996).

7. Brooks 1979, 1981, 1985, 1986, 1987, 1988b, 1989c, 1990; Mitter and Brooks 1983; Brooks and Mitter 1984; and Brooks and McLennan 1993b and references therein.

the evolutionary forces operating on any biological association (Futuyma and Slatkin 1983). The synthesis of information from all of these approaches has indicated that coevolutionary associations are structured in much the same way that communities are structured:

1. Mutual association without mutual modification. This is the null hypothesis for coevolution.[8] It is characterized by congruence between the phylogeny of the host and the phylogeny of its associate, presumably reflecting episodes of simultaneous vicariant speciation in the two lineages (fig. 8.1). *Support for this hypothesis offers relatively weak explanatory power.* For example, discovering that a particular set of associations has resulted from a long history of mutual association eliminates coevolutionary explanations based on host switching, but does not allow us to distinguish the effects of a fortuitous historical correlation (the "California Model," or allopatric cospeciation: Brooks and McLennan 1991) from the effects of some mutual interaction that maintains or promotes the association and its diversification. And even if we assume the latter, the macroevolutionary patterns on their own do not allow us to differentiate among a variety of ways in which the associated species might be causally, rather than casually, intertwined.

2. Colonization. One of four things can happen to an associate species when it colonizes a new host species. First, the colonizer may go extinct, in which case we will have no record of the attempt. Second, the new host species may possess the same resource as the colonizer's ancestral host (Jermy 1976, 1984). As far as the colonizer is concerned, the new host is simply more of the same habitat (Janzen 1985b). To the unwary researcher the expansion of the host range will appear to indicate that the associate species is becoming a generalist, when in fact it has not changed its resource preference at all (appearances can be misleading). We make this kind of mistake in our interpretation of colonization scenarios for a very simple reason: we are focusing our attention on the interaction between particular species, while the associate species is focusing its attention on finding a particular resource. The specific status of the vessel carrying that resource is immaterial to the associate. Third, the associate may speciate on the colonized host species, but still be tracking the same resource. In this case, we may be fooled into thinking a specialist on a different resource has evolved when it has not. Such a dynamic produces clades of specialists on the same resource, regardless of whether the ancestral resource is *plesiomorphically* (e.g., Taber and

8. Janzen 1968, 1980, 1981, 1983, 1985a,b; Brooks 1979; Mitter and Brooks 1983; Brooks and Mitter 1984; and Brooks and McLennan 1991, 1993b.

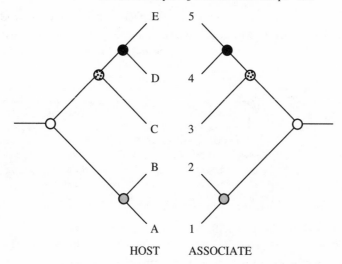

Figure 8.1. The California model of coevolution. Complete congruence between the phylogeny for the hosts (taxa represented by *letters*) and the phylogeny for their associates (taxa represented by *numbers*) is due solely to simultaneous episodes of vicariant speciation (identified by circles on each node). The host and associate are simply hanging out together, sharing space and time. They may also be interacting with each other, but we cannot tell that from the patterns alone.

Pease 1990; Berenbaum and Passoa 1999) or *convergently* (e.g., Becerra 1997; Köpf et al. 1998) widespread.

Finally, the associate species may colonize a host species with a novel resource. If the colonizer already possesses a trait that allows it to access the novel resource (plesiomorphic trait, apomorphic function), it may establish itself on the new species. If the colonizer does not speciate, it has expanded its resource base. In this case, appearances will not be misleading, because the colonizing species really has added a novel resource to its list (has become more of a generalist). If the colonist speciates, then we have the emergence a different specialist. A special case of the colonization plus novel resource scenario is called "resource or phylogenetic tracking" (Kethley and Johnston 1975). In this scenario, the evolutionary sequence of events is as follows: a new host species evolves, characterized in part by an evolutionarily modified (novel) form of the required resource. This new species of host is then colonized by individuals from the species associated with the ancestral host. Some of these individuals adapt to the new form of the resource, eventually producing, in their turn, a new species of associate (Crespi and Sandoval 2000). And so the cycle continues. In order for this to occur, the ancestral host species *must* persist after the speciation event that produces descendants bearing the modified resource; otherwise, the ancestral associate species would have no resource base to support it through the colonization phase.

 The take home message from this section is simple: There are no predicted phylogenetic patterns that can be used on their own either to distinguish one colonization scenario from another, or to distinguish particular scenarios (like phylogenetic tracking) from the null hypothesis. The only way to differentiate among these possibilities is to add details about the nature of the resource and its evolution to the phylogenetic information for the hosts and associates.

3. Mutual association with mutual modification (evolutionary arms race). This is the classical coevolution model, originally proposed to explain the evolution of insect-plant systems.[9] The primary assumption in the arms-race model is that coevolving associations are maintained by mutual adaptive responses (Sperling and Feeny 1995; Armbruster 1997). In a nutshell, as applied to the original insect-plant systems, the argument goes something like this: phytophagous insects reduce the fitness of their hosts. Plants which, by chance, evolve traits that make them unpalatable to these insects (defense mechanisms) will increase their fitness relative to their undefended brethren, and the new defense mechanism will spread throughout the species (anagenesis). However, some insects will, in their turn, evolve a counterdefense allowing them to feed on the previously protected plants. If this trait confers a fitness advantage to those individuals (e.g., through access to an unused resource base), the counterdefense mechanism will spread throughout the insect species (more anagenesis). The insect species will now enjoy unhampered dining on the previously protected members of the plant species, and the cycle will begin anew. Notice that the focus is shifted away from the actual resource that the associate is using (the "food") to the way in which the host attempts to prevent the associate from gaining access to that resource (the defense traits).

As Ehrlich and Raven suggested, an arms race can produce many different phylogenetic patterns. If, for example, the defense and the counterdefense traits arise relatively quickly, we might see congruent portions of the host-associate phylogenies (for a possible example, see Vane-Wright et al. 1992). On the other hand, if the time scale on which the

---

9. Mode 1958; Ehrlich and Raven 1964; Feeny 1976; Dawkins and Krebs 1979; Berenbaum 1983; Abrams 1986; Thompson 1986; Vermeij 1994; Sperling and Feeny 1995; and Becerra 1997. See also "exclusion model" in Ronquist and Nylin 1990.

defense and counterdefense traits originate is longer than the time between speciation events, we might expect to find that the associate group is missing from most members of the host clade characterized by possession of the defense trait. A third possible pattern results when one or more relatively old members of a host clade are colonized by more recently derived members of the associate group bearing the counterdefense trait. In this case we might find evidence that some associates have "back colonized" hosts (Mitter and Brooks 1983; Brooks and Mitter 1984).

Now, let us suppose that both the resource (food) and the trait involved in protecting that resource (defense) are taxonomically widespread. And let us further suppose that associate group ABCidae, coevolving quite happily with host group NOPidae, eventually evolves a counterdefense. Within the context of the ABCidae-NOPidae interaction, this fits our arms race scenario. Now, stretch your imagination and suppose further that some members of species B colonize a novel host species from a different clade (species X from XYZidae) bearing the plesiomorphic resource and the plesiomorphic defense trait. If you only look at the relationship between host species X and associate species B, you could conclude that you have evidence for an arms race between the two. However, this is not an ancient association; in fact, these two species did not interact at all before the colonization, and their association is the result of colonization without modification on either side. In the absence of data pertaining to both the *origin of the association* and the *origin/modification of the characters involved in the association*, you (once again) cannot tell whether the apparent congruence between defense and counterdefense is coincidental or causal (e.g., Sperling and Feeny 1995; Vollrath and Parker 1997; Weller et al. 1999; Turner et al. 2000). Because of this, you might miss the opportunity to study the origin of a new arms race dynamic between the colonizer and the new host species (or the opportunity to document the effects of a lethal invader).

 Once again, there is no one macroevolutionary pattern that, on its own, can identify an arms-race scenario unambiguously.

Part of the confusion about the relative roles of different models in explaining coevolutionary processes has arisen from attempts to derive phylogenetic patterns that could unambiguously identify each model.[10] Those initial attempts to simplify nature encouraged researchers to begin

---

10. Ronquist and Nylin 1990. We ourselves made this attempt: see Brooks and McLennan 1991.

collecting data from a variety of associations, data which are now revealing that the original assumptions were too simple. For example, the term "cospeciation" was originally coined to describe congruent portions of host and associate phylogenies and to differentiate between those portions and incongruencies due to host switching (Brooks 1979). This was an important distinction for researchers attempting to understand the conditions under which speciation was initiated. "Cospeciation" has more recently been used by some researchers to indicate the rate at which speciation of hosts and associates is completed (Hafner and Nadler 1988, 1990). These researchers argue that the two lineages may diverge together by the same amount (**synchronous cospeciation**), the associate may diverge less (i.e., complete speciation later) than the host (**delayed cospeciation of associates**), or the host may diverge less than the associate (**delayed cospeciation of hosts**). This research provided an important insight for the study of coevolution: using congruent portions of host and associate phylogenies is a way to calibrate studies of divergence rates, much as we use sister groups to calibrate estimates of relative species richness (chapter 6).

Observing patterns of congruence between host and associate phylogenies on its own, however, tells us nothing about the underlying mechanisms producing that congruence. Based upon the coevolutionary processes we have been discussing so far, there are at least three ways to explain the observation that you have perfect congruence between host and associate phylogenies (fig. 8.2: more explanations will occur to you). By now you should recognize the falsification of simplifying assumptions as a general theme in this book—"one gene, one selection vector, one character modification" is no longer considered a valid way to study adaptation; "all of nature is organized in one general vicariance pattern" has been falsified in historical biogeography; "all species in a community evolved elsewhere and dispersed into the community" has been falsified by comparative phylogeneticists; and "each coevolutionary process leaves a distinct macroevolutionary signature" does not hold for coevolutionary studies.

How Do We Begin?

In order to study coevolution, we need two pieces of information. First, phylogenetic trees for the host and associate clades are necessary to pinpoint when the association originated and how it has been evolving over time. For example, is every instance of host speciation coupled with an instance of associate speciation? What role has colonization played in the history of the particular coevolving clades (and the further evolution

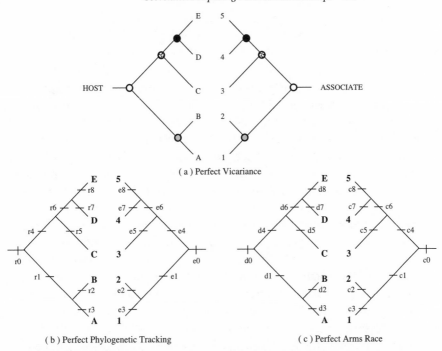

Figure 8.2. So many ways to explain the observation that you have perfect congruence between host and associate phylogenies: (*a*) Perfect vicariant speciation: both lineages have responded in the same manner (speciated) to a series of vicariance events. We know nothing about the interaction (if any) beyond this point. (*b*) Perfect phylogenetic tracking: every speciation event in the host lineage has been accompanied by a modification of the resource, r. Members from the ancestral association colonize the new host and speciate, accompanied by a modification in their ability, e, to access the new resource. (*c*) Perfect arms race: The origins of defense, d, and counterdefense, c, traits are coupled with the speciation of host and associate in a manner directly analogous to (b).

of the associate in either expanding its resource base or forming new associations)? Second, we must understand the particular resource (and defense trait where applicable) that is involved in the coevolutionary interaction. What is that resource? How is it distributed (e.g., Will et al. 2000)? Are colonizers tracking widespread, plesiomorphic resources, or are they evolving to meet the demands of a novel resource? Does the evolution of defense-counterdefense traits appear coupled or staggered in a particular association? Neither of these pieces of information is sufficient to explain coevolution; both are necessary.

SOME GENERAL EXAMPLES

Although complex, coevolutionary systems are subject to the same effects as those we discussed in chapter 7 for communities. For example,

Benz (cited in O'Grady 1989) demonstrated substantial diversification in habitat preference (site selection) within the copepod communities on sharks' gills. Interestingly, the phylogenetic structure of these gill niches was retained even though the specific shark and copepod species varied from community to community (fig. 8.3). In almost every case, species of *Pandarus* inhabit the gill arch, *Eudactylinodes* and *Gangliopus* attach themselves to the secondary lamellae, *Nemesis* burrow into the efferent arterioles, *Phyllothyreus* inhabit the superficial portions of the inter-branchial septum, *Paeon* embed in the interbranchial septum, and *Kroyeria* are found in the water channels of the secondary lamellae or embedded in the interbranchial septum. Simkova et al. (2000) reported similar microhabitat partitioning among nine monogeneans of the genus *Dactylogyrus* living on the gills of the roach (*Rutilus rutilus*). Pérez and Atyeo (1984) reported an even more spectacular example involving four species of Mexican parrots each inhabited by at least 15 species of feather mites, each of which lives on or in a particular portion of the feathers found in a particular part of the bird. As a result, there is virtually no overlap among mite species. As with the copepods, this site specificity is phylogenetically conservative among mite congeners living on different bird species. And finally, Taber and Pease (1990) showed that different species of paramyxoviruses (causal agents of measles, mumps, Newcastle disease, rinderpest, canine distemper, and human parinfluenza) occurred in homologous tissues of nonsister groups more often than they occurred in sister groups. Once again, it is the host resource and not the host species that holds the key.

Swallowtails and Umbels, Caterpillars and Chemicals. Swallowtail butterflies (Papilionidae) can be found just about anywhere on the planet (except at the poles, of course), feeding on a wide variety of host plants (summarized in Aubert et al. 1999). Within this large and diverse family, one clade, the *Papilio machaon* species group, has attracted the attention of researchers because some of its members display an unusual ability to feed (as caterpillars) upon umbellifers (Apiaceae: think of Queen Anne's lace), plants that are spurned by most other swallowtails (Scriber 1984). The reason that most swallowtails eschew umbellifers is simple; many of those plant species produce toxic substances called linear furanocoumarins (LFCs), and a few produce a variant called angular furanocoumarins (AFCs), which are evidently even more toxic to would-be grazers. Experimental investigation using one member of the *Papilio machaon* species group, *P. polyxenes* (the black swallowtail butterfly), as the test subject demonstrated that caterpillars respond to ingesting an LFC with an enhanced growth rate (Berenbaum 1981, 1983; Cohen et al. 1990), but

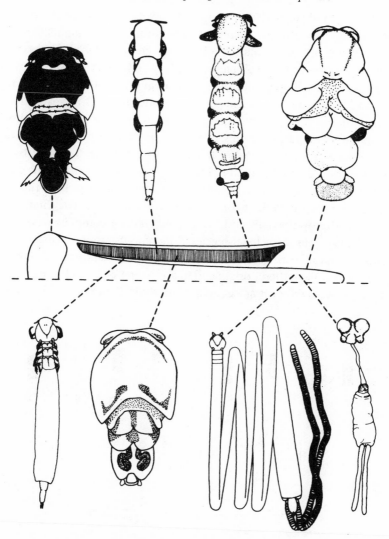

Figure 8.3. Distribution of eight species of copepods on the gill of a shark. The copepods in this community are (*clockwise, from top left*) *Pandarus cranchii, Eudactylinodes uncinata, Nemesis lamna, Phyllothyreus cornutus, Paeon vaissieri, Kroyeria caseyi, Gangliopus pyriformis,* and *K. lineata.* (From O'Grady 1989.)

show a reduced growth rate when fed plants containing an AFC. Evidently the black swallowtail has been able to evade the first line of plant defense, only to fall before the second. Another member of the group, *P. brevicauda,* appears to feed quite happily on LFC- and AFC-containing hosts, implying that it at least has evolved a counterdefense against both chemical types (Berenbaum and Feeny 1981).

Sperling and Feeny (1995) wondered how the defense-counterdefense interaction between the *Papilio machaon* group and the umbellifers had been played out in the macroevolutionary arena. For example, was the counterdefense to the LFCs plesiomorphic? Did *P. brevicauda* display an autapomorphic change that gave it an advantage in the arms race over both its hosts and other members of its clade? In order to answer these (and other) questions, they mapped the presence or absence of hosts containing the various forms of coumarin (not to be confused with cumin) onto the phylogeny for the butterfly group (fig. 8.4). This initial investigation is based upon the simplifying assumption that if the caterpillars are feasting on plants with a particular coumarin, then they are relatively resistant to the toxic effects of the chemicals. The interaction between the swallowtails and the umbellifers is a complicated one, in part because one of the basal species, *P. indra,* is not found in association with coumarin-bearing hosts. There are two optimizations for the ability to tolerate LFCs: either it originated in ancestor 1 and was lost in *P. indra* and *P. m. oregonius* or it arose independently in *P. alexanor* and ancestor

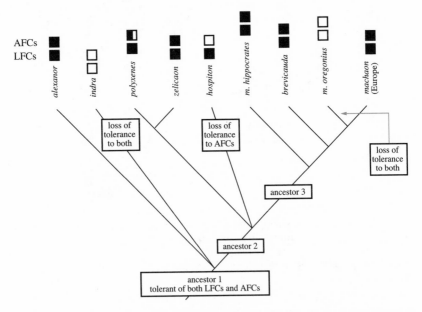

Figure 8.4. Swallowtails and umbellifers. Phylogenetic tree for the *Papilio machaon* species group with tolerance to LFC (linear furanocoumarins) and AFC (angular furanocoumarins) mapped above each species. *Black box* = caterpillars feed on host plants with the chemical defense; *white box* = caterpillars feed on host plants without the chemical defense; *m. = machaon*. There are several different optimizations for the evolution of LFC and AFC tolerance, only one of which is shown in this figure.

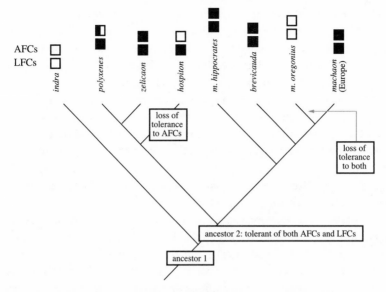

Figure 8.5. Another look at swallowtails and umbellifers. Tolerances to furanocoumarins mapped above an updated phylogeny for the group (consensus of two equally parsimonious trees: Caterino and Sperling 1999). *Black box* = caterpillars feed on host plants with the chemical defense; *white box* = caterpillars feed on host plants without the chemical defense; *m.* = *machaon*. There are two different optimizations for the evolution of AFC tolerance, only one of which is shown in this figure.

2 and was lost in *P. m. oregonius.* There are three optimizations for the ability to tolerate AFCs (involving origins in either ancestor 1, 2, or 3 with corresponding changes in gains and losses: you can work this out for yourself).

Two recent phylogenetic studies have helped resolve the basal ambiguity by removing *P. alexanor* from the *P. machaon* group (Aubert et al. 1999; Caterino and Sperling 1999). This removal resolves the evolution of LFC tolerance in the group (fig. 8.5): it unambiguously originated in ancestor 2 and was lost in *P. m. oregonius.* The point of origin for AFC tolerance is still ambiguous, but the number of equally parsimonious optimizations is reduced from three to two. At the moment, there is no way to choose between those two options. Five things, however, are clear even with this ambiguity: (1) *P. indra* displays the plesiomorphic condition (absence of tolerance to furanocoumarins); (2) the ability to tolerate LFCs is very old within the clade (ancestor 2); (3) the counterdefense to the origin of AFCs is not an autapomorphy for *A. brevicauda* (many members of the group have out-maneuvered the umbellifers); (4) the ability to tolerate both LFCs and AFCs may not be due to the same underlying mechanism (see *P. hospiton*); and (5) *P. m. oregonius* appears to have lost its tolerance for LFCs and AFCs. We say "appears to have lost its tolerance" for umbellifer-based chemicals because the butterfly has shifted its host preference from umbellifers to composites (which do not produce furanocoumarins). This host shift is interesting in itself because even fewer swallowtail species feed on composites than feed on umbellifers. The question of how the new association became established awaits resolution of the relationships among the *P. machaon* subspecies (there are at least four others that feed only on composites) and answers to (yet more) questions: Why don't more swallowtail caterpillars feed on composites? Are the *P. machaon* subspecies that feed on composites still capable of tolerating FCs even though they no longer feed on umbellifers? If so, does resistance to FCs confer some sort of resistance to composites, assuming that they are also producing a chemical that is toxic to swallowtails (and thus permit the host switch to composites)?

This study is fascinating because it demonstrates that defense/counterdefense traits may be quite evolutionarily labile, contrary to our more simplistic picture of one defense, sweeping through the hosts, followed by the counterdefense, sweeping through the associates. The study also highlights many directions for future research. For example, just because a swallowtail species associates with host plants that do not produce furanocoumarins, does this mean that its caterpillars are incapable of tolerating those chemicals? Perhaps the caterpillar simply does not encounter umbellifers containing FCs in its habitat (recall that not all species of umbellifer produce these toxins), but would be quite capable of feasting upon those plants if it did. Or, perhaps FCs really are toxic to the caterpillars. Clearly we need to move to the next stage of the study, feeding experiments in the lab, in order to differentiate between these alternate explanations for the "absence" of furanocoumarin-bearing hosts in the repertoire of *P. hospiton* (AFCs at least) and populations of *P. machaon*.

We also need to determine how the butterflies are managing to tolerate the chemicals. Cohen et al. (1990) demonstrated that at least one enzyme, cytochrome $P450$, is involved in detoxifying FCs in the black swallowtail. Is this enzyme widespread throughout the group? Are different enzymes used to detoxify linear and angular FCs (and does this explain the anomalous *P. hospiton*)? Do these enzymes serve a different function in the sister group to, and other close relatives of, the *Papio machaon* species group? Does "lack of tolerance" mean "absence of enzyme," "deactivation of enzyme," or "change in enzymatic function"? Is the same or a different mechanism responsible for the ability of *P. alexanor* to feed on FC-producing umbellifers, given that the butterfly has now been removed from the *P. machaon* species group (although its exact placement with respect to that group is still in flux: see Aubert et al. 1999 and Caterino and Sperling 1999)?

Finally, Sperling and Feeny noted that population levels of these swallowtail butterflies were generally quite low relative to their abundant host plants. Given this, they proposed that the evolution of the toxic chemicals in the plants (the defense move) was probably most strongly influenced by an arms race between umbellifers and other, more population-rich diners, although the interaction between swallowtails and umbellifers would reinforce that dynamic. This would make the system an excellent example of diffuse coevolution. Of course, the observation begs one last question: Are the low levels of these butterfly populations unusual among swallowtails? If so, is there a relationship between the unusual host preference and the unusual population demographics of the butterflies?

Birches and the Midges That Gall Them. Dipteran insects of the genus *Semudobia* (family Cecidomyiidae, subfamily Cecidomyiinae) are commonly called gall midges. The association between gall midges and their hosts is more intimate than a simple dinner-diner relationship. Females lay their eggs in bracts or fruits of various species of birches (genus *Betula*) and the larvae develop *in situ*, drawing both sustenance and shelter from their host, inducing a thickening in the plant tissue in return (gall formation).[11] Research on members of other gall midge tribes indicates that characteristics of the host plant have considerable influence on larval development (Åhman 1981, 1985; Skuhravy et al. 1983), which may provide a barrier to speciation via host switching in these insects (Roskam 1985). Although as many as three species of *Semudobia* may co-occur in the same host, none of the five species that make up the genus is found on plants other than birches. This specificity led Roskam (1985, 1992) to investigate the coevolutionary aspects of the association between birches and their gall midges.

The birch genus *Betula* comprises four sections: *Costatae, Humiles, Acuminatae,* and *Exelsae* (fig. 8.6b). Members of the sections *Costatae* and *Humiles* bear erect pendulous flowers, called catkins, and retain their fruits over the winter, while their relatives, the *Acuminatae* and *Exelsae,* display pendulous catkins and drop their fruits in the autumn. There are pronounced differences in host preferences among the gall midges. *Semudobia skuhravae* (species 1: fig. 8.6a) occurs commonly in association

---

11. For studies on gall formation in a variety of systems, see Stern 1995; Crespi and Worobey 1998; Nyman et al. 1998, 2000; and Stone and Cook 1998.

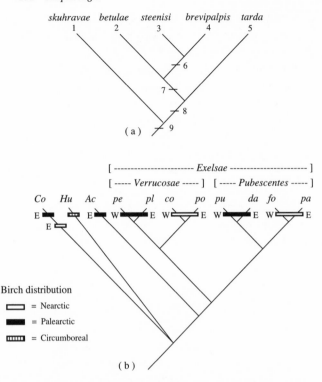

Figure 8.6.   Phylogenetic trees for the birches and their gall midges. (*a*) The gall midges, genus *Semudobia*, coded for coevolutionary analysis. (*b*) The host birches, genus *Betula*. Sections: *Co* = *Costatae*; *Hu* = *Humiles*; *Ac* = *Acuminatae*. Species: *pe* = *Verrucosae pendula*; *pl* = *V. platyphylla*; *co* = *V. coerulea*; *po* = *V. populifolia*; *pu* = *Pubescentes pubescens*; *da* = *P. davurica*; *fo* = *P. fontinalis*; *pa* = *P. papyrifera*. The geographic distributions of the birches are mapped onto their phylogenetic tree: W = west; E = east.

with members of sections *Costatae* and *Humiles* and less so with members of *Exelsae*. Within the Palearctic *Exelsae*, *S. betulae* (species 2) is regularly found on species in the series *Verrucosae* and is only occasionally associated with the series *Pubescentes*. The reverse situation occurs for *S. tarda* (species 5), a gall midge displaying a marked preference for *Pubescentes* and a secondary preference for *Verrucosae*. Relationships between the insects and their host plants produce a similar pattern of preferences in the Nearctic *Exelsae*, where *S. steenisi* (species 3) and *S. brevipalpis* (species 4) occur commonly in association with birches within *Verrucosae* and less commonly among those within *Pubescentes*. Gall midges are never found on birches in the section *Acuminatae*. Mapping the host preferences of the five contemporaneous species of *Semudobia* onto the host phylogeny (fig. 8.7a) reveals that no two species of *Semudobia* show

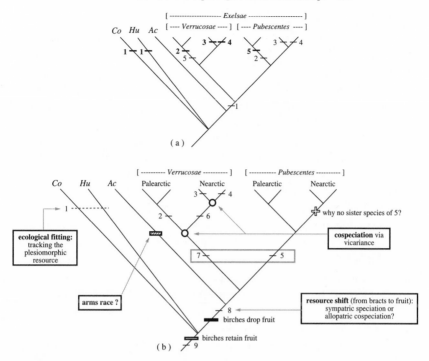

Figure 8.7.   Of birches and gall midges. *Co = Costatae; Hu = Humiles; Ac = Acuminatae.* 1 = *Semudobia skuhravae;* 2 = *S. betulae;* 3 = *S. steenisi;* 4 = *S. brevipalpis;* 5 = *S. tarda.* (*a*) The entire range of host preferences for the five gall midge species mapped onto the host phylogeny. *Bold numbers* = preferred birch host for the particular species; *nonbold numbers* = secondarily preferred hosts. (*b*) Phylogeny for the gall midges, mapped onto the phylogeny for the host birches. *Dotted box* = possible lineage duplication in the birches and the midges; *cross* = possible extinction.

their greatest preference for the same birches. Based on these distributions, and the assumption that the associations of greatest preference are those of longest evolutionary duration, Roskam postulated that the majority of the birch–gall midge associations could be explained by simultaneous episodes of vicariant speciation (allopatric cospeciation).

Let's examine this proposition further, based upon some of our new ideas about coevolutionary dynamics (fig. 8.7b). First, we need to expand the scope of our study to include the sister groups of the midges and birches in order to determine how the association between the two clades began. Past that point, the association between *Semudobia skuhravae* (species 1) and members of sections *Costatae, Humiles,* and *Exelsae* represents an example of ecological fitting as the gall midge tracks the plesiomorphic resource (birch bracts). By the time ancestor 8 speciated, producing *S. tarda* and ancestor 7, a *resource* switch had occurred, so that all descendants of ancestor 8 obligately oviposit in fruit. You cannot tell, based upon these data, whether that switch was associated with (1) simultaneous vicariance of the birch and the gall midge clade, coupled with modifications in both the fruit-producing abilities of the birch (origin of dropped fruit) and the egg-laying preferences of the midge (prefer the fruit), or (2) sympatric speciation of ancestor 8 associated with a resource switch to fruit. In either event, the fact that these fruit are now "dropped" allowed species 1 (which is widespread) to co-occur with the descendants of ancestor 8, because the two midge lineages are now spatially separated on the same host. In order to study this particular modification further, we need to determine exactly what the trait is that we are calling "preference for bract versus fruit." Perhaps this trait is a response by the midge to a particular chemical or visual cue (or a combination of cues). Details of the cue(s) aside, studies have demonstrated that individual midges that overwinter in the soil (which you must do if you hatch in a fruit lying on the ground) have an advantage over midges that remain within the birch tree itself (Möhn 1961). This advantage explains, in part, why the resource switch was originally successful. Now all we need to know is, What trait was modified?

The remaining speciation events occurred within a common ecological context (eggs laid in dropped fruit). The second speciation event was associated with a lineage duplication in the hosts that could be a reflection of either an old vicariance event (allopatric cospeciation; the explanation favored by Roskam and Van Uffelen [1981]) or simultaneous episodes of sympatric speciation producing the ancestors of the *Verrucosae* and *Pubescentes* series and ancestor 7 + *S. tarda* (species 5). Next, two vicariance events occurred. The first isolated *S. betulae* (species 2) in the Palearctic and the ancestor of *S. steenisi* and *S. brevipalpis* in the Nearctic (ancestor 6), and the second isolated *S. steenisi* (species 3) and *S. brevipalpis* (species 4) in the west and east Nearctic, respectively (see fig. 8.6b for distributions). Additional examples of ecological fitting occurred as *S. steenisi* (species 3) and *S. brevipalpis* (species 4) in the Nearctic and *S. betulae* (species 2) in the Palearctic colonized members of the *Pubescentes* series, while *S. tarda* (species 5) in the Palearctic colonized sympatric members of the *Verrucosae* series. None of the midges speciated following these colonizations; they simply expanded their host range. In all cases, the preferences for the colonized hosts are weak, indicating that there are factors other than simply "birch fruit" influencing the midge-birch interaction. Perhaps, for example, the birch fruits have themselves changed and so are not completely equivalent across birch species.

The patterns shown in figure 8.7b highlight two additional areas for investigation. First, why is *S. tarda* (species 5) not found in the Nearctic

Pubescentes series? *Semudobia steenisi* (species 3) and *S. brevipalpis* (species 4) are capable of utilizing those birches, which suggests that they provide adequate resources for *Semudobia* development. This implies that the explanation for the absence of *S. tarda* (species 5) or a sister species of 5 may be geographic; that is, a chance extinction event that had nothing to do with the coevolutionary association. Or perhaps, as discussed previously, there really is something unusual about the fruit of Nearctic *Pubescentes*. Second, how do we explain the absence of gall midges in the bracts or fruit of any *Acuminatae* species? This appears to be an excellent test case for an evolutionary arms-race, which the plants appear to have won, at least for the moment. If this is true, members of *Acuminatae* may produce an apomorphic, unrecognizable cue, a substance that repels *Semudobia* species, or a substance that adversely affects development of gall midge larvae regardless of whether the midges develop in the bracts or the fruit.

Coevolution from the Host Perspective: *Dalechampia*. *Dalechampia* is a genus of approximately 120 species of viny euphorbs distributed throughout the tropical regions of the world. The origins of this genus lie in western Gondwana, what we now call the Neotropics. The earliest members of the group apparently were dispersed throughout Gondwana by the end of the Cretaceous, when the supercontinent broke up. This resulted in substantial *in situ* speciation. A second wave of dispersal from the Neotropics to Africa and Madagascar occurred sometime during the Tertiary, resulting in a secondary radiation of the group in those regions (Armbruster 1994). Members of the genus are distinctive in having complex pseudanthial inflorescences (blossoms), each made up of a few (usually 3) pistillate flowers and several (4–16) staminate flowers acting together as a single pollination unit.

   Dalechampia species host both mutualist and herbivore insect associates. The herbivores include nymphalid butterfly larvae that feed on the leaves, numerous nocturnal insects that feed on the flowers, and insects and small vertebrates that feed on the seeds (Armbruster 1997). The mutualists (all bees except for a few species in Madagascar, where everything is different and beetles stand in for the bees) include (1) male euglossines (Apidae), which use the monoterpene fragrance secretions of the flowers as precursors in their biosynthesis of sex pheromones (Dressler 1982; Williams and Whitten 1983; Whitten et al. 1986; Armbruster et al. 1989); (2) female euglossines (*Euglossa, Eulaema,* and *Eufriesia*), anthidiines (Megachilidae: *Hypanthidium*), and worker meliponines (Apidae: *Trigona*), which use oxidated triterpene resin secretions as nest-building material; and (3) worker *Trigona* species, which use the pollen for provisioning larvae. The bees are called mutualists rather than herbivores because, in the course of collecting resin, fragrances, or pollen, they pollinate the host plant's blossoms. Scott Armbruster and coworkers have spent nearly 20 years investigating the complex evolutionary history of *Dalechampia,* concentrating primarily on species living in the Neotropics. This research has produced information about the phylogeny, historical biogeography, mating systems, pollinator systems, herbivores, and morphological evolution for 42 species (Armbruster 1988, 1992, 1993, 1994, 1996a,b, 1997; Armbruster and Mziray 1987; Armbruster et al. 1989, 1997). Field biologists will tell you what a tremendous accomplishment this represents. That information gives us an excellent opportunity to examine coevolution from the perspective of the host rather than the associate.

The story of *Dalechampia* is one of ecological fitting in a complex diffuse coevolutionary context, involving eight traits associated with evolutionary trade-offs between attracting pollinators and deterring nonmutualistic herbivores (Armbruster 1996a,b, 1997). The optimizations for seven of those traits (fig. 8.8) support the following complex scenario:

1. The plants are pollinated ancestrally by male euglossine bees seeking fragrances (terpenes).

2. Resin appeared, secreted in small, diffuse amounts by the bractlets of the staminate inflorescences. Bear in mind that the plants are still pollinated by fragrance-seeking male euglossines, so resin did not serve a function in the plant–pollinator dynamic at its point of origin. Experimental analyses, however, indicated that this was indeed a significant evolutionary event from the plant's perspective, because those resins protect the nutritious flowers from florivorous insects (Armbruster et al. 1997). This step in the story, then, is driven by a plant-herbivore interaction akin to an arms race.

3. Large, brightly colored involucral bracts evolved. At this point on the tree, it is difficult to determine what was pollinating the plants because field observations are absent for the critical species *D. attenuistylus* and *D. heterobractea.* Armbruster (1993) hypothesized that they were pollinated by male euglossines based upon stigma morphology. From this, we would hypothesize that the male bees also use visual cues (in additional to the volatile fragrances) to locate appropriate flowers. It is possible that the showy bracts also attract florivores, so there may be an intricate relationship between costs (be eaten) and benefits (be pollinated). Depending upon the answers to two questions—What pollinates *D. attenuistylus* and *D. heterobractea*? and, Are those pollinators (and any florivores) attracted to the showy bracts?—we may (tentatively) give two

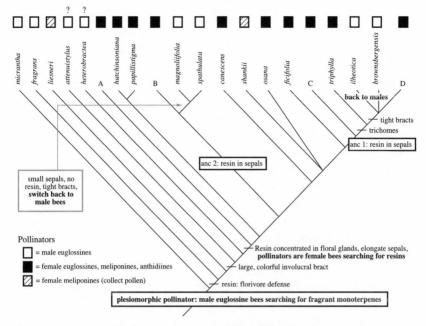

Figure 8.8. Coevolution from the plants' perspective: the dynamic interaction among pollinators, herbivores, and *Dalechampia.* Type of pollinator is mapped above the simplified tree for *Dalechampia.* A = *pentaphylla* + *leutzelburgii* + *megacarpa;* B= *websteri* + *juruana* + *schottii* + *dioscoreifolia* + *karsteniana* + *"bella"* + *aristolochiifolia;* C = *caperonioides* + *schippii;* D = *arenalensis* + *tenuiramea* + *heteromorpha* + *cissifolia* + *stipulacea* + *stipulacea* var. *minor* + *variifolia* + *grenadilla* + *tiliifolia* + *affinis* + *magnistipulata* + *armbrusteri* + *pernambucensis* + *scandens* var. A + *scandens* var. B + *scandens* var. C.

thumbs up for the importance of pollinator mutualisms and two thumbs down for herbivore defense at this point in evolution.

4. The next step in the evolutionary game involved two important changes: a pollinator shift from male euglossines to female euglossines, meliponines, and anthidiines and the ability of *Dalechampia* species to concentrate resin in a large gland in the anterior part of the flower (Armbruster 1996a,b). Female bees are very attracted to this stash of concentrated resin because it is much easier to collect (they take advantage of the waterproof, slow-hardening qualities of resin in nest construction). Concentrating the resin in this manner, however, forces the bee to assume a "head into the gland, body over the flower interior" position, which maximizes the contact between the pollinator and the pollen producing-accepting parts of the flower. This change in the structural deployment of resin co-opted the function of resins, emphasizing reward

(to the bee) over defense (resin is acting as an exaptation: Armbruster 1996a,b, 1997). Once again, then, we have a balance between increasing the effectiveness of pollination and decreasing the effectiveness of antiflorivore function (the resin is no longer dispersed around the floral components). Interestingly, a novel counterdefense on the part of the plants also originated at this time: the sepals, which are plesiomorphically small, became quite large, partially or completely covering the ovary.

5. Showy bracts and elongate sepals set the stage for a change in the use of resin; specifically, the elongated sepals now began to secrete the same resin as the floral glands. This novel ability deployed the resins around the flower once more, increasing the effectiveness of their antiflorivore function, providing protection for the ovary and developing seeds. At this point, the pollinator has already served its function, and the plant's "attention" is on protecting the outcome of that pollination. Resin-secreting sepals appear to have originated at least twice and possibly three times in this genus (see Armbruster 1997 for details). It is tempting to consign this trait to the "key innovation" category because the clade comprising ancestor 1 and its descendants is more species rich than its sister group (approximately 19 species versus one species). The second origin of resin-producing sepals in ancestor 2, however, is not associated with obvious asymmetry in species number. This is not surprising, given the observation that the trait originated from a different background in each lineage.

6. Showy bracts and elongate sepals also set the stage for at least six independent origins of nocturnal floral closing. Like resin glands in the sepals, this new character also seems to serve a purely florivore defense function. For example, Armbruster and Mziray (1987) demonstrated that *Dalechampia* species with nocturnal flower closure exhibited almost 90 percent less damage to their flowers than those lacking the ability. Notice that no change in floral structure was required here. In fact, the large showy bracts have been co-opted to serve two functions: advertisement by day and protection by night.

7. The next change to occur involved the evolution of long, sharp trichomes on the sepals covering the ovary, then persistent bracts that close around the ovary (developing fruit plus seeds: two origins). Both of these traits have been postulated to play an important role in deterring herbivores (seed and fruit eaters). The appearance of trichomes is not as mysterious as it might seem. All *Dalechampia* species have small, stinging trichomes on their leaves, and sepals are simply modified leaves (Armbruster 1983). Like the preceding two characters (resin producing sepals and elongated sepals), these traits do not affect the plant-pollinator interaction because they come into play after pollination has occurred.

8. Finally, a small clade of three species in group D (*D. stipulacea* + *D. variifolia* + *D. grenadilla*) exhibits an elaboration of resin production, with glands on the leaves and stipules, suggesting further herbivore defense, in this case against folivores.

These eight evolutionary innovations are centered around the maintenance of the resin secreting glands, which integrate pollinator attraction and antiherbivore defense. This amounts to a "coevolutionary constraint architecture" that explains the 35 species of *Dalechampia* in this example that are pollinated by resin-gathering bees. At the same time, the phylogenetic range of those pollinating bees is relatively small (members of three tribes in two families), but there is little indication of cospeciation. If this system is so constrained historically, why do we not see cospeciation between bees and *Dalechampia* species? Armbruster noted that *Dalechampia* species with large deposits of resin have large distances between the resin glands and the reproductive parts of the flowers, while those with small deposits of resin have small distances. He further noted that small, resin-collecting bees visit *Dalechampia* species with large and small resin deposits, but are only able to pollinate those with small deposits, whereas large bees visit only those *Dalechampia* species with large deposits. Thus, ecological fitting explains why we see no cospeciation between *Dalechampia* species and their resin-collecting pollinators; any resin-collecting bee will do, so long as it is the proper size to do the job. The pollinators are specialists tracking a resource that is so widespread that there is no effective coevolutionary pressure on body size.

Only ten species (22 percent) in this example are not pollinated by resin-collecting bees. The basal two members of *Dalechampia*, *D. micrantha* and *D. fragrans*, are pollinated by fragrance-collecting male euglossine bees, which appears to be the plesiomorphic association for the group. *Dalechampia micrantha* does not have resin-secreting bractlets, so it is understandable that it would attract no resin-collecting pollinators. *Dalechampia fragrans* has only small resin-secreting glands, and thus also may not produce enough resin to attract resin-collecting female bees. Four other species of *Dalechampia* are pollinated by fragrance-gathering male euglossines. Two of those species, *D. magnoliifolia* and *D. spathulata*, are sister species that have secondarily lost resin secretion in bracts by a peculiar mechanism—their biosynthetic pathway to producing triterpenes apparently has been "short-circuited"—leaving them able to produce only monoterpenes in their bractlet glands. Monoterpenes are the fragrance chemicals that attract male euglossine bees. These species are thus functionally in the same condition as *D. micrantha*, though for different evolutionary reasons. The other two male-pollinated species are *D. ilheotica* and *D. brownsbergensis,* distantly related to *D. magnolii-folia* and *D. spathulata.* Although not each other's closest relative, these two species are members of the clade exhibiting the highest number of herbivore defense characteristics. Both species secrete fragrances attractive to male euglossine bees from the stigmatic surface of the pistillate flowers; in addition, *D. ilheotica* has vestigial but functional resin-secreting glands, while *D. brownsbergensis* has vestigial and nonfunctional resin-secreting glands.

This is one of the first studies to examine diffuse coevolution within an explicitly phylogenetic framework. As you can tell from the preceding discussion, that examination is based upon integrating a substantial amount of information from phylogenetic analysis, experimental studies, and detailed field observations. The years it took to collect such a database have been amply rewarded (and not just by fragrance and resin): the research highlights a fascinating (and complicated) interaction among plant-pollinator and plant-herbivore systems. Most of the changes involving the plant-pollinator system appear to have occurred relatively early in the history of the clade (e.g., the origin of monoterpene fragrances, the origin of colorful bracts, the origin of glands concentrating resins: see fig. 8.8). The putative arms race between the plants and their herbivores occurred in two general bursts. The earliest characteristics were involved in protecting the flowers *before* pollination (the origin of resin), while later changes were focused upon protecting the ovary + developing seeds complex *after* pollination. There are at least two points throughout this complex history when the need to attract pollinators was balanced against the need to deter herbivores, with emphasis placed on the pollinators (the origin of the showy bracts and the concentration of resin in a gland). Finally, although there is relatively little cospeciation between the plants and their pollinators, there is still substantial phylogenetic conservatism in the system. Female bees collecting resin rewards and males tracking fragrances both appear to be tracking plesiomorphic resources. In order to address this issue, we need to incorporate information from the pollinator's perspective into the study. Are all female euglossines attracted to resins? Male euglossines are also important pollinators of orchids in these areas; are they responding to the same or similar chemicals when they seek out *Dalechampia* fragrances? (One study suggests this is the case: Whitten et al. 1986.) And finally, what is the dynamic between the plants and the appearance of the (rare) pollen-eating pollinators?

Recent studies of tropical figs, their pollinator wasps, and nematodes parasitizing the wasps suggest similar evolutionary complexities (Herre

1993, 1995, 1997; West and Herre 1994; Herre et al. 1996) as researchers strive to expand the scope of coevolutionary studies beyond the realm of "single host species : single associate species."

Integrating Experimental and Phylogenetic Studies of Coevolution: Understanding Host Specificity

Researchers began thinking about coevolutionary dynamics because they noticed that particular groups tended to be associated with one another. So, for example, swallowtail butterflies in the *Papilio machaon* group are found predominantly with umbellifers and *Semudobia* midges are found only on birches. These associations are rarely, however, absolute; sometimes an associate species switches hosts and a new kind of association begins. The existence of host switching is an interesting phenomenon because it begs two general questions: Why don't species host switch more frequently (are they bound to their association by lack of genetic variability in traits involved in the interaction, or is stabilizing selection holding the association together)? and, When host switching does occur, is it just a random phenomenon (based, say, upon spatial proximity), or are there more stringent guidelines concerning what is and what is not an acceptable new host? Although the quick answer to the latter question is, There must be guidelines, otherwise we would have no evidence for coevolution, closer inspection reveals this answer to be rather superficial. What are the guidelines? Are they the same for every association?

HOST SPECIFICITY AND RESOURCE SPECIALIZATION

The concept of resource specialization has intrigued biologists for decades. The fascination stems, in part, from the observation that logically constructed arguments about the evolution of specialization always seem to lead us into one paradox or another. So let's begin with a thought experiment: assume that specialists are better at utilizing their specific host than are any generalists that might also use that host (Price 1980; Kuris and Norton 1985; Futuyma and Moreno 1988). When the host is abundant, the specialist will flourish. But, because we live in a Darwinian rather than Lamarckian world, if that host goes extinct or its abundance is reduced severely, then the specialist may not be able to find another suitable host in time to avoid extinction itself. This kind of scenario is called the "gambler's ruin." That is, it does not matter how much you win, you can always lose it all if you remain in the game. Any short-term benefits of pronounced host specificity thus appear to be offset by a severe cost in the long term: increasing specialization may push lineages into evolutionary "blind alleys"[12] because the risk of extinction increases as a species becomes less capable of responding to any decrease in host abundance (by finding something new to eat or somewhere new to lay its eggs, etc.). Given this cost versus benefit argument, we might expect to find the world populated by roughly an equal number of specialists and generalists, if not by a majority of generalists. Instead, pronounced host specificity appears to predominate.[13]

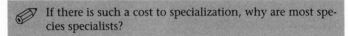

If there is such a cost to specialization, why are most species specialists?

There are two ways to answer this question, both of which depend upon the way we identify a specialist. First, the "cost" of specialization may be highly context dependent.[14] For example, although resource specialists may exhibit specialized morphologies that enhance their abilities to utilize a preferred host, those same morphologies can sometimes be used to exploit other hosts when the preferred one is in short supply (Robinson and Wilson 1998; "survival of the adequate": Brooks and Wiley 1986, 1988). In this case, we have mistakenly conferred the title "specialist" upon a species that has a narrow host preference but a wider host range. This mistake stems from incomplete descriptions of the specialist's biology.

Second, evolutionary history can "lower the cost" of specialization. Consider an insect that specializes on a resource associated with a single host-plant species in a given area. Now suppose environmental changes reduce the abundance of the host, so the survival of the insect depends on establishing a new association with a new host plant. If the resource preferred by the insect is widespread, the same environmental changes that reduce the abundance of the original host may increase the abundance of another plant species possessing the same resource. The insect species thus may be able to establish a "new" specialized association *without the cost of evolving novel abilities*. In this case, although the insect is a specialist, the resource it is tracking is not limited to a particular host species, which decreases the chances that host extinction will drive the

12. Mayr 1963; Flessa et al. 1975; Doyle 1976; Futuyma and Moreno 1988; Moran 1988; Thompson 1994; Thompson et al. 1997; Kelley and Farrell 1998; Conti et al. 1999; Kniskern and Rausher 2001; and Rausher 2001.

13. For example, Janzen 1988; Jaenike 1990; Futuyma 1991; Bernays and Chapman 1994; Thompson 1994; Carriere 1998; Janz and Nylin 1998; Norton and Carpenter 1998; Nylin and Janz 1999; Ashen and Goff 2000; Caillaud and Via 2000; and D'Haese 2000.

14. Kuris and Norton 1985; Futuyma and Moreno 1988; and Timms and Read 1999. See Johnson and Steiner 2000 for reviews of theories about the evolution of specialization.

insect out of the game (because the insect is capable, in effect, of placing bets on different numbers on the roulette wheel).

This observation, however, leads us to another seeming paradox:

 If the resource is widespread, why does the associate not colonize all available sources (and look like a generalist)?

There are two reasons why a widespread resource may not be accessible to an associate. First, the geographic distribution of a given associate may not coincide with the total geographic distribution of all suitable hosts. For example, allopatric speciation in species that share a widespread resource (plesiomorphic or homoplasious) will tend to produce a wide but patchy distribution of the suitable resource that could account for observations of pronounced host specificity (Pellmyr 1992a,b). If the associate is not distributed as widely as all of the hosts having the resource, there may be a limited number of suitable hosts species available in the geographic regions where the associate actually lives.

Second, aspects of host biology other than the possession of the appropriate resource may make some host species "invisible" or "inaccessible" to the associate even though they are sympatric. For example, if species A bearing resource "x" is highly abundant in a community, then the less abundant species B and C, which also bear resource "x," may not be "apparent" to a specialist on that resource (Feeny 1976; Courtney 1981, 1983, 1985; Wiklund 1984; Dobler et al. 1996). Such density-dependent factors would provide the appearance of close ecological tracking between the specialist and species A. Now, let us say that some environmental stressor decreases the abundance of species A, and C begins to proliferate rapidly. From the perspective of our specialist, nothing too untoward has happened because there is still plenty of resource "x" available (it just happens to come packaged as species C now). From the perspective of a biologist studying the interactions, the specialist has rapidly established a new ecological association (shifted from species A to species C), indicating an apparent breakdown in, then reestablishment of, specialization on a new resource (species C). In fact, there has been no breakdown in specialization at all in this case. This manifestation of ecological fitting could explain apparent rapid and virtually unconstrained evolution of novel specialized host associations even when the associate retains the ability to utilize more than one host.[15] The misunderstanding between the specialist and the biologist arises because *the*

15. Singer et al. 1971, 1992, 1993; Singer and Parmesan 1993; Singer 1994; Radtkey and Singer 1995; and Janz and Nylin 1998.

specialist is tracking the resource provided by the host species and not the species per se, while the biologist is tracking the host species, not the resource.
If a widespread resource is coupled with a counterdefense by some (but not all) hosts, or if the resource is present in different concentrations in different hosts, this may establish a further wrinkle in the coevolutionary fabric; that is, the associate may have a hierarchy of preferences even though it is tracking the same resource (**host rank order**).[16] The hierarchy arises because the costs of accessing the resource may not be identical across all host species. Such costs, in turn, will depend upon many different factors, including concentration of the resource, host density, and difficulty in extracting the resource. The sequence of host rank order may even vary among individuals in the same species (Singer et al. 1992). Once again, although the process is more complicated, the outcome may be the same as the simple case in which the resource is evenly distributed among different hosts; that is, the associates will be capable of demonstrating "conservatism in host plant choice in primary habitat and also innovative change in host plant utilization in novel habitats where the ancestral host plants are missing" (Wiklund 1981: 170). In other words, the associates will be "faux generalists," true specialists on a plesiomorphically (or, less often, homoplasiously) distributed resource.

HOST SPECIFICITY AND INFORMATION PROCESSING ABILITIES

Within the past decade, some biologists have begun investigating the relationship between host specificity and the ability of the associate to find a host. Courtney et al. (1989) presented a model based on the discovery that female butterflies have a hierarchy of preferences in plant species used for oviposition founded on different resources rather than on different availability of the same resource. They postulated that the butterflies would tend to specialize on the highest ranked and most abundant of the hosts in their hierarchy. For example, if an associate evolves the ability to utilize a host representing a novel resource, it may prefer that host, but may also *retain sufficient information to use the plesiomorphic resource.* In such cases, if preferred hosts are patchily distributed, then a generalist may appear to be a specialist. (Note that this condition is really just a case of incomplete knowledge about the host preferences of a given associate.)

Expanding on an earlier suggestion by Levins and MacArthur (1969), Bernays and Wcislo (1994; also Bernays 1996, 1998) extended the model

---

16. Singer 1971; Wiklund 1973, 1974, 1975, 1981; Janz and Nylin 1997; and Carriere 1998. For an excellent and extensive discussion, see Janz 1999.

by Courtney et al. to include two constraints based on butterfly biology. First, butterflies do not have very large brains, so Bernays and Wcislo argued that butterflies choosing among hosts make their decisions based on a relatively small number of key signals in their environment. This implies that if the information processing system of the butterfly is geared toward discriminating a hierarchy of host species, it will not have the capacity to discriminate accurately among plants within a species (host quality). There may in fact be a trade-off between number of species in the preference hierarchy and the butterfly's ability to determine host quality. Experimental tests with Nymphalini butterflies have confirmed this prediction: generalists are not as good as specialists at discriminating among different quality conspecific hosts (Janz and Nylin 1997; Nylin et al. 2000; see also Bernays and Funk 1999).

The second constraint arises because female butterflies do not have unlimited time to oviposit. If anything, time is pressing. Given this, females should become less choosy as time passes and they have not found a suitable (preferred) oviposition site. Logically, they should eventually stop seeking the best species and settle for ovipositing on the first available acceptable plant. Instead, "giving up" time appears to vary among females, even among those displaying an almost identical host-rank hierarchy (Nylin and Janz 1993). This scenario mirrors the situation in which females that have a limited amount of time to find an acceptable mate vary in their willingness to accept a less preferred male as time goes by.

The existence of this second type of host-rank hierarchy provides an additional way in which history can lower the cost of specialization. In this case, the historical accumulation of host-utilization abilities permits a species to be a specialist at any given time, showing conservatism in host choice in the given habitat, while retaining the ability to change host utilization when habitat change eliminates the preferred host species (as anticipated by Wiklund [1981]).

European Butterflies and Host Plants: A Scandinavian View. A large research group at Stockholm University has been investigating the question of host-preference evolution in papilionoid butterflies for almost 30 years (e.g., Wiklund 1973; Janz et al. 2001). Thousands of observation hours and mark recapture data from the field, combined with elegant experimental manipulations of adults and larvae in both the laboratory and the field, are finally beginning to reveal the secret lives of these butterflies. For the purposes of this example, we will focus on the Nymphalini, the group including the stunning Golden Angelwing, Painted Lady, Indian Admiral, and Comma butterflies. A search through published reports in the literature, combined with data collected by the research group itself, indicated that an association with a clade of plants called rosid 1b (Chase et al. 1993) was plesiomorphic for the entire superorder of butterflies (Janz and Nylin 1998). How would response to that plesiomorphic preference fare when the search moved from the familial to the species level?

In order to address this question, Niklas Janz (1999; Janz et al. 2001) reared 23 nymphaline species on the nine most commonly encountered families of plants throughout the butterflies' range. These experiments provided the researchers with a relative measure of host suitability from the perspective of the caterpillar. The caterpillar's view was added to that of the adult female (oviposition preference) and the results optimized onto a total evidence (morphology + molecular) phylogenetic tree for the Nymphalini (Nylin et al. 2001). The optimization unambiguously indicated that preference for the Urticaceae (stinging nettles: rosid 1b) was plesiomorphic for the clade (fig. 8.9), mirroring the results from the higher-level analysis. Ancestor 1 expanded its host range through the addition of the Salicaceae, a member of the sister group to the plesiomorphic host clade. Ancestor 2 underwent a reduction in host range by losing the Salicaceae and reverting back to the plesiomorphic host. Although the addition of hosts to the repertoire within the descendants of ancestor 1 appears quite exuberant, in most cases those additions involve other members of the plesiomorphic host clade (rosid 1b: potentially "faux generalists") or close relatives of that clade (other rosids). Most unexpectedly, however, all but two of the butterfly species retained the ability to survive on Urticaceae, indicating that changes in host preference did not involve one host supplanting another in the butterfly's favor. Three true specialists appear within the dynamic ancestor 1 group: *N. californica*, associated with a novel member of the plesiomorphic clade (the Rhamnaceae); *P. egea*, reduced to the original host Urticaceae; and *K. canace*, which shows the only true shift to a novel and very distantly related host group, the monocot Smilacaceae.

Overall, then, the preference hierarchy in these butterflies appears to have been built by the addition of plant groups either within the plesiomorphically preferred clade or within successively more distantly related groups. Even most of those "more distantly related" hosts, however, fall within the asterid + rosid clade, so it is tempting to hypothesize that the resource being tracked by the butterflies has not changed substantially over time. This might explain why most species that prefer plants other than Urticaceae are still capable of utilizing the plesiomorphic host if forced to do so. The dynamic waxing and waning of specialization and generalization displayed by this group is thus not so chaotic as it might initially have seemed. Indeed, it appears that substantial hierarchical structure can result from a rather simple process involving (hypo-

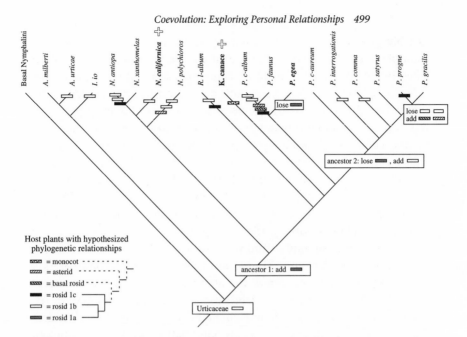

Figure 8.9. Coevolution from the butterflies' perspective. Preferred host clades are mapped onto a simplified phylogenetic tree for the Nymphalini butterflies (strict consensus of two equally parsimonious trees, CI = 0.402: Nylin et al. 2001). Basal Nymphalini = *Araschnia, Vanessa, Bassaris,* and *Cynthia* (additional host switching occurs here: see Janz et al. 2001 for discussion). *Cross* = loss of ability to survive on all previous hosts except the indicated group; *A.* = *Aglais, I.* = *Inachis; K.* = *Kaniska; N.* = *Nymphalis; P.* = *Polygonia; R.* = *Roddia.* The phylogenetic relationships among the plant groups are indicated to the right of the key (see discussion in Janz and Nylin 1998). *Solid lines* = hypothesized sister groups; *dotted lines* = more distant relationships. The plant clades include the following groups of relevance to the butterflies: rosid 1a = Salicaceae; rosid 1b = Cannabidaceae, Rhamnaceae, Rosaceae, Ulmaceae, Urticaceae; rosid 1c = Betulaceae, Fagaceae; basal rosid = Grossulariaceae; asterid = Ericaceae; monocot = Smilacaceae.

thetically) a response to a widespread resource, coupled with minor evolutionary changes in the resource that permit host expansion (by retaining enough chemical similarity) without requiring the origin of dramatically different associate abilities.[17]

HOST SPECIFICITY AND INDIRECT EFFECTS

Naturally, the story does not end here. Bernays and Graham added another layer of complexity to the issue when they proposed that observed host range in the wild might be narrow, not because of the direct interaction between hosts and associates, but rather because of indirect effects,

17. See Wahlberg 2001 for a discussion about the conservative nature of the chemical cue tracked by another group of Nymphaline butterflies, the Melitaeine.

notably predation by natural enemies of the associate.[18] Consider, for example, a mollusk that feeds on a single species of algae and is itself eaten by many different predators. The mollusk might eat the algae because of a long history of coevolution between the two lineages. The mollusk might also eat the algae because it grows in the most predator-free part of the habitat (Abrams 2000a,b). More likely, there is some combination of interactions between diner 1 (the mollusk) and dinner 1 (the algae) and between diner 2 (various fish, other mollusks, perhaps an echinoderm or two) and dinner 2 (the mollusk) that shape the "coevolutionary relationship" between the mollusk and the algae. This appears to be exactly the kind of dynamic for which diffuse coevolution has been developed (Barbosa 1988; Fox 1988).

Indirect effects provide a third way history can lower the costs of specialization. To understand this, we must return to the second paradox above: Why do associates not utilize all hosts having suitable resources? The answer is that resource availability is not the same as resource distribution. Trophic segregation and allopatry are two powerful "indirect effects" that may limit actual (versus potential) host utilization (the observed host range) at any given time. Oddly, these limiting factors actually lower the cost of specialization by increasing the chances that the associate will come into contact with appropriate "new" hosts if there is a change in the environment, leading to altered trophic interactions or to geographic dispersal.

HOST SPECIFICITY AND EVOLUTIONARY COSTS REVISITED

We have now discovered that various manifestations of phylogenetic history can lower the cost of specialization in host utilization in three fundamental ways. The first of these has to do with the nature of the resource (part of the nature of the [host] organism). The second has to do with the nature of the associate's abilities to find and utilize hosts (part of the nature of the [associate] organism). The third has to do with the distribution (apparency and availability) of host resources in the environment (the nature of the conditions). If (co)evolution is the result of interactions between the nature of the organism (host and associate) and the nature of the conditions, specialization is not as costly as we have previously thought. Being a specialist simply means optimizing any local situation and letting historical effects hedge your bets against future (and therefore unpredictable) changes.

18. Bernays and Graham 1988 and references therein; Rausher 1988; Bernays 1989; Bernays and Chapman 1994; and Abrams 2000a,b—but see also Courtney 1988; Jermy 1988; Schultz 1988; and Thompson 1988a.

History lowers the cost of specialization and host specificity in both preference and ability, eliminating the paradox presented by the existence of so many specialists.

The findings of the past decade or so thus turn our traditional question on its ear: if the cost of being a specialist is much lower than we thought, perhaps we should be asking, Is there a cost to being a generalist? The answer to this question is yes. In fact there are many such costs. A generalist may utilize a subpar host if search time is limited—remember that generalists cannot distinguish quality of hosts as well as specialists. The chance of making a mistake also increases as the range of possibilities increases (Bernays 1996, 1998). These two problems have direct ramifications for the fitness of the associate; experiments indicate that larval life history traits may be influenced greatly by host type. Given this, when is it beneficial to be a generalist? Sören Nylin (1988) examined this question using the butterfly species *Polygonia c-album* living along a north-south gradient. In the north, the environment is highly predictable with respect to butterfly reproduction: each year there will be more than enough time for a single clutch, but never enough time for two clutches. In the south, the environment is less predictable: in some years there is enough time for two clutches, in other years there is not. Nylin postulated that in environments sometimes favorable to two clutches per season, there will be increased specialization on hosts that support the fastest growth rate or shortest development time (1988; see also Scribner and Lederhouse 1992). This will limit the chances of wasting energy on a second clutch that does not survive, but will also limit the host range and select against "trying out" new hosts. In areas where there is never enough time for a second clutch, but plenty of time for a single clutch, the field is open to alternative oviposition strategies and the inclusion of additional species into the host repertoire. It appears that the answer to, When is it beneficial to be a generalist? is not, When you need to switch hosts in order to survive, but rather, When life is good and you can afford to experiment with new hosts. Experimental studies have corroborated this hypothesis.[19]

HOST SPECIFICITY AND MACROEVOLUTIONARY TRENDS

The effects of history on coevolutionary associations manifest themselves in more ways than just simple cospeciation. In fact, we would expect episodes of cospeciation to be predominantly byproducts of con-

19. Nylin and Janz 1993; Hodkinson 1997; Janz and Nylin 1997; Nylin et al. 2000; and Janz et al. 2001.

comitant vicariant speciation in host and associate (allopatric cospeciation). Furthermore, if the resource an associate needs is widespread across clade "A," it doesn't matter which member of that clade the associate colonizes first, it can always survive on any other members of the group if and when it comes into contact with them. Consequently, there are no predicted patterns of cospeciation/host switching *with respect to the host species or host phylogeny*. There may, however, be macroevolutionary patterns associated with the tracking of the *host resource*. There are three major considerations:

1. Trends in specificity. This could appear as (1) progressive specialization in coevolving lineages, which could be manifested as increasing host specificity (e.g., Smiley 1978); repeated production of specialists from generalists; or increasing specialization that limits host range (e.g., Crespi and Sandoval 2000) or (2) progressive generalization in coevolving lineages, which could be manifested as decreasing host specificity; repeated production of generalists from specialists; or decreasing specialization that expands host range. To date it appears there are no general "macroevolutionary trends" in the degree of specialization that can be applied to all clades. Some researchers have documented the evolution of specialists in clades of generalists (e.g., Kelley and Farrell 1998; Vermeij and Carlson 2000), while others have uncovered the evolution of generalists in clades of specialists (D'Haese 2000; Scheffer and Wiegmann 2000; Janz et al. 2001). Once again, it appears to be the details of each particular system that are important. If being a generalist is a relatively uncommon and transient phenomenon, then the perception that there is a trend toward or away from specialization may depend on the particular time slice you are studying.

2. Patterns of host switching. As discussed above, this can involve either switching to hosts representing novel resources or switching to novel hosts representing the same resource. The first possibility should be less common than the second, since it requires the evolution of new capabilities. (This scenario is analogous to the "colonizer + apomorphic trait" versus the "colonizer + plesiomorphic trait" discussed in chapter 7.) In addition, if the first kind of shift is associated with good times, it should not be correlated with particular episodes of environmental change, while the second may often be correlated with such episodes.

3. Congruence and incongruence between host and associate phylogenies at different taxonomic levels. If the host species represents the resource being tracked, we would expect to see congruence between host and associate phylogenies on all levels of comparison (assuming that we always use monophyletic higher taxa for comparisons). Any apparent

disagreement between the two would be due to either undetected extinctions or sampling error. By contrast, if the associate is tracking a widespread plesiomorphic resource but is capable of occasionally evolving novel resource preferences, then we expect to find higher-level congruence between the host and associate phylogenies (phylogenetic conservatism) that progressively disappears as we move toward species-level investigations.[20]

The next example touches on all three of the above categories.

Leaf Eaters and Leaf Providers: A North American View. Douglas Futuyma and coworkers have been addressing fundamental questions about coevolutionary mechanisms for almost two decades (see, e.g., Futuyma 1983, 1991; Futuyma and Kim 1987). They have focused their attention on a small (and thus tractable) group of insects, the North American leaf beetles in the genus *Ophraella,* which contains 14 species that dine on the foliage of composites in the Asteraceae from northern Canada to Mexico. Members of this research group have combined the power of phylogenetic and experimental approaches to search for the mechanisms underlying the ability to host switch. They have been particularly intrigued by the idea that successful host switching might be constrained by internal (genetic) factors. Their studies have followed a two-step cycle that is a continual feedback loop between the phylogenetic and experimental research:

1. What are the macroevolutionary patterns of host switching in the group? A total evidence phylogenetic analysis of morphological, allozyme, and mitochondria DNA sequences for 12 of the 14 leaf beetle species produced a single phylogenetic tree (Funk et al. 1995a,b). Optimizing general host groups onto that tree provided evidence for four episodes of host switching (fig. 8.10): from Astereae to Anthemideae, Eupatorieae, and Heliantheae independently, and from Heliantheae back to Astereae (Futuyma and McCafferty 1990; Funk et al. 1995c). This relatively conservative perspective was altered somewhat when genera rather than tribes of host plants were considered (fig. 8.11), mainly due to the paraphyly of *O. communa* (according to the phylogenetic tree, there are at least three macrospecies in this complex) and host switching within the Heliantheae (mainly by members of the *O. communa* complex). In fact, there are only two cases of possible cospeciation, that is, sister beetles existing on sister genera of plants: *O. notulata* + *O. slobod-*

---

20. For an early review of this pattern, see Mitter and Brooks 1983. See also Van Klinken 2000; and Janz et al. 2001.

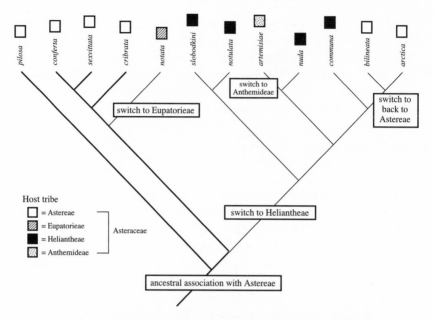

Figure 8.10. Leaf beetles and the Asteraceae they eat. Host tribe preferences mapped onto a total evidence tree for *Ophraella*. (Modified from Funk et al. 1995b.)

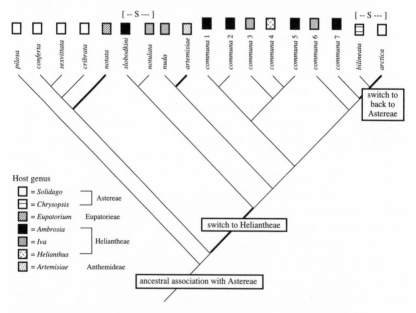

Figure 8.11. Leaf beetles and their associated plants: expanding the host resolution. S = host plants are sister groups; bold line = major switch to a different host tribe. (For details of major host switches, see figure 8.10.)

kini and *O. bilineata* + *O. arctica*. The rest of the insect associations exhibit rampant, but not random, host switching. For example, while there is substantial movement in the *O. communa* group, the switching is always within the Heliantheae. Even the two largest jumps, from *Solidago* to *Eupatorium* by *O. notata* and from *Ambrosia* to *Artemisiae* by *O. artemisiae*, still occur within the larger context of the Asteraceae. In other words, the host switching is not randomly distributed across all possible plants in the beetles' environment; most sister species of *Ophraella* feed on closely related host plants.

2. How much variability in feeding responses to, and survivorship on, new hosts does each beetle species display? This question was initially investigated using three different populations of *O. communa* (*communa* 2, 5, and 7 on fig. 8.11), all of which feed naturally on *Ambrosia* (Futuyma et al. 1993). Feeding and rearing experiments indicated a three-tier system of response by these beetles to nonpreferred hosts from the entire *Ophraella* range: (1) larvae refused to eat *Solidago* and *Artemisia Carruthii*; (2) larvae fed moderately on *Chrysopsis*, *Artemisia vulgaris*, and *Eupatorium*, but mortality on these hosts was almost absolute; (3) larval feeding, growth, and survival was highest on *Iva*. Genetic variation leads to differential feeding and survival, depending upon the host plant species, so some host switches are more likely than others even though all of the proffered hosts fall within the total range of *Ophraella*. In other words, if you are an *O. communa* beetle, you are not going to host shift to the *Solidago* plant growing beside your *Ambrosia* hosts, even if all of your host plants die. You simply do not have the capability to shift to that plant. If, however, you are lucky enough to live in an area where *Iva* species are also growing, then you may be able to find a new food source should something catastrophic happen to your *Ambrosia* hosts. The research also indicated that the probability of surviving a host shift is not random; of novel host plants offered, *O. communa* larvae fed and grew the best on the sister group (*Iva*) to their normal host (*Ambrosia*). Not surprisingly the two groups are more similar to one another in secondary chemistry (Heywood et al. 1977; Seaman 1982; Futuyma and McCafferty 1990) than they are to the other Asteraceae offered in the tests. This would appear to be an example of an associate species tracking a plesiomorphic resource when it moves to a novel host.

The feeding experiments were repeated for three other members of the leaf beetle clade, *O. conferta*, *O. artemisiae* (Futuyma et al. 1994), and *O. notulata* (Futuyma et al. 1995). The results of these experiments paralleled the study discussed above. Eighteen out of 39 tests of larval and adult feeding responses and 14 out of 16 tests of larval survival showed no evidence of genetic variation (that is, the leaf beetle species could not survive on the novel host plant). The amount of genetic variation that was present could be arranged in the following sequence: larval consumption > adult consumption >> larval survivorship (as predicted by Wiklund 1975), indicating that the bottleneck to a successful host shift occurred prior to adulthood. Once again, degree of variation was structured phylogenetically; that is, beetles were more likely to feed and survive on plants that were closely related to their normal host and, less strongly, to respond to the host plant their own sister species was feeding on. It is important to note that in no case did a leaf beetle show as strong a feeding response or as high a survival rate on an alternate host as it did on its normal host, so, although host switching may in some cases be possible, it is not a likely event when the preferred host is available.

There thus appear to be phylogenetically based host preferences and tolerances that constrain but do not prohibit the establishment of new associations through host switching. In other words, *Ophraella* species are as opportunistic as they can be given the abilities they inherited and the circumstances in which they find themselves. Occasionally an evolutionary novelty arises, permitting the colonization of a new type of host plant, but most evolution in the group has been variations on an ancestral (host) theme, even when there has been speciation by host switching. (Sound familiar? Remember the Nymphalini butterflies?) Other studies of chrysomelids are beginning to show much the same picture.[21] This suggests that the features documented by Futuyma et al. for *Ophraella* may be general for all chrysomelids. But the phenomenon may be even more general than that. For example, all but one member of the *Phycomyza ilicis* species group of agromyzid dipterans inhabit various species of holly (hence their common name, holly leaf miners), but there is little congruence between the phylogeny of the insects and of their holly hosts (Scheffer and Wiegmann 2000). Similarly, at least some of the behavioral and life history traits critical to the establishment and maintenance of the complex mutualism between yuccas and their obligate yucca moth pollinators are plesiomorphic for the moth family.[22] This system, which has persisted over many species and a wide geographical area for at least 40 million years (Pellmyr and Leebens-Mack 1999, 2000), involves substantial host switching by yucca moths, but only among species of yucca.

---

21. Farrell and Mitter 1990; Dobler et al. 1996; Mardulyn et al. 1997; Dobler and Farrell 1999; Garin et al. 1999; Lingafelter and Konstantinov 1999; Gomez-Zurita, Juan, and Petitpierre 2000; and Gomez-Zurita, Petitpierre, and Juan 2000.

22. Pellmyr and Thompson 1992; Pellmyr and Huth 1994; Pellmyr, Leebens-Mack, and Huth 1996; Pellmyr, Thompson, et al. 1996; Pellmyr et al. 1997; Leebens-Mack et al. 1998; and Groman and Pellmyr 2000.

Speciation Again

In the previous four examples, we have identified numerous cases of speciation by host switching, but we have not yet discussed what mode of speciation accounts for those events. Speciation by host switching has traditionally been assigned to sympatric speciation. The new host is assumed to represent a new type of habitat and speciation is driven by ecological segregation.[23] This mode of speciation is problematic because it is at once theoretically attractive and perplexingly paradoxical. The attraction lies in the model's invocation of adaptive processes to drive speciation. The paradox is twofold. First, once colonization of a new type of resource (habitat or host) within the ancestral species range has occurred, the probability that the new resource will exert strong directional selection pressure on the colonizing population should be high for species displaying pronounced habitat specificity. These species are more tightly coupled to their resource bases and thus should be more sensitive evolutionarily than their generalist counterparts to changes in that component of their environment. However, the likelihood of a habitat change occurring in the first place is decreased for species that only respond to a small number of cues. Therefore, the species least likely to successfully colonize new habitats in the first place (specialists) are the ones most likely to speciate as a result of any such switch; while the species most likely to successfully colonize new habitats (generalists) are the least likely to speciate as a result of the interaction.

 Is there a way out of this problem that does not involve abandoning host switching as a potential influence on the speciation of coevolving lineages?

We believe there is, but it requires that we reexamine our concept of speciation by host switching, this time from the associate's perspective rather than the host's perspective (for an extensive discussion, see Brooks and McLennan 1993b). A new host may or may not represent a new kind of resource, but it definitely represents new geography. Thinking of hosts as geographical areas eliminates the host specificity paradox because, once a species has moved into a new area (a new host species), the potential for speciation is created by the effects of geographical isolation on gene flow, even if the host does not represent a new kind of environment that provokes an adaptive response by the colonizers (although such responses can accelerate the speciation process). *Speciation by host switching is thus an example of peripheral isolates allopatric specia-*

---

23. See, e.g., Diehl and Bush 1989; Grant and Grant 1989; Tauber and Tauber 1989; Adamson and Caira 1994; and Jousson et al. 2000.

tion, not sympatric speciation (Brooks and McLennan 1993b; Funk et al. 1995b; Funk 1998).

The confusion about modes of speciation in coevolution was created in the first place by taking the "bird's eye view," or the geographic perspective of the hosts. When you take the "worm's (or beetle's) eye view," or the geographic perspective of the associates (which, after all, are the ones doing the switching), it is easier to see a new host species as a new piece of real estate, rather like a new island archipelago.[24] We believe it is time for a new view of coevolution and the origin of associations. This view must be based on the notion that associated species (e.g., phytophagous insects, parasitoids, parasites of all kinds) have their own individual evolutionary tendencies and fates (nature of the organism) and thus may exhibit geographic distribution patterns that can be determined, evaluated, and explained without reference to a priori theories about host evolution or the influence of hosts on the evolution of their associates.

So, as we discussed above, *host switching involves the movement of a small subset of a species into a new geographical area.* Once there, the colonizers will either simply add the new host to the species range of preferred habitats, or they will speciate (via a peripheral isolates mode). The interactions depicted in table 8.1 highlight two interesting, and unique, aspects of speciation by host switching: (1) the associate is not speciating in association with its "normal" host and (2) speciation is initiated by an active movement of the associate. New associations can also arise without the movement of an associate population onto a new host species. For example, an ancestral associate species inhabiting an ancestral host species can be subdivided geographically, resulting in the evolution of new associations (allopatric cospeciation: Brooks 1979, 1985; Brooks and McLennan 1991). There are three possible outcomes of such separation: (1) the associate will speciate and the host will not; (2) the host will speciate and the associate will not; or (3) both the associate and the host will speciate. Speciation can also be initiated by character changes within one associate deme that result in reproductive isolation of the new phenotype from another sympatric deme without involving any changes around or to the host; this is sympatric speciation from the associate's perspective. This type of speciation can be recognized by (1) the presence of sister species utilizing the same host species that (2) differ in some apomorphic ecological or genetic characteristics that could, in themselves, produce independent species. Both allopatric cospeciation

---

24. See also Janzen (1968, 1973a), who first suggested that hosts could be construed as islands on both ecological and evolutionary time scales.

Table 8.1. Modes of Speciation for the Associate in a Coevolutionary Interaction

Associate Sister-Species Distribution	What Initiates Speciation from the Associate's Perspective	Host Distribution	Speciation Mode of Associate
Allopatric	geographical changes subdivide an ancestral species (passive)	allopatric (same host species or sister species)	VICARIANCE
Allopatric	host switching (active)	sympatric (different host species)	PERIPHERAL ISOLATES
Allopatric	behavioral or ecological segregation of hosts via sympatric speciation (passive)	sympatric (sister species)	VICARIANCE (?)
Sympatric	interactions between two diverging populations (no movement)	sympatric (same host species)	SYMPATRIC

and sympatric speciation have one thing in common: new associations do not originate as a result of active colonization by the associate because both the host and associate maintain the ancestral association throughout the speciation process.

Although the hypothesis that host switching represents peripheral isolates, rather than sympatric, speciation was originally based upon examples from host-parasite associations (Brooks and McLennan 1993b), free living systems also fit the scenario. For example, Funk et al. (1995b) identified at least three speciation events following host shifts by members of *Ophraella* (see fig. 8.11) due to peripheral isolates speciation. They identified putative peripheral isolates as being species that had (1) host shifted and (2) displayed a much smaller distribution than their sister species without overlapping them (*O. artemisiae* versus *O. nuda*; *O. communa* versus the *O. bilineata* + *O. arctica* pair [tentative, pending the resolution of the *O. communa* species complex]; and *O. arctica* versus *O. bilineata*). Knowles et al. (1999) suggested that these episodes of speciation by host switching in *Ophraella* occurred in a burst during the Plio-Pleistocene, a period of tremendous environmental change, when many previously allopatric species would have been forced into the same locations by advancing glaciers, creating new opportunities for host switching onto hosts possessing the same resource as the host with which the species originally evolved (for additional examples, see Ward 1991; Brower 1996; Jiggins et al. 1997; Funk 1998). In a complementary fashion, Kerdelhue et al. (1999) found that each species of the hymenopteran genus *Ceratosolena* pollinated only a single species of *Ficus* (figs), except in areas of overlap between two or more species (of insects and figs). From the parasitological side, each eucestode species in the genus *Alcataenia* inhabits only one species of avian host in the family Alcidae (e.g., puffins, auks, and murres), but the phylogeny of *Alcataenia* does not match the phylogeny of the alcids. The biogeographic distributions of *Alcataenia* species, however, do conform to a documented dispersal corridor through the Arctic during the past 3–5 million years, which also seems to explain the evolution of some of the parasites of pinniped mammals occurring in the same area.[25] These all appear to be examples of host switching involving novel hosts possessing a plesiomorphic resource occurring in conjunction with environmental disturbance, as discussed above.

Studying More Than One Clade of Associates at a Time

As you recall from chapter 4, we use the power of phylogenetic systematic approaches in historical biogeography to identify episodes of simultaneous vicariant speciation (the historical backbone), peripheral isolates or sympatric speciation, postspeciation dispersal, and extinction. Proponents of both the ontology of simplicity (e.g., Page 1990, 1994; Charleston 1998; Page and Charleston 1998) and the ontology of complexity (Brooks 1985, 1990; Wiley 1988a,b; Ronquist 1995, 1997b, 1998; Van Veller and Brooks 2001) agree that the methodology of historical biogeography can be applied to studies of coevolving lineages. The same methodological principles should be applicable to the history of individual clades or associated clades in either a biogeographic or coevolutionary context because the explanation for homology, common history, is the same at all levels of biological organization. The important difference between phylogenetic systematic reconstruction of genealogies, associations between lineages and areas, and associations among lineages is that the explanations for homoplasy on those different levels invoke different processes (Brooks 1990). For example, when we speak of homoplasious traits for individual taxa, we postulate multiple origins of traits mediated by the effects of natural selection, common developmental pathways, or both. Homoplasy in a coevolutionary analysis, on the other hand, can indicate episodes of either (1) host switching, when the homoplasies are parallelisms or convergences (including multiple species in association with the same host and widespread taxa) or (2) extinction, when the homoplasies are reversals (missing taxa). So long as we are aware of the level at which we are analyzing evolution, we should

25. Hoberg 1986, 1992, 1995, 1996; Hoberg and Adams 1992, 2000; and Hoberg et al. 1997. See also Brooks and McLennan 1993b.

be able to invoke appropriate mechanisms to explain departures from common history.

When we study interspecific ecological associations involving two or more different clades, we expect that the history of association between coevolving lineages might be a complex one. In the following pages we will discuss a relatively simple example, which nonetheless will point the way to substantial amounts of interesting complexity. You should recognize that we are using primary and secondary BPA.

Do Parasitic Worms Ape Their Host Phylogeny? The classification of the great ape species has been the subject of often heated debate ever since 1758, when Linnaeus proposed that human beings be classified with other apes. Parasitologists have long attempted to decipher what, if anything, parasites could tell them about human origins and genealogical relationships. As noted in the introduction of this chapter, Fahrenholz (1913) suggested that relationships among biting lice supported the hypothesis that humans were more closely related to great apes than to any other primate group. A distinguished list of parasitologists provided additional evidence for this hypothesis (especially for the relationship between humans and the African great apes, *Pan* and *Gorilla*) from helminths and from malaria (*Plasmodium*).[26]

Brooks and Glen (1982) presented the first phylogenetic systematic analysis examining the relationship between great apes and their parasites based on the pinworm (nematodes) genus *Enterobius* studied by Cameron (1929), Sandosham (1950), and Inglis (1961). Subsequently, a group of hookworms (*Oesophagostomum* [*Conoweberia*]: Glen and Brooks 1985) and a new phylogenetic analysis of *Enterobius* and relatives (Hugot 1999) were added to the database. This gives us two groups with which to perform a BPA.

1. Because we are primarily interested in the relationships between the worms and their great ape hosts, we have simplified the phylogenetic trees for the parasites by collapsing each monophyletic group that inhabits other primates into a single lineage. The phylogenetic tree for the pinworms (fig. 8.12) depicts three such lineages: *E. brevicauda* + *E. bipapillatus* + *E. macaci* (6), which inhabit cercopithecines; the *Colobenterobius* clade (7), the members of which live in colobines; the *Trypanoxyuris* clade (8), the members of which plesiomorphically inhabit New World monkeys. The phylogenetic tree for the hookworms (fig. 8.13) de-

---

26. Helminths: Cameron 1929; Sandosham 1950; Inglis 1961; Dunn 1966; and Kuhn 1967. Malaria: Garnham 1973; and Peters et al. 1976.

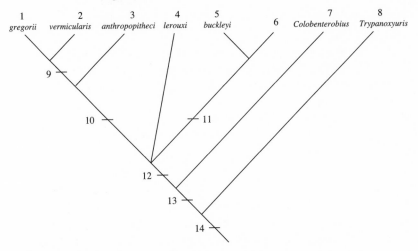

Figure 8.12. Simplified phylogenetic tree for the pinworm genus *Enterobius* and relatives (consensus of three trees) coded for coevolutionary analysis. 6 = the monophyletic group *Enterobius brevicauda* + *E. bipapillatus* + *E. macaci.* (Modified from Hugot 1999.)

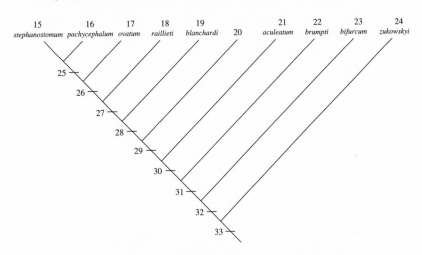

Figure 8.13. The phylogenetic tree for the hookworm genus *Oesophagostomum* (*Conoweberia*). 20 = *O. xeri* and *O. susannae.* (Modified from Glen and Brooks 1985.)

picts one such lineage: the rodent-inhabiting clade of *O. xeri* + *O. susannae* (20).

2. Table 8.2 is the matrix for primary BPA.

3. The current estimate of anthropoid phylogeny is shown in figure 8.14. When we are studying coevolution, we use the phylogenetic relationships of the associates to reconstruct a *host cladogram* rather than

Table 8.2. Primary Matrix Listing the Distribution of Pinworm and Hookworm Clades among Their Primate Hosts, along with the Binary Codes Representing the Phylogenetic Relationships for Both Clades

Host	Species	Binary Code			
New World monkeys	8, 24	0000000100	0001000000	0001000000	001
Cercopithecines	6, 16, 21, 23	0000010000	1111010000	1010111111	111
Colobines	7, 21, 22, 23	0000001000	0011000000	1110000001	111
Hylobates	17, 18	??????????	????001100	0000011111	111
Pongo	5, 19	0000100000	1111000010	0000000111	111
Gorilla	4, 15	0001000000	0111100000	0000111111	111
Pan	3, 15, 23	0010000001	0111100000	0010111111	111
Homo	1, 2, 15, 23	1100000011	0111100000	0010111111	111
Rous	20	??????????	????000001	0000000011	111

NOTE: Species 1–8 = *Enterobius;* Species 15–24 = *Oesophagostomum;* Rous = rodents; ? = species missing from the host group.

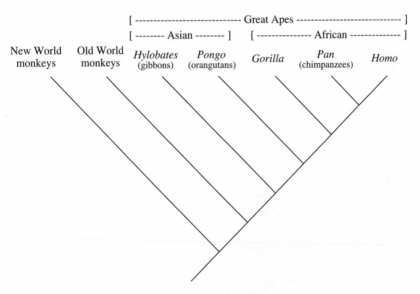

Figure 8.14. Phylogenetic tree for the great apes (Hominidae) and their relatives. The Old World monkey clade includes the Cercopithecinae (e.g., baboons, macaques) and the Colobinae (e.g., colobines, langurs). The New World monkeys include such famous representatives as the howler monkeys and the tamarins. For recent phylogenetic analyses, see Shoshani et al. 1996, Barriel 1997, and Goodman et al. 1998 (whose hierarchical classification scheme we follow in this example).

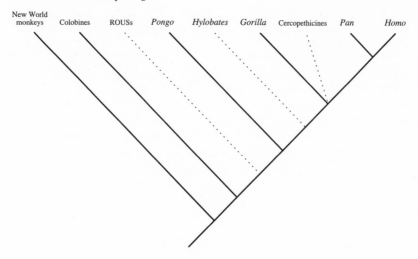

Figure 8.15. Primary BPA of the parasite data. *Bold lines* = historical backbone of association between the primates and their pinworms and hookworms; *dotted lines* = host misplaced on the tree, indicating interactions other than cospeciation. ROUSs = rodents.

an *area cladogram.* Primary BPA produces two equally parsimonious host cladograms (the consensus is shown in fig. 8.15), which depict substantial, but not complete, congruence with the phylogenetic relationships among the great apes and between the great apes and the Old and New World monkeys (compare the host cladogram with the host phylogeny shown in fig. 8.14). Although this suggests a long history of association between these parasites and the great apes, it also indicates "there are more things in heaven and earth" than just cospeciation. In particular, the cercopithecines (considered a single taxon for purposes of this study) should be the sister group of the colobines, *Hylobates* (gibbons) should be the sister group of the rest of the great apes (*Pongo + Gorilla + Pan + Homo*), and to date no one has suggested that any rodent is, in fact, a primate.

4. Computer-assisted secondary BPA (table 8.3) produces one most parsimonious host cladogram with a CI of 97%, but that host cladogram places *Oesophagostomum bifurcum* (species 23) as the ancestor to species 31, rather than as its sister. This misplacement occurs because *O. bifurcum* has a widespread distribution (it is the only species to inhabit *Homo, Pan,* and the cercopithecines), which, as you recall from chapter 4, short-circuits the computer programs. Figure 8.16 depicts the host cladogram derived by maintaining the phylogenetic relationships of *O. bifurcum* in accordance with the requirements of assumption 0. It differs from the machine-generated host cladogram only in the placement of *Homo₂* and *Pan₂* (the computer pulls these two taxa into a polytomy at the base

Table 8.3. Secondary Matrix Listing the Distribution of Pinworm and Hookworm Clades among Their Primate Hosts, along with the Binary Codes Representing the Phylogenetic Relationships for Both Clades

Host	Species	Binary Code			
New World monkeys	8, 24	0000000100	0001000000	0001000000	001
Cercopithecines$_1$	21, 23	??????????	????000000	1010000001	111
Cercopithecines$_2$	6	0000010000	1111??????	??????????	???
Cercopithecines$_3$	16	??????????	????010000	0000111111	111
Colobines	7, 21, 22, 23	0000001000	0011000000	1110000001	111
Hylobates$_1$	18	??????????	????000100	0000001111	111
Hylobates$_2$	17	??????????	????001000	0000011111	111
Pongo	5, 19	0000100000	1111000010	0000000111	111
Gorilla	4, 15	0001000000	0111100000	0000111111	111
Pan$_1$	3, 15	0010000001	0111100000	0000111111	111
Pan$_2$	23	??????????	????000000	0010000000	011
Homo$_1$	1, 2, 15	1100000011	0111100000	0000111111	111
Homo$_2$	23	??????????	????000000	0010000000	011
Rous	20	??????????	????000001	0000000011	111

NOTE: The Cercopithecines are listed as three separate taxa, while *Hylobates, Pan,* and *Homo* are each listed as two separate taxa. Species 1–8 = *Enterobius;* species 15–24 = *Oesophagostomum;* Rous = rodents; ? = species missing from the host group.

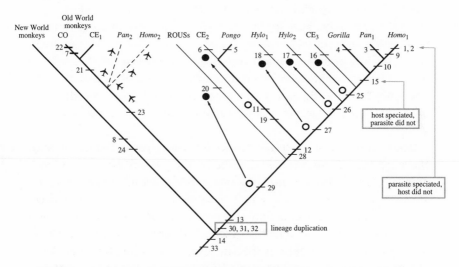

Figure 8.16. Secondary BPA of the parasite data. *Bold lines* = historical backbone of association between the primates and their pinworms and hookworms; *dotted lines and airplanes* = postspeciation dispersal by the parasite into new hosts; *open circle* (ancestor) + *solid circle* (peripheral isolate) = instance of peripheral isolates speciation via host switching; *numbers accompanying slash marks* = species codes (from figs. 8.12 and 8.13); CE = cercopithecines; CO = colobines; *Hylo* = *Hylobates;* ROUSs = rodents.

of CO + CE$_1$) and in having a CI of 100%, which indicates the most parsimonious representation of the data.

The secondary BPA provides a much fuller account of the evolutionary associations between the two parasite groups and their hosts than does primary BPA. It appears that hookworms and pinworms were already jointly associated in the common ancestor of the anthropoid primates (ancestors 14 and 33 on fig. 8.16). In order to pinpoint the exact origin of those associations, we need to expand the phylogenetic scope of the analysis. There is a strong historical backbone of cospeciation among these organisms (bold lines on fig. 8.16). This includes three apparent cases of lineage duplication in the ancestor of the Old World monkeys + great apes (denoted by the joint occurrence of ancestors 30, 31, and 32). As with all lineage duplications, these may represent ancient cases of sympatric speciation. The secondary BPA suggests that there have been six episodes of host switching. The most obvious of these is the occurrence of the hookworms *Oesophagostomum xeri* + *O. susannae* (20) in South African rodents, which represents an episode of *speciation by host switching* from apes to rodents. This type of exchange is evidently not completely unique; Hugot (1999) documented instances of other pinworms (*Trypanoxyuris*) speciating after they had transferred from New World monkeys to squirrels. The occurrences of *O. pachycephalum* (species 16) in cercopithecines and *O. ovatum* (species 17) and *O. raillieti* (species 18) in *Hylobates* also appear to represent episodes of speciation by host switching from one or more species ancestral to the African great apes. The only pinworm to host switch was species 6 (recall that "6" actually refers to the ancestor of a monophyletic group of three species in cercopithecines; see fig. 8. 12). Ancestor 6 colonized the cercopithecines from the *Pongo* lineage, where it subsequently diversified in the new host group.

The occurrence of *Oesophagostomum bifurcum* (species 23) in *Pan* and *Homo* represents *postspeciation dispersal* by the parasite from Old World monkeys (dotted lines + airplanes on fig. 8.16). Because *Homo* and *Pan* are sister groups, it is actually most parsimonious to postulate that *O. bifurcum* colonized the common ancestor of *Pan* + *Homo*, in which case a single host switch explains the observed increase in host range. Once there, the hookworm did not speciate again, even though its new host did, producing the chimpanzee and *Homo* lineages. The presence of another hookworm, *O. stephanostomum* (species 15), in *Gorilla, Pan,* and *Homo* also represents a case of the parasite failing to speciate with its host. In other words, the expanded host ranges of *O. stephanostomum* and *O. bifurcum* were not gained by rampant host switching but rather by the parasite species *not speciating when its host did. Oesophagostomum bifurcum, O. brumpti* (species 22), and *O. aculeatum* (species 21) inhabit various Old World monkeys. At this level of analysis we cannot tell if this large host range was achieved through association by descent (hosts continued to speciate while the parasites did not) or widespread colonization. In order to investigate this problem further, we need a detailed phylogeny for the Old World monkeys and intensive inventories of the parasites of all primate species.

One advantageous feature of using BPA to study the evolution of host and associate relationships is that you can add historical biogeographical information to help explain your findings. Geographic distribution data for these two parasite groups suggest complex, and as yet incompletely resolved, movements of primates throughout the tropical regions of the Old World. *Oesophagostomum bifurcum* and *O. brumpti* occur in both Africa and Asia, while *O. aculeatum* is endemic to Asia, possibly suggesting dispersal from Africa to Asia (although *O. xeri* + *O. susannae* are endemic to South Africa). *Oesophagostomum blanchardi* (species 19), *O. raillieti* (species 18), and *O. ovatum* (species 17) are Asian endemics, while *O. pachycephalum* (species 16) and *O. stephanostomum* (species 15) are African endemics, supporting an explanation of dispersal from Asia to Africa by the common ancestor(s) of, not surprisingly, the African great apes (*Gorilla, Pan,* and *Homo*). And what about us? Stepping out of the African forest into the African savannah, our ancestors may have added *O. bifurcum* (remember, *O. bifurcum* may have been acquired by the ancestor of humans and chimpanzees) to their parasite repertoire through host switches from Old World monkeys. The parasites then indicate that *Homo* moved to Asia, leaving *O. stephanostomum* behind in Africa (humans in Asia do not have this hookworm). The occurrence of *O. bifurcum* in humans in Africa and Asia is not surprising; the host cladogram suggests that *O. bifurcum* occurred in Old World monkeys on both continents long before *Homo* arrived in Asia. This period may also have seen the differentiation of the pinworms *E. vermicularis* (species 2) and *E. gregorii* (species 1), sister species in the same host (us) produced in the absence of host speciation. From the parasites' perspective, this is either a case of sympatric speciation or an interesting form of peripheral isolates speciation, in which the geographic dispersal of the hosts acted like the drifting of an island arc. In any event, once again we find evidence of host switching associated with environmental change (in this case mediated by host dispersal).

This apparently is not an isolated case in the evolution of human parasites. Hoberg et al. (2000) recently discovered evidence that two cases of speciation by host switching, correlated with the shift from scavenging to predation (hunting) more than a million years ago in Africa, explained the association between humans and tapeworms in the genus *Taenia* that infect humans (including the beef tapeworm, *T. saginata,* and the pork tapeworm, *T. solium*). Furthermore, they found that *T. saginata* and *T. asiatica,* sister species occurring in Africa and Asia, respectively, diverged from each other between 750,000 and 1,700,000 years ago, perhaps at the same time *E. vermicularis* and *E. gregorii* were differentiating (but in different parts of the host).

Finally, we must ask, Why do cercopithecines appear not to have cospeciated with pinworms and gibbons appear not to have cospeciated with hookworms, when the evidence of speciation by colonization shows they are perfectly good hosts? Why do gibbons appear to have no pinworms at all in the wild, when they can acquire various species in zoos, including *E. vermicularis* from their human handlers and visitors? These questions await further research, including better inventories of gibbon parasites in the wild.

Are Parasite/Vertebrate Systems Different from Insect/Plant Systems, and If So, Are Those Differences Qualitative or Quantitative?

Some researchers believe that there are qualitative differences between the systems, because parasites of vertebrates appear to be more dependent on their hosts than are phytophagous insects, so much so that they are actually incapable of host switching. As discussed previously, this perspective stems from orthogenetic thinking about the nature and evolutionary significance of parasitism. It continues today in a number of forms, ranging from the proposal of "phylogenetic specificity" (Burt and Jarecka 1982; Humphery-Smith 1989) through assertions that parasites are the most extreme resource specialists on the planet (Adamson and Caira 1994) to the maximum cospeciation approach. The default for this latter approach forbids host switching, explaining apparent episodes of host switching as being due to unrecognized lineage duplication and extinction (Hafner and Nadler 1988, 1990; Page 1990, 1994; Charleston 1998; Page and Charleston 1998). If we use a method (BPA) that does not forbid any evolutionary processes a priori, however, we routinely uncover evidence for lineage duplication, extinction, *and* host switching in a variety of parasite/vertebrate systems.[27] Despite numerous contributions to the issue,[28] there is still no agreement among researchers about

27. See examples in Brooks and McLennan 1993b. For recent examples involving mange mites of ungulates and various groups of African ticks and their vertebrate hosts, see Ramey et al. 2000; Cumming 2000; and Smith 2001, respectively.

28. For example, Futuyma and Slatkin 1983; Nitecki 1983; Wheeler and Blackwell 1984; Kim 1985; Whitfield 1990; Toft et al. 1991; Brooks and McLennan 1993b; and Clayton and Moore 1997.

the best way to uncover macro[co]evolutionary patterns. We believe we cannot resolve the "it's all cospeciation" versus the "coevolutionary mosaic" issue by recourse solely to the macroevolutionary patterns. *The only way to resolve the debate is by experimental investigations demonstrating whether parasite species are, or are not, capable of host switching.*

We have already indicated that gibbons in zoos acquire human pinworm species from their handlers and visitors. Studies performed in more rigorous laboratory settings have also demonstrated that at least some parasite species are capable of surviving in many different hosts, including hosts they never associate with in nature (Watertor 1967; Blankespoor 1974; O'Grady 1987). We believe the explanation for some episodes of host switching is thus the same as the explanations we have discussed for phytophagous insects: parasite species are tracking plesiomorphic resources. If there is a difference between the two systems, it is more likely to be one of degree, not kind. Parasites may ultimately show more cospeciation and less host switching than has been discovered in insect/plant systems. This, of course, is a testable prediction. *But* it can only be examined if we use (1) a method that does not eliminate the possibility of host switching a priori (BPA), followed by (2) experimental demonstration that the episodes of host switching identified by the BPA could actually have occurred. For example, if the colonizer is tracking a plesiomorphic resource, both it and its sister species should be able to survive in each other's hosts, just as has been documented for insect/plant systems.[29]

SUMMARY

> No generalization is worth a damn, including this one.
>
> Mark Twain

Like a good murder mystery, unraveling the secrets of coevolution requires that one know both the means (the nature of the organism) and the opportunity (the nature of the conditions) before proposing an explanation for the historical and ecological context of host specificity. The mechanisms underlying observed specialization depend upon whether the specificity originates from relatively fixed aspects of the biology of the associate or of the host (means), or a spatio-temporally unique combination of many factors affecting the host and the associate (opportunity). To carry the analogy further, every good murder mystery requires motive, in this case, the resource. A detailed description of the

29. For example, studies by the Stockholm Group and by Futuyma and coworkers discussed above. For an example using vertebrate parasites, see O'Grady 1987.

resource being utilized by the "specialist" has been the missing piece in our mystery for decades, one that has caused substantial confusion in discussions about the evolution of generalists and specialists because *the host is so often confused with the resource.* With your new understanding of character evolution in a phylogenetic context, however, you can now see that there are really three categories of associates: (1) true generalists (many different resources, many different hosts); (2) true specialists (one resource, one host); and (3) faux generalists (one resource, many different hosts). It is the species that fall into category 3 that have created the nightmare for theoretical discussions, because they are capable of such apparently rapid changes in host association (remember our discussion of widespread resources), which you could not detect if you focused on the hosts rather than the resource. These species are also, presumably, the type of specialist that are the most buffered from the orthogenetic evolutionary blind alley, the ones most likely to survive a catastrophic loss of a host or two. The obvious question now is, How many species fall into these three categories? In particular, How many "true specialists" really exist in nature?

Our summary for this chapter thus mirrors our summary for the previous chapter. First, it is possible to combine phylogenetic, biogeographic, ecological, and behavioral data to ask questions about the relative contributions of historical and nonhistorical influences on the spatial and resource allocation patterns in coevolving associations. Second, historical influences on coevolution are ubiquitous. And third, patterns of coevolution are generally not the result of a single influence, most certainly not of simple cospeciation.[30] We expect the interaction between motive, means, and opportunity to produce dynamic and complex coevolutionary patterns. For example, Thompson (1988b, 1994, 1997, 1999a,b,c) proposed an elegant view of coevolution in which an associate species is in reality a mosaic of coadapted populations. Although each population is relatively specialized as a result of local adaptive responses (see also Williams 1992; Pratt 1994; Gomulkiewicz et al. 2000), the entire species is more generalized than any one population

30. For additional examples, see mites and vertebrates/insects: O'Connor 1984, 1985, 1988; rusts and plants: Hart 1988; various plant-pollinator systems: Armbruster 1992; Chase and Hills 1992; Farrell et al. 1992; McDade 1992b; Miller 1992; and Stein 1992; various eukaryotic groups parasitizing vertebrates: Brooks and McLennan 1993b; figs and their pollinating wasps: Machado et al. 1996; protists and termites/cockroaches: Grandcolas and Deleporte 1996; haemosporinid microbes and their vertebrate/dipteran hosts: Carreno et al. 1997; cleaner wrasses and their client fish: Grutter and Poulin 1998; brood parasitism in finches: Klein and Payne 1998; and Nyman et al. 1998, 2000; sawflies and willows: Nyman et al. 1998; feather mites and birds: Dabert and Mironov 1999; and Mironov and Dabert 1999; and gall-inducing and inquiliine thrips of the subfamily Phlaeothripiinae and their plants: Morris et al. 1999.

due to the exchange of information by reproduction. This coevolutionary deployment of microspecies of hosts and associates is analogous to the deployment of cichlid microspecies in African Rift Lakes (chapter 3). We think the bulk of macroevolutionary changes in coevolving systems fit Thompson's scenario, which can be summarized as "mosaic micro-coevolution leads to mosaic macro-coevolution" (see also Brown et al. 1997; Dilley et al. 2000).

We must therefore study each system on its own merits, and that requires both a phylogenetic framework and the use of experimental studies to delineate the complexity. The only thing we expect to find is that, given the means and the opportunity, coevolutionary systems will be as opportunistic as possible. In other words, most coevolution will be diffuse. The importance of diffuse coevolution should not be surprising to you. We have already seen its effects in communities, for example, influencing the assemblage of snakes and anoles on Hispaniola Island (chapter 7). Overall, we expect diffuse coevolution to provide substantial indirect effects on all biological assemblages, whether we call them coevolutionary modules, communities, or biotas.

The Frontiers of Discovery in Coevolution
THE COEVOLUTIONARY MOSAIC MEETS THE MAJOR TRANSITIONS

The greatest challenge for our voyage of discovery is to explain host switching. Such switching can range from relatively simple moves (which is always a good place to start) to coevolutionary "heroic episodes" or "major transitions" (sensu Maynard Smith and Szathmary 1995) in the history of life on this planet. There are (at least) three categories of such transitions:

1. Taking advantage of opportunities permitted by the rules of the game. In general evolutionary terms (Maynard Smith and Szathmary 1995), such coevolutionary transitions are the easiest (least costly in time and genetic change) to achieve but provide the least evolutionary payoff when successful. They require only altered environmental conditions (e.g., bringing associate species into contact with novel hosts bearing the plesiomorphic resource). Or, to return to Darwin, such transitions require only a change in the nature of the conditions, and not the nature of the organism.

2. Changing the rules of the game. The evolution of new abilities may permit an associate species to adopt a new type of host association, that is, to move to a host that possesses apomorphic resources. This kind of coevolutionary change is more difficult to achieve than simply switching among different species of hosts representing the same resource, be-

cause it requires a change in the nature of the organism. If successful, such transitions may provide a greater evolutionary payoff than the ones discussed above, because a new type of host establishes a new set of rules for the game; for example, the new hosts may differ ecologically or behaviorally from the ancestral hosts, or the new resource may be distributed differently among the members of the new host group.

Most comparative phylogenetic studies have focused on this kind of evolutionary or coevolutionary transition. Ross (1972a) set the stage for these studies with his proposal that only approximately one out of every 30 speciation events in a variety of insect clades was correlated with geographic dispersal from the primitive climatic zone to a derived one, or with shifts from the plesiomorphic condition to any apomorphic state in life cycles, adaptive syndromes, or host preferences. He concluded that such shifts were consistent with, but much less frequent than, phylogenetic diversification. Furthermore, he felt there were no predictable patterns explaining the shifts that did occur and suggested such transitions constituted a biological "uncertainty principle."

3. Changing the game. This type of coevolutionary transition represents our greatest challenge. It involves explanations for the origins of major modes of life that depend on specialized associations with other species, such as herbivory, parasitism, or mutualism. Such transitions are the most difficult to achieve because they require the origin of least two novel traits on the part of the associate (i.e., two changes in the nature of the organism), one providing the means of utilizing a dramatically novel resource and the other providing the means of locating that resource. The conjunction of two such improbable innovations would tend to make such transitions either permanently irreversible or reversible on such long time scales that they would serve as constraints on future evolution in the lineage (Brooks and Wiley 1988; Maynard Smith and Szathmary 1995; Brooks 1998a). As great as the cost is, the evolutionary payoff for such transitions can be very large, which may be why they are often characteristic of species-rich groups. For example, initial assessments of the codiversification between flowering plants and phytophagous insects suggest that phytophagy has originated relatively rarely throughout the millions of years that insects have inhabited this planet, despite the fact that the vast majority of insects are phytophagous.[31] And although comparable studies of parasite-vertebrate systems are only just beginning (see Brooks and McLennan 1993b; Poulin 1998), similar patterns have begun to emerge. For example, a single origin of

31. Mitter et al. 1988; Mitter and Farrell 1991; Wiegmann et al. 1993; Farrell and Mitter 1994; and Farrell 1998.

parasitism produced the flukes, monogeneans, and tapeworms (the majority of platyhelminth species), while the vast majority of the nematodes parasitizing vertebrates also arose from a common ancestor.

Let us examine these three categories of transitions by returning to the leaf beetles (*Ophraella*) (fig. 8.17). It appears that much of the speciation in the group took place within the context of the plesiomorphic resource (host switching among members of the same tribe: category 1). Many of those events evidently occurred in a burst during the Plio-Pleistocene, a time of substantial environmental change (Knowles et al. 1999: see chapters 3 and 6 for other studies suggesting periodic bursts of speciation). This implies that rates of speciation may have increased during that period of environmental crisis because the beetles were given the opportunity to host switch (through, e.g., increased sympatry of previously allopatric hosts bearing the plesiomorphic resource as organisms were forced together in rapidly shrinking refuge areas). Colonization of more distantly related host plants or movement between tribes (category 2), encompasses fewer events (see fig. 8.10 versus fig. 8.11), and as yet there is no indication that these events were correlated with each other

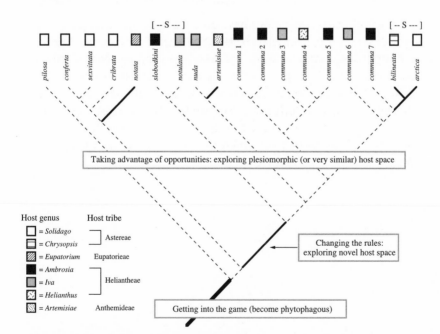

Figure 8.17. Three different types of transitions in coevolution mapped onto the phylogeny for *Ophraella*. *Dotted lines* = taking advantage of an opportunity; *bold lines* = changing the rules of the game; *thick line at base of phylogenetic tree* = changing the game (this shift happened prior to the origin of *Ophraella*). S = host plants are sister groups.

or with any particular episode of environmental change. These host switches are more difficult to understand. When and how did they happen? For example, this may be the place to examine the generality of Nylin's proposal that host switching to a novel resource is more likely to happen during "good times" (environmentally speaking).

And finally, the fact that leaf beetles are herbivores in the first place is even older and more phylogenetically conservative, at least for members of the clade comprising the Chrysomelidae (all of which are herbivores). Eventually, we will want to know when and how herbivory arose in the first place (category 3); was it a feature of the first coleopteran ancestor, or is it even older?

 Even though only two pairs of recently evolved sister species of *Ophraella* show possible cospeciation patterns, this system nonetheless has evolved in a context highly influenced by phylogenetic conservatism. There are two critical take-home messages from this and other studies. First, *history does not equal cospeciation*. Second, *historical effects in coevolution need not be simple in order to be important*.

Coevolutionary associations have been studied with increasing intensity ever since Ehrlich and Raven (1964) published their pioneering work on butterfly–plant interactions. The debate concerning the evolutionary processes underlying extant patterns of association has been particularly vigorous, but until recently the discussions generally have not incorporated phylogenetic information. The studies discussed in this chapter clearly demonstrate how beneficial a dialogue between "pattern" and "process" can be, particularly when that dialogue occurs among members of the same general research group.

Chapter 9

Biodiversity:
Exploring the Future

[O]ur understanding of the application of conservation biology to the problem of management of the natural world cannot be considered apart from an attempt to understand the
Human Condition in all its richness and contradiction.

Kiester 1997b:163

One of the most significant scientific contributions to our
understanding of the natural world in the last generation of the twentieth century was the discovery that many species are disappearing from
this planet, and that *Homo sapiens* is primarily responsible for those
disappearances. The magnitude of this "biodiversity crisis" is hotly debated (see, e.g., Ghilarov 1996), but no one questions the underlying
premise that the state of the biosphere should be a profound concern
for science and society.[1] The survival of our species and the quality of
life for future generations depends on our interactions with other species
on this planet. Although most countries have identified biodiversity as
a critical natural resource, the term "resource" is often equated solely
with the economic bottom line. This in itself would not be such a problem if the bottom line were served by strategies that are sustainable in
the long term. Instead, the lure of the quick return has prevailed, and
few cultures have adopted measures to ensure that natural resources are
monitored against overexploitation. People committed to "conservation" and those committed to "development" thus generally find themselves on either side of a widening and increasingly hostile rift, passionately coveting seemingly incompatible goals. Compromises proposed to
satisfy both parties seldom satisfy either, not to mention the wildlife involved.

1. Raven 1980, 1988, 1992; Wilson 1985, 1988, 1990, 1992; Ehrlich and Wilson 1991;
Janzen 1991, 1992, 1993, 1994a,b, 1996, 1997a,b,c, 1998, 1999, 2000a,b; Pimm 1991;
Soulé 1991; Mayden 1992b; Smith et al. 1993; Janzen and Hallwachs 1994; Marini-Bottolo
1994; Savage 1995; Epstein 1997, 1999; Pimentel et al. 1997; Reaka-Kudla et al. 1997;
Ponder and Lunney 1999; and Williams 2000.

More unfortunately, those compromises, as well as policies mandated to forestall the overall crisis, have often been based upon too little scientific information about the organisms in question to be considered educated decisions for their conservation. For example, until the final decade of the twentieth century, conservation biologists generally did not incorporate information about the historical origins of either species or species characters into their models. Proposals such as the theory of island biogeography (MacArthur and Wilson 1967; Wilson 1985, 1988) and discussions about speciation and extinction rates were based upon the assumption that communities progress toward an equilibrium number of species (e.g., Van Valen 1973; Raup 1981, 1985; Raup and Sepkoski 1982). Once at equilibrium, community membership was expected to turn over on a regular basis. Any sudden drop in the number of species within a community would thus be balanced by a sudden increase in the number of species colonizing the community. Simple extrapolation from this local and nonhistorical perspective to evolution might indicate that increased extinction rates would be rapidly matched by speciation rates of equal magnitude. Without evidence that this actually occurs, however, this is a dangerous assumption on which to base conservation policies.

Ricklefs (1987) was among the first researchers to question this perspective seriously. He made three significant observations. First, the equilibrium number of species in any local community is a subset of a regional pool of species, which are produced and maintained by evolutionary processes occurring on larger spatial scales and longer temporal scales than the dynamics explained by constructs such as the MacArthur-Wilson model. This means we cannot assume that increased rates of local extinction will affect evolutionary rates in any compensatory manner. Second, so long as the regional pool of species is maintained, local extinctions can be reversed. This means that increased rates of local extinction will be matched by increased rates of colonization only to the extent that local extinctions do not cause any of the species in the regional pool to become extinct. And finally, depletion of the regional pool of species can have a strongly detrimental effect on local diversity, since local diversity is a subset of that regional diversity. This was a clear mandate for biologists to focus their attention on the historical, that is, macroevolutionary, components of biodiversity.

Evolution binds all organisms in both a common hierarchy of life on this planet and a common hierarchy of processes shaping the universe. This web of existence is strongly influenced by the irreversible and indelible effects of time. Because of this, the potential of the future is hidden within the constraints of the past in biological systems. Biodiversity thus represents a continuum across a variety of scales, encompassing ecological, and phylogenetic components within a temporal/spatial framework or fabric.[2] We need to buy time by gathering more information faster and putting it in a widely accessible predictive framework earlier so we can begin making rational, long-range decisions informed by all available data. We have stated throughout this book our belief that the predictable parts of biology are those slowly evolving components of biodiversity embedded in phylogenetic history. We have also stated our belief that as a historical science, biology comprises both discovery modes and evaluation modes. We have tried to show the usefulness of this approach in establishing protocols for studying evolutionary mechanisms. For biodiversity, discovery is embedded in basic taxonomic inventories and the integration of taxonomic identity and all elements of natural history summarized in predictive, that is, phylogenetic, classifications. In this chapter we discuss some ways the research program we have discussed in the previous chapters can contribute to biodiversity science.

SYSTEMATICS AND THE BIODIVERSITY CRISIS

> The systematics community first called attention to the impending biodiversity crisis and has played the primary role in documenting it.
>
> Savage 1995:673

Each day we learn more about the importance of the documented portions of the biosphere. Against this backdrop, it is sobering to reflect on the fact that we have not documented, and thus do not understand, the overwhelming majority of the biosphere's total diversity (Barnes 1989; Cotterill and Dangerfield 1997; Lunney 1998). Consequently, we often have no idea about what we have already lost or might be losing. At best, we have only incomplete information on how to preserve what remains. This puts biologists in the weak position of having to advocate extreme caution in the advent of human development projects linked to loss of habitat and diversity on the basis of what we do not know, and on the fear that our ignorance will lead us to make mistakes both in the short and the long terms. Caution cannot become a synonym for stasis or inaction. We cannot be satisfied with simply slowing the rate at which species are lost or habitats are destroyed, because extinction is an

---

2. Ricklefs 1987; Barrowclough 1992; Eldredge 1992a,b; Platnick 1993; Hoberg 1997a; and Ricklefs and Miller 1999.

irreversible process. We can never bring back an extinct species, and its potential to play a role in the survival of our species is lost forever. Each species lost thus represents an irreversible loss of evolutionary potential, the very potential that has been the source of biotic recovery from ecological perturbations and environmental disasters on a global scale in the past (Jablonski 1991). Over the last decade, therefore, biologists have begun to take a stronger, proactive stand, advocating immediate action to document the world's biodiversity. Sadly, and paradoxically, this push for documentation has occurred at a time when the people who have the special skills required to collect, describe, and classify the world's biodiversity (systematists) are themselves an endangered group.

The Convention on Biological Diversity (CBD: Glowka et al. 1994) designated ecosystems management and sustainable development as the fundamental organizing principles for managing global biodiversity. Biologists and managers quickly realized that the current inventory of the world's species is far too limited to implement this mandate properly and that a critical shortage of trained taxonomists prohibits any ready solution to the problem (e.g., Gallagher 1989). The United Nations Environment Program (UNEP) in biodiversity, called DIVERSITAS (Loreau and Olivieri 1999), coined the term "taxonomic impediment" to refer to this critical lack of global taxonomic expertise that prevents initiation and completion of biodiversity research programs.[3] This concern led to Systematics Agenda 2000 (SA2000), an intensive professional inventory of the value of taxonomic expertise to this planet, and a set of recommendations for revitalizing systematic biology and justifying the allocation of resources necessary to carry out such a revitalization.[4] In 1998 and 2000, the Conference of the Parties (COP) to the Convention on Biological Diversity endorsed a Global Taxonomy Initiative (GTI) to improve taxonomic knowledge and capacity to further national needs and activities for the conservation, sustainable use, and equitable sharing of benefits and knowledge of biodiversity (GTI 1999; Cresswell 2000). The GTI has three major components: (1) Taxonomic Inventory; (2) Storage of Information about Species in Predictive Classifications; and (3) Electronic Management of Systematic Knowledge Bases, including museum collections.

---

3. SA2000 1994; Blackmore 1996; Hoagland 1996; PCAST 1998; and Ponder and Lunney 1999.

4. Eldredge 1992a,b; Forey et al. 1994; SA2000 1994; Brooks et al. 1995; Claridge 1995; Cotterill 1995; Cracraft 1995; Davis 1995; Eshbaugh 1995; Jones 1995; Lauder et al. 1995; McNeely 1995; Miller and Rossman 1995; Prance 1995; Savage 1995; Simpson and Cracraft 1995; Wheeler 1995; Balick 1996; Blackmore 1996; Monson 1996; Oliver 1996; Richardson 1996; Rossman and Miller 1996; Vane-Wright 1996; Veccione and Collette 1996; and Ponder and Lunney 1999.

The Importance of Basic Taxonomy and Taxonomic Inventories

The inventory and preservation of global biotic resources is an enormous challenge (Stork and Gaston 1990). The most recent estimates of species diversity on this planet have ranged from 5 to 30 million, only 1.7 million of which are currently known (Wilson 1988; Hawksworth 1995). Many of these species will never be known because humans are changing natural ecosystems 1,000–10,000 times faster than evolutionary processes have replaced extinct species in the past (Wilson 1988). Inventory is central to sustainable management of any resource; running a successful business requires an inventory of the goods and services provided to customers as well as a management plan to ensure that the business can always provide customers with the goods and services they need. This is as true for long-lasting conservation of biodiversity resources as it is for any major corporation.[5] Policies directed at the preservation of natural resources must be cognizant of the patterns, types, and magnitude of biodiversity. The inventory of species and their ecosystems is thus a necessary and critical phase in achieving the objectives of the CBD. There is not, of course, universal agreement that taxonomic inventories are desirable, but such objections typically arise from academic turf wars far removed from the day-to-day struggles of biologists and biodiversity managers working at ground zero. For example, Renner and Ricklefs concluded that "the current fad of biodiversity inventories will pass rather soon as a central focus of conservation. This having happened, conservationists will be able to concentrate on more practical issues concerning the preservation and rational development of natural ecosystems, and systematists will be more free to work on the intellectual development of their discipline and its reconciliation with biology as a whole" (1994:78). Although Renner and Ricklefs were expressing a sentiment shared by others (e.g., Hopper 2000), the systematics community saw their statement as extremely aggressive. Minelli (1994), Beattie (1994), Cotterill et al. (1994), Rejmanuk et al. (1994) and Pine (1994) disagreed on two counts. First, the intellectual development of systematics, especially with respect to phylogenetic analysis, and its reconciliation with biology as a whole has been one of, if not *the,* major themes of evolutionary biology during the past 25 years. We suggest this reconciliation has been largely accomplished, illustrated in part by the studies described in this book. Second,

---

5. For example, Janzen 1991, 1992, 1994a,b, 1997a,b, 1999; Sandlund et al. 1992; Sandlund and Schei 1993; Janzen and Hallwachs 1994; Brooks 1998b; and Softing et al. 1998.

> Ŷ The lack of taxonomic inventories is largely responsible
> for our failure to preserve biodiversity: you cannot pre-
> serve, rationally develop, understand, or legally protect
> species whose existence has not been documented.

A common sports adage states, One should never change a winning game and always change a losing game. So far, we have been engaged in a losing game. The traditional hands-off management approach to conservation of species and systems has produced few instances of success. The North American fish fauna serves as an excellent example, but the problems are not unique to this fauna or the associated agencies. Deacon et al. (1979) compiled an inventory of 251 rare species (not including any recently extinct species). A decade later, 26 of the species originally included were absent from the updated list, but 16 had been removed because of enhanced inventory information pertaining to status or taxonomy and 10 because they had become extinct. "[N]ot a single fish warranted removal from the list because of recovery efforts" (Williams et al. 1989:2). In fact, the updated list included 139 *new* taxa. "Comparison of the 1979 and 1989 lists indicates that recovery efforts have been locally effective for some species, but are clearly lagging behind deterioration of the overall fish fauna" (Williams et al. 1989:2). During the last century, the extinctions of 3 genera, 27 species, and 13 subspecies have been documented for North American freshwater fish alone (Miller et al. 1989). Under the existing policies of conservation, this degradation of aquatic biodiversity will continue. On a global basis, the situation is similar; people's lives are not improving, and we continue to lose biodiversity on a daily basis.

The Importance of Storing and Managing Biodiversity Information in Predictive Classifications

A crucial element in preserving biodiversity is managing information about the known and the yet to be discovered and described species on this planet. To do so requires knowledge of phylogenetic relationships; phylogenetic classification systems are the most predictive information systems about organisms and their places in the biosphere.[6] Why should this be? The most persistent units of biodiversity are species, genealogical systems that store and transmit information leading to the emer-

---

6. For example, Wheeler 1990, 1993; Erwin 1991; Brooks and McLennan 1991, 1993a; Faith 1991, 1992, 1996; Brooks et al. 1992a,b, 2000, 2001; Sanderson et al. 1993; Stiassny 1993; Forey et al. 1994; SA2000 1994; Humphries et al. 1995; Moritz 1995; Simpson and Cracraft 1995; Faith and Walker 1996; and Janzen and Gamez 1997.

gence of ecosystems with their complex interactions. Species are thus vessels of future potential as well as living legacies of past modifications and stasis, shaped by millennia of biotic and abiotic interactions partitioned into ecologically coherent units. They are history embodied. The predictable parts of biological systems are the stable elements, form and function, autecological and synecological, that have persisted through evolutionary time. Shared history allows us to make predictions, and this saves time and money, two resources that are in short supply in battling the biodiversity crisis. For example, one of the most significant recent contributions to cancer therapy was the discovery of a substance called Taxol, produced by a tree species called the American yew. When the therapeutic properties of Taxol became known, plant taxonomists suggested that a close relative of the American yew, the European yew, might well produce a compound similar to Taxol; in short order, the anti-cancer agent Taxotene was discovered. Urbatsch et al. (2000) recently utilized a phylogenetic analysis of coneflowers and their relatives to suggest which species are most closely related to, and thus most likely to possess therapeutic properties similar to, Echinacea. The power of systematics to "buy time" is embodied in phylogenetic classifications (SA2000 1994; Simpson and Cracraft 1995; GTI 1999).

A Plea for Preserving Systematic Biology

Systematics analyzes the historical thread that binds life together.

Simpson and Cracraft 1995:670

In a few years, there are likely to be too few or no experts who can identify arachnid, insect and molluscan vectors of human disease.

Davis 1995:705

Systematic biology and the biosphere share one important characteristic: both may be facing imminent extinction. We have insufficient taxonomic expertise across all components of diversity to address global needs for survey and inventory in a timely manner (Davis 1995; Goodfellow et al. 1999). Systematists are changing from their traditional status as "collectors of things" to being "managers of information." It is no longer important, or even relevant, to have more specimens of a particular species in one collection than are found in any other collection; rather, it is important to know how much information is available about each species. It appears to us that, at the moment, society at large is not willing to invest huge sums of money in classical set-aside conservation projects, nor to invest in ever-expanding museum collections serving only as repositories of material accessible to an ever-decreasing number of specialists, especially if their only concern is esoteric research (e.g., Davis 1996; Cotterill 1997a,b, 1999; Brooke 2000).

The ongoing internal debate among systematic biologists about whether systematics should be a service discipline or an independent research discipline threatens to weaken systematic biology in its efforts to make a significant contribution to society and thereby ensure its own survival. Already too many ecologists and biodiversity managers believe that systematists are only good for providing names. The fundamental importance of systematic biology stems from its being *both* an essential service profession and an essential element of virtually every area of modern evolutionary biology (Ferris and Ferris 1989). Systematic biologists must understand that biodiversity preservation is based on such issues as economic development, conservation, solving major environmental challenges, and limiting the impacts of emerging pathogens and parasites. Inventories that seek to identify the critical components of biodiversity are requisite in achieving these goals, which are mutually beneficial to the scientific community and the general population. They are an essential part of helping to save the biosphere by helping to improve the socioeconomic status of as many people as possible. Ecologists are also grappling with these issues at the moment (Lubchenco et al. 1991; Benson 2000).

The call for more inventory work in biodiversity represents a tremendous opportunity and challenge for systematic biology. Taxonomic identifications based on morphology, ecology, and behavior provide the basis for assessing biodiversity. As yet, we have no cheap, convenient, exhaustive technology for using molecular markers to identify species and describe their natural history in the field—this is not Star Trek, and we have no tricorders. Consequently, all assessments of the basic units of biodiversity, including assessments of their molecular genetic status, are derived ultimately from morphological determinations made by those few surviving taxonomists and biologists with taxonomic expertise. All the expensive technology in the world will not help if you do not know whether you have collected a snake or a caecilian. And to exacerbate matters, the world of the small number of new systematists who are being trained is progressively a world of ultracold freezers, cryotubes, automated sequencers, and computers in a rapidly expanding flood of G's, C's, A's and T's. It is less and less a world of the natural history of each species, which is the essential information that biodiversity inventories need to provide.[7] We fully endorse Donoghue and Alverson's (2000) call

7. Greene 1986b, 1994b; Greene and Losos 1988; Janzen 1993, 1994a,b, 1996; New 1999; Nielsen 1999; Brooks and Hoberg 2000; and Donoghue and Alverson 2000.

for a new Age of Discovery to document this planet's biodiversity, complementing the call for an International Biodiversity Observation Year (Wall et al. 2001).

Systematic biology has long been charged with the responsibility of naming and classifying the species on this planet, and modern phylogenetic methods have produced maximally efficient modes of storing and transmitting information through classifications. The Internet gives us a powerful mechanism for disseminating enormous amounts of information quickly and widely. Consequently, all those interested in preserving, managing, and sustainably using biodiversity should have a vested interest in supporting a strong systematic biology. Within the scientific community, however, taxonomists do not have a history of close and cordial interactions with other specialists, not to mention with each other. There are many reasons for this (e.g., Hull 1988; Funk 2001), but we must be able to overcome the historical constraints of sectarian competition for academic positions and prestige. The extent to which the scientific community can help preserve biodiversity is the extent to which each participant can give up something of his or her own immediate personal agenda to help achieve a greater good.

COMPARATIVE PHYLOGENETIC STUDIES IN THE SERVICE OF BIODIVERSITY SCIENCE
What Are We to Protect, and How Do We Know?

The Convention on Biological Diversity states that species are the basic units of biodiversity, and a growing number of countries have enacted legislation to protect endangered species. It is thus not surprising that the controversy about species concepts (see chapter 3) has spilled over into biodiversity and conservation biology,[8] where management efforts have recently focused on protocols for designating **Evolutionarily Significant Units** (ESUs).[9]

Conservation agencies and especially governments have the greatest influence in dictating the fate of our biodiversity. They view wildlife species and ecosystems as part of a regional, national, or global patrimony that has value in and of itself for aesthetic, educational, historical, economic, medical, ecological, and scientific reasons. They aim to prevent wildlife species from becoming extirpated or extinct and to provide for

8. For example, Karr 1990; O'Brien and Mayr 1991; Brooks et al. 1992; Mayden and Wood 1995; Mayden and Kuhajda 1996; Sites and Crandall 1997; Zamudio and Greene 1997; Walker et al. 1998; Austin 1999; and Brooks and McLennan 1999.

9. Ryder 1986; Moritz 1995; Bowen 1998; King et al. 1998; Nammack 1998; Roe and Lydeard 1998; Riddle and Hafner 1999; Taylor 1999; and Crandall et al. 2000.

the recovery of those that are extirpated, endangered, or threatened as a result of human activity. Often, however, they have no concept of exactly what they are protecting. A tacit assumption that species must be real entities underlies most of these conservation measures. For example, consider a species of fish widely distributed throughout the streams and ditches in your area. If one population goes extinct because the Department of Highways comes through and scours a ditch, the entire species obviously does not go extinct. In fact, because conservationists believe that the collection of populations representing a species exhibits more geographical and ecological cohesion and temporal persistence than the populations themselves, they believe it is reasonable to recolonize that ditch with representatives of the same species from another area, and to expect those new colonizers to function in the same manner as their resident predecessors. It is for this reason that species are recognized as the basic unit of biodiversity, and species names are used in ecological and biogeographical analyses (Kelt and Brown 2000).

Efforts to preserve and sustain biodiversity must therefore include a view of species that will allow all conservation agencies and governments to apply the same objective standards when making policy decisions with regard to individual species. We believe the hierarchical view of discovery and evaluation presented in chapter 3 can be adapted to serve as a template for making policy decisions on a case by case basis. That protocol will produce a decisive answer to the question of whether or not a population or collection of populations under consideration for special conservation status is, or is not, a distinct evolutionary species (macrospecies). Implementing this kind of decision-making process calls for government, industry, and the academic sectors to work together toward a common goal. That is, while either side in any dispute about species identity can present evidence directed at the same issue, *the type of evidence required must be clearly spelled out in advance.* This is a crucial step. Without a clearly defined way to assess the evidence, discussions (hearings, trials) can be held hostage to the "my expert is better known, older, comes from a more powerful institution, etc. than your expert" ploy, which reduces the decision to one based upon personalities and money rather than data. We offer three principles to help guide the assessment of specieshood presented in chapter 3 within the context of a dispute about "conservation versus development."

> **Precautionary Principle (akin to the medical dictum "do no harm"):** Make no decisions that will knowingly result in destruction of biodiversity, so always opt for preserving evolutionary (macro)species.

Macrospecies are self-contained information systems irreversibly split from all others. This status can be confirmed unambiguously with a combination of population genetic, phylogenetic, and natural history (primarily ecology and reproductive biology) studies (see also Mayden and Kuhajda 1996; Lydeard and Roe 1998; Parkinson et al. 2000; Szalanski et al. 2000). First, establish whether the population(s) in question is diagnosably distinct, on the basis of fixed genetic differences, from its sister group (Has speciation been initiated?); then, ask if those differences have been translated into irreversible lineage splitting (Has speciation been completed?).

How Many Cantils Are There, Anyhow?

> Walking in their habitats fosters a special kind of respect for nature, an urgency that both humbles and enlivens us. Perhaps most important, with venomous snakes we contemplate violence and mortality without implications of real evil, devoid of anthropocentric traps laid by fur, feather and facial expression; with pitvipers and their unusual infrared vision, we arrive at the very cusp of mystery, the illusive but tantalizing limits on empathy. And when we know snakes as living creatures in nature, venomous or otherwise, they can take on grander meaning. Sidewinders *(Crotalus cerastes)* and Peringuey's Adders *(Bitis peringueyi)* epitomize the demands of life on shifting desert sands, and flying-snakes *(Chrysopelea)* reflect the unique botanical characteristics of Asian rain forests. Each of the more than 2,700 species of snakes embodies special relationships with its environment, and the earth would be poorer without them.
>
> Greene 1997:305

There are approximately 157 species of pit vipers inhabiting temperate to tropical areas around the world. Some of these species will be familiar to anyone who has walked in the deserts of Arizona (western diamondback rattlesnake) or the rain forests of Cost Rica (eyelash pit viper). Avid hiker or not, almost everyone is familiar (if only by name) with snakes in the genus *Agkistrodon,* the copperhead *(A. contortrix)* and the cottonmouth or water moccasin *(A. piscivorus)* of the southeastern United States, and the cantil or castellana *(A. bilineatus)* of Mexico and Central America. The cantil has traditionally been divided into four allopatric subspecies, one widely distributed along the west coast of Mexico, Guatemala, and El Salvador *(A. b. bilineatus),* and three with more restricted ranges *(A. b. taylori* in northeastern Mexico, *A. b. russeolus* farther south along the coast of the Yucatan peninsula, and *A. b. howardgloydi* along the Pacific coast of Honduras, Nicaragua, and northern Costa Rica). All *Agkistrodon* species, even the "widespread" ones, are threatened with fragmentation and destruction of their preferred habitats. Needless to say, habitat loss is of particular concern to species with small, restricted ranges. Bearing this in mind, Parkinson et al. (2000) investigated the question of how many macrospecies, masquerading taxonomically as "subspecies," were contained within the genus.

Phylogenetic analysis of sequences from three mitochondrial genes and two tRNAs produced strong support for the tree shown in figure 9.1, recognizing *Agkistrodon contortrix* (copperhead) as the sister species of *A. piscivorus* (cotton mouth) + the four *A. bilineatus* "subspecies" and *A. b taylori* (Taylor's cantil) as the basal member of the *A. bilineatus* clade. The analysis indicated that Taylor's cantil is as genetically distinct from the other three *A. bilineatus* subspecies as it is from the cottonmouth, supporting previous morphological assessments that *A. b. taylori* is "completely distinct" from all other members of the genus (Burger and Robertson 1951; Gloyd and Conant 1990; Knight et al. 1992). All lines of evidence, molecular, morphological, and the allopatric distribution, convinced Parkinson et al. that Taylor's cantil should be elevated from a subspecies to a distinct (macro)species *(A. taylori)*. What about the remaining three subspecies of *A. bilineatus*? The current disjunct distributions of these cantils cannot be unambiguously explained by the data at hand. It is possible that the ancestor of *A. bilineatus* dispersed across Middle America, only to be fragmented at a later time (e.g., via climatic changes

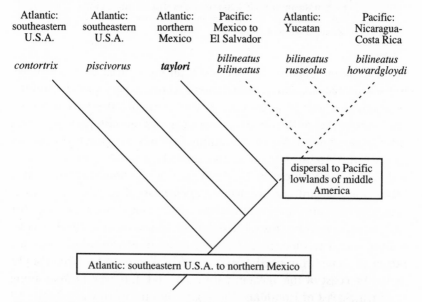

Figure 9.1. Searching for macrospecies. Phylogenetic tree based upon molecular characters for snakes in the pitviper genus *Agkistrodon*. Dotted lines = subspecies of *A. bilineatus*. The newly elevated macrospecies is highlighted in bold (Modified from Parkinson et al 2000).

that eliminated suitable interconnected habitats). It is also possible that the disjunct distributions represent movements by founder populations tracking the opening and closing of dispersal corridors (e.g., Stuart 1954; Wilson and McCranie 1998). Whatever the scenario, these subspecies have not been isolated from one another for a long enough period to produce either morphological or molecular differentiation. For example, a population in Guatemala displays morphological characteristics of all three subspecies! *Agkistrodon b. russeolus* and *A. b. howardgloydi* show a sequence divergence of only 0.003. The divergence between these two subspecies and *A. b. bilineatus* is much greater, 0.018, but is still only about half of the divergence between *A. taylori* and *A. bilineatus*. This does, however, raise the intriguing possibility that the subgroups *A. b. bilineatus* and *A. b. russeolus* + *A. b. howardgloydi* are in the process of speciating right before our eyes. If we cannot argue for the protection of each subspecies individually because they are not distinct macrospecies, we could perhaps argue that the two groups have unique research value. They are, in fact, living laboratories, offering us the opportunity to study the process of speciation *in situ*, rather than just extrapolating backward from the end product. Such opportunities are indeed rare.

 We believe that following this portion of the "assessing specieshood" protocol discussed in chapter 3 (see fig. 3.25) will show that many of the presumptive microspecies are actually good evolutionary (macro)species.[10]

Macrospecies may comprise many microspecies, any of which may become macrospecies if they experience an irreversible splitting event. It may thus be argued that the fewer microspecies we protect, the fewer options for continued evolutionary diversification there will be (Crandall et al. 2000). But that is not necessarily true. Evolutionary potential is in the dynamics of the evolutionary process over extended periods of time, not in current diversity at any point in time. This is what makes macrospecies more than the sum of their parts (Avise 1989a,b, 1992, 2000). It is true that microspecies are the source of macrospecies, but it is equally true that most microspecies never become macrospecies; in fact, many of them go extinct over time scales relevant to evolution. Suppose our ditch population represented the only source of a particular version of an enzyme that might enhance the survivability of the species should environmental changes occur in the future. Should it have been

10. For example, leopard frogs, birds of paradise, sturgeon, silky pocket mice, various passerines: Hillis 1988; Cracraft 1992; Mayden and Kuhajda 1996; Dimmick et al. 1999; and Zink et al. 2000, respectively.

given special status as an ESU (Moritz 1995; Tagliaro et al. 1997)? We believe not. If the ditch population did not show any fixed genetic differences from the rest of its species, then we can assume the same ancestral genetic architecture that gave rise to the mutant enzyme still exists in the rest of the species (for an empirical example, see Szalanski et al. 2000). Opting to isolate a particular microspecies for conservation, based on contemporaneous genetic or ecological considerations (Crandall et al. 2000), will always be ambiguous and will possibly be shortsighted from the standpoint of maintaining evolutionary potential (Brooks and McLennan 1991, 1993a; Brooks et al. 1992).

Note also that the mode of initiating speciation is irrelevant to determining species status. At the same time, it is important to recognize that allopatric speciation is the predominant mode of speciation, and that allopatric speciation requires range expansion, either to set the stage for the process (allopatric mode I) or to initiate it (allopatric mode II). Recognition of this leads us to assert that we should always maintain at least some microspecies within any given macrospecies, connected by dispersal corridors. This will help ensure that additional diversification will be possible when and if conditions permit.

> **"Species are evolutionary, not political units" Principle:** If a species' range is bisected asymmetrically by a political boundary, the species does not automatically become an ESU in the political area where its fewest members live. Such status must be argued on the basis of evidence other than population size in particular areas.

This principle reasserts the primacy of biology over politics. We must think of a species' conservation in terms of the species' biology rather than political boundaries. Species, after all, do not recognize our political borders. You do not get to be an ESU just because you represent a geographically apparent yet evolutionarily ephemeral microspecies.

Finally, who is going to pay for all this work by systematists and evolutionary biologists? This leads us to the third guideline.

> **"The Polluter (or Developer or Importer) Pays" Principle:** Any protocol for making a management decision should place the responsibility and financial burden of demonstrating no harm will be done upon those who wish to engage in behavior that might place a species in jeopardy. In legal terms, "the disputed biological entity is an ESU until proven otherwise."

The Agony of Choice

Even when we have an good understanding of the macrospecies in question and of their more or less well-differentiated parts, we are often still confronted with conflicting socioeconomic agendas that threaten biodiversity. In most cases, we simply do not have enough time to document the natural history of each member of a clade well enough to make fully informed decisions about conservation measures. Ecologists have long used diversity indices to produce coarse-grained comparative "snapshots" of relative species richness in communities and ecosystems, which can be used to help make priority decisions about the conservation of ecosystems.[11] During the past decade, a number of systematists have developed diversity indices that attempt to highlight certain species or groups with respect to various criteria that might be useful to biodiversity and conservation managers. These indices are not meant to suggest that once we have identified some magical "minimum" number of species necessary to maintain the biosphere, all the rest can go extinct. Rather, they are based upon the "agony of choice" (Vane-Wright et al. 1991; Crozier 1992)—a realistic assessment that we cannot save everything and we do not have enough time to collect all the information we need to make fully informed decisions about what to save.

Recognizing that the biodiversity crisis is not only upon us, it is overwhelming us, it follows that evolutionary biologists should at least make an attempt to provide a priority list of species. For example, we might decide to focus our conservation attempts on the old, relictual endemic species, on the assumption that they are least likely to continue to evolve and adapt to changing conditions, and thus more in need of help (Nixon and Wheeler 1990, 1992a,b; Vane-Wright et al. 1991). We could decide to save the youngest species, on the assumption that they are the most likely to continue to evolve and adapt to changing conditions (Erwin 1991; Faith 1991, 1992, 1996; Humphries and Williams 1994). Or, we could decide to save that subset of species that contains the most genetic diversity (Faith 1992, 1996; Moritz 1995; Tagliaro et al. 1997), perhaps the ultimate bet-hedging protocol. All of these ideas have been proposed (for a review, see Humphries et al. 1995), and all have potential drawbacks (May 1990, 1994; Krajewski 1994; Faith and Walker 1996; Nee and May 1997).

There are two fundamental issues at stake. The first is the question of whether or not we can tell which species are likely to produce more species and which are likely to be evolutionary dead-ends. We believe the answer to this problem is equivocal. The development of comparative biology during the past generation suggests strongly that while there are many clade-specific generalizations, there are few, if any, global

11. For example, Jaccard 1908; Whittaker 1965, 1972; McIntosh 1967; Hurlbert 1971; Putnam and Wratten 1984; Ricklefs 1987; Losos et al. 1989; Cousins 1991; Ricklefs and Miller 1999; and Izsak and Papp 2000.

generalizations. Consequently, the utility of any such indices, and the choice of which one to use, may depend on the clade being assessed and the questions being asked (e.g,. by biodiversity managers) about the future of its members. The second issue involves the assumption that species exist solely as independent genetic information systems. For example, Nee and May (1997) suggested that a loss of 50 percent of the world's species would result in only a 5 percent loss of overall genetic diversity, and a loss of 95 percent of the world's species would still leave us with 80 percent of the total genetic diversity of the biosphere. This proposal, although mathematically sophisticated, was based in part on the archaic assumption that species are interchangeable units. If extinction is not phylogenetically random (Bininda-Emonds et al. 2000; Purvis et al. 2000), and if ecological interactions among species are causally important in maintaining biosphere viability (e.g., Nijs and Impens 2000), then Nee and May's position is naïve at best and a recipe for disaster at worst. We must consider the possibility that some species may be more important than others in terms of maintaining ecosystem integrity, even if they are not obvious repositories of special genetic significance. Loss of that genetic information would certainly affect the ways in which and the extent to which the remaining biological diversity could survive and new diversity could evolve. We must also admit that we have no real understanding yet of how to measure the value of "genetic information." A complete description of the genotype does not guarantee absolute understanding of the phenotype, let alone the complicated network of interactions among phenotypes in particular environments. Genetics is important, but no more so than behavior, development, ecology, physiology, and morphology. Remember from chapter 6 that Raup (1996) found that each of the relatively few mass extinction spikes for marine Phanerozoic species was always closely preceded by a another major spike, extending the time of faunal die-offs. These coordinated double spikes might well indicate initial extinction of some taxa, followed by extinctions of larger numbers of taxa as a result of ecological cascade effects.

Functional Diversification

The majority of characters displayed by an ancestral species do not change when the ancestor speciates. If every character changed with each speciation event, we would expect to find phylogenetic analyses consistently producing large polytomies. We find instead substantial phylogenetic resolution in most cases, resolution based upon shared special characters that stay the same through one or more speciation events (synapomorphies). The assertion that some adaptive features are conservative evolutionarily may at first sound counterintuitive. We believe, however, that the assertion is consistent with MacArthur's proposal that we should search for evidence of competitive interactions among sympatric *congeners*. Macroevolutionary conservatism in ecology and behavior leads to the potential for niche overlap in closely related species. This can only occur if diversification in ecology and behavior are conservative on a macroevolutionary scale. We can take advantage of this conservatism in *predicting resource requirements of threatened and endangered species based on the requirements of their close relatives.*

Less than 10 percent of the species on this planet have been identified and named. It gets worse, because we know very little about the ecology and behavior of most species that we have named, making the job of protecting them frustrating and potentially disastrous (Kleiman 1980). We can use the power of our research program to help focus conservation responses in a number of ways. For example, Gittleman (1994) suggested that giant and red pandas *differ* from their closest relatives (bears and raccoons, respectively) in ways that are relevant to their threatened existence. We can also use the tools of our voyage to make predictions about the probable characteristics of poorly studied, rare, or endangered species; that is, we can concentrate on potential *similarities* between threatened species and their relatives. Consider the elusive, and endangered, boulder darter, *Etheostoma wapiti*. *Etheostoma wapiti* is a member of a clade comprising six species, five of which display the derived character of depositing their eggs beneath or behind large slab rocks (Etnier and Williams 1989). Because *E. wapiti* is also found in habitats containing slab rocks, just like its relatives (hence the name boulder darter), Brooks et al. (1992a) predicted that it would display the same breeding strategy. They proposed that this was an example in which both a rare species and its widely distributed relatives shared a common ecology involving resources that were severely limited in the areas inhabited by the rare species. A subsequent study confirmed both the prediction and the proposal (see Brooks et al. 1992b): the boulder darter is endangered because water control through hydroelectric dams upstream leads to fluctuating water levels. Drops in those levels during the breeding season expose the boulders and the underlying egg masses. If more information about the behavior and ecology of the boulder darter is required to formulate conservation policies, studies could be conducted without disturbing the already endangered populations of *E. wapiti* by using its widespread and relatively common sister species, *E. vulneratum*.

Let us consider some more fish. Many native species (e.g., the shiners *Notropis simus*, *N. orca*, *N. bairdi*, *N. girardi*, and *N. buccala*) are disappearing from river drainages throughout the central Plains of North America. Why? In order to answer this question, we need to know something about the evolutionary history of these fish. The river drainages they inhabit are typified by extreme seasonal fluctuations in water levels, shifting substrates, and variable temperatures. The fish species that historically dominated those rivers possess morphological, ecological, and behavioral attributes in tune with these variable conditions. For example, cyclical droughts and floods were a fact of river life for centuries. The species listed above and others also at risk are those that require the inundation of floodplains during their spring and summer spawnings to ensure successful reproduction. Given this ecological constraint, it is not surprising that these fish should be disappearing in rivers that have been dammed to control the annual cycle of flooding and to reroute water through irrigation channels. Destroy the conditions under which the species evolved, destroy the species. More depressingly, the species listed above are all part of a clade sharing similar feeding and reproductive ecologies (Mayden 1989). Environmental alterations that impact negatively on that shared portion of their evolutionary history will cause not just one species, but all species, to go extinct. Fortunately, there is a positive side to this shared history: what is good for one is good for all. In this case, we would need to implement only a single management policy aimed at maintaining regions of the rivers with their natural, highly variable flow regimes to help preserve all members of the shiner clade (for more examples, see Byrne et al. 1999; Jennings et al. 1999; Parkinson et al. 2000).

 Predicting the requirements of endangered or rare species based upon information from close relatives is not foolproof because, after all, each species has its own suite of autapomorphies. It does, however, provide a starting point based on information other than inspired guesswork.

Priority Areas: What Are the Options?

The research program we have outlined in this book provides a means of integrating smoothly the CBD's two-pronged position that species are the basic units of biodiversity and ecosystems are the basic units of biodiversity management. The core of resident species exhibiting plesiomorphic interactions will serve as a stabilizing force with respect to each other's evolution and to the addition of colonizers. Paleontological data indicate that the structure and composition of ecological associations have always been highly conservative with episodic evolutionary restructuring associated with moderate- to large-scale environmental (climatic, tectonic, and astronomic) changes.[12] The problem therefore becomes, How do we protect (or manage) entire chunks of the biosphere?

Less than a decade after its publication, ecologists began proposing design criteria for biodiversity reserves based on the MacArthur-Wilson model of Island Biogeography (e.g., Diamond 1975; Wilcox 1980; Murphy and Ehrlich 1989). It took only a decade to discover that conserving biodiversity was not going to be that easy (e.g., Noss 1986; Bieregaard et al. 1992). Workers quickly divided into two camps, those advocating several small reserves and those favoring single large reserves.[13] It soon became apparent that both sides of the SLOSS debates were missing an important point: biological diversity is *evolved* diversity, and evolution does not occur within human-defined spatial boundaries, large or small. Scientists working on both macroevolutionary (e.g., Erwin 1991) and microevolutionary (e.g., Pimm and Gittleman 1992; Avise 2000) scales stated that preservation of biological diversity required not only conserved areas, but also biotic connections among those areas. Often called "dispersal corridors," these connections permit biotic expansion and contraction, setting the stage for the production of new species and new ecological associations. They are also hedges against local extinction becoming regional extinction (Ricklefs 1987). *If we do not provide space for species to spread out and find their own futures, building biodiversity reserves is tantamount to creating zoos and herbaria, betting that we can maintain standing diversity while halting evolution.* Our own history argues against that bet. More important, recent studies on what happens to biological diversity when species are forced into small areas on evolutionary time scales also appears to argue against the establishment of small, restricted conservation areas.

FOREST REFUGIA AS MODELS OF BIODIVERSITY RESERVES

Natural environmental disturbances occur on a variety of temporal and spatial scales. Most such disturbances are relatively ephemeral; consequently, their effects on biodiversity are often modeled using immigration/emigration phenomena, assuming a constant, or static, pool of species with access to the area in which disturbance occurs. In a few cases, however, environmental disturbances are thought to have played a sig-

12. Boucot 1975a,b, 1978, 1981, 1982, 1983, 1990; Brooks and Wiley 1988; Behrensmeyer et al. 1992; DiMichele 1994; and Conway Morris 1998.

13. For a discussion of the history of the "SLOSS" [Single Large Or Several Small] debates, see Murphy 1989; and Ambrose and Bratton 1990.

nificant role in the production (speciation) or elimination (extinction) of species. As we discussed in chapter 4, one class of such disturbances is the formation of "refugia" in the western hemisphere as a result of Pliocene/Pleistocene glaciation. From the Arctic to southern South America, widespread glaciation is thought to have dissected habitat into "islands" (the refugia) representing remnants of preglaciation habitat and biotas embedded in "seas" of new habitat created by the changing climatic conditions. In the Northern Hemisphere, the new habitat represented vast sheets of glacial ice and debris. In the Southern Hemisphere, glaciation in the mountains reduced precipitation in Amazonia, resulting in replacement of forests with grasslands. In the mountains of South America themselves, the glaciation dissected previously widespread biotas into mountain valleys, where the climate remained relatively more equable during the Ice Age.

During the 1970s and 1980s, refugia were viewed as biodiversity "arks," places where standing diversity is sequestered during the invasion of inhospitable environmental conditions, and as biodiversity "pumps," sources of species formation (Lynch 1989). Both of these perspectives emphasized the benefits of refugia and provided optimistic models for the establishment of nature reserves. One of the great advances in biology over the past 25 years, however, has been the recognition that evolutionary systems exist in a delicate balance between benefits and costs. There is no such thing as a free lunch. So, what cost do species pay when they retreat to refugia? In a novel approach to the problem, Amorim (1991) combined data from the number of refugia per refugium cycle and the number of cycles to construct a simple but elegant simulation of expected versus actual diversity in the Neotropics. He came to the startling conclusion that there is in fact less diversity in the Neotropics than would be expected if refugia served primarily as centers of diversification. This means that either little speciation, or more extinction than speciation, has occurred there. His analysis uncovered the missing cost in the scenario: extinction.

When we consider refugia from this more dynamic and nonequilibrium perspective, we see both speciation and extinction at work, with no guarantee that one will balance the other. These simulation results are congruent with emerging macroevolutionary findings. Whether produced episodically or continually, biological diversity is produced slowly and over relatively large geographic expanses, especially with respect to the rate of human encroachment on wildlands. We must therefore rein in the hopeful metaphor of nature reserves as places where old diversity survives forever and new diversity emerges quickly. *In its place, we must implement policies that mimic, insofar as possible, the biotic expan-*

sion that sets the stage for speciation, the real mechanism of biodiversity renewal.

Avoiding the Traps

A comparative phylogenetic perspective highlights a number of potential pitfalls that must be avoided when establishing protected areas. First, if we choose to preserve areas of high species richness, we must understand where that diversity came from. An area may be species rich because it contains a large number of endemics. It may also be species rich because it represents a region of overlap between two biotas. In the latter case it is possible that such a region is composed of marginal habitat that has limited the expansion of both biotas in the past. Congeneric species are found together less often than expected given the number of species that have access to any given community (Bowers and Brown 1982). Or, in historical biogeographical terms, most sister species tend to be allopatric. Add this to the observation that closely related species tend to utilize the same resources (MacArthur 1958, 1972; Root 1967), and this means that congeners rarely occur together often enough to allow adaptations for competitive exclusion to evolve (Maurer 1985; Graham 1986). Hence, artificially preserving the region of overlap, to the exclusion of the larger areas that contain most of the preferred habitat for the species in each biota, could be counterproductive for the survival of any of them. Because evolution is a slow and unpredictable process (in a Darwinian world), such forced competition could lead to a rapidly cascading series of regional extinctions, the exact opposite of our intention. We would thus violate the precautionary principle by trading evolutionary potential for current diversity, and possibly harming the current diversity in the process.

Comparative phylogenetic studies can also provide information that will complement current conservation/management practices based on theories about the relationship between species number and area size or number. For example, historical biogeographical studies (chapter 4) can identify areas that have been associated with hot spots of evolutionary activity in the past. Such areas of endemism are interesting because they do not always encompass the greatest number of extant species, nor are they always extremely large or centrally located.[14] Whether this means such areas will be "hot spots" of evolution in the future is debatable and debated (Erwin 1991; Brooks et al. 1992a,b; Mayden 1992b). Regardless, ecosystems in areas of endemism contain substantial numbers of species

---

14. For example, Erwin 1991; Brooks et al. 1992a,b; Emberton 1995; Linder 1995; Savage 1995; Amorim and Pires 1996; Williams et al. 1996; and Crandall 1998.

that have been associated with each other for a long time. For example, the North American Central Highland freshwater fish communities contain sister species with historically replicated composition and biogeographic patterns suggesting origins that substantially predate the Pleistocene Ice Age (Mayden 1988; chapter 4). More recently, a similar study using 33 clades of insects, fish, amphibians, and reptiles distributed throughout nine areas of endemism in Mesoamerica (Marshall and Liebherr 2000) produced a single most parsimonious area cladogram (using primary BPA only) suggesting that approximately five out of every eight species in those areas have long-term historical associations with each other.[15]

As we pointed out in chapters 7 and 8, ecological associations encompass more than just different species living together in isolation. Making decisions based solely on species occurrences, even within a phylogenetic context, thus neglects a truism of community and ecosystem ecology: *some species are more important than others in maintaining ecosystem stability, resilience and integrity.* For example, the existence of some species with key food linkages can change the energy flow dynamics of both terrestrial and aquatic ecosystems (Power 1990). Species playing these fundamental roles must be identified and protected in order to more effectively preserve whole ecosystems (Walker et al. 1999; Crandall et al. 2000). We believe strongly that the core of old species and old interactions (especially redundant ones resulting from phylogenetic conservatism) in a community is what gives it a degree of stability and resilience in the face of perturbations (including anthropomorphic ones). The same phylogenetic conservatism that makes communities stable and predictable may, however, also make them slow to respond to drastic environmental changes. Our ability to predict the behavior of a community under a variety of disturbances thus depends less on the overall diversity of the ecosystem (Hodgson et al. 1998) than on our understanding of the community's evolutionary legacy (Meffe 1990; Abrams 1996; Foster 2000; Kaiser 2000). As a consequence, if a choice must be made about areas requiring protection, phylogeneticists can provide information about regions that have been important in the origin and long-term maintenance of biological diversity (Coates 2000). Equiarte et al. (1999) have recently proposed a management protocol based on identifying the minimum set of areas that encompasses the highest proportion of total phylogenetic diversity for any given indicator taxon. Connecting these areas via dispersal corridors is the next logical step in a plan to conserve

15. See Crother and Guyer 1996 for a similar study of Caribbean historical biogeography.

biotas within an ecological and geographical context that permits the greatest survival of standing diversity and greatest possible expression of evolutionary potential (Hoctor et al. 2000).

> *A critical issue:* We need a global decision about what we want, a series of natural history parks where pieces of nature are held in stasis (Do we want to turn the planet into a global zoo/botanical garden?) or an evolving biosphere? Only after we have made this choice can we make rational policies. We, of course, support the second proposal.

Using Phylogenies in Human Affairs
INTRODUCED SPECIES

The introduction of nonnative species into ecosystems, or dispersal, is a widespread ecological and evolutionary phenomenon. Megadiverse regions, such as Mexico, Costa Rica, and Brazil, owe a portion of their diversity to dispersal in some form. (Remember the Neotropical snakes, cotton-stainer bugs, and eutherian mammals?) Evolutionary agents of dispersal include continental drift and other tectonic processes, watershed capture, and biotic expansion and contraction, all of which may affect multiple species simultaneously, and rare dispersal events affecting single species (chapter 4). Although many mammals and birds are vectors for dispersal of plant, animal, and microbial diversity, human introductions differ by not necessarily being evolutionarily based, affecting species from ecosystems that are not geographically adjacent or closely related. As a consequence, the effects of anthropogenically introduced species need not be tempered by common evolutionary history and thus may be difficult to predict (Soulé 1990).

The threat of deliberate and accidental introductions of nonnative species is probably greater today than in any time in human history and will increase in the future (Soulé 1990). Many countries with developing economies are attracted to the economic benefits of certain introduced species (in which case they are called "beneficial species"), and most of these countries have rich native biodiversity that could be affected adversely by those introductions. Those effects are usually negative, ranging from reduction in abundance of one or more native species (through direct or indirect action of the introduced species) to catastrophic collapse of diverse, complex ecosystems into less diverse, simpler, and often less productive states. Because each ecosystem is historically unique, we must always be cautious about making policy decisions by extrapolation from one ecosystem to another. For example, the assumption that each ecosystem has a similar number of equivalent niches may lead an un-wary researcher to conclude that a foreign species can be safely intro-duced because it is filling an "unoccupied" niche. Consider an aquatic ecosystem filled with many species of small fish. Is this community "in-complete," containing open "top-line piscivore niches," or does it owe its diversity to an absence of top-line piscivores? This question has, un-fortunately, been answered several times. Aquatic ecosystems with top-line piscivores typically contain a low diversity of prey species, and eco-systems where such predators have been introduced usually experience a drop in species richness to levels approaching those of the native eco-system for the predator (for sobering discussions, see Kaufman 1992 and Chapleau et al. 1997).

Comparative phylogenetic studies can contribute important informa-tion to the debate about introduced species. First, in the absence of writ-ten records, the only way to distinguish introduced species from rare native species is by reference to the *geographical origin of the species*. This requires historical biogeographical information, which requires the ser-vices of a professional systematist. Second, understanding a species' phy-logenetic legacy may give us an insight into its likely impact on the new ecosystem once it is introduced (e.g., Dobson 1988; Daehler 1998; Huckins et al. 2000; Kaiser 2000), or help us develop strategies minimiz-ing any postintroduction negative effects. In other words, incorporating evolutionary information into the decision-making process allows us to be both proactive and reactive. This kind of information is particularly critical when the deliberate introduction of nonnative species for bio-control purposes is at issue (a practice generally used in agriculture). As you know from reading chapters 7 and 8, we need to consider the evolu-tionary "pedigree" of any proposed biocontrol agent, especially ones that are thought to have specialized coevolutionary relationships.[16] Re-member ecological fitting: a "specialist" (chapter 8) may provide un-pleasant surprises if it is a member of a generalist clade, or if it prefers a plesiomorphic resource that is more abundant, more accessible, or more palatable in a beneficial species than in the one it is meant to control. Conversely, insects beneficial to a given group of wild plants may also be beneficial to their domestic relatives. Agricultural operations situated close to conserved wildlands where such species reside will reap substan-tial economic benefits, both directly through the decreased use of pes-ticides and indirectly through positive public relations (Janzen 1998, 1999, 2000a,b).

16. Hanson 1990; Holt and Hochberg 1997; Jervis 1997; Liu et al. 2000; and Martinez-Ghersa et al. 2000.

The first person to provide explicit phylogenetically based insights into crop-disease relationships was Virginia Ferris. Ferris (1979) used a historical biogeographical approach to evaluate hypotheses about the origin and evolution (and control) of cyst nematodes (Heteroderidae). She was particularly interested in tracing the roots of four heteroderid species. Two of those species, *Globodera rostochiensis* and *G. pallida,* are by far the most important pests of potatoes in North America. Nematologists first thought that these nematodes originated in Europe, and then later hypothesized that they originated in the Andes and were transported to Europe. In either case, there was little doubt that the nematodes eventually found their way to North America with European settlers (Stone 1977; Stone et al. 1977). The cereal-cyst nematode *Heterodera avenae,* was also thought to have originated in Europe, then moved with the Europeans and their grain as they colonized other parts of the world, including North America (Meagher 1977). The fourth species, the soybean-cyst nematode *Heterodera glycine,* was supposedly introduced to North America in some bulbs from Japan, its place of origin (Riggs 1977).

Ferris's analysis uncovered Pangaean distribution patterns for the major clades within the family, indicating that the heteroderids are very old. Her study suggested that the common ancestor of *Globodera* probably infected native Solanaceae in North America as a result of biotic connections between Eurasia and North America while both areas were joined as Laurasia. Tuberous Solanaceae (precursors of potatoes) are thought to have originated in what is today Mexico and the southern United States during the Cretaceous to early Tertiary (Hawkes 1972). They had arrived in the Andes by the Pliocene (Raven and Axelrod 1974), presumably with a full complement of associates, including cyst nematodes. (See also Pellmyr et al. 1998 for a similar case involving an introduced umbel, *Bowlesia incana,* and a native lepidopteran pollinator, *Greya powelli.*) In other words, the agricultural "problem" was created by superimposing introduced plant species (crops) on endemic nematode communities in North America.

Ferris felt her conclusions were startling only because so few inventories of the cyst nematodes associated with native plants other than crop species had been conducted. She noted, for example, that Rigg's (1977) report listed many wild hosts for *H. glycine.* She further postulated that many of the "species" of cyst nematodes recognized at that time would be found, upon closer study, to be groups of species. Finally, and most significantly, Ferris pointed out that if the various cyst nematodes were in North America before potatoes, sugar beets, cereal grains, and soybeans were introduced, then we could only hope to manage, not eradicate, cyst nematodes by letting infected fields lie fallow for a year. What

was needed was a whole new way of thinking about the nematode problem. For example, Are there native species that the cyst nematodes prefer
to the introduced species? If so, then perhaps we can plant rows of native species between crop rows, producing fields in which crops and cyst
nematodes coexist spatially but do not actually interact. This might result in lowered per-hectare production, which would be offset by (1) not
having to let the fields lie fallow, thereby periodically losing an entire
year's production, and (2) reduced use of nematicides, which kill beneficial as well as pest species. The exuberant use of nematicides also tends
to create nematicide-resistant strains of pest nematodes, leading to a
double spiral of higher production costs for farmers and more anthropogenic habitat degradation.

Overall then, global management of cyst nematodes (you can replace
"cyst nematode" with any pest organism) in an evolutionary framework
requires (1) a large-scale inventory of all cyst nematodes in a given region, such as North America and (2) phylogenetic, historical biogeographical, and coevolutionary analyses of those species to provide the historical platform for (3) asking questions about the true host preferences of
cyst nematode species infecting introduced crop species and their closest
relatives that do not.

We have emphasized throughout this book that one of the effects of
phylogenetic history is to "lower the cost of evolution." Some researchers see socioeconomic development involving sustainable use of biodiversity as the critical element in preserving that biodiversity.[17] Taking
advantage of evolutionary conservatism can lower the costs of that development. Human beings have always understood that the predictability of genealogy pays off. This is why whenever a group of humans
moves into a new locality, they are likely to seek plants and animals
similar to (i.e., related to) the ones they found useful in the old area (or
to avoid plants and animals similar to ones that were dangerous in the
old area). In the past, such decisions had a direct impact on survivorship.
Today, we are more sophisticated and believe we can engineer virtually
any species to produce what we want. Engineering still must take place
within a genealogical matrix, so it is not surprising that many corporations have recently been quietly extolling the importance of evolution.
In this case, rather than direct effects on survivorship (although indirect
effects certainly apply if the research is attempting to modify cancer-
causing genes, rather than genes that cause unsightly wrinkles), the pre-

17. For example, Janzen 1991, 1992, 1993, 1994a,b, 1997a,b, 1998, 1999, 2000a,b;
Sandlund et al. 1992; Sandlund and Shei 1993; Janzen and Hallwachs 1994; Brooks 1998b;
and Softing et al. 1998.

dictability of phylogeny offers substantial savings in research and development time and expense. For example, there is growing interest in the notion that wild stocks of any domestic species and their closest relatives should be conserved as sources of genes for resistance to disease and as a hedge against genetic bottlenecks and inbreeding problems. In this regard, we think it should be a high priority to determine the wild sister groups of all our domestic species and institute plans to conserve those species *and* the areas in which they reside.

Such studies are well underway; recent examples involve tracing the roots of apples (Zhou and Li 2000), citrus (Nicolosi et al. 2000), pistachios (Kafkas and Perl-Treves 2001), buckwheat (Tsuji and Ohnishi 2001), and wheat (von Buren 2001). Happily, a side effect of this recognition is that it enhances the "value" of an increasing number of species, making it more likely that they, and the areas in which they reside, will be preserved. This is the "use it or lose it" criterion for biodiversity conservation (Janzen 1988, 1991, 1992, 1999, 2000a,b). For example, suppose you are searching through soil microbes, looking for novel genetic products. You find a new microbe that produces something interesting. How can you maximize the benefits of this finding? First, you might reconstruct the phylogenetic relationships of the new microbe then ask what is known about the properties of its closest relative(s)—this will enable you to focus your research beam on the most likely avenues for development. Second, you might shift your search away from a general inventory of the microbes in one habitat to a focused search for more relatives of your novel microbe in other habitats. In both cases, research and development costs (in time and money) will be lowered. How does this help the cause of biodiversity? Remember the Taxol and Taxotene story discussed above, involving the American yew and the European yew. A close relative of these trees, the Himalayan yew, is one of the world's ten most endangered species. This gives us a commercial reason to save and propagate that species (something that speaks directly to managers concentrating on the bottom line) *in addition to* bolstering arguments for the intrinsic biological importance of each species.

EMERGING DISEASES

Phylogenetic information can also give us some insights into origins and epidemiology of emerging diseases.[18] Parasitologists, for example, have traditionally taken advantage of the phylogenetically conservative na-

18. For example, Choo et al. 1989; Gorman et al. 1990a,b; Choo et al. 1991; Hendriks et al. 1991; Bowman et al. 1992; Chan, Bernard, et al. 1992; Chan, McCormish et al. 1992; Davis 1995, 1996; Holmes et al. 1995; Bowen et al. 1997; Epstein 1997, 1999; Hoberg 1997b; and Daszak et al. 2000.

ture of life cycles to guide eradication and control programs. Once we understood that it was possible to control one species of malaria by eradicating mosquito populations, we were able to attack all species of malaria in the same way. All species of *Schistosoma* are transmitted by aquatic larvae that penetrate exposed skin and take up residence in the circulatory or lymphatic system. This epidemiology is very old—all members of the Schistosomidae and its two immediate sister groups exhibit the same pattern, whether they live in a mammal, bird, crocodilian, turtle, fish, or elasomobranch.[19] In addition, because many agents of emerging diseases do not cause disease in their native ecological context, phylogenetic information about both a pathogen and the immune system of its hosts (native and novel) may provide insight into the evolution of both susceptibility and resistance to disease.[20]

The enthusiasm for using phylogenetic information in basic and applied medical and public health research has outrun taxonomic resources. "Thousands of disease organisms and their vectors need to be identified and distinguished accurately" (Davis 1995:705). It has also exceeded the availability of systematic expertise. Typically, phylogenetic analyses will be performed by expert geneticists and molecular biologists who have access to a user-friendly computer program for generating phylogenetic trees from sequence data. As you will remember from chapter 2, generating sequences is barely the beginning of systematic character analysis, and producing a tree is barely the beginning of interpreting and using the results properly. The next three examples highlight the importance of bringing phylogenetic systematic expertise to bear on such issues.

The Scourge of Malaria: Whither Falciparum? Millions of people around the world are infected with members of at least four different species of malaria *(Plasmodium ovale, P. malariae, P. vivax, and P. falciparum)*, all of which include drug resistant strains. Of these, *P. falciparum* is the most virulent and exhibits the most unusual symptomology. Since virulence is often taken as evidence for a recent host shift by a pathogen, Boyd (1949) suggested that *P. falciparum* colonized human beings from birds. He hypothesized that the colonization was facilitated by the origins of agriculture; increasing concentrated patches of edible plant material attracted birds, while building reservoirs and canals for irrigation increased the suitable breeding habitat for mosquitoes. Waters et al. (1991) were

19. For a summary of the degree of phylogenetic conservatism in parasite life cycles, see Brooks and McLennan 1993b.

20. See also Dobson and Hudson 1986; Dobson and May 1986a,b; Dobson and Carper 1992; Davis et al. 1995; and Ewald 1995.

the first authors to investigate this hypothesis within a phylogenetic context. They mapped host type onto a phylogenetic tree for ten *Plasmodium* species (there are more than 170 species of *Plasmodium* parasitizing amniotes: Levine 1988) and concluded that their results supported Boyd's agriculture hypothesis (fig. 9.2a). If you try the optimization yourself, you will discover that the phylogenetic patterns actually falsify the hypothesis about the sequence of parasite switching: an ancestral population of *Plasmodium* appears to have jumped *from* human beings *to* birds, and not vice versa (Brooks and McLennan 1992b). This may well have happened as a result of the advent of agriculture, as suggested by Boyd. Although the patterns on the reduced tree constructed by Waters et al. suggest that *P. falciparum* is part of a clade of species all associated with *Homo sapiens,* this still does not answer the question, Why is *P. falciparum* so virulent in human beings?

Escalante and Ayala (1994) presented a genetic distance analysis of *Plasmodium* species (fig. 9.2b) and concluded that *P. falciparum* is actually

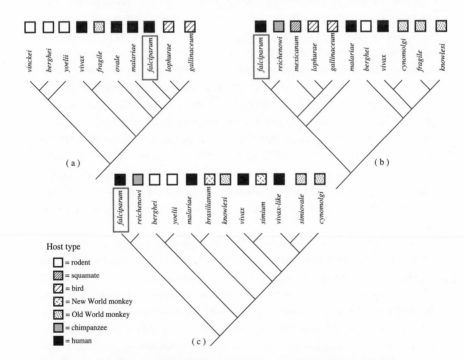

Figure 9.2. Coevolution of *Plasmodium* and their hosts mapped above different tree topologies. (*a*) phylogenetic tree reconstructed using mitochondrial DNA (Waters et al. 1991). (*b*) Dendrogram based on genetic distances for 11 species of *Plasmodium* (Escalante and Ayala 1994). (*c*) Dendrogram based on genetic distances for 12 species of *Plasmodium* (Escalante et al. 1995). We leave the optimizations of host switching on each tree up to you.

the sister species of *P. reichenowi*, which infects chimpanzees *(Pan troglodytes)*. They also concluded that *P. falciparum* is only distantly related to the other malaria species in humans, concurring with Brooks and McLennan's assessment that the data do not support an origin of *P. falciparum* via a host switch from birds. In another genetic distance analysis (fig. 9.2c), Escalante et al. (1995) found that while *P. falciparum* and *P. reichenowi* are highly distinct from one another (molecular clock estimates suggesting a divergence time of 6–10 million years), *Plasmodium malariae* is indistinguishable from *P. brasilianum* and *P. vivax* is indistinguishable from *P. simium*. Both *P. brasilianum* and *P. simium* parasitize New World monkeys. A fourth form, which Escalante et al. called "vivax-like," is indistinguishable from *P. simiovale*, which parasitizes macaques. The authors suggested that species of *Plasmodium* other than *P. falciparum* have colonized humans rather recently, perhaps in the past 15,000 years. This scenario turns the assumptions about length of association and degree of parasite pathogenicity on its head by suggesting that the species with which human beings have been coevolving for millions of years is the most pathogenic, while species that have only recently colonized us are less virulent. In other words, specific details of the parasite's biology, and specific details of the parasite-host interaction, will determine the degree of pathogenicity in any association: there are no general rules (Anderson and May 1985; Koella and Restif 2001). Finally, Escalante and Ayala (1995) assessed the degree of similarity between *Plasmodium* and other members of the protist phylum Apicomplexa, which includes a number of important agents of parasitic disease in humans, pets, and livestock (such as *Eimeria* and *Toxoplasma*). They concluded that their genetic data indicated great antiquity for *Plasmodium*, the lineage giving rise to human malarias being about 130 million years old, old enough to have been associated with some of the earliest mammals.

At the moment, we are still in the dark about the evolutionary roots of the malaria species inhabiting humans. Unfortunately the different studies investigating the problem did not include all the same taxa in their analyses. As you know, missing data can dramatically alter our hypotheses about cospeciation and host switching. In this particular case, the absence of data from malaria species inhabiting other primates, specifically the remaining members of the Hominidae (*Gorilla, Pongo,* and *Hylobates*), substantially weakens any hypothesis about directions and frequency of host switching. Two things are desperately needed here if we are ever to understand the relationship between coevolution and disease. First, researchers need to use phylogenetic systematic methodology to address the issue. As you recall from chapter 2, distance-based (phenetic) methods do not distinguish among plesiomorphies, apomorphies, and homoplasies, so they are not reliable general methods for determining phylogenetic relatedness. This does not mean that genetic distance data are useless. Quite the contrary. If such data are placed within a phylogenetic framework, they can give us valuable information about the age of lineage splitting events and the amount of change that has taken place in two lineages since they diverged. Second, those phylogenetic systematic analyses must include, minimally, all the primate and avian malaria species (and eventually all the species; analyses by researchers investigating the relationships of flowering plants and anoles have demonstrated that it is possible, albeit difficult, to analyze such large data sets). Escalante et al.'s proposal that *P. falciparum* is a longtime associate of *Homo sapiens,* and that the other human malarias, *P. vivax* and *P. malariae,* are more recent colonizers is fascinating and raises numerous questions that can only be answered with a robust phylogenetic analysis. At what point did the association between *Plasmodium* and the Hominidae begin? What characters have changed throughout the evolutionary history of the different human malarias (if indeed they are paraphyletic) that might be responsible for either the reduced virulence of *P. vivax* and *P. malariae* or the increased virulence of *P. falciparum*? The most obvious place to look for these answers, as you know by now, is by comparing the epidemiology of the *Plasmodium* species inhabiting sister species, then comparing those data on a larger phylogenetic scale.

AIDS: Are We Source or Receiver? Problems similar to those discussed above have manifested themselves during the quest to uncover the origin(s) of HIV (Human Immunodeficiency Virus).[21] Myers et al. (1993) published a phylogenetic study that established the most widespread and commonly accepted story about the origin of the virus and the disease it causes: the two strains of HIV (HIV-1 and HIV-2) are each derived independently from different groups of African primates.[22] Once again, the virulence of the disease was taken as de facto evidence that the virus is a recent colonizer of humans. Mindell et al. (1995) noted that Myers et al. had used only one strain of HIV-1 and two strains of HIV-2 in their analysis, compared with multiple strains of various SIVs (Simian Immunodeficiency

21. For the first use of phylogenetic analysis as a forensics tool, see Ou et al. 1992.

22. See also Guyader et al. 1987; Huet et al. 1990; Gao et al. 1992; and Leigh Brown and Holmes 1995.

Viruses). When Mindel et al. reanalyzed the same data, using similar numbers of strains for HIV-1 and HIV-2 as for the other viruses, they discovered quite a different story. The new analysis suggested a monophyletic origin of both HIV strains, one that predates the current AIDS pandemic. It also suggested relatively recent host transfers from humans to nonhuman primates (for an excellent discussion, see Packer 1997).

Influenza: Chickens to Pigs to Us, Oh No! Finally, phylogenetic studies by Gorman, Bean, et al. (1990) and Gorman, Donis, et al. (1990; see also Kroll and Moxon 1990) support the hypothesis that the Influenza A virus originated in birds. Subsequently, swine acquired the virus, which differentiated into a swine form. Human Influenza A is apparently a derivative of the swine form, showing that domestication can have costs as well as benefits. Swine remain suitable hosts for all three forms of the virus. When swine are infected with avian and human or swine and human Influenza A at the same time, there is a possibility of genetic exchange that can produce viral strains highly pathogenic to humans. Evolutionary biologists believe that such highly pathogenic strains are self-limiting on time scales that are short indeed for evolution, but that can be quite long if your family lives in an infected area. Evolutionary biologists are also intrigued by the number of times such pathogenic hybrid strains can be generated and the diversity of viral genetic backgrounds that can be co-opted. Once again, what intrigues the evolutionist in Toronto terrifies the epidemiologist in rural China.

These three examples should give serious pause both to evolutionary biologists and to medical and public health researchers (see also Janzen 1997c; Demanche et al. 2001; and Pavesi 2001). The quality of life and even the survival of hundreds of millions of people depend on the decisions of those few people who are responsible for determining the direction and funding of research. Because all biological diversity is evolved diversity, all applied research must be informed by the most current evolutionary theory and information. Physicists lost their status as impartial observers, absolved from responsibility for the social applications of their research, with the development of nuclear weapons. The biodiversity crisis has likewise stripped away that status from biologists. Whether we like it or not, contemporary biologists are either part of the solution or they are part of the problem. Remember our conclusion in chapter 8 that host switching should increase during periods of environmental disruption. From this it follows that the more humans encroach on novel habitats, the more they will create prime conditions for emerging diseases. Or, as the cartoon character Pogo once stated, "We have met the enemy and they is us."

DENOUEMENT

> Without a living past, we have only an inert present, and a
> dead future.
>
> Carlos Fuentes

We have for some time been concerned that human activities are changing the ecoscape faster than systems can adapt to those changes. As a result of research efforts outlined in this book, we now believe that ecological structure in biotas is much more conservative evolutionarily than evolutionary biologists once thought, raising questions about the adaptability of ecosystems on time scales relevant to human activities. The more phylogenetically conservative the critical elements of ecosystem structure and function, the slower ecosystems will be in their responses to perturbations. And if it is not bad enough that current ecological structure is more conservative than we thought, paleontological studies indicate that this may be the way it has always been. As a consequence, our concern about the survivability of any given ecosystem should be directly proportional to the degree of phylogenetic conservatism in its structure and composition. Those ecosystems that have been around for the longest time and contain the largest number of endemic species may be the most in need of immediate conservation. Similarly, those species that have been part of any given ecosystem for the longest periods of time may be those most in need of protection against exploitation and removal.

Humans study and preserve their history because it has had a great impact on their present existence. These histories include heroes and heroic events. The successful preservation of biodiversity requires that we find, explain, and preserve the heroic episodes in the history of life on this planet. We can make a difference, but our voyage of discovery will be long and difficult because biological diversity has been shaped by billions of years of interactions among a variety of evolutionary processes, none of which can be assigned the dominant role in evolution. Every clade and every community is a unique mosaic of phylogenetic conservatism and innovation, of species that evolved *in situ* and species that have dispersed into the area from elsewhere. Neglecting the complexity of that historical process, which is represented in phylogenetic patterns, weakens our understanding of the way biodiversity has been produced and replaced in the past (Brooks et al. 1992; Mayden 1992b; Brooks and Hoberg 2000). The research program outlined in this book offers a framework for integrating information from systematics and natural history in a predictive manner that will serve the professional agendas of biologists involved in biodiversity initiatives while giving efficient data to conservation managers. Systematists are the keepers of genealogical information. Evolutionary biologists, whether they call themselves ecologists, ethologists, or some other name, are the interpreters of how those genealogical information units interact with one another. Together, all bring important insights to questions about the production and maintenance of biodiversity, and that includes a full appreciation for one biological reality of *Homo sapiens:*

> ⚠ Historical and contemporary deterioration of biodiversity is highly correlated with human population growth and the associated demands on the biosphere. Without a doubt, until overpopulation is controlled the battle to save biodiversity will be formidable; even sustainable use of biodiversity cannot be maintained in the face of unbridled population increase.[23]

Although this sounds grim, there is always hope. Humans save what they value. For most of this chapter, we have stressed ways in which preserving biodiversity can enhance our material quality of life, certainly a major component of human values (see Janzen 1997b, 1998, 1999, 2000a; Brooks 1998b). Biodiversity also affects us in quite a different way because we do more than just use it. We are part of the biosphere and thus have a natural affinity for all living things ("biophilia": Wilson 1984, 1988, 1989, 1992; Takacs 1996; or "Kantian aesthetics": Kiester 1997a,b). This natural affinity expresses itself in many different ways, depending upon the culture and the century. One subtle outcome of the relationship is that it helps us understand ourselves and find our place in a seemingly incomprehensible universe. Imagine how little you would understand about your own behavior if you had never known anyone in your family. Now imagine how even that small amount of knowledge would be diminished if you had been displaced from your cultural and deeper genealogical roots. Our connection to the biosphere is part of that continuous relationship, from parent to culture to all human beings to all primates to all life. Perhaps understanding that connection will lead to a more universal commitment to save ourselves along with as much of the biosphere as we can.

> In the end we will conserve only what we love. We love only what we understand. We will understand only what we are taught.
>
>   Baba Dioum

23. For an excellent discussion of the consequences of human overpopulation, see Nentwig 1999.

Within biological science, this calls for renewed emphasis on the value and relevance of natural history, in our teaching, in our research, and in our dialogue with the public (Greene and Losos 1988; Greene 1994b, 1996, 1997, 1998, 1999). For example, what can we possibly gain from studying ancient species, the ones we call "living fossils"? For some people, these species are the focus of vain efforts to delay, reverse, or avoid human mortality and its consequences. For us, species such as coelacanths are shining examples of survival in the face of local, regional, and global changes on a magnitude we can still only dimly imagine. They provide us with hope that, no matter how bad things appear to be, life will find a way. Likewise, clades of species arrayed around the globe bear witness to the endless surprise and wonder embodied in the process we call evolution. Every species on this planet has a story to tell, both of itself and its associations with other species, and of its ancestry and ancestral associations. This is the source of our existence, and the sheer beauty of it all makes human beings feel good to be near it, to be in it, to be part of it. To paraphrase Harry Greene, one of the most eloquent advocates for biodiversity, we are all "standing on the edge of mystery." Throughout this book we have shown you how to begin your own personal voyage from the edge into the heart of that mystery.

Your Responsibility to Ensure That
the Voyage Continues
Begins Now.

References

Abé, H. 1998. *Rhombognathine mites: Taxonomy, phylogeny, and biogeography.* Hokkaido: Hokkaido Univ. Press.

Abrams, P. A. 1986. Adaptive responses of predators to prey and prey to predators: Failure of the arms race analogy. *Evolution* 40:1229–1247.

———. 1996. Evolution and the consequences of species introductions and deletions. *Ecology* 77:1321–1328.

———. 2000a. Character shifts of prey species that share predators. *Am Nat* 156:S45–S61.

———. 2000b. The evolution of predator-prey interactions: Theory and evidence. *Annu Rev Ecol Syst* 31:79–105.

Ackerly, D. D. 2000. Taxon sampling, correlated evolution, and independent contrasts. *Evolution* 54:1480–1492.

Ackerly, D. D., and M. J. Donoghue. 1995. Phylogeny and ecology reconsidered. *J Ecol* 83:730–733.

———. 1998. Leaf size, sampling allometry, and Corner's rules: Phylogeny and correlated evolution in maples *(Acer). Am Nat* 152:767–791.

Acot, P. 1997. The Lamarckian cradle of scientific ecology. *Acta Biotheor* 45:185–193.

Adams, B. J. 1999. Species concepts and the evolutionary paradigm in modern nematology. *J Nematol* 30:1–21.

Adams, E. N., III. 1972. Consensus techniques and the comparison of taxonomic trees. *Syst Zool* 21:390–397.

———. 1986. N-trees as nestings: Complexity, similarity, and consensus. *J Classif* 3:299–317.

Adamson, M. L., and J. N. Caira. 1994. Evolutionary factors influencing the nature of parasite specificity. *Parasitology* 109:S85–S95.

Aerts, P., R. Van Damme, B. Vanhooydonck, B. A. Zaaf, and A. Herrel. 2000. Lizard locomotion: How morphology meets ecology. *Neth J Zool* 50:261–277.

Ah-King, M., and B. S. Tullberg. 2000. Phylogenetic analysis of twinning in callitrichinae. *Am J Primatol* 51:135–146.

Ahlberg, P. E., and A. R. Milner. 1994. The origin and early diversification of tetrapods. *Nature* 368:507–513.

Ahmad, I., and C. W. Schaefer. 1987. Food plants and feeding biology of the Pyrrhocoroidea (Hemiptera). *Phytophaga* 1:92.

Åhman, I. 1981. Några kålväxters potential som värdväxter för skidgallmyggen. *Ent Tidskr* 102:111–119.

———. 1985. Larval feeding period and growth of *Dasineura brassicae* Winn. (Dipt., Cecidomyiidae) on brassica host plants. *Oikos* 44:191–194.

Alberch, P. 1980. Ontogenesis and morphological diversification. *Am Zool* 20:653–667.

———. 1982. Developmental constraints in evolutionary processes. In *Evolution and Development,* ed. J. T. Bonner, 313–332. New York: Springer Verlag.

———. 1985. Problems with the interpretation of developmental sequences. *Syst Zool* 34:46–58.

Alberch, P., and J. Alberch. 1981. Heterochronic mechanisms of morphological diversification and evolutionary change in the neotropical salamander, *Bolitoglossa occidentalis* (Amphibia: Plethodontidae). *J Morphol* 167:249–264.

Alberch, P., S. J. Gould, G. F. Oster, and D. B. Wake. 1979. Size and shape in ontogeny and phylogeny. *Paleobiology* 5:296–315.

Albert, J. S., M. J. Lannoo, and T. Yuri. 1998. Testing hypotheses of neural evolution in gymnotiform electric fishes using phylogenetic character data. *Evolution* 52:1760–1780.

Albertson, R. C., J. A. Markert, P. D. Danley, and T. D. Kocher. 1999. Phylogeny of a rapidly evolving clade: The cichlid fishes of Lake Malawi, East Africa. *Proc Natl Acad Sci* (USA) 96:5107–5110.

Albrecht, H. 1966. Zur Stammesgeschichte einiger Bewegungsweisen bei Fischen, untersucht am Verhalten von *Haplochromis* (Pisces, Cichlidae). *Z Tierpsychol* 23:270–302.

Alcock, J. 1984. *Animal behavior.* 3d ed. Sunderland, MA: Sinauer.

Alexander, R. D. 1962. The role of behavioral study in cricket classification. *Syst Zool* 11:53–72.

Allard, M. F., and J. M. Carpenter. 1996. On weighting and congruence. *Cladistics* 12: 183–198.

Allmon, W. D. 1994. Taxic evolutionary paleoecology and the ecological context of macroevolutionary change. *Evol Ecol* 8:95–112.

Altaba, C. R. 1991. The importance of ecological and historical factors in the production of benzaldehyde by tiger beetles. *Syst Zool* 40:101–105.

Ambrose, J. P., and S. B. Bratton. 1990. Trends in landscape heterogeneity along the borders of Great Smoky Mountains National Park. *Conserv Biol* 4:135–143.

Amorim, D. S. 1991. Refuge model simulations: Testing the theory. *Rev Bras Entomol* 35: 803–812.

———. 1994. *Elementos Basicos de Sistematica Filogenetica.* 1st ed. Riberao Preto, Brazil: Sociedade Brasileira de Entomologia.

———. 1997. *Elementos Basicos de Sistematica Filogenetica.* 2d ed. Riberao Preto, Brazil: Sociedade Brasileira de Entomologia.

Amorim, D. S., and M. R. S. Pires. 1996. Neotropical biogeography and a method for maximum biodiversity estimation. In *Biodiversity in Brazil: A first approach,* ed. C. E. M. Buicudo and N. A. Menezes, 183–219. São Paulo: CNPq.

Amundson, R. 1996. Historical development of the concept of adaptation. In *Adaptation,* ed. M. R. Rose and G. V. Lauder, 11–53. New York: Academic Press.

Anderberg, A. A., and A. Tehler. 1990. Consensus trees: A necessity in taxonomic practice. *Cladistics* 6:399–402.

Anderberg, A. A., I. Trift, and M. Källersjö. 2000. Phylogeny of *Cyclamen* L. (Primulaceae): Evidence from morphology and sequence data from the internal transcribed spacers of nuclear ribosomal DNA. *Plant Syst Evol* 220:147–160.

Andersen, N. M. 1982. The semi-aquatic bugs (Hemiptera, Gerromorpha): Phylogeny, adaptations, biogeography, and classification. *Entomograph* 3:1–455.

———. 1991. Marine insects: Genital morphology, phylogeny, and evolution of sea skaters, genus *Halobates* (Hemiptera: Gerridae). *Zool J Linn Soc* 103:21–60.

———. 1993. The evolution of wing polymorphism in water striders (Gerridae): A phylogenetic approach. *Oikos* 67:433–443.

———. 1994. The evolution of sexual size dimorphism and mating systems in water striders (Hemiptera: Gerridae): A phylogenetic approach. *Ecoscience* 1:208–214.

———. 1995. Cladistic inference and evolutionary scenarios: Locomotory structure, function, and performance in water striders. *Cladistics* 11:279–295.

———. 1996. Ecological phylogenetics of mating systems and sexual dimorphism in water striders (Hemiptera: Gerridae). *Vie Milieu* 46:103–114.

———. 1997a. Phylogenetic tests of evolutionary scenarios: The evolution of flightlessness and wing polymorphism in insects. *Mem Mus Natl Hist Nat* (Paris) 173:91–108.

———. 1997b. A phylogenetic analysis of the evolution of sexual dimorphism and mating systems in water striders (Hemiptera: Gerridae). *Biol J Linn Soc* 61:345–368.

———. 1999. The evolution of marine insects: Phylogenetic, ecological, and geographical aspects of species diversity in marine water striders. *Ecography* 22:98–111.

———. 2001. The impact of W. Hennig's "phylogenetic systematics" on contemporary entomology. *Eur J Entomol* 98:133–150.

Anderson, F. E. 2000. Phylogenetic relationships among loliginid squids (Cephalopoda: Myopsida) based on analyses of multiple data sets. *Zool J Linn Soc* 130:603–633.

Anderson, R. M., and R. M. May. 1982. Coevolution of hosts and parasites. *Parasitology* 85:411–426.

———. 1985. Epidemiology and genetics in the coevolution of parasites and hosts. *Proc Roy Soc Lond* B 219:281–283.

Anderson, R. S. 1993. Weevils and plants: Phylogenetic versus ecological mediation of the evolution of host plant associations in Curculioninae (Coleoptera: Curculionidae). *Mem Entomol Soc Can* 165:197–232.

Anderson, S. 1974. Patterns of faunal evolution. *Annu Rev Biol* 49:1–15.

Anderson, S., and C. S. Anderson. 1975. Three Monte Carlo models of faunal evolution. *Am Mus Novit* 2563:1–6.

Andersson, L. 1996. An ontological dilemma: Epistemology and methodology of historical biogeography. *J Biogeogr* 23:269–277.

Andersson, M. 1999. Phylogeny, behaviour, plumage evolution, and neoteny in skuas, Stercorariidae. *J Avian Biol* 30:205–215.

Andrews, J. M. 1927. Host-parasite specificity in the coccidia of mammals. *J Parasitol* 13:183–194.

Antonovics, J., and P. H. van Tierderen. 1991. Ontoecogenophyloconstraints? The chaos of constraint terminology. *Trends in Ecol Evol* 6:166–167.

Apakupakul, K., M. E. Siddall, and E. M. Burreson. 1999. Higher level relationships of leeches (Annelida: Clitellata: Euhirudinea) based on morphology and gene sequences. *Mol Phylogenet Evol* 12:350–359.

Arata, A. A. 1964. The anatomy and taxonomic significance of the male accessory reproductive glands of muroid rodents. *Bull Florida State Mus* 9:1–42.

Archibald, J. D. 1993. The importance of phylogenetic analysis for the assessment of species turnover: A case history of Paleocene mammals in North America. *Paleobiology* 19:1–27.

Archie, J. W. 1985. Methods for coding variable morphological features for numerical taxonomic analysis. *Syst Zool* 34:326–345.

———. 1989. Phylogenies of plant families: A demonstration of phylogenetic randomness in DNA sequence data derived from proteins. *Evolution* 43:1796–1799.

Armbruster, W. S. 1983. *Dalechampia scandens*. In *Costa Rican natural history,* ed. D. H. Janzen, 230–233. Chicago: Univ. Chicago Press.

———. 1988. Multilevel comparative analysis of the morphology, function, and evolution of *Dalechampia* blossoms. *Ecology* 69:1746–1761.

———. 1992. Phylogeny and the evolution of plant-animal interactions. *BioScience* 42: 12–20.

———. 1993. Evolution of plant pollination systems: Hypotheses and tests with the neotropical vine *Dalechampia*. *Evolution* 47:1480–1503.

———. 1994. Early evolution of *Dalechampia* (Euphorbiaceae): Insights from phylogeny, biogeography, and comparative biology. *Ann Missouri Bot Gard* 81:302–316.

———. 1996a. Cladistic analysis and revision of *Dalechampia* sections *Rhopalostylus* and *Brevicolumnae* (Euphorbiaceae). *Syst Bot* 212:209–235.

———. 1996b. Exaptation, adaptation, and homoplasy: Evolution of ecological traits in *Dalechampia* vines. In *Homoplasy: The recurrence of similarity in evolution,* ed. M. J. Sanderson and L. Hufford, 227–243. New York: Academic Press.

———. 1997. Exaptations link evolution of plant-herbivore and plant-pollinator interactions: A phylogenetic inquiry. *Ecology* 78:1661–1672.

Armbruster, W. S., J. J. Howard, T. P. Clausen, E. M. Debevec, J. Loquvam, M. Matsuki, B. Cerendolo, and F. Andel. 1997. Do biochemical exaptations link evolution of plant defense and pollination systems?: Historical hypotheses and experimental test with *Dalechampia* vines. *Am Nat* 148:461–484.

Armbruster, W. S., S. Keller, M. Matsuki, and T. P. Clausen. 1989. Pollination of *Dalechampia magnoliifolia* by male euglossine bees (Apidae: Euglossini). *Am J Bot* 76:1279–1285.

Armbruster, W. S., and W. S. Mziray. 1987. Pollination and herbivore ecology of an African *Dalechampia* (Euphorbiaceae): Comparisons with New World species. *Biotropica* 19:64–73.

Armstrong, D. P., and M. Westoby. 1993. Seedlings from large seeds tolerate defoliation better: A test using phylogenetically independent contrasts. *Ecology* 74:1092–1100.

Arnold, E. N. 1990. Why do morphological phylogenies vary in quality? An investigation based on the comparative history of lizard clades. *Proc Roy Soc Lond* B 240: 135–172.

———. 1994. Investigating the origins of performance advantage: Adaptation, exaptation, and lineage effects. In *Phylogenetics and Ecology,* ed. P. Eggleton and R. I. Vane-Wright, 123–168. London: Academic Press.

———. 1995. Identifying the effects of history on adaptation: Origins of different sand-diving techniques in lizards. *J Zool* 235:351–388.

Arnold, S. J. 1987. The comparative ethology of courtship in salamandrid salamanders: 1. *Salamandra* and *Chioglossa*. *Ethology* 74:133–145.

———. 1992. Constraints on phenotypic evolution. *Am Nat* 140:S85–S107.

Arnold, S. J., P. A. Verrell, and S. G. Tilley. 1996. The evolution of asymmetry in sexual isolation: A model and test case. *Evolution* 50:1024–1033.

Arnqvist, G., M. Edvardsson, U. Friberg, and T. Nilsson. 2000. Sexual conflict promotes speciation in insects. *Proc Nat Acad Science* (USA) 97:10460–10464.

Arntzen, J. W., and M. Sparreboom. 1989. A phylogeny for Old World newts, genus *Triturus*: Biochemical and behavioral data. *J Zool* (London) 219:645–664.

Arthur, W., and C. Kettle. 2001. Geographic patterning of variation in segment number in geophilomorph centipedes: Clines and speciation. *Evol Dev* 3:34–40.

Ashe, J. S. 2000. Mouthpart structure of *Stylogymnusa subantarctica* Hammond, 1975 (Coleoptera: Staphylindae: Aleocharinae) with a reanalysis of the phylogenetic position of the genus. *Zool J Linn Soc* 130:471–498.

Ashen, J. B., and L. J. Goff. 2000. Molecular and ecological evidence for species specificity and coevolution in a group of marine algal bacterial symbioses. *Appl Envir Microbiol* 66:3024–3030.

Ashlock, P. D. 1974. The uses of cladistics. *Annu Rev Ecol Syst* 5:81–99.

Atz, J. W. 1970. The application of the idea of homology to behavior. In *Development and evolution of behavior*, ed. L. R. Aronson, E. Tobach, D. S. Lehrman, and J. S. Rosenblatt, 53–74. San Francisco: W. H. Freeman.

Aubert, J., L. Legal, H. Descimon, and F. Michel. 1999. Molecular phylogeny of Swallowtail butterflies of the tribe Papilionini (Papilionidae, Lepidoptera). *Mol Phylogenet Evol* 12:156–167.

Austin, M. P. 1999. The potential contribution of vegetation ecology to biodiversity research. *Ecography* 22:465–484.

Autumn, K., D. Jindrich, D. DeNardo, and R. Mueller. 1999. Locomotor performance at low temperature and the evolution of nocturnality in geckos. *Evolution* 53:580–599.

Avers, C. 1989. *Process and pattern in evolution*. Oxford: Oxford Univ. Press.

Avise, J. C. 1989a. Gene trees and organismal histories: A phylogenetic approach to population biology. *Evolution* 43:1192–1208.

———. 1989b. A role for molecular genetics in the recognition and conservation of endangered species. *Trends Ecol Evol* 4:279–281.

———. 1992. Molecular population structure and the biogeographic history of a regional fauna: A case history with lessons for conservation biology. *Oikos* 63:62–76.

———. 2000. *Phylogeography: The history and formation of species*. Cambridge, MA: Harvard Univ. Press.

Avise, J. C., J. Arnold, R. M. Ball, E. Bermingham, T. Lamb, J. E. Neigel, C. A. Reed, and N. C. Saunders. 1987. Intraspecific phylogeography: The mitochondrial DNA bridge between population genetics and systematics. *Annu Rev Ecol Syst* 18:489–522.

Axelius, B. 1991. The genus *Xanthophytum* (Rubiaceae): Taxonomy, phylogeny, and biogeography. *Blumea* 34:425–497.

Axelrod, D. I. 1979. Age and origin of Sonoran desert vegetation. *California Acad Sci Occas Pap* 132:1–74.

———. 1983. Paleobotanical history of the western deserts. In *Origin and evolution of deserts*, ed. S. G. Wells and D. R. Haragan, 113–129. Albuquerque: Univ. of New Mexico Press.

Ayala, F. J. 1982. The genetic structure of species. In *Perspectives on evolution*, ed. R. Milkman, 3–81. Sunderland, MA: Sinauer.

Baas, P., K. Kalkman, and R. Geesink, eds. 1990. *The plant diversity of Malesia*. Dordrecht: Kluwer Academic.

Backlund, M., B. Oxelman, and B. Bremer. 2000. Phylogenetic relationships within the Gentianales based on ndhF and rbcL sequences, with particular reference to the Loganiaceae. *Am J Bot* 87:1029–1043.

Baer, J.-G. 1948. Les helminthes, parasites des vertèbres: Relations phylogénétique entre leur évolution et celle de leurs hotes. *Ann Sci Franche Comte* 2:99–113.

Baerends, G. P. 1958. Comparative methods and the concept of homology in the study of behaviour. *Arch Neerl Zool* 13, suppl. no. 1:401–417.

Baker, F. C. 1927. Molluscan associations of White Lake, Michigan: A study of a small inland lake from an ecological and systematic viewpoint. *Ecology* 8:353–370.

Baker, R. H., and R. DeSalle. 1997. Multiple sources of character information and the phylogeny of Hawaiian drosophilids. *Syst Biol* 46:654–673.

Baker, R. J., C. S. Hood, and R. L. Honeycutt. 1989. Phylogenetic relationships and classification of the higher categories of the New World bat family Phyllostomidae. *Syst Zool* 38:228–238.

Baker, R. J., M. B. Qumsiyeh, and C. S. Hood. 1987. Role of chromosomal banding patterns in understanding mammalian evolution. In *Current Mammalogy*, vol. 2, ed. H. H. Genoways, 67–96. New York: Plenum Press.

Baker, W. J., J. Dransfield, and T. A. Hedderson. 2000. Phylogeny, character evolution, and a new classification of the calamoid palms. *Syst Bot* 25:297–322.

Bakker, R. T. 1983. The deer flees, the wolf pursues: Incongruencies in predator-prey coevolution. In *Coevolution*, ed. D. J. Futuyma and M. Slatkin, 350–382. Sunderland, MA: Sinauer.

———. 1986. *The dinosaur heresies*. New York: Morrow.

Bakker T. C. M. 1993. Positive genetic correlation between female preference and preferred male ornament in sticklebacks. *Nature* 363:255–257.

———. 1999. The study of intersexual selection using quantitative genetics. *Behaviour* 136:1237–1265.

Bakker, T. C. M., and M. Milinski. 1991. Sequential female choice and the previous male effect in sticklebacks. *Behav Ecol Sociobiol* 29:205–210.

Bakker, T. C. M., and B. Mundwiler. 1994. Female mate choice and male red coloration in a natural three-spined stickleback *(Gasterosteus aculeatus)* population. *Behav Ecol* 5:74–80.

Baldwin, B. G., D. W. Kyhos, and J. Dvorak. 1990. Chloroplast DNA evolution and adaptive radiation in the Hawaiian silversword alliance (Asteraceae: Madiinae). *Ann Missouri Bot Gard* 77:96–109.

Balick, M. J. 1996. Transforming ethnobotany for the new millennium. *Ann Missouri Bot Gard* 83:58–66.

Bandoni, S. M., and D. R. Brooks. 1987. Revision and phylogenetic analysis of the Amphilinidea Poche, 1922 (Platyhelminthes: Cercomeria: Cercomeromorpha). *Can J Zool* 65:1110–1128.

Baptista, L. F., and P. W. Trail. 1992. The role of song in the evolution of passerine diversity. *Syst Biol* 41:242–247.

Barbosa, P. 1988. Some thoughts on the evolution of "host range." *Ecology* 69:912–915.

Barclay, R. M. R. 1994. Constraints on reproduction by flying vertebrates: Energy and calcium. *Am Nat* 144:1021–1031.

Bargelloni, L., L. Zane, G. Lecointre, and T. Patarnello. 2000. Molecular zoogeography of Antarctic euphausiids and notothenoids: From species phylogenies to intraspecific patterns of genetic variation. *Antarct Sci* 12:259–268.

Barker, G. 1993. Models of biological change: Implications of three studies of "Lamarckian" change. *Perspect Ethol* 10:229–248.

Barkman, T. J. 2001. Character coding of secondary chemical variation for use in phylogenetic analyses. *Biochem Syst Ecol* 29:1–20.

Barkman, T. J., G. Chenery, J. R. McNeal, J. Lyons-Weiler, W. J. Ellisens, G. Moore, A. D. Wolfe, and C. W. dePamphilis. 2000. Independent and combined analyses of sequences from all three genomic compartments converge on the root of flowering plant phylogeny. *Proc Natl Acad Sci* (USA) 97:13166–13171.

Barnes, R. D. 1989. Diversity of organisms: How much do we know? *Am Zool* 29:1075–1084.

Barr, T. C., Jr. 1968. Cave ecology and evolution of troglobites. *Evol Biol* 2:35–102.

Barraclough, T. G., P. H. Harvey, and S. Nee. 1995. Sexual selection and taxonomic diversity in passerine birds. *Proc Roy Soc Lond* B 259:211–215.

Barrett, M, M., J. Donoghue, and E. Sober. 1991. Against consensus. *Syst Zool* 40:486–493.

Barrett, S. C. H. 1989. Mating system evolution and speciation in heterostylous plants.

In *Speciation and its consequences,* ed. D. Otte and J. A. Endler, 257–283. Sunderland, MA.: Sinauer.

———. 1995. Mating-system evolution in flowering plants: Micro- and macroevolutionary approaches. *Acta Bot Neerl* 44:385–402.

Barrier, M., B. G. Baldwin, R. H. Robichaux, and M. D. Purugganan. 1999. Interspecific hybrid ancestry of a plant adaptive radiation: Allopolyploidy of the Hawaiian Silversword alliance (Asteraceae) inferred from floral homeotic gene duplications. *Mol Biol Evol* 16:1105–1113.

Barrowclough, G. F. 1992. Systematics, biodiversity, and conservation biology. In *Systematics, ecology, and the biodiversity crisis,* ed. N. Eldredge, 121–143. New York: Columbia Univ. Press.

Barrowclough, G. F., R. J. Gutierrez, and G. G. Groth. 1999. Phylogeography of spotted owl *(Strix occidentalis)* populations based on mitochondrial DNA sequences: Gene flow, genetic structure, and a novel biogeographic pattern. *Evolution* 53: 919–931.

Barton, N. H. 1989. Founder effect speciation. In *Speciation and its consequences,* ed. D. Otte and J. A. Endler, 229–256. Sunderland, MA: Sinauer.

———. 1996. Natural selection and random genetic drift as causes of evolution in islands. *Phil Trans Roy Soc Lond* B 351:785–795.

Barton, N. H., and B. Charlesworth. 1984. Genetic revolutions, founder effects, and speciation. *Annu Rev Ecol Syst* 15:133–164.

Basibuyuk, H. H., and D. L. J. Quicke. 1999. Grooming behaviours in the Hymenoptera (Insecta): Potential phylogenetic significance. *Zool J Linn Soc* 125:349–382.

Bateman, R. M., P. R. Crane, W. A. DiMichele, P. R. Kenrick, N. P. Rowe, T. Speck, and W. E. Stein. 1998. Early evolution of land plants: Phylogeny, physiology, and ecology of the primary terrestrial radiation. *Annu Rev Ecol Syst* 29:263–292.

Bates, H. W. 1862. Contributions to an insect fauna of the Amazon Valley, Lepidoptera: Heliconidae. *Trans Linn Soc Lond* 23:495–566.

Baube, C. L., W. J. Rowland, and J. B. Fowler. 1995. The mechanisms of colour-based mate choice in female threespines: Hue, contrast, and configurational cues. *Behaviour* 132:979–996.

Baum, D. A. 1992a. Combining trees as a way of combining data sets for phylogenetic inference, and the desirability of combining gene trees. *Taxon* 41:1–10.

———. 1992b. Phylogenetic species concepts. *Trends Ecol Evol* 7:1–2.

Baum, D. A., and M. J. Donoghue. 1995. Choosing among alternative "phylogenetic" species concepts. *Syst Bot* 20:560–573.

Baum, D. A., and A. Larson. 1991. Adaptation reviewed: A phylogenetic methodology for studying character macroevolution. *Syst Zool* 40:1–18.

Baum, D. A., and K. L. Shaw. 1995. Genealogical perspectives on the species problem. In *Experimental and molecular approaches to plant biosystematics,* ed. P. C. Hoch and A. G. Stephenson, 289–303. St. Louis: Missouri Botanical Garden.

Beattie, A. J. 1994. Systematics and biodiversity. *Trends Ecol Evol* 9:227–228.

Beatty, J. 1982. Classes and cladists. *Syst Zool* 31:25–34.

———. 1994. Theoretical pluralism in biology. In *Interpreting the hierarchy of nature: From systematic patterns to evolutionary process theories,* ed. L. Grande and O. Rieppel, 33–57. London: Academic Press.

Becerra, J. X. 1997. Insects on plants: Macroevolutionary chemical trends in host use. *Science* 276:253–256.

Begle, D. P. 1991. Relationships of the osmeroid fishes and the use of reductive characters in phylogenetic analysis. *Syst Zool* 40:33–53.

Begun, D., C. Ward, and M. Rose, eds. 1997. *Function, phylogeny, and fossils: Miocene hominoid evolution and adaptations.* New York: Plenum Press.

Behrensmeyer, A. K., J. D. Damuth, W. A. DiMichele, R. Potts, H.-D. Sues, and S. L. Wing. eds. 1992. *Terrestrial ecosystems through time.* Chicago: Univ. Chicago Press.

Bell, G. 1989. A comparative method. *Am Nat* 133:553–571.

Bell, M. A., G. Orti, J. A. Walker, and J. P. Koenings. 1993. Evolution of pelvic reduction in threespine stickleback fish: A test of competing hypotheses. *Evolution* 47:906–914.

Bell, M. A., and S. A. Foster. 1994. *The evolutionary biology of the threespine stickleback.* Oxford: Oxford Univ. Press.

Belyea, L. R., and J. Lancaster. 1999. Assembly rules within a contingent ecology. *Oikos* 86:402–416.

Bemis, W. E. 1987. Feeding systems of living Dipnoi: Anatomy and function. In *The biology and evolution of lungfishes*, ed. W. E. Bemis, W. W. Burggren, and N. E. Kemp, 249–275. New York: Alan Liss.

Bemis, W. E., and Lauder, G. V. 1986. Morphology and function of the feeding apparatus of the lungfish, *Lepidosiren paradoxa*. *J Morphol* 187:81–108.

Bengtsson, J. 1989. Interspecific competition increases local extinction rate in a metapopulation system. *Nature* 340:713–715.

Benson, K. R. 2000. The emergence of ecology from natural history. *Endeavour* 24:59–62.

Benson, W. W., K. S. Brown Jr., and L. E. Gilbert. 1975. Coevolution of plants and herbivores: Passion flower butterflies. *Evolution* 29:659–680.

Benton, M. J. 1990. Evolution of large size. In *Palaeobiology: A synthesis*, ed. D. E. G. Briggs and P. R. Crowther, 147–152. Oxford: Blackwell.

———. 1996. On the nonprevalence of competitive replacement in the evolution of tetrapods. In *Evolutionary paleobiology*, ed. D. Jablonski, D. H. Erwin, and J. H. Lipps, 185–210. Chicago: Univ. Chicago Press.

———. 2000. Stem nodes, crown clades, and rank-free lists: Is Linnaeus dead? *Biol Rev* 75:633–648.

Bentzen, P., and J. D. McPhail. 1984. Ecology and evolution of sympatric sticklebacks *(Gasterosteus)*: Specialization for alternate trophic niches in the Enos Lake species pair. *Can J Zool* 52:2280–2286.

Berenbaum, M. R. 1981. Effects of linear furanocoumarins on an adapted specialist insect *(Papilio polyxenes)*. *Ecol Entomol* 6:345–351.

———. 1983. Coumarins and caterpillars: A case for coevolution. *Evolution* 37:163–179.

Berenbaum, M. R., and P. Feeny. 1981. Toxicity of angular furanocoumarins to swallowtail butterflies: Escalation in a coevolutionary arms race? *Science* 212:927–929.

Berenbaum, M. R., and S. Passoa. 1999. Generic phylogeny of North American Depressariinae (Lepidoptera: Elachistidae) and hypotheses about coevolution. *Ann Entomol Soc Am* 92:971–986.

Bernays, E. A. 1989. Host range in phytophagous insects: The potential role of generalist predators. *Evol Ecol* 3:299–311.

———. 1996. Selective attention and host-plant specialization. *Entomol Exp Appl* 80:125–131.

———. 1998. The value of being a resource specialist: Behavioral support for a neural hypothesis. *Am Nat* 151:451–464.

Bernays, E. A., and M. Chapman. 1994. *Host-plant selection by phytophagous insects*. New York: Chapman and Hall.

Bernays, E. A., and D. J. Funk. 1999. Specialists make faster decisions than generalists: Experiments with aphids. *Proc Roy Soc Lond* B 266:151–156.

Bernays, E. A., and M. Graham. 1988. On the evolution of host specificity in phytophagous arthropods. *Ecology* 69:8865–8892.

Bernays, E. A., and W. T. Wcislo. 1994. Sensory capabilities, information processing, and resource specialization. *Quart Rev Biol* 69:187–204.

Beuttell, K., and J. B. Losos. 1999. Ecological morphology of Caribbean anoles. *Herpetol Monogr* 13:1–28.

Bieregaard, R. O., Jr., T. E. Lovejoy, V. Kapos, A. Augusto dos Santos, and R. W. Hutchings. 1992. The biological dynamics of tropical rainforest fragments. *BioScience* 42:859–866.

Bininda-Emonds, O. R. P., and H. N. Bryant. 1998. Properties of matrix representation with parsimony analyses. *Syst Biol* 47:497–508.

Bininda-Emonds, O. R. P., D. P. Vázquez, and L. L. Manne. 2000. The calculus of biodiversity: Integrating phylogeny and conservation. *Trends Ecol Evol* 15:92–93.

Björklund, M. 1990. A phylogenetic interpretation of sexual dimorphism in body size and ornament in relation to mating system in birds. *J Evol Biol* 3:171–183.

———. 1994. The independent contrast method in comparative biology. *Cladistics* 10:425–433.

Björklund, M., and J. Merilä. 1993. Morphological differentiation in *Carduelis* finches: Adaptive vs. constraint models. *J Evol Biol* 6:359–373.

Blackmore, S. 1996. Knowing the Earth's biodiversity: Challenges for the infrastructure of systematic biology. *Science* 274:63–64.

Blackmore, S., C. A. McConchie, and R. B. Knox. 1987. Phylogenetic analysis of the male ontogenetic program in aquatic and terrestrial monocotyledons. *Cladistics* 3:333–347.

Blankespoor, H. D. 1974. Host-induced variation in *Plagiorchis noblei* Park, 1936 (Plagiorchiidae; Trematoda). *Am Midl Nat* 91:415–433.

Bledsoe, A. H., and R. J. Raikow. 1990. A quantitative assessment of congruence between molecular and morphological estimates of phylogeny. *J Mol Evol* 30:247–259.

Block, B. A. 1991a. Endothermy in fish: Thermogenesis, ecology, and evolution. In *Biochemistry and molecular biology of fishes*, vol. 1, ed. P. W. Hochachka and T. Mommsen, 269–311. Amsterdam: Elsevier Science.

———. 1991b. Evolutionary novelties: How fish have built a heater out of muscle. *Am Zool* 31:726–742.

Blouw, D. M., and G. J. Boyd. 1992. Inheritance of reduction, loss, and asymmetry of the pelvis of *Pungitius pungitius* (ninespine stickleback). *Heredity* 68:33–42.

Boas, F. 1896. The limitations of the comparative method in anthropology. *Science* 4:901–908.

———. 1898. A precise criterion of species. *Science* 7:860–861.

Bonaccorso, F. J. 1979. Foraging and reproductive ecology in a Panamanian bat community. *Bull Florida State Mus Biol Sci* 24:359–408.

Bond, J. E., and B. D. Opell. 1998. Testing adaptive radiation and key innovation hypotheses in spiders. *Evolution* 52:403–414.

Bond, W. J. 1989. The tortoise and the hare: Ecology of angiosperm dominance and gymnosperm persistence. *Biol J Linn Soc* 36:227–249.

Bonner, J. T., ed. 1982a. *Heterochrony in evolution: A multidisciplinary approach.* New York: Plenum Press.

———. 1982b. *Evolution and development.* New York: Springer Verlag.

Bookstein, F. L., B. Chernoff, R. L. Elder, J. M. Humphries, Jr., G. R. Smith, and R. E. Strauss. 1985. *Morphometrics in evolutionary biology.* Acad Nat Sci Spec Publ 15.

Borgia, G. 1985a. Bower destruction and sexual competition in the male satin bowerbird *(Ptilonorhynchus violaceus). Behav Ecol Sociobiol* 18:91–100.

———. 1985b. Bower quality, number of decorations, and mating success of male satin bowerbirds *(Ptilonorhynchus violaceus):* An experimental analysis. *Anim Behav* 33:266–271.

———. 1986. Sexual selection in bowerbirds. *Sci Am* 254:70–79.

———. 1995a. Complex male display and female choice in the spotted bowerbird: Specialized functions for different bower decorations. *Anim Behav* 49:1291–1301.

———. 1995b. Threat reduction as a cause of differences in bower architecture, bower decoration, and male display in two closely related bowerbirds, *Chlamydera nuchalis* and *C. maculata. Emu* 95:1–12.

———. 1995c. Why do bowerbirds build bowers? *Am Sci* 83:542–547.

Borgia, G., and S. W. Coleman. 2000. Co-option of male courtship signals from aggressive display in bowerbirds. *Proc Roy Soc Lond* B 267:1735–1740.

Borgia, G., and D.C. Presgraves. 1998. Coevolution of elaborated male display traits in the spotted bowerbird: An experimental test of the threat reduction hypothesis. *Anim Behav* 56:1121–1128.

Borowik, O. A. 1995. Coding chromosomal data for phylogenetic analysis: phylogenetic resolution of the *Pan-Homo-Gorilla* trichotomy. *Syst Biol* 44:563–570.

Bostwick, K. S. 2000. Display behaviors, mechanical sounds, and evolutionary relationships of the club-winged manakin *(Machaeropterus deliciosus). Auk* 117:465–478.

Boucot, A. J. 1975a. Standing diversity of fossil groups in successive intervals of geologic time viewed in the light of changing levels of provincialism. *J Paleontol* 49:1105–1111.

———. 1975b. *Evolution and extinction rate controls.* New York: Elsevier.

———. 1978. Community evolution and rates of cladogenesis. In *Evolutionary biology*, ed. M. K. Hecht, W. C. Steere, and B. Wallace, 11:545–655. New York: Plenum Press.

———. 1981. *Principles of benthic marine paleoecology.* New York: Academic Press.

———. 1982. *Paleobiologic evidence of behavioral evolution and coevolution.* Corvallis, OR: by the author.

———. 1983. Does evolution take place in an ecological vacuum? II. *J Paleontol* 57:1–30.

———. 1990. Community evolution: Its evolutionary and biostratigraphic significance. In *Paleocommunity temporal dynamics: The long-term development of multi-species assemblies,* ed. W. Miller, III., Paleontol Soc Spec Publ 5:48–70.

Bouton, N. 2000. Progressive invasion and allopatric speciation can also explain distribution patterns of rock-dwelling cichlids from southern Lake Victoria: A comment on Seehausen and van Alphen (1999). *Ecol Let* 3:166–171.

Bouton, N., O. Seehausen, and J. J. M. van Alphen. 1997. Resource partitioning among rock-dwelling haplochromines (Pisces: Cichlidae) from Lake Victoria. *Ecol Freshwater Fish* 6:225–240.

Bowen, B. W. 1998. What is wrong with ESUs? The gap between evolutionary theory and conservation genetics. *J Shellfish Res* 17:1355–1358.

Bowen, M. D., C. J. Peters, and S. T. Nichol. 1997. Phylogenetic analysis of the Arenaviridae: Patterns of virus evolution and evidence for cospeciation between arenaviruses and their rodent hosts. *Mol Phylogenet Evol* 8:301–316.

Bowers, M. A., and J. H. Brown. 1982. Body size and coexistence in desert rodents: Chance or community structure? *Ecology* 63:391–400.

Bowler, P. J. 1983. *The eclipse of Darwinism.* Baltimore, MD: Johns Hopkins Univ. Press.

———. 1996. *Life's splendid drama: Evolutionary biology and the reconstruction of life's ancestry, 1860–1940.* Chicago: Univ. Chicago Press.

Bowman, B. H., J. W. Taylor, and T. J. White. 1992. Molecular evolution of the fungi: Human pathogens. *Mol Biol Evol* 9:893–904.

Boyd, M. F. 1949. Historical review. In *Malariology,* vol. 1, ed. M. F. Boyd, 3–25. Philadelphia: Saunders.

Boyden, A. 1947. Homology and analogy: A critical review of the meanings and implication of these concepts in biology. *Am Midl Nat* 37:648–660.

Bragg, A. N., and C. C. Smith. 1943. Observations on the ecology and natural history of anura, IV. The ecological distribution of toads in Oklahoma. *Ecology* 24:285–309.

Bramble, D. M., and D. B. Wake 1985. Feeding mechanisms of lower tetrapods. In *Functional vertebrate morphology,* ed. M. Hildebrand, D. M. Bramble, K. F. Liem, and D. B. Wake, 230–261. Cambridge, MA: Harvard Univ. Press.

Brandt, A. 2000. Hypotheses on southern ocean peracarid evolution and radiation (Crustacea, Malacostraca). *Antarct Sci* 12:269–275.

Bremer, K. 1990. Combinable component consensus. *Cladistics* 6:369–372.

———. 1992. Ancestral areas: A cladistic reinterpretation of the center of origin concept. *Syst Biol* 41:436–445.

———. 1994. Branch support and tree stability. *Cladistics* 10:295–304.

Bremer, K., and H.-E. Wanntorp. 1977. Phylogenetic systematics in botany. *Taxon* 27:317–329.

———. 1979. Hierarchy and reticulation in systematics. *Syst Zool* 28:624–627.

Bretsky, P. W. 1973. Evolutionary patterns in the Paleozoic Bivalvia: Documentation and some theoretical considerations. *Bull Geol Soc Am* 84:2079–2096.

Bridle, J. R., and C. D. Jiggins. 2000. Adaptive dynamics: is speciation too easy? *Trends Ecol Evol* 15:225–226.

Brochu, C. A. 1997. Fossils, morphology, divergence timing, and the phylogenetic relationships of *Gavialis. Syst Biol* 46:479–522.

Brokaw, N., and R. T. Busing. 2000. Niche versus chance and tree diversity in forest gaps. *Trends Ecol Evol* 15:183–187.

Brooke, M. de L. 2000. Why museums matter. *Trends Ecol Evol* 15:136–137.

Brooks, D. R. 1979. Testing the context and extent of host-parasite coevolution. *Syst Zool* 28:299–307.

———. 1980. Allopatric speciation and non-interactive parasite community structure. *Syst Zool* 29:192–203.

———. 1981. Hennig's parasitological method: A proposed solution. *Syst Zool* 30:229–249.

———. 1985. Historical ecology: A new approach to studying the evolution of ecological associations. *Ann Missouri Bot Gard* 72:660–680.

———. 1986. Analysis of host-parasite coevolution. In *Parasitology—Quo Vadit? Proceedings of the 6th International Congress of Parasitology,* ed. M. J. Howell, 291–300. Canberra: Australian Academy of Science.

————. 1987. Analysis of host-parasite coevolution. *Int J Parasitol* 17:291–300.

————. 1988a. Scaling effects in historical biogeography: A new view of space, time and form. *Syst Zool* 38:237–244.

————. 1988b. Macroevolutionary comparisons of host and parasite phylogenies. *Annu Rev Ecol Syst* 19:235–259.

————. 1989a. A summary of the database pertaining to the phylogeny of the major groups of parasitic platyhelminths, with a revised classification. *Can J Zool* 67: 714–720.

————. 1989b. The phylogeny of the Cercomeria (Platyhelminthes: Rhabdocoela) and general evolutionary principles. *J Parasitol* 75:606–616.

————. 1989c. Coevolution of helminth parasites and vertebrates. In *Current concepts in parasitology,* ed. R. A. Ko, 255–267. Hong Kong: Univ. Hong Kong Press.

————. 1990. Parsimony analysis in historical biogeography and coevolution: Methodological and theoretical update. *Syst Zool* 39:14–30.

————. 1995. What is "cladistics"? *Semiotic Rev Books* 6.3:1–2.

————. 1996. Explanations of homoplasy at different levels of biological organization. In *Homoplasy: The recurrence of similarity in evolution,* ed. R. J. Sanderson and L. Hufford, 3–36. London: Academic Press.

————. 1998a. The unified theory of evolution and selection processes. In *Evolutionary systems: Biological and epistemological perspectives on selection and self-organization,* ed. G. van de Vijver, S. N. Salthe, and M. Delpos, 113–128. Dordrecht: Kluwer Academic.

————. 1998b. Triage for the biosphere. In *The Brundtland Commission's report—10 Years,* ed. G. B. Softing, G. Benneh, K. Hindar, and A. Wijkman, 71–80. Oslo: Scandinavian Univ. Press.

————. 1999. Review of *Phylogenies and the comparative method in animal behavior,* ed. E. P. Martins. *Behav Processes* 47:135–136.

————. 2000. The nature of the organism: Life has a life of its own. *Proc NY Acad Sci* 901:257–265.

Brooks, D. R., and S. M. Bandoni. 1988. Coevolution and relicts. *Syst Zool* 37:19–33.

Brooks, D. R., S. M. Bandoni, C. M. Macdonald, and R. T. O'Grady. 1989. Aspects of the phylogeny of the Trematoda Rudolphi, 1808 (Platyhelminthes: Cercomeria). *Can J Zool* 67:2609–2624.

Brooks, D. R., and D. R. Glen. 1982. Pinworms and primates: A case study in coevolution. *Proc Helminthol Soc Wash* 49:76–85.

Brooks, D. R., and E. P. Hoberg. 2000. Triage for the biosphere: The need and rationale for taxonomic inventories and phylogenetic studies of parasites. *Comp Parasitol* 67:1–25.

Brooks, D. R., V. León-Règagnon, and G. Pérez Ponce de León. 2000. Parasitos y la biodiversidad. In *Enfoques contemporaneos para el estudio de la biodiversidad,* ed. M. H. Hernandez, A. N. Garcia, M. Ulloa, and F. Alvarez, 245–289. Mexico City: UNAM.

Brooks, D. R., and D. A. McLennan. 1991. *Phylogeny, ecology, and behavior: A research program in comparative biology.* Chicago: Univ. Chicago Press.

————. 1992a. Historical ecology as a research program in macroevolution. In *Systematics, historical ecology, and North American freshwater fishes,* ed. R. L. Mayden, 76–113. Stanford: Stanford Univ. Press.

————. 1992b. The evolutionary origin of *Plasmodium falciparum. J Parasitol* 78:564–566.

————. 1993a. Historical ecology: Examining phylogenetic components of community evolution. In *Species diversity in ecological communities,* ed. R. E. Ricklefs and D. Schluter, 267–280. Chicago: Univ. Chicago Press.

————. 1993b. *Parascript: Parasites and the language of evolution.* Washington, DC: Smithsonian Institution Press.

————. 1993c. Macroevolutionary trends in the morphological diversification among the parasitic flatworms (Platyhelminthes: Cercomeria). *Evolution* 47:495–509.

————. 1993d. Comparative studies of adaptive radiations with an example using parasitic flatworms (Platyhelminthes: Cercomeria). *Am Nat* 142:755–778.

————. 1994. Historical ecology as a research programme: Scope, limitations, and the future. In *Phylogenetics and ecology,* ed. Eggleton and R. Vane-Wright, Linnaean Society Symposium Series no. 17, pp. 1–27. London: Academic Press.

———. 1999. Species: Turning a conundrum into a research program. *J Nematol* 31:117–133.

———. 2001. A comparison of a discovery-based and an event-based method of historical biogeography. *J Biogeogr* 28:757–767.

Brooks, D. R., D. A. McLennan, J. M. Carpenter, S. G. Weller, and J .A. Coddington. 1995. Systematics, ecology, and behavior. *BioScience* 45:687–695.

Brooks, D. R., R. L. Mayden, and D. A. McLennan. 1992a. Phylogeny and biodiversity: Conserving our evolutionary legacy. *Trends Ecol Evol* 7:55–59.

———. 1992b. Phylogenetics and conservation. *Trends Ecol Evol* 7:353.

Brooks, D. R., and C. Mitter. 1984. Analytical basis of coevolution. In *Fungus-insect relationships: Perspectives in ecology and evolution,* ed. Q. Wheeler and M. Blackwell, 42–53. New York: Columbia Univ. Press.

Brooks, D. R., R. T. O'Grady, and D. R. Glen. 1985a. Phylogenetic analysis of the Digenea (Platyhelminthes: Cercomeria) with comments on their adaptive radiation. *Can J Zool* 63:411–443.

———. 1985b. The phylogeny of the Cercomeria Brooks, 1982 (Platyhelminthes). *Proc Helminthol Soc Wash* 52:1–20.

Brooks, D. R., M. Van Veller, and D. A. McLennan. 2001. How to do BPA, Really. *J Biogeogr* 28:345–358.

Brooks, D. R., and E. O. Wiley. 1985. Theories and methods in different approaches to systematics. *Cladistics* 1:1–14.

———. 1986. *Evolution as entropy: Toward a unified theory of biology.* 1st ed. Chicago: Univ. Chicago Press.

———. 1988. *Evolution as entropy: Toward a unified theory of biology.* 2d ed. Chicago: Univ. Chicago Press.

Broughton, R. E., S. E. Stanley, and R. T. Durrett. 2000. Quantification of homoplasy for nucleotide transitions and transversions and a reexamination of assumptions in weighted phylogenetic analysis. *Syst Biol* 49:617–627.

Brower, A. V. Z. 1996. Parallel race formation and the evolution of mimicry in *Heliconius* butterflies: A phylogenetic hypothesis from mitochondrial DNA sequences. *Evolution* 50:195–221.

———. 1997. The evolution of ecologically important characters in *Heliconius* butterflies (Lepidoptera: Nymphalidae): A cladistic review. *Zool J Linn Soc* 119:457–472.

Brower, A. V. Z., and R. DeSalle. 1994. Practical and theoretical considerations for choice of DNA sequence region in insect molecular systematics and a short review of published studies using nuclear gene regions. *Ann Entomol Soc Am* 87:702–716.

Brower, A. V. Z., R. DeSalle, and A. Vogler. 1996. Gene trees, species trees, and systematics: A cladistic perspective. *Annu Rev Ecol Syst* 27:423–450.

Brower, A. V. Z., and V. Schawaroch. 1996. Three steps in homology assessment. *Cladistics* 12:265–272.

Brown, J. H. 1971. Mechanisms of competitive exclusion between two species of chipmunks. *Ecology* 52:305–311.

———. 1973. Species diversity of seed-eating desert rodents in sand dune habitats. *Ecology* 54:775–787.

———. 1981. Two decades of homage to Santa Rosalia: Toward a general theory of diversity. *Am Zool* 21:877–888.

———. 1984. On the relationship between abundance and distribution of species. *Am Nat* 124:255–279.

———. 1995. *Macroecology.* Chicago: Univ. Chicago Press.

Brown, J. H., B. J. Fox, and D. A. Kelt. 2000. Assembly rules: Desert rodent communities are structured at scales from local to continental. *Am Nat* 156:314–321.

Brown, J. H., and A. C. Gibson. 1983. *Biogeography.* 1st ed. St. Louis, MO: Mosby.

———. 1998. *Biogeography.* 2d ed. St. Louis, MO: Mosby.

Brown, J. H., and B. A. Maurer. 1987. Evolution of species assemblages: Effects of energetic constraints and species dynamics on the diversification of the North American avifauna. *Am Nat* 130:1–17.

———. 1989. Macroecology: The division of food and space among species on continents. *Science* 243:1145–1150.

Brown, J. H., and Z. Zeng. 1989. Comparative population ecology of eleven species of rodents in the Chihuahuan desert. *Ecology* 70:1507–1525.

Brown, J. M., J. H. Leebens-Mack, J. N. Thompson, Ø. Pellmyr, and R. G. Harrison. 1997. Phylogeography and host association in a pollinating seed parasite *Greya politella* (Lepidoptera: Prodoxidae). *Mol Ecol* 6:215–224.

Brown, K. S., J. R. G. Turner, and P. M. Shepard. 1974. Quaternary forest refugia in Tropical America: Evidence from race formation in *Heliconius* butterflies. *Proc Roy Soc Lond* B 187:369–378.

Brown, W. L., Jr. 1957. Centrifugal speciation. *Quart Rev Biol* 32:247–277.

Brown, W. L., Jr., and E. O. Wilson. 1956. Character displacement. *Syst Zool* 5:49–64.

Brues, C. T. 1920. The selection of food-plants by insects, with special reference to lepidopterous larvae. *Am Nat* 54:313–332.

———. 1924. The specificity of food-plants in the evolution of phytophagous insects. *Am Nat* 58:127–144.

Brumfield, R. T., and A. P. Capparella. 1996. Historical diversification of birds in northwestern South America: A molecular perspective on the role of vicariant events. *Evolution* 50:1607–1624.

Brundin, L. 1966. Transantarctic relationships and their significance, as evidenced by chironomid midges. *Kungl Svenska Vetenskap Handl* 11:1–472.

———. 1972. Evolution, causal biology, and classification. *Zool Scripta* 1:107–120.

Bruneau, A. 1997. Evolution and homology of bird pollination syndromes in *Erythrina* (Leguminosae). *Am J Bot* 84:54–71.

Bryant, H. N. 1989. An evaluation of cladistic and character analyses as hypotheticodeductive procedures, and the consequences for character weighting. *Syst Zool* 38:214–237.

Bull, J. J., J. P. Huelsenbeck, C. W. Cunningham, D. L. Swofford, and P. J. Waddell. 1993. Partitioning and combining data in phylogenetic analysis. *Syst Biol* 42:384–397.

Bürger, R., and M. Lynch. 1995. Evolution and extinction in a changing environment: A quantitative-genetic analysis. *Evolution* 49:151–163.

Burger, W. L., and W. B. Robertson. 1951. A new subspecies of the Mexican moccasin, *Agkistrodon bilineatus*. *Univ Kansas Sci Bull* 34:213–218.

Burghardt, G. M., and J. L. Gittleman. 1990. Comparative behavior and phylogenetic analysis. In *Interpretation and explanation in the study of behavior: Comparative perspectives*, ed. M. Bekoff and D. Jamieson, 192–225. Boulder, CO: Westview Press.

Burns, K. J. 1998. A phylogenetic perspective on the evolution of sexual dichromatism in tanagers (Thraupidae): The role of female versus male plumage. *Evolution* 52:1219–1224.

Burt, A. 1989. Comparative methods using phylogenetically independent contrasts. *Oxford Surv Evol Biol* 6:33–53.

Burt, M. D. B., and L. Jarecka. 1982. Phylogenetic host specificity of cestodes. In *Deuxieme symposium sur la spécificité parasitaire des parasites de vertébrés. Mem Mus Natl Hist Nat* A 123:47–51.

Buscalioni, A. D., F. Ortega, D. Rasskin-Guttman, and B. Perez-Moreno. 1997. Loss of carpal elements in crocodilian limb evolution: Morphogenetic model corroborated by paleontological data. *Biol J Linn Soc* 62:133–144.

Buschbeck, E. K. 2000. Neurobiological constraints and fly systematics: How different types of neural characters can contribute to a higher level dipteran phylogeny. *Evolution* 54:888–898.

Bush, G. L. 1966. The taxonomy, cytology, and evolution of the genus *Rhagoletis* in North America (Diptera, Tephritidae). *Bull Mus Comp Zool* 134:431–562.

———. 1969. Sympatric host race formation and speciation in frugivorous flies of the genus *Rhagoletis* (Diptera, Tephritidae). *Evolution* 23:237–251.

———. 1974. The mechanism of sympatric host race formation in the true fruit flies (Tephrytidae). In *Genetic mechanisms of speciation in insects*, ed. M. J. D. White, 3–23. Sydney: Australia and New Zealand Book Co.

———. 1975a. Modes of animal speciation. *Annu Rev Ecol Syst* 6:339–364.

———. 1975b. Sympatric speciation in phytophagous parasitic insects. In *Evolutionary strategies of parasitic insects and mites*, ed. P. W. Price, 187–206. New York: Plenum Press.

———. 1982. What do we really know about speciation? In *Perspectives on evolution*, ed. R. Milkman, 119–128. Sunderland, MA: Sinauer.

———. 1994. Sympatric speciation in animals: New wine in old bottles. *Trends Ecol Evol* 9:285–288.

Bush, G. L., and D. J. Howard. 1986. Allopatric and non-allopatric speciation: Assumptions and evidence. In *Evolutionary process and theory,* ed. S. Karlin and E. Nevo, 411–438. New York: Academic Press.

Buth, D. G. 1984. The application of electrophoretic data in systematics. *Annu Rev Ecol Syst* 15:501–522.

Butler, P. M. 1982. Directions of evolution in the mammalian dentition. In *Problems of phylogenetic reconstruction,* ed. K. A. Joysey and A. E. Friday, Syst Assoc Spec Vol 21:235–244. London: Academic Press.

Butlin, R. K. 1987a. A new approach to sympatric speciation. *Trends Ecol Evol* 2:310–311.

———. 1987b. Speciation by reinforcement. *Trends Ecol Evol* 2:8–13.

———. 1989. Reinforcement of premating isolation. In *Speciation and its consequences,* ed. D. Otte and J. A. Endler, 158–179. Sunderland, MA.: Sinauer.

Butlin, R. K., and M. G. Ritchie. 1989. Genetic coupling in mate recognition systems: What is the evidence? *Biol J Linn Soc* 37:237–246.

———. 1994. Mating behaviour and speciation. In *Behaviour and evolution,* ed. P. J. B. Slater and T. R. Halliday, 43–79. Cambridge: Cambridge Univ. Press.

Butterfield, N.J. 1997. Plankton ecology and the Proterozoic-Phanerozoic transition. *Paleobiology* 23:247–262.

Byrne, M., A. Cerra, M. Hart, and M. Smith. 1999. Life history diversity and molecular phylogeny in the Australian sea star genus *Patiriella.* In *The other 99%: The conservation and biodiversity of invertebrates,* ed. W. Ponder and D. Lunney, 188–195. *Trans Roy Soc NSW.* Mosman, NSW: Royal Society of New South Wales.

Cade, T. J. 1963. Observations of torpidity in captive chipmunks of the genus *Eutamias. Ecology* 44:255–261.

———. 1984a. Molecular systematics of neotropical xenodontine colubrid snakes: I. South American xenodontines. *Herpetologica* 40:8–20.

———. 1984b. Molecular systematics of neotropical xenodontine colubrid snakes: II. Central American xenodontines. *Herpetologica* 40:21–30.

———. 1984c. Molecular systematics of neotropical xenodontine colubrid snakes: III. Overview of xenodontine phylogeny and the history of New World snakes. *Copeia* 1984:641–652.

———. 1985. The neotropical colubrid snake fauna (Serpentes: Colubridae): Lineage components and biogeography. *Syst Zool* 34:1–20.

Cadle, J. E., and H. W. Greene. 1993. Phylogenetic patterns, biogeography, and the ecological structure of neotropical snake assemblages. In *Species diversity in ecological communities,* ed. R. E. Ricklefs and D. Schluter, 281–293. Chicago: Univ. Chicago Press.

Caillaud, M. C., and S. Via. 2000. Specialized feeding behavior influences both ecological specialization and assortative mating in sympatric host races of pea aphids. *Am Nat* 156:606–621.

Cain, S. A. 1944. *Foundations of plant biogeography.* New York: Harper and Row.

Caldwell, J. P. 1996. The evolution of myrmecophagy and its correlates in poison frogs (Family Dendrobatidae). *J Zool* 240:75–101.

Caldwell, M. W. 1999. Squamate phylogeny and the relationships of snakes and mosasauroids. *Zool J Linnaean Soc* 125:115–147.

Cameron, T. W. M. 1929. The species of *Enterobius* in primates. *J Helminthol* 7:161–182.

Camp, W. H. 1947. Distributional patterns in modern plants and the problems of ancient dispersals. *Ecol Mono* 17:159–183.

Campbell, K. E., Jr., and D. Frailey. 1984. Holocene flooding and species diversity in southeastern Amazonia. *Quatern Res* 21:369–375.

Candolle, A.-P. de. 1844. *Théorie élémentaire de la botanique.* 3d ed. Paris: Roret.

Cannatella, D.C., and R. O. de Sá. 1993. *Xenopus laevis* as a model organism. *Syst Biol* 42:476–507.

Cannatella, D.C., D. M. Hillis, P. T. Chippendale, L. Weigt, A. S. Rand, and M. J. Ryan. 1998. Phylogeny of frogs in the *Physalaemus pustulosus* species group, with an examination of data incongruence. *Syst Biol* 47:311–335.

Carefoot, T. H. 1987. *Aplysia:* Its biology and ecology. *Annu Rev Oceanogr Mar Biol* 25: 167–284.

Carleton, M. D. 1980. Phylogenetic relationships in neotomine-peromyscine rodents (Muroidea) and a reappraisal of the dichotomy within New World Cricetinae. *Misc Publ Mus Zool Univ Mich* 157:1–146.

Carpenter, C. C. 1956. Body temperatures of three species of *Thamnophis*. *Ecology* 37:732–735.

Carpenter, J. M. 1982. The phylogenetic relationships and natural classification of the Vespoidea (Hymenoptera). *Syst Entomol* 7:11–38.

———. 1987. Phylogenetic relationships and classification of the Vespinae (Hymenoptera: Vespidae). *Syst Entomol* 12:413–431.

———. 1988a. Choosing among equally parsimonious trees. *Cladistics* 4:291–296.

———. 1988b. The phylogenetic system of the Stenogastrinae (Hymenoptera: Vespidae). *J NY Entomol Soc* 96:140–175.

———. 1989. Testing scenarios: Wasp social behavior. *Cladistics* 5:131–144.

———. 1991. Phylogenetic relationships and the evolution of social behavior in the Vespidae. In *The social biology of wasps*, ed. K. G. Ross and R. W. Matthews, 7–32. Ithaca, NY: Cornell Univ. Press.

Carpenter, J. M., and J. M. Cumming. 1985. A character analysis of the North American potter wasps (Hymenoptera: Vespidae; Eumeninae). *J Nat Hist* 19:877–916.

Carpenter, K. E. 1990. A phylogenetic analysis of the Caesionidae (Perciformes: Lutjanidae). *Copeia* 1990:692–717.

Carreno, R. A., J. C. Kissinger, T. F. McCutchan, and J. R. Barta. 1997. Phylogenetic analysis of haemosporinid parasites (Apicomplexa: Haemosporina) and their coevolution with vectors and intermediate hosts. *Arch Protistenkd* 148:245–252.

Carreno, R. A., L. A. Durden, D. R. Brooks., A. Abrams, and E. P. Hoberg. 2001. *Parelaphostrongylus tenuis* and other parasites of white-tailed deer (*Odocoileus virginianus*) in Costa Rica. *Comp Parasitol* 68:177–184.

Carrier, D. R. 1991. Conflict in the hypaxial musculo-skeletal system: documenting an evolutionary constraint. *Am Zool* 31:644–654.

Carriere, Y. 1998. Constraints on the evolution of host choice by phytophagous insects. *Oikos* 82:401–406.

Carson, H. L. 1975. The genetics of speciation at the diploid level. *Am Nat* 109:83–92.

———. 1982. Speciation as a major reorganization of polygenic balances. In *Mechanisms of speciation*, ed. C. Barigozzi, 411–433. New York: Alan R. Liss.

———. 1983. Chromosomal sequences and interisland colonizations in Hawaiian *Drosophila*. *Genetics* 103:465–482.

Carson, H. L., and D. A. Clague. 1995. Geology and Biogeography of the Hawaiian Islands. In *Hawaiian biogeography: Evolution on a hot spot archipelago*, ed. W. L. Wagner and V. A. Funk, 14–29. Washington, DC: Smithsonian Institution Press.

Carson, H. L., and K. Y. Kaneshiro. 1976. *Drosophila* of Hawaii: Systematics and evolutionary genetics. *Annu Rev Ecol Syst* 7:311–346.

Carson, H. L., and A. R. Templeton. 1984. Genetic revolutions in relation to speciation phenomena: The founding of new populations. *Annu Rev Ecol Syst* 15:97–131.

Castresana, J. 2000. Selection of conserved blocks from multiple alignments for their use in phylogenetic analysis. *Mol Biol Evol* 17:540–552.

Caterino, M. S., S. Cho, and F. A. H. Sperling. 2000. The current state of insect molecular systematics: A thriving Tower of Babel. *Annu Rev Entomol* 45:1–54.

Caterino, M. S., and F. A. H. Sperling. 1999. *Papilio* phylogeny based on mitochondrial cytochrome oxidase I and II genes. *Mol Phylogenet Evol* 11:122–137.

Chacón, P., and W. G. Eberhard. 1980. Factors affecting numbers and kinds of prey caught in artificial spider webs with considerations of how orb-webs trap prey. *Bull Brit Arachnol Soc* 5:29–38.

Chan, K., and B. Moore. 1999. Accounting for modes of speciation increases power and realism of tests of phylogenetic asymmetry. *Am Nat* 153:332–346.

Chan, S.-Y., H.-U. Bernard, C.-K. Ong, S.-P. Chan, B. Hoffman, and H. Delius. 1992. Phylogenetic analysis of 48 papillomavirus types and 28 subtypes and variants: A showcase for the molecular evolution of DNA viruses. *J Virol* 66:5714–5725.

Chan, S.-W., F. McOrmish, E. C. Holmes, B. Dow, J. F. Peutherer, E. Follett, P. L. Yap, and P. Simmonds. 1992. Analysis of a new hepatitis C virus type and its phylogenetic relationships to existing variants. *J Virol* 73:1131–1141.

Chapin, J. P. 1917. The classification of the weaver-birds. *Bull Am Mus Nat Hist* 37:243–280.

———. 1932. The birds of the Belgian Congo, part I. *Bull Am Mus Nat Hist* 65:1–391.

Chapleau, F., C. S. Findlay, and E. Szenasy. 1997. Impact of piscivorous fish introductions on fish species richness of small lakes in Gatineau Park, Quebec. *Ecoscience* 4:259–268.

Chapman, F. M. 1917. The distribution of bird-life in Colombia. *Bull Am Mus Nat Hist* 36:1–729.

———. 1926. The distribution of bird-life in Ecuador. *Bull Am Mus Nat Hist* 55:1–784.

Charleston, M. A. 1998. Jungles: A new solution to the host/parasite phylogeny reconciliation problem. *Math Biosci* 149:191–223.

Charlesworth, B., and S. Rouhani. 1988. The probability of peak shifts in a founder population: II. An additive polygenic trait. *Evolution* 42:1129–1145.

Chase, M. W., and H. G. Hills. 1992. Orchid phylogeny, flower sexuality, and fragrance-seeking. *BioScience* 42:43–49.

Chase, M. W., D. E. Soltis, R. G. Olmstead, D. Morgan, D. H. Les, B. D. Mishler, M. R. Duvall, R. A. Price, H. G. Hills, Y.-L. Qiu, K. A. Kron, J. H. Rettig, E. Conti, J. D. Palmer, J. R. Manhart, K. J. Sytsma, H. J. Michaels, W. J. Kress, K. G. Karol, W. D. Clark, M. Hedrén, B. S. Gaut, R. K. Jansen, K-J Kim, C. F. Wimpee, J. F. Smith, G. R. Furnier, S. H. Strauss, Q.-Y. Xiang, G. M. Plunkett, P. S. Soltis, S. M. Swensen, S. E. Williams, P. A. Gadek, C. J. Quinn, L. E. Equiarte, E. Golenberg, G. H. Learn, S. W. Graham, S. C. H. Barrett, S. Dayanandan, and V. A. Albert. 1993. Phylogenetics of seed plants: An analysis of nucleotide sequences from the plastid gene rbcL. *Ann Missouri Bot Gard* 80:528–580.

Cheetham, A. H. 1986. Tempo of evolution in Neogene bryozoans: Rates of morphologic change within and across species boundaries. *Paleobiology* 12:190–202.

Chernoff, B. 1982. Character variation among populations and the analysis of biogeography. *Am Zool* 22:425–439.

Chesser, R. T., and R. M. Zink. 1994. Modes of speciation in birds: A test of Lynch's method. *Evolution* 48:490–497.

Cheverud, J. M. 1996. Developmental integration and the evolution of pleiotropy. *Am Zool* 36:44–50.

Cheverud, J. M., M. M. Dow, and W. Leutenegger. 1985. The quantitative assessment of phylogenetic constraints in comparative analyses: Sexual dimorphism in body weight among primates. *Evolution* 39:1335–1351.

Chippindale, P. T., and J. J. Wiens. 1994. Weighting, partitioning, and combining characters in phylogenetic analysis. *Syst Biol* 43:278–287.

Choo, Q. L., G. Kuo, A. J. Weiner, L. R. Overby, D. W. Bradley, and M. Houghton. 1989. Isolation of cDNA derived from a blood-borne non-A, non-B hepatitis genome. *Science* 244:359–362.

Choo, Q. L., K. H. Richman, J. H. Han, K. Berger, C. Lee, C. Dong, C. Gallegos, D. Coit, R. Medina Selby, P. J. Barr, A. J. Weiner, D. W. Bradley, G. Kuo, and M. Houghton. 1991. Genetic organization and diversity of the hepatitis C virus. *Proc Natl Acad Sci* (USA) 88:2451–2455.

Chor, S., M. D. Hendy, B. R. Holland, and D. Penny. 2000. Multiple maxima of likelihood in phylogenetic trees: An analytic approach. *Mol Biol Evol* 17:1529–1541.

Christie, D. M., R. A. Duncan, A. R. McBirney, M. A. Richards, W. M. White, K. S. Harpp, and C. C. Fox. 1992. Drowned islands downstream from the Galapagos hotspot imply extended speciation times. *Nature* 355:246–248.

Christoffersen, M. L. 1994. An overview of cladistic applications. *Rev Nordestina Biol* 9:133–141.

Civetta, A., and R. S. Singh. 1998. Sex and speciation: Genetic architecture and evolutionary potential of sexual versus nonsexual traits in the sibling species of the *Drosophila melanogaster* complex. *Evolution* 52:1080–1092.

Clague, J. J. 1983. Glacio-isostatic effects of the Cordilleran Ice Sheet, British Columbia, Canada. In *Shorelines and isostasy*, ed. D. E. Smith and A. G. Dawson, 321–343. London: Academic Press.

Clague, J. J., J. R. Harper, R. J. Hebba, and D. E. Howes. 1982. Late Quaternary sea levels and crustal movements, coastal British Columbia. *Can J Earth Sci* 19:597–618.

Claridge, M. F. 1995. Introducing Systematics Agenda 2000. *Biodiv Conserv* 4:451–454.

Clark, C., and D. J. Curran. 1986. Outgroup analysis, homoplasy, and global parsimony: A response to Maddison, Donoghue, and Maddison. *Syst Zool* 35:422–426.

Clarke, A. 2000. Evolution in the cold. *Antarct Sci* 12:257.

Clayton, D., and J. Moore, eds. 1997. *Host-parasite evolution: General principles and avian models.* Oxford: Oxford Univ. Press.

Clutton-Brock, T. H., and P. H. Harvey. 1977. Primate ecology and social organization. *J Zool* (London) 183:1–39.

———. 1984. Comparative approaches to investigating adaptation. In *Behavioural ecology: An evolutionary approach,* 2d ed, ed. J. R. Krebs and N. B. Davies, 7–29. Sunderland, MA: Sinauer.

Coates, D. J. 2000. Defining conservation units in a rich and fragmented flora: Implications for the management of genetic resources and evolutionary processes in south west Australia. *Austral J Bot* 48:329–339.

Coddington, J. A. 1985. Review of *The explanation of organic diversity: The comparative method and adaptations for mating,* by M. Ridley. *Cladistics* 1:102–107.

———. 1986. The monophyletic origin of the orb web. In *Spiders: Webs, behavior, and evolution,* ed. W. A. Shear, 319–363. Stanford, CA: Stanford Univ. Press.

———. 1988. Cladistic tests of adaptational hypotheses. *Cladistics* 4:3–22.

———. 1989. Spinneret silk spigot morphology: Evidence for the monophyly of orb-weaving spiders, Cyrtophorinae (Araneidae), and the group Theridiidae plus Nesticidae. *J Arachnol* 17:71–95.

———. 1990. Bridges between evolutionary pattern and process. *Cladistics* 6:379–386.

———. 1992. Avoiding phylogenetic bias. *Trends Ecol Evol* 7:68–69.

———. 1994. The roles of homology and convergence in studies of adaptation. In *Phylogenetics and ecology,* ed. P. Eggleton and R. Vane-Wright, 53–78. London: Academic Press.

Coddington, J. A., G. Hormiga, and N. Scharff. 1997. Giant female or dwarf male spiders? *Nature* 285:687–688.

Coddington, J. A., and H. W. Levi. 1991. Systematics and evolution of spiders (Araneae). *Annu Rev Ecol Syst* 22:565–592.

Cody, M. L. 1973. Character convergence. *Annu Rev Ecol Syst* 4:189–211.

Cogger, H. G., and R. G. Zweifel. 1998. *Encyclopedia of reptiles and amphibians.* San Diego, CA: Academic Press.

Cognato, A. I., S. J. Seybold, D. L. Wood, and S. A. Teale. 1997. A cladistic analysis of pheromone evolution in *Ips* bark beetles (Coleoptera: Scolytidae). *Evolution* 51: 313–318.

Cohen, A. S., L. Kaufman, and R. Ogutu-Ohwayo. 1996. Anthropogenic threats, impacts, and conservation strategies in the African Great Lakes: A review. In *The limnology, climatology, and paleoclimatology of the East African Lakes,* ed. T. C. Johnson and E. O. Odada, 575–624. Amsterdam: Gordon and Breach.

Cohen, M. B., M. R. Berenbaum, and M. A. Schuler. 1990. Immunochemical analysis of cytochrome $P450$ monooxygenase diversity in the black swallowtail caterpillar, *Papilio polyxenes. Insect Biochem* 20:777–783.

Colbourne, J. K. 1998. A documentary on the evolutionary history of *Daphnia.* Ph.D. diss., Univ. of Guelph, Guelph, Ontario.

Colbourne, J. K., T. J. Crease, L. J. Weider, P. D. N. Herbert, F. Dufresne, and A. Hobaek. 1998. Phylogenetics and evolution of a circumarctic species complex (Cladocera: *Daphnia pulex*). *Biol J Linn Soc* 65:347–366.

Colbourne, J. K., and P. D. Hebert. 1996. The systematics of North American *Daphnia* (Crustacea: Anomopoda): A molecular phylogenetic approach. *Phil Trans Roy Soc Lond* B 351:349–360.

Collazo, A. 2000. Developmental variation, homology, and the pharyngula stage. *Syst Biol* 49:3–18.

Collette, B. B. 1961. Correlations between ecology and morphology in anoline lizards from Havana, Cuba, and southern California. *Bull Mus Comp Zool* 125:137–162.

Collins, J. P. 1986. Evolutionary ecology and the use of natural selection in ecological theory. *J Hist Biol* 19:257–288.

Connell, J. H. 1978. Diversity in tropical rain forests and coral reefs. *Science* 199:1302–1309.

———. 1980. Diversity and the coevolution of competitors, or the ghost of competition past. *Oikos* 35:131–138.

Connell, J. H., and E. Orias. 1964. The ecological regulation of species diversity. *Am Nat* 98:399–414.

Connor, E. F., and E. D. McCoy. 1979. The statistics and biology of the species-area relationship. *Am Nat* 113:791–833.

Connor, E. F., and M. P. Taverner. 1997. The evolution and adaptive significance of the leaf-mining habit. *Oikos* 79:6–25.

Constance, L. 1953. The role of plant ecology in biosystematics. *Ecology* 34:642–649.

Constantinescu, M., and D. Sankoff. 1986. Tree enumeration modulo a consensus. *J Classif* 3:349–356.

———. 1995. An efficient algorithm for supertrees. *J Classif* 12:101–112.

Conti, E., D. E. Soltis, T. M. Hardig, and J. Schneider. 1999. Phylogenetic relationships of the silver saxifrages (*Saxifraga,* sect. Ligulatae Haworth): Implications for the evolution of substrate specificity, life histories, and biogeography. *Mol Phylogenet Evol* 13:536–555.

Conway Morris, S. 1995. Ecology in deep time. *Trends Ecol Evol* 10:290–294.

———. 1998. The evolution of diversity in ancient ecosystems: A review. *Phil Trans Roy Soc Bio Sci* 353:327–345.

Cook, J. A., and E. P. Lessa. 1998. Are rates of diversification in subterranean South American tuco-tucos (genus *Ctenomys,* Rodentia: Octodontidae) unusually high? *Evolution* 52:1521–1527.

Cooper, W. E., Jr. 1989a. Absence of prey odor discrimination by iguanid and agamid lizards in applicator tests. *Copeia* 1989:472–478.

———. 1989b. Prey odor discrimination in the varanoid lizards *Heloderma suspectum* and *Varanus exanthematicus. Ethology* 81:250–258.

———. 1994. Chemical discrimination by tongue-flicking in lizards: A review with hypotheses on its origin and its ecological and phylogenetic relationships. *J Chem Ecol* 20:439–487.

———. 1995. Foraging mode, prey chemical discrimination, and phylogeny in lizards. *Anim Behav* 50:973–985.

———. 1997. Correlated evolution of prey chemical discrimination with foraging, lingual morphology, and vomeronasal chemoreceptor abundance in lizards. *Behav Ecol Sociobiol* 41:257–265.

———. 1999. Supplementation of phylogenetically correct data by two-species comparison: Support for correlated evolution of foraging mode and prey chemical discrimination in lizards extended by first intrageneric evidence. *Oikos* 87:97–104.

———. 2000. An adaptive difference in the relationship between foraging mode and responses to prey chemicals in two congeneric scincid lizards. *Ethology* 106:193–206.

Cope, E. D. 1885. On the evolution of the Vertebrata, progressive and retrogressive. *Am Nat* 19:140–148, 234–247, 341–353.

———. 1896. *The primary factors of organic evolution.* Chicago: Open Court.

Cotgreave, P. 1994. The relation between body size and abundance in a bird community: The effects of phylogeny and competition. *Proc Roy Soc Lond* B 205:147–149.

Cotterill, F. P. D. 1995. Systematics, biological knowledge, and environmental conservation. *Biodiv Conserv* 4:183–205.

———. 1997a. The second Alexandrian tragedy, and the fundamental relationship between biological collections and scientific knowledge. In *The value and valuation of natural science collections,* ed. J. R. Nudds and C. W. Pettitt, 227–241. London: Geological Society of London.

———. 1997b. The growth of the WCCR or the extinction of biosystematic resources? *Nat Hist Conserv* 11:7–11.

———. 1999. Toward exorcism of the ghost of W. T. Thisleton-Dyer: A comment on "overduplication" and the scientific properties, uses, and values of natural science specimens. *Taxon* 48:35–39.

Cotterill, F. P. D., and J. M. Dangerfield. 1997. The state of biological knowledge. *Trends Ecol Evol* 12:206.

Cotterill, F. P. D., C. W. Hustler, and D. G. Broadley. 1994. Systematics and biodiversity. *Trends Ecol Evol* 9:228.

Courtney, S. P. 1981. Coevolution of pierid butterflies and their cruciferous food plants. IV. Crucifer apparency and *Anthocharis cardamines* (L.) oviposition. *Oecologia* 51:91–96.

————. 1983. Coevolution of pierid butterflies and their cruciferous food plants. V. habitat selection, community structure, and speciation. *Oecologia* 54:101–107.

————. 1985. Apparency in coevolving relationships. *Oikos* 44:91–98.

————. 1988. If it's not coevolution, it must be predation? *Ecology* 69:910–911.

Courtney, S. P., G. K. Chen, and A. Gardner. 1989. A general model for individual host selection. *Oikos* 55:55–65.

Cousins, S. H. 1991. Species diversity measurement: Choosing the right index. *Trends Ecol Evol* 6:190–192.

Coyne, J. A. 1990. Endless forms most beautiful. *Nature* 344:30.

————. 1994. Mayr, Ernst, and the origin of species. *Evolution* 48:19–30.

Coyne, J. A., and M. Kreitman. 1986. Evolutionary genetics of two sibling species, *Drosophila simulans* and *D. sechella*. *Evolution* 40:673–691.

Coyne, J. A., and H. A. Orr. 1989. Patterns of speciation in *Drosophila*. *Evolution* 43:362–381.

————. 1997a. The evolutionary genetics of speciation. *Phil Trans Roy Soc* B 353:287–305.

————. 1997b. "Patterns of speciation in *Drosophila*" revisited. *Evolution* 51:295–303.

Coyne, J. A., H. A. Orr, and D. J. Futuyma. 1988. Do we need a new species concept? *Syst Zool* 37:190–200.

Cracraft, J. 1974. Continental drift and vertebrate distribution. *Annu Rev Ecol Syst* 5:215–261.

————. 1978. Science, philosophy, and systematics. *Syst Zool* 27:213–215.

————. 1979. Phylogenetic analysis, evolutionary models, and paleontology. In *Phylogenetic analysis and palentology*, ed. J. Cracraft and N. Eldredge, 7–39. New York: Columbia Univ. Press.

————. 1981a. The use of functional and adaptive criteria in phylogenetic systematics. *Am Zool* 21:21–38.

————. 1981b. Pattern and process in paleobiology: The role of cladistic analysis in systematic paleontology. *Paleobiology* 7:456–468.

————. 1982a. A nonequilibrium theory for the rate-control of speciation and extinction and the origin of macroevolutionary patterns. *Syst Zool* 31:348–365.

————. 1982b. Geographic differentiation, cladistics, and vicariance biogeography: Reconstructing the tempo and mode of evolution. *Am Zool* 22:411–424.

————. 1983a. Species concepts and speciation analysis. In *Current Ornithology*, vol. 1, ed. R. F. Johnston, 159–187. New York: Plenum Press.

————. 1983b. Cladistic analysis and vicariance biogeography. *Am Sci* 7:273–281.

————. 1984. Conceptual and methodological aspects of the study of evolutionary rates, with some comments on bradytely in birds. In *Living fossils*, ed. N. Eldredge and S. M. Stanley, 95–104. New York: Springer Verlag.

————. 1985a. Biological diversification and its causes. *Ann Missouri Bot Gard* 72:794–822.

————. 1985b. Species selection, macroevolutionary analysis, and the "hierarchical theory of evolution." *Syst Zool* 34:222–229.

————. 1986. Origin and evolution of continental biotas: Speciation and historical congruence within the Australian avifauna. *Evolution* 40:977–996.

————. 1987. Species concepts and the ontology of evolution. *Biol Philos* 2:329–346.

————. 1988. Deep-history biogeography: Retrieving the historical pattern of evolving continental biotas. *Syst Zool* 37:221–236.

————. 1989a. Speciation and its ontology: The empirical consequences of alternative species concepts for understanding patterns and processes of speciation. In *Speciation and its consequences*, ed. D. Otte and J. A. Endler, 28–59. Sunderland, MA: Sinauer.

————. 1989b. Species as entities in biological theory. In *What the philosophy of biology is: Essays for David Hull*. ed. M. Ruse, 33–54. New York: Kluwer Academic.

————. 1990. The origin of evolutionary novelties: Pattern and process at different hierarchical levels. In *Evolutionary innovation*, ed. M. Nitecki, 21–44. Chicago: Univ. Chicago Press.

————. 1992. The species of the birds-of-paradise (Paradisaeidae): Applying the phylogenetic species concept to a complex pattern of diversification. *Cladistics* 8:1–43.

————. 1994. Species diversity, biogeography, and the evolution of biotas. *Am Zool* 34:33–47.

———. 1995. The urgency of building global capacity for biodiversity science. *Biodiv Conserv* 4:463–475.

Cracraft, J., and K. Helm-Bychowski. 1991. Parsimony and phylogenetic inference using DNA sequences: Some methodological strategies. In *Phylogenetic analysis of DNA sequences*, ed. M. M. Miyamoto and J. Cracraft, 184–220. New York: Oxford Univ. Press.

Cracraft, J., and D. P. Mindell. 1989. The early history of modern birds: A comparison of molecular and morphological evidence. In *The hierarchy of life: Molecules and morphology in phylogenetic analysis*, ed. B. Fernhom, K. Bremer, and H. Jornvall, 389–403. Amsterdam: Elsevier.

Cracraft, J., and R. O. Prum. 1988. Patterns and processes of diversification: Speciation and historical congruence in some Neotropical birds. *Evolution* 42:603–620.

Craig, C. L., and G. D. Bernard. 1990. Insect attraction to ultra-violet reflecting spiders and web decorations. *Ecology* 71:616–623.

Craig, C. L., G. D. Bernard, and J. A. Coddington. 1994. Evolutionary shifts in the spectral properties of spider silks. *Evolution* 48:287–296.

Craig, T. P., J. K. Itami, W. G. Abrahamson, and J. D. Horner. 1993. Behavioral evidence for host-race formation in *Eurostoa solidaginis*. *Evolution* 47:1696–1710.

Crampton, G. C. 1929. The terminal abdominal structures of female insects compared throughout the orders from the standpoint of phylogeny. *J NY Ent Soc* 37:453–496.

Crandall, K. A. 1998. Conservation phylogenetics of Ozark crayfishes: Assigning priorities for aquatic habitat protection. *Biol Conserv* 84:107–117.

Crandall, K. A., O. R. P. Bininda-Emonds, G. M. Mace, and R. K. Wayne. 2000. Considering evolutionary processes in conservation biology. *Trends Ecol Evol* 15:290–295.

Crane, P. R. 1985. Phylogenetic relationships in seed plants. *Cladistics* 1:329–348.

Crane, P. R., E. M. Friis, and K. R. Pedersen. 1995. The origin and early diversification of angiosperms. *Nature* 374:27–33.

Crane, P. R., and S. Lidgard. 1989. Angiosperm diversification and paleolatitudinal gradients in Cretaceous floristic diversity. *Science* 246:675–678.

Crapon de Caprona, M.-D. 1986. The use of fertile hybrids for the study of the accuracy of species recognition in cichlids. *Ann Mus Roy Afr Centr Sc Zool* 251:117–120.

Crapon de Caprona, M.-D., and B. Fritzsch. 1984. Interspecific fertile hybrids of Haplochromine Cichlidae (Teleostei) and their possible importance for speciation. *Neth J Zool* 34:504–538.

Crepet, W. L. 1984. Advanced (constant) insect pollination systems: Pattern of evolution and implication vis-à-vis angiosperm diversity. *Ann Missouri Bot Gard* 71:607–630.

Crespi, B. J. 1996. Comparative analysis of the origins and losses of eusociality: Causal mosaics and historical uniqueness. In *Phylogenies and the comparative method in animal behavior*, ed. E. P. Martins, 253–287. New York: Oxford Univ. Press.

Crespi, B. J., and C. P. Sandoval. 2000. Phylogenetic evidence for the evolution of ecological specialization in *Timema* walking sticks. *J Evol Biol* 13:249–262.

Crespi, B. J., and M. Worobey. 1998. Comparative analysis of gall morphology in Australian gall thrips: The evolution of extended phenotypes. *Evolution* 52:1686–1696.

Cressey, R. F., B. Collette, and J. Russo. 1983. Copepods and scombrid fishes: A study in host-parasite relationships. *Fish Bull* 81:227–265.

Cresswell, I. D. 2000. The Global Taxonomy Initiative: *Quo vadis? Biol Int* 38:12–16.

Crisci, J. V., M. M. Cigliano, J. J. Morrone, and S. Roig-Juñent. 1991. Historical biogeography of southern South America. *Syst Zool* 40:152–171.

Crisci, J. V., and T. F. Stuessy. 1980. Determining primitive character states for phylogenetic reconstructions. *Syst Bot* 5:112–135.

Crisp, M. D., H. P. Linder, and P. H. Weston. 1995. Cladistic biogeography of plants in Australia and New Guinea: Congruent pattern reveals two endemic tropical tracks. *Syst Biol* 44:457–473.

Croizat, L. 1952. *Manual of phytogeography*. The Hague: Junk.

———. 1958. *Panbiogeography*, vol. 1: *The New World*. Deventer, Netherlands: N. V. Drukkerij Salland.

———. 1964. *Space, time, and form: The biological synthesis*. Deventer, Netherlands: N. V. Drukkerij Salland.

Croizat, L., G. Nelson, and D. E. Rosen. 1974. Centers of origin and related concepts. *Syst Zool* 23:265–287.

Crompton, D. W. T., and S. M. Joyner. 1980. *Parasitic worms.* London: Wykeham.

Crother, B. I., and C. Guyer. 1996. Caribbean historical biogeography: Was the dispersal-vicariance debate eliminated by an extraterrestrial bolide? *Herpetologica* 52:440–465.

Crother, B. I., and W. F. Presch. 1992. The phylogeny of xantusiid lizards: The concern for analysis in the search for the best estimate of phylogeny. *Mol Phylogenet Evol* 1:289–294.

———. 1994. Xantusiid lizards, concern for analysis, and the search for the best estimate of phylogeny: Further comments. *Mol Phylogenet Evol* 3:272–275.

Crowe, T. M. 1994. Morphometrics, phylogenetic models, and cladistics: Means to an end or much ado about nothing? *Cladistics* 10:77–84.

Crozier, R. H. 1992. Genetic diversity and the agony of choice. *Conserv Biol* 61:11–15.

Cumming, G. S. 2000. Host use does not clarify the evolutionary history of African ticks (Acari: Ixodidae). *Afr Zool* 35:43–50.

Cunningham, C. W., H. Zhu, and D. M. Hillis. 1998. Best-fit maximum likelihood models for phylogenetic inference: Empirical tests with known phylogenies. *Evolution* 52:978–987.

Cunningham, S. A. 1995. Problems with null models in the study of phylogenetic radiation. *Evolution* 49:1292–1294.

Dabert, J., and S. V Mironov. 1999. Origin and evolution of feather mites. *Exper Appl Acarol* 23:437–454.

Daehler, C. C. 1998. The taxonomic distribution of invasive angiosperm plants: Ecological insights and comparison to agricultural weeds. *Biol Conserv* 84:167–180.

Damgaard, J., N. M. Andersen, L. Cheng, and F. A. H. Sperling. 2000. Phylogeny of sea skaters, *Halobates* Eschscholtz (Hemiptera, Gerridae), based on mtDNA sequence and morphology. *Zool J Linn Soc.* 130:511–526.

Darlington, P. J., Jr. 1938. The origin of the fauna of the Greater Antilles, with a discussion of dispersal of animals over water and through air. *Quart Rev Biol* 13:274–300.

———. 1957. *Zoogeography: The geographical distribution of animals.* New York: Wiley.

Darwin, C. 1837. On certain areas of elevation and subsidence in the Pacific and Indian oceans, as deduced form the study of coral formations. *Proc Geol Soc Lond* 2:552–554.

———. 1859. *The origin of species.* 1st ed. London: John Murray.

———. 1871. *The descent of man, and selection in relation to sex.* London: John Murray.

———. 1872. *The origin of species.* 6th ed. London: John Murray.

Daszak, P., A. A. Cunningham, and A. D. Hyatt. 2000. Wildlife ecology—Emerging infectious diseases of wildlife—Threats to biodiversity and human health. *Science* 287:443–449.

Davidson, D. W., and S. R. Morton. 1984. Dispersal adaptations of some *Acacia* species in the Australian arid zone. *Ecology* 65:1038–1051.

Davidson, J. F. 1952. The use of taxonomy in ecology. *Ecology* 33:297–299.

Davis, G. M. 1995. Systematics and public health. *BioScience* 45:705–714.

———. 1996. Collections of biological specimens essential for science and society. *Assoc Syst Coll Newsletter* 24:77–78, 88–90.

Davis, G. M., C. Spolsky, and Z. Yi. 1995. Malacology, biotechnology and snail-borne diseases. *Acta Med Philipp* 31:45–60.

Davis, J. I. 1995. A phylogenetic structure of the monocotyledons, as inferred from chloroplast DNA restriction site variation, and a comparison of measures of clade support. *Syst Bot* 20:503–527.

———. 1996. Phylogenetics, molecular variation, and species concepts. *BioScience* 46:502–511.

Davis, J. I., M. W. Frohlich, and R. J. Soreng. 1993. Cladistic characters and cladogram stability. *Syst Bot* 18:188–196.

Davis, J. I., and K. C. Nixon. 1992. Populations, genetic variation, and the delimitation of phylogenetic species. *Syst Bot* 41:421–435.

Dawkins, R., and J. R. Krebs. 1979. Arms races between and within species. *Proc Roy Soc Lond* B 205:489–511.

Day, T. 2000. Competition and the effect of spatial resource heterogeneity on evolutionary diversification. *Am Nat* 155:790–803.

Deacon, J. E., G. Kobetich, J. D. Williams, and S. Contreras-Balderas. 1979. Fishes of North America endangered, threatened or of special concern: 1979. *Fisheries* 4:29–44.

DeBach, P., and R. A. Sundby. 1963. Competitive displacement between ecological homologues. *Hilgardia* 34:105–166.

DeBry, R. W., and N. A. Slade. 1985. Cladistic analysis of restriction endonuclease cleavage maps within a maximum-likelihood framework. *Syst Zool* 34:21–34.

De Jong, H. 1998. In search of historical biogeographic patterns in the western Mediterranean terrestrial fauna. *Biol J Linn Soc* 65:99–164.

De Jong, R. 1980. Some tools for evolutionary and phylogenetic studies. *Z Zool Syst Evolutionsforsch* 18:1–23.

Deleporte, P. 1993. Characters, attributes, and tests of evolutionary scenarios. *Cladistics* 9:427–432.

Demanche, C., M. Berthelemy, T. Petit, B. Polack, A. E. Wakefield, E. Dei-Cas, and J. Guillot. 2001. Phylogeny of *Pneumocystis carinii* from 18 primate species confirms host specificity and suggests coevolution. *J Clin Microbiol* 39:2126–2133.

De Pinna, M. C. C. 1999. Species concepts and phylogenetics. *Rev Fish Biol Fisheries* 9: 353–373.

De Sá, R. O., and D. M. Hillis. 1990. Phylogenetic relationships of the pipid frogs *Xenopus* and *Silurana:* An integration of ribosomal DNA and morphology. *Mol Biol Evol* 7:365–376.

DeSalle, R., and D. A. Grimaldi. 1991. Morphological and molecular systematics in the Drosophilidae. *Annu Rev Ecol Syst* 22:447–475.

DeSalle, R., and A. R. Templeton. 1987. Comments on "The significance of asymmetrical sexual isolation." *Evol Biol* 21:21–27.

Desutter-Grandcolas, L. 1993. The cricket fauna of Chiapanecan caves (Mexico): Systematics, phylogeny, and the evolution of troglobitic life (Orthoptera, Grylloidea, Phalangopsidae, Luzarinae). *Int J Speleol* 22:1–82.

———. 1994. Test phylogénétique de l'adaptation à la vie troglobie chez des grillons (Insecta, Orthoptera, Grylloidea). *CR Acad Sci* (Paris) 317:907–912.

———. 1995. Toward the knowledge of the evolutionary biology of phalangopsid crickets (Orthoptera, Grylloidea, Phalangopsidae): Data, questions, and scenarios. *J Orth Res* 4:163–175.

———. 1997. A phylogenetic analysis of the evolution of the stridulatory apparatus of the true crickets (Orthoptera, Grylloidea). *Cladistics* 13:101–108.

Deutsch, J. C. 1997. Colour and diversification in Malawi cichlids: Evidence for adaptation, reinforcement, or sexual selection? *Biol J Linn Soc* 62:1–14.

Dewsbury, D. A. 1972. Patterns of copulatory behavior in male mammals. *Quart Rev Biol* 47:1–33.

———. 1975. Diversity and adaptation in rodent copulatory behavior. *Science* 190:947–954.

———. 1988. A test for the role of copulatory plugs in sperm competition in deer mice *(Peromyscus maniculatus). J Mammal* 69:854–857.

D'Haese, C. 2000. Is psammophily an evolutionary dead end? A phylogenetic test in the genus *Willemia* (Collembola: Hypogastruridae). *Cladistics* 16:255–273.

Dial, B. E., and L. L. Grismer. 1992. A phylogenetic analysis of physiological-ecological character evolution in the lizard genus *Coleonyx* and its implications for historical biogeographic reconstruction. *Syst Biol* 41:178–195.

Dial, K. P., and J. M. Marzluff. 1989. Nonrandom diversification within taxonomic assemblages. *Syst Zool* 38:26–37.

Diamond, J. M. 1975. The island dilemma: Lessons of modern geographical studies for the design of nature reserves. *Biol Conserv* 7:129–146.

Diaz-Uriarte, R., and T. Garland, Jr. 1996. Testing hypotheses of correlated evolution using phylogenetically independent contrasts: Sensitivity to deviations from Brownian motion. *Syst Biol* 45:27–47.

Dickinson, H., and J. Antonovics. 1973. Theoretical consideration of sympatric divergence. *Am Nat* 107:256–274.

Dieckmann, U., and M. Doebeli. 1999. On the origin of species by sympatric speciation. *Nature* 400:354–357.

Diehl, S. R., and G. L. Bush. 1984. An evolutionary and applied perspective of insect biotypes. *Annu Rev Entomol* 29:471–504.

———. 1989. The role of habitat preference in adaptation and speciation. In *Speciation and its consequences,* ed. D. Otte and J. A. Endler, 345–365. Sunderland, MA: Sinauer.

Dietz, R. S., and J. C. Holden. 1966. Miogeoclines (Miogeosynclines) in space and time. *J Geol* 74:566–583.

———. 1970. The breakup of Pangaea. *Sci Am* 223:30–41.

Dilley, J. D., P. Wilson, and M. R. Mesler. 2000. The radiation of *Calochortus:* Generalist flowers moving through a mosaic of potential pollinators. *Oikos* 89:209–222.

DiMichele, W. A. 1994. Ecological patterns in time and space. *Paleobiology* 20:89–92.

Dimmick, W. W., M. J. Ghedotti, M. J. Grose, A. M. Maglia, D. J. Meinhardt, and D. S. Pennock. 1999. The importance of systematic biology in defining units of conservation. *Conserv Biol* 13:653–660.

Dobler, S., and B. D. Farrell. 1999. Host use evolution in *Chrysochus* milkweed beetles: Evidence from behaviour, population genetics, and phylogeny. *Mol Ecol* 8:1297–1307.

Dobler, S., P. Mardulyn, J. M. Pasteels, and M. Rowell-Rahier. 1996. Host-plant switches and the evolution of chemical defense and life history in the leaf beetle genus *Oreina. Evolution* 50:2373–2386.

Dobson, A. P. 1988. Restoring island ecosystems: The potential of parasites to control introduced animals. *Conserv Biol* 2:31–39.

Dobson, A. P., and R. Carper. 1992. Global warming and potential changes in host-parasite and disease-vector relationships. In *Global warming and biological diversity,* ed. R. L. Peters and T. E. Lovejoy, 201–217. New Haven, CT: Yale Univ. Press.

Dobson, A. P., and P. J. Hudson. 1986. Parasites, disease and the structure of ecological communities. *Trends Ecol Evol* 1:11–15.

Dobson, A. P., and R. M. May. 1986a. Disease and conservation. In *Conservation biology: The science of scarcity and diversity,* ed. M. E. Soulé, 345–365. Sunderland, MA: Sinauer.

———. 1986b. Patterns of invasions by pathogens and parasites. In *Ecology of biological invasions of North America and Hawaii,* ed. H. A. Moore and J. A. Drake, 5876. Ecological Studies vol. 58. New York: Springer Verlag.

Dobson, F. S. 1985. The use of phylogeny in behavior and ecology. *Evolution* 39:1384–1388.

Dobzhansky, T. 1937. *Genetics and the origin of species.* 1st ed. New York: Columbia Univ. Press.

———. 1940. Speciation as a stage in evolutionary divergence. *Am Nat* 74:312–321.

———. 1951. *Genetics and the origin of species.* 3d ed. New York: Columbia Univ. Press.

———. 1970. *Genetics of the evolutionary process.* New York: Columbia Univ. Press.

———. 1976. Organismic and molecular aspects of species formation. In *Molecular evolution,* ed. F. J. Ayala, 95–105. Sunderland, MA: Sinauer.

Dobzhansky, T., and A. B. da Cunha. 1955. Differentiation of nutritional preferences in Brazilian species of *Drosophila. Ecology* 36:34–39.

Dodd, M. E., J. Silvertown, and M. W. Chase. 1999. Phylogenetic analysis of trait evolution and species diversity variation among angiosperm families. *Evolution* 53:732–744.

Dodson, E. O., and P. Dodson. 1985. *Evolution: Process and product.* Boston: Prindle, Weber, and Schmidt.

Doebeli, M., and U. Dieckmann. 2000. Evolutionary branching and sympatric speciation caused by different types of ecological interactions. *Am Nat* 156:S77–S101.

Doesburg, P. H. van. 1968. A revision of the New World species of *Dysdercus* Guérin Méneville (Heteroptera, Pyrrhocoridae). *Zool Verh* 97:215.

Dollo, L. 1893. Les lois de l'évolution. *Bull Soc Belg Geol Paleontol Hydrol* 7:164–166.

———. 1922. Les Céphalopodes déroulés et l'irréversibilité de l'évolution. *Bijdr Dierk* 1922:215–227.

Dominey, W. J. 1984. Effects of sexual selection and life histories on speciation: Species flocks in African cichlids and Hawaiian *Drosophila.* In *Evolution of fish species flocks,* ed. A. A. Echelle and I. Kornfield, 231–249. Orono, ME: Univ. Maine Press.

Donoghue, M. J. 1985. A critique of the biological species concept, and recommendation for a phylogenetic alternative. *The Bryologist* 88:172–181.

———. 1989. Phylogenies and the analysis of evolutionary sequences, with examples from seed plants. *Evolution* 43:1137–1156.

———. 1990. Why parsimony? *Evolution* 44:1121–1123.

Donoghue, M. J., and W. S. Alverson. 2000. A new age of discovery. *Ann Missouri Bot Gard* 87:110–126.

Donoghue, M. J., and P. D. Cantino. 1984. The logic and limitations of the outgroup substitution approach to cladistic analysis. *Syst Bot* 9:192–202.

Donoghue, M. J., J. A. Doyle, J. Gauthier, A. G. Kluge, and T. Rowe. 1989. The importance of fossils in phylogeny reconstruction. *Annu Rev Ecol Syst* 20:431–460.

Donoghue, M. J., and M. J. Sanderson. 1992. The suitability of molecular and morphological evidence in reconstructing plant phylogeny. In *Molecular systematics of plants*, ed. P. S. Soltis, D. E. Soltis, and J. A. Doyle, 340–368. New York: Chapman and Hall.

Doolittle, W. F. 2000. Uprooting the tree of life. *Sci Am*, February, 90–95.

Dougherty, E. C. 1949. The phylogeny of the nematode family Metastrongylidae Leiper (1909): A correlation of host and symbiote evolution. *Parasitology* 39:222–234.

Doyle, J. A. 1998. Phylogeny of vascular plants. *Annu Rev Ecol Syst* 29:567–599.

Doyle, J. A., and M. J. Donoghue. 1986. Seed plant phylogeny and the origin of angiosperms: An experimental cladistic approach. *Bot Rev* 52:321–431.

———. 1987. The importance of fossils in elucidating seed plant phylogeny and macroevolution. *Rev Palaeobot Palynol* 50:63–95.

———. 1993. Phylogenies and angiosperm diversification. *Paleobiology* 19:141–167.

Doyle, J. A., and P. K. Endress. 2000. Morphological phylogenetic analysis of basal angiosperms: Comparison and combination with molecular data. *Int J Plant Sci* 161:S121–S153.

Doyle, R. W. 1976. Analysis of habitat loyalty and habitat preference in the settlement behavior of planktonic marine larvae. *Am Nat* 110:719–730.

Drake, J. A. 1990. Communities as assembled structures: Do rules govern pattern? *Trends Ecol Evol* 5:159–163.

Dressler, R. L. 1982. Biology of the orchid bees (Euglossini). *Annu Rev Ecol Syst* 13:73–94.

Drummond, H. 1981. The nature and description of behaviour patterns. In *Perspectives in ethology*, vol. 4, ed. P. P. G. Bateson and P. Klopfer, 1–33. New York: Plenum Press.

Ducke, A. 1913. Über Phylogenie und Klassification der sozialen Vespiden. *Zool Jahrb Abt Syst Oekol Geogr Tiere* 36:303–330.

Duellman, W. E. 1985a. Systematic zoology: Slicing the Gordian knot with Ockham's razor. *Am. Zool* 25:751–762.

———. 1985b. Reproductive modes in anuran amphibians: Phylogenetic significance of adaptive strategies. *S Afr J Sci* 81:174–178.

———. 1993. Amphibians in Africa and South America: Evolutionary history and ecological comparisons. In *Biological relationships between Africa and South America*, ed. P. Goldblatt, 200–243. New Haven: Yale Univ. Press.

Duffy, J. E., C. L. Morrison, and R. Ríos. 2000. Multiple origins of eusociality among sponge-dwelling shrimps (*Synalpheus*). *Evolution* 54:503–516.

Dufresne, F., and P. D. N. Hebert. 1994. Hybridization and origins of polyploidy *Proc Royal Soc Lond* B 258:141–146.

Dunford, C., and R. Davis. 1975. Cliff chipmunk vocalizations and their relevance to the taxonomy of coastal sonoran chipmunks. *J Mammal* 56:207–212.

Dunham, A. E., and D. B. Miles. 1985. Patterns of covariation in the life history traits of squamate reptiles: The effects of size and phylogeny reconsidered. *Am Nat* 126:231–257.

Dunn, F. L. 1966. Patterns of parasitism in primates: Phylogenetic and ecological interpretations, with particular reference to the Hominoidea. *Folia Primatologia* 4:329–345.

Dupré, J. 1992. Species: Theoretical contexts. In *Keywords in evolutionary biology*, ed. E. F. Keller and E. A. Lloyd, 312–317. Cambridge, MA: Harvard Univ. Press.

Dupuis, C. 1978. Permanence et actualité de la systématique: La "Systématique Phylogénétique" de W. Hennig (Historique, Discussion, choix de références). *Cah Nat* 34:1–69.

————. 1984. Willi Hennig's impact on taxonomic thought. *Annu Rev Ecol Syst* 15:1–24.

Eastman, J. T. 2000. Antarctic notothenoid fishes as subjects for research in evolutionary biology. *Antarct Sci* 12:276–287.

Eastman, J. T., and A. R. McCune. 2000. Fishes of the Antarctic continental shelf: Evolution of a marine species flock? *J Fish Biol* 57:84–102.

Eberhard, W. G. 1985. *Sexual selection and animal genitalia*. Cambridge, MA: Harvard Univ. Press.

————. 1989. Effects of orb-web orientation and spider size on prey retention. *Bull Brit Arachnol Soc* 8:45–48.

Edwards, A. W. F. 1972. *Likelihood*. Cambridge: Cambridge Univ. Press.

————. 1996. The origin and early development of the method of minimum evolution for the reconstruction of phylogenetic trees. *Syst Biol* 45:79–91.

Eernisse, D. J., and A. G. Kluge. 1993. Taxonomic congruence versus total evidence, and amniote phylogeny inferred from fossils, molecules and morphology. *Mol Biol Evol* 10:1170–1195.

Eggleton, P., and R. Belshaw. 1993. Comparisons of dipteran, hymenopteran, and coleopteran parasitoids: Provisional phylogenetic explanations. *Biol J Linn Soc* 48:213–226.

Eggleton, P., and R. Vane-Wright, eds. 1994. *Phylogenetics and ecology*. Linnaean Society Symposium Series no. 17. London: Academic Press.

Ehrlich, P. R., and P. H. Raven. 1964. Butterflies and plants: A study in coevolution. *Evolution* 18:586–608.

————. 1969. Differentiation of populations. *Science* 165:1228–1232.

Ehrlich, P. R., and E. O. Wilson. 1991. Biodiversity studies: Science and policy. *Science* 253:758–762.

Ehrman, L., and M. Wasserman. 1987. The significance of asymmetrical sexual isolation. *Evo Biol* 21:1–20.

Eibl-Eibesfeldt, I. 1975. *Ethology: The biology of behavior*. 2d ed. New York: Holt, Rhinehart, and Winston.

Eichler, W. 1941a. Wirtsspezifitat und stammesgeschichtliche Gleichlaufigkeit (Fahrenholz'sche Regel) bei Parasiten im allgemeinen und bei Mallophagen im besonderen. *Zool Anz* 132:254–262.

————. 1941b. Korrelation in der Stammesentwicklung von Wirten und Parasiten. *Z Parasitenkd* (Berlin) 12:94.

————. 1942. Die Entfaltungsregel und andere Gesetzmassigkeiten in den parasitologischen Beziehungen der Mallophagen und anderer standiger Parasiten zu ihrer Wirten. *Zool Anz* 137:77–83.

————. 1948a. Evolutionsfragender Wirtsspezifitat. *Biol Zentralbl* 67:373–406.

————. 1948b. Some rules in ectoparasitism. *Ann Mag Nat Hist* 1:588–598.

————. 1966. Two new evolutionary terms for speciation in parasitic animals. *Syst Zool* 15:216–218.

————. 1973. Neuere überlegungen zu den parasitophyletischen Regeln. *Helminthologia* 14:1–4.

————. 1982. Les règles parasitophyletiques comme phénomène de la théorie de l'évolution. 2d Symposium on Host Specificity among Parasite of Vertebrates. *Mem Mus Natl Hist Nat*, n. s. 123:69–71.

Eldredge, N. 1972. Systematics and evolution of *Phacops rana* (Green, 1932) and *Phacops iowensis* Delo, 1935 (Trilobita) from the Middle-Devonian of North America. *Bull Am Mus Nat Hist* 147:45–114.

————. 1973. Systematics of Lower and Middle Devonian species of the trilobite *Phacops* Emmrich in North America. *Bull Am Mus Nat Hist* 151:285–338.

————. 1976. Differential evolutionary rates. *Paleobiology* 2:174–177.

————. 1979. Cladism and common sense. In *Phylogenetic analysis and paleontology*, ed. J. Cracraft and N. Eldredge, 165–198. New York: Columbia Univ. Press.

————. 1985. The ontology of species. In *Species and speciation*, ed. E. Vrba, 17–20. Transvaal Museum monograph no. 4. Pretoria, South Africa: Transvaal Museum.

————. ed. 1992a. *Systematics, ecology, and the biodiversity crisis*. New York: Columbia Univ. Press.

————. 1992b. Where the twain meet: Causal intersections and the genealogical and ecological realms. In *Systematics, ecology, and the biodiversity crisis,* ed. N. Eldredge, 1–14. New York: Columbia Univ. Press.

Eldredge, N., and L. Branisa. 1980. Calmonid trilobites of the Lower Devonian *Scaphioceolia* zone of Bolivia, with remarks on related species. *Bull Am Mus Nat Hist* 165:181–290.

Eldredge, N., and J. Cracraft. 1980. *Phylogenetic patterns and the evolutionary process.* New York: Columbia Univ. Press.

Eldredge, N., and S. J. Gould. 1972. Punctuated equilibria: An alternative to phyletic gradualism. In *Models in paleobiology,* ed. T. J. M. Schopf, 82–115. San Francisco: W. H. Freeman.

Ellsworth, D. L., L. R. Honeycutt, N. J. Silvy, J. W. Bickham, and W. D. Klimstra. 1994. Historical biogeography and contemporary patterns of mitochondrial DNA variation in white-tailed deer from the southeastern United States. *Evolution* 48:122–136.

Emberton, K. C. 1995. When shells do not tell: 145 million years of evolution in North American polygyrid land snails, with a revision and conservation priorities. *Malacologia* 37:69–110.

Emerson, A. E. 1938. Termite nests: A study of the phylogeny of behavior. *Ecol Monogr* 8:247–284.

Emerson, S. B. 1986. Heterochrony and frogs: The relationship of a life history trait to morphological form. *Am Nat* 127:167–183.

Endler, J. A. 1977. *Geographic variation, speciation, and clines.* Monographs in Population Biology no. 10. Princeton, NJ: Princeton Univ. Press.

———. 1982a. Problems in distinguishing historical from ecological factors in biogeography. *Am Zool* 22:441–452.

———. 1982b. Pleistocene forest refuges: Fact or fancy? In *Biological diversification in the tropics,* ed. G. Prance, 641–657. New York: Columbia Univ. Press.

———. 1989. Conceptual and other problems in speciation. In *Speciation and its consequences,* ed. D. Otte and J. A. Endler, 625–648. Sunderland, MA: Sinauer.

Epstein, P. R. 1997. Climate, ecology, and human health. *Consequences* 3:3–19.

———. 1999. Climate and health. *Science* 285:347–348.

Equiarte, L. E., J. Larson-Guerra, J. Núñez-Farfán, A. Martínez-Palacios, K. S. Del Prado, and H. T. Arita. 1999. Phylogenetic diversity and conservation: Examples at different scales and a population level proposal for *Agave victoriae reginae* in the Mexican Chihuahuan Desert. *Rev Chil Hist Nat* 72:475–492.

Ereshevsky. M. 1992. *The units of evolution: Essays on the nature of species.* Cambridge, MA: MIT Press.

Eriksson, O., and B. Bremer. 1991. Fruit characteristics, life forms, and species richness in the plant family Rubiaceae. *Am Nat* 138:751–761.

———. 1992. Pollination systems, dispersal modes, life forms, and diversification rates in angiosperm families. *Evolution* 46:258–266.

Erwin, D. H. 1989a. Regional paleoecology of Permian gastropod genera, south-western United States, and the end-Permian mass extinction. *Palaios* 4:424–438.

———. 1989b. The end-Permian mass extinction: What really happened and did it matter? *Trends Ecol Evol* 4:225–229.

———. 1990. Carboniferous-Triassic gastropod diversity patterns and the Permo-Triassic mass extinction. *Paleobiology* 16:187–203.

———. 1993. *The great Paleozoic crisis.* New York: Columbia Univ. Press.

———. 1996. Understanding biotic recoveries: Extinction, survival, and preservation during the end-Permian mass extinction. In *Evolutionary paleobiology,* ed. D. Jablonski, D. H. Erwin, and J. H. Lipps, 398–418. Chicago: Univ. Chicago Press.

Erwin, T. L. 1985. The taxon pulse: A general pattern of lineage radiation and extinction among carabid beetles. In *Taxonomy, phylogeny, and zoogeography of beetles and ants,* ed. G. E. Ball, 437–472. Dordrecht: W. Junk.

———. 1991. An evolutionary basis for conservation strategies. *Science* 253:750–752.

Escalante, A. A., and F. J. Ayala. 1994. Phylogeny of the malarial genus *Plasmodium,* derived from rRNA gene sequences. *Proc Natl Acad Sci* (USA) 91:11373–11377.

———. 1995. Evolutionary origin of *Plasmodium* and other Apicomplexa based on rRNA genes. *Proc Natl Acad Sci* (USA) 92:5793–5797.

Escalante, A. A., E. Barrio, and F. J. Ayala. 1995. Evolutionary origin of human and primate malarias: Evidence from the circumsporozoite protein gene. *Mol Biol Evol* 12:616–626.

Eshbaugh, W. H. 1995. Systematics Agenda 2000: An historical perspective. *Biodiv Conserv* 4:455–462.

Estrada, A. R., and A. Silva Rodríguez. 1984. Análisis de la ecomorfología de 23 especies de lagartos cubanos del género *Anolis*. *Ciencias Biológicas* 12:91–104.

Etheridge, R. 1959. The relationships of the anoles (Reptilia: Sauria: Iguanidae): An interpretation based on skeletal morphology. Ph.D. diss., Univ. of Michigan, Ann Arbor.

Etnier, D. A., and Williams, J. D. 1989. *Etheostoma (Nothonotus) wapiti* (Osteichthyes: Percidae): A new darter from the southern bend of the Tennessee River system in Alabama and Tennessee. *Proc Biol Soc Wash* 102:987–1000.

Ewald, P. W. 1995. The evolution of virulence: A unifying link between parasitology and ecology. *J Parasit* 81:659–669.

Ewens, W. J., P. J. Brockwell, J. M. Gani, and S. I. Resnick. 1987. Minimum viable population sizes in the presence of catastrophes. In *Viable populations for conservation*, ed. M. E. Soulé, 59–68. New York: Cambridge Univ. Press.

Facelli, J. M., and S. T. A. Pickett. 1990. Markovian chains and the role of history in succession. *Trends Ecol Evol* 5:27–30.

Fagerstrom, J. A. 1978. Modes of evolution and their chronostratigraphic significance: Evidence from Devonian invertebrates in the Michigan Basin. *Paleobiology* 4:381–393.

Fahrenholz, H. 1913. Ectoparasiten unde abstammungslehre. *Zool Anz* (Leipzig) 41:371–374.

Faith, D. P. 1989. Homoplasy as pattern: Multivariate analysis of morphological convergence in Anseriformes. *Cladistics* 5:235–258.

———. 1991. Conservation evaluation and phylogenetic diversity. *Biol Conserv* 61:1–10.

———. 1992. Systematics and conservation: On predicting the feature subsets of taxa. *Cladistics* 8:361–373.

———. 1996. Conservation priorities and phylogenetic pattern. *Conserv Biol* 10:1286–1289.

Faith, D. P., and J. W. H. Trueman. 2001. Towards an inclusive philosophy for phylogenetic inference. *Syst Biol* 50:331–350.

Faith, D. P., and P. A. Walker. 1996. How do indicator groups provide information about the relative biodiversity of different sets of areas? On hotspots, complementarity, and pattern-based approaches. *Biodiv Lett* 3:18–25.

Fang, Q. Q., A. Mitchell, J. C. Regier, C. Mitter, T. P. Friedlander, and R. W. Poole. 2000. Phylogenetic utility of the nuclear gene dopa decarboxylase in noctuoid moths (Insecta : Lepidoptera : Noctuoidea). *Mol Phylogenet Evol* 15:473–486.

Farías, I. P., G. Ortí, and A. Meyer. 2000. Total evidence: Molecules, morphology, and the phylogenetics of Cichlid fishes. *J Exp Zool* 288:76–92.

Farías, I. P., G. Ortí, I. Sampaio, H. Schneider, and A. Meyer. 1999. Mitochondrial DNA phylogeny of the family Cichlidae: Monophyly and fast molecular evolution of the Neotropical assemblage. *J Mol Evol* 48:703–711.

Fariña, R. A. 1996. Trophic relationships among Lujanian mammals. *Evol Theory* 11:125–134.

Farine, J. P., O. Bonnard, R. Brossut, and J. L. LeQuere. 1992. Chemistry of pheromonal and defensive secretions in nymphs and adults of *Dysdercus cingulatus* Fabr. (Heteroptera, Pyrrhocoridae). *J Chem Ecol* 18:65–76.

Farrell, B. D. 1998. "Inordinate fondness" explained: Why are there so many beetles? *Science* 281:555–559.

Farrell, B. D., and C. Mitter. 1990. Phylogenesis of insect/plant interactions: Have *Phyllobrotica* leaf beetles (Chrysomelidae) and the Lamiales diversified in parallel? *Evolution* 44:1389–1403.

———. 1994. Adaptive radiation in insects and plants: Time and opportunity. *Am Zool* 34:57–69.

Farrell, B. D., C. Mitter, and D. J. Futuyma. 1992. Diversification at the insect-plant interface. *BioScience* 42:34–42.

Farris, J. S. 1969. A successive approximations approach to character weighting. *Syst Zool* 18:374–385.

———. 1970. Methods for computing Wagner trees. *Syst Zool* 19:83–92.

———. 1973. A probability model for inferring phylogenetic trees. *Syst. Zool* 22:250–256.

———. 1976. Expected asymmetry of phylogenetic trees. *Syst Zool* 25:196–198.

———. 1978. Inferring phylogenetic trees from chromosome inversion data. *Syst Zool* 27:275–284.

———. 1982. Outgroups and parsimony. *Syst Zool* 31:328–334.

———. 1983. The logical basis of phylogenetic analysis. In *Advances in cladistics: Proceedings of the second meeting of the Willi Hennig Society,* vol. 2, ed. N. I. Platnick and V. A. Funk, 7–36. New York: Columbia Univ. Press.

———. 1986. Distances and statistics. *Cladistics* 2:144–157.

———. 1988. Hennig86. Ver. 1. 5. Port Jefferson Station, NY: by the author.

———. 1989a. The retention index and homoplasy excess. *Syst Zool* 38:406–407.

———. 1989b. The retention index and the rescaled consistency index. *Cladistics* 5:417–419.

———. 1990. Phenetics in camouflage. *Cladistics* 6:91–100.

Farris, J. S., A. G. Kluge, and M. J. Eckardt. 1970. A numerical approach to phylogenetic systematics. *Syst Zool* 19:172–189.

Feeny, P. P. 1976. Plant apparency and chemical defense. In *Recent advances in phytochemistry: Biochemical interactions between plants and insects,* ed. J. Wallace and R. Mansell, 10:1–40.

Felsenstein, J. 1973. Maximum likelihood and minimum-steps methods for estimating evolutionary trees from data on discrete characters. *Syst Zool* 22:240–249.

———. 1978. Cases in which parsimony or compatibility can be positively misleading. *Syst Zool* 27:401–410.

———. 1981a. Evolutionary trees from DNA sequences: A maximum likelihood approach. *J Mol Evol* 17:368–376.

———. 1981b. A likelihood approach to character weighting and what it tells us about parsimony and compatibility. *Biol J Linn Soc* 16:183–196.

———. 1981c. Skepticism towards Santa Rosalia, or why are there so few kinds of animals? *Evolution* 35:124–138.

———. 1982. Numerical methods for inferring phylogenetic trees. *Quart Rev Biol* 57:379–404.

———. 1983a. Parsimony in systematics: Biological and statistical issues. *Annu Rev Ecol Syst* 14:313–333.

———. 1983b. Methods for inferring phylogenetic trees: A statistical view. In *Numerical taxonomy,* ed. J. Felsenstein, 315–334. New York: Springer Verlag.

———. 1984. The statistical approach to inferring phylogeny and what it tells us about parsimony and compatibility. In *Cladistics: Perspectives on the reconstruction of evolutionary history,* ed. T. Duncan and T. F. Stuessy, 169–191. New York: Columbia Univ. Press.

———. 1985a. Phylogenies and the comparative method. *Am Nat* 125:1–15.

———. 1985b. Confidence limits on phylogenies: An approach using the bootstrap. *Evolution* 39:783–791.

———. 1988a. Phylogenies from molecular sequences: Inference and reliability. *Annu Rev Genet* 22:521–565.

———. 1988b. Phylogenies and quantitative characters. *Annu Rev Ecol Syst* 19:445–471.

———. 1988c. The detection of phylogeny. In *Prospects in systematics,* ed. D. L. Hawksworth, 112–127. Oxford: Systematics Association, Clarendon Press.

Ferraguti, M., and C. Erseus. 1999. Sperm types and their use for a phylogenetic analysis of aquatic clitellates. *Hydrobiologia* 402:225–237.

Ferrarezzi, H., and E. do Amaral Gimenez. 1996. Systematic patterns and the evolution of feeding habits in Chiroptera (Archonta: Mammalia). *J Comp Biol* 1:75–94.

Ferris, V. R. 1979. Cladistic approaches in the study of soil and plant parasitic nematodes. *Am Zool* 19:1195–1215.

———. 1994. The future of nematode systematics. *Fundam Appl Nematol* 17:97–101.

Ferris, V. R., and J. M. Ferris. 1989. Why ecologists need systematists: Importance of systematics to ecological research. *J Nematol* 21:308–314.

Fialkowski, K. R. 1988. Lottery of sympatric speciation computer model. *J Theor Biol* 130:379–390.

———. 1992. Sympatric speciation: A simulation model of imperfect assortative mating. *J Theor Biol* 157:9–30.

Fink, W. L. 1982. The conceptual relationship between ontogeny and phylogeny. *Paleobiology* 8:254–264.

Fink, W. L., and. M. L. Zelditch. 1995. Phylogenetic analysis of ontogenetic shape transformations: A reassessment of the piranha genus *Pygocentrus* (Teleostei). *Syst Biol* 44:343–360.

Fischer, C., M. Mahner, and E. Wachmann. 2000. The rhabdom structure in the ommatidia of the Heteroptera (Insecta), and its phylogenetic significance. *Zoomorphology* 120:1–13.

Fisher, R. A. 1930. *The genetical theory of natural selection.* Oxford: Clarendon Press.

———. 1958. *The genetical theory of natural selection.* 2d ed. New York: Dover.

Fitter, A. H. 1995. Interpreting quantitative and qualitative characteristics in comparative analyses. *J Ecol* 83:730.

Flessa, K. W., and D. Jablonski. 1996. The geography of evolutionary turnover: A global analysis of extant bivalves. In *Evolutionary paleobiology,* ed. D. Jablonski, D. H. Erwin, and J. H. Lipps, 376–397. Chicago: Univ. Chicago Press.

Flessa, K. W., and J. S. Levinton. 1975. Phanerozoic diversity patterns: Tests for randomness. *J Geol* 83:239–248.

Flessa, K. W., K. V. Powers, and J. L. Cisne. 1975. Specialization and evolutionary longevity in the Arthropoda. *Paleobiology* 1:71–81.

Flores-Villela, O. A. 1991. Análisis de la distribución de la herpetofauna de México. Ph.D. diss., Facultad de Ciencias, UNAM, Mexico City.

Flores-Villela, O. A., K. M. Kjer, M. Benabib, and J. W. Sites. 2000. Multiple data sets, congruence, and hypothesis testing for the phylogeny of basal groups of the lizard genus *Sceloporus* (Squamata, Phrynosomatidae). *Syst Biol* 49:713–739.

Floyd, S. K., and W. E. Friedman. 2000. Evolution of endosperm developmental patterns among basal flowering plants. *Int J Plant Sci* 161:S57–S81.

Forbes, E. 1846. On the connexion between the distribution of the existing fauna and flora of the British Isles, and the geological changes which have affected their area, especially during the epoch of northern drift. *Mem Geol Surv Gr Brit* 1:336–432.

Ford, E. B. 1971. *Ecological genetics.* 3d ed. London: Chapman and Hall.

Foreman, R. E., A. Gorbman, J. M. Dodd, and R. Olsson, eds. 1985. *Evolutionary biology of primitive fishes.* New York: Plenum Press.

Forey, P. L., C. J. Humphries, I. L. Kitching, R. W. Scotland, D. J. Siebert, and D. M. Williams. 1992. *Cladistics: A practical course in systematics.* Oxford: Clarendon Press.

Forey, P. L., C. J. Humphries, and R. I. Vane-Wright, eds. 1994. *Systematics and conservation evaluation.* Oxford: Clarendon Press.

Foster, D. R. 2000. From bobolinks to bears: Interjecting geographical history into ecological studies, environmental interpretation, and conservation planning. *J Biogeogr* 27:27–30.

Foster, D. R., P. K. Schoonmaker, and S. T. A. Pickett. 1990. Insights from paleoecology to community ecology. *Trends Ecol Evol* 5:119–122.

Fox, B. J. 1987. Species assembly and the evolution of community structure. *Evol Ecol* 1:201–213.

Fox, L. R. 1988. Diffuse coevolution within complex communities. *Ecology* 69:906–907.

Fraser, D. F. 1976. Coexistence of salamanders in the genus *Plethodon:* A variation of the Santa Rosalia theme. *Ecology* 57:238–251.

Fredericq, S., D. W. Freshwater, and M. H. Hommersand. 1999. Observations on the phylogenetic systematics and biogeography of the Solieriaceae (Gigartinales, Rhodophyta) inferred from rbcL sequences and morphological evidence. *Hydrobiologia* 399:25–38.

Freeman, P. W. 1995. Nectarivorous feeding mechanisms in bats. *Biol J Linn Soc* 56:439–463.

———. 1998. Form, function, and evolution in skulls and teeth of bats. In *Bat biology and conservation,* ed. T. H. Kunz and P. A. Racey, 140–156. Washington, DC: Smithsonian Institution Press.

———. 2000. Macroevolution in Microchiroptera: Recoupling morphology and ecology with phylogeny. *Evol Ecol Res* 2:317–335.

Frey, D. G. 1987. The taxonomy and biogeography of the Cladocera. *Hydrobiologia* 145: 5–17.

Frey, J. K. 1993. Modes of peripheral isolate formation and speciation. *Syst Biol* 42:373–381.

Friedlander, T. P., J. C. Regier, and C. Mitter. 1994. Phylogenetic information content of five nuclear gene sequences in animals: Initial assessment of character sets from concordance and divergence studies. *Syst Biol* 43:511–525.

Friedman, W. E., and J. S. Carmichael. 1998. Heterochrony and developmental innovation: Evolution on female gametophyte ontogeny in *Gnetum,* a highly apomorphic seed plant. *Evolution* 52:1016–1030.

Friedmann, H. 1929. *The cowbirds.* Springfield, IL: Charles C. Thomas.

Fritsch, P. W. 1999. Phylogeny of *Styrax* based on morphological characters, with implications for biogeography and infrageneric classification. *Syst Bot* 24:356–378.

Frost, D. R., and A. G. Kluge. 1994. A consideration of epistemology in systematic biology, with special reference to species. *Cladistics* 10:259–293.

Frumhoff, P. C., and H. K. Reeve. 1994. Using phylogenies to test hypotheses of adaptation: A critique of some current proposals. *Evolution* 48:172–180.

Fryer, G. 1959. The trophic interrelationships and ecology of some littoral communities of Lake Nyasa with especial reference to the fishes, and a discussion of the evolution of a group of rock-frequenting Cichlidae. *Proc Zool Soc Lond* 132:153–281.

———. 1991. A daphniid *Ephippium* (Branchiopoda: Anomopoda) of Cretaceous age. *Zool J Linn Soc* 102:163–167.

Fryer, G., P. H. Greenwood, and J. F. Peake. 1983. Punctuated equilibria, morphological stasis, and the palaeontological documentation of speciation: A biological appraisal of a case history in an African Lake. *Biol J Linn Soc* 20:195–205.

Fryer, G., and T. D. Iles. 1972. *The cichlid fishes of the Great Lakes of Africa: Their biology and evolution.* Edinburgh: Oliver and Boyd.

Funk, D. J. 1998. Isolating a role for natural selection in speciation: Host adaptation and sexual isolation in *Neochlamisus bebbianae* leaf beetles. *Evolution* 52:1744–1759.

Funk, D. J., D. J. Futuyma, G. Ortí, and A. Meyer. 1995a. A history of host associations and evolutionary diversification for *Ophraella* (Coleoptera: Chrysomelidae): New evidence from mitochondrial DNA. *Evolution* 49:1017–1022.

———. 1995b. Mitochondrial DNA sequences and multiple data sets: A phylogenetic study of phytophagous insects (Chrysomelidae: *Ophraella*). *Mol Biol Evol* 12:627–640.

———. 1995c. A history of host associations and evolutionary diversification for *Ophraella* (Coleoptera: Chrysomelidae): New evidence from mitochondrial DNA. *Evolution* 49:1008–1017.

Funk, V. A. 1981. Special concerns in estimating plant phylogenies. In *Advances in cladistics,* vol. 1, ed. V. A. Funk and D. R. Brooks, 73–86. New York: New York Botanical Garden.

———. 1982. Systematics of *Montanoa* (Asteraceae: Heliantheae). *Mem NY Bot Gard* 36:1–135.

———. 1985. Phylogenetic patterns and hybridization. *Ann Missouri Bot Gard* 72:681–715.

Funk, V. A., and D. R. Brooks. 1990. Phylogenetic systematics as the basis of comparative biology. *Smithson Contr Bot* 73:1–45.

Funk, V. A., and T. F. Stuessy. 1978. Cladistics for the practicing plant taxonomist. *Syst Bot* 3:159–178.

Funk, V. A., and W. L. Wagner. 1995. Biogeographic patterns in the Hawaiian Islands. In *Hawaiian biogeography: Evolution on a hot spot archipelago,* ed. W. L. Wagner and V. A. Funk, 379–419. Washington, D.C.: Smithsonian Institution Press.

Futuyma, D. J. 1983. Evolutionary interactions among herbivorous insects and plants. In *Coevolution,* ed. D. J. Futuyma and M. Slatkin, 207–231. Sunderland, MA: Sinauer.

———. 1986. *Evolutionary biology.* 2d ed. Sunderland, MA: Sinauer.

———. 1989. Speciational trends and the role of species in macroevolution. *Am Nat* 134:318–321.

———. 1991. Evolution of host specificity in herbivorous insects: Genetic, ecological, and phylogenetic aspects. In *Plant-animal interactions: Evolutionary ecology in tropical and temperate regions,* ed. P. W. Price, T. M. Lewinsohn, G. W. Fernandes, and W. W. Benson, 431–454. New York: Wiley.

———. 1998. *Evolutionary biology.* 3d ed. Sunderland, MA: Sinauer.

Futuyma, D. J., M. C. Keese, and D. J. Funk. 1995. Genetic constraints on macroevolution: The evolution of host affiliation in the leaf-beetle genus *Ophraella*. *Evolution* 49:797–809.

Futuyma, D. J., M. C. Keese, and S. J. Scheffer. 1993. Genetic constraints and the phylogeny of insect-plant associations: Responses of *Ophraella communa* (Coleoptera: Chrysomelidae) to host-plants of their congeners. *Evolution* 47:888–905.

Futuyma, D. J., and J. Kim. 1987. Phylogeny and coevolution. *Science* 237:441–447.

Futuyma, D. J., and G. C. Mayer. 1980. Non-allopatric speciation in animals. *Syst Zool* 29:254–271.

Futuyma, D. J., and S. S. McCafferty. 1990. Phylogeny and the evolution of host plant associations in the leaf beetle genus *Ophraella* (Coleoptera, Chrysomelidae). *Evolution* 44:885–913.

Futuyma, D. J., and C. Mitter. 1996. Insect-plant interactions: The evolution of component communities. *Phil Trans Roy Soc Lond* B 351:1361–1366.

Futuyma, D. J., and G. Moreno. 1988. The evolution of ecological specialization. *Annu Rev Ecol Syst* 19:207–233.

Futuyma, D. J., and M. Slatkin, eds. 1983. *Coevolution*. Sunderland, MA: Sinauer.

Futuyma, D. J., J. Walsh, T. Morton, D. J. Funk, and M. C. Keese. 1994. Genetic variation in a phylogenetic context: Responses of two specialized leaf beetles (Coleoptera: Chrysomelidae) to host-plants of their congeners. *J Evol Biol* 7:127–146.

Gaffney, E. S. 1979. An introduction to the logic of phylogeny reconstruction. In *Phylogenetic analysis and paleontology*, ed. J. Cracraft and N. Eldredge, 79–111. New York: Columbia Univ. Press.

Gaffney, E. S., L. Dingus, and M. K. Smith. 1995. Why cladistics. *J Nat Hist* 104:33–35.

Galis, F., and J. A. J. Metz. 1998. Why are there so many cichlid species? *Trends Ecol Evol* 13:2–3.

Gallagher, J. C. 1989. Toward a cladistic view of phytoplankton physiological ecology. *Biol Oceanogr* 6:279–289.

Gamberale, G., and B. S. Tullberg. 1996a. Evidence for a peak-shift in predator generalization among aposematic prey. *Proc Roy Soc Lond* B 263:1329–1334.

———. 1996b. Evidence for a more effective signal in aggregated aposematic prey. *Anim Behav* 52:597–601.

Gao, F., L. Yue, A. T. White, P. G. Pappas, J. Barchue, A. P. Hanson, B. M. Greene, P. M. Sharp, G. M. Shaw, and B. H. Hahn. 1992. Human infection by genetically diverse SIVsm-related HIV-2 in West Africa. *Nature* 358:495–499.

García-Ramos, G., and M. Kirkpatrick. 1997. Genetic models of adaptation and gene flow in peripheral populations. *Evolution* 51:21–28.

Garin, C. F., C. Juan, and E. Petitpierre. 1999. Mitochondrial DNA phylogeny and the evolution of host plant use in Palearctic *Chrysolina* (Coleoptera, Chrysomelidae) leaf beetles. *J Mol Evol* 48:435–444.

Garland, T., Jr. 1992. Rate tests for phenotypic evolution using phylogenetically independent contrasts. *Am Nat* 140:509–519.

Garland, T., Jr., P. H. Harvey, and A. R. Ives. 1992. Procedures for the analysis of comparative data using phylogenetically independent contrasts. *Syst Biol* 41:18–32.

Garland, T., Jr., R. B. Huey, and A. F. Bennett. 1991. Phylogeny and coadaptation of thermal physiology in lizards: A reanalysis. *Evolution* 45:1969–1975.

Garland, T., Jr., and A. R. Ives. 2000. Using the past to predict the present: Confidence intervals for regression equations in phylogenetic comparative methods. *Amer Nat* 155:346–364.

Garland, T., Jr., P. E. Midford, and A. R. Ives. 1999. An introduction to phylogenetically based statistical methods, with a new method for confidence intervals on ancestral values. *Amer Zool* 39:374–388.

Garnham, P. C. C. 1973. Distribution of malaria parasites in primates, insectivores, and bats. *Symp Zool Soc Lond* 33:377–404.

Gatesy, J. 2000. Linked branch support and tree stability. *Syst Biol* 49:800–897.

Gatesy, J., and P. Arctander. 2000. Hidden morphological support for the phylogenetic placement of *Pseudoryx nghetinhensis* with bovine bovids: A combined analysis of gross anatomical evidence and DNA sequences from five genes. *Syst Biol* 49:515–538.

Gatesy, J., R. DeSalle, and W. C. Wheeler. 1993. Alignment-ambiguous nucleotide sites and the exclusion of data. *Mol Phylogenet Evol* 2:152–157.

Gauld, I. D., and D. B. Wahl. 2000. The Labeninae (Hymenoptera: Ichneumonidae): A study in phylogenetic reconstruction and evolutionary biology. *Zool J Linn Soc* 129:271–347.

Gauthier, J., A. G. Kluge, and T. Rowe. 1988. Amniote phylogeny and the importance of fossils. *Cladistics* 4:105–209.

Gavrilets, S. 1996. On phase three of the shifting-balance theory. *Evolution* 50:1034–1041.

———. 1997a. Evolution and speciation on holey adaptive landscapes. *Trends Ecol Evol* 12:307–312.

———. 1997b. Hybrid zones with Dobzhansky-type epistatic evolution. *Evolution* 51:1027–1035.

———. 1999. A dynamical theory of speciation on holey adaptive landscapes. *Am Nat* 154:1–22.

———. 2000. Waiting time to parapatric speciation. *Proc Roy Soc Lond* B 267:2483–2492.

Gavrilets, S., R. Acton, and J. Gravner. 2000. Dynamics of speciation and diversification in a metapopulation. *Evolution* 54:1493–1501.

Gavrilets, S., and C. R. B. Boake. 1998. On the evolution of premating isolation after a founder event. *Am Nat* 152:706–716.

Gavrilets, S., and A. Hastings. 1996. Founder effect speciation: A theoretical reassessment. *Am Nat* 147:466–491.

Gavrilets, S., H. Li, and M. D. Vose. 1998. Rapid parapatric speciation on holey adaptive landscapes. *Proc Roy Soc Lond* B 265:1483–1489.

———. 2000. Patterns of parapatric speciation. *Evolution* 54:1126–1134.

Geel, B. van, and T. van Der Hammen. 1973. Upper Quaternary vegetation and climatic sequence of the Fuquene area (eastern cordillera, Colombia). *Palaeogeogr Palaeoclimatol Palaeoecol* 14:9–92.

Genner, M. J., G. F. Turner, S. Barker, and S. J. Hawkins. 1999. Niche segregation among Lake Malawi cichlid fishes? Evidence from stable isotope signatures. *Ecol Let* 2:185–190.

Genner, M. J., G. F. Turner, and S. J. Hawkins. 1999. Foraging of rocky habitat cichlid fishes in Lake Malawi: Coexistence through niche partitioning? *Oecologia* 121:283–292.

Getty, T. 2000. A constrained view of constraints. *Trends Ecol Evol* 16:249.

Ghebrehiwet, M., B. Bremer, and M. Thulin. 2000. Phylogeny of the tribe Antirrhineae (Scrophulariaceae) based on morphological and ndhF sequence data. *Plant Syst Evol* 220:223–239.

Ghilarov, A. 1996. What does "biodiversity" mean—scientific problem or convenient myth? *Trends Ecol Evol* 11:304–306.

Ghiselin, M. T. 1974. A radical solution to the species problem. *Syst Zool* 23:536–544.

Giddings, L. V., K. Y. Kaneshiro, and W. W. Anderson, eds. 1989 *Genetics, speciation, and the founder principle.* New York: Oxford Univ. Press.

Giddings, L. V., and A. R. Templeton. 1983. Behavioral phylogenies and the direction of evolution. *Science* 220:372–378.

Gilbert, C. R. 1976. Composition and derivation of the North American freshwater fish fauna. *Fla Sci* 39:104–111.

Gilbert, L. E. 1983. Coevolution and mimicry. In *Coevolution,* ed. D. J. Futuyma and M. Slatkin, 263–281. Sunderland, MA: Sinauer.

———. 1991. Biodiversity of a Central American *Heliconius* community: Pattern, process and problems. In *Plant-animal interactions: Evolutionary ecology in tropical and temperate regions,* ed. P. W. Price, T. M. Lewinsohn, G. W. Fernandes, and W. W. Benson, 403–427. New York: John Wiley and Sons.

Gilinsky, N. L. 1981. Stabilizing selection in the Archaeogastropoda. *Paleobiology* 7:316–331.

Gillespie, L. J., L. L. Consaul, and S. G. Aiken. 1997. Hybridization and the origin of the arctic grass *Poa hartzii* (Poaceae): Evidence from the morphology and chloroplast DNA restriction site data. *Can J Bot* 75:1978–1997.

Gilmour, 1940. Taxonomy and philosophy. In *The new systematics,* ed. J. Huxley, 461–474. Oxford: Oxford Univ. Press.

Gimenez, E. A., H. Ferrarezzi, and V. A. Taddei. 1996. Lingual morphology and cladistic analysis of the New World nectar-feeding bats (Chiroptera: Phyllostomidae). *J Comp Biol* 1:41–64.

Giribet, G., D. L. Distel, M. Polz, W. Sterrer, and W. C. Wheeler. 2000. Triplobastic relationships with emphasis on the acoelomates and the position of Gnathostomulida, Cycliophora, Platyhelminthes, and Chaetognatha: A combined approach of 18S rDNA sequences and morphology. *Syst Biol* 49:539–562.

Gissi, C., A. Reyes, G. Pesole, C. Saccone. 2000. Lineage-specific evolutionary rate in mammalian mtDNA. *Mol Biol Evol* 17:1022–1031.

Gittleman, J. L. 1981. The phylogeny of parental care in fishes. *Anim Behav* 29:936–941.

———. 1994. Are the pandas successful specialists or evolutionary failures? *BioScience* 44:456–464.

Gittleman, J. L., and M. Kot. 1990. Adaptation: Statistics and a null model for estimating phylogenetic effects. *Syst Zool* 39:227–241.

Gittleman, J. L., and H.-K. Luh. 1992. On comparing comparative methods. *Annu Rev Ecol Syst* 23:383–404.

Givnish, T. J., K. J. Sytsma, J. F. Smith, and W. J. Hahn. 1994. Thorn-like prickles and heterophylly in *Cyanea:* Adaptations to extinction avian browsers on Hawaii? *Proc Natl Acad Sci* (USA) 91:2810–2814.

Glazier, D. S. 1987. Energetics and taxonomic patterns of species diversity. *Syst Zool* 36:62–71.

Gleason, J. M., and M. G. Ritchie. 1998. Evolution of courtship song and reproductive isolation in the *Drosophila willistoni* species complex: Do sexual signals diverge the most quickly? *Evolution* 52:1493–1500.

Glen, D. R., and D. R. Brooks. 1985. Phylogenetic relationships of some strongylate nematodes of primates. *Proc Helminthol Soc Wash* 52:227–236.

Global Taxonomy Initiative. *See* GTI.

Glossip, D., and J. B. Losos. 1997. Ecological correlates of number of subdigital lamellae in anoles. *Herpetologica* 53:192–199.

Glowka, L., F. Burhenne-Guilmin, H. Synge, J. A. McNeely, and L. Gündling. 1994. *A guide to the Convention on Biological Diversity.* Cambridge: IUCN.

Gloyd, H. K., and R. Conant. 1990. Snakes of the *Agkistrodon* complex: A monographic review. Contr Herpetol 6. St. Louis, MO: Soc Study Amphib Rept.

Goldman, N. 1988. Methods for discrete coding of morphological characters for numerical analysis. *Cladistics* 4:59–71.

———. 1990. Maximum likelihood inference of phylogenetic trees, with special reference to a Poisson Process Model of DNA substitution and to parsimony analyses. *Syst Zool* 39:345–361.

———. 1993a. Statistical tests of models of DNA substitution. *J Mol Evol* 36:182–198.

———. 1993b. Simple diagnostic statistical tests of models for DNA substitution. *J Mol Evol* 37:345–361.

Goldman, N., J. P. Anderson, and A. G. Rodrigo. 2000. Likelihood tests of topologies in phylogenetics. *Syst Biol* 49:652–670.

Goldman, N., and S. Whelan. 2000. Statistical tests of gamma distributed rate heterogeneity in models of sequence evolution in phylogenetics. *Mol Biol Evol* 17:975–978.

Goldstein, P. Z., and R. DeSalle. 2000. Phylogenetic species, nested hierarchies, and character fixation. *Cladistics* 16:364–384.

Gomez-Zurita, J., C. Juan, and E. Petitpierre. 2000. The evolutionary history of the genus *Timarcha* (Coleoptera, Chrysomelidae) inferred from mitochondrial COII gene and partial 16S rDNA sequences. *Mol Phylogenet Evol* 14:304–317.

Gomez-Zurita, J., E. Petitpierre, and C. Juan. 2000. Nested cladistic analysis, phylogeography, and speciation in the *Timarcha goettingensis* complex (Coleoptera, Chrysomelidae). *Mol Ecol* 9:557–570.

Gomulkiewicz, R., J. N. Thompson, R. D. Holt, S. L. Nuismer, and M. E. Hochberg. 2000. Hot spots, cold spots, and the geographic mosaic theory of coevolution. *Am Nat* 156:156–174.

Goodfellow, R., G. Scudder, and I. Smith. 1999. Living capital. *Biotech Focus,* November/December, 22–23.

Goodman, D. 1987. Consideration of stochastic demography in the design and management of biological reserves. *Nat Res Mod* 1:205–234.

Goodnight, C. J. 1987. On the effect of founder events on epistatic genetic variance. *Evolution* 41:80–91.

Gordon, A. D. 1986. Consensus supertrees: The synthesis of rooted trees containing overlapping sets of labeled leaves. *J Classif* 3:335–348.

Gorman, G. C., D. Buth, M. Soulé, and S. Y. Yang. 1983. The relationships of the Puerto Rican *Anolis*: Electrophoretic and karyotypic studies. In *Advances in herpetology and evolutionary biology: Essays in honor of Ernest E. Williams*, ed. A. G. J. Rhodin and K. Miyata, 626–642. Cambridge: Museum of Comparative Zoology.

Gorman, O. 1992. Evolutionary ecology and historical ecology: Assembly, structure, and organization of stream fish communities. In *Systematics, historical ecology, and North American freshwater fishes*, ed. R. Mayden, 659–688. Stanford, CA: Stanford Univ. Press.

Gorman, O., W. J. Bean, Y. Kawaoka, and R. G. Webster. 1990. Evolution of the nucleoprotein gene of Influenza A virus. *J Virol* 64:1487–1497.

Gorman, O., R. O. Donis, Y. Kawaoka, and R. G. Webster. 1990. Evolution of Influenza A virus PB2 genes: implications for evolution of the ribonucleoprotein complex and origin of human Influenza A virus. *J Virol* 64:4893–4902.

Gosline, J. M., M. E. Demont, and M. Denny. 1992. The structure and properties of spider silk. *Endeavour* 10:37–43.

Gosline, J. M., M. Denny, and M. E. Demont. 1984. Spider silk as rubber. *Nature* 309: 551–552.

Gould, G. C. 1995. Hedgehog phylogeny (Mammalia, Erinaceidae): The reciprocal illumination of the quick and the dead. *Am Mus Novit* 3131:1–45.

Gould, S. J. 1970. Dollo on Dollo's law: Irreversibility and the status of evolutionary laws. *J Hist Biol* 3:189–212.

———. 1977. *Ontogeny and phylogeny*. Cambridge, MA: Harvard Univ. Press.

———. 1980. Is a new and general theory of evolution emerging? *Paleobiology* 6:119–120.

———. 1982a. Change in developmental timing as a mechanism of macroevolution. In *Evolution and development*, ed. J. T. Bonner, 333–346. New York: Springer Verlag.

———. 1982b. The meaning of punctuated equilibrium and its role in validating a hierarchical approach to macroevolution. In *Perspectives on evolution*, ed. R. Milkman, 83–104. Sunderland, MA: Sinauer.

———. 1985. The paradox of the first tier: An agenda for paleobiology. *Paleobiology* 11:2–12.

———. 1986. Evolution and the triumph of homology, or why history matters. *Am Sci* 74:60–69.

———. 1989. A developmental constraint in *Cerion*, with comments on the definition and interpretation of constraint in evolution. *Evolution* 43:516–539.

———. 1994. Mayr, Ernst, and the centrality of species. *Evolution* 48:31–35.

———. 1995. Tempo and mode in the macroevolutionary reconstruction of Darwinism. In *Tempo and mode in evolution: Genetics and paleontology 50 years after Simpson*, ed. W. M. Fitch and F. J. Ayala, 125–144, Washington, DC: National Academy of Sciences Press.

Gould, S. J., and N. Eldredge. 1977. Punctuated equilibria: The tempo and mode of evolution reconsidered. *Paleobiology* 3:115–151.

———. 1993. Punctuated equilibrium comes of age. *Nature* 366:223–227.

Gould, S. J., and R. C. Lewontin. 1979. The spandrels of San Marco and the Panglossian paradigm: A critique of the adaptationist programme. *Proc Roy Soc Lond* B 205: 581–598.

Gould, S. J., D. M. Raup, J. J. Sepkoski, T. J. M. Schopf, and D. S. Simberloff. 1977. The shape of evolution: A comparison of real and random clades. *Paleobiology* 3:23–40.

Gould, S. J., and E. S. Vrba. 1982. Exaptation: A missing term in the science of form. *Paleobiology* 8:4–15.

Grady, J. M., and W. H. LeGrande. 1992. Phylogenetic relationships, modes of speciation, and historical biogeography of the madtom catfishes, genus *Noturus* Rafinesque (Siluriformes: Ictaluridae). In *Systematics, historical ecology, and North American freshwater fishes*, ed. R. L. Mayden, 747–777. Stanford: Stanford Univ. Press.

Graham, R. W. 1986. Response of mammalian communities to environmental changes during the late Quaternary. In *Community ecology*, ed. J. Diamond and T. J. Case, 300–313. New York: Harper and Row.

Graham, S. W., and S. C. H. Barrett. 1995. Phylogenetic systematics of Pontederiales: Implications for breeding-system evolution. In *Monocotyledons: Systematics and evolution*, ed. P. J. Rudall, P. J. Cribb, D. F. Cutler, and C. J. Humphries, 415–441. Kew: Royal Botanic Gardens.

Graham, S. W., and R. G. Olmstead. 2000. Utility of 17 chloroplast genes for inferring the phylogeny of the basal angiosperms. *Am J Bot* 87:1712–1730.

Graham, S. W., P. A. Reeves, A. C. E. Burns, and R. G. Olmstead. 2000. Microstructural changes in noncoding chloroplast DNA: Interpretation, evolution, and utility of indels and inversions in basal angiosperm phylogenetic inference. *Int J Plant Sci* 161:S83–S96.

Grandcolas, P. 1993. The origin of biological diversity in a tropical cockroach lineage: A phylogenetic analysis of habitat choice and biome occupancy. *Acta Oecologica* 14:259–270.

———. 1998. Phylogenetic analysis and the study of community structure. *Oikos* 82: 397–400.

Grandcolas, P., and P. Deleporte. 1996. The origin of protistan symbionts in termites and cockroaches: A phylogenetic perspective. *Cladistics* 12:93–98.

Grandcolas, P., P. Deleporte, and L. Desutter-Grandcolas. 1994. Why to use phylogeny in evolutionary ecology? *Acta Oecologica* 15:661–673.

Grandcolas, P., P. Deleporte, L. Desutter-Grandcolas, and C. Daugeron. 2001. Phylogenetics and ecology: As many characters as possible should be included in cladistic analysis. *Cladistics* 17:104–110.

Grande, L., and O. Rieppel, eds. 1994. *Interpreting the hierarchy of nature: From systematic patterns to evolutionary process theories.* New York: Academic Press.

Grant, P. R. 2000. R. C. L. Perkins and evolutionary radiations on islands. *Oikos* 89:195–201.

Grant, P. R., and B. R. Grant. 1989. Sympatric speciation and Darwin's finches. In *Speciation and its consequences*, ed. D. Otte and J. A. Endler, 433–457. Sunderland, MA: Sinauer.

Grant, V. 1949. Pollination systems as isolating mechanisms in angiosperms. *Evolution* 3:82–97.

———. 1963. *The origin of adaptations.* New York: Columbia Univ. Press.

———. 1977. *Organismic evolution.* San Francisco: W. H. Freeman.

———. 1981. *Plant speciation.* 2d ed. New York: Columbia Univ. Press.

———. 1985. *The evolutionary process: A critical review of evolutionary theory.* New York: Columbia Univ. Press.

Graybeal, A. 1994. Evaluating the phylogenetic utility of genes: A search for genes informative about deep divergence among vertebrates. *Syst Biol* 43:174–193.

Green, M. D., M. G. P. Van Veller, and D. R. Brooks. Forthcoming. Assessing modes of speciation: Range asymmetry and biogeographical congruence. *J Biogeogr* 00:000–000.

Greene, H. W. 1983. Dietary correlates of the origin and radiation of snakes. *Am Zool* 23:431–441.

———. 1986a. Diet and arboreality in the emerald monitor, *Varanus prasinus,* with comments on the study of adaptation. *Fieldiana Zool N Series* 31:1–12.

———. 1986b. Natural history and evolutionary biology. In *Predator-prey relationships: Perspective and approaches from the study of lower vertebrates*, ed. M. E. Feder and G. V. Lauder, 99–108. Chicago: Univ. Chicago Press.

———. 1988. Antipredator mechanisms in reptiles. In *Biology of the reptilia*, vol. 16: *Ecology B, defense, and life history*, ed. C. Gans and R. B. Huey, 1–152. New York: Alan Liss.

———. 1994a. Homology and behavioral repertoires. In *Homology: The hierarchical basis of comparative biology*, ed. B. K. Hall, 369–391. San Diego: Academic Press.

———. 1994b. Systematics and natural history: Foundations for understanding and conserving biodiversity. *Am Zool* 34:48–56.

———. 1996. Natural history and evolutionary biology. In *Predator-prey relationships: Perspectives and approaches from the study of lower vertebrates*, ed. M. E. Feder and G. V Lauder, 99–108. Chicago: Univ. Chicago Press.

———. 1997. *Snakes: The evolution of mystery in nature.* Berkeley, CA: Univ. California Press.

———. 1998. We are primates and we are fish: Teaching monophyletic organismal biology. *Integr Biol* 3:108–111.

————. 1999. Natural history and behavioural homology. In *Novartis foundation symposium 222: Homology,* ed. G. R. Bock and G. Cardew, 173–188. New York: John Wiley and Sons.

Greene, H. W., and G. M. Burghardt. 1978. Behavior and phylogeny: Constriction in ancient and modern snakes. *Science* 200:74–77.

Greene, H. W., and J. B. Losos. 1988. Systematics, natural history, and conservation. *BioScience* 38:458–462.

Greenwood, P. H. 1964. Explosive speciation in African lakes. *Proc Roy Inst Gr Brit* 40:256–269.

————. 1965. The cichlid fishes of Lake Nabugabo, Uganda. *Bull Brit Mus Nat Hist* (Zool.) 12:315–357.

————. 1973. Morphology, endemism, and speciation in African cichlid fishes. *Verh Dtsch Zool Ges* 1973:115–124.

————. 1974. The cichlid fishes of Lake Victoria, East Africa: The biology and evolution of a species flock. *Bull Brit Mus Nat Hist* (Zool.), suppl. no. 6:1–134.

————. 1984. African cichlid fishes and evolutionary theories. In *Evolution of fish species flocks,* ed. A. A. Echelle and I. Kornfield, 141–154. Orono, ME: Univ. Maine Press.

Griffiths, T. A., A. Truckenbrod, and P. J. Sponholz. 1992. Systematics of megadermatid bats (Chiroptera, Megadermatidae) based on hyoid morphology. *Am Mus Novit* 3041:1–21.

Griswold, C. E., J. A. Coddington, G. Hormiga, and N. Scharff. 1998. Phylogeny of the orb-web building spiders (Araneae, Orbiculariae: Deinopoidea, Araneoidea). *Zool J Linn Soc* 123:1–99.

Groman, J. D., and Ø. Pellmyr. 2000. Rapid evolution and specialization following host colonization in a yucca moth. *J Evol Biol* 13:223–236.

Grundy, W. N., and G. J. P. Naylor. 1999. Phylogenetic inference from conserved sites alignments. *J Exper Zool* 285:128–139.

Grutter, A. S., and R. Poulin. 1998. Cleaning of coral reef fishes by the Wrasse *Labroides dimidiatus:* Influence of client body size and phylogeny. *Copeia* 1998:120–127.

GTI (Global Taxonomy Initiative). 1999. *Using systematic inventories to meet country and regional needs: A report of the DIVERSITAS/ Systematics Agenda 2000 International Workshop, September 17–19, 1998, American Museum of Natural History.* New York: The Center for Biodiversity and Conservation, American Museum of Natural History.

Guilford, T. 1990. The evolution of aposematism. In *Insect defenses: Adaptive mechanisms and strategies of prey and predators,* ed. D. L. Evans and J. O. Schmidt, 23–62. Albany, NY: SUNY Press.

Guralnick, R. P., and D. R. Lindberg. 1999. Integrating developmental evolutionary patterns and mechanisms: A case study using the gastropod radula. *Evolution* 53:447–459.

Guyader, M., M. Emerman, P. Sonigo, F. Clavel, L. Montagnier, and M. Alizon. 1987. Genome organization and transactivation of the human immunodeficiency virus type 2. *Nature* 326:662–669.

Guyer, C., and J. B. Slowinski. 1991. Comparisons of observed phylogenetic topologies with null expectations among three monophyletic lineages. *Evolution* 45:340–350.

————. 1993. Adaptive radiation and the topology of large phylogenies. *Evolution* 47:253–263.

————. 1995. Reply to Cunningham. *Evolution* 49:1294–1295.

Haffer, J. 1967. Speciation in Colombian forest birds west of the Andes. *Am Mus Novit* 294:1–57.

————. 1969. Speciation in Amazonian forest birds. *Science* 165:131–137.

————. 1974. *Avian speciation in tropical South America.* Cambridge, MA: Nuttall Ornithological Club.

————. 1987. Biogeography of Neotropical birds. In *Biogeography and quaternary history in tropical America,* ed. C. T. Whitmore and G. T. Prance, 105–150. New York: Oxford Univ. Press.

————. 1993. Time's cycle and time's arrow in the history of Amazonia. *Biogeographica* 69:15–45.

Hafner, M. S., and S. A. Nadler. 1988. Phylogenetic trees support the coevolution of parasites and their hosts. *Nature* 332:258–260.

—. 1990. Cospeciation in host-parasite assemblages: Comparative analysis of rates of evolution and timing of cospeciation events. *Syst Zool* 39:192–204.

Hagen, D. W., and L. G. Gilbertson. 1972. Geographic variation and environmental selection in *Gasterosteus aculeatus* L. in the Pacific northwest, America. *Evolution* 26:32–51.

Hagen, D. W., and J. D. McPhail. 1970. The species problem within *Gasterosteus aculeatus* on the Pacific coast of North America. *J Fish Res Board* Canada 27:147–155.

Hairston, N. G. 1951. Interspecies competition and its probable influence upon the vertical distribution of Appalachian salamanders of the genus *Plethodon. Ecology* 32:266–274.

—. 1981. An experimental test of a guild: Salamander competition. *Ecology* 62:65–72.

Hairston, N. G., F. E. Smith, and L. B. Slobodkin. 1960. Community structure, population control, and competition. *Am Nat* 94:421–425.

Haldane, J. B. S. 1932. *The causes of evolution.* London: Longman.

—. 1935. Darwinism under revision. *Rationalist Ann* (1934):19–29.

—. 1937. The effect of variation on fitness. *Am Nat* 71:337–349.

Hall, B. K. 1992. *Evolutionary developmental biology.* London: Chapman and Hall.

—, ed. 1994. *Homology: The hierarchical basis of comparative biology.* New York: Academic Press.

Hall, B. P., and R. E. Moreau. 1970. *An atlas of speciation in African passerine birds.* London: Natural History Museum.

Hall, G. S. 1996. Modern approaches to species concepts in downy mildews. *Plant Pathol* 45:1009–1026.

Hallam, A. 1989. What can the fossil record tell us about macroevolution? In *Evolutionary biology at the crossroads,* ed. M. K. Hecht, 59–73. Flushing, NY: Queen's College Press.

Halloy, M., R. Etheridge, and G. M. Burghardt. 1998. To bury in sand: Phylogenetic relationships among lizard species of the *boulengeri* group, *Liolaemus* (Reptilia: Squamata: Tropiduridae), based on behavioral characters. *Herpetol Monogr* 12:1–37.

Hamilton, A. C. 1976. The significance of patterns of distribution shown by forest plants and animals in tropical Africa for the reconstruction of upper Pleistocene paleoenvironments: A review. *Paleoecol Afr* 9:63–97.

Hammen, T. van Der. 1972. Changes in vegetation and climate in the Amazon Basin and surrounding areas during the Pleistocene. *Geol Mijnb* 51:641–643.

Hammen, T. van Der, and E. Gonzales. 1960. Upper Pleistocene and Holocene climate and vegetation of the "Sabana de Bogota" (Colombia, South America). *Leidse Geol Meded* 25:261–315.

Hancock, J. M., and A. P. Vogler. 2000. How slippage derived sequences are incorporated into rRNA variable region secondary structure: Implications for phylogeny reconstruction. *Mol Phylogenet Evol* 14:366–374.

Hansen, T. F. 1997. Stabilizing selection and the comparative analysis of adaptation. *Evolution* 51:1341–1351.

Hansen, T. F., W. S. Armbruster, and L. Antonsen. 2000. Comparative analysis of character displacement and spatial adaptations as illustrated by the evolution of *Dalechampia* blossoms. *Am Nat* 156:S17–S34.

Hanson, P. 1990. La sistematic aplicada al estudio de la biologia de los parasitoides. *Manejo Integrado de Plagas* (Costa Rica) 15:53–66.

Harlan, J. R., and J. M. J. DeWet 1963. The compilospecies concept. *Evolution* 17:497–501.

—. 1975. On a wing and a prayer: The origins of polyploidy. *Bot Rev* 41:361–390.

Harlin, M. 1996. Biogeographic patterns and the evolution of eureptantic nemerteans. *Biol J Linn Soc* 58:325–342.

Harper, E. M. 1994. Are conchiolin sheets in corbulid bivalves primarily defensive? *Palaeontology* 37:551–578.

Harris, D. J., E. N. Arnold, and R. H. Thomas. 1998. Rapid speciation, morphological evolution, and adaptation to extreme environments in South African sand lizards *(Meroles)* as revealed by mitochondrial gene sequences. *Mol Phylogenet Evol* 10:37–48.

Harrison, L. 1914. The Mallophaga as a possible clue to bird phylogeny. *Austral Zool* 1:7–11.

—. 1915a. Mallophaga from *Apteryx* and their significance: With a note on *Rallicola. Parasitology* 8:88–100.

———. 1915b. The relationship of the phylogeny of the parasite to that of the host. *Rep Brit Assoc Adv Sci* 85:476–477.

———. 1916. Bird-parasites and bird-phylogeny. *Ibis* 10:254–263.

———. 1922. On the Mallophagan family Trimenoponidae, with a description of a new genus and species from an Australian marsupial. *Austral Zool* 2:154–159.

———. 1924. The migration route of the Australian marsupial fauna. *Austral Zool* 3:247–263.

———. 1926. Crucial evidence for Antarctic radiation. *Am Nat* 60:374–383.

———. 1928a. On the genus *Stratiodrilus* (Archiannelida: Histriobdellidae), with a description of a new species from Madagascar. *Rec Austral Mus* 16:116–121.

———. 1928b. Host and parasite. *Proc Linn Soc NSW* 53:ix–xxxi.

———. 1929. The composition and origins of the Australian fauna, with special reference to the Wegener hypothesis. *Rep Mtgs Australas Assoc Adv Sci* (Perth) 1926:332–396.

Harrison, R. G., and D. M. Rand. 1989. Mosaic hybrid zones and the nature of species boundaries. In *Speciation and its consequences*, ed. D. Otte and J. A. Endler, 111–133. Sunderland, MA: Sinauer.

Harry, M., M. Solignac, and D. Lachaise. 1996. Adaptive radiation in the Afrotropical region of the Paleotropical genus *Lissocephala* (Drosophilidae) on the pantropical genus *Ficus* (Moraceae). *J Biogeogr* 23:543–552.

———. 1998. Molecular evidence for parallel evolution of adaptive syndromes in fig-breeding *Lissocephala* (Drosophilidae). *Mol Phylogenet Evol* 9:542–551.

Hart, J. A. 1985a. Peripheral isolation and the origin of diversity in *Lepechinia* sect. *Parviflorae* (Lamiaceae). *Syst Bot* 10:134–146.

———. 1985b. Evolution of dioecism in *Lepechinia* willd. sect. *Parviflorae* (Lamiaceae). *Syst Bot* 10:147–154.

———. 1988. Rust fungi and host plant coevolution: Do primitive hosts harbor primitive parasites? *Cladistics* 4:339–366.

Hart, M. W., M. Byrne, and M. J. Smith. 1997. Molecular phylogenetic analysis of life-history evolution in asterinid starfish. *Evolution* 51:1848–1861.

Harvey, P. H. 1996. Phylogenies for ecologists. *J Anim Ecol* 65:255–263.

Harvey, P. H., and T. Clutton-Brock. 1985. Life history variation in primates. *Evolution* 39:559–581.

Harvey, P. H., A. J. Leigh Brown, and J. Maynard Smith. 1995. New uses for new phylogenies: Editors' introduction. *Philos Trans Roy Soc Lond* 349:3–4.

Harvey, P. H., and G. M. Mace. 1982. Comparisons between taxa and adaptive trends. In *Current problems in sociobiology*, ed. King's College Sociobiology Group, 343–361. Cambridge: Cambridge Univ. Press.

Harvey, P. H., and M. Pagel. 1991. *The comparative method in evolutionary biology*. Oxford: Oxford Univ. Press.

Harvey, P. H., and A. Purvis. 1991. Comparative methods for explaining adaptations. *Nature* 351:619–624.

Harvey, P. H., A. F. Read, and S. Nee. 1995a. Why ecologists need to be phylogenetically challenged. *J Ecol* 83:535–536.

———. 1995b. Further remarks on the role of phylogeny in comparative ecology. *J Ecol* 83:733–734.

Håstad, O., and M. Björklund. 1998. Nucleotide substitution models and estimation of phylogeny. *Mol Biol Evol* 15:1381–1389.

Hastenrath, S. 1971a. On the Pleistocene snowline depression in the arid regions of the South American Andes. *J Glaciol* 10:255–267.

———. 1971b. On snowline depression and atmospheric circulation in the tropical Americas during the Pleistocene. *S Afr Geogr J* 53:53–69.

Haszprunar, G. 1992. The types of homology and their significance for evolutionary biology and phylogenetics. *J Evol Biol* 5:13–24.

Hatfield, T., and D. Schluter. 1996. A test for sexual selection on hybrids of two sympatric sticklebacks. *Evolution* 50:2429–2434.

———. 1999. Ecological speciation in sticklebacks: Environment-dependent hybrid fitness. *Evolution* 53:866–873.

Hawkes, J. G. 1972. Evolutionary relationships in wild tuber-bearing *Solanum* species. *Symp Biol Hung* 12:65–69.

Hawksworth, D. L. 1995. Magnitude and distribution of biodiversity. In *Global biodiversity assessment,* ed. V. H. Heywood and R. T. Watson, 107–192. Cambridge: Cambridge Univ. Press.

Hay, D. E., and J. D. McPhail. 1975. Mate selection in threespine sticklebacks *(Gasterosteus). Can J Zool* 53:441–450.

Hayami, I. 1978. Notes on the rates and patterns of size change in evolution. *Paleobiology* 4:252–260.

Haydon, D., R. R. Radtkey, and E. R. Pianka. 1993. Experimental biogeography: Interactions between stochastic, historical, and ecological processes in a model archipelago. In *Species diversity in ecological communities,* ed. R. E. Ricklefs and D. Schluter, 117–130. Chicago: Univ. Chicago Press.

Hayes, T. B. 1997. Hormonal mechanisms as potential constraints on evolution: Examples from the Anura. *Am Zool* 37:482–490.

Heard, S. B., and D. L. Hauser. 1995. Key evolutionary innovations and their ecological mechanisms. *Hist Biol* 10:151–173.

Hebert, P. D. N. 1987. Genotypic characteristics of the Cladocera. *Hydrobiologia* 145:183–193.

Hedenäs, L., and A. Kooijman. 1996. Phylogeny and habitat adaptations within a monophyletic group of wetland moss genera. *Plant Syst Evol* 199:33–52.

Hedges, S. B. 1996a. Historical biogeography of West Indian vertebrates. *Annu Rev Ecol Syst* 27:163–196.

———. 1996b. Vicariance and dispersal in Caribbean biogeography. *Herpetologica* 52:466–473.

Hedges, S. B., and K. L. Burnell. 1990. The Jamaican radiation of *Anolis* (Sauria: Iguanidae): An analysis of relationships and biogeography using sequential electrophoresis. *Carib J Sci* 26:31–44.

Hedin, M. C. 1997. Speciational history in a diverse clade of habitat-specialized spiders (Araneae: Nesticidae: *Nesticus*): Inferences from geographic-based sampling. *Evolution* 51:1929–1945.

Heinroth, O. 1911. Beiträge zur Biologie, namentlich Ethologie und Psychologie der Anatiden. *Verh Ver Int Ornithol Kongr* (Berlin) 1910:589–702.

Hellmayr, C. E. 1910. The birds of the Rio Madeira. *Novit Zoologicae* 17:257–428.

Helm-Bychowski, K., and J. Cracraft. 1993. Recovering phylogenetic signal from DNA sequences: Relationships within the corvine assemblage (class Aves) as inferred from complete sequences of the mitochondrial cytochrome-b gene. *Mol Biol Evol* 10:1196–1214.

Henderson, R. W. 1982. Trophic relationships and foraging strategies of some New World tree snakes *(Leptophis, Oxybelis, Uromacer). Amphibia-Reptilia* 3:71–80.

Henderson, R. W., T. A. Noeske-Hallin, B. I. Crother, and A. Schwartz. 1988. The diets of Hispaniolan colubrid snakes II: Prey species, prey size, and phylogeny. *Herpetologica* 44:55–70.

Hendriks, L., A. Goris, Y. van de Peer, J.-M. Neefs, M. Vancanneyt, K. Kersters, G. L. Hennebert, and R. de Wachter. 1991. Phylogenetic analysis of five medically important *Candida* species as deduced on the basis of small ribosomal subunit RNA sequences. *J Genet Micro* 137:1223–1230.

Hendy, M. D., and D. Penny. 1989. A framework for the quantitative study of evolutionary trees. *Syst Zool* 38:297–309.

Hendy, M. D., M. A. Steel, D. Penny, and I. M. Henderson. 1988. Families of trees and consensus. In *Classification and related methods of data analysis,* ed. H. H. Bock, 355–362. Amsterdam: Elsevier.

Hennig, W. 1950. *Grundzüge einer Theory der phylogenetischen Systematik.* Berlin: Deutscher Zentralverlag.

———. 1965. Phylogenetic systematics. *Annu Rev Entomol* 10:97–116.

———. 1966a. *Phylogenetic Systematics.* Urbana: Univ. Illinois Press.

———. 1966b. The Diptera fauna of New Zealand as a problem in systematics and zoogeography. Trans. P. Wygodzinsky. *Pacif Insects Monogr* 9:1–81.

Herberstein, M. E., C. L. Craig, J. A. Coddington, and M. A. Elgar. 2000. The functional significance of silk decorations of orb-web spiders: A critical review of the empirical evidence. *Biol Rev* 75:649–669.

Herendeen, P. S., and R. B. Miller. 2000. Utility of wood anatomical characters in cladistic analysis. *IAWA J* 21:247–276.

Hernandez, M. H., A. N. Garcia, M. Ulloa, and F. Alvarez N., eds. 2000. *Enfoques contemporaneos para el estudio de la biodiversidad.* Mexico City: Inst. Biol., UNAM.

Herre, E. A. 1993. Population structure and the evolution of virulence in nematode parasites of fig wasps. *Science* 259:1442–1445.

———. 1995. Factors affecting the evolution of virulence: Nematode parasites of fig wasps as a case study. *Parasitology* 111:S179–S191.

———. 1997. An overview of studies on a community of Panamanian figs. *J Biogeogr* 23:593–607.

Herre, E. A., C. A. Machado, E. Bermingham, J. D. Nason, D. M. Windsor, S. S. McCafferty, W. Van Houten, and K. Bachmann. 1996. Molecular phylogenies of figs and their pollinator wasps. *J Biogeogr* 23:521–530.

Herrera, C. M. 1992. Historical effects and sorting processes as explanations for contemporary ecological patterns: Character syndromes in Mediterranean plants. *Am Nat* 140:421–446.

Herrick, F. H. 1911. Nest and nest-building in birds. *J Anim Behav* 1:159–192, 244–277, 336–373.

Hewitt, G. M. 1989. The subdivision of species by hybrid zones. In *Speciation and its consequences,* ed. D. Otte and J. A. Endler, 85–110. Sunderland, MA: Sinauer.

———. 1999. Post-glacial re-colonization of European biota. *Biol J Linn Soc* 68:87–112.

Hey, J. 1992. Using phylogenetic trees to study speciation and extinction. *Evolution* 46: 627–640.

Heywood, V. H., J. R. Harborne, and B. L. Turner, eds. 1977. *The biology and chemistry of the Compositae.* London: Academic Press.

Hickey, L. J., and J. A. Wolfe. 1975. The bases of angiosperm phylogeny: Vegetative morphology. *Ann Missouri Bot Gard* 62:583–589.

Higashi, M., G. Takimoto, and N. Yamamura. 1999. Sympatric speciation by sexual selection. *Nature* 402:523–526.

Higgs, P. G., and B. Derrida. 1991. Stochastic model for species formation in evolving populations. *J Phys* A: *Math Gen* 24:L985–L991.

———. 1992. Genetic distance and species formation in evolving populations. *J Mol Evol* 35:454–465.

Hill, J. E., and J. D. Smith. 1984. *Bats: A natural history.* London: British Museum (Natural History).

Hillis, D. M. 1985. Evolutionary genetics of the Andean lizard genus *Pholidobolus* (Sauria: Gymnopthalmidae): Phylogeny, biogeography, and a comparison of tree construction techniques. *Syst Zool* 34:109–126.

———. 1987. Molecular versus morphological approaches to systematics. *Annu Rev Ecol Syst* 18:23–42.

———. 1988. Systematics of the *Rana pipiens* complex: Puzzle and paradigm. *Annu Rev Ecol Syst* 19:39–63.

———. 1991. Discriminating between phylogenetic signal and random noise in DNA sequences. In *Phylogenetic analysis of DNA sequences,* ed. M. M. Miyamoto and J. Cracraft, 278–294. New York: Oxford Univ. Press.

———. 1994. Homology in molecular biology. In *Homology: The hierarchical basis of comparative biology,* ed. B. K. Hall, 339–368. New York: Academic Press.

———. 1995. Approaches for assessing phylogenetic accuracy. *Syst Biol* 44:3–16.

Hillis, D. M., M. W. Allard, and M. M. Miyamoto. 1993. Analysis of DNA sequence data: Phylogenetic inference. *Methods in Enzymol* 224:456–487.

Hillis, D. M., and J. J. Bull. 1993. An empirical test of bootstrapping as a method for assessing confidence in phylogenetic analysis. *Syst Biol* 42:182–192.

Hillis, D. M., J. J. Bull, M. E. White, M. R. Badgett, and I. J. Molineux. 1992. Experimental phylogenetics: Generation of a known phylogeny. *Science* 255:589–592.

Hillis, D. M., and S. K. Davis. 1986. Evolution of ribosomal DNA: Fifty million years of recorded history in the frog genus *Rana. Evolution* 40:1275–1288.

Hillis, D. M., J. S. Frost, and D. A. Wright. 1983. Phylogeny and biogeography of the *Rana pipiens* complex: A biochemical evaluation. *Syst Zool* 32:132–143.

Hillis, D. M., and J. P. Huelsenbeck. 1992. Signal, noise, and reliability in molecular phylogenetic analyses. *J Heredity* 83:189–195.

Hillis, D. M., J. P. Huelsenbeck, and D. L. Swofford. 1994. Hobgoblin of phylogenetics? *Nature* 369:363–364.

Hillis, D. M., and C. Moritz, eds. 1990. *Molecular systematics*. 1st ed. Sunderland, MA: Sinauer.

Hillis, D. M., C. Moritz, and B. K. Mable, eds. 1996. *Molecular Systematics*. 2d ed. Sunderland, MA.: Sinauer.

Hixon, M. A. 1980. Competitive interactions between California reef fishes of the genus *Embiotoca*. *Ecology* 61:918–931.

Hoagland, K. E. 1996. The taxonomic impediment and the Convention on Biodiversity. *Assoc Syst Coll Newsletter* 24:62–62, 66–67.

Hoberg, E. P. 1986. Evolution and historical biogeography of a parasite-host assemblage: *Alcataenia* spp. (Cyclophyllidea: Dilepididae) in Alcidae (Charadriiformes). *Can J Zool* 64:2576–2589.

———. 1987. Recognition of larvae of the Tetrabothriidae (Eucestoda): Implications for the origins of tapeworms in marine homeotherms. *Can J Zool* 65:997–1000.

———. 1992. Congruent and synchronic patterns in biogeography and speciation among seabirds, pinnipeds, and cestodes. *J Parasitol* 78:601–615.

———. 1995. Historical biogeography and modes of speciation across high latitude seas of the Holarctic: Concepts for host-parasite coevolution among the Phocini (Phocidae) and Tetrabothriidae (Eucestoda). *Can J Zool* 73:45–57.

———. 1996. Faunal diversity among avian parasite assemblages: The interaction of history, ecology, and biogeography in marine systems. *Bull Scand Soc Parasitol* 6:65–89.

———. 1997a. Phylogeny and historical reconstruction: Host-parasite systems as keystones in biogeography and ecology. In *Biodiversity II: Understanding and protecting our biological resources*, ed. M. L. Reaka-Kudla, D. E. Wilson, and E. O. Wilson, 243–261. Washington, DC: Joseph Henry Press.

———. 1997b. Parasite biodiversity and emerging pathogens: A role for systematics in limiting impacts on genetic resources. In *Global genetic resources: Access, ownership, and intellectual property rights*, ed. K. E. Hoagland and A. Y. Rossman., 71–83. Washington, DC: Association of Systematics Collections.

Hoberg, E. P., and A. M. Adams. 1992. Phylogeny, historical biogeography, and ecology of *Anophryocephalus* spp. (Eucestoda: Tetrabothriidae) among pinnipeds of the Holarctic during the late Tertiary and Pleistocene. *Can J Zool* 70:703–719.

———. 2000. Phylogeny, history, and biodiversity: Understanding faunal structure and biogeography in the marine realm. *Bull Scand Soc Parasitol* 10:19–37.

Hoberg, E. P., N. L. Alkire, A. de Queiroz, and A. Jones. 2000. Out of Africa: Origins of the *Taenia* tapeworms in humans. *Proc Roy Soc Lond* B 268:781–787.

Hoberg, E. P., D. R. Brooks, and D. Siegel-Causey. 1997. Host-parasite co-speciation: History, principles, and prospects. In *Host-parasite evolution: General principles and avian models*, ed. D. H. Clayton and J. Moore, 212–235. Oxford: Oxford Univ. Press.

Hoberg, E. P., J. Mariaux, and D. R. Brooks. 2001. Phylogeny among orders of the Eucestoda (Cercomeromorphae): Integrating morphology, molecules, and total evidence. In *The Platyhelminthes: Phylogenetic perspectives*, ed. D. T. J. Littlewood and R. A. Bray, 112–126. London: Taylor and Francis.

Hobson, E. S. 1991. Trophic relationships of fishes specialized to feed on zooplankters above coral reefs. In *The ecology of fishes on coral reefs*, ed. P. F. Sale, 69–95. New York: Academic Press.

Hoctor, T. S., M. H. Carr, and P. D. Zwick. 2000. Identifying a linked reserve system using a regional landscape approach: The Florida ecological network. *Conserv Biol* 14: 984–1000.

Hodgson, J. G., K. Thompson, and P. J. Wilson. 1998. Does biodiversity determine ecosystem function? The ecotron experiment reconsidered. *Funct Ecol* 12:843–848.

Hodkinson, I. D., 1997. Progressive restriction of host plant exploitation along a climatic gradient: The willow psyllid *Cacopsylla groenlandica* in Greenland. *Ecol Entomol* 22:47–54.

Hoelzer, G. A., and D. J. Melnick. 1994. Patterns of speciation and limits to phylogenetic resolution. *Trends Ecol Evol* 9:104–107.

Hoffman, A. A., and M. J. Hercus. 2000. Environmental stress as an evolutionary force. *BioScience* 50:217–226.

Hoffmeister, D. F. 1986. *Mammals of Arizona*. Tucson: Univ. Arizona Press and Arizona Game and Fish Dept.

Höglund, J., and B. Sillén-Tullberg, 1994. Does lekking promote the evolution of male-biased size dimorphism in birds? On the use of comparative approaches. *Am Nat* 144:881–889.

Hollocher, H., C.-T. Ting, M.-L. Wu, and C.-I. Wu. 1997. Incipient speciation by sexual isolation in *Drosophila melanogaster:* Extensive genetic divergence without re-inforcement. *Genetics* 147:1191–1201.

Holland, P. W. H. 1999. The future of evolutionary developmental biology. *Nature* 402, suppl. (December 2):C41–C44.

Holmes, E. C., S. Nee, A. Rambaut, G. P. Garnett, and P. H. Harvey. 1995. Revealing the history of infectious disease epidemics through phylogenetic trees. *Phil Trans Roy Soc Lond* B 349:33–40.

Holt, R. D. 1996. Demographic constraints in evolution: Towards unifying the evolution-ary theories of senescence and niche conservatism. *Evol Ecol* 10:1–11.

Holt, R. D., and M. E. Hochberg. 1997. When is biological control evolutionarily stable? *Ecology* 78:1673–1683.

Honeycutt, R. L., and V. M. Sarich. 1987. Albumin evolution and subfamilial relation-ships among New World leaf-nosed bats (Family Phyllostomidae). *J Mammal* 62:805–811.

Hopper, S. D. 2000. How well do phylogenetic studies inform the conservation of Austra-lian plants? *Austral J Bot* 48:321–328.

Hori, M. 1983. Feeding ecology of thirteen species of *Lamprologus* (Teleostei; Cichlidae) coexisting at a rocky shore of Lake Tanganyika. *Physiol Ecol* (Japan) 20:129–149.

Hormiga, G., N. Scharff, and J. Coddington. 2000. The phylogenetic basis of sexual dimorphism in orb weaving spiders (Araneae, Orbiculariae). *Syst Biol* 49:435–462.

Hormiga, G., W. G. Eberhard, and J. A. Coddington. 1995. Web-construction behaviour in Australian *Phonognatha* and the phylogeny of Nephiline and Tetragnathid spi-ders (Araneae: Tetragnathidae). *Aust J Zool* 43:313–364.

Horton, C. C., and D. H. Wise. 1983. The experimental analysis of competition between two syntopic species of orb-web spiders (Araneae: Araneidae). *Ecology* 64:929–944.

Hostert, E. E. 1997. Reinforcement: A new perspective on an old controversy. *Evolution* 51:697–702.

Houle, A. 1997. The role of phylogeny and behavioral competition in the evolution of coexistence among primates. *Can J Zool* 75:827–846.

Hovenkamp, P. 1997. Vicariance events, not areas, should be used in biogeographical analysis. *Cladistics* 13:67–79.

Howard, D. J. 1993. Reinforcement: Origin, dynamics, and fate of an evolutionary hy-pothesis. In *Hybrid zones and the evolutionary process*, ed. R. G. Harrison, 46–69. New York: Oxford Univ. Press.

Howarth, F. G. 1983. Ecology of cave arthropods. *Annu Rev Entomol* 28:365–389.

———. 1987. The evolution of non-relictual tropical troglobites. *Int J Speleol* 16:1–16.

———. 1991. Hawaiian cave fauna: Macroevolution on young islands. In *The unity of evo-lutionary biology*, ed. E. C. Dudley, 286–295. Portland, OR: Discorides Press.

Hubbell, S. P., and L. K. Johnson. 1978. Comparative foraging behavior of six stingless bee species exploiting a standardized resource. *Ecology* 59:1123–1136.

Huckins, C. J. F., C. W. Osenberg, and G. G. Mittelbach. 2000. Species introductions and their consequences: An example with congeneric sunfish. *Ecol Appl* 10:612–625.

Huelsenbeck, J. P. 1995. Performance of phylogenetic methods in simulation. *Syst Biol* 44:1748.

Huelsenbeck, J. P., J. J. Bull, and C. W. Cunningham. 1996. Combining data in phyloge-netic analysis. *Trends Ecol Evol* 11:152–158.

Huelsenbeck, J. P., and K. A. Crandall. 1997. Phylogeny estimation and hypothesis test-ing using maximum likelihood. *Annu Rev Ecol Syst* 28:437–466.

Huelsenbeck, J. P., and D. M. Hillis. 1993. Success of phylogenetic methods in the four-taxon case. *Syst Biol* 42:247–264.

Huelsenbeck, J. P., and B. Rannala. 1997. Phylogenetic methods come of age: Testing hypotheses in an evolutionary context. *Science* 276:227–232.

Huet, T., A. Cheynier, A. Meyerhans, G. Roelants, and S. Wain-Hobson. 1990. Genetic organization of a chimpanzee lentivirus related to HIV-1. *Nature* 345:356–359.

Huey, R. B. 1987. Phylogeny, history, and the comparative method. In *New directions in ecological physiology*, ed. M. E. Feder, A. F. Bennett, W. Burggren, and R. B. Huey, 76–98. Cambridge: Cambridge Univ. Press.

Huey, R. B., and A. F. Bennett. 1987. Phylogenetic studies of coadaptation: Preferred temperatures versus optimal performance temperatures of lizards. *Evolution* 41:1098–1115.

Huey, R. B., and J. G. Kingsolver. 1993. Evolution of resistance to high temperature in ectotherms. *Am Nat* 142:S21–S46.

Huey, R. B., and E. R. Pianka. 1977. Patterns of niche overlap among broadly sympatric versus narrowly sympatric Kalahari lizards (Scincidae: *Mabuya*). *Ecology* 58:119–128.

Huey, R. B., and T. P. Webster. 1976. Thermal biology of *Anolis* lizards in a complex fauna: The *cristatellus* group on Puerto Rico. *Ecology* 57:985–994.

Hufford, L. 1995. Patterns of ontogenetic development in perianth diversification of *Besseya* (Scrophulariaceae). *Am J Bot* 82:655–680.

———. 1996. Ontogenetic evolution, clade diversification, and homoplasy. In *Homoplasy: The recurrence of similarity in evolution*, ed. M. J. Sanderson and L. Hufford, 271–301. San Diego: Academic Press.

Hughes, J. M. 1996. Phylogenetic analysis of the Cuculidae (Aves, Cuculiformes) using behavioral and ecological characters. *Auk* 113:10–22.

Hughes, T. P. 1989. Community structure and the diversity of corals reefs: The role of history. *Ecology* 70:275–279.

Hugot, J. P. 1999. Primates and their pinworm parasites: The Cameron hypothesis revisited. *Syst Biol* 48:523–546.

Hull, D. L. 1965. The effect of essentialism on taxonomy: Two thousand years of stasis (II). *Br J Philos Sci* 16:1–18.

———. 1976. Are species really individuals? *Syst Zool* 25:174–191.

———. 1978. A matter of individuality. *Philos Sci* 45:335–360.

———. 1980. Individuality and selection. *Annu Rev Ecol Syst* 11:311–332.

———. 1988. *Science as a process*. Chicago: Univ. Chicago Press.

———. 1999. On the plurality of species: Questioning the party line. In *Species: New interdisciplinary essays*, ed. R. A. Wilson, 23–48. Cambridge, MA: MIT Press.

Humphery-Smith, I. 1989. The evolution of phylogenetic specificity among parasitic organisms. *Parasitol Today* 5:385–387.

Humphries, C. J. 1983. Primary data in hybrid analysis. In *Advances in cladistics*, vol. 2, ed. N. I. Platnick and V. A. Funk, 89–103. New York: Columbia Univ. Press.

Humphries, C. J., and P. H. Williams. 1994. Cladograms and trees in biodiversity. In *Models in phylogeny reconstruction*, ed. R. W. Scotland, D. J. Siebert, and D. M. Williams, 335–352, Oxford: Clarendon Press.

Humphries, C. J., P. H. Williams, and R. I. Vane-Wright. 1995. Measuring biodiversity value for conservation. *Annu Rev Ecol Syst* 26:93–111.

Humphries, S., and G. D. Ruxton. 1999. Bower-building: Coevolution of display traits in response to the costs of female choice? *Ecology Letters* 2:404–413.

Hunt, J. H. 1999. Trait mapping and salience in the evolution of eusocial vespid wasps. *Evolution* 53:225–237.

Hunt, R. M. J., X. Xiang-Xu, and J. Kaufan. 1983. Miocene burrows of extinct bear dogs: Indication of early denning behavior of large mammalian carnivores. *Science* 221:364–366.

Hunter, J. P. 1998. Key innovations and the ecology of macroevolution. *Trends Ecol Evol* 13:31–36.

Hurlbert, S. H. 1971. The nonconcept of species diversity: A critique and alternative parameters. *Ecology* 52:577–586.

Hutchinson, G. E. 1959. Homage to Santa Rosalia, or why are there so many kinds of animals? *Am Nat* 93:145–159.

Huxley, J. S. 1938. Darwin's theory of sexual selection and the data subsumed by it, in light of recent research. *Am Nat* 72:416–433.

———, ed. 1940. *The new systematics*. Oxford: Oxford Univ. Press.

———. 1942. *Evolution, the modern synthesis*. London: Allen and Unwin.

————. 1953. *Evolution in action.* New York: Harper.

Hwang, D. 1992. [A discussion on outgroup analysis in cladistics.] *Sinozoologia* 4:149–157. (In Chinese.)

Ihering, H. von. 1891. On the ancient relations between New Zealand and South America. *Trans Proc NZ Inst* 24:431–445.

————. 1902. Die Helminthen als Hilfsmittel der zoogeographischen Forschung. *Zool Anz* (Leipzig) 26:42–51.

Inglis, W. G. 1961. The oxyurid parasites (Nematodes) of primates. *Proc Zool Soc Lond* 136:103–122.

Irschick, D. J., and J. B. Losos. 1998. A comparative analysis of the ecological significance of maximal locomotor performance in Caribbean *Anolis* lizards. *Evolution* 52:219–226.

Irschick, D. J., L. J. Vitt, P. A. Zani, and J. B. Losos. 1997. A comparison of evolutionary radiations in mainland and Caribbean *Anolis* lizards. *Ecology* 78:2191–2203.

Iwaniuk, A. N., S. M. Pellis, and I. Q. Whishaw. 2000. The relative importance of body size, phylogeny, locomotion, and diet in the evolution of forelimb dexterity in fissiped carnivores (Carnivora). *Can J Zool* 78:1110–1125.

Izsak, J., and L. Papp. 2000. A link between ecological diversity indices and measures of biodiversity. *Ecol Model* 130:151–156.

Jablonski, D. 1986a. Causes and consequences of mass extinctions: A comparative approach. In *Dynamics of extinction,* ed. D. K. Elliott, 183–229. New York: Wiley.

————.1986b. Background and mass extinctions: The alternation of macroevolutionary regimes. *Science* 231:129–133.

————.1986c. Larval ecology and macroevolution in marine invertebrates. *Bull Mar Sci* 39:565–587.

————. 1989. The biology of mass extinction: A palaeontological view. *Phil Trans Roy Soc Lond* B 325:357–368.

————. 1991. Extinctions: A paleontological perspective. *Science* 253:754–757.

————. 1995. Extinctions in the fossil record. In *Extinction rates,* ed. J. H. Lawton and R. M. May, 25–44. Oxford: Oxford Univ. Press.

Jablonski, D., and D. J. Bottjer. 1990a. The origin and diversification of major groups: Environmental patterns and macroevolutionary lags. In *Major evolutionary radiations,* ed. P. D. Taylor and G. P. Larwood, 17–57. Oxford: Clarendon Press.

————. 1990b. The ecology of evolutionary innovation: The fossil record. In *Evolutionary innovations,* ed. M. H. Nitecki, 253–288. Chicago: Univ. Chicago Press.

Jablonski, D., D. H. Erwin, and J. H. Lipps, eds. 1996. *Evolutionary paleobiology.* Chicago: Univ. Chicago Press.

Jablonski, D., and J. J. Sepkoski, Jr. 1996. Paleobiology, community ecology, and scales of ecological pattern. *Ecology* 77:1367–1378.

Jaccard, P. 1908. Nouvelle recherches sur la distribution florale. *Bull Soc Vaud Sci Nat.* 44: 223–270.

Jackman, T. R., A. Larson, K. de Quieroz, and J. B. Losos. 1999. Phylogenetic relationships and tempos of early diversification in *Anolis* lizards. *Syst Biol* 48:254–285.

Jackson, J. B. C., and A. H. Cheetham. 1990. Evolutionary significance of morphospecies: A test with cheilostome Bryozoa. *Science* 248:579–583.

————. 1994. Phylogeny reconstruction and the tempo of speciation in cheilostome Bryozoa. *Paleobiology* 20:407–423.

Jackson, K., D. G. Butler, and D. R. Brooks. 1996. Habitat and phylogeny influence salinity discrimination in crocodilians: Implications for osmoregulatory physiology and historical biogeography. *Biol J Linn Soc* 58:371–383.

Jacobs, D. K., and C. G. Wray. 1995. Developmental gene trees and phylogenies that include developmental gene-sequences: Phylogeny character tracing and functional evolution. *Dev Biol* 170:743–743.

Jaenike, J. 1990. Host specialization in phytophagous insects. *Annu Rev Ecol Syst* 21:243–273.

Jaksic, F. M., H. Núñez, and F. P. Ojeda. 1980. Body proportions, microhabitat selection, and adaptive radiation of *Liolaemus* lizards in central Chile. *Oecologia* 45:178–181.

Janz, N. 1999. *Ecology and evolution of butterfly host plant range.* Stockholm: Jannes Snabbtryck AB.

Janz, N., and S. Nylin. 1997. The role of female search behaviour in determining host plant range in plant feeding insects: A test of the information processing hypothesis. *Proc Roy Soc Lond* B 264:701–707.

———. 1998. Butterflies and plants: A phylogenetic study. *Evolution* 52:486–502.

Janz, N., S. Nylin, and K. Bybom. 2001. Evolutionary dynamics of host plant specialization: A case study. *Evolution* 55:783–796.

Janzen, D. H. 1968. Host plants as islands in evolutionary and contemporary time. *Am Nat* 102:592–595.

———. 1973a. Host plants as islands: Competition in evolutionary and contemporary time. *Am Nat* 107:786–790.

———. 1973b. Comments on host-specificity of tropical herbivores and its relevance to species richness. In *Taxonomy and ecology,* ed. V. H. Heywood, 201–211. New York: Academic Press.

———. 1980. When is it coevolution? *Evolution* 34:611–612.

———. 1981. Patterns of herbivory in a tropical deciduous forest. *Biotropics* 13:271–282.

———. 1983. Dispersal of seeds by vertebrate guts. In *Coevolution,* ed. D. J. Futuyma and M. Slatkin, 232–262. Sunderland, MA: Sinauer.

———. 1985a. Coevolution as a process: What parasites of animals and plants do NOT have in common. In *Coevolution of parasitic arthropods and mammals,* ed. K. C. Kim, 83–99. New York: Wiley and Sons.

———. 1985b. On ecological fitting. *Oikos* 45:308–310.

———. 1985c. A host plant is more than its chemistry. *Bull IL Nat Hist Surv* 33:141–174.

———. 1988. Ecological characterization of a Costa Rican dry forest caterpillar fauna. *Biotropica* 20:120–135.

———. 1991. How to save tropical biodiversity. *Am Entomol* 37:159–171.

———. 1992. A south-north perspective on science in the management, use, and economic development of biodiversity. In *Conservation of biodiversity for sustainable development,* ed. O. T. Sandlund, K. Hindar, and A. H. D. Brown, 27–52. Oslo: Scandinavian Univ. Press.

———. 1993. Taxonomy: Universal and essential infrastructure for development and management of tropical wildland biodiversity. In *Proceedings of the Norway/UNEP Expert Conference on Biodiversity, Trondheim, Norway,* ed. O. T. Sandlund and P. J. Schei, 100–113. Trondheim, Norway: NINA.

———. 1994a. Priorities in tropical biology. *Trends Ecol Evol* 9:365–367.

———. 1994b. Wildland biodiversity management in the tropics: Where are we now and where are we going? *Vida Silvestre Neotrop* 3:3–15.

———. 1996. On the importance of systematic biology in biodiversity development. *Assoc Syst Coll Newsletter* 24:17, 23–28.

———. 1997a. Wildland biodiversity management in the tropics. In *Biodiversity II: Understanding and protecting our biological resources,* ed. M. L. Reaka-Kudla, D. E. Wilson, and E. O. Wilson, 411–434. Washington, DC: Joseph Henry Press.

———. 1997b. How to grow a wildland: The gardenification of nature. *Insect Sci Appl* 17:269–276.

———. 1997c. Causes and consequences of biodiversity loss: Liquidation of natural capital and biodiversity resource development in Costa Rica. In *Biodiversity and human health,* eds. F. Grifo and J. Rosenthal, 302–311. Washington, DC: Island Press.

———. 1998. Gardenification of wildland nature and the human footprint. *Science* 279:1312–1313.

———. 1999. Gardenification of tropical conserved wildlands: Multitasking, multicropping, and multiusers. *Proc US Natl Acad Sci* 96:5987–5994.

———. 2000a. How to grow a wildland: The gardenification of nature. In *Nature and human society,* ed. P. H. Raven and T. Williams, 521–529. Washington, DC: National Academy of Science Press.

———. 2000b. Costa Rica's Area de Conservación Guanacaste: A long march to survival through non-damaging bio-development. *Biodiversity* 1:7–20.

Janzen, D. H., and R. Gamez. 1997. Assessing information needs for sustainable use and conservation of biodiversity. In *Biodiversity information: Needs and options,* ed. D. I. Hawksworth, P. M. Kirk, and S. D. Clarke, 21–29. Wallingford, Oxon, UK: Oxford, CAB International.

Janzen, D. H., and W. Hallwachs. 1994. Ethical aspects of the impact of humans on biodiversity. In *Man and his environment: Tropical forests and the conservation of species*, ed. G. B. Marini-Bottolo, 227–255. Pontificiae Academiae Scientarum Scriptum Varia no. 84. Vatican: Pontificiae Academiae Scientarum.

Janzen, D. H., and P. S. Martin. 1982. Neotropical anachronisms: The fruits the Gomphotheres ate. *Science* 215:19–27.

Jennings, S., J. D. Reynolds, and N. V. C. Polunin. 1999. Predicting the vulnerability of tropical reef fishes to exploitation with phylogenies and life histories. *Conserv Biol* 13:1466–1475.

Jensen, J. S. 1990. Plausibility and testability: Assessing the consequences of evolutionary innovation. In *Evolutionary innovations,* ed. M. H. Nitecki, 171–190. Chicago: Univ. Chicago Press.

Jensen, K. R. 1997. Evolution of the *Sacoglossa* (Mollusca: Opishtobranchia) and the ecological associations with their food plants. *Evol Ecol* 11:301–335.

Jermiin, L. S., G. J. Olsen, K. L. Mengersen, and S. Easteal. 1997. Majority-rule consensus of phylogenetic trees obtained by maximum-likelihood analysis. *Mol Biol Evol* 14:1296–1302.

Jermy, T. 1976. Insect-host plant relationships: Coevolution or sequential evolution? *Symp Biol Hung* 16:109–113.

———. 1984. Evolution of insect/host plant relationships. *Am Nat* 124:609–630.

———. 1988. Can predation lead to narrow food specialization in phytophagous insects? *Ecology* 69:902–904.

Jervis, M. A. 1997. Parasitoids as limiting and selective factors: Can biological control be evolutionarily stable? *Trends Ecol Evol* 12:378–380.

Jiggins, C. D., and J. Mallet. 2000. Bimodal hybrid zones and speciation. *Trends Ecol Evol* 15:250–255.

Jiggins, C. D., and W. O. McMillan. 1997. The genetic basis of an adaptive radiation: Warning colour in two *Heliconius* species. *Proc Roy Soc Lond* B 264:1167–1175.

Jiggins, C. D., W. O. McMillan, and J. Mallet. 1997. Host plant adaptation has not played a role in the recent speciation of *Heliconius himera* and *Heliconius erato. Ecol Entomol* 22:361–365.

Johns, G. C., and J. C. Avise. 1998. Tests for ancient species flocks based on molecular phylogenetic appraisals of *Sebastes* rockfishes and other marine fishes. *Evolution* 52:1135–1146.

Johnson, C. N. 1998. Species extinction and the relationship between distribution and abundance. *Nature* 394:272–274.

Johnson, K. G., A. F. Budd, and T. A. Stemann. 1995. Extinction selectivity and ecology of Neogene Caribbean reef corals. *Paleobiology* 21:52–73.

Johnson, K. P., F. McKinney, R. Wilson, and M. D. Sorenson. 2000. The evolution of postcopulatory displays in dabbling ducks (Anatini): A phylogenetic perspective. *Anim Behav* 59:953–963.

Johnson, P. A., F. C. Hoppensteadt, J. L. Smith, and G. L. Bush. 1996. Conditions for sympatric speciation: A diploid model incorporating habitat fidelity and non-habitat assortative mating. *Evol Ecol* 10:187–205.

Johnson, S. D., H. D. Linder, and K. E. Steiner. 1998. Phylogeny and radiation of pollination systems in *Disa* (Orchidaceae). *Am J Bot* 85:402–411.

Johnson, S. D., and K. E. Steiner. 2000. Generalization versus specialization in plant pollination systems. *Trends Ecol Evol* 15:140–143.

Johnson, T. C., K. Kelts, and E. Odada. 2000. The Holocene history of Lake Victoria. *Ambio* 29:2–11.

Johnston, M. R., R. E. Torbachnik, and F. L. Bookstein. 1991. Landmark-based morphometric of spiral accretionary growth. *Paleobiology* 17:19–36.

Johnston, S. J. 1912. On some trematode parasites of Australian frogs. *Proc Linn Soc NSW* 37:285–362.

———. 1914. Australian trematodes and cestodes. *Med J Austral* 1:243–244.

———. 1916. On the trematodes of Australian birds. *J Proc Roy Soc NSW* 50:187–261.

Johst, K., and R. Brandl. 1997. Body size and extinction risk in a stochastic environment. *Oikos* 78:612–617.

Jones, R., D. C. Culver, and T. C. Kane. 1992. Are parallel morphologies of cave organisms the result of similar selection pressures? *Evolution* 46:353–365.

Jones, R. S. 1968. Ecological relationships in Hawaiian and Johnston Island Acanthuridae (Surgeonfishes). *Micronesica* 4:310–361.

Jones, T. 1995. Down in the woods they have no names: BioNet-International—Strengthening systematics in developing countries. *Biodiv Conserv* 4:501–509.

Jordal, B. H., B. B. Normark, and B. D. Farrell. 2000. Evolutionary radiation of an inbreeding haplodiploid beetle lineage (Curculionidae, Scolytinae). *Biol J Linn Soc* 71:483–499.

Jordan, D. S., and V. Kellogg. 1908. The law of geminate species. *Am Nat* 42:73–80.

Jordano, P. 1995. Angiosperm fleshy fruits and seed dispersers: A comparative analysis of adaptation and constraints in plant-animal interactions. *Am Nat* 145:163–191.

Joron, M., and J. L. B. Mallet. 1998. Diversity in mimicry: Paradox or paradigm? *Trends Ecol Evol* 13:461–466.

Jousson, O., P. Bartoli, and J. Pawlowski. 2000. Cryptic speciation among intestinal parasites (Trematoda: Digenea) infecting sympatric host fishes (Sparidae). *J Evol Biol* 13:778–785.

Joysey, K. A., and A. E. Friday, eds. 1992. *Problems of phylogenetic reconstruction*. London: Academic Press.

Juberthie, C. 1984. La colonisation du milieu souterrain: Théories et méthodes, relations avec la spéciation et l'évolution souterraine. *Mem Biospeleol* 11:65–102.

Kafkas, S., and R. Perl-Treves. 2001. Morphological and molecular phylogeny of *Pistachia* species in Turkey. *Theor Appl Genet* 102:908–915.

Kaiser, J. 2000. Ecology: Does biodiversity help fend off invaders? *Science* 288:785–786.

Kalko, E. K. V., and M. A. Condon. 1998. Echolocation, olfaction, and fruit display: How bats find fruit of flagellichorous cucurbits. *Funct Ecol* 12:364–372.

Kampny, C. M., and N. G. Dengler. 1997. Evolution of flower shape in Veroniceae (Scrophulariaceae). *Plant Syst Evol* 205:1–25.

Kaneshiro, K. Y. 1976. Ethological isolation and phylogeny in the *plantibia* subgroup of Hawaiian *Drosophila*. *Evolution* 30:740–745.

———. 1980. Sexual isolation, speciation, and the direction of evolution. *Evolution* 34:437–444.

Kaneshiro, K. Y., and L. V. Giddings. 1987. The significance of asymmetrical sexual isolation and the formation of new species. *Evol Biol* 21:29–43.

Kardon, G. 1998. Evidence from the fossil record of an antipredatory exaptation: Conchiolin layers in corbulid bivalves. *Evolution* 52:68–79.

Karlsson, B. 1995. Resource allocation and mating systems in butterflies. *Evolution* 49:955–961.

Karr, J. R. 1990. Biological integrity and the goal of environmental legislation: Lessons for conservation biology. *Conserv Biol* 4:244–250.

Katinas, L., and J. V. Crisci. 2000. Cladistic and biogeographic analyses of the genera *Moscharia* and *Polyachyrus* (Asteraceae, Mutisieae). *Syst Bot* 25:33–46.

Kaufman, L. 1992. Catastrophic change in species-rich freshwater ecosystems: The lessons of Lake Victoria. *BioScience* 42:846–858.

Kavanaugh, D. H. 1972. Hennig's principles and methods of phylogenetic systematics. *Biologist* 54:115–127.

Kawecki, T. J. 1997. Sympatric speciation via habitat specialization driven by deleterious mutations. *Evolution* 51:1751–1763.

Kearn, G. C. 1986. The eggs of monogeneans. *Adv Parasitol* 25:175–273.

Keast, A. 1961. Bird speciation on the Australian continent. *Bull Mus Comp Zool* 123:305–495.

Keen, W. H. 1982. Habitat selection and interspecific competition in two species of plethodontid salamanders. *Ecology* 63:94–102.

Kéler, S. 1938. Zur Geschichte der Mallophagen-forschung. *Z Parasitenkd* 10:31–66.

———. 1939. Zur Kenntnis der Mallophagen-Fauna Polens: 2. Beitrag. *Z Parasitenkd* 11: 47–57.

Kelley, S. T., and B. D. Farrell. 1998. Is specialization a dead end? The phylogeny of host use in *Dendroctonus* bark beetles (Scolytidae). *Evolution* 52:1731–1743.

Kellogg, E. A. 2000. The grasses: A study in macroevolution. *Annu Rev Ecol Syst* 31:217–238.

Kellogg, V. L. 1896. New Mallophaga, 1. With special reference to a collection from maritime birds of the Bay of Monterey, California. *Proc Cal Acad Sci* 6:31–168.

———. 1906. Notes and literature: Notes on evolution. *Am Nat* 43:379–383.

———. 1913. Distribution and species-forming of ectoparasites. *Am Nat* 47:129–158.

Kelly, C. K., and A. Purvis. 1993. Seed size and establishment in tropical trees: On the use of taxonomic relatedness in determining ecological patterns. *Oecologia* 94:356–360.

Kelly, J. K. 1999. Response to selection in partially self-fertilizing populations: II. Selection on multiple traits. *Evolution* 53:350–357.

Kelt, D. A. 1999. On the relative importance of history and ecology in structuring communities of desert small mammals. *Ecography* 22:123–137.

Kelt, D. A., and J. H. Brown. 2000. Species as units of analysis in ecology and biogeography: Are the blind leading the blind? *Global Ecol Biogeogr* 9:213–217.

Kendrick, P., and P. R. Crane. 1997. *The origin and early diversification of land plants: A cladistic study.* Washington, DC: Smithsonian Institution Press.

Kennedy, M., H. G. Spencer, and R. D. Gray. 1996. Hop, step, and gape: Do the social displays of the Pelecaniformes reflect phylogeny? *Anim Behav* 51:273–291.

Kerdelhue, C., I. La Clainche, and J. Y. Rasplus. 1999. Molecular phylogeny of the *Ceratosolen* species pollinating *Ficus* of the subgenus *Sycomorus sensu stricto:* Biogeographical history and origins of the species specificity breakdown cases. *Mol Phylogenet Evol* 11:401–414.

Kethley, J. B., and D. E. Johnston. 1975. Resource tracking patterns in bird and mammal ectoparasites. *Misc Publ Entomol Soc Am* 9:231–236.

Ketterson, E. D., and V. Nolan Jr. 1999. Adaptation, exaptation, and constraint: A hormonal perspective. *Am Nat* 154:S4–S25.

Key, K. H. L. 1968. The concept of stasipatric speciation. *Syst Zool* 17:14–22.

Kiester, A. R. 1997a. Aesthetics of biological diversity. *Hum Ecol Forum* 3:151–157.

———. 1997b. Response to Ribe and Levine. *Hum Ecol Forum* 3:163.

Kiltie, R. A. 1984. Size ratios among sympatric neotropical cats. *Oecologia* 61:411–416.

———. 1988. Interspecific size regularities in tropical felid assemblages. *Oecologia* 76:97–105.

Kim, J. 1993. Improving the accuracy of phylogenetic estimation by combining different methods. *Syst Biol* 42:331–340.

Kim, K. C., ed. 1985. *Coevolution of parasitic arthropods and mammals.* New York: Wiley InterScience.

Kime, N. M., A. S. Rand, M. Kapfer, and M. J. Ryan. 1998. Consistency of female choice in the túngara frog: A permissive preference for complex characters. *Anim Behav* 55:641–649.

Kimura, M. 1980. A simple model for estimating evolutionary rate of base substitutions through comparative studies of nucleotide sequences. *J Mol Evol* 16:111–120.

King, C. E. 1964. Relative abundance of species and MacArthur's model. *Ecology* 45:716–727.

King, D. G. 1991. The origin of an organ: Phylogenetic analysis of evolutionary innovation in the digestive tract of flies (Insecta: Diptera). *Evolution* 45:568–588.

King, T. L., E. C. Pendleton, and R. F. Villela. 1998. Gene conservation: Management and evolutionary units in freshwater bivalve management: Introduction to the proceedings. *J Shellfish Res* 17:1351–1353.

Kingsland, S. 1985. *Modelling nature.* Chicago: Univ. Chicago Press.

Kingsolver, J. G. 1983. Thermoregulation and flight in *Colias* butterflies: Elevational patterns and mechanistic limitations. *Ecology* 64:534–545.

Kirkpatrick, M. 1982. Sexual selection and the evolution of female choice. *Evolution* 36:1–12.

———. 2000. Reinforcement and divergence under assortative mating. *Proc Roy Soc Lond* B 267:1649–1655.

———. 2001. Reinforcement during ecological speciation. *Proc Roy Soc Lond* B 268:1259–1263.

Kirkpatrick, M., and D. Lofsvold. 1992. Measuring selection and constraint in the evolution of growth. *Evolution* 46:954–971.

Kirkpatrick, M., and M. Slatkin. 1993. Searching for evolutionary patterns in the shape of phylogenetic trees. *Evolution* 47:1171–1181.

Kishino, H., and M. Kasegawa. 1989. Evaluation of the maximum likelihood estimates of the evolutionary tree topologies from sequence data, and the branching order in Hominoidea. *J Mol Evol* 29:170–179.

Kitchell, J. A., D. L. Clark, and A. M. Gombos, Jr. 1986. Biological selectivity of extinction: A link between background and mass extinction. *Palaios* 1:504–511.

Klassen, G. J., R. D. Mooi, and A. Locke. 1991. Consistency indices and random data. *Syst Zool* 40:446–457.

Kleiman, D. G. 1980. The sociobiology of captive propagation. In *Conservation biology*, ed. M. E. Soulé and B. A. Wilcox, 243–261. Sunderland, MA: Sinauer.

Klein, N. K., and W. M. Brown. 1994. Intraspecific molecular phylogeny in the yellow warbler (*Dendroica petechia*), and implications for avian biogeography in the West Indies. *Evolution* 48:1914–1932.

Klein, N. K., and R. B. Payne. 1998. Evolutionary associations of brood parasitic finches (*Vidua*) and their host species: Analyses of mitochondrial DNA restriction sites. *Evolution* 52:566–582.

Klicka, J., and R. M. Zink. 1997. The importance of recent ice ages in speciation: A failed paradigm. *Science* 277:1666–1669.

Klompen, J. S. H., and B. M. O'Connor. 1989. Ontogenetic patterns and phylogenetic analysis in Acari. In *Le concept de stase et l'ontogenèse des Arthropodes*, ed. H. M. André and J.-Cl. Lions, 91–103. Wavre, Belgium: AGAR.

Kluge, A. G. 1985. Ontogeny and phylogenetic systematics. *Cladistics* 1:13–27.

———. 1988. Parsimony in vicariance biogeography: A quantitative method and a Greater Antillean example. *Syst Zool* 37:315–328.

———. 1989. A concern for evidence and a phylogenetic hypothesis of relationships among *Epicrates* (Boidae, Serpentes). *Syst Zool* 38:315–328.

———. 1990. Species as historical individuals. *Biol Philos* 5:417–431.

———. 1991. Boine snake phylogeny and research cycles. *Misc Publ Mus Zool Univ Mich* 178:1–58.

———. 1997. Testability and the refutation and corroboration of cladistic hypotheses. *Cladistics* 13:81–96.

———. 1998a. Total evidence or taxonomic congruence: Cladistics or consensus classification. *Cladistics* 14:151–158.

———. 1998b. Sophisticated falsification and research cycles: Consequences for differential character weighting in phylogenetic systematics. *Zool Scr* 26:349–360.

———. 1999. The science of phylogenetic systematics: Explanation, prediction, and test. *Cladistics* 15:429–436.

———. 2001. Philosophical conjectures and their refutation. *Syst Biol* 50:322–330.

Kluge, A. G., and J. S. Farris. 1969. Quantitative phyletics and the evolution of anurans. *Syst Zool* 18:1–32.

Kluge, A. G., and A. J. Wolf. 1993. Cladistics: What's in a word? *Cladistics* 9:183–199.

Knight, A., L. D. Densmore, and E. D. Rael. 1992. Molecular systematics of the *Agkistrodon* complex. In *Biology of the pitvipers*, ed. J. A. Campbell and E. D. Brodie Jr., 49–70. Tyler, TX: Selva.

Knight, J. B., L. R. Cox, A. M. Keen, A. G. Smith, R. L. Batten, E. L. Yochelson, N. H. Ludbrook, R. Robertson, C. M. Yonge, and R. C. Moore. 1960. Mollusca I. In *Treatise on invertebrate paleontology*, ed. R. C. Moore and C. W. Pitrat, 12–1351. Boulder, CO: Geological Society of America; Lawrence: Univ. Kansas Press.

Knight, M. E., G. F. Turner, C. Rico, M. J. H. van Oppen, and G. M. Hewitt. 1998. Microsatellite paternity analysis on captive Lake Malawi cichlids supports reproductive isolation by direct mate choice. *Mol Ecol* 7:1605–1610.

Kniskern, J., and M. D. Rausher. 2001. Two modes of host enemy coevolution. *Pop Ecol* 43:3–14.

Knowles. L. L., D. J. Futuyma, W. F. Eanes, and B. Rannala. 1999. Insight into speciation from historical demography in the phytophagous beetle genus *Ophraella*. *Evolution* 53:1846–1856.

Kodric-Brown, A., and J. H. Brown. 1984. Truth in advertising: The kinds of traits favored by sexual selection. *Am Nat* 124:309–323.

Koella, J. C., and O. Restif. 2001. Coevolution of parasite virulence and host life history. *Ecol Lett* 4:207–214.

Köhler, T., and F. Vollrath. 1995. Thread biomechanics in the two orb-weaving spiders, *Aranues diadematus* (Araneae, Araneidae) and *Uloborus walckenaerius* (Araneae, Uloboridae). *J Exp Zool* 271:1–17.

Kohn, A. J. 1959. The ecology of *Conus* in Hawaii. *Ecol Monogr* 29:47–90.

Kondrashov, A. S., and M. V. Mina. 1986. Sympatric speciation: When is it possible? *Biol J Linn Soc* 27:201–223.

Kool, S. P. 1987. Significance of radular characters in reconstruction of thaidid phylogeny (Neogastropoda: Muricacea). *Nautilus* 101:117–132.

Köpf, A., N. E. Rank, H. Roininen, R. Julkunen-Tiitto, J. M. Pasteels, and J. Tahvanainen. 1998. The evolution of host-plant use and sequestration in the leaf beetle *Phratora* (Coleoptera: Chrysomelidae). *Evolution* 52:517–528.

Kornet, D. J. 1993a. Permanent splits as speciation events: A formal reconstruction of the internodal species concept. *J Theor Biol* 164:407–435.

———. 1993b. Reconstructing species: Demarcations in genealogical networks. Ph.D. diss., Leiden Univ., Leiden.

Kornet, D. J., and J. W. McAllister. 1993. The composite species concept. In *Reconstructing species: Demarcations in genealogical networks*, by D. J. Kornet, 61–89. Ph.D. diss., Leiden Univ., Leiden.

Kornet, D. J., J. A. J. Metz, and H. A. J. M. Schellinx. 1995. Internodons as equivalence classes in the genealogical network: Building-blocks for a rigorous species concept. *J Math Biol* 34:110–122.

Kornet, D. J., and H. Turner. 1999. Coding polymorphism for phylogeny reconstruction. *Syst Biol* 48:365–379.

Kornfield, I., and P. E. Smith. 2000. African Cichlid fishes: Model systems for evolutionary biology. *Annu Rev Ecol Syst* 31:163–196.

Kraak, S. B. M., T. C. M. Bakker, and B. Mundwiler. 1999. Sexual selection in sticklebacks in the field: Correlates of reproductive, mating, and paternal success. *Behav Ecol* 10:696–706.

Kraak, S. B. H., B. Mundwiler, and P. J. B. Hart. 2001. Increased number of hybrids between benthic and limnetic three-spined sticklebacks in Enos Lake, Canada; the collapse of a species pair? *J Fish Biol* 58:1458–1464.

Krajewski, C. 1994. Phylogenetic measures of biodiversity: A comparison and critique. *Biol. Conserv* 69:33–39.

Kramer, E. M., and V. F. Irish. 2000. Evolution of the petal and stamen developmental programs: Evidence from comparative studies of the lower eudicots and basal angiosperms. *Int J Plant Sci* 161:S29–S40.

Krenn, H. W., and N. P. Kristensen. 2000. Early evolution of the proboscis of Lepidoptera (Insecta): External morphology of the galea in basal glossatan moth lineages, with remarks on the origin of the pilifers. *Zool Anz* 239:179–196.

Kristoffersen, M. L. 1994. An overview of cladistic applications. *Rev Nordestina Biol* 9:133–141.

Kroll, J. S., and E. R. Moxon. 1990. Capsulation in distantly related strains of *Haemophilus* influenzae type b: Genetic drift and gene transfer at the capsulation locus. *J Bact* 172:1374–1379.

Kuhn, H. J. 1967. Parasites and phylogeny of the catarrhine primates. In *Taxonomy and phylogeny of the Old World primates with references to the origin of man*, ed. B. Chiarelli, 187–195. Torino: Rosenberg and Sellier.

Kuris, A. M., and S. F. Norton. 1985. Evolutionary importance of overspecialization: Insect parasitoids as an example. *Am Nat* 126:387–391.

Kusmierski, R., G. Borgia, R. H. Crozier, and B. H. Y. Chan. 1993. Molecular information on bowerbird phylogeny and the evolution of exaggerated male characteristics. *J Evol Biol* 6:737–752.

Kusmierski, R., G. Borgia, A. Uy, and R. H. Crozier. 1997. Labile evolution of display traits in bowerbirds indicates reduced effects of phylogenetic constraints. *Proc Royal Soc Lond* B 264:307–313.

Lack, D. 1947. *Darwin's finches*. Cambridge: Cambridge Univ. Press.

Laerm, J. 1974. A functional analysis of morphological variation and differential niche utilization in basilisk lizards. *Ecology* 55:404–411.

Lake, J. A. 1991. The order of sequence alignment can bias the selection of tree topology. *Mol Biol Evol* 8:378–385.

Lamarck, J.-B. 1809. *Philosophie zoologique, ou exposition des considérations relatives à l'histoire naturelle des animaux: A la diversité de leur organization et des facultés qu'ils en ob-*

tiennent; aux causes physiques qui maintiennnent en eux la vie et donnent lieu aux mouve-ments qu'ils exécutent; enfin, à celles produisent, les unes le sentiment, et les autres l'intelligence de ceux qui en sont doués. Paris: Dentu.

Lande, R. 1978. Evolutionary mechanisms of limb loss in tetrapods. *Evolution* 32:73–92.

———. 1980. Genetic variation and phenotypic evolution during allopatric speciation. *Am Nat* 116:463–479.

———. 1981. Models of speciation by sexual selection on polygenic traits. *Proc Natl Acad Sci* (USA) 78:3721–3725.

———. 1982. Rapid origin of sexual isolation and character divergence in a cline. *Evolution* 36:213–223.

———. 1986. The dynamics of peak shifts and the pattern of morphological evolution. *Paleobiology* 12:343–354.

———. 1993. Risks of population extinction from demographic and environmental stochasticity, and random catastrophes. *Am Nat* 142:911–927.

Lande, R., and M. Kirkpatrick. 1988. Ecological speciation by sexual selection. *J Theor Biol* 133:85–98.

Langtimm, C. A., and D. A. Dewsbury. 1991. Phylogeny and evolution of rodent copulatory behaviour. *Anim Behav* 41:217–225.

Lanyon, S. M. 1994. Polyphyly of the blackbird genus *Agelaius* and the importance of assumptions of monophyly in comparative studies. *Evolution* 48:679–693.

LaPointe, F.-J., and G. Cucumel. 1997. The average consensus procedure: Combination of weighted trees containing identical or overlapping sets of taxa. *Syst Biol* 46: 306–312.

Lara, M. C., and J. L. Patton. 2000. Evolutionary diversification of spiny rats (genus *Triomys*, Rodentia: Echimyidae) in the Atlantic Forest of Brazil. *Zool J Linn Soc* 130: 661–686.

Larson, A. 1994. The comparison of morphological and molecular data in phylogenetic systematics. In *Molecular ecology and evolution: Approaches and applications*, ed. B. Schierwater, B. Streit, G. P. Wagner, and R. DeSalle, 371–390. Basel: Burkhäuser Verlag.

Larson, A., D. B. Wake, L. R. Maxson, and R. Highton. 1981. A molecular phylogenetic perspective on the origins of morphological novelties in the salamanders of the Tribe Plethodontini (Amphibia, Plethodontidae). *Evolution* 35:405–422.

Larson, G. A. 1976. Social behaviour and feeding ability of two phenotypes of *Gasterosteus aculeatus* in relation to their spatial and trophic segregation in a temperate lake. *Can J Zool* 54:107–121.

Latta, R. G., and J. B. Mitton. 1999. Historical separation and present gene flow through a zone of secondary contact in Ponderosa pine. *Evolution* 53:769–776.

Lauder, G. V. 1980. Evolution of the feeding mechanism in primitive actinopterygian fishes: A functional anatomical analysis of *Polypterus, Lepisosteus,* and *Amia. J Morphol* 163:283–317.

———. 1981. Form and function: Structural analysis in evolutionary biology. *Paleobiology* 7:430–442.

———. 1982a. Historical biology and the problem of design. *J Theor Biol* 97:57–67.

———. 1982b. Patterns of evolution in the feeding mechanism of actinopterygian fishes. *Am Zool* 22:275–285.

———. 1983a. Neuromuscular patterns and the origin of trophic specialization in fishes. *Science* 219:1235–1237.

———. 1983b. Functional and morphological bases of trophic specialization in sunfishes (Teleostei: Centrarchidae). *J Morphol* 178:1–21.

———. 1985a. Aquatic feeding in lower vertebrates. In *Functional vertebrate morphology*, ed. M. Hildebrand, D. Bramble, K. F. Liem, and D. B. Wake, 210–229. Cambridge, MA: Harvard Univ. Press.

———. 1985b. Functional morphology of the feeding mechanism in lower vertebrates. In *Functional morphology of vertebrates*, ed. H.-R. Duncker and G. Fleischer, 179–188. New York: Springer Verlag.

———. 1986. Homology, analogy, and the evolution of behavior. In *Evolution of animal behavior*, ed. M. H. Nitecki and J. A. Kitchell, 9–40. Oxford: Oxford Univ. Press.

———. 1988. Phylogeny and physiology. *Evolution* 42:1113–1114.

———. 1989a. Caudal fin locomotion in ray-finned fishes: Historical and functional analyses. *Am Zool* 29:85–102.

———. 1989b. Review of *Genetics, paleontology, and macroevolution*, by J. S. Levinton. *J Vert Paleo* 9:122–123.

———. 1990. Functional morphology and systematics: Studying functional patterns in an historical context. *Annu Rev Ecol Syst* 21:317–340.

———. 1991a. An evolutionary perspective on the concept of efficiency: How does function evolve? In *Efficiency and economy in animal physiology*, ed. R. W. Blake, 169–184. New York: Cambridge Univ. Press.

———. 1991b. Biomechanics and evolution: Integrating physical and historical biology in the study of complex systems. In *Biomechanics in evolution*, ed. J. M. V. Rayner and R. J. Wootton, 1–19. Cambridge: Cambridge Univ. Press.

———. 1996. The argument from design. In *Adaptation*, ed. M. R. Rose and G. V. Lauder, 55–91. New York: Academic Press.

———. 2000. Function of the caudal fin during locomotion in fishes: Kinematics, flow visualization, and evolutionary patterns. *Am Zool* 40:101–122.

Lauder, G. V., A. W. Crompton, C. Gans, J. Hanken, K. F. Liem, W. O. Maier, A. Meyer, R. Presley, O. C. Rieppel, G. Roth, D. Schluter, and G. A. Zweers. 1989. Group report. How are feeding systems integrated and how have evolutionary innovations been introduced? In *Complex organismal functions: Integration and evolution in vertebrates*, ed. D. B. Wake and G. Roth, 97–115. New York: Wiley.

Lauder, G. V., R. B. Huey, R. K. Monson, and R. J. Jensen. 1995. Systematics and the study of organismal form and function. *BioScience* 45:696–704.

Lauder, G. V., and K. F. Liem. 1989. The role of historical factors in the evolution of complex organismal functions. In *Complex organismal functions: Integration and evolution in vertebrates*, ed. D. B. Wake and G. Roth., 63–78. Dahlem Conferenzen: S. Bernhard.

Lauder, G. V., and H. B. Shaffer. 1985. Functional morphology of the feeding mechanism in aquatic ambystomatid salamanders. *J Morphol* 185:297–326.

———. 1993. Design of feeding systems in aquatic vertebrates: Major patterns and their evolutionary implications. In *The skull*, vol. 3: *Functional and evolutionary mechanisms*, ed. J. Hanken, and B. K. Hall, 113–149. Chicago: Univ. Chicago Press.

Lauder, G. V., and P. C. Wainwright. 1992. Functions and history: The pharyngeal jaw apparatus in primitive ray-finned fishes. In *Systematics, historical ecology, and North American freshwater fishes*, ed. R. Mayden, 455–471. Stanford, CA: Stanford Univ. Press.

LaVal, R. K., and H. S. Fitch. 1977. Structure, movements and reproduction in three Costa Rica bat communities. *Occas Pap Mus Nat Hist Univ Kansas* 69:1–28.

Lavin, P. A., and J. D. McPhail. 1985. The evolution of freshwater diversity in the threespine stickleback, *Gasterosteus aculeatus:* Site-specific differentiation of trophic morphology. *Can J Zool* 63:2632–2638.

———. 1986. Adaptive divergence of trophic phenotype among freshwater populations of the threespine stickleback *(Gasterosteus aculeatus)*. *Can J Fish Aquat Sci* 43:2455–2463.

———. 1987. Morphological divergence and the optimization of trophic characters among lacustrine populations of the threespine stickleback *(Gasterosteus aculeatus)*. *Can J Fish Aquat Sci* 44:1820–1829.

Lawlor, L. R., and J. Maynard Smith. 1976. The coevolution and stability of competing species. *Am Nat* 110:79–99.

Lawton, J. H., and D. R. Strong. 1981. Community patterns and competition in folivorous insects. *Am Nat* 118:317–338.

Lecointre, G., H. Philippe, H. L. Van Le, and H. LeGuayader. 1994. Species sampling has a major impact on phylogenetic inference. *Mol Phylogenet Evol* 2:205–224.

Lee, M. S. Y. 2000a. Tree robustness and clade significance. *Syst Biol* 49:829–836.

———. 2000b. Soft anatomy, diffuse homoplasy, and the relationships of lizards and snakes. *Zool Scripta* 29:101–130.

Lee, M. S. Y., and P. Doughty. 1997. The relationship between evolutionary theory and phylogenetic analysis. *Biol Rev* 72:471–495.

Lee, M. S. Y., and R. Shine. 1998. Reptilian viviparity and Dollo's Law. *Evolution* 52:1441–1450.

Leebens-Mack, J., Ø. Pellmyr, and M. Brock. 1998. Host specificity and the genetic structure of two Yucca moth species in a Yucca hybrid zone. *Evolution* 52:1376–1382.

Leigh, E. G., Jr. 1981. The average lifetime of a population in a varying environment.
J Theor Biol 90:213–239.

Leigh Brown, A. J., and E. C. Holmes. 1995. Evolutionary biology of human immunodeficiency virus. *Annu Rev Ecol Syst* 25:127–165.

Leishman, M. R., M. Westoby, and E. Jurado. 1995. Correlates of seed size variation: A
comparison of five temperate floras. *J Ecol* 83:517–530.

Lenski, R. E., and M. Travisano. 1994. Dynamics of adaptation and diversification: A
10,000 generation experiment with bacterial populations. *Proc Natl Acad Sci* (USA)
91:6808–6814.

Les, D. H., E. L. Schneider, D. J. Padgett, P. S. Soltis, D. E. Soltis, and M. Zanis. 1999. Phylogeny, classification, and floral evolution of water lilies (Nymphaeaceae; Nym-
phaeales): A synthesis of non-molecular, rbcL, matK, and 18SrDNA data. *Syst Bot*
24:28–46.

Leschen, R. A. B. 2000. Beetles feeding on bugs (Coleoptera, Hemiptera): Repeated shifts
from mycophagous ancestors. *Invert Taxon* 14:917–929.

Lessa, E. P., and R. A. Fariña. 1996. Reassessment of extinction patterns among the late
Pleistocene mammals of South America. *Palaeontology* 39:651–662.

Lessios, H. A., B. D. Kessing, D. R. Roberston, and G. Paulay. 1999. Phylogeography of
the pantropical sea urchin *Eucidaris* in relation to land barriers and ocean cur-
rents. *Evolution* 53:806–817.

Levin, D. A. 1979. The nature of plant species. *Science* 204:381–384.

———. 1984. Immigration in plants: An exercise in the subjunctive. In *Perspectives in
plant population ecology*, ed. R. Dirzo and J. Sarukhan, 242–260. Sunderland, MA:
Sinauer.

———. 1993. Local speciation in plants: The rule not the exception. *Syst Bot* 18:197–208.

Levine, N. D. 1988. *The protozoan phylum Apicomplexa*. Boca Raton, FL: CRC Press.

Levins, R., and R. H. MacArthur. 1969. A hypothesis to explain the incidence of monoph-
agy. *Ecology* 50:910–911.

Levinton, J. S. 1988. *Genetics, paleontology, and macroevolution*. Cambridge, UK: Cambridge Univ. Press.

Levitan, D. R. 2000. Optimal egg size in marine invertebrates: Theory and phylogenetic
analysis of the critical relationship between egg size and development time in
echinoids. *Am Nat* 156:175–192.

Lewis, P.O. 2001. Phylogenetics turns over a new leaf. *Trends Ecol Evol* 16:30–36.

Lewontin, R. C. 1978. Adaptation. *Sci Am* 239:212–230.

———. 1983. Gene, organism, and environment. In *Evolution from molecules to men*, ed.
D. S. Bendall, 273–285. Cambridge: Cambridge Univ. Press.

Lewy, Z., and C. Samtleben. 1979. Functional morphology and paleontological significance of the conchiolin layers in corbulid pelecypods. *Lethaia* 12:341–351.

Li, W.-H. 1997. *Molecular evolution*. Sunderland, MA: Sinauer.

Lieberman, B. S. 1993. Systematics and biogeography of the *"Metacryphaeus* Group" Calmoniidae (Trilobita, Devonian), with comments on adaptive radiations and the
geological history of the Malvinokaffric realm. *J Paleontol* 67:549–570.

———. 1994. Evolution of the trilobite subfamily Proetinae Salter, 1864, and the origin
diversification, evolutionary affinity, and extinction of the middle Devonian proe-
tid fauna of eastern North America. *Bull Am Mus Nat Hist* 223:1–176.

———. 1997. Early Cambrian paleogeography and tectonic history: A biogeographic approach. *Geology* 25:1039–1042.

———. 1999. Systematic revision of the Olenelloidea (Trilobita, Cambrian). *Bull Peabody
Mus* 45:1–150.

———. 2000. *Paleobiogeography: Using fossils to study global changes, plate tectonics, and evolution*. Topics in geobiology no. 16. New York: Kluwer Academic/Plenum Press.

———. 2001. Phylogenetic analysis of the Olenellina Walcott, 1890 (Trilobita, Cambrian). *J Paleontol* 75:96–115.

Lieberman, B. S., G. D. Edgecombe, and N. Eldredge. 1997. Systematics and biogeography of the *"Malvinella* Group," Calmoniidae (Trilobita, Devonian). *J Paleontol* 65:
824–843.

Lieberman, B. S., and N. Eldredge. 1996. Trilobite biogeography in the Middle Devonian: Geological processes and analytical methods. *Paleobiology* 22:66–79.

Lieberman, B. S., and G. J. Kloc. 1997. Evolutionary and biogeographic patterns in the Asteropyginae (Trilobita, Devonian) Delo, 1935. *Bull Am Mus Nat Hist* 232:1–127.

Liebherr, J. K. 1988. General patterns in West Indian insects, and graphical biogeographic analysis of some circum-Caribbean *Platynus* beetles (Carabidae). *Syst Zool* 38:385–409.

Liem, K. F. 1973. Evolutionary strategies and morphological innovations: Cichlid pharyngeal jaws. *Syst Zool* 22:424–441.

———. 1980. Adaptive significance of intra- and interspecific differences in the feeding repertoires of cichlid fishes. *Am Zool* 20:295–314.

Liem, K. F., and J. W. M. Osse. 1975. Biological versatility, evolution, and food resource exploitation in African cichlid fishes. *Am Zool* 15:427–454.

Liem, K. F., and A. P. Summers. 2000. Integration of versatile functional design, population ecology, ontogeny, and phylogeny. *Neth J Zool* 50:245–259.

Liem, K. F., and D. B. Wake. 1985. Morphology: Current approaches and concepts. In *Functional vertebrate morphology*, ed. M. Hildebrand, D. M. Bramble, K. F. Liem, and D. B. Wake, 366–377. Cambridge: Harvard Univ. Press.

Light, J. E., and M. E. Siddall. 1999. Phylogeny of the leech family Glossiphonidae based on mitochondrial gene sequences and morphological data. *J Parasitol* 85:815–823.

Lim, B. K. 1993. Cladistic re-appraisal of neotropical stenodermatine bat phylogeny. *Cladistics* 9:147–165.

Lin, L. H., D. T. Edmonds, and F. Vollrath. 1995. Structural engineering of an orb-spider's web. *Nature* 373:146–148.

Lindeman, P. V. 2000. Resource use of five sympatric turtle species: Effects of competition, phylogeny, and morphology. *Can J Zool* 78:992–1008.

Lindeman, R. L. 1942. The trophic-dynamic aspect of ecology. *Ecology* 23:399–418.

Lindenfors, P., and B. S. Tullberg. 1998. Phylogenetic analyses of primate size evolution: The consequences of sexual selection. *Biol J Linn Soc* 64:413–447.

Linder, H. P. 1995. Setting conservation priorities: The importance of endemism and phylogeny in the southern African orchid genus *Herschelia*. *Conserv Biol* 9:585–595.

Lingafelter, S. W., and A. S. Konstantinov. 1999. The monophyly and relative rank of alticine and gelerucine leaf beetles: A cladistic analysis using adult morphological characters (Coleoptera: Chrysomelidae). *Entomol Scand* 30:397–416.

Liou, L. W., and T. D. Price. 1994. Speciation by reinforcement of premating isolation. *Evolution* 48:1451–1459.

Lippitsch, E. 1998. Phylogenetic study of cichlid fishes in Lake Tanganyika: A lepidological approach. *J. Fish Biol* 53:752–766.

Liu, J., G. O. Poinar, and R. E. Berry. 2000. Control of insect pests with entomopathogenic nematodes: The impact of molecular biology and phylogenetic reconstruction. *Annu Rev Entomol* 45:287–306.

Liu, K., and P. A. Colinvaux. 1985. Forest changes in the Amazon Basin during the last glacial maximum. *Nature* 318:556–557.

Loehle, C., and J. H. K. Pechmann. 1988. Evolution: The missing ingredient in systems ecology. *Am Nat* 132:884–899.

Lomnicki, A. 1974. Evolution of the herbivore-plant, predator-prey, and parasite-host systems: A theoretical model. *Am Nat* 108:167–180.

Lord, J. M., M. Westoby, and M. R. Leishman. 1995. Seed size and phylogeny in six temperate floras: Constraints, niche conservation, and adaptation. *Am Nat* 146:349–364.

Loreau, M., and I. Olivieri. 1999. Diversitas: An international programme of biodiversity science. *Trend Ecol Evol* 14:2–3.

Lorenz, K. 1941. Comparative studies on the behaviour of Anatinae [Vergleichende Bewegungstudien an Anatien], trans. C. H. D. Clarke. *J Ornithol* 89:194–294.

———. 1950. The comparative method in studying innate behaviour patterns. *Symp Soc Exp Biol* 4:221–268.

———. 1958. The evolution of behaviour. *Sci Am* 199:67–78.

Losos, J. B. 1990a. A phylogenetic analysis of character displacement in Caribbean *Anolis* lizards. *Evolution* 44:558–569.

———. 1990b. Ecomorphology, performance capability, and scaling of West Indian *Anolis* lizards: An evolutionary analysis. *Ecol Monogr* 60:369–388.

———. 1990c. The evolution of form and function: Morphology and locomotor performances in West Indian *Anolis* lizards. *Evolution* 44:1189–1203.

———. 1990d. Concordant evolution of locomotor behaviour, display rate, and morphology in *Anolis* lizards. *Anim Behav* 39:879–890.

———. 1992. The evolution of convergent structure in Caribbean *Anolis* communities. *Syst Biol* 41:403–420.

———. 1994. Integrative approaches to evolutionary ecology: *Anolis* lizards as model systems. *Annu Rev Ecol Syst* 25:467–493.

———. 1996. Phylogenetic perspectives on community ecology. *Ecology* 77:1344–1354.

Losos, J. B., and F. R. Adler. 1995. Stumped by trees? A generalized model for patterns of organismal diversity. *Am Nat* 145:329–342.

Losos, J. B., and D. J. Irschick. 1996. The effect of perch diameter on escape behaviour in *Anolis* lizards: Laboratory predictions and field tests. *Anim Behav* 51:593–602.

Losos, J. B., T. R. Jackman, A. Larson, K. de Queiroz, and L. Rodríguez-Schettino. 1998. Contingency and determinism in replicated adaptive radiations of island lizards. *Science* 279:2115–2118.

Losos, J. B., and D. B. Miles. 1994. Adaptation, constraint, and the comparative method: Phylogenetic issues and methods. In *Ecological morphology,* ed. P. C. Wainwright and S. M. Reilly, 13–41. Chicago: Univ. Chicago Press.

Losos, J. B., S. Naeem, and R. K. Colwell. 1989. Hutchinsonian ratios and statistical power. *Evolution* 43:1820–1826.

Losos, J. B., and K. de Queiroz. 1997. Evolutionary consequences of ecological release in Caribbean *Anolis* lizards. *Biol J Linn Soc* 61:459–483.

Losos, J. B., and D. Schluter. 2000. Analysis of an evolutionary species-area relationship. *Nature* 408:847–850.

Losos, J. B., and B. Sinervo. 1989. The effect of morphology and perch diameter on sprint performance of *Anolis* lizards. *J Exp Biol* 145:23–30.

Losos, J. B., K. I. Warheit, and T. W. Schoener. 1997. Adaptive differentiation following experimental island colonization in *Anolis* lizards. *Nature* 387:70–73.

Löther, R. 1990. Species and monophyletic taxa as individual substantial systems. In *The plant diversity of Malesia,* ed. P. Baas et al., 371–378. The Hague: Kluwer Academic.

Lubchenco, J., A. M. Olson, L. B. Brubaker, S. R. Carpenter, M. M. Holland, S. P. Hubbell, S. A. Levin, J. A. MacMahon, P. A. Matson, J. M. Melillo, H. A. Mooney, C. H. Peterson, H. R. Pulliam, L. A. Real, P. Regal, and P. G. Risser. 1991. The sustainable biosphere initiative: An ecological research agenda. *Ecology* 72:371–412.

Lucas, G., P. Kaufman, and L. Kasdan. 1981. *Raiders of the lost ark.* Hollywood, CA: Paramount Pictures.

Lucchitta, I. 1990. History of the grand Canyon and of the Colorado River in Arizona. In *Grand Canyon geology,* ed. S. S. Beus and M. Morales, 311–332. New York: Oxford Univ. Press.

Luckow, M., and A. Bruneau. 1997. Circularity and independence in phylogenetic tests of ecological hypotheses. *Cladistics* 13:145–151.

Ludwig, D. 1976. A singular perturbation problem in the theory of population extinction. *Soc Ind App Math—Am Math Soc Proc* 10:87–104.

Lundin, K. 2000. Phylogeny of the Nemertodermatida (Acoelomorpha, Platyhelminthes): A cladistic analysis. *Zool Scr* 29:65–74.

Lunney, D. 1998. Lack of awareness and knowledge about biodiversity. *Austral Zool* 30:381–382.

Luter, C. 2000.The origin of the coelom in Brachiopoda and its phylogenetic significance. *Zoomorphology* 120:15–28.

Lydeard, C. 1993. Phylogenetic analysis of species richness: Has viviparity increased diversification of actinopterygian fishes? *Copeia* 1993:514–518.

Lydeard C., and K. J. Roe. 1998. Phylogenetic systematics: The missing ingredient in the conservation of freshwater unionid bivalves. *Fisheries* 23:16–17.

Lynch, J. D. 1986. Origins of the high Andean Herpetological fauna. In *High altitude tropical biogeography,* ed. F. Vuilleumier and M. Monasterio, 478–499. London: Oxford Univ. Press.

———. 1988. Refugia. In *Analytical biogeography: An integrated approach to the study of ani-*

mal and plant distributions, ed. A. A. Myers and P. S. Giller, 311–342. London: Chapman and Hall.

———. 1989. The gauge of speciation: On the frequencies of modes of speciation. In *Speciation and its consequences,* ed. D. Otte and J. Endler, 527–553. Sunderland, MA: Sinauer.

———. 1999. Ranas pequeñas, la geometría de evolución, y la especiación en los Andes colombianos. *Rev Acad Colombiana Cien Exa Fis Nat* 23:143–159.

Lynch, M. 1991. Methods for the analysis of comparative data in evolutionary biology. *Evolution* 45:1065–1080.

Lynch, M., and R. Lande. 1993. Evolution and extinction in response to environmental change. In *Biotic interactions and global change,* ed. P. M. Kareiva, J. G. Kingsolver, and R. B. Huey, 234–250. Sunderland, MA: Sinauer.

Lyons-Weiler, J., and G. A. Hoelzer. 1997. Escaping from the Felsenstein zone by detecting long branches in phylogenetic data. *Mol Phylogenet Evol* 8:375–384.

———. 1999. Null model selection, compositional bias, character state bias, and the limits of phylogenetic information. *Mol Biol Evol* 16:1400–1405.

Lyons-Weiler, J., G. A. Hoelzer, and R. J. Tansch. 1998. Optimal outgroup analysis. *Biol J Linn Soc* 64:493–511.

Lyons-Weiler, J., and K. Takahashi. 1999. Branch length heterogeneity leads to non-independent branch length estimates and can decrease the efficiency of methods of phylogenetic inference. *J Mol Evol* 49:392–405.

Mabee, P. M. 1993. Phylogenetic interpretation of ontogenetic change: Sorting out the actual and artefactual in an empirical case study of centrarchid fishes. *Biol J Linn Soc* 107:175–291.

———. 2000. Developmental data and phylogenetic systematics: Evolution of the vertebrate limb. *Am Zool* 40:789–800.

Mabee, P. M., and J. Humphries. 1993. Coding polymorphic data: Examples from allozymes and ontogeny. *Syst Biol* 42:166–181.

MacArthur, R. H. 1957. On the relative abundance of bird species. *Proc Natl Acad Sci* 43: 293–295.

———. 1958. Population ecology of some warblers of northeastern coniferous forests. *Ecology* 39:599–619.

———. 1960. On the relative abundance of species. *Am Nat* 94:25–36.

———. 1965. Patterns of species diversity. *Biol Rev* 40:510–533.

———. 1969. Patterns of communities in the tropics. *Biol J Linn Soc* 1:9–30.

———. 1972. *Geographical ecology.* New York: Harper and Row.

MacArthur, R. H., and R. Levins. 1967. The limiting similarity, convergence, and divergence of co-existing species. *Am Nat* 101:377–385.

MacArthur, R. H., and E. O. Wilson. 1963. An equilibrium theory of insular zoogeography. *Evolution* 17:373–387.

———. 1967. *The theory of island biogeography.* Princeton: Princeton Univ. Press.

Macedonia, J. M., and K. F. Stanger. 1994. Phylogeny of the Lemuridae revisited: Evidence from communication signals. *Folia Primatol* 63:1–43.

Machado, C. A., E. A. Herre, S. McCafferty, and E. Bermingham. 1996. Molecular phylogenies of fig pollinating and non-pollinating wasps and the implications for the origin and evolution of the fig-fig wasp mutualism. *J Biogeogr* 23:531–542.

Maddison, D. R. 1994. Phylogenetic methods for inferring the evolutionary history and processes of change in discretely valued characters. *Annu Rev Entomol* 39:267–292.

Maddison, W. P. 1989. Reconstructing character evolution on polytomous cladograms. *Cladistics* 5:365–377.

———. 1990. A method for testing the correlated evolution of two binary characters: Are gains and losses concentrated on certain branches of a phylogenetic tree? *Evolution* 44:539–557.

———. 1991. Squared-change parsimony reconstructions of ancestral states for continuous valued characters on a phylogenetic tree. *Syst Zool* 40:304–314.

———. 1993. Missing data versus missing characters in phylogenetic analysis. *Syst Biol* 42:581–587.

Maddison, W. P., M. J. Donoghue, and D. R. Maddison. 1984. Outgroup analysis and parsimony. *Syst Zool* 33:83–103.

Maddison, W. P., and D. R. Maddison. 2000. MacClade. Analysis of phylogeny and character evolution. Ver. 4. Sunderland, MA: Sinauer.

Maddison, W. P., and M. McMahon. 2000. Divergence and reticulation among montane populations of a jumping spider (*Habronattus pugillis* Griswold). *Syst Biol* 49:400–421.

Maddison, W. P., and M. Slatkin. 1991. Null models for the number of evolutionary steps in a character on a phylogenetic tree. *Evolution* 45:1184–1197.

Maderson, P. F. A. 1982. The role of development in macroevolutionary change: Group report. In *Evolution and development*, ed. J. T. Bonner, 279–312. New York: Springer Verlag.

Maier, W. 1999. On the evolutionary biology of early mammals: With morphological remarks on the interaction between ontogenetic adaptation and phylogenetic transformation. *Zool Anz.* 238:55–74.

Mallet, J., and M. Joron. 1999. Evolution of diversity in warning color and mimicry: Polymorphisms, shifting balance, and speciation. *Annu Rev Ecol Syst* 30:201–233.

Malthus, T. R. 1826. *An essay on the principle of population.* 6th ed. London.

Manter, H. W. 1966. Parasites of fishes as biological indicators of recent and ancient conditions. In *Host-parasite relationships,* ed. J. E. McCauley, 59–71. Corvallis: Oregon State Univ. Press.

Mardulyn, P., M. C. Milinkovitch, and J. M. Pasteels. 1997. Phylogenetic analysis of DNA and allozyme data suggest that *Gonioctena* leaf beetles (Coleoptera; Chrysomelidae) experienced convergent evolution in their history of host-plant family shifts. *Syst Biol* 46:722–747.

Margulis, L. 1981. *Symbiosis in cell evolution.* San Francisco: W. H. Freeman.

Margulis, L., M. Dolan, and R. Guerrero. 1999. The molecular tangled bank: Not seeing the phylogenies for the trees. *Biol Bull* 196:413–414.

Margush, T., and F. R. McMorris. 1981. Consensus n-trees. *Bull Math Biol* 43:239–244.

Marini-Bottolo, G. B., ed. 1994. *Man and his environment: Tropical forests and the conservation of species.* Pontificiae Academiae Scientarum Scriptum Varia no. 84. Vatican: Pontificiae Academiae Scientarum.

Markowicz, Y., and S. Loiseaux-De Goer. 1991. Plastid genomes of the Rhodophyta and Chromophyta constitute a distinct lineage which differs from that of the Chlorophyta and have a composite phylogenetic origin, perhaps like that of the Euglenophyta. *Curr Genet* 20:427–430.

Marler, C. A., and M. J. Ryan. 1997. Origin and maintenance of a female mating preference. *Evolution* 51:1244–1248.

Marlier, G. 1959. Observations sur la biologie littorale du lac Tanganyika. *Rev Zool Bot Africaine* 59:164–183.

Marshall, C. J., and J. K. Liebherr. 2000. Cladistic biogeography of the Mexican transition zone. *J Biogeogr* 27:203–216.

Marshall, C. R., E. C. Raff, and R. A. Raff. 1994. Dollo's law and the death and resurrection of genes. *Proc Natl Acad Sci* (USA) 91:12283–12287.

Marshall, D. J., and L. Coetzee. 2000. Historical biogeography and ecology of a continental Antarctic mite genus, *Maudheimia* (Acari: Oribatidae): Evidence for a Gondwanian origin and Plio-Pleistocene speciation. *Zool J Linn Soc* 129:111–128.

Martens, K. 1997. Speciation in ancient lakes. *Trends Ecol Evol* 12:177–182.

Martens, P. 1971. *Les gnétophytes.* Berlin: Gebrüder Borntraeger.

Martin, P., I. Kaygorodova, D. Y. Sherbakov, and E. Verheyen. 2000. Rapidly evolving lineages impede the resolution of phylogenetic relationships among Clitellata (Annelida). *Mol Phylogenet Evol* 15:355–368.

Martin, P. S., and R. G. Klein. 1984. *Quaternary extinctions: A prehistoric revolution.* Tuscan: Univ. of Arizona Press.

Martinez-Ghersa, M. A., C. M. Ghersa, and E. H. Satorre. 2000. Coevolution of agricultural systems and their weed companions: Implications for research. *Field Crops Res* 67:181–190.

Martins, E. P. 1993. A comparative study of the evolution of *Sceloporus* push-up displays. *Am Nat* 142:994–1018.

———, ed. 1996. *Phylogenies and the comparative method in animal behavior.* New York: Oxford Univ. Press.

------. 2000. Adaptation and the comparative method. *Trends Ecol Evol* 15:296–299.

Martins, E. P., and T. Garland Jr. 1991. Phylogenetic analysis of the correlated evolution of continuous characters: A simulation study. *Evolution* 45:534–557.

Martins, E. P., and T. F. Hansen. 1997. Phylogenies and the comparative method: A general approach to incorporating phylogenetic information into the analysis of interspecific data. *Am Nat* 149:646–667.

Masters, J. C., and R. J. Rayner. 1993. Competition and macroevolution: The ghost of competition yet to come? *Biol J Linn Soc* 49:87–98.

------. 1996. The recognition concept and the fossil record: Putting the genetics back into phylogenetic species. *S Afr J Sci* 92:225–231.

Masters, J. C., and H. G. Spencer. 1989. Why we need a new genetic species concept. *Syst Zool* 38:270–279.

Mathews, S., and M. J. Donoghue. 2000. Basal angiosperm phylogeny inferred from duplicate phytochromes A and C. *Int J Plant Sci* 161:S41–S55.

Mathews, W. H., J. G. Fyles, and H. W. Nasmith. 1970. Postglacial crustal movements in southwestern British Columbia and adjacent Washington state. *Can J Earth Sci* 62:1402–1408.

Matile, L. 1970. L'origine des Diptères cavernicoles. Académie de la République socialiste roumaine, livre du centenaire Emile G. Racovitza 1868–1968, 307–311.

Mattern, M., and D. A. McLennan. 2000. Phylogeny and speciation of felids. *Cladistics* 16:232–253.

Matthew, W. D. 1915. *Climate and evolution. Ann NY Acad Sci* 24:171–318.

------. 1939. *Climate and evolution.* NY Acad Sci Spec Publ 1.

Matthews, E. G. 2000. Origins of Australian arid-zone tenebrionid beetles. *Invert Taxon* 14:941–951.

Maurer, B. A. 1985. On the ecological and evolutionary roles of interspecific competition. *Oikos* 45:300–302.

------. 1989. Diversity dependent species dynamics: Incorporating the effects of population level processes on species dynamics. *Paleobiology* 15:133–146.

------. 1991. Ecological aspects of macroevolution. *Adv Ecol Res* 1:29–39.

------. 1999. *Untangling ecological complexity: The macroscopic perspective.* Chicago: Univ. of Chicago Press.

------. 2000. Macroecology and consilience. *Global Ecol Biogeogr* 9:275–280.

Maurer, B. A., and J. H. Brown. 1988. Distribution of energy use and biomass among species of North American terrestrial birds. *Ecology* 69:1923–1932.

Maurer, B. A., J. H. Brown, and R. D. Rusler. 1992. The micro and macro in body size evolution. *Evolution* 46:939–953.

May, M. L. 1977. Thermoregulation and reproductive activity in tropical dragonflies of the genus *Micrathyria. Ecology* 58:787–798.

May, R. M. 1990. Taxonomy as destiny. *Nature* 347:129–130.

------. 1994. Conceptual aspects of the quantification of the extent of biological diversity. *Phil Trans Roy Soc Lond* B 345:13–20.

Mayden, R. L. 1985. Biogeography of the Ouachita Highland fishes. *Southwest Nat* 30:195–211.

------. 1986. Speciose and depauperate phylads and tests of punctuated and gradual evolution: Fact or artifact? *Syst Zool* 35:591–602.

------. 1987a. Historical ecology and North American highland fishes: A research program in community ecology. In *Community and evolutionary ecology of North American stream fishes,* ed. W. J. Matthews and D.C. Heins, 210–222. Norman: Univ. Oklahoma Press.

------. 1987b. Pleistocene glaciation and historical biogeography of North American central highland fishes. In *Quaternary environments of Kansas,* ed. W. C. Johnson, 141–151. Guidebook Series 5. Lawrence: Kansas Geological Survey.

------. 1988. Biogeography, parsimony, and evolution in North American freshwater fishes. *Syst Zool* 37:329–355.

------. 1989. Phylogenetic studies of North American minnows, with emphasis on the genus *Cyprinella* (Teleostei: Cypriniformes). *Univ. Kansas Mus Nat Hist Misc Publ* 80:1–189.

------. 1991. The wilderness of Panbiogeography: A synthesis of space, time and form? *Syst Zool* 40:503–519.

———, ed. 1992a. *Systematics, historical ecology, and North American freshwater fishes.* Palo Alto, CA: Stanford Univ. Press.

———. 1992b. An emerging revolution in comparative biology and the evolution of North American freshwater fishes. In *Systematics, historical ecology, and North American freshwater fishes,* ed. R. Mayden, 864–890. Stanford, CA: Stanford Univ. Press.

———. 1997. A hierarchy of species concepts: The denouement in the saga of the species problem. In *Species: The units of biodiversity,* ed. M. F. Claridge, H. A. Dawah, and M. R. Wilson, 381–424. London: Chapman and Hall.

———. 1999. Consilience and a hierarchy of species concepts: Advances toward closure on the species puzzle. *J Nematol* 31:95–116.

Mayden, R. L., and B. R. Kuhajda. 1996. Systematics, taxonomy, and conservation status of the endangered Alabama sturgeon, *Scaphirhynchus suttkusi* Williams and Clemmer (Actinopterygii, Acipenseridae). *Copeia* 1996:241–273.

Mayden, R. L., and R. H. Matson. 1992. Systematics and biogeography of the Tennessee Shiner, *Notropis leuciodus* (Cope) (Teleostei: Cyprinidae). *Copeia* 1992:954–968.

Mayden, R. L., and E. O. Wiley. 1992. The fundamentals of phylogenetic systematics. In *Systematics, historical ecology, and North American freshwater fishes,* ed. R. Mayden, 114–185. Stanford, CA: Stanford Univ. Press.

Mayden, R. L., and R. M. Wood. 1995. Systematics, species concepts, and the evolutionarily significant unit in biodiversity and conservation biology. *Am Fisheries Soc Symp* 17:58–113.

Mayer, W. E., H. Tichy, and J. Klein. 1998. Phylogeny of African cichlid fishes as revealed by molecular markers. *Heredity* 80:702–714.

Maynard Smith, J. 1966. Sympatric speciation. *Am Nat* 100:637–650.

———. 1976. What determines the rate of evolution? *Am Nat* 110:331–338.

———. 1982. Evolution and the theory of games. Cambridge, MA: Cambridge Univ. Press.

Maynard Smith, J., R. Burian, S. Kauffman, P. Alberch, J. Campbell, B. Goodwin, R. Lande, D. Raup, and L. Wolpert. 1985. Developmental constraints and evolution. *Quart Rev Biol* 60:265–287.

Maynard Smith, J., and E. Szathmary. 1995. *The major transitions in evolution.* Oxford: W. H. Freeman.

———. 1999. *The origins of life.* Oxford: W. H. Freeman.

Mayr, E. 1940. Speciation phenomena in birds. *Am Nat* 74:249–278.

———. 1942. *Systematics and the origin of species.* New York: Columbia Univ. Press.

———. 1954. Change of genetic environment and evolution. In *Evolution as a process,* ed. J. Huxley, A. C. Hardy, and E. B. Ford, 157–180. London: Allen and Unwin.

———. 1957. Evolutionary aspects of host specificity among parasites of vertebrates. In *Premier symposium sur la spécificité parasitaire des parasites de vertébrés,* ed J. G. Baer, 7–14. Neuchâtel: International Union of Biological Sciences and Université Neuchâtel; Paul Attinger.

———. 1958. Behavior and systematics. In *Behavior and evolution,* ed. A. Roe and G. G. Simpson, 341–366. New Haven, CT: Yale Univ. Press.

———. 1960. The emergence of evolutionary novelties. In *Evolution of life,* ed. S. Tax, 349–380. Chicago: Univ. Chicago Press.

———. 1963. *Animal species and evolution.* Cambridge, MA: Harvard Univ. Press.

———. 1969. *Principles of systematic zoology.* New York: McGraw-Hill.

———. 1970. *Populations, species, and evolution.* Cambridge, MA: Belknap Press.

———. 1976. Is the species a class or an individual? *Syst Zool* 25:192.

———. 1982. Processes of speciation in animals. In *Mechanisms of speciation,* ed. C. Barigozzi, 1–19. New York: Alan R. Liss.

———. 1988. *Toward a new philosophy of biology: Observations of an evolutionist.* Cambridge, MA: Belknap Press.

Mayr, E., and R. J. O'Hara. 1986. The biogeographic evidence supporting the Pleistocene forest refuge hypothesis. *Evolution* 40:55–67.

McClure, M. S., and P. W. Price. 1975. Competition among sympatric *Erythroneura* leafhoppers (Homoptera: Cicadellidae) on American Sycamore. *Ecology* 56:1388–1397.

McCoy, E. D., and K. L. Heck, Jr. 1976. Biogeography of corals, seagrasses, and mangroves: An alternative to the center of origin concept. *Syst Zool* 25:201–210.

McDade, L. A. 1990. Hybrids and phylogenetic systematics I. Patterns of character expression in hybrids and their implications for cladistic analysis. *Evolution* 44:1685–1700.

———. 1992a. Hybrids and phylogenetic systematics II. The impact of hybrids on cladistic analysis. *Evolution* 46:1329–1346.

———. 1992b. Pollinator relationships, biogeography, and phylogenetics. *BioScience* 42:21–26.

———. 1997. Hybrids and phylogenetic systematics III. Comparison with distance methods. *Syst Bot* 22:669–683.

McDougall, I. 1964. Potassium-argon ages from lavas of the Hawaiian Islands. *Geol Soc Am Bull* 75:107–128.

McElroy, D. M., and I. Kornfield. 1990. Sexual selection, reproductive behavior, and speciation in the Mbuna species flock of Lake Malawi (Pisces: Cichlidae). *Envir Biol Fish* 28:273–284.

McIntosh, R. P. 1967. An index of diversity and the relation of certain concepts to diversity. *Ecology* 48:392–404.

———. 1985. *The background of ecology: Concept and theory.* Cambridge: Cambridge Univ. Press.

———. 1987. Pluralism in ecology. *Annu Rev Ecol Syst* 18:321–341.

McKaye, K. R. 1991. Sexual selection and the evolution of the cichlid fishes of Lake Malawi, Africa. In *Cichlid fishes: Behaviour, ecology, and evolution,* ed. M. H. A. Keenleyside, 241–257. London: Chapman and Hall.

McKaye, K. R., J. H. Howard, J. R. Stauffer, Jr., R. P. Morgan, and F. Shonhiwa. 1993. Sexual selection and genetic relationships of a sibling species complex of bower building cichlids in Lake Malawi, Africa. *Jap J Ichthyol* 40:15–21.

McKinney, M. L. 1986. Ecological causation of heterochrony: A test and implications for evolutionary theory. *Paleobiology* 12:282–289.

———, ed. 1988. *Evolution and development.* Dahlem Conference. New York: Springer Verlag.

———. 1993. *Evolution of life: Processes, patterns, and prospects.* New York: Prentice Hall.

McKinnon, J. S. 1995. Video mate preferences of female three-spined sticklebacks from populations with divergent male coloration. *Anim Behav* 50:1645–1655.

McKitrick, M. C. 1985. Monophyly of the Tyrannidae (Aves): Comparison of morphology and DNA. *Syst Zool* 34:35–45.

———. 1993. Phylogenetic constraint in evolutionary theory: Has it any explanatory power? *Annu Rev Ecol Syst* 24:307–330.

———. 1994. On homology and ontological relationship of parts. *Syst Biol* 43:1–10.

McKitrick, M. C., and R. M. Zink. 1988. Species concepts in ornithology. *The Condor* 90:1–14.

McKnight, M. L. 1995. Mitochondrial DNA phylogeography of *Perognathus amplus* and *Perognathus longimembris* (Rodentia: Heteromidae): A possible mammalian ring species. *Evolution* 49:816–826.

McKnight, M. L., and M. R. Lee. 1992. Karyotypic variation in the pocket mice, *Perognathus amplus* and *P. longimembris.* *J Mammal* 73:625–629.

McLennan, D. A. 1991. Integrating phylogeny and experimental ethology: From pattern to process. *Evolution* 45:1773–1789.

———. 1993a. Phylogenetic relationships in the Gasterosteidae: An updated tree based on behavioral characters with a discussion of homoplasy. *Copeia* 1993:318–326.

———. 1993b. Temporal changes in the structure of the male nuptial signal in the brook stickleback, *Culaea inconstans* (Kirtland). *Can J Zool* 71:1111–1119.

———. 1994. A phylogenetic approach to the evolution of fish behaviour. *Fish Biol and Fisheries* 4:430–460.

———. 1996. Integrating phylogenetic and experimental analyses: The evolution of male and female nuptial coloration in the Gasterosteidae. *Syst Biol* 45:261–277.

———. 2000. The macroevolutionary diversification of female and male components of the stickleback breeding system. *Behaviour* 137:1029–1045.

McLennan, D. A., and D. R. Brooks. 1991. Parasites and sexual selection: A macroevolutionary perspective. *Quart Rev Biol* 66:255–286.

———. 1993. The phylogenetic component of cooperative breeding in perching birds: A commentary. *Am Nat* 141:790–795.

————. 2001. Phylogenetic systematics: Five steps to enlightenment. In *Fossils, phylogeny, and form: An analytical approach*, ed. J. Adrian, J. Edgecombe, and B. Lieberman, 7–28. London: Kluwer Academic/Plenum.

McLennan, D. A., D. R. Brooks, and J. D. McPhail. 1988. The benefits of communication between comparative ethology and phylogenetic systematics: A case study using gasterosteid fishes. *Can J Zool* 66:2177–2190.

McLennan, D. A., and M. Y. Mattern. 2001. The phylogeny of the Gasterosteidae: Combining behavioral and morphological data sets. *Cladistics* 17:11–27.

McLennan, D. A., and J. D. McPhail. 1989. Experimental investigations of the evolutionary significance of sexually dimorphic nuptial colouration in *Gasterosteus aculeatus* (L.): Temporal changes in the structure of the male mosaic signal. *Can J Zool* 67:1767–1777.

————. 1990. Experimental investigations of the evolutionary significance of sexually dimorphic nuptial colouration in *Gasterosteus aculeatus* (L.): The relationships between male colour and female behaviour. *Can J Zool* 68:482–492.

McLennan, D. A., and M. J. Ryan. 1997. Responses to conspecific and heterospecific olfactory cues in the swordtail *Xiphophorus cortezi*. *Anim Behav* 54:1077–1088.

————. 1999. Interspecific recognition and discrimination based upon olfactory cues in northern swordtails. *Evolution* 53:880–888.

McMillan, C. 1954. Parallelisms between ecology and systematics. *Ecology* 35:92–94.

McMillan, W. O., C. D. Jiggins, and J. Mallet. 1997. What initiates speciation in passion vine butterflies? *Proc Natl Acad Sci* (USA) 94:8628–8633.

McNamara, K. J. 1982. Heterochrony and phylogenetic trends. *Paleobiology* 8:130–142.

————. 1986. A guide to the nomenclature of heterochrony. *J Paleontol* 60:4–13.

McNeely, J. A. 1995. Keep all the pieces: Systematics Agenda 2000 and world conservation. *Biodiv Conserv* 4:510–519.

McPeek, M. A. 1995. Testing hypotheses about evolutionary change on single branches of a phylogeny using evolutionary contrasts. *Am Nat* 145:686–703.

————. 1996. Linking local species interactions to rates of speciation in communities. *Ecology* 77:1355–1366.

McPeek, M. A., and J. M Brown. 2000. Building a regional species pool: Diversification of the *Enflame* damselflies in Eastern North American waters. *Ecology* 81:904–920.

McPhail, J. D. 1984. Ecology and evolution of sympatric sticklebacks *(Gasterosteus):* Morphological and genetic evidence for a species pair in Enos Lake, British Columbia. *Can J Zool* 62:1402–1408.

————. 1992. Ecology and evolution of sympatric sticklebacks *(Gasterosteus):* Evidence for a species pair in Paxton Lake, Texada Island, British Columbia. *Can J Zool* 70:361–369.

————. 1993. Ecology and evolution of sympatric sticklebacks *(Gasterosteus):* Origin of the species pairs. *Can J Zool* 71:515–523.

————. 1994. Speciation and the evolution of reproductive isolation in sticklebacks *(Gasterosteus)* of southwestern British Columbia. In *Evolutionary biology of the threespine stickleback*, ed. M. A. Bell and S. A. Foster, 399–437. Oxford: Oxford Univ. Press.

Meagher, J. W. 1977. World dissemination of the cereal-cyst nematode (*Heterodera avenae*) and its potential as a pathogen of wheat. *J Nematol* 9:9–15.

Medina, M., and P. J. Walsh. 2000. Molecular systematics of the order Anaspidea based on mitochondrial DNA sequences (12S, 16S, and COI). *Mol Phylogenet Evol* 15:41–58.

Meffe, G. K. 1990. Post-defaunation recovery of fish assemblages in southeastern blackwater streams. *Ecology* 71:637–667.

Meier, R., P. Kores, and S. Darwin. 1991. Homoplasy slope ratio: A better measurement of observed homoplasy in cladistic analyses. *Syst Zool* 40:74–88.

Menge, B. A., and A. M. Olsen. 1990. Role of scale and environmental factors in regulation of community structure. *Trends Ecol Evol* 5:52–57.

Metcalf, M. M. 1920. Upon an important method of studying problems of relationship and geographical distribution. *Proc Natl Acad Sci* (USA) 6:432–433.

————. 1922. The host parasite method of investigation and some problems to which it gives approach. *Anat Rec* 23:117.

————. 1923a. The opalinid ciliate infusorians. *Bull US Natl Mus* 120:1–484.

————. 1923b. The origin and distribution of the Anura. *Am Nat* 57:385–411.

———. 1929. Parasites and the aid they give in problems of taxonomy, geographic distribution, and paleogeography. *Smith Misc Coll* 81:1–36.

———. 1940. Further studies on the opalinid ciliate infusorians. *Proc US Nat Mus* 87:465–634.

Meyer, A., C. M. Montero, and A. Spreinat. 1996. Molecular phylogenetic inferences about the evolutionary history of East African cichlid fish radiations. In *The limnology, climatology, and paleoclimatology of the East African Lakes*, ed. T. C. Johnson and E. O. Odada, 303–323. Amsterdam: Gordon and Breach.

Miadlikowski, J., and F. Lutzoni. 2000. Phylogenetic revision of the genus *Peltigera* (lichen forming Ascomycota) based on morphological, chemical and large subunit nuclear ribosomal DNA. *Int J Plant Sci* 161:925–958.

Michener, C. D. 1953. Life-history studies in insect systematics. *Syst Zool* 2:112–118.

———. 1967. Diverse approaches to systematics. In *Evolutionary biology*, ed. T. Dobzhansky, 1–38. New York: Appleton-Century-Crofts.

Mickevich, M. F., and S. Weller. 1990. Evolutionary character analysis: Tracing character evolution on a cladogram. *Cladistics* 6:137–170.

Miles, D. B., and A. E. Dunham. 1992. Comparative analyses of phylogenetic effects in the life history patterns of iguanid reptiles. *Am Nat* 139:848–869.

———. 1993. Historical perspectives in ecology and evolutionary biology: The use of phylogenetic comparative analyses. *Annu Rev Ecol Syst* 24:587–619.

Milinkovitch, M. C., R. G. LeDuc, J. Adachi, F. Farnir, M. Georges, and M. Hasegawa. 1996. Effects of character weighting and species sampling on phylogeny reconstruction: A case study based on DNA sequence data in cetaceans. *Genetics* 144:1817–1833.

Milinski, M., and T. C. M. Bakker. 1990. Female sticklebacks use male coloration in mate choice and hence avoid parasitized males. *Nature* 344:330–333.

———. 1992. Costs influence sequential mate choice in sticklebacks, *Gasterosteus aculeatus. Proc Roy Soc Lond* B. 250:229–233.

Miller, A. H. 1949. Some ecologic and morphologic considerations in the evolution of higher taxonomic categories. In *Ornithologie als Biologische Wissenschaft*, ed. E. Mayr and E. Schüz, 84–88. Heidelberg: Carl Winter.

Miller, D. 1999. Being an absolute skeptic. *Science* 284:1625–1626.

Miller, D. R., and A. Y. Rossman. 1995. Systematics, biodiversity, and agriculture. *BioScience* 45:680–686.

Miller, E. H. 1988. Description of bird behavior for comparative purposes. *Curr Ornithol* 5:347–394.

Miller, J. S. 1992. Host-plant associations among prominent moths. *BioScience* 42:50–57.

Miller, J. S., and J. W. Wenzel. 1995. Ecological characters and phylogeny. *Annu Rev Entomol* 40:389–415.

Miller, R. R., J. D. Williams, and J. E. Williams. 1989. Extinctions of North American fishes during the past century. *Fisheries* 14:22–38.

Miller, T. E., and J. Travis. 1996. The evolutionary role of indirect effects in communities. *Ecology* 77:1329–1335.

Milne, M. J., and L. J. Milne. 1939. Evolutionary trends in caddis-worm case construction. *Ann Entomol Soc Am* 32:533–542.

Mindell, D. P. 1992. Phylogenetic consequences of symbioses: Eukarya and Eubacteria are not monophyletic taxa. *BioSystems* 27:53–62.

———. 1993. Merger of taxa and definition of monophyly. *BioSystems* 31:130–133.

Mindell, D. P., and R. L. Honeycutt. 1990. Ribosomal RNA in vertebrates: Evolution and phylogenetic applications. *Annu Rev Ecol Syst* 21:541–566.

Mindell, D. P., J. W. Shultz, and P. W. Ewald. 1995. The aids pandemic is new, but is HIV new? *Syst Biol* 44:77–92.

Mindell, D. P., J. W. Sites Jr., and D. Graur. 1989. Speciation evolution: A phylogenetic test with allozymes in *Sceloporus* (Reptilia). *Cladistics* 5:49–61.

Minelli, A. 1994. Systematics and biodiversity. *Trends Ecol Evol* 9:227.

Mironov, S. V., and J. Dabert. 1999. Phylogeny and cospeciation in feather mites of the subfamily Avenzoariinae (Analgoidea: Avenzoariidae). *Exper Appl Acarol* 23:525–549.

Mishler, B. D. 1994. Cladistic analysis of molecular and morphological data. *Am J Phys Anthropol* 94:143–156.

———. 2000. Deep phylogenetic relationships among "plants" and their implications for classification. *Taxon* 49:661–683.

Mishler, B. D., and R. N. Brandon. 1987. Individuality, pluralism, and the phylogenetic species concept. *Biol Phil* 2:397–414.

Mishler, B. D., and S. P. Churchill. 1984. A cladistic approach to the phylogeny of the "Bryophytes." *Brittonia* 36:406–424.

Mishler, B. D., and M. J. Donoghue. 1982. Species concepts: A case for pluralism. *Syst Zool* 31:491–503.

Mitchell, A., C. Mitter, and J. C. Regier. 2000. More taxa or more characters revisited: Combining data from nuclear protein coding genes for phylogenetic analyses of Noctuoidea (Insecta: Lepidoptera) *Syst Biol* 49:202–224.

Mitter, C., and D. R. Brooks. 1983. Phylogenetic aspects of coevolution. In *Coevolution,* ed. D. J. Futuyma and M. Slatkin, 65–98. Sunderland, MA: Sinauer.

Mitter, C., and B. Farrell. 1991. Macroevolutionary aspects of insect-plant relationships. In *Insect-plant interactions,* vol. 3, ed. E. Bernays, 35–78. Boca Raton, FL: CRC Press.

Mitter, C., B. Farrell, and D. J. Futuyma. 1991. Phylogenetic studies of insect-plant interactions: Insights into the genesis of diversity. *Trends Ecol Evol* 6:290–293.

Mitter, C., B. Farrell, and B. Wiegmann. 1988. The phylogenetic study of adaptive zones: Has phytophagy promoted insect diversification? *Am Nat* 132:107–128.

Miyamoto, M. M. 1981. Congruence among character sets in phylogenetic studies of the frog genus *Eleutherodactylus*. *Syst Zool* 30:281–290.

———. 1983. Frogs of the *Eleutherodactylus rugulosus* group: A cladistic study of allozyme, morphological, and karyological data. *Syst Zool* 32:109–124.

———. 1984. Central American frogs allied to *Eleutherodactylus cruentus:* Allozyme and morphological data. *J Herpetol* 18:256–263.

———. 1985. Consensus cladograms and general classifications. *Cladistics* 1:186–189.

Miyamoto, M. M., M. W. Allard, R. M. Adkins, L. L. Janecek, and R. L. Honeycutt. 1994. A congruence test of reliability using linked mitochondrial DNA sequences. *Syst Biol* 43:236–249.

Miyamoto, M. M., and J. Cracraft, eds. 1991. *Phylogenetic analysis of DNA sequences.* New York: Oxford Univ. Press.

Miyamoto, M. M., and W. M. Fitch. 1995. Testing species phylogenies and phylogenetic methods with congruence. *Syst Biol* 44:64–76.

Miyatake, T., and T. Shimizu. 1999. Genetic correlations between life-history and behavioral traits can cause reproductive isolation. *Evolution* 53:201–208.

Mode, C. J. 1958. A mathematical model for the co-evolution of obligate parasites and their hosts. *Evolution* 12:158–165.

Moermond, T. C. 1979a. Habitat constraints on the behavior, morphology, and community structure of *Anolis* lizards. *Ecology* 60:152–164.

———. 1979b. The influence of habitat structure on *Anolis* foraging behaviour. *Behaviour* 70:141–167.

Möhn, E. 1961. Gallmücken (Diptera, Itonidae) aus El Salvador. 4. Zur Phylogenie der neotropischen und holarktischen Region. *Senck Biol* 42:131–330.

Møller, A. P. 1993. Developmental stability, sexual selection, and speciation. *J Evol Biol* 6:493–509.

Møller, A. P., and T. R. Birkhead. 1992. A pairwise comparative method as illustrated by copulation frequency in birds. *Am Nat* 139:644–656.

Møller, A. P., and J. J. Cuervo. 1998. Speciation and feather ornamentation in birds. *Evolution* 52:859–869.

Monson, R. K. 1996. The use of phylogenetic perspective in comparative plant physiology and developmental biology. *Ann Missouri Bot Gard* 83:3–16.

Mooi, R. 1990. Paedomorphosis, Aristotle's Lantern, and the origin of the sand dollars. *Paleobiology* 16:25–48.

Mooi, R., P. F. Cannell, V. A. Funk, P. M. Mabee, R. T. O'Grady, and C. K. Starr. 1989. Historical perspectives, ecology, and tiger beetles: An alternative discussion. *Syst Zool* 38:191–195.

Moore, B. 1920. The scope of ecology. *Ecology* 1:3–5.

Moore, J. 1981. Asexual reproduction and environmental predictability in cestodes (Cyclophyllidea: Taeniidae). *Evolution* 35:723–741.

Moore, R., C. Laliker, and A. Fisher. 1952. *Invertebrate fossils.* New York: McGraw-Hill.

Morales, E. 2000. Estimating phylogenetic inertia in *Tithonia* (Asteraceae): A comparative approach. *Evolution* 54:475–484.

Moran, N. A. 1988. The evolution of host-plant alternation in aphids: Evidence for specialization as a dead end. *Am Nat* 132:681–706.

———. 1994. Adaptation and constraint in the complex life cycles of animals. *Annu Rev Ecol Syst* 25:573–600.

Moreau, R. E. 1952. Africa since the Mesozoic: With particular reference to certain biological problems. *Proc Zool Soc Lond* 121:869–913.

Morgan, D. R. 1997. Decay analysis of large sets of phylogenetic data. *Taxon* 46:509–517.

Morgan, T. H. 1932. *The scientific basis of evolution.* New York: W. W. Norton.

Morin, L. 2000. Long branch attraction effects and the status of "basal eukaryotes": Phylogeny and structural analysis of the ribosomal RNA gene cluster of the free living diplomonad *Trepomonas agilis. J Eukaryot Microbiol* 47:167–177.

Moritz, C. 1995. Use of molecular phylogenies for conservation. *Phil Trans Roy Soc Lond* B 349:113–118.

Moritz, C., T. E. Dowling, and W. M. Brown. 1987. Evolution of animal mitochondrial DNA: Relevance for population biology and systematics. *Annu Rev Ecol Syst* 18:269–292.

Morris, D.C., L. A. Mound, M. P. Schwartz, and B. J. Crespi. 1999. Morphological phylogenetics of Australian gall inducing thrips and their allies: The evolution of host plant affiliations, domicile use, and social behaviour. *Syst Entomol* 24:289–299.

Morris, P. J., L. C. Ivany, K. M. Schopf, and C. E. Brett. 1995. The challenge of paleoecological stasis: Reassessing sources of evolutionary stability. *Proc Natl Acad Sci* (USA) 92:11269–11273.

Mort, M. E., P. S. Soltis, D. E. Soltis, and M. L. Mabry. 2000. Comparison of three methods for estimating internal support on phylogenetic trees. *Syst Biol* 49:160–171.

Motta, P. J. 1988. Functional morphology of the feeding apparatus of ten species of Pacific butterflyfishes (Perciformes, Chaetodontidae): An ecomorphological approach. *Environ Biol Fishes* 22:39–67.

———. 1989. Dentition patterns among Pacific and western Atlantic butterflyfishes (Perciformes, Chaetodontidae): Relationship to feeding ecology and evolutionary history. *Environ Biol Fishes* 25:159–170.

Mottishaw, P. D., S. J. Thornton, and P. W. Hochachka. 1999. The diving response mechanism and its surprising evolutionary path in seals and sea lions. *Am Zool* 39:434–450.

Moyle, P. B., and J. J. Cech Jr. 1982. *Fishes: An introduction to ichthyology.* Englewood Cliffs, NJ: Prentice Hall.

Müller, F. 1879. *Ituna* and *Thyridia:* A remarkable case of mimicry in butterflies. *Proc Entomol Soc Lond* 1879:20–29.

Muller, H. J. 1939. Reversibility in evolution considered from the standpoint of genetics. *Biol Rev* 14:261–280.

———. 1942. Isolating mechanisms, evolution, and temperature. *Biol Symp* 6:71–125.

Muñoz-Chápuli, A. V. De Andrés, and G. Dingerkus. 1994. Coronary artery anatomy and elasmobranch phylogeny. *Acta Zool* 75:249–254.

Murphy, D. D. 1989. Conservation and confusion: Wrong species, wrong scale, wrong conclusions. *Conserv Biol* 3:82–84.

Murphy, D. D., and P. R. Ehrlich. 1989. The conservation biology of California's remnant native grasslands. In *Grassland structure and function: California annual grassland,* 201–211, ed. L. F. Huenneke and H. A. Mooney. Dordrecht, Netherlands: Kluwer Academic.

Murphy, R. W. 1988. The problematic phylogenetic analysis of interlocus heteropolymer isozyme characters: A case study from sea snakes and cobras. *Can J Zool* 66:2628–2633.

———. 1993. The phylogenetic analysis of allozyme data: Invalidity of coding alleles by presence/absence and recommended procedures. *Biochem Syst Ecol* 21:25–38.

Myers, G., K. MacInnes, and L. Myers. 1993. Phylogenetic moments in the AIDS epidemic. In *Emerging viruses,* ed. S. S. Morse, 120–137. Oxford: Oxford Univ. Press.

Nagel, L., and D. Schluter. 1998. Body size, natural selection, and speciation in sticklebacks. *Evolution* 52:209–218.

Nammack, M. 1998. National Marine Fisheries Service and the evolutionarily significant unit: Implications for management of freshwater mussels. *J Shellfish Res* 17:1415–1418.

Nandi, O. I., M. W. Chase, and P. K. Endress. 1998. A combined cladistic analysis of angiosperms using *rbcL* and non-molecular data sets. *Ann Missouri Bot Gard* 85:137–212.

Navidi, W. C., G. A. Churchill, and A. V. von Haeseler. 1991. Methods for inferring phylogenies from nucleic acid sequence data using maximum likelihood and linear invariants. *Mol Biol Evol* 8:128–143.

Naylor, G., and F. Kraus. 1995. The relationship between *s* and *m* and the retention index. *Syst Biol* 44:559–562.

Near, T. J., J. C. Porterfield, and L. M. Page. 2000. Evolution of cytochrome *b* and the molecular systematics of *Ammocrypta* (Percidae: Etheostomatinae). *Copeia* 2000:701–711.

Nee, S., T. G. Barraclough, and P. H. Harvey. 1996. Temporal changes in biodiversity: Detecting patterns and identifying causes. In *Biodiversity: A biology of numbers and difference*, ed. K. J. Gaston, 230–252. Oxford: Blackwell Science.

Nee, S., and R. M. May. 1997. Extinction and the loss of evolutionary history. *Science* 278:692–694.

Nelson, G. 1969. Infraorbital bones and their bearing on the phylogeny and geography of osteoglossomorph fishes. *Am Mus Nat Hist Nov* 2394:1–37.

———. 1978. Ontogeny, phylogeny, paleontology, and the Biogenetic Law. *Syst Zool* 21:324–345.

———. 1983. Reticulation in cladograms. In *Advances in cladistics: Proceedings of the second meeting of the Willi Hennig Society*, vol. 2, ed. N. I. Platnick and V. A. Funk, 105–111. New York: Columbia Univ. Press.

Nelson, G., and P. Y. Ladiges. 1991a. Standard assumptions for biogeographic analysis. *Austral Syst Bot* 4:41–58.

———. 1991b. Three-area statements: Standard assumptions for biogeographic analysis. *Syst Biol* 40:470–485.

Nelson, G., and N. Platnick. 1981. *Systematics and biogeography: Cladistics and vicariance.* New York: Columbia Univ. Press.

Nelson, G., and D. E. Rosen, eds. 1980. *Vicariance biogeography: A critique.* New York: Columbia Univ. Press.

Nelson, J. S., and F. M. Atton. 1971. Geographical and morphological variation in the presence and absence of the pelvic skeleton in the brook stickleback, *Culaea inconstans* (Kirkland), in Alberta and Saskatchewan. *Can J Zool* 49:343–352.

Nentwig, W. 1999. The importance of human ecology at the threshold of the next millennium: How can population growth be stopped? *Naturwissenschaften* 86:411–421.

New, T. R. 1999. Descriptive taxonomy as a facilitating discipline in invertebrate conservation. In *The other 99%: The conservation and biodiversity of invertebrates*, ed. W. Ponder and D. Lunney, 154–158. *Trans Roy Soc NSW*. Mosman: Royal Society of New South Wales.

Newman, C. M., J. E. Cohen, and C. Kipnis. 1985. Neo-Darwinian evolution implies punctuated equilibria. *Nature* 315:400–401.

Newton, A. E., C. J. Cox, J. G. Duckett, J. A. Wheeler, B. Goffinet, T. A. J. Hedderson, and B. D. Mishler. 2000. Evolution of the major moss lineages: Phylogenetic analyses based on multiple gene sequences and morphology. *Bryologist* 103:187–211.

Newton, A. E., and E. de Luna. 1999. A survey of morphological characters for phylogenetic study of the transition to pleurocarpy. *Bryologist* 102:651–682.

Nicholls, J. A., M. C. Double, D. M. Rowell, and R. D. Magrath. 2000. The evolution of cooperative and pair breeding in thornbills *Acanthiza* (Pardalotidae). *J Avian Biol* 31:165–176.

Nicolosi, E., Z. N. Deng, A. Gentile, S. La Malfa, G. Continella, and E. Tribulato. 2000. Citrus phylogeny and genetic origin of important species as investigated by molecular markers. *Theor Appl Genet* 100:1155–1166.

Nielsen, E. S. 1999. Systematics and conservation. In *The other 99%: The conservation and biodiversity of invertebrates*, ed. W. Ponder and D. Lunney, 166–173. *Trans Roy Soc NSW*. Mosman: Royal Society of New South Wales.

Nijhout, H. F. 1991. *The development and evolution of butterfly wing patterns.* Washington DC: Smithsonian Institution Press.

——. 1994. Developmental perspectives on evolution of butterfly mimicry. *BioScience* 44:148–157.

Nijs, I., and I. Impens. 2000. Biological diversity and probability of local extinction of ecosystems. *Funct Ecol* 14:46–54.

Nishida, M. 1991. Lake Tanganyika as an evolutionary *reservoir* of old lineages of East African cichlid fishes: Inferences from allozyme data. *Experientia* 47:974–979.

Nitecki, M. H., ed. 1983. *Coevolution.* Chicago: Univ. Chicago Press.

Nixon, K. C., and J. M. Carpenter. 1996. On simultaneous analysis. *Cladistics* 12:221–241.

——. 2000. On the other "phylogenetic systematics." *Cladistics* 16:298–318.

Nixon, K. C., and J. I. Davis. 1991. Polymorphic taxa, missing values and cladistic analysis. *Cladistics* 7:233–241.

Nixon, K. C., and Q. D. Wheeler. 1990. An amplification of the phylogenetic species concept. *Cladistics* 6:211–223.

——. 1992a. Extinction and the origin of species. In *Extinction and phylogeny,* ed. M. J. Novacek and Q. D. Wheeler, 216–234. New York: Columbia Univ. Press.

——. 1992b. Measures of phylogenetic diversity. In *Extinction and phylogeny,* ed. M. J. Novacek and Q. D. Wheeler, 216–234. New York: Columbia Univ. Press.

Norell, M. A., and M. J. Novacek. 1992. Congruence between superpositional and phylogenetic patterns: Comparing cladistic patterns with fossil records. *Cladistics* 8:319–337.

Norris, R. D. 1991. Biased extinction and evolutionary trends. *Paleobiology* 17:388–399.

Norton, D. A., and M. A. Carpenter. 1998. Mistletoes as parasites: Host specificity and speciation. *Trends Ecol Evol* 13:101–105.

Noss, R. F. 1986. Dangerous simplification in conservation biology. *Bull Ecol Soc Am* 67: 278–279.

Novacek, M. J. 1984. Evolutionary stasis in the elephant-shrew, *Rhynchocyon.* In *Living fossils,* ed. N. Eldredge and S. M. Stanley, 4–22. New York: Springer Verlag.

——. 1992a. Fossils as critical data for phylogeny. In *Extinction and phylogeny,* ed. M. J. Novacek and Q. D. Wheeler, 46–88. New York: Columbia Univ. Press.

——. 1992b. Fossils, topologies, missing data, and the higher level phylogeny of eutherian mammals. *Syst Biol* 41:58–73.

Novacek, M. J., and M. A. Norell. 1982. Fossils, phylogeny, and taxonomic rates of evolution. *Syst Zool* 31:366–375.

Novacek, M. J., and Q. D. Wheeler, eds. 1992. *Extinction and phylogeny.* New York: Columbia Univ. Press.

Núñez-Farfán, J., and C. Cordero, eds. 1993. Topicos de biologia evolutiva: Diversidad y adaptación. Mexico City: Centro de Ecología, Universidad Nacional Autónoma de México.

Nylin, S. 1988. Host plant specialisation and seasonality in a polyphagous butterfly, *Poygonia c-album. Oikos* 53:381–386.

Nylin, S., A. Bergström, and N. Janz. 2000. Butterfly host plant choice in the face of possible confusion. *J Insect Behav* 13:469–482.

Nylin, S., and N. Janz. 1993. Oviposition preference and larval performance in *Polygonia c-album* (Lepidoptera: Nymphalidae): The choice between bad and worse. *Ecol Entomol* 18:394–398.

——. 1999. Ecology and evolution of host plant range: Butterflies as a model group. In *Herbivores: Between plants and predators,* ed. H. Olff, V. K. Brown, and R. H. Drent, 31–54. Oxford: Blackwell.

Nylin, S., K. Nyblom, F. Ronquist, N. Janz, J. Belicek, and M. Källersjö. 2001. Phylogeny of *Polygonia, Nymphalis* and related butterflies (Lepidoptera: Nymphalidae): A total evidence analysis. *Zool J Linn Soc* 132:441–468.

Nyman, T., H. Roininen, and J. A. Vuorinen. 1998. Evolution of different gall types in willow-feeding sawflies (Hymenoptera: Tenthredinidae). *Evolution* 52:465–474.

Nyman, T., A. Widmer, and H. Roininen. 2000. Evolution of gall morphology and host-plant relationships in willow-feeding sawflies (Hymenoptera: Tenthredinidae). *Evolution* 54:526–533.

Oakley, T. H., and C. W. Cunningham. 2000. Independent contrasts succeed where ancestor reconstruction fails in a known bacteriophage phylogeny. *Evolution* 54:397–405.

Oberdorff, T., S. Lek, and J-F. Guegan. 1999. Patterns of endemism in riverine fish of the northern hemisphere. *Ecol Let* 2:75–81.

O'Brien, S. J., and E. Mayr. 1991. Bureaucratic mischief: Recognizing endangered species and subspecies. *Science* 251:1187–1188.

O'Connor, B. M. 1984. Co-evolutionary patterns between astigmatid mites and primates. In *Acarology VI*, vol. 1., ed. D. A. Griffiths and C. E. Bowman, 19–28. Chichester: Horwood.

———. 1985. Hypoderatid mites (Acari) associated with cormorants (Aves: Phalacrocoracidae), with description of a new species. *J Med Entomol* 22:324–331.

———. 1988. Host associations and coevolutionary relationships of astigmatid mite parasites of New World primates: I. Families Psoroptidae and Audycoptidae. *Fieldiana* 39:245–260.

———. 1992. Ontogeny and systematics of the genus *Cerophagus* (Acari: Gaudiellidae), mites associated with bumblebees. *The Great Lakes Nat* 25:173–189.

Odling-Schmee, F. J., K. N. Laland, and M. W. Feldman. 1996. Niche construction. *Am Nat* 86:309–326.

O'Donald, P. 1962. The theory of sexual selection. *Heredity* 17:541–552.

———. 1967. A general model of sexual and natural selection. *Heredity* (London) 22:499–518.

Odum, E. P. 1969. The strategy of ecosystem development. *Science* 164:262–270.

O'Grady, R. T. 1987. Phylogenetic systematics and the evolutionary history of some intestinal flatworm parasites (Trematoda: Digenea: Plagiorchioidea) of anurans. Ph.D. diss., Univ. British Columbia.

———. 1989. Parasite:host specificity. In *Parasitology: The biology of animal parasites*, 6th ed., ed. E. R. Noble and G. A. Noble, 495–511. Philadelphia: Lea and Febiger.

O'Grady, R. T., and G. B. Deets. 1987. Coding multi-state characters, with special reference to the use of parasites as characters of their hosts. *Syst Zool* 36:1–21.

O'Hara, R. J. 1993. Systematic generalization, historical fate, and the species problem. *Syst Biol* 42:231–246.

———. 1994. Evolutionary history and the species problem. *Am Zool* 34:12–22.

Oksanen, L. 1988. Ecosystem organization: Mutualism and cybernetics or plain Darwinian struggle for existence? *Am Nat* 131:424–444.

Oliver, J. H., Jr. 1996. Importance of systematics to public health: Ticks, microbes, and disease. *Ann Missouri Bot Gard* 83:37–46.

Olmland, K. E. 1994. Character congruence between a molecular and a morphological phylogeny for dabbling ducks (*Anas*). *Syst Biol* 43:369–386.

———. 1997. Correlated rates of molecular and morphological evolution. *Evolution* 51:1381–1393.

Olmstead, R. G., and J. A. Sweere. 1994. Combining data in phylogenetic systematics: An empirical approach using three molecular data sets in the Solenaceae. *Syst Biol* 43:467–481.

O'Neil, P., and J. Schmitt. 1993. Genetic constraints on the independent evolution of male and female reproductive characters in the tristylous plant *Lythrum salicaria*. *Evolution* 47:1457–1471.

Opell, B. D. 1997a. A comparison of capture thread and architectural features of deinopoid and araneoid orb-webs. *J Arachnol* 25:295–306.

———. 1997b. The material cost and stickiness of capture threads and the evolution of orb-weaving spiders. *Biol J Linn Soc* 62:443–458.

———. 1998. Economics of spider orb-webs: The benefits of producing adhesive capture threads and of recycling silk. *Funct Ecol* 12:613–624.

———. 1999. Changes in spinning anatomy and thread stickiness associated with the origin of orb-weaving spiders. *Biol J Linn Soc* 68:593–612.

Opell, B. D., and J. E. Bond. 2001. Changes in the mechanical properties of capture threads and the evolution of modern orb weaving spiders. *Evol Ecol Res* 3:567–581.

Orr, H. A. 1990. "Why polyploidy is rarer in animals than in plants" revisited. *Am Nat* 136:759–770.

Orr, H. A., and J. A. Coyne. 1992. The genetics of adaptation: A reassessment. *Am Nat* 140:725–742.

Orr, H. A., and L. H. Orr. 1996. Waiting for speciation: The effect of population subdivision on the time to speciation. *Evolution* 50:1742–1749.

Orr, M. R., and T. B. Smith. 1998. Ecology and speciation. *Trends Ecol Evol* 13:502–506.

Ortí, G., M. A. Bell, T. E. Reimchen, and A. Meyer. 1994. Global survey of mitochondrial DNA sequences in the threespine stickleback: Evidence for recent migrations. *Evolution* 48:608–622.

Ortolani, A. 1999. Spots, stripes, tail tips, and dark eyes: Predicting the function of carnivore colour patterns using the comparative method. *Biol J Linn Soc* 67:433–476.

Osborn, H. F. 1910. *The age of mammals in Europe, Asia and North America.* New York: MacMillan.

———. 1934. Aristogenesis, the creative principle in the origin of species. *Am Nat* 68:193–235.

Osche, G. 1958. Beiträge zur Morphologie, Ökologie, und Phylogenie der Ascaridoidea. Parallelen in der Evolution von Parasit und Wirt. *Z Parasitenkd* 18:479–572.

———. 1960. Systematische, morphologische und parasitophyletische studien an parasitischen Oxyuroidea (Nematoda) exotischer Diplopoden (ein Beitrag zur Morphologie des Sexualdimorphismus). *Zool Jahrb* 87:395–440.

———. 1963. Morphological, biological, and ecological considerations in the phylogeny of parasitic nematodes. In *The lower metazoa, comparative biology, and phylogeny,* ed. E. C. Dougherty, 283–302. Berkeley: Univ. California Press.

Ospovat, D. 1978. Perfect adaptation and teleological explanation. *Stud Hist Biol* 2:33–56.

———. 1981. *The development of Darwin's theory.* Cambridge: Cambridge Univ. Press.

Ota, R., P. J. Waddell, M. Hasegawa, H. Shimodaira, and H. Kishino. 2000. Appropriate likelihood ratio tests and marginal distributions for evolutionary tree models with constraints on parameters. *Mol Biol Evol* 17:798–803.

Otte, D. 1989. Speciation in Hawaiian crickets. In *Speciation and its consequences,* ed. D. Otte and J. A. Endler, 482–526. Sunderland, MA: Sinauer.

Ou, C.-Y., C. A. Ciesielski, G. Myers, C. I. Bandea, C.-C. Luo, B. T. M. Korber, J. I. Mullins, G. Schochetam, R. L. Berkelman, A. N. Economou, J. J. Witte, L. J. Furman, G. A. Satten, K. A. McInnes, J. W. Curran, H. W. Jaffe, Laboratory Investigation Group, Epidemiologic Investigation Group. 1992. Molecular epidemiology of HIV transmission in a dental practice. *Science* 256:1165–1171.

Owen, D. F., D. A. S. Smith, I. J. Gordon, and A. M. Owiny. 1993. Polymorphic Müllerian mimicry in a group of African butterflies: A re-assessment of the relationships between *Danaus chryippus, Acraea encedon,* and *Acraea encedana* (Lepidoptera: Nymphalidae). *J Zool* 231:93–108.

Owen, R. B., R. Crossley, T. C. Johnson, D. Tweddle, I. Kornfield, S. Davison, D. H. Eccles, and D. E. Endstrom. 1990. Major low levels of Lake Malawi and their implications for speciation rates in cichlid fishes. *Proc Roy Soc Lond* B 240:519–553.

Packer, L. 1991. The evolution of social behavior and nest architecture in sweat bees of the subgenus *Evylaeus* (Hymenoptera: Halictidae): A phylogenetic approach. *Behav Ecol Sociobiol* 29:153–160.

———. 1997. The utility of phylogenetic systematics in biology: Examples from medicine and behavioural ecology. *Mem Mus Natl Hist Nat* 173:11–29.

———. 1998. A phylogenetic analysis of western European species of the *Lasioglossum leucozonium* species-group (Hymenoptera: Halictidae): Sociobiological and taxonomic implications. *Can J Zool* 76:1611–1621.

Packer, L., and J. S. Taylor. 1997. How many hidden species are there? An application of the phylogenetic species concept to genetic data for some comparatively well known bee "species." *Can Entomol* 129:587–594.

Page, R. D. M. 1987. Graphs and generalized tracks: Quantifying Croizat's panbiogeography. *Syst Zool* 36:1–17.

———. 1988. Quantitative cladistic biogeography: Constructing and comparing area cladograms. *Syst Zool* 37:254–270.

———. 1990. Temporal congruence and cladistic analysis of biogeography and cospeciation. *Syst Zool* 39:205–226.

———. 1993. *COMPONENT: Tree comparison software for Microsoft Windows.* Vers. 2. 0. London: Natural History Museum.

———. 1994. Parallel phylogenies: Reconstructing the history of host-parasite assemblages. *Cladistics* 10:155–173.

————. 1996. On consensus, confidence, and "total evidence." *Cladistics* 12:83–92.

Page, R. D. M., and M. A. Charleston. 1998. Trees within trees: Phylogeny and historical associations. *Trends Ecol Evol* 13:356–359.

Pagel, M. D. 1992. A method for the analysis of comparative data. *J Theor Biol* 156:431–442.

————. 1994a. The adaptationist wager. In *Phylogenetics and ecology*, ed. P. Eggleton and R. Vane-Wright, 29–51. Linn Soc Symp Series 17. London: Academic Press.

————. 1994b. Detecting correlated evolution on phylogenies: A general method for the comparative analysis of discrete characters. *Proc Roy Soc Lond* B 255:37–45.

————. 1999. Inferring the historical patterns of biological evolution. *Nature* 401:877–884.

Pagel, M. D., and P. H. Harvey. 1992. On solving the correct problem: Wishing does not make it so. *J Theor Biol* 156:425–430.

Paine, R. T. 1966. Food web complexity and species diversity. *Am Nat* 100:65–75.

Paley, W. 1802. *Natural theology, or evidences of the existence and attributes of the Deity collected from the appearance of Nature*. By the author.

Pandolfi, J. M. 1992. Successive isolation rather than evolutionary centres for the origination of Indo-Pacific reef corals. *J Biogeogr* 19:593–609.

Pannell, J. R., and B. Charlesworth. 1999. Neutral genetic diversity in a metapopulation with recurrent local extinction and recolonization. *Evolution* 53:664–676.

Papp, L. 1982. Cavernicolous Diptera of the Genova Museum. *Rev Suisse Zool* 89:7–22.

Park, O. 1945. Observations concerning the future of ecology. *Ecology* 26:1–9.

Park, T., and M. B. Frank. 1948. The fecundity and development of the flour beetles, *Tribolium confusum* and *Tribolium castaneum*, at three constant temperatures. *Ecology* 29:368–374.

Parker, J. S. 1930. Some effects of temperature and moisture upon *Melanoplus mexicanus* Saussure and *Camnula pellucida* Scudder (Orthoptera). *Univ Montana Agric Exp Sta Bull* 223.

Parkinson, C. L., K. R. Zamudio, and H. W. Greene. 2000. Phylogeography of the pit-viper clade *Agkistrodon:* Historical ecology, species status, and conservation of cantils. *Mol Ecol* 9:411–420.

Parrish, J. D., and S. B. Saila. 1970. Interspecific competition, predation, and species diversity. *J Theor Biol* 27:207–220.

Parsons, P. A. 1993. Stress, extinctions, and evolutionary change: From living organisms to fossils. *Biol Rev* 68:313–333.

Paterson, A. M., R. D. Gray, and G. P. Wallis. 1995. Penguins, petrels, and parsimony: Does cladistic analysis of behaviour reflect seabird phylogeny? *Evolution* 49:974–989.

Paterson, H. E. H. 1978. More evidence against speciation by reinforcement. *S Afr J Sci* 74:369–371.

————. 1981. The continuing search for the unknown and the unknowable: A critique of contemporary ideas on speciation. *S Afr J Sci* 77:119–133.

————. 1982. Perspective on speciation by reinforcement. *S Afr J Sci* 78:53–57.

————. 1985. The recognition concept of species. In *Species and speciation*, ed. E. S. Vrba, 21–29. Transvaal Mus Monogr 4. Pretoria: Transvaal Museum.

————. 1987. A view of species. *Riv Biol—Biology Forum* 80:211–215.

Patten, B. C. 1978. Systems approach to the concept of environment. *Ohio J Sci* 78:206–222.

————. 1982. Environs: Relativistic elementary particles for ecology. *Am Nat* 119:179–219.

Patterson, C. 1981a. Significance of fossils in determining evolutionary relationships *Annu Rev Ecol Syst* 12:195–223.

————. 1982. Morphological characters and homology. In *Problems of phylogeny reconstruction,* ed. K. A. Joysey and A. E. Friday, 21–74. London: Academic Press.

————, ed. 1987. *Molecules and morphology in evolution: Conflict or compromise?* Cambridge: Cambridge Univ. Press.

Patterson, C., D. M. Williams, and C. J. Humphries. 1993. Congruence between molecular and morphological phylogenies. *Annu Rev Ecol Syst* 24:153–188.

Patton, J. L., M. N. F. da Silva, and J. R. Malcolm. 2000. Mammals of the Rio Jurua and the evolutionary and ecological diversification of Amazonia. *Bull Am Mus Nat Hist* 244:3–306.

Patton, J. L., and S. W. Sherwood. 1983. Chromosome evolution and speciation in rodents. *Annu Rev Ecol Syst* 14:139–158.

Patton, J. L., and M. F. Smith. 1989. Population structure and the genetic and morphologic divergence among pocket gopher species (genus *Thonomys*). In *Speciation and its consequences*, ed. D. Otte and J. A. Endler, 284–304. Sunderland, MA: Sinauer.

Pavan, C., T. Dobzhansky, and H. Burla. 1950. Diurnal behavior of some neotropical species of *Drosophila*. *Ecology* 31:36–43.

Pavesi, A. 2001. Origin and evolution of GBV C/Hepatitis G virus and relationships with ancient human migrations. *J Mol Evol* 53:104–113.

PCAST (President's Committee of Advisors on Science and Technology). 1998. *Teaming with life: Investing in science to understand and use America's living capital*. PCAST Panel on Biodiversity and Ecosystems. Washington, DC: U.S. Govt. Printing Office.

Pearson, D. L. 1990. The evolution of multi anti-predator characteristics as illustrated by tiger beetles. *Florida Entomol* 73:67–70.

Pearson, D. L., M. S. Blum, T. H. Jones, H. M. Fales, E. Gonda, and B. R. Witte. 1988. Historical perspective and the interpretation of ecological patterns: Defensive compounds of tiger beetles (Coleoptera: Cicindelidae). *Am Nat* 132:404–416.

Pease, C. M., R. Lande, and J. J. Bull. 1989. A model of population growth, dispersal, and evolution in a changing environment. *Ecology* 70:1657–1664.

Peat, H. J., and A. H. Fitter. 1994. Comparative analyses of ecological characteristics of British angiosperms. *Biol Rev* 69:95–115.

Peck, S. B. 1978. New montane *Ptomophagus* beetles from New Mexico and zoogeography of southwestern caves (Coleoptera; Leiodidae; Catopinae). *Southwest Nat* 23:227–238.

———. 1990. Eyeless arthropods of the Galapagos Islands, Ecuador: Composition and origin of the Cryptozoic fauna of a young, tropical, oceanic archipelago. *Biotropica* 22:366–381.

Pellmyr, Ø. 1992a. Evolution of insect pollination and angiosperm diversification. *Trends Ecol Evol* 7:46–49.

———. 1992b. The phylogeny of a mutualism: Evolution and coadaptation between *Trollius* and its seed-parasitic pollinators. *Biol J Linn Soc* 47:337–365.

Pellmyr, Ø., and C. J. Huth. 1994. Evolutionary stability of mutualism between yuccas and yucca moths. *Nature* 372:257–260.

Pellmyr, Ø., and J. Leebens-Mack. 1999. Forty million years of mutualism: Evidence for Eocene origin of the yucca-yucca moth association. *Proc Natl Acad Sci* (USA) 96:9178–9183.

———. 2000. Reversal of mutualism as a mechanism for adaptive radiation in yucca moths. *Am Nat* 156:S62–S76.

Pellmyr, Ø., J. Leebens-Mack, and C. J. Huth. 1996. Non-mutualistic yucca moths and their evolutionary consequences. *Nature* 380:155–156.

Pellmyr, Ø., J. Leebens-Mack, and J. N. Thompson. 1998. Herbivores and molecular clocks as tools in plant biogeography. *Biol J Linn Soc* 63:367–378.

Pellmyr, Ø., L. K. Massey, J. L. Hamrick, and M. A. Feist. 1997. Genetic consequences of specialization: Yucca moth behavior and self-pollination in yuccas. *Oecologia* 109:273–278.

Pellmyr, Ø., and J. N. Thompson. 1992. Multiple occurrences of mutualism in the yucca moth lineage. *Proc Natl Acad Sci* (USA) 89:2927–2929.

Pellmyr, Ø., J. N. Thompson, J. M. Brown, and R. G. Harrison. 1996. Evolution of pollination and mutualism in the yucca moth lineage. *Am Nat* 148:827–847.

Pennings, S. C. 1990. Multiple factors promoting narrow host range in the sea hare, *Aplysia californica*. *Oecologia* 82:192–200.

———. 1994. Interspecific variation in chemical defenses in the sea hares (Opisthobranchia: Anaspidea). *J Exper Marine Biol Ecol* 180:203–219.

Pérez, T. M., and W. T. Atyeo. 1984. Site selection of the feather and quill mites of Mexican parrots. In *Acarology IV*, vol. 1, ed. D. A. Griffiths and C. E. Bowman, 563–570. Chichester: Horwood.

Pérez-Ponce de León, G. 1997. La taxonomia en México: El papel de la sistematica filogenetica. *Ciencia* 48:33–39.

Pérez-Ponce de León, G., V. León-Règagnon, and L. García-Prieto. 1997. ¿Que es la sistematica filogenetica? *Ciencia y desarollo* 135:61–65.

Perkins, R. C. L. 1903. Vertebrata. In *Fauna hawaiiensis,* ed. D. Sharp, 365–466. Cambridge: Cambridge Univ. Press.

———. 1913. Introduction to *Fauna hawaiiensis,* ed. D. Sharp, xv–ccxxviii, Cambridge: Cambridge Univ. Press.

Perrin, N., and J. Travis. 1992. On the use of constraints in evolutionary biology and some allergic reactions to them. *Funct Ecol* 6:361–363.

———. 1994. Reply to TALMUD. *Funct Ecol* 8:140.

Peters, W., P. C. C. Garnham, R. Killick-Kendrick, N. Rajapaksa, W. Cheong, and F. Cadigan. 1976. Malaria of the orang-utan (*Pongo pygmaeus*) in Borneo. *Proc Roy Soc Lond* B 275:439–482.

Peterson, A. T. 1992. Phylogeny and rates of molecular evolution in the *Aphelocoma* jays (Corvidea). *Auk* 109:133–147.

Peterson, C. H., and R. Black. 1988. Density-dependent mortality caused by stress interacting with biotic history. *Am Nat* 131:257–270.

Petren, K., and T. J. Case. 1997. A phylogenetic analysis of body size evolution and biogeography in chuckwallas (*Sauromalus*) and other iguanines. *Evolution* 51:206–219.

Petrunkevitch, A. 1926. The value of instinct as a taxonomic character in spiders. *Biol Bull* 50:427–432.

Pfefferkorn, H. W. 1999. Recuperation from mass extinctions. *Proc Natl Acad Sci* (USA) 96:13597–13599.

Phelps, S. M. 2001. History's lessons: A neural network approach to receiver biases and the evolution of communication. In *Recent advances in anuran communication,* ed. M. J. Ryan, 167–180. Washington, DC: Smithsonian Institution Press.

Phelps, S. M., and M. J. Ryan. 2000. History influences signal recognition: Neural network models of túngara frogs. *Proc Roy Soc Lond* B 267:1633–1639.

———. 1998. Neural networks predict the response biases of female túngara frogs. *Proc Roy Soc Lond* B 265:279–285.

Philips, A., D. Janies, and W. Wheeler. 2000. Multiple sequence alignment in phylogenetic analysis. *Mol Phylogenet Evol* 16:317–330.

Phillips, C. A. 1994. Geographic distribution of mitochondrial DNA variants and the historical biogeography of the spotted salamander, *Ambystoma maculatum. Evolution* 48:597–607.

Phillips, R. B., M. P. Matsuoka, I. Konon, and K. M. Reed. 2000. Phylogenetic analysis of mitochondrial and nuclear sequences supports inclusion of *Acantholingua ohridana* in the genus *Salmo. Copeia* 2000:546–550.

Pianka, E. R. 1975. Niche relations of desert lizards. In *Ecology and evolution of communities,* ed. M. L. Cody and J. M. Diamond, 292–314. Cambridge, MA: Harvard Univ. Press.

———. 1980. Guild structure in desert lizards. *Oikos* 35:194–201.

Pigliucci, M., and J. Kaplan. 2000a. The fall and rise of Dr. Pangloss: Adaptationism and the Spandrels paper 20 years later. *Trends Ecol Evol* 15:66–70.

———. 2000b. Reply from M. Pigliucci and J. Kaplan. *Trends Ecol Evol* 16:249.

Pigliucci, M., G. A. Tyler, and C. D. Schlichting. 1998. Mutational effects on constraints on character evolution and phenotypic plasticity in *Arabidopsis thaliana. J Genet* 77:95–103.

Pimentel, D., and R. Al-Hafidh. 1965. Ecological control of a parasite population by genetic evolution in the host-parasite system. *Ann Entomol Soc Am* 58:1–6.

Pimentel, D., and A. B. Soans. 1972. Animal population regulated to carrying capacity of plant host by genetic feedback. In *Dynamics of populations: Proceedings of the Advanced Study Institute on Dynamics of Numbers in Populations, Oosterbeek, the Netherlands, 7–8 September, 1970,* ed. P. J. den Boer and G. R. Gradwell, 313–326. Wageningen: Pudoc.

Pimentel, D., and F. A. Stone, 1968. Evolution and population ecology of parasite-host systems. *Can Entomol* 100:455–468.

Pimentel, D., C. Wilson, C. McCullum, R. Huang, P. Owen, J. Flack, Q. Tran, T. Saltman, and B. Cliff. 1997. Economic and environmental benefits of biodiversity. *BioScience* 47:747–756.

Pimm, S. L. 1991. *The balance of nature?* Chicago: Univ. Chicago Press.

Pimm, S. L., and J. L. Gittleman. 1992. Biological diversity: Where is it? *Science* 255:940.

Pimm, S. L., H. L. Jones, and J. Diamond. 1988. On the risk of extinction. *Am Nat* 132: 757–785.

Pine, R. H. 1994. Systematics and biodiversity. *Trends Ecol Evol* 9:229.

Pisano, E., and C. Ozouf-Costaz. 2000. Chromosome change and evolution in the Antarctic fish order Notothenoidei. *Antarct Sci* 12:334–342.

Pitelka, L. F. 1977. Energy allocation in annual and perennial lupines (*Lupinus:* Leguminosae). *Ecology* 58:1055–1065.

Plath, O. E. 1934. *Bumblebees and their ways.* New York: Macmillan.

Platnick, N. I. 1989. *Advances in spider taxonomy 1981–1987.* New York: Manchester Univ. Press.

———. 1993. Patterns of biodiversity. In *Systematics, ecology, and the biodiversity crisis,* ed. N. Eldredge, 15–24. New York: Columbia Univ. Press.

Platnick, N. I., and E. S. Gaffney. 1977. Systematics: A Popperian perspective. [Reviews of "The logic of scientific discovery" and "Conjectures and refutations," by Karl R. Popper.] *Syst Zool* 26:360–365.

———. 1978a. Evolutionary biology: A Popperian perspective. *Syst Zool* 27:137–141.

———. 1978b. Systematics and the Popperian paradigm. *Syst Zool* 27:381–388.

Platnick, N. I., C. E. Griswold, and J. A. Coddington. 1991. On missing entries in cladistic analysis. *Cladistics* 7:337–343.

Platnick, N. I., C. J. Humphries, G. Nelson, and D. M. Williams. 1996. Is Farris optimization perfect?: Three-taxon statements and multiple branching. *Cladistics* 12:243–252.

Platnick, N. I., and G. Nelson. 1978. A method of analysis for historical biogeography. *Syst Zool* 27:1–16.

Polhemus, D. A. 1996. Two new species of *Rhagovelia* from the Philippines, with a discussion of zoogeographic relationships between the Philippines and New Guinea (Heteroptera: Veliidae). *J NY Ent Soc* 103:55–68.

Ponder, W., and D. Lunney, eds. 1999. *The other 99%: The conservation and biodiversity of invertebrates. Trans Roy Soc NSW.* Mosman: Royal Society of New South Wales.

Popper, K. R. 1968. *The logic of scientific discovery.* New York: Harper and Row.

———. 1992. *Realism and the aim of science.* London: Routledge.

Pough, F. H., and J. D. Groves. 1983. Specializations of the body form and food habits of snakes. *Am Zool* 23:443–454.

Poulin, R. 1998. *Evolutionary ecology of parasites.* New York: Chapman and Hall.

Pounds, J. A. 1988. Ecomorphology, locomotion, and microhabitat structure: Patterns in a tropical mainland *Anolis* community. *Ecol Monogr* 58:299–320.

Power, M. E. 1990. Effects of fish in river food webs. *Science* 250:811–814.

Prance, G. T. 1973. Phytogeographic support for the theory of Pleistocene forest refuges in the Amazon Basin based on evidence from distribution patterns in Caryocaraceae, Chrysobalanaceae, Dichapetalaceae, and Lecythidaceae. *Acta Amazonica* 3:5–28.

———, ed. 1982a. *Biological diversification in the tropics.* New York: Columbia Univ. Press.

———. 1982b. Forest refuges: Evidence from woody gymnosperms. In *Biological diversification in the tropics,* ed. G. T. Prance, 137–157. New York: Columbia Univ. Press.

———. 1995. Systematics, conservation, and sustainable development. *Biodiv Conserv* 4:490–500.

Pratt, G. F. 1994. Evolution by *Euphilotes* (Lepidoptera: Lycaenidae) by seasonal and host shifts. *Biol J Linn Soc* 51:387–416.

Presch, W. 1989. Systematics and science: A comment. *Syst Zool* 38:181–189.

President's Committee of Advisors on Science and Technology. *See* PCAST.

Preston, E. M. 1973. A computer simulation of competition among five sympatric congeneric species of xanthid crabs. *Ecology* 54:469–483.

Price, P. W. 1980. *Evolutionary biology of parasites.* Princeton, NJ: Princeton Univ. Press.

Price, P. W., and T. G. Carr. 2000. Comparative ecology of membracids and tenthredinids in a macroevolutionary context. *Evol Ecol Res* 2:645–665.

Price, T. 1998. Sexual selection and natural selection in bird speciation. *Phil Trans Roy Soc Lond* B 353:251–260.

Price, T., I. J. Lovette, E. Bermingham, H. Lisle Gibbs, and A. D. Richman. 2000. The imprint of history on communities of North American and Asian warblers. *Am Nat* 156:354–367.

Prum, R. O. 1990. Phylogenetic analysis of the evolution of display behaviour in the Neotropical manikins (Aves: Pipridae). *Ethology* 84:202–231.

——. 1993. Phylogeny, biogeography, and evolution of the broadbills (Eurylaimidae) and asities (Phileptittidae) based on morphology. *Auk* 110:304–324.

——. 1994. Phylogenetic analysis of the evolution of alternative social behavior in the manakins (Aves: Pipridae). *Evolution* 48:1657–1675.

Ptacek, M. B., and J. Travis. 1998. Hierarchical patterns of covariance between morphological and behavioural traits. *Anim Behav* 56:1044–1048.

Ptacek, M. B., H. C. Gerhardt, and R. D. Sage. 1994. Speciation by polyploidy in tree frogs: Multiple origins of the tetraploid, *Hyla versicolor. Evolution* 48:898–908.

Purvis, A. 1995. A modification of Baum and Ragan's method for combining phylogenetic trees. *Syst Biol* 44:251–255.

——. 1996. Using interspecies phylogenies to test macroevolutionary hypotheses. In *New uses for new phylogenies,* ed. P. H. Harvey, A. J. Leigh Brown, J. Maynard Smith, and S. Nee, 153–168. Oxford: Oxford Univ. Press.

Purvis, A., P. M. Agapow, J. L. Gittleman, and G. M. Mace. 2000. Nonrandom extinction and the loss of evolutionary history. *Science* 288:328–330.

Purvis, A., S. Nee, and P. H. Harvey. 1995. Macroevolutionary inferences from primate phylogeny. *Proc Roy Soc Lond* B 260:329–333.

Purvis, A., and A. Rambaut. 1995. Comparative analysis by independent contrasts (CAIC): An Apple Macintosh application for analysing comparative data. *Comp Appl Biosciences* 11:247–251.

Putnam, R. J., and S. D. Wratten. 1984. *Principles of ecology.* London: Croom Helm.

Qiu, Y. L., J. Lee, D. E. Soltis, P. S. Soltis, M. Zanis, E. A. Zimmer, Z. Chen, V. Savolainen, and M. W. Chase. 2000. Phylogeny of basal angiosperms: Analyses of five genes from three genomes. *Int J Plant Sci* 161:S3–S27.

Queiroz, A. de. 1999. Do image-forming eyes promote evolutionary diversification? *Evolution* 53:1654–1664.

Queiroz, A. de, and P. H. Wimberger. 1993. The usefulness of behavior for phylogeny estimation: Levels of homoplasy in behavioral and morphological characters. *Evolution* 47:46–60.

Queiroz, K. de. 1985. The ontogenetic method for determining character polarity and its relevance to phylogenetic systematics. *Syst Zool* 34:280–299.

——. 1988. Systematics and the Darwinian revolution. *Philos Sci* 55:238–259.

——. 1996. Including the characters of interest during tree reconstruction and the problems of circularity and bias in studies of character evolution. *Am Nat* 148:700–708.

Queiroz, K. de, and M. J. Donoghue. 1990. Phylogenetic systematics or Nelson's version of cladistics? *Cladistics* 4:317–338.

Queiroz, K. de, and M. J. Donoghue, and J. Kim. 1995. Separate versus combined analysis of phylogenetic data. *Annu Rev Ecol Syst* 26:657–681.

Radinsky, L. 1975. Primate brain evolution. *Am Sci* 63:656–663.

Radtkey, R. R., S. M. Fallon, and T. J. Case. 1997. Character displacement in some *Cnemidophorus* lizards revisited: A phylogenetic analysis. *Proc Natl Acad Sci* (USA) 94: 9740–9745.

Radtkey, R. R., and M. C. Singer. 1995. Repeated reversals of host-preference evolution in a specialist insect herbivore. *Evolution* 49:351–359.

Raff, R. A. 1996. *The shape of life: Genes, development, and the evolution of animal form.* Chicago: Univ. Chicago Press.

Raff, R. A., and E. C. Raff, eds. 1987. *Development as an evolutionary process: Proceedings of a meeting held at the Marine Biological Laboratory in Woods Hole, Massachusetts, August 23 and 24, 1985.* New York: A. R. Liss.

Ragan, M. A. 1992. Phylogenetic inference based on matrix representation of trees. *Mol Phylogenet Evol* 1:53–58.

Raikow, R. J. 1988. An analysis of evolutionary success. *Syst Zool* 37:76–79.

——. 1990. Individual variation in avian muscles and its significance for the reconstruction of phylogeny. *Syst Zool* 39:362–370.

Raikow, R. J., and A. H. Bledsoe. 2000. Phylogeny and evolution of the passerine birds. *BioScience* 50:487–499.

Rainey, P. B., and M. Travisano. 1998. Adaptive radiation in a heterogeneous environment. *Nature* 394:69–72.

Ramey, R. R., S. T. Kelley, W. M. Boyce, and B. D. Farrell. 2000. Phylogeny and host specificity of psoroptic mange mites (Acarina: Psoroptidae) as indicated by TTS sequence data. *J Med Entomol* 37:791–796.

Rand, A. S. 1964. Ecological distribution in anoline lizards of Puerto Rico. *Ecology* 45: 745–757.

Rand, A. S., and M. J. Ryan. 1981. The adaptive significance of a complex vocal repertoire in a Neotropical frog *Physalaemus pustulosus. Z Tierpsychol* 57:209–214.

Rand, A. S., and E. E. Williams. 1969. The anoles of La Palma: Aspects of their ecological relationships. *Breviora* 327:1–19.

Räsänen, M. E., J. S. Salo, and R. J. Kalliola. 1987. Fluvial perturbance in the western Amazonian Basin: Regulation by long-term sub-Andean tectonics. *Science* 238:1398–1401.

Rau, P. 1929. The habitat and dissemination of four species of *Polistes* wasps. *Ecology* 10:191–200.

———. 1931. Polistes wasps and their use of water. *Ecology* 12:690–693.

Raup, D. M. 1977. Stochastic models in evolutionary palaeontology. In *Patterns of evolution as illustrated by the fossil record,* ed. A. Hallum, 59–78. Amsterdam: Elsevier.

———. 1981. Extinction: Bad genes or bad luck? *Acta Geol Hisp* 16:25–33.

———. 1985. Mathematical models of cladogenesis. *Paleobiology* 11:42–52.

———. 1991. A kill curve for Phanerozoic marine species. *Paleobiology* 17:37–48.

———. 1993. *Extinction: Bad genes or bad luck?* Oxford: Oxford Univ. Press.

———. 1996. Extinction models. In *Evolutionary paleobiology,* ed. D. Jablonski, D. H. Erwin, and J. H. Lipps, 419–433. Chicago: Univ. Chicago Press.

Raup, D. M., and S. J. Gould. 1974. Stochastic simulation and evolution of morphology: Towards a nomothetic paleontology. *Syst Zool* 23:305–322.

Raup, D. M., S. J. Gould, T. J. M. Schopf, and D. S. Simberloff. 1973. Stochastic models of phylogeny and the evolution of diversity. *J Geol* 81:525–542.

Raup, D. M., and J. J. Sepkoski Jr. 1982. Mass extinctions in the marine fossil record. *Science* 215:1501–1503.

Rausher, M. D. 1988. Is coevolution dead? *Ecology* 69:898–901.

———. 2001. Coevolution and plant resistance to natural enemies. *Nature* 411:857–864.

Raven, P. H. 1977. A suggestion concerning the Cretaceous rise to dominance of the angiosperms. *Evolution* 31:451–452.

———, ed. 1980. *Research priorities in tropical biology.* Washington, DC: National Academy of Sciences Press.

———. 1988. Biological resources and global stability. In *Evolution and coadaptation in biotic communities,* ed. S. Kawano, J. H. Connell, and T. Hidaka, 3–27. Tokyo: Univ. Tokyo Press.

———. 1992. The nature and value of biodiversity. In *Global biodiversity strategy,* 1–5. New York: World Resources Institute, World Conservation Union, United National Environmental Programme.

Raven, P. H., and D. I. Axelrod. 1974. Angiosperm biogeography and past continental movements. *Ann Missouri Bot Gard* 68:539–673.

Read, A. F., and S. Nee. 1995. Inference from binary comparative data. *J Theor Biol* 173: 99–108.

Reaka-Kudla, M. L., D. O. Wilson, and E. O. Wilson, eds. 1997 *Biodiversity II: Understanding and protecting our biological resources.* Washington, DC: Joseph Henry Press.

Reeve, H. K., and P. W. Sherman. 1993. Adaptation and the goals of evolutionary research. *Quart Rev Biol* 68:2–31.

Regan, C. T. 1926. Organic Evolution. *Report of the British Association for the Advancement of Science* 1925:75–86.

Reilly, S. M., and G. V. Lauder. 1992. Morphology, behavior, and evolution: Comparative kinematics of aquatic feeding in salamanders. *Brain Behav Ecol* 40:182–196.

Reilly, S. M., E. O. Wiley, and D. J. Meinhardt. 1997. An integrative approach to heterochrony: The distinction between interspecific and intraspecific phenomena. *Biol J Linn Soc* 60:119–143.

Reimchen, T. E. 1989. Loss of nuptial color in threespine sticklebacks (*Gasterosteus aculeatus*). *Evolution* 43:450–460.

Reiss, J. O. 1989. The meaning of developmental time: A metric for comparative embryology. *Am Nat* 134:170–189.

Reist, J. D. 1980. Selective predation upon pelvic phenotypes of brook stickleback, *Culaea inconstans*, by selected invertebrates. *Can J Zool* 58:1253–1258.

Reisz, R., and H. Sues. 1998. Origins and early evolution of herbivory in tetrapods. *Trends Ecol Evol* 13:141–145.

Rejmanuk, M., P. S. Ward, G. L. Webster, and J. M. Randall. 1994. Systematics and biodiversity. *Trends Ecol Evol* 9:228–229.

Remane, A. 1956. *Die Gründlagen des naturlichen System der vergleichenden Anatomie und Phylogenetik*. 2. Leipzig: Geest und Portig, KG.

———. 1961. Gedanken zum Problem: Homologie und Analogie, Preadaptation und Parallelität. *Zool Anz* 166:447–470.

Renner, S. S. 1999. Circumscription and phylogeny of the Laurales: Evidence from molecular and morphological data. *Am J Bot* 86:1301–1315.

Renner, S. S., and R. E. Ricklefs. 1994. Systematics and biodiversity. *Trends Ecol Evol* 9:78.

Renzaglia, K. S., R. J. Duff, D. L. Nickrent, and D. J. Garbary. 2000. Vegetative and reproductive innovations of early land plants: Implications for a unified phylogeny. *Phil Trans Roy Soc Lond* B 355:769–793.

Reydon, T. 2000. Monists, pluralists, and biologists. In *Global village: Acta 22e Nederlands-Vlaamse Filosofiedag*. http://www.leidenuniv.nl/philosophy/publicaties/filosophidag/index.html.

Reynolds, R. P., and N.J. Scott. 1982. Use of a mammalian resource by a Chihuahuan snake community. *U.S. Dep Inter, Fish Wildl Res Rep* 13:99–118.

Ribbink, A. J. 1994. Alternative perspectives on some controversial aspects of cichlid fish speciation. *Arch Hydriobiol Beih Ergebn Limnol* 44:101–125.

Rice, R. R., and E. E. Hostert. 1993. Laboratory experiments on speciation: What have we learned in 40 years? *Evolution* 47:1637–1653.

Rice, S. H. 1997. The analysis of ontogenetic trajectories: When a change in size or shape is not heterochrony. *Proc Natl Acad Sci* (USA) 94:907–912.

Rice, W. R. 1984. Disruptive selection on habitat preference and the evolution of reproductive isolation: A simulation study. *Evolution* 38:1251–1260.

———. 1985. Disruptive selection on habitat preference and the evolution of reproductive isolation: An exploratory experiment. *Evolution* 39:1645–1656.

———. 1987. Speciation via habitat specialization. *Evol Ecol* 1:301–314.

Rice, W. R., and G. W. Salt. 1988. Speciation via disruptive selection on habitat preference: Experimental evidence. *Am Nat* 129:183–187.

———. 1990. The evolution of reproductive isolation as a correlated character under sympatric conditions: Experimental evidence. *Evolution* 44:1140–1152.

Richardson, P. M. 1996. The Systematics Agenda 2000 Symposium: Introduction. *Ann Missouri Bot Gard* 83:1–2.

Richman, A. D. 1996. Ecological diversification and community structure in the Old World leaf warblers (genus *Phylloscopus*): A phylogenetic perspective. *Evolution* 50:2461–2470.

Ricklefs, R. E. 1987. Community diversity: Relative roles of local and regional processes. *Science* 235:167–171.

———. 1989. Speciation and diversity: The integration of local and regional processes. In *Speciation and its consequences*, ed. D. Otte and J. A. Endler, 599–622. Sunderland, MA: Sinauer.

———. 1990. *Ecology*. 3d ed. San Francisco: W. H. Freeman.

Ricklefs, R. E., and G. W. Cox. 1972. Taxon cycle of the West Indian avifauna. *Am Nat* 106:195–219.

———. 1978. Stage of taxon cycle, habitat distribution, and population density in the avifauna of the West Indies. *Am Nat* 112:875–895.

Ricklefs, R. E., and G. L. Miller. 1999. *Ecology*. 4th ed. San Francisco: W. H. Freeman.

Ricklefs, R. E., and S. S. Renner. 1994. Species richness within families of flowering plants. *Evolution* 48:1619–1636.

Ricklefs, R. E., and D. Schluter, eds. 1993. *Species diversity in ecological communities* Chicago: Univ. Chicago Press.

Ricklefs, R. E., and J. M. Starck. 1996. Application of phylogenetically independent contrasts: A mixed progress report. *Oikos* 77:167–173.

Ridder-Numan, J. 1995. Phylogeny and biogeography of *Spatholobus, Butea, Meizotropis*, and *Kunsteria* (Leguminosae-Papilionoideae). In *Advances in legume systematics,* vol. 7, ed. M. Crisp and J. J. Doyle, 133–139. Kew: Royal Botanical Gardens.

————. 1996. Historical biogeography of the Southeast Asian genus *Spatholobus* (Legum.-Papilionoideae) and its allies. *Blumea,* suppl. no. 10:1–144.

Riddle, B. R. 1996. The molecular phylogeographic bridge between deep and shallow history in continental biotas. *Trends Ecol Evol* 11:207–211.

Riddle, B. R., and D. J. Hafner. 1999. Species as units of analysis in ecology and biogeography: Time to take the blinders off. *Global Ecol Biogeogr* 8:433–441.

Riddle, B. R., D. J. Hafner, L. F. Alexander, and J. R. Jaeger. 2000. Cryptic vicariance in the historical assembly of a Baja California peninsular desert biota. *Proc Natl Acad Sci* (USA) 97:14438–14443.

Ridgeway, M. S., and J. D. McPhail. 1984. Ecology and evolution of sympatric sticklebacks (*Gasterosteus*): Mate choice and reproductive isolation in the Enos Lake species pair. *Can J Zool* 62:1813–1818.

Ridley, M. 1983. *The explanation of organic diversity: The comparative method and adaptations for mating.* Oxford: Clarendon Press.

————. 1986. *Evolution and classification.* London: Longman.

————. 1989. The cladistic solution to the species problem. *Biology and Philosophy* 4:1–16.

Riedl, R. 1978. *Order in living organisms.* New York: John Wiley and Sons.

Rieppel, O. 1992. Homology and logical fallacy. *J Evol Biol* 5:701–715.

Rieseberg, L. H., and J. M. Burke. 2001. The biological reality of species: Gene flow, selection, and collective evolution. *Taxon* 50:47–67.

Riggs, R. D. 1977. Worldwide distribution of soybean-cyst nematode and its economic importance. *J Nematol* 9:34–39.

Ringo, J. M. 1977. Why 300 species of Hawaiian *Drosophila*? *Evolution* 31:695–754.

Ritchie, P. A., S. Lavoué, and G Lecointre. 1997. Molecular phylogenetics and the evolution of Antarctic Notothenoid fishes. *Comp Biochem Physiol* 118A:1009–1025.

Robinson, B. W., and D. S. Wilson. 1998. Optimal foraging, specialization, and a solution to Liem's paradox. *Am Nat* 151:223–235.

Robinson, J. V., and J. E. Dickerson Jr. 1987. Does invasion sequence affect community structure? *Ecology* 68:587–595.

Rodhouse, P. G., G. di Prisco, and J. C. Hureau. 2000. Special issue: Evolutionary biology of Antarctic organisms. *Antarct Sci* 12:258.

Roe, K. J., and C. Lydeard. 1998. Species delineation and the identification of evolutionarily significant units: Lessons from the freshwater mussel genus *Potamilus* (Bivalvia: Unionidae). *J Shellfish Res* 17:1359–1363.

Rogers, W. P. 1962. *The nature of parasitism: The relation of some metazoan parasites to their hosts.* New York: Academic Press.

Rognes, K. 1997. The Calliphoridae (Blowflies) (Diptera: Oestridae) are not a monophyletic group. *Cladistics* 13:27–66.

Rohde, K. 1989. At least eight types of sense receptors in an endoparasitic flatworm: A counter-trend to sacculinization. *Naturwissenschaften* 76:383–385.

Ronquist, F. 1994. Ancestral areas and parsimony. *Syst Biol* 43:267–274.

————. 1995. Reconstructing the history of host-parasite associations using generalised parsimony. *Cladistics* 11:73–89.

————. 1996. Matrix representation of trees, redundancy, and weighting. *Syst Biol* 45: 247–253.

————. 1997a. Dispersal-vicariance analysis: A new approach to the quantification of historical biogeography. *Syst Biol* 46:195–203.

————. 1997b. Phylogenetic approaches in coevolution and biogeography. *Zool Scr* 26: 313–322.

————. 1998. Three-dimensional cost-matrix optimization and maximum cospeciation. *Cladistics* 14:167–172.

Ronquist, F., and S. Nylin. 1990. Process and pattern in the evolution of species associations. *Syst Zool* 39:323–344.

Root, R. B. 1967. The niche exploitation theory of the blue-grey gnatcatcher. *Ecol Monogr* 37:317–350.

Rosa, D. 1918. *Ologenesi.* Florence: R. Bemparad.

Rosa, D. 1931. *L'ologenèse: Nouvelle théorie de l'évolution des êtres vivants.* Paris: F. Alcan.

Rose, M. R., and G. V. Lauder, eds. 1996. *Adaptation.* New York: Academic Press.

Rosen, D. E. 1975. A vicariance model of Caribbean biogeography. *Syst Zool* 24:431–464.

———. 1978. Vicariant patterns and historical explanation in biogeography. *Syst Zool* 27:159–188.

———. 1979. Fishes from the uplands and intermontane basins of Guatemala: Revisionary studies and comparative biogeography. *Bull Am Mus Nat Hist* 162: 267–376.

———. 1985. Geological hierarchies and biogeographic congruence in the Caribbean. *Ann Missouri Bot Gard* 72:636–659.

Rosenberg, G. 1996. Independent evolution of terrestriality in Atlantic truncatellid gastropods. *Evolution* 50:682–693.

Rosenzweig, M. L. 1978. Competitive speciation. *Biol J Linn Soc* 10:275–289.

———. 1997. Tempo and mode of speciation. *Science* 277:1622–1623.

Roskam, J. C. 1985. Evolutionary patterns in gall midge-host plant associations (Diptera, Cecidomyiidae). *Tijdschr Entomol* 128:193–213.

———. 1992. Evolution of the gall-inducing guild. In *Biology of insect-induced galls,* ed. J. D. Shorthouse and O. Rohfritsch, 34–49. Oxford: Oxford Univ. Press.

Roskam, J. C., and G. van Uffelen. 1981. Biosystematics of insects living in female birch catkins: III. Plant-insect relations between white birches, *Betula* L., section *Excelsae* (Koch) and gall midges of the genus *Semudobia* Kieffer (Diptera, Cecidomyiidae). *Neth J Zool* 31:533–553.

Ross, H. H. 1972a. The origin of species diversity in ecological communities. *Taxon* 21:253–259.

———. 1972b. An uncertainty principle in ecological evolution. In *A symposium on ecosystematics,* ed. R. T. Allen and F. C. James, 133–157. *Univ Arkansas Mus Occas Pap* 4.

———. 1974. *Biological systematics.* Reading, MA: Addison-Wesley.

Ross, K. G., and J. M. Carpenter. 1991a. Phylogenetic analysis and the evolution of queen number in eusocial Hymenoptera. *J Evol Biol* 4:117–130.

———. 1991b. Population genetic structure, relatedness, and breeding systems. In *The social biology of wasps,* ed. K. G. Ross and R. W. Matthews, 451–479. Ithaca, NY: Cornell Univ. Press.

Ross, R. M., and W. D. Allmon. 1990. *Causes of evolution: A paleontological perspective.* Chicago: Univ. Chicago Press.

Ross, S. T. 1986. Resource partitioning in fish assemblages: A review of field studies. *Copeia* 1986:352–388.

Rossiter, A. 1995. The cichlid fish assemblages of Lake Tanganyika: Ecology, behaviour, and evolution of its species flock. *Adv Ecol Res* 26:187–252.

Rossman, A. Y., and D. R. Miller. 1996. Systematics solves problems in agriculture and forestry. *Ann Missouri Bot Gard* 83:17–28.

Roth, V. L. 1984. On homology. *Biol J Linn Soc* 22:13–29.

———. 1988. The biological basis of homology. In *Ontogeny and systematics,* ed. C. J. Humphries, 1–26. New York: Columbia Univ. Press.

———. 1991. Homology and hierarchies: Problems solved and unresolved. *J Evol Biol* 4:167–194.

———. 1994. Within and between organisms: Replicators, lineages, and homologues. In *Homology: The hierarchical basis of comparative biology,* ed. B. K. Hall, 301–337. New York: Academic Press.

Rotheray, G., and F. Gilbert. 1999. Phylogeny of palaearctic syrphidae (Diptera): Evidence from larval stages. *Zool J Linn Soc* 127:1–112.

Roughgarden, J. 1974. Niche width: Biogeographic patterns among *Anolis* lizard populations. *Am Nat* 108:429–442.

———. 1992. Comments on the paper by Losos: Character displacement versus taxon loop. *Copeia* 1992:288–295.

Roughgarden, J., and S. Pacala. 1989. Taxon cycle among *Anolis* lizard populations: Review of evidence. In *Speciation and its consequences,* ed. D. Otte and J. A. Endler, 403–431. Sunderland, MA: Sinauer.

Roulston, T. H., and S. L. Buchman. 2000. A phylogenetic reconsideration of the pollen starch–pollination correlation. *Evol Ecol Res* 2:627–643.

Rouse, G. W. 1999. Polychaete sperm: Phylogenetic and functional considerations. *Hydrobiologia* 402:215–224.

Rowland, W. J. 1982. The effects of male nuptial coloration on stickleback aggression: A reexamination. *Behaviour* 80:118–126.

———. 1984. The relationships among nuptial coloration, aggression, and courtship of male three-spined sticklebacks, *Gasterosteus aculeatus. Can J Zool* 62:999–1004.

———. 1988. Aggression versus courtship in threespine sticklebacks and the role of habituation to neighbours. *Anim Behav* 36:348–357.

———. 1989. The effects of body size, aggression, and nuptial coloration on competition for territories in male threespine sticklebacks, *Gasterosteus aculeatus. Anim Behav* 37:282–289.

Rowland, W. J., K. J. Bolyard, and A. D. Halpern. 1995a. The dual effect of stickleback nuptial colouration on rivals: Manipulation of a graded signal using video playback. *Anim Behav* 50:267–272.

Rowland, W. J., K. J. Bolyard, J. J. Jenkins, and J. Fowler. 1995b. Video playback experiments on stickleback mate choice: Female motivation and attentiveness to male colour cues. *Anim Behav* 49:1559–1567.

Rüber, L., E. Verheyens, and A. Meyer. 1999. Replicated evolution of trophic specializations in an endemic cichlid fish lineage from Lake Tanganyika. *Proc Natl Acad Sci* (USA) 96:10230–10235.

Ruedi, M. 1996. Phylogenetic evolution and biogeography of Southeast Asian shrews (genus *Crocidura:* Soricidae). *Biol J Linn Soc* 58:197–219.

Rundle, H. D., and D. Schluter. 1998. Reinforcement of stickleback mate preferences: Sympatry breeds contempt. *Evolution* 52:200–208.

Rundle, H. D., A. O. Mooers, and M. C. Whitlock. 1998. Single founder-flush events and the evolution of reproductive evolution. *Evolution* 52:1850–1855.

Ruse, M. 1979. *The Darwinian revolution.* Chicago: Univ. of Chicago Press.

Russell, E. S. 1916. *Form and function.* London: John Murray.

Ryan, M. J. 1983. Sexual selection and communication in a Neotropical frog, *Physalaemus pustulosus. Evolution* 37:261–272.

———. 1985. *The túngara frog: A study in sexual selection and communication.* Chicago: Univ. Chicago Press.

———. 1990a. Sexual selection, sensory systems, and sensory exploitation In *Oxford surveys in evolutionary biology,* vol. 7, ed. D. J. Futuyma and J. Antonovics, 157–195. Oxford: Oxford Univ. Press.

———. 1990b. Signals, species, and sexual selection. *Am Sci* 78:46–52.

———. 1991. Sexual selection and communication in frogs. *Trends Ecol Evol* 6:351–355.

———. 1997. Sexual selection and mate choice. In *Behavioral ecology: An evolutionary approach,* ed. J. R. Krebs and N. B. Davies, 179–202. Oxford: Blackwell Science.

Ryan, M. J., J. H. Fox, W. Wilczynski, and A. S. Rand. 1990. Sexual selection for sensory exploitation in the frog *Physalaemus pustulosus. Nature* 343:66–67.

Ryan, M. J., and A. Keddy-Hector. 1992. Directional patterns of female mate choice and the role of sensory biases. *Am Nat* 139:S4–S35.

Ryan, M. J., and A. S. Rand. 1990. The sensory basis of sexual selection for complex calls in the túngara frog, *Physalaemus pustulosus* (sexual selection for sensory exploitation). *Evolution* 44:305–314.

———. 1993a. Species recognition and sexual selection as a unitary problem in animal communication. *Evolution* 47:647–657.

———. 1993b. Sexual selection and signal evolution: The ghost of biases past. *Phil Trans Roy Soc Lond* B 340:187–195.

———. 1993c. Phylogenetic patterns of behavioural mate recognition systems in the *Physalaemus pustulosus* species group (Anura: Leptodactylidae): The role of ancestral and derived characters and sensory exploitation. In *Evolutionary patterns and processes,* ed. D. R. Lees and D. Edwards, 251–267. London: Academic Press.

———. 1995. Female responses to ancestral advertisement calls in the túngara frog. *Science* 269:390–392.

———. 1999. Phylogenetic influence on mating call preferences in female túngara frogs, *Physalaemus pustulosus. Anim Behav* 57:945–956.

Ryan, M. J., A. S. Rand, and L. A. Weigt. 1996. Allozyme and advertisement call variation in the túngara frog, *Physalaemus pustulosus. Evolution* 50:2435–2453.

Ryan, M. J., M. D. Tuttle, and A. S. Rand. 1982. Bat predation and sexual advertisement in a Neotropical anuran. *Am Nat* 119:136–139.

Ryder, O. A. 1986. Species, conservation, and systematics: The dilemma of subspecies. *Trends Ecol Evol* 1:9–10.

SA2000. 1994. Systematics Agenda 2000: Charting the biosphere. Technical Report. New York: American Museum of Natural History.

Sabrosky, C. W. 1950. Taxonomy and ecology. *Ecology* 31:151–152.

Saether, O. A. 1977. Female genitalia in Chironomidae and other Nematocera: Morphology, phylogenies, keys. *Bull Fish Res Board Can* 197:1–211.

———. 1979a. Hierarchy of the Chironomidae with special emphasis on the female genitalia. In *Recent developments in chironomid studies (Diptera: Chironomidae)*, ed. O. A. Saether, 17–26. *Entomol Scand*, suppl. no. 10.

———. 1979b. Underlying synapomorphies and anagenetic analysis. *Zool Scr* 8:305–312.

———. 1979c. Underliggende synapomorfi enestående innvendig parallelisme belyst ved eksempler fra Chironomidae og Chaoboridae (Diptera). *Entomol Tidskr* 100:173–180.

———. 1983. The canalized evolutionary potential: Inconsistencies in phylogenetic reasoning *Syst Zool* 32:343–359.

Sallum, M. A. M., T. R. Schultz, and R. C. Wilkinson. 2000. Phylogeny of the Anophelinae (Diptera: Culicidae) based on morphological characters. *Ann Entomol Soc Am* 93:745–775.

Salter, L. A., and D. K. Pearl. 2001. Stochastic search strategy for estimation of maximum likelihood phylogenetic trees. *Syst Biol* 50:7–17.

Sanderson, M. J. 1989. Confidence limits on phylogenies: The bootstrap revisited. *Cladistics* 5:113–129.

———. 1990. Estimating rates of speciation and evolution: A bias due to homoplasy. *Evolution* 43:1781–1795.

———. 1991. In search of homoplastic tendencies: Statistical inference of topological patterns in homoplasy. *Evolution* 45:351–358.

———. 1994. Reconstructing the history of evolutionary processes using maximum likelihood. In *Molecular evolution of physiological processes*, ed. D. M. Frambrough, 13–26. New York: Rockefeller Univ. Press.

———. 1995. Objections to boot-strapping: A critique. *Syst Biol* 44:299–320.

Sanderson, M. J., B. G. Baldwin, G. Bharathan, C. S. Campbell, C. von Dohlen, D. Ferguson, J. M. Porter, M. F. Wojciechowski, and M. J. Donoghue. 1993. The growth of phylogenetic information and the need for a phylogenetic data base. *Syst Biol* 42:562–568.

Sanderson, M. J., and G. Bharathan. 1993. Does cladistic information affect inferences about branching rates? *Syst Biol* 42:1–17.

Sanderson, M. J., and M. J. Donoghue. 1989. Patterns of variation in levels of homoplasy. *Evolution* 43:1781–1795.

———. 1994. Shifts in diversification rate with the origin of angiosperms. *Science* 264: 1590–1593.

———. 1996. Reconstructing shifts in diversification rates on phylogenetic trees. *Trends Ecol Evol* 11:15–20.

Sanderson, M. J., and L. Hufford, eds. 1996 *Homoplasy: The recurrence of similarity in evolution*. New York: Academic Press.

Sanderson, M. J., A. Purvis, and C. Henze. 1998. Phylogenetic supertrees: Assembling the trees of life. *Trends in Ecol Evol* 13:105–109.

Sandlund, O. T., K. Hindar, and A. H. D. Brown, eds. 1992. *Conservation of biodiversity for sustainable development*. Oslo: Scandinavian Univ. Press.

Sandlund, O. T., and P. J. Schei, eds. 1993. *Proceedings of the Norway/UNEP expert conference on biodiversity, Trondheim, Norway*. Trondheim: NINA.

Sandosham, A. A. 1950. On *Enterobius vermicularis* (Linnaeus, 1758) and some related species from primates and rodents. *J Helminthol* 24:171–204.

Sang, T., and D. M. Zhang. 1999. Reconstructing hybrid speciation using sequences of low copy nuclear genes: Hybrid origins of five *Paeonia* species based on Adh gene phylogenies. *Syst Bot* 24:148–163.

Sang, T., and Y. Zhong. 2000. Testing hybridization hypotheses based on incongruent gene trees. *Syst Biol* 49:422–434.

Sangster, G., C. J. Hazevoet, A. B. Van den Berg, C. S. Roselaar, and R. Sluys. 1999. Dutch avifaunal list: Species concepts, instability, and taxonomic changes in 1977–1998. *Ardea* 87:139–165.

Savage, J. M. 1982. The enigma of the Central American herpetofauna: Dispersals or vicariance? *Ann Missouri Bot Gard* 69:464–547.

———. 1995. Systematics and the biodiversity crisis. *BioScience* 45:673–679.

Savill, N.J., D. C. Hoyle, and P. G. Higgs. 2001. RNA sequence evolution with secondary structure constraints: Comparison of substitution rate models using maximum likelihood methods. *Genetics* 157:399–411.

Savolainen, V., S. B. Hoot, C. M. Morton, D. E. Soltis, C. Bayer, M. F. Fay, A. Y. De Bruijn, S. Sullivan, and Y. L. Qiu. 2000. Phylogenetics of flowering plants based on combined analysis of plastid atpB and rbcL gene sequences. *Syst Biol* 49:306–362.

Sawyer, R. T. 1971. The phylogenetic development of brooding behaviour in the Hirudinea. *Hydrobiologia* 37:197–204.

———. 1986. *Leech biology and behaviour.* Oxford: Oxford Univ. Press.

Schaefer, S. A., and G. V. Lauder. 1986. Historical transformation of functional design: Evolutionary morphology of feeding mechanisms in loricarioid catfishes. *Syst Zool* 35:489–508.

———. 1996. Testing historical hypotheses of morphological change: Biomechanical decoupling in loricarioid catfishes. *Evolution* 50:1661–1675.

Scharff, N., and J. A. Coddington. 1997. A phylogenetic analysis of the orb-weaving spider family Araneidae (Arachnida, Araneae) *Zool J Linn Soc* 120:355–434.

Scheffer, S. J., and B. M. Wiegmann. 2000. Molecular phylogenetics of the holly leaf miners (Diptera: Agromyzidae: *Phytomyza*): Species limits, speciation, and dietary specialization. *Mol Phylogenet Evol* 17:244–255.

Scheltema, A. H., and C. Schander. 2000. Discrimination and phylogeny of *Solenogaster* species through the morphology of hard parts (Mollusca, Aplacophora, Neomeniomorpha). *Biol Bull* 198:121–151.

Schluter, D. 1993. Adaptive radiation in sticklebacks: Size, shape, and habitat use efficiency. *Ecology* 74:699–709.

———. 1994. Experimental evidence that competition promotes divergence in adaptive radiation. *Science* 266:798–801.

———. 1995. Adaptive radiation in sticklebacks: Trade-offs in feeding performance and growth. *Ecology* 76:82–90.

———. 1996a. Ecological speciation in postglacial fishes *Phil Trans Roy Soc Lond* B 351: 807–814.

———. 1996b. Ecological causes of adaptive radiation. *Am Nat* 148:S40–S64.

———. 1998. Ecological causes of speciation. In *Endless forms: Species and speciation,* ed. D. J. Howard and S. H. Berlocher, 114–129. New York: Oxford Univ. Press.

———. 2000. Ecological character displacement in adaptive radiation. *Am Nat* 156:S4–S16.

Schluter, D., and J. D. McPhail. 1992. Ecological character displacement and speciation in sticklebacks. *Am Nat* 140:85–108.

Schluter, D., and L. M. Nagel. 1995. Parallel speciation by natural selection. *Am Nat* 146:292–301.

Schluter, D., and T. Price. 1993. Honesty, perception and population divergence in sexually selected traits. *Proc Roy Soc Lond* B 253:117–122.

Schluter, D., and R. E. Ricklefs. 1993. Convergence and the regional component of species diversity. In *Species diversity in ecological communities: Historical and geographical perspectives,* ed. R. E. Ricklefs and D. Schluter, 230–240. Chicago: Univ. Chicago Press.

Schmitz, J., and R. F. A. Moritz. 1998. Molecular phylogeny of Vespidae (Hymenoptera) and the evolution of sociality in wasps. *Mol Phylogenet Evol* 9:183–191.

Schneirla, T. C. 1952. A consideration of some conceptual trends in comparative psychology. *Psychol Bull* 49:559–597.

Schoch, R. M. 1986. *Phylogeny reconstruction in paleontology.* New York: Van Nostrand Reinhold.

Schoener, T. W. 1965. The evolution of bill size differences among sympatric congeneric species of birds. *Evolution* 19:189–213.

———. 1968a. Sizes of feeding territories among birds. *Ecology* 49:123–141.

———. 1968b. The *Anolis* lizards of Bimini: Resource partitioning in a complex fauna. *Ecology* 49:704–726.

Schoener, T. W., and G. C. Gorman. 1968. Some niche differences in three Lesser Antillean lizards of the genus *Anolis. Ecology* 49:819–830.

Scholz, C. A., and B. Rosendahl. 1988. Low lake stands in Lakes Malawi and Tanganyika, delineated with multifold seismic data. *Science* 240:1645–1648.

Schopf, T. J. M. 1979. Evolving palaeontological views on deterministic and stochastic approaches. *Paleobiology* 5:337–352.

Schopf, T. J. M., D. M. Raup, S. J. Gould, and D. S. Simberloff. 1975. Genomic versus morphologic rates of evolution: Influence of morphologic complexity. *Paleobiology* 1:63–70.

Schroder, G. D. 1987. Mechanisms for coexistences among three species of *Dipdomys:* Habitat selection and an alternative. *Ecology* 68:1071–1083.

Schultz, J. C. 1988. Many factors influence the evolution of herbivore diets, but plant chemistry is central. *Ecology* 69:896–897.

Schwartz, A., and R. W. Henderson. 1991. *Amphibians and reptiles of the West Indies: Descriptions, distributions, and natural history.* Gainesville, FL: Univ. Florida Press.

Schwartz, R. M., and M. O. Dayhoff. 1978. Origins of prokaryotes, eukaryotes, mitochondria, and chloroplasts. *Science* 199:395–403.

Schwenk, K. 1988. Comparative morphology of the Lepidosaur tongue and its relevance to squamate phylogeny. In *Phylogenetic relationships of the lizard families Essays commemorating Charles L Camp,* ed. R. Estes and G. Pregill, 569–598. Stanford, CA: Stanford Univ. Press.

———. 1993. The evolution of chemoreception in squamate reptiles: A phylogenetic approach. *Brain Behav Evol* 41:124–137.

———. 1994. Why snakes have forked tongues. *Science* 263:1573–1577.

———. 1995a. A utilitarian approach to evolutionary constraint. *Zoology* 98:251–262.

———. 1995b. Of tongues and noses: Chemoreception in lizards and snakes. *Trends Ecol Evol* 10:7–12.

Schwenk, K., and D. A. Bell. 1988. A cryptic intermediate in the evolution of chameleon tongue projection. *Experientia* 44:697–700.

Schwenk, K., and G. S. Throckmorton. 1989. Functional and evolutionary morphology of lingual feeding in squamate reptiles: Phylogenetics and kinematics. *J Zool* 219:153–175.

Sclater, P. L., and O. Salvin. 1867. List of birds collected by Mr. Wallace on the lower Amazons and Rio Negro. *Proc Zool Soc Lond* 1867:556–596.

Scriber, J. M. 1984. Larval foodplant utilization by the world Papilionidae (Lep): Latitudinal gradients reappraised. *Tokurana (Acta Rhopalocerologica)* 6/7:1–50.

Scriber, J. M., and R. C. Lederhouse. 1992. The thermal environment as a resource dictating geographic patterns of feeding specialization of insect herbivores. In *Effects of resource distribution on animal-plant interactions,* ed. M. R. Hunter, T. Ohgushi, and P. W. Price, 429–466. New York: Academic Press.

Scudder, G. G. E., L. V. Moore, and M. B. Isman. 1986. Sequestration of cardenolides in *Oncopeltus fasciatus:* Morphological and physiological adaptations. *J Chem Ecol* 12:1171–1187.

Seaman, F. C. 1982. Sesquiterpene lactones as taxonomic characters of the Asteraceae. *Bot Rev* 48:121–595.

Seaman, F. C., and V. A. Funk. 1983. Cladistic analysis of complex natural products: Developing transformation series from sesquiterpene lactone data. *Taxon* 32:1–27.

Seehausen, O. 1996. *Lake Victoria rock cichlids, taxonomy, ecology and distribution.* Zevenhuizen: Verduijn Cichlids.

———. 2000. Explosive speciation rates and unusual species richness in haplochromine cichlid fishes: Effects of sexual selection. *Adv Ecol Res* 31:237–274.

Seehausen, O., P. J. Mayhew, and J. J. M. van Alphen. 1999. Evolution of colour patterns in East African cichlid fish. *J Evol Biol* 12:514–534.

Seehausen, O., and J. J. M. van Alphen. 1998. The effects of male coloration on female mate choice in closely related Lake Victoria cichlids (*Haplochromis nyererei* complex). *Behav Ecol Sociobiol* 42:1–8.

———. 1999. Can sympatric speciation by disruptive sexual selection explain rapid evolution of cichlid diversity in Lake Victoria? *Ecol Let* 2:262–271.

Seehausen, O., J. J. M. van Alphen, and R. Lande. 1999. Color polymorphism and sex ratio distortion in a cichlid fish as an incipient stage in sympatric speciation by sexual selection. *Ecol Let* 2:367–378.

Seehausen, O., J. J. M. van Alphen, and F. Witte. 1997. Cichlid fish diversity threatened
by eutrophication that curbs sexual selection. *Science* 277:1808–1811.

Seehausen, O., F. Witte, J. J. M. van Alphen, and N. Bouton. 1998. Direct mate choice
maintains diversity among sympatric cichlids in Lake Victoria. *J Fish Biol* 53,
suppl. A:37–55.

Sennblad, B., and B. Bremer. 2000. Is there a justification for differential a priori
weighting in coding sequences? A case study from rbcL and Apocynaceae s. l. *Syst
Biol* 49:101–113.

Sepkoski, J. J., Jr. 1978. A kinetic model of Phanerozoic taxonomic diversity: I. Analysis
of marine orders. *Paleobiology* 4:223–251.

Seutin, G., N. K. Klein, R. E. Ricklefs, and E. Bermingham. 1994. Historical biogeography
of the bananaquit (*Coereba flaveola*) in the Caribbean region: A mitochondrial
DNA assessment. *Evolution* 48:1041–1061.

Shaffer, H. B., and M. L. McKnight. 1996. The polytypic species revisited: Genetic differentiation and molecular phylogenetics of the tiger salamander *Ambystoma tigri-
num* (Amphibia: Caudata) complex. *Evolution:* 50:417–433.

Shaffer, M. L. 1981. Minimum population sizes for species conservation. *BioScience*
31:131–134.

———. 1987. Minimum viable populations: Coping with uncertainty. In *Viable populations for conservation,* ed. M. E. Soulé, 69–86. New York: Cambridge Univ. Press.

Shaw, A. J., and B. Goffinet. 2000. Molecular evidence of reticulate evolution in the peat-
mosses (*Sphagnum*), including *S. ehyalinum* sp. nov. *Bryologist* 103:357–374.

Shaw, K. L. 1995a. Biogeographic patterns of two independent Hawaiian cricket radiations (*Laupala* and *Prognathogryllus*). In *Hawaiian biogeography: Evolution on a hot
spot archipelago,* ed. W. L. Wagner and V. A. Funk, 39–56. Washington, DC: Smithsonian Institution Press.

———. 1995b. Phylogenetic tests of the sensory exploitation model of sexual selection.
Trends Ecol Evol 10:117–120.

———. 1996a. Sequential radiations and patterns of speciation in the Hawaiian cricket
genus *Laupala* inferred from DNA sequences. *Evolution* 50:237–255.

———. 1996b. Polygenic inheritance of a behavioral phenotype: Interspecific genetics of
song in the Hawaiian cricket genus *Laupala. Evolution* 50:256–266.

———. 1999. A nested analysis of song groups and species boundaries in the Hawaiian
cricket genus *Laupala. Mol Phylogenet Evol* 11:332–341.

Shea, B. T. 1983. Allometry and heterochrony in the African apes. *Am J Phys Anthropol*
62:275–289.

Sheehan, P. M. 1985. Reefs are not so different—they follow the evolutionary pattern of
level-bottom communities. *Geology* 13:46–49.

Shettleworth, S. J. 1993. Where is the comparison in comparative cognition? *Psychol Sci*
4:179–184.

Shine, R. 1988. Parental care in reptiles. In *Biology of the reptilia,* vol. 16: *Ecology B defense
and life history,* ed. C. Gans and R. B. Huey, 275–329. New York: Alan Liss.

Shubin, N., D. B. Wake, and A. J. Crawford. 1995. Morphological variation in the limbs
of *Taricha granulosa* (Caudata: Salamandridae): Evolutionary and phylogenetic implications. *Evolution* 49:874–884.

Sick, H. 1967. *Ornitologia Brasileira.* Vol. 2. Brasilia: Editora Universidade de Brasilia.

Siddall, M. E., and E. M. Burreson. 1995. Phylogeny of the Euhirudinea: Independent
evolution of blood feeding by leeches? *Can J Zool* 73:1048–1064.

———. 1996. Leeches (Oligochaeta?: Euhirudinea), their phylogeny, and the evolution
of life-history strategies. *Hydrobiologia* 334:277–285.

Siddall, M. E., and A. G. Kluge. 1997. Probabilism and phylogenetic inference. *Cladistics*
13:313–336.

Sidor, C. A., and J. A. Hopson. 1998. Ghost lineages and "mammalness": Assessing the
temporal pattern of character acquisition in the Synapsida. *Paleobiology* 24:254–
273.

Siegel-Causey, D. 1991. Systematics and biogeography of North pacific shags, with a description of a new species. *Occas Pap Mus Nat Hist Univ Kansas* 140:1–17.

Sillén-Tullberg, B. 1985. The higher survival of an aposematic than of a cryptic form of a distasteful bug. *Oecologia* 67:411–415.

———. 1988. Evolution of gregariousness in aposematic butterfly larvae: A phylogenetic analysis. *Evolution* 42:293–305.

———. 1993. The effect of biased inclusion of taxa on the correlation between discrete characters in phylogenetic trees. *Evolution* 47:1182–1191.

Sillén-Tullberg, B., and A. P. Møller. 1993. The relationship between concealed ovulation and mating systems in anthropoid primates: A phylogenetic analysis. *Am Nat* 141:1–25.

Sillén-Tullberg, B., and H. Temrin. 1994. On the use of discrete characters in phylogenetic trees with special reference to the evolution of avian mating systems. In *Phylogenetics and ecology*, ed. P. Eggleton and R. Vane-Wright, 311–322. London: Academic Press.

Sillén-Tullberg, B., C. Wiklund, and T. Järvi. 1982. Aposematic coloration in adults and larvae of *Lygaeus equestris* and its bearing on Müllerian mimicry: An experimental study on predation on living bugs by the great tit *Parus major*. *Oikos* 39:131–136.

Silvain, J.-F., and A. Delobel. 1998. Phylogeny of West African *Caryedon* (Coleoptera: Bruchidae) congruence between molecular and morphological data. *Mol Phylogenet Evol* 9:533–541.

Silvertown, J., M. Dodd, and D. Gowing. 2001. Phylogeny and niche structure of meadow plant communities. *J Ecol* 89: 428–435.

Simberloff, D. S. 1971. Population sizes of congeneric bird species on islands. *Am Nat* 105:190–193.

———. 1987. Calculating probabilities that cladograms match: A method of biogeographical inference. *Syst Zool* 36:175–195.

———. 1988. Effects of drift and selection on detecting similarities between large cladograms. *Syst Zool* 37:56–59.

Simberloff, D., K. L. Heck, E. D. McCoy, and E. F. Connor. 1980. There have been no statistical tests of cladistic biogeographic hypotheses. In *Vicariance biogeography: A critique*, ed. G. Nelson and D. E. Rosen, 40–63. New York: Columbia Univ. Press.

Simkova, A., Y. Desdevises, M. Gelnar, and S. Morand. 2000. Co-existence of nine gill parasites (*Dactylogyrus:* Monogenea) parasitising the roach (*Rutilus rutilus*): History and present ecology. *Int J Parasitol* 30:1077–1088.

Simons, A. M. 1992. Phylogenetic relationships of the *Boleosoma* species group (Percidae: Etheostoma). In *Systematics, historical ecology, and North American freshwater fishes*, ed. R. L. Mayden, 268–292. Stanford, CA: Stanford Univ. Press.

Simpson, B. B. 1975. Pleistocene changes in the flora of the high tropical Andes. *Paleobiology* 1:273–294.

———. 1983. An historical phytogeography of the high Andean flora. *Rev Chilena Hist Nat* 56:109–122.

Simpson, B. B., and J. Cracraft. 1995. Systematics: The science of biodiversity. *BioScience* 45:670–672.

Simpson, B. B., and J. Haffer. 1978. Speciation patterns in the Amazonian forest biota. *Annu Rev Ecol Syst* 9:497–518.

Simpson, B. B., and C. A. Todzia. 1990. Patterns and processes in the development of the high Andean flora. *Am J Bot* 77:1419–1432.

Simpson, G. G. 1944. *Tempo and mode in evolution*. New York: Columbia Univ. Press.

———. 1950. History of the fauna of Latin America. *Am J Sci* 38:361–389.

———. 1951. The species concept. *Evolution* 5:285–298.

———. 1953. *The major features of evolution*. New York: Columbia Univ. Press.

———. 1960. *The meaning of evolution*. New Haven, CT: Yale Univ. Press.

———. 1961. *Principles of animal taxonomy*. New York: Columbia Univ. Press.

———. 1980. *Splendid isolation: The curious history of South American mammals*. New Haven, CT: Yale Univ. Press.

Singer, M. C. 1971. Evolution of food-plant preference in the butterfly *Euphydryas eitha*. *Evolution* 25:383–389.

———. 1994. Behavioral constraints on the evolutionary expansion of insect diet: A case history from checkerspot butterflies. In *Behavioral mechanisms in evolutionary ecology*, ed. L. Real, 279–296. Chicago: Univ. Chicago Press.

Singer, M. C., P. R. Ehrlich, and L. E. Gilbert. 1971. Butterfly feeding on lycopsid. *Science* 172:1341–1342.

Singer, M. C., D. Ng, and C. D. Thomas. 1992. Rapidly evolving associations among oviposition preferences fail to constrain evolution of insect diet. *Am Nat* 139:9–20.

Singer, M. C., and C. Parmesan. 1993. Sources of variations in patterns of plant-insect associations. *Nature* 361:251–253.

Singer, M. C., C. D. Thomas, and C. Parmesan. 1993. Rapid human-induced evolution of insect host associations. *Nature* 366:681–683.

Sites, J. W., and K. A. Crandall. 1997. Testing species boundaries in biodiversity studies. *Conserv Biol* 11:1289–1297.

Sites, J. W., and K. M. Reed. 1994. Chromosomal evolution, speciation, and systematics: Some relevant issues. *Herpetologica* 50:237–249.

Sitte, P. 1993. Symbiogenetic evolution of complex cells and complex plastids. *Eur J Protistol* 29:131–143.

Skala, Z., and J. Zrzavy. 1994. Phylogenetic reticulations and cladistics: Discussion of methodological concepts. *Cladistics* 10:305–313.

Skelton, P. W., ed. 1993a. *Evolution: A biological and paleontological approach.* NY: Addison-Wesley.

Skelton, P. W. 1993b. Adaptive radiation: Definition and diagnostic tests. In *Evolutionary patterns and processes,* ed. D. R. Lees and D. Edwards, 40–45. Linn Soc Symp, vol. 14. London: Academic Press.

Skuhravy, V., M. Skuhrava, and J. W. Brewer. 1983. Ecology of the saddle gall midge *Haplodiplosis marginata* (Von Roser) (Diptera, Cecidomyiidae). *Z Angew Entomol* 96:476–490.

Slikas, B. 1998. Recognizing and testing homology of courtship displays in storks (Aves: Ciconiiformes: Ciconiidae). *Evolution* 52:884–893.

Slobodkin, L. B., and H. L. Sanders. 1969. On the contribution of environmental predictability to species diversity. In *Diversity and stability in ecological systems,* ed. G. M. Woodwell and H. H. Smith, 82–95. Upton, NY: Brookhaven National Laboratory.

Slowinski, J. B. 1990. Probabilities of n-trees under two models: A demonstration that asymmetrical interior nodes are not improbable. *Syst Zool* 39:89–94.

Slowinski, J. B., and C. B. Guyer. 1989. Testing the stochasticity of patterns of organismal diversity: An improved null model. *Am Nat* 134:907–921.

———. 1993. Testing whether certain traits have caused amplified diversification: An improved method based on a model of random speciation and extinction. *Am Nat* 142:1019–1024.

Sluys, R. 1989. Rampant parallelism: An appraisal of the use of nonuniversal derived character states in phylogenetic reconstructions. *Syst Zool* 38:350–370.

———. 1991. Species concepts, process analysis, and the hierarchy of nature. *Experientia* 47:1162–1170.

Sluys, R., and C. J. Hazevoet. 1999. Pluralism in species concepts: Dividing nature at its diverse joints. *Species Diversity* 4:243–256.

Smiley, J. 1978. Plant chemistry and the evolution of host specificity: New evidence from *Heliconius* and *Passiflora. Science* 201:745–746.

Smith, C. C., and A. N. Bragg. 1949. Observations on the ecology and natural history of anura: VII. Food and feeding habits of the common species of toads in Oklahoma. *Ecology* 30:333–349.

Smith, F. D. M., R. M. May, R. Pellew, Y. H. Johnson, and K. R. Walker. 1993. How much do we know about the current extinction rate? *Trends Ecol Evol* 8:375–378.

Smith, H. M. 1965. More evolutionary terms. *Syst Zool* 14:57–58.

———. 1969. Parapatry: Sympatry or allopatry? *Syst Zool* 18:254–255.

Smith, J. F. 2000. Phylogenetic signal common to three data sets: Combining data which initially appear heterogeneous. *Plant Syst Evol* 221:179–198.

Smith, K. K. 2001. Heterochrony revisited: The evolution of developmental sequences. *Biol J Linn Soc* 73:169–186.

Smith, V. S. 2001. Avian louse phylogeny (Phthiraptera: Ischnocera): A cladistic study based on morphology. *Zool J Linn Soc* 132:81–144.

Sneath, P. H. A., and R. R. Sokal. 1973. *Numerical taxonomy.* San Francisco: W. H. Freeman.

Snethlage, E. 1913. Über der Verbreitung der Vogelarten in Unteramazonien. *J Ornithol* 61:469–539.

Snow, D. W., ed. 1978. *An atlas of speciation in African non-passerine birds*. London: Natural History Museum.

———. 1981. Tropical frugivorous birds and their food plants. A world survey. *Biotropica* 13:1–14.

Sober, E. 1983. Parsimony in systematics: Philosophical issues. *Annu Rev Ecol Syst* 14:335–357.

———. 1988a. *Reconstructing the past: Parsimony, evolution, and inference*. Cambridge, MA: MIT Press.

———. 1988b. The conceptual relationship of cladistic phylogenetics and vicariance biogeography. *Syst Zool* 37:245–253.

Soest, R. W. M. van, and E. Hajdu. 1997. Marine area relationships from twenty sponge phylogenies. A comparison of methods and coding strategies. *Cladistics* 13:1–20.

Softing, G. B., G. Benneh, K. Hindar, and A. Wijkman, eds. 1998. *The Brundtland Commission's Report—10 Years*. Oslo: Scandinavian Univ. Press.

Sokal, R. R., and T. J. Crovello. 1970. The biological species concept: A critical evaluation. *Am Nat* 104:127–153.

Sokal, R. R., and P. H. A. Sneath. 1963. *The principles of numerical taxonomy*. San Francisco: W. H. Freeman.

Soltis, D. E., and P. S. Soltis. 1999. Polyploidy: Recurrent formation and genome evolution. *Trends Ecol Evol* 14:348–352.

Soltis, P. S., D. E. Soltis, and J. J. Doyle, eds. 1992. *Molecular systematics in plants*. New York: Chapman and Hall.

Soltis, D. E., P. S. Soltis, D. L. Nickrent, L. A. Johnson, W. J. Hahn, S. B. Hoot, J. A. Sweere, R. K. Kuzoff, K. A. Kron, M. W. Chase, S. M. Swensen, E. A. Zimmer, S. Chow, L. J. Gillespie, W. J. Kress, and K. J. Sytsma. 1997. Angiosperm phylogeny inferred from 18S ribosomal DNA sequences. *Ann Missouri Bot Gard* 84:1–49.

Soltis, P. S., D. E. Soltis, S. G. Weller, A. K. Sakai, and W. L. Wagner. 1996. Molecular phylogenetic analysis of the Hawaiian endemics *Schiedea* and *Alsinidendron* (Caryophyllaceae). *Syst Bot* 21:365–379.

Soltis, P. S., D. E. Soltis, M. J. Zanis, and S. Kim. 2000. Basal lineages of angiosperms: Relationships and implications for floral evolution. *Int J Plant Sci* 161:S97–S107.

Sorenson, M. D. 1999. TreeRot. Ver. 2. Boston, MA: Boston Univ.

Sosa, V., and E. De Luna. 1998. Morphometrics and character state recognition for cladistic analyses in the *Bletia reflexa* complex (Orchidaceae). *Plant Syst Evol* 212:185–213.

Soulé, M. E. 1990. The onslaught of alien species, and other challenges in the coming decades. *Conserv Biol* 4:233–239.

———. 1991. Conservation: Tactics for a constant crisis. *Science* 253:744–750.

Sperling, F. A. H., and P. Feeny. 1995. *Ubmellifer* and composite feeding in *Papilio*: Phylogenetic frameworks and constraints on caterpillars. In *Swallowtail butterflies: Their ecology and evolutionary biology*, ed. J. M. Scribner, Y. Tsubaki, and R. C. Lederhouse, 299–306. Gainesville, FL: Scientific Publishers.

Spieth, H. T. 1974. Mating behavior and evolution of the Hawaiian *Drosophila*. In *Genetic mechanisms of speciation in insects*, ed. M. J. D. White, 94–101. Sydney: Australia and New Zealand Book Co.

Stager, J. C., and T. C. Johnson. 2000. A 12,4000 C-14 yr offshore diatom record from east central Lake Victoria, East Africa. *J Paleolimnol* 23:373–383.

Stammer, H. J. 1955. Ökologische Wechselsbeziehungen zwischen Insekten und anderen Tiergruppen. *7 Wand Versamml Dtsch Entomol*, 12–61.

———. 1957. Gedanken zu den parasitophyletischen Regeln und zur Evolution der Parasiten. *Zool Anz* 159:255–267.

Stamps, J. A. 1977. Social behaviour and spacing patterns in lizards. In *Biology of the reptilia*, vol. 7A: *Ecology*, ed. C. Gans and D. W. Tinkle, 265–335. London: Academic Press.

Staniczek, A. H. 2000. The mandible of silverfish (Insecta: Zygentoma) and mayflies (Ephemeroptera): Its morphology and phylogenetic significance. *Zool Anz* 239:147–178.

Stanley, S. M. 1973. An explanation for Cope's Rule. *Evolution* 27:1–26.

——. 1979. *Macroevolution: Pattern and process.* San Francisco: Freeman.

——. 1987. *Extinction.* New York: Scientific American Library.

Stanley, S. M., P. W. Signor, S. Lidgard, and A. F. Karr. 1981. Natural clades differ from "random" clades: Simulations and analysis. *Paleobiology* 7:115–127.

Stauffer, J. R., N. J. Bowers, T. D. Kocher, and K. R. McKaye. 1996. Evidence of hybridization between *Cynotilapia afra* and *Pseudotropheus zebra* (Teleostei: Cichlidae) following intralacustrine translocation in Lake Malawi. *Copeia* 1996:203–208.

Stebbins, G. L. 1971. Adaptive radiation of reproductive characteristics in angiosperms: II. Seeds and seedlings. *Annu Rev Ecol Syst* 2:237–260.

——. 1981. Why are there so many species of flowering plants? *BioScience* 31:573–577.

——. 1989. Plant speciation and the founder principle. In *Genetics, speciation, and the founder principle,* ed. L. V. Giddings, K. Y. Kaneshiro, and W. W. Anderson, 113–125. New York: Oxford Univ. Press.

Steel, M. 1992. The complexity of reconstructing trees from qualitative characters and subtrees. *J Classif* 9:91–116.

Steel, M., D. Huson, and P. J. Lockhart. 2000. Invariable sites models and their use in phylogeny reconstruction. *Syst Biol* 49:225–232.

Steel, M., and D. Penny. 2000. Parsimony, likelihood, and the role of models in molecular phylogenetics. *Mol Biol Evol* 17:839–850.

Stein, B. A. 1992. Sicklebill hummingbirds, ants, and flowers. *BioScience* 42:27–33.

Stenseth, N. C. 1984. Why mathematical models in evolutionary ecology? In *Trends for ecological research for the 1980's,* ed. J. H. Cooley and F. B. Golley, 239–287. New York: Plenum Press.

Stenseth, N. C., and J. Maynard Smith. 1984. Coevolution in ecosystems: Red queen evolution or stasis? *Evolution* 38:870–880.

Stern, D. L. 1995. Phylogenetic evidence that aphids, rather than plants, determine gall morphology. *Proc Roy Soc Lond* B 260:85–89.

Stevens, P. F. 1980. Evolutionary polarity of character states. *Annu Rev Ecol Syst* 11:333–358.

——. 1981. On ends and means, or how polarity criteria can be assessed. *Syst Zool* 11:333–358.

——. 1992. Species: Historical perspectives. In *Keywords in evolutionary biology,* ed. E. F. Keller and E. A. Lloyd, 302–311. Cambridge, MA: Harvard Univ. Press.

Stiassny, M. L. 1987. Cichlid familial relationships and the placement of the Neotropical genus *Cichla* (Perciformes, Labroidei). *J Nat Hist* 21:1311–1331.

——. 1990. *Tylochromis,* relationships and the phylogenetic status of the African Cichlidae. *Am Mus Nov* 2993:1–14.

——. 1991. Phylogenetic interrelationships of the family Cichlidae: An overview. In *Cichlid fishes: Behaviour, ecology, and evolution,* ed. M. H. A. Keenleyside, 1–35. London: Chapman and Hall.

——. 1992. Atavisms, phylogenetic character reversals, and the origin of evolutionary novelties. *Neth J Zool* 42:260–276.

——. 1993. Phylogenetic analysis and the role of systematics in the biodiversity crisis. In *Systematics, ecology, and the biodiversity crisis,* ed. N. Eldredge, 109–120. New York: Columbia Univ. Press.

Stiller, J. W., and B. D. Hall. 1999. Long branch attraction and the rDNA model of early eukaryotic evolution. *Mol Biol Evol* 16:1270–1279.

Stone, A. R. 1977. Recent developments and some problems in the taxonomy of cyst nematodes, with a classification of the Heteroderoidea. *Nematologica* 23:273–288.

Stone, A. R., P. R. Thompson, and B. E. Hopper. 1977. *Globodera pallida* present in Newfoundland. *Plant Dis Rep* 61:590–591.

Stone, G. N., and J. M. Cook. 1998. The structure of cynipid oak galls: Patterns in the evolution of an extended phenotype. *Proc Roy Soc Lond* B 265:979–988.

Stone, J. R. 1997. Using shell parameters as complementary data in phylogenetic systematic analysis: Evolution of form in five species of littorinids (Mollusca: Gastropoda). *The Veliger* 40:12–22.

——. 1998. Ontogenic tracks and evolutionary vestiges in morphospace. *Biol J Linn Soc* 64:223–238.

Stonor, C. R. 1936. The evolution and mutual relationships of some members of the Paradisaeidae. *Proc Zool Soc Lond* 1936:1177–1185.

Storey, M., J. J. Mahoney, A. D. Saunders, R. A. Duncan, S. P. Kelley, and M. F. Coffin. 1995. Timing of hot spot-related volcanism and breakup of Madagascar and India. *Science* 267:852–855.

Stork, N. E., and K. J. Gaston. 1990. Counting species one by one. *New Scientist* 11:43–47.

Strauss, R. E. 1990. Heterochronic variation in the developmental timing of cranial ossifications in poeciliid fishes (Cyprinodontiformes). *Evolution* 44:1558–1567.

Streelman, J. T., R. Zardoya, A. Meyer, and S. A. Karl. 1998. Multilocus phylogeny of cichlid fishes (Pisces: Perciformes): Evolutionary comparison of microsatellite and single-copy nuclear loci. *Mol Biol Evol* 15:798–808.

Strong, D. R., Jr., D. Simberloff, L. G. Abele, and A. B. Thistle, eds. 1984. *Ecological communities: Conceptual issues and the evidence.* Princeton, NJ: Princeton Univ. Press.

Stuart, A. E., and F. F. Hunter. 1998. End-products of behaviour versus behavioural characters: A phylogenetic investigation of pupal cocoon construction and form in some North American black flies (Diptera: Simuliidae). *Syst Entomol* 23:387–398.

Stuart, L. C. 1954. Description of a subhumid corridor across northern Central America, with comments on its herpetofaunal indicators. *Contr Lab Vert Biol Univ Mich* 65:1–26.

Sturmbauer, C. 1998. Explosive speciation in cichlid fishes of the African Great Lakes: A dynamic model of adaptive radiation. *J Fish Biol* 53, suppl. A: 18–36.

Sturmbauer, C., W. Mark, and R. Dallinger. 1992. Ecophysiology of aufwuchs eating cichlids in Lake Tanganyika: Niche separation by trophic specialisation. *Environ Biol Fishes* 35:283–290.

Sturmbauer, C., and A. Meyer. 1993. Mitochondrial phylogeny of the endemic mouthbrooding lineages of cichlid fishes from Lake Tanganyika. *Mol Biol Evol* 10:751–768.

Sültmann, H., W. E. Mayer, F. Figueroa, H. Tichy, and J. Klein. 1995. Phylogenetic analysis of cichlid fishes using nuclear DNA markers. *Mol Biol Evol* 12:1033–1047.

Sues, H.-D., ed. 2000. *Evolution of herbivory in terrestrial vertebrates: Perspectives from the fossil record.* Cambridge: Cambridge Univ. Press.

Sundberg, P. 1989a. Phylogeny and cladistic classification of the paramonostiliferous family Plectonemertidae (phylum Nemertea). *Cladistics* 5:87–100.

———. 1989b. Phylogeny and cladistic classification of terrestrial nemerteans: The genera *Pantinonemertes* Moore & Gibson and *Geonemertes* Semper. *Zool J Linn Soc* 95:363–372.

———. 1993. Phylogeny, natural groups, and nemertean classification. *Hydrobiologia* 266:103–113.

Swiderski, D. L., M. L. Zelditch, and W. L. Fink. 1998. Why morphometrics is not special: Coding quantitative data for phylogenetic analysis. *Syst Biol* 47:508–519.

Swofford, D. L. 1991. When are phylogeny estimates from morphological and molecular data incongruent? In *Phylogenetic analysis of DNA sequences,* ed. M. M. Miyamoto and J. Cracraft, 295–333. New York: Oxford Univ. Press.

Swofford, D. L., and S. H. Berlocher. 1987. Inferring evolutionary trees from gene frequency data under the principle of maximum parsimony. *Syst Zool* 36:293–325.

Swofford, D. L., and W. P. Maddison. 1987. Reconstructing ancestral character states under Wagner parsimony. *Math Biosci* 87:199–229.

———. 1992. Parsimony, character-state reconstructions, and evolutionary inferences. In *Systematics, historical ecology, and North American freshwater fishes,* ed. R. L. Mayden, 186–223. Stanford, CA: Stanford Univ. Press.

Swofford, D. L., and G. J. Olsen. 1990. Phylogeny reconstruction. In *Molecular Systematics,* ed. D. M. Hillis and C. Moritz, 411–501. Sunderland, MA: Sinauer.

Swofford, D. L., G. J. Olsen, P. J. Waddell, and D. M. Hillis. 1996. Phylogenetic inference. In *Molecular systematics,* ed. D. M. Hillis, C. Mortiz, and B. K. Mable, 407–514. Sunderland, MA: Sinauer.

Systematics Agenda 2000. *See* SA2000.

Szalanski, A. L., D. S. Sikes, R. Bischof, and M. Fritz. 2000. Population genetics and phylogenetics of the endangered American burying beetle, *Nicrophorus americanus* (Coleoptera: Silphidae). *Ann Entomol Soc Am* 93:589–594.

Taber, S. W., and C. M. Pease. 1990. Paramyxovirus phylogeny: Tissue tropism evolves slower than host specificity. *Evolution* 44:435–438.

Tagliaro, C. H., M. P. C. Schneider, H. Schneider, I. C. Sampaio, and M. J. Stanhope.

1997. Marmoset phylogenetics, conservation perspectives, and the evolution of mtDNA control region. *Mol Biol Evol* 14:674–684.

Takacs, D. 1996. *The idea of biodiversity: Philosophies of paradise.* Baltimore, MD: Johns Hopkins Univ. Press.

Talbot, M. 1934. Distribution of ant species in the Chicago region with reference to ecological factors and physiological toleration. *Ecology* 15:416–439.

———. 1945. A comparison of flights of four species of ants. *Am Midl Nat* 34:504–510.

———. 1948. A comparison of two ants of the genus *Formica. Ecology* 29:316–325.

Talbot, M. R., and T. Laerdal. 2000. The late Pleistocene-Holocene palaeolimnology of Lake Victoria, based upon elemental and isotopic analyses of sedimentary organic matter. *J Paleolimnol* 23:141–164.

Tauber, C. A., and M. J. Tauber. 1977a. Sympatric speciation based on allelic changes at three loci: Evidence from natural populations in two habitats. *Science* 197:1298–1299.

———. 1977b. A genetic model for sympatric speciation through habitat diversification and seasonal isolation. *Nature* 268:702–705.

———. 1989. Sympatric speciation in insects: Perception and perspective. In *Speciation and its consequences,* ed. D. Otte and J. A. Endler, 307–344. Sunderland, MA: Sinauer.

Taylor, D. J., P. D. N. Herbert, and J. K. Colbourne. 1996. Phylogenetics and evolution of the *Daphnia longispina* group (Crustacea) based on 12S rDNA sequence and allozyme variation. *Mol Phylogenet Evol* 5:495–510.

Taylor, E. B. 1999. Species pairs of north temperate freshwater fishes: Evolution, taxonomy, and conservation. *Rev Fish Biol Fisheries* 9:299–324.

Taylor, E. B., and J. D. McPhail. 1999. Evolutionary history of an adaptive radiation in species pairs of threespine sticklebacks (*Gasterosteus*): Insights from mitochondrial DNA. *Biol J Linn Soc* 66:271–291.

———. 2000. Historical contingency and ecological determinism interact to prime speciation in sticklebacks, *Gasterosteus. Proc Roy Soc Lond* B 267:2375–2384.

Tehler, A. 1995. Morphological data, molecular data, and total evidence in phylogenetic analyses. *Can J Bot* 73:S667–S676.

Templeton, A. R. 1979. Once again, why 300 species of Hawaiian *Drosophila? Evolution* 33:513–517.

———. 1980. The theory of speciation by the founder principle. *Genetics* 92:1011–1038.

———. 1981. Mechanisms of speciation: A population genetic approach. *Annu Rev Ecol Syst* 12:23–48.

———. 1982. Genetic architectures of speciation. In *Mechanisms of speciation,* ed. C. Barigozzi, 105–121. New York: Alan R. Liss.

———. 1989a. The meaning of species and speciation: A genetic perspective. In *Speciation and its consequences.* ed. D. Otte and J. Endler, 3–27. Sunderland, MA: Sinauer.

———. 1989b. Founder effects and the evolution of reproductive isolation. In *Genetics, speciation, and the founder principle,* ed. L. V. Giddings, K. Y. Kaneshiro, and W. W. Anderson, 329–344. Oxford: Oxford Univ. Press.

———. 1996. Experimental evidence for the genetic-transilience model of speciation. *Evolution* 50:909–915.

Temrin, H., and B. Sillén-Tullberg. 1994. The evolution of avian mating systems: A phylogenetic analysis of male and female polygamy and length of pair bond. *Biol J Linn Soc* 52:121–149.

———. 1995. A phylogenetic analysis of the evolution of avian mating systems in relation to altricial and precocial young. *Behav Ecol* 6:296–307.

Thien, L. B., H. Azuma, and S. Kawano. 2000. New perspectives on the pollination biology of basal angiosperms. *Int J Plant Sci* 161:S225–S235.

Thierry, B., A. N. Iwaniuk, and S. M. Pellis. 2000. The influence of phylogeny on the social behaviour of macaques (Primates: Cercopithecidae, genus *Macaca*). *Ethology* 106:713–728.

Thompson, J. N. 1986. Constraints on arms races in coevolution. *Trends Ecol Evol* 1:105–107.

———. 1988a. Evolutionary genetics of oviposition preference in swallowtail butterflies. *Evolution* 42:1223–1234.

————. 1988b. Coevolution and alternative hypotheses on insect/plant interactions. *Ecology* 69:893–895.

————. 1994. *The coevolutionary process.* Chicago: Univ. Chicago Press.

————. 1997. Evaluating the dynamics of coevolution among geographically structured populations. *Ecology* 78:1619–1623.

————. 1999a. Specific hypotheses on the geographic mosaic of coevolution. *Am Nat* 153:S1–S14.

————. 1999b. Coevolution and escalation: Are ongoing coevolutionary meanderings important? *Am Nat* 153:S92–S93.

————. 1999c. The evolution of species interactions. *Science* 284:2116–2118.

Thompson, J. N., B. M. Cunningham, K. A. Segraves, D. M. Althoff, and D. Wagner. 1997. Plant polyploidy and insect/plant interactions. *Am Nat* 150:73–743.

Thornton, J. W., and R. DeSalle. 2000. A new method to localize and test the significance of incongruence: Detecting domain shuffling in the nuclear receptor superfamily. *Syst Biol* 49:183–201.

Tier, C., and F. B. Hanson. 1981. Persistence in density-dependent stochastic populations. *Math Biosci* 53:89–117.

Tierderen, P. H. van, and J. Antonovics. 1994. Constraints in evolution: On the baby and the bath water. *Funct Ecol* 8:139–140.

Tiffney, B. H. 1984. Seed size, dispersal syndromes, and the rise of angiosperms: Evidence and hypothesis. *Ann Missouri Bot Gard* 71:551–576.

————. 1986. Fruit and seed dispersal and the evolution of the Hammamelidae. *Ann Missouri Bot Gard* 73:394–416.

Timm, R. M., D. E. Wilson, B. L. Clauson, R. K. LaVal, and C. S. Vaughan. 1989. Mammals of the La Selva-Braulio Carrillo complex, Costa Rica. *N Am Fauna* 75:1–162.

Timms, R., and A. F. Read. 1999. What makes a specialist special? *Trends Ecol Evol* 14:333–334.

Tinbergen, N. 1951. *The study of instinct.* London: Oxford Univ. Press.

————. 1953. *Social behaviour in animals: With special reference to vertebrates.* London: Methuen.

————. 1964. On aims and methods of ethology. *Z Tierpsychol* 20:410–433.

Tinsley, R. C. 1983. Ovoviviparity in platyhelminth life-cycles. *Parasitology* 86:161–196.

Todd, P. M., and G. F. Miller. 1991. On the sympatric origin of species: Mercurial mating in the quicksilver model. In *Proceedings of the fourth international conference on genetic algorithms,* ed. R. K. Belew and L. B. Booker, 547–554. San Mateo, CA: Morgan Kauffman.

Toft, C. A. 1985. Resource partitioning in amphibians and reptiles. *Copeia* 1985:1–21.

Toft, C. A., A. Aeschlimann, and L. Bolis, eds. 1991 *Parasite-host associations: Co-existence or conflict?* Oxford: Oxford Scientific Publications.

Travisano, M., and P. B. Rainey. 2000. Studies of adaptive radiation using model microbial systems. *Am Nat* 156:S35–S44.

Trivers, R. L. 1972. Parental investment and sexual selection. In *Sexual selection and the descent of man, 1871–1971,* ed. B. Campbell, 136–179. Chicago: Aldine.

Trontelj, P., B. Sket, and G. Steinbruck. 1999. Molecular phylogeny of leeches: Congruence of nuclear and mitochondrial rDNA data sets and the origin of bloodsucking. *J Zool Syst Evol Res* 37:141–147.

Trouvé, S., P. Sasal, J. Jourdans, F. N. Renaud, and S. Morand. 1998. The evolution of life-history traits in parasitic and free-living platyhelminthes: A new perspective. *Oecologia* 115:370–378.

Trueb, L., and R. Cloutier. 1991. A phylogenetic investigation of the inter- and intrarelationships of the Lissamphibia (Amphibia: Temnospondyli). In *Origins of the higher groups of tetrapods: Controversy and consensus,* ed. H.-P. Schultze and L. Trueb, 223–313. Ithaca, NY: Cornell Univ. Press.

Tsuji, K., and O. Ohnishi. 2001. Phylogenetic position of east Tibetan natural populations in Tartary buckwheat (*Fagopyrum tartaricum* Gaert.) revealed by RAPD analysis. *Genet Res Crop Evol* 48:63–67.

Tullberg, B. S., and A. F. Hunter. 1996. Evolution of larval gregariousness in relation to repellent defences and warning coloration in tree-feeding Macrolepidoptera: A phylogenetic analysis based on independent contrasts. *Biol J Linn Soc* 57:253–276.

Tuomikoski, R. 1967. Notes on some principles of phylogenetic systematics. *Ann Entomol Fenn* 33:137–147.

Tuffley, C., and M. Steel. 1997. Links between maximum likelihood and maximum parsimony under a simple model of site substitution. *Bull Math Biol* 59:581–607.

Turner, A. M., R. J. Bernot, and C. M. Boes. 2000. Chemical cues modify species interactions: The ecological consequences of predator avoidance by freshwater snails. *Oikos* 88:148–158.

Turner, G. F., and M. T. Burrows. 1995. A model of sympatric speciation by sexual selection. *Proc Roy Soc Lond* B 260:287–292.

Turner, H. 1995. Cladistic and biogeographic analyses of *Arytera* Blume and *Mischarytera* gen. nov. (Sapindaceae), with notes on methodology and a full taxonomic revision. *Blumea*, suppl. no. 9:1–230.

———. 1996. Sapindaceae and the biogeography of eastern Australia. *Aust Syst Bot* 8:133–167.

Turner, J. R. G. 1971. Studies of Müllerian mimicry and its evolution in burnet moths and heliconid butterflies. In *Ecological genetics and evolution,* ed. R. Creed, 224–260. Oxford: Blackwell.

———. 1977. Butterfly mimicry: The genetical evolution of an adaptation. *Evol Biol* 10:163–206.

———. 1983. Mimetic butterflies and punctuated equilibria: Some old light on a new paradigm. *Biol J Linn Soc* 20:277–300.

———. 1984. Mimicry: The palatability spectrum and its consequences. In *The biology of butterflies,* ed. R. I. Vane-Wright and P. R. Ackery, 141–161. London: Academic Press.

———. 1987. The evolutionary dynamics of Batesian and Müllerian mimicry: Similarities and differences. *Ecol Entomol* 12:81–95.

Tuttle, M. D., and M. J. Ryan. 1981. Bat predation and·the evolution of frog vocalizations in the Neotropics. *Science* 214:677–678.

Tyler, J. C., G. D. Johnson, I. Nakamura, and B. B. Colette. 1989. Morphology of *Luvarus imperialis* (Luvaridae), with a phylogenetic analysis of the Acanthuroidei (Pisces). *Smithsonian Contr Zool* 485:1–78.

Udvardy, M. D. F. 1969. *Dynamic zoogeography with special reference to land animals.* New York: Van Nostrand Reinhold.

Ulanowicz, R. E. 1986. *Growth and development: Ecosystems phenomenology.* New York: Springer Verlag.

———. 1997. *Ecology: The ascendent perspective.* New York: Columbia Univ. Press.

Urbatsch, L. E., B. G. Baldwin, and M. J. Donoghue. 2000. Phylogeny of the coneflowers and relatives (Heliantheae: Asteraceae) based on nuclear rDNA internal transcribed spacer (ITS) sequences and DNA restriction site data. *Syst Bot* 25:539–565.

Uy, J. A. C., and G. Borgia. 2000. Sexual selection drives rapid divergence in bowerbird display traits. *Evolution* 54:273–278.

Valentine, J. W., and D. Jablonski. 1991. Biotic effects of sea level change: The Pleistocene test. *J Geophys Res* 96:6873–6878.

Vamosi, S. M., T. Hatfield, and D. Schluter. 2000. A test of ecological selection against young-of-the-year hybrids of sympatric sticklebacks. *J Fish Biol* 57:109–121.

Vandel, A. 1964. *Biospéologie: La biologie des animaux cavernicoles.* Paris: Gauthier-Villars.

Vane-Wright, R. I. 1996. Systematics and the conservation of biological diversity *Ann Missouri Bot Gard* 83:47–57.

Vane-Wright, R. I., C. J. Humphries, and P. H. Williams. 1991. What to protect? Systematics and the agony of choice. *Biol Conserv* 55:235–254.

Vane-Wright, R. I., S. Schulz, and M. Boppré. 1992. The cladistics of *Amauris* butterflies: Congruence, consensus, and total evidence. *Cladistics* 8:125–138.

Vane-Wright, R. I., and C. R. Smith. 1991. Phylogenetic relationships of three Africa swallowtail butterflies, *Papilio dardanus, P. phorcas,* and *P. constantinus:* A cladistic analysis (Lepidoptera: Papilionidae). *Syst Entomol* 16:275–291.

Vanhooydonck, B., and R. van Damme. 1999. Evolutionary relationships between body shape and habitat use in lacertid lizards. *Evol Ecol Res* 1:785–805.

Van Klinken, R. D. 2000. Host specificity constrains evolutionary host change in the psyllid *Prosopidopsylla flava. Ecol Entomol* 25:413–422.

Van Leeuwen, J. L., J. H. De Groot, and W. M. Kier. 2000. Evolutionary mechanics of protrusible tentacles and tongues. *Neth J Zool* 50:113–139.

Van Valen, L. M. 1971. Adaptive zones and the orders of mammals. *Evolution* 25:420–428.

———. 1973. A new evolutionary law. *Evol Theory* 1:1–30.

———. 1976. Ecological species, multispecies, and oaks. *Taxon* 25:233–239.

Van Valkenburgh, B. 1994. Ecomorphological analysis of fossil vertebrates and their paleocommunities. In *Ecological morphology,* ed. P. C. Wainwright and S. M. Reilly, 140–168. Chicago: Univ. Chicago Press.

Van Veller, M. G. P., and D. R. Brooks. 2001. When simplicity is not parsimonious: *A priori* and *a posteriori* approaches in historical biogeography. *J Biogeogr* 28:1–11.

Van Veller, M. G. P., D. J. Kornet, and M. Zandee. 2000. Methods in vicariance biogeography: Assessment of the implementation of assumptions zero, 1 and 2. *Cladistics* 16:319–345.

———. 2001a. *A posteriori* and *a priori* methodologies for testing hypotheses of causal processes in vicariance biogeography. *Cladistics* 00:000–000.

———. 2001b. Testing hypotheses regarding general patterns in vicariance biogeography with *a posteriori* and *a priori* methods. *Cladistics* 00:000–000.

Van Veller, M. G. P., M. Zandee, and D. J. Kornet. 1999. Two requirements for obtaining valid common patterns under different assumptions in vicariance biogeography. *Cladistics* 15:393–406.

Vanzolini, P. E., and E. E. Williams. 1970. South American anoles: The geographic differentiation and evolution of the *Anolis chrysolepis* species group (Sauria: Iguanidae) *Arq Zool São Paulo* 19:1–298.

Vargas, P., B. G. Baldwin, and L. Constance. 1999. A phylogenetic study of *Sanicula* sect. *Sanicoria* and S. sect. *Sandwichenses* (Apiaceae) based on nuclear rDNA and morphological data. *Syst Bot* 24:228–248.

Vargas, P., H. A. McAllister, S. L. Jury, and M. J. Wilkinson. 1999. Polyploid speciation and *Hedera* (Araliaceae): Phylogenetic and biogeographic insights based on chromosome counts and ITS sequences. *Plant Syst Evol* 219:165–179.

Veccione, M., and B. B. Collette. 1996. Fisheries agencies and marine biodiversity. *Ann Missouri Bot Gard* 83:29–36.

Vermeij, G. J. 1994. The evolutionary interaction among species: Selection, escalation, and coevolution. *Annu Rev Ecol Syst* 25:219–236.

Vermeij, G. J., and S. J. Carlson. 2000. The muricid gastropod subfamily Rapaninae: Phylogeny and ecological history. *Paleobiology* 26:19–46.

Vermeij, G. J., and A. P. Covich. 1978. Coevolution of freshwater gastropods and their predators. *Am Nat* 112:833–843.

Verschaffelt, E. 1910. The cause determining the selection of food in some herbivorous insects. *Proc Acad Sci Amsterdam* 13:536–542.

Via, S., A. Bouck, and S. Skillman. 2000. Reproductive isolation between divergent races of pea aphids on two hosts: II. Selection against migrants and hybrids in the parental environments. *Evolution* 54:1626–1637.

Vickery, R. K. 1995. Speciation by aneuploidy and polyploidy in *Mimulus* (Scrophulariaceae). *Great Basin Nat* 55:174–176.

Vidal, N., S. G. Kindl, A. Wong, and B. Hedges. 2000. Phylogenetic relationships of Xenodontine snakes inferred from 12S and 16S ribosomal RNA sequences. *Mol Phylogenet Evol* 14:389–402.

Vilhelmsen, L. 2000. Cervical and prothoracic skeletomusculature in the basal Hymenoptera (Insecta): Comparative anatomy and phylogenetic implications. *Zool Anz* 239:105–138.

Vincent, W. F. 2000. Evolutionary origins of Antarctic microbiota: Invasion, selection and endemism. *Antarct Sci* 12:374–385.

Vitt, L. J., P. A. Zani, and M. C. Esposito. 1999. Historical ecology of Amazonian lizards: Implications for community ecology. *Oikos* 87:286–294.

Vogler, A. P., and R. DeSalle. 1993. Phylogeographic patterns in coastal North American tiger beetles (*Cicindela dorsalis* Say) inferred from mitochondrial DNA sequences. *Evolution* 47:1192–1202.

Vogler, A. P., and K. C. Kelley. 1996. At the interface of phylogenetics and ecology: The case of chemical defenses in *Cicindela. Acta Zool Fenn* 33:39–47.

————. 1998. Covariation of defensive traits in tiger beetles (genus *Cicindella*): A phylogenetic approach using mtDNA. *Evolution* 52:529–538.

Vollrath, F., and D. T. Edmonds. 1989. Modulation of the mechanical properties of spider silk by coating with water. *Nature* 340:305–307.

Vollrath, F., and G. A. Parker. 1992. Sexual dimorphism and distorted sex ratios in spiders. *Nature* 360:156–159.

————. 1997. Reply to Coddington et al.: Giant female or dwarf male spiders? *Nature* 385:688.

von Buren, M. 2001. Polymorphisms in two homologous gamma gliadin genes and the evolution of cultivated wheat. *Genet Res Crop Evol* 48:205–220.

Voorzanger, B., and W. J. van der Steen. 1982. New perspectives on the Biogenetic Law? *Syst Zool* 31:202–205.

Voss, R. 1979. Male accessory glands and the evolution of copulatory plugs in rodents. *Occas Pap Mus Zool Univ Michigan* 689:1–27.

Vrba, E. S. 1980. Evolution, species, and fossils: How does life evolve? *S Afr J Sci* 76:61–84.

————. 1984a. What is species selection? *Syst Zool* 33:318–328.

————. 1984b. Evolutionary pattern and process in the sister-group Alcelaphini-Aepycerotini (Mammalia: Bovidae). In *Living fossils*, ed. N. Eldredge and S. M. Stanley, 62–79. New York: Springer Verlag.

————. 1989. Levels of selection and sorting with special reference to the species level. In *Oxford Surveys in Evolutionary Biology*, vol. 6, ed. P. H. Harvey and L. Partridge, 111–168. Oxford: Oxford University Press.

Vrba, E. S., and S. J. Gould, 1986. The hierarchical expansion of sorting and selection: Sorting and selection cannot be equated. *Paleobiology* 12:217–228.

Vuilleumier, B. S. 1971. Pleistocene changes in the flora and fauna of South America. *Science* 173:771–780.

Wade, M. J., and C. J. Goodnight. 1998. Perspective: The theories of Fisher and Wright in the context of metapopulations: When nature does many small experiments. *Evolution* 52:1537–1553.

Wägele, J. W. 1996a. First principles of phylogenetic systematics, a basis for numerical methods used for morphological and molecular characters. *Vie Milieu* 46:125–138.

————. 1996b. Identification of apomorphies and the role of groundplan patterns in molecular systematics. *J Zool Syst Evol Res* 34:313–339.

————. 1999. Major sources of error in phylogenetic systematics. *Zool Anz* 238:329–337.

Wägele, J. W., and F. Rödding. 1998. A priori estimation of phylogenetic information conserved in aligned sequences. *Mol Phylogenet Evol* 9:358–265.

Wägele, J. W., and G. Stanjek 1995. Arthropod phylogeny inferred from partial 12s rRNA revisited: Monophyly of the Tracheata depends on sequence alignment. *J Zool Syst Evol Res* 33:75–80.

Wagner, G. P., and K. Schwenk. 1999. Evolutionary stable configurations: Functional integration and the evolution of phenotypic stability. *Evol Biol* 31:155–217.

Wagner, M. 1868. *Die Darwin'sche Theorie und das Migrationsgesetz der Organismen*. Leipzig: Duncker and Humboldt.

Wagner, W. H., Jr. 1983. Reticulistics: The recognition of hybrids and their role in cladistics and classification. In *Advances in cladistics: Proceedings of the second meeting of the Willi Hennig Society*, vol. 2, ed. N. I. Platnick and V. A. Funk, 63–79. New York: Columbia Univ. Press.

Wagner, W. L., and V. A. Funk, eds. 1995 *Hawaiian biogeography: Evolution on a hot spot archipelago*. Washington, DC: Smithsonian Institution Press.

Wahlberg, N. 2001. The phylogenetics and biochemistry of host plant specialization in Melitaeine butterflies (Lepidoptera: Nymphalidae). *Evolution* 55:522–537.

Wainwright, P. C. 1989. Prey processing in haemulid fishes: Patterns of variation in pharyngeal jaw muscle activity. *J Exp Biol* 141:359–375.

Wainwright, P. C. 1996. Ecological explanation through functional morphology: The feeding biology of sunfishes. *Ecology* 77:1336–1343.

Wainwright, P. C., and G. V. Lauder. 1986. Feeding biology of sunfishes: Patterns of variation in prey capture. *Zool J Linn Soc* 88:217–228.

————. 1992. The evolution of feeding biology in sunfishes (Centrarchidae). In *Systematics, historical ecology, and North American freshwater fishes*, ed. R. Mayden, 472–491. Stanford, CA: Stanford Univ. Press.

Wainwright, P. C., and S. M. Reilly, eds. 1994. *Ecological morphology: Integrative organismal biology.* Chicago: Univ. Chicago Press.

Wainwright, P. C., C. P. Sanford, S. M. Reilly, and G. V. Lauder. 1989. Evolution of motor patterns: Aquatic feeding in salamanders and ray-finned fishes. *Brain Beh Evol* 34:329–341.

Wake, D. B. 1982. Functional and developmental constraints and opportunities in the evolution of feeding systems in urodeles. In *Environmental adaptation and evolution,* ed. D. Mossakowski and G. Royth, 51–66. New York: G. Fischer.

———. 1991. Homoplasy: The result of natural selection or evidence of design limitations? *Am Nat* 138:543–567.

Wake, D. B., and G. Roth, eds. 1989 *Complex organismal functions: Integration and evolution in vertebrates.* New York: John Wiley and Sons.

Wake, D. B., K. P. Yanev, and M. M. Frelow. 1989. Sympatry and hybridization in a "ring species": The plethodontid salamander *Ensatina eschscholtzii.* In *Speciation and its consequences,* ed. D. Otte and J. A. Endler, 134–157. Sunderland, MA: Sinauer.

Wake, M. H. 1990. The evolution of integration of biological systems: An evolutionary perspective through study of cells, tissues, and organs. *Am Zool* 30:897–906.

———. 1992. Morphology, the study of form and function. In *Oxford surveys in evolutionary biology,* vol. 9: *Modern evolutionary biology,* ed. D. J. Futuyma and J. Antonovics, 289–346. Oxford: Oxford Univ. Press.

———. 1993. Non-traditional characters in the assessment of caecilian phylogenetic relationships. *Herpetol Monogr* 7:42–55.

———. 1994. The use of unconventional morphological characters in analysis of systematic patterns and evolutionary processes. In *Interpreting the hierarchy of nature: From systematic patterns to evolutionary process theories,* ed. L. Grande and O. Rieppel, 173–200. London: Academic Press.

Walker, B., A. Kinzig, and J. Langridge. 1999. Plant attribute diversity, resilience, and ecosystem function: The nature and significance of dominant and minor species. *Ecosystems* 2:95–113.

Walker, D., G. Orti, and J. C. Avise. 1998. Phylogenetic distinctiveness of a threatened aquatic turtle *(Sternotherus depressus). Conserv Biol* 12:639–645.

Wall, D., H. Mooney, G. Adams, G. Boxshall, A. Dobson, T. Nakashizuku, J. Seyani, C. Samper, and S. Sarukhan. 2001. An international biodiversity observation year. *Trends Ecol Evol* 16:52–54.

Wallace, A. R. 1855. On the law which has regulated the introduction of new species. *Ann Mag Nat Hist,* 2d ser. 16:184–196.

———. 1878. *Tropical nature and other essays.* London: Macmillan.

———. 1889a. *Travels on the Amazon.* New York: Harper and Row.

———. 1889b. *Darwinism.* 1st ed. London: Macmillan.

———. 1905. *Darwinism.* 3d ed. London: Macmillan.

Wallace, B. 1955. Inter-population hybrids in *Drosophila melanogaster. Evolution* 9:302–316.

Walter, G. H., and R. Hengeveld. 2000. The structure of the two ecological paradigms. *Acta Biotheor* 48:15–46.

Wang, X. M., R. H. Tedford, and B. E. Taylor. 1999. Phylogenetic systematics of the Borophaginae (Carnivora: Canidae). *Bull Am Mus Nat Hist* 243:1–391.

Wanntorp, H.-E. 1983a. Reticulated cladograms and the identification of hybrid taxa. In *Advances in cladistics: Proceedings of the second meeting of the Willi Hennig Society,* vol. 2, ed. N. I. Platnick and V. A. Funk, 81–88. New York: Columbia Univ. Press.

———. 1983b. Historical constraints in adaptation theory: Traits and non-traits. *Oikos* 41:157–160.

Wanntorp, H.-E., D. R. Brooks, T. Nilsson, S. Nylin, F. Ronquist, S. C. Stearns, and N. Weddell. 1990. Phylogenetic approaches in ecology. *Oikos* 57:119–132.

Ward, D., and M. K. Seely. 1996. Adaptation and constraint in the evolution of the physiology and behavior of the Namib desert tenebrionid beetle genus *Onymacris. Evolution* 50:1231–1240.

Ward, S. A. 1991. Theoretical perspectives on sympatric speciation in aphids (Homoptera: Aphidinea: Aphididae). *Entomol Gener* 16:177–192.

Waters, A. P., D. G. Higgins, and T. F. McCutchan. 1991. *Plasmodium falciparum* appears to have arisen as a result of lateral transfer between avian and human hosts. *Proc Natl Acad Sci* (USA) 88:3140–3144.

Watertor, J. L. 1967. Intraspecific variation of adult *Telorchis bonnerensis* (Trematoda: Telorchiidae) in amphibian and reptilian hosts. *J Parasitol* 53:962–968.

Watkins, J. 1984. *Science and skepticism*. Princeton: Princeton Univ. Press.

Watrous, L. E., and Q. D. Wheeler. 1981. The outgroup comparison method of character analysis. *Syst Zool* 30:1–11.

Webb, C. O. 2000. Exploring the phylogenetic structure of ecological communities: An example for rain forest trees. *Am Nat* 156:145–155.

Webb, J. K., and R. Shine. 1994. Feeding habits and reproductive biology of Australian pygopodid lizards of the genus *Aprasia*. *Copeia* 1994:390–398.

Webb, S. D. 1976. Mammalian faunal dynamics of the Great American interchange. *Paleobiology* 2:216–234.

———. 1985. Late Cenozoic mammal dispersals between the Americas. In *The great American biotic interchange*, ed. F. G. Stehli and S. D. Webb, 357–386. New York: Plenum Press.

Wegener, A. 1912. *Die Entstehung der Kontinente*. *Petermanns Mitt* 58:185–195, 253–256, 305–309.

———. 1924. *The origin of continents and ocean basins*. London: Methuen.

Weiblen, G. D., R. K. Oyama, and M. J. Donoghue. 2000. Phylogenetic analysis of dioecy in monocotyledons. *Am Nat* 155:46–58.

Weider, L. J., A. Hobaek, J. K. Colbourne, T. J. Crease, F. Dufresne, and P. D. N. Hebert. 1999. Holarctic phylogeography of an asexual species complex: I. Mitochondrial DNA variation in Arctic *Daphnia*. *Evolution* 53:777–792.

Weinberg, J. R., V. R. Starczak, and D. Jörg. 1992. Evidence for rapid speciation following a founder event in the laboratory. *Evolution* 46:1214–1220.

Weisrock, D. W., and F. J. Janzen. 2000. Comparative molecular phylogeography of North American softshell turtles (*Apalone*): Implications for regional and wide scale historical evolutionary forces. *Mol Phylogenet Evol* 14:152–164.

Weitzman, S. H., and M. Weitzman. 1982. Biogeography and evolutionary diversification in Neotropical freshwater fishes, with comments on the refuge theory. In *Biological diversification in the tropics*, ed. G. I. Prance, 403–422. New York: Columbia Univ. Press.

Weller, S. G., and A. K. Sakai. 1999. Using phylogenetic approaches for the analysis of plant breeding system evolution. *Annu Rev Ecol Syst* 30:167–199.

Weller, S. G., W. L. Wagner, and A. K. Sakai. 1995. A phylogenetic analysis of *Schiedea* and *Alsinidendron* (Caryophyllaceae: Alsinoideae): Implications for the evolution of breeding systems. *Syst Bot* 20:315–337.

Weller, S. J., N. L. Jacobson, and W. E. Conner. 1999. The evolution of chemical defences and mating systems in tiger moths (Lepidoptera: Arctiidae). *Biol J Linn Soc* 68:557–578.

Welzen, P. C. van. 1990. *Guioa Cav (Sapindaceae): Taxonomy, phylogeny, and historical biogeography*. Leiden: Rijksherbarium/Hortus Botanicus.

Wenzel, J. W. 1991. Evolution of nest architecture. In *The social biology of wasps*, ed. K. G. Ross and R. W. Matthews, 480–519. Ithaca, NY: Cornell Univ. Press.

———. 1992. Behavioral homology and phylogeny. *Annu Rev Ecol Syst* 23:361–381.

———. 1993. Application of the biogenetic law to behavioral ontogeny: A test using nest architecture in paper wasps. *J Evol Biol* 6:229–247.

Wenzel, J. W., and J. M. Carpenter. 1994. Comparing methods: Adaptive traits and tests of adaptation. In *Phylogenetics and ecology*, ed. P. Eggleton and R. I. Vane-Wright, 79–101. London: Academic Press.

Wenzel, J. W., and M. E. Siddall. 1999. Noise. *Cladistics* 15:51–64.

Werdelin, L., and B. Tullberg. 1995. A comparison of two methods to study correlated discrete characters on phylogenetic trees. *Cladistics* 11:265–277.

Werner, Y. L., A. Bouskila, S. J. J. F. Davies, and N. Werner. 1997a. Observations and comments on active foraging in geckos. *Russian J Herpetol* 4:34–39.

Werner, Y. L., S. Okada, H. Ota, G. Perry, and S. Tokunaga. 1997b. Varied and fluctuating foraging modes in nocturnal lizards of the family Gekkonidae. *Asiat Herpetol Res* 7:153–165.

Werth, C. R., and M. D. Windham. 1991. A model for divergent, allopatric speciation of polyploid Pteridophytes resulting from silencing of duplicate-gene expression. *Am Nat* 137:515–526.

West, S. A., and E. A. Herre. 1994. The ecology of the New World fig-parasitizing wasps *Idarnes* and implications for the evolution of the fig-pollinator mutualism. *Proc Roy Soc Lond* B 258:67–72.

West-Eberhard, M. J. 1978. Polygyny and the evolution of social behavior in wasps. *J Kansas Entomol Soc* 51:832–856.

———. 1983. Sexual selection, social competition, and speciation. *Quart Rev Biol* 58:155–183.

———. 1989. Phenotypic plasticity and the origins of diversity. *Annu Rev Ecol Syst* 20:249–278.

Westneat, M. W. 1994. Transmission force and velocity in the feeding mechanisms of labrid fishes (Teleostei, Perciformes). *Zoomorphology* 114:103–118.

———. 1995a. Phylogenetic systematics and biomechanics in ecomorphology. *Environ Biol Fishes* 44:263–282.

———. 1995b. Feeding, function, and phylogeny: Analysis of historical biomechanics in labrid fishes using comparative methods. *Syst Biol* 44:361–383.

Westneat, M. W., and P. C. Wainwright. 1989. Feeding mechanism of *Epibulus insidiator* (Labridae; Teleostei): Evolution of a novel functional system. *J Morphol* 202:129–150.

Westoby, M. 1988. Comparing Australian ecosystems to those elsewhere. *BioScience* 38:549–556.

Westoby, M., M. R. Leishman, and J. M. Lord. 1995a. On misinterpreting the "phylogenetic correction." *J Ecol* 83:531–534.

———. 1995b. Further remarks on phylogenetic correction. *J Ecol* 83:727–730.

Wetterer, A. L., M. V. Rockmand, and N. B. Simmons. 2000. Phylogeny of phyllostomid bats (Mammalia: Chiroptera): Data from diverse morphological systems, sex chromosomes, and restriction sites. *Bull Am Mus Nat Hist* 248:4–200.

Wheeler, Q. D. 1990. Insect diversity and cladistic constraints. *Ann Entomol Soc Am* 83:1031–1047.

———. 1993. A crisis of biodiversity, systematics, and ecology. *BioScience* 43:578–580.

———. 1995. Systematics, the scientific basis for inventories of biodiversity. *Biodiv Conserv* 4:476–489.

———. 1999. Why the phylogenetic species concept?—Elementary. *J Nematol* 31:134–141.

Wheeler, Q. D., and M. Blackwell, eds. 1984. *Fungus-insect relationships: Perspectives in ecology and evolution.* New York: Columbia Univ. Press.

Wheeler, W. C. 1992. Extinction, sampling, and molecular phylogenetics. In *Extinction and phylogeny,* ed. M. J. Novacek and Q. D. Wheeler, 205–215. New York: Columbia Univ. Press.

Wheeler, W. C., R. T. Schuh, and R. Bang. 1993. Cladistic relationships among higher groups of heteroptera: Congruence between morphological and molecular data sets. *Entomol Scand* 24:121–137.

Wheeler, W. M. 1919. The parasitic *Aculeata:* A study in evolution. *Proc Am Philos Soc* 58:1–40.

———. 1928. *The social insects: Their origin and evolution.* New York: Harcourt, Brace.

Whelan, S., and N. Goldman. 1999. Distribution of statistics used for the comparison of models of sequence evolution in phylogenetics. *Mol Biol Evol* 16:1292–1299.

White, M. J. D. 1978. *Modes of speciation.* San Francisco: W. H. Freeman.

Whitfield, J. B. 1990. Parasitoids, polyadenoviruses, and endosymbiosis. *Parasitol Today* 6:381–384.

Whitlock, M. C. 1997. Founder effects and peak shifts without genetic drift: Adaptive peak shifts occur easily when environments fluctuate slightly. *Evolution* 51:1044–1048.

Whitlock, M. C., and M. J. Wade. 1995. Speciation: Founder events and their effects on X-linked and autosomal genes. *Am Nat* 145:676–685.

Whitman, C. O. 1899. Animal behavior. In *Biological lectures, Wood's Hole,* ed. C. O. Whitman, 285–338. Boston: Ginn.

Whittaker, R. H. 1965. Dominance and diversity in land plant communities. *Science* 147:250–260.

———. 1972. Evolution and measurement of species diversity. *Taxon* 21:213–251.

Whitten, W. M., N. H. Williams, W. S. Armbruster, M. A. Battiste, L. Strekowski, and

N. Lindquist. 1986. Carvone oxide: An example of convergent evolution in euglossine-pollinated plants. *Syst Bot* 11:222–228.

Wickler, W. 1961. Ökologie und Stammesgeschichte von Verhaltensweisen. *Fortschr Zool* 13:303–365.

Wickman, P.-O. 1992. Sexual selection and butterfly design: A comparative study. *Evolution* 46:1525–1536.

Wiegmann, B. M., C. Mitter, and B. Farrell. 1993. Diversification of carnivorous insects: Extraordinary radiation or specialized dead end? *Am Nat* 142:737–754.

Wiens, J. J. 1995. Polymorphic characters in phylogenetic analysis. *Syst Biol* 44:482–500.

Wiens, J. J., and B. D. Hollingsworth. 2000. War of the iguanas: Conflicting molecular and morphological phylogenies and long-branch attraction in iguanid lizards. *Syst Biol* 49:143–159.

Wiens, J. J., and T. W. Reeder. 1995. Combining data sets with different numbers of taxa for phylogenetic analysis. *Syst Biol* 44:548–558.

Wiklund, C. 1973. Host plant suitability and the mechanism of host selection in larvae of *Papilio machaon*. *Ent Exp Appl* 16:232–242.

———. 1974. Oviposition preference in *Papilio machaon* in relation to the host plants of the larvae. *Ent Exp Appl* 17:189–198.

———. 1975. The evolutionary relationship between adult oviposition preferences and larval host plant range in *Papilio machaon*. *Oecologia* 18:185–197.

———. 1981. Generalist vs. specialist oviposition behaviour in *Papilio machaon* (Lepidoptera) and functional aspects on the hierarchy of oviposition preferences. *Oikos* 36:163–170.

———. 1984. Egg-laying patterns in butterflies in relation to their phenology and the visual apparency and abundance of their host plants. *Oecologia* 63:23–29.

Wilcox, B. A. 1980. Insular ecology and conservation. In *Conservation biology: An evolutionary-ecology perspective*, ed. M. E. Soulé and B. A. Wilcox, 95–117. Sunderland, MA: Sinauer.

Wiley. E. O. 1975. Karl R. Popper, systematics, and classification: A reply to Walter Bock and other evolutionary taxonomists. *Syst Zool* 24:233–242.

———. 1976. The phylogeny and biogeography of fossil and recent gars (Actinopterygii: Lepisosteidae). *Misc Pub Univ Kansas Mus Nat Hist* 64:1–111.

———. 1978. The evolutionary species concept reconsidered. *Syst Zool* 27:17–26.

———. 1980a. Is the evolutionary species concept fiction? A consideration of classes, individuals, and historical entities. *Syst Zool* 29:76–80.

———. 1980b. The metaphysics of individuality and its consequences for systematic biology. *Brain Behav Sci* 4:302–303.

———. 1981. *Phylogenetics: The theory and practice of phylogenetic systematics*. New York: Wiley InterScience.

———. 1986a. The evolutionary basis for phylogenetic classification. In *Systematics and evolution: A matter of diversity*, ed. P. Hovenkamp, 55–64. Utrecht: Univ. Utrecht Press.

———. 1986b. Approaches to outgroup comparison. In *Systematics and evolution: A matter of diversity*, ed. P. Hovenkamp, 173–191. Utrecht: Univ. Utrecht Press.

———. 1986c. Process and pattern: Cladograms and trees. In *Systematics and evolution: A matter of diversity*, ed. P. Hovenkamp, 233–247. Utrecht: Univ. Utrecht Press.

———. 1986d. Methods in vicariance biogeography. In *Systematics and evolution: A matter of diversity*, ed. P. Hovenkamp, 283–306. Utrecht: Univ. Utrecht Press.

———. 1986e. Historical ecology and coevolution. In *Systematics and evolution: A matter of diversity*, ed. P. Hovenkamp, 331–341. Utrecht: Univ. Utrecht Press.

———. 1986f. La sistematica en la revolución Darwiniana. *Ann Mus Hist Nat Valparaiso* 17:25–31.

———. 1988a. Vicariance biogeography. *Annu Rev Ecol Syst* 19:513–542.

———. 1988b. Parsimony analysis and vicariance biogeography. *Syst Zool* 37:271–290.

———. 1989. Kinds, individuals, and theories. In *What the philosophy of biology is: Essays for David Hull*, ed. M. Ruse, 289–300. Dordrecht: Kluwer Academic.

Wiley, E. O., and D. R. Brooks. 1982. Victims of history: A nonequilibrium approach to evolution. *Syst Zool* 31:1–24.

Wiley, E. O., G. D. Johnson, and W. W. Dimmick. 1998. The phylogenetic relationships of lampridiform fishes (Teleostei: Acanthomorpha), based on a total-evidence analysis of morphological and molecular data. *Mol Phylogenet Evol* 10:417–425.

Wiley, E. O., and R. L. Mayden. 1985. Species and speciation in phylogenetic systematics, with examples from the North American fish fauna. *Ann Missouri Bot Gard* 72:596–635.

———. 2000a. The evolutionary species concept. In *Species concepts and phylogenetic theory: A debate,* ed. Q. D. Wheeler and R. Meier, 70–89. New York: Columbia Univ. Press.

———. 2000b. Comments on alternative species concepts. In *Species concepts and phylogenetic theory: A debate,* ed. Q. D. Wheeler and R. Meier, 146–158. New York: Columbia Univ. Press.

———. 2000c. A reply to our critics. In *Species concepts and phylogenetic theory: A debate,* ed. Q. D. Wheeler and R. Meier, 198–208 New York: Columbia Univ. Press.

Wiley, E. O., D. J. Siegel-Causey, D. R. Brooks, and V. A. Funk. 1991. *The compleat cladist: A primer of phylogenetic procedures.* Mus Nat Hist Univ Kansas Spec Publ 19.

———. Forthcoming. *The compleat cladist: A primer of phylogenetic procedures.* 2d ed. Mus Nat Hist Univ Kansas Spec Publ.

Wilga, C. D., P. C. Wainwright, and P. J. Motta. 2000. Evolution of jaw depression mechanics in aquatic vertebrates: Insights from Chondrichthyes. *Biol J Linn Soc* 71:165–185.

Wilkinson, M. 1994. Common cladistic information and its consensus representation: Reduced Adams and reduced cladistic consensus trees and profiles. *Syst Biol* 43:343–368.

———. 1995. Coping with abundant missing entries in phylogenetic inference using parsimony. *Syst Biol* 44:501–514.

———. 1996. Majority-rule consensus trees and their use in bootstrapping. *Mol Biol Evol* 13:437–444.

Wilkinson, M., and R. A. Nussbaum. 1999. Evolutionary relationships of the lungless caecilian *Atretochoana eiselti* (Amphibia: Gymnophiona: Typhlonectidae). *Zool J Linn Soc* 126:191–223.

Wilkinson, M., J. L. Thorley, and P. Upchurch. 2000. A chain is no stronger than its weakest link: Double decay analysis of phylogenetic hypotheses. *Syst Biol* 49:754–776.

Will, K. W., A. B. Attygalle, and K. Herath. 2000. New defensive chemical data for ground beetles (Coleoptera: Carabidae): Interpretations in a phylogenetic framework. *Biol J Linn Soc* 71:459–481.

Williams, C. B. 1964. *Patterns in the balance of nature.* London: Academic Press.

Williams, E. E. 1972. The origin of faunas. Evolution of lizard congeners in a complex island fauna: A trial analysis. *Evol Biol* 6:47–89.

———. 1983. Ecomorphs, faunas, island size, and diverse end points in island radiations of *Anolis*. In *Lizard ecology: Studies of a model organism,* ed. R. B. Huey, E. R. Pianka, and T. W. Schoener, 326–370. Cambridge, MA: Harvard Univ. Press.

Williams, G. C. 1992. *Natural selection: Domains, levels, and challenges.* Oxford: Oxford Univ. Press.

Williams, J. E. 2000. The biodiversity crisis and adaptation to climate change: A case study from Australia's forests. *Environ Monit Assess* 61:65–74.

Williams, J. E., J. E. Johnson, D. A. Hendrickson, S. Contreras-Balderas, J. D. Williams, M. Navarro-Mendoza, D. E. McAllister, and J. E. Deacon. 1989. Fishes of North America endangered, threatened, or of special concern: 1989. *Fisheries* 14:2–20.

Williams, N. H., and W. M. Whitten. 1983. Orchid floral fragrances and euglossine bees: Methods and advances in the last sesquidecade. *Biol Bull* 164:355–395.

Williams, P. H., G. T. Prance, C. G. Humphries, and K. S. Edwards. 1996. Promise and problems in applying quantitative complementary areas for representing the diversity of some Neotropical plants (families Dichapetalaceae, Lecythidaceae, Caryocaraceae, Chrysobalanaceae, and Proteaceae). *Biol J Linn Soc* 58:125–157.

Wilson, A. C., G. L. Bush, S. M. Case, and M. C. King. 1975. Social structuring of mammalian populations and rate of chromosomal evolution. *Proc Natl Acad Sci (USA)* 72:5061–5065.

Wilson, D. S. 1980. *The natural selection of populations and communities.* Menlo Park: Benjamin Cummings.

———. 1983. The group selection controversy: History and current status. *Annu Rev Ecol Syst* 14:159–117.

Wilson, D. E., and D. M. Reeder. 1993. Introduction to *Mammal species of the world: A tax-*

onomic and geographic reference, ed. D. E. Wilson and D. M. Reeder, 1–12. Washington, DC: Smithsonian Institution Press.

Wilson, E. O. 1971. The plight of taxonomy. *Ecology* 52:741.

———. 1984. *Biophilia*. Cambridge, MA: Harvard Univ. Press.

———. 1985. The biodiversity crisis: A challenge to science. *Issues Sci Technol* (Fall):20–29.

———. 1988. The current state of biological diversity. In *Biodiversity*, ed. E. O. Wilson, 1–18. Washington DC: National Academy of Sciences Press.

———. 1989. The coming pluralization of biology and the stewardship of systematics. *BioScience* 39:242–245.

———. 1990. Threats to biodiversity. In *Managing planet Earth: Readings from Scientific American*, 49–59. San Francisco: Freeman.

———. 1992. *The diversity of life*. Cambridge, MA: Harvard Univ. Press.

Wilson, L. D., and J. R. McCranie. 1998. The biogeography of the herpetofauna of the subhumid forest of Middle America (Isthmus of Tehuantepec to northwestern Costa Rica). *Royal Ontario Museum of Life Science Contributions* 163:1–50.

Wilson, G. D. F., and R. T. Johnson. 1999. Ancient endemism among freshwater isopods (Crustacea, Phreatoicidea). In *The other 99%: The conservation and biodiversity of invertebrates*, ed. W. Ponder and D. Lunney, 264–268. *Trans Royal Soc NSW*. Mosman: Royal Society of New South Wales.

Wilson, R. A., G. Smith, and M. R. Thomas. 1982. Fascioliasis. In *Population dynamics of infectious diseases: Theory and applications*, ed. R. M. Anderson, 262–319. London: Chapman and Hall.

Wimberger, P. H., and A. de Queiroz. 1996. Comparing behavioral and morphological characters as indicators of phylogeny. In *Phylogenies and the comparative method in animal behavior*, ed. E. P. Martins, 206–233. Oxford: Oxford Univ. Press.

Windley, B. F. 1986. *The evolving continents*. New York: Wiley.

Wing, S. L., and L. D. Boucher. 1998. Ecological aspects of the Cretaceous flowering plant radiation. *Annu Rev Earth Planet Sci* 26:379–421.

Winkler, D. W. 2000. The phylogenetic approach to avian life histories: An important complement to within population studies. *Condor* 102:52–59.

Winnepenninckx, B., and T. Backeljau. 1996. 18S rRNA alignments derived from different secondary structure models can produce alternative phylogenies. *J Zool Syst Evol Res* 34:135–143.

Winterbottom, R. 1990. The *Trimmatom nanus* species complex (Actinopterygii, Gobiidae): Phylogeny and progenetic heterochrony. *Syst Zool* 39:253–265.

———. 1993. Morphological evidence for the phylogeny of recent genera of surgeonfishes (Percomorpha, Acanthuridae), with comments on the Acanthuroidei. *Copeia* 1993:21–39.

Winterbottom, R., and D. A. McLennan. 1993. Cladogram versatility-evolution and biogeography of acanthuroid fishes. *Evolution* 47:1557–1571.

Witt, J. D. S., and P. D. N. Hebert. 2000. Cryptic species diversity and evolution in the amphipod genus *Hyallela* within central glaciated North America: A molecular phylogenetic approach. *Can J Fish Aquat Sci* 57:687–698.

Witt, P. N., C. F. Reed, and D. B. Peakall. 1968. A spider's web: Problems in regulatory biology. Berlin: Springer Verlag.

Witt, P. N., M. B. Scarboro, R. Daniels, D. B. Peakall, and R. L. Gause. 1976. Spider webbuilding in outer space: Evaluation of records from the Skylab spider experiment. *J Arachnol* 4:115–124.

Witte, H., and D. Döring. 1999. Canalized pathways of change and constraints in the evolution of reproductive modes of microarthropods. *Exp App Acarol* 23:181–216.

Wong, A. H. K., D. D. Williams, D. J. McQueen, E. Demers, and C. W. Ramcharan. 1998. Macroinvertebrate abundance in two lakes with contrasting fish communities. *Arch Hydrobiol* 141:283–302.

Wootton, R. J. 1976. *The biology of the sticklebacks*. London: Academic Press.

———. 1984. *A functional biology of sticklebacks*. Berkeley: Univ. of California Press.

Wray, G. A. 1996. Parallel evolution of nonfeeding larvae in echinoids. *Syst Biol* 45:308–322.

Wray, G. A., J. S. Levinton, L. H. Shapiro. 1996. Molecular evidence for deep Precambrian divergences among Metazoan phyla. *Science* 274:568–573.

Wray, G. A., and C. J. Lowe. 2000. Developmental regulatory genes and echinoderm evolution. *Syst Biol* 49:28–51.

Wright, D. H. 1983. Species-energy theory: An extension of species-area theory. *Oikos* 41:496–506.

Wright, S. 1931. Evolution in Mendelian populations. *Genetics* 16:97–159.

———. 1940. Breeding structure of populations in relation to speciation. *Am Nat* 74:232–248.

———. 1949. Adaptation and selection. In *Genetics, paleontology, and evolution,* ed. G. L. Jepson, G. G. Simpson, and E. Mayr, 365–389. Princeton, NJ: Princeton Univ. Press.

———. 1978a. Modes of speciation. *Paleobiology* 4:373–379.

———. 1978b. *Evolution and the genetics of populations,* vol. 4: *Variability within and among populations.* Chicago: Univ. Chicago Press.

Wright, W. G. 1998. Evolution of nonassociative learning: Behavioral analysis of a phylogenetic lesion. *Neurobiol Learn Mem* 69:326–337.

———. 2000. Neuronal and behavioral plasticity in evolution: Experiments in a model lineage. *BioScience* 50:883–894.

Wright, W. G., D. Kirchman, D. Rozen, and B. Maynard. 1996. Phylogenetic analysis of learning-related neuromodulation in molluscan mechanosensory neurons. *Evolution* 50:2248–2263.

Wright, W. G., E. F. McCance, and T. J. Carew. 1996. Developmental emergence of long-term memory for sensitization in *Aplysia. Neurobiol Learn Mem* 65:261–268.

Wroe, S., M. Ebach, S. Ahyong, C. de Muizon, and J. Muirhead. 2000. Cladistic analysis of dasyuromorphian (Marsupalia) phylogeny using cranial and dental characters. *J Mammal* 81:1008–1024.

Wu, C.-I. 1985. A stochastic simulation study on speciation by sexual selection. *Evolution* 39:66–82.

Wulff, E. V. 1950. *An introduction to historical plant biogeography.* Waltham, MA: Chronica Botanica. (Translation from Russian.)

Wyner, Y., R. DeSalle, and R. Absher. 2000. Phylogeny and character behavior in the family Lemuridae. *Mol Phylogenet Evol* 15:124–134.

Xu, S. Z. 2000. Phylogenetic analysis under reticulate evolution. *Mol Biol Evol* 17:897–907.

Yang, L. L., B. M. Wiegmann, D. K. Yeates, and M. E. Irwin. 2000. Higher level phylogeny of the Therevidae (Diptera: Insecta) based on 28S ribosomal and elongation factor-1 alpha gene sequences. *Mol Phylogenet Evol* 15:440–451.

Yang, Z. 1994. Estimating the pattern of nucleotide substitution. *J Mol Evol* 39:105–111.

———. 1996. Maximum-likelihood models for combined analyses of multiple sequence data. *J Mol Evol* 42:587–596.

Yang, Z., and J. P. Bielawski. 2000. Statistical methods for detecting molecular adaptation. *Trends Ecol Evol* 15:496–503.

Young, B. A. 1993. Evaluating hypotheses for the transfer of stimulus particles to Jacobson organ in snakes. *Brain Behav Evol* 41:203–209.

Zahavi, A. 1975. Mate selection: A selection for a handicap. *J Theor Biol* 53:205–214.

Zamparo, D., D. R. Brooks, E. P Hoberg, and D. A. McLennan. 2001. Phylogenetic analysis of the Rhabdocoela (Platyhelminthes) with emphasis on the Neodermata and relatives. *Zool Scr* 30:59–77.

Zamudio, K. 1998. The evolution of female-biased sexual size dimorphism: A population-level comparative study in horned lizards (*Phrynosoma*). *Evolution* 52:1821–1833.

Zamudio, K. R., and H. W. Greene. 1997. Phylogeography of the bushmaster (*Lachesis muta:* Viperidae): Implications of neotropical biogeography, systematics, and conservation. *Biol J Linn Soc* 62:421–442.

Zandee, M., and M. C. Roos. 1987. Component-compatibility in historical biogeography. *Cladistics* 3:305–332.

Zelditch, M. L., and W. L. Fink. 1996. Heterochrony and heterotopy: Stability and innovation in the evolution of form. *Paleobiology* 22:241–254.

———. 1998. Partial warps, phylogeny, and ontogeny: A reply to Adams and Rosenberg. *Syst Biol* 47:345–348.

Zelditch, M. L., W. L. Fink, D. L. Swiderski, and B. L. Lundrigan. 1998. On applications of geometric morphometrics to studies of ontogeny and phylogeny: A reply to Rohlf. *Syst Biol* 47:159–167.

Zelditch, M. L., H. D. Sheets, and W. L. Fink. 2000. Spatiotemporal reorganization of growth rates in the evolution of ontogeny. *Evolution* 54:1363–1371.

Zhou, Z. Q., and Y. N. Li. 2000. The RAPD evidence for the phylogenetic relationship of the closely related species of cultivated apple. *Genet Res Crop Evol* 47:353–357.

Zimmerman, B. L., and D. Simberloff. 1996. An historical interpretation of habitat use by frogs in a Central Amazonian forest. *J Biogeogr* 23:27–46.

Zink, R. M. 1994. The geography of mitochondrial DNA variation, population structure, hybridization, and species limits in the fox sparrow *(Passerella iliaca)*. *Evolution* 48:96–111.

———. 1996. Comparative phylogeography in North American birds. *Evolution* 50:308–317.

———. 1997. Species concepts. *Bull Br Ornith Club* 117:97–109.

Zink, R. M., R. C. Blackwell-Rago, and F. Ronquist. 2000. The shifting roles of dispersal and vicariance in biogeography. *Proc Roy Soc Lond* B 267:497–503.

Zink, R. M., and M. C. McKitrick. 1995. The debate over species concepts and its implications for ornithology. *Auk* 112:701–719.

Zrzavy, J. 1992. Evolution of antennae and historical ecology of the hemipteran insects (Paraneoptera). *Acta Entomol Bohemoslov* 89:77–86.

———. 1994. Red bugs and the origin of mimetic complexes (Heteroptera: Pyrrhocoridae: *Dysdercus* spp.). *Oikos* 69:346–352.

———. 1997. Phylogenetics and ecology: All characters should be included in the cladistic analysis. *Oikos* 80:186–192.

Zrzavy, J., and O. Nedved. 1997. Phylogeny of the New World Dysdercus (Insecta: Hemiptera: Pyrrhocoridae) and evolution of their colour patterns. *Cladistics* 13:109–123.

———. 1999. Evolution of mimicry in the New World *Dysdercus* (Hemiptera: Pyrrhocoridae). *J Evol Biol* 12:956–969.

Zrzavy, J., and Z. Skala. 1993. Holobionts, hybrids, and cladistics. *BioSystems* 31:127–130.

Zuiganov, V. V. G., G. J. Golovatjuk, K. A. Savvaitova, and V. F. Bugaev. 1987. Genetically isolated sympatric forms of threespine stickleback, *Gasterosteus aculeatus*, in Lake Azabachije (Kamchatka peninsula). *Envir Biol Fishes* 18:241–247.

Zweers, G. A., and J. C. Vanden Berge. 1997. Evolutionary transitions in the trophic system of the wader-waterfowl complex. *Neth J Zool* 47:255–287.

Zyskowski, K., and R. O. Prum. 1999. Phylogenetic analysis of the nest architecture of Neotropical ovenbirds (Furnariidae). *Auk* 116:891–911.

Index

DATE DUE

DEC 2 0 2001

REC'D SEP 1 7 2002

REC'D JAN 27 2004

Demco, Inc. 38-293